$P(A \text{ and } B)$	Joint probability of events A and B
$P(A \text{ or } B)$	Probability of the union of events A and B
$P(A \mid B)$	Probability that event A will occur given event B has occurred
q	$1 - p$, probability of failure for a binomial experiment
$\hat{q}$	$1 - \hat{p}$ where $\hat{p}$ is the sample proportion
$\bar{q}$	$1 - \bar{p}$ where $\bar{p}$ is the pooled sample proportion for two samples
Q_1, Q_2, Q_3	First, second, and third quartiles, respectively
R	Number of rows in a contingency table
r	Linear correlation coefficient for sample data
r^2	Coefficient of determination
ρ	(Greek letter *rho*) Linear correlation coefficient for population data
S	Sample space
s	Sample standard deviation
s^2	Sample variance
s_b	Estimator of σ_b
s_d	Standard deviation of the paired differences for a sample
$s_{\bar{d}}$	Estimator of $\sigma_{\bar{d}}$
s_e	Standard deviation of errors for the sample regression model
s_p	Pooled standard deviation
$s_{\hat{p}}$	Estimator of $\sigma_{\hat{p}}$
$s_{\hat{p}_1 - \hat{p}_2}$	Estimator of $\sigma_{\hat{p}_1 - \hat{p}_2}$
$s_{\bar{x}}$	Estimator of $\sigma_{\bar{x}}$
$s_{\bar{x}_1 - \bar{x}_2}$	Estimator of $\sigma_{\bar{x}_1 - \bar{x}_2}$
$s_{\hat{y}_m}$	Estimator of $\sigma_{\hat{y}_m}$
$s_{\hat{y}_p}$	Estimator of $\sigma_{\hat{y}_p}$
Σ	(Greek letter capital *sigma*) summation notation
σ	(Greek letter lower-case *sigma*) Population standard deviation
σ^2	Population variance
σ_b	Standard deviation of the sampling distribution of b
σ_d	Standard deviation of the paired differences for the population
$\sigma_{\bar{d}}$	Standard deviation of the sampling distribution of $\bar{d}$
σ_ϵ	Standard deviation of errors for the population regression model
$\sigma_{\hat{p}}$	Standard deviation of the sampling distribution of $\hat{p}$
$\sigma_{\hat{p}_1 - \hat{p}_2}$	Standard deviation of the sampling distribution of $\hat{p}_1 - \hat{p}_2$
$\sigma_{\bar{x}}$	Standard deviation of the sampling distribution of $\bar{x}$
$\sigma_{\bar{x}_1 - \bar{x}_2}$	Standard deviation of the sampling distribution of $\bar{x}_1 - \bar{x}_2$
$\sigma_{\hat{y}_m}$	Standard deviation of $\hat{y}$ when estimating $\mu_{y \mid x}$
$\sigma_{\hat{y}_p}$	Standard deviation of $\hat{y}$ when predicting y_p
t	The t distribution
T_i	Sum of the values included in sample i in one-way ANOVA
x	(1) Variable; (2) random variable; (3) independent variable in a regression model
$\bar{x}$	Sample mean
y	(1) Variable; (2) dependent variable in a regression model
$\hat{y}$	Estimated or predicted value of y using a regression model
z	Units of the standard normal distribution

CHAPTER 5 · DISCRETE RANDOM VARIABLES AND THEIR PROBABILITY DISTRIBUTIONS

- Mean of a discrete random variable x: $\mu = \Sigma x P(x)$
- Standard deviation of a discrete random variable x:
$$\sigma = \sqrt{\Sigma x^2 P(x) - \mu^2}$$
- n factorial: $n! = n\,(n-1)\,(n-2)\ldots 3\cdot 2\cdot 1$
- Number of combinations of n items selected x at a time:
$$\binom{n}{x} = \frac{n!}{x!\,(n-x)!}$$
- Binomial probability formula:
$$P(x) = \binom{n}{x} p^x q^{n-x}$$
- Mean and standard deviation of the binomial distribution:
$$\mu = np \quad \text{and} \quad \sigma = \sqrt{npq}$$
- Poisson probability formula:
$$P(x) = \frac{\lambda^x\, e^{-\lambda}}{x!}$$
- Mean, variance, and standard deviation of the Poisson probability distribution:
$$\mu = \lambda, \quad \sigma^2 = \lambda, \quad \text{and} \quad \sigma = \sqrt{\lambda}$$

CHAPTER 6 · CONTINUOUS RANDOM VARIABLES AND THE NORMAL DISTRIBUTION

- z value for an x value:
$$z = \frac{x - \mu}{\sigma}$$
- Value of x when μ, σ, and z are known: $x = \mu + z\sigma$

CHAPTER 7 · SAMPLING DISTRIBUTIONS

- Mean of $\bar{x}$: $\mu_{\bar{x}} = \mu$
- Standard deviation of $\bar{x}$ when $n/N \le .05$: $\sigma_{\bar{x}} = \sigma/\sqrt{n}$
- z value for $\bar{x}$:
$$z = \frac{\bar{x} - \mu}{\sigma_{\bar{x}}}$$
- Population proportion: $p = X/N$
- Sample proportion: $\hat{p} = x/n$
- Mean of $\hat{p}$: $\mu_{\hat{p}} = p$
- Standard deviation of $\hat{p}$ when $n/N \le .05$: $\sigma_{\hat{p}} = \sqrt{pq/n}$
- z value for $\hat{p}$:
$$z = \frac{\hat{p} - p}{\sigma_{\hat{p}}}$$

CHAPTER 8 · ESTIMATION OF THE MEAN AND PROPORTION

- Margin of error for the point estimation of μ:
$$\pm 1.96\,\sigma_{\bar{x}} \quad \text{or} \quad \pm 1.96\,s_{\bar{x}}$$
where $\sigma_{\bar{x}} = \sigma/\sqrt{n}$ and $s_{\bar{x}} = s/\sqrt{n}$

- Confidence interval for μ for a large sample:
$$\bar{x} \pm z\,\sigma_{\bar{x}} \quad \text{if } \sigma \text{ is known}$$
$$\bar{x} \pm z\,s_{\bar{x}} \quad \text{if } \sigma \text{ is not known}$$
where $\sigma_{\bar{x}} = \sigma/\sqrt{n}$ and $s_{\bar{x}} = s/\sqrt{n}$
- Confidence interval for μ for a small sample:
$$\bar{x} \pm t\,s_{\bar{x}} \quad \text{where} \quad s_{\bar{x}} = s/\sqrt{n}$$
- Margin of error for the point estimation of p:
$$\pm 1.96\,s_{\hat{p}} \quad \text{where} \quad s_{\hat{p}} = \sqrt{\hat{p}\hat{q}/n}$$
- Confidence interval for p for a large sample:
$$\hat{p} \pm z\,s_{\hat{p}} \quad \text{where} \quad s_{\hat{p}} = \sqrt{\hat{p}\hat{q}/n}$$
- Maximum error of the estimate for μ:
$$E = z\,\sigma_{\bar{x}} \quad \text{or} \quad z\,s_{\bar{x}}$$
- Determining sample size for estimating μ: $n = z^2\,\sigma^2/E$
- Maximum error of the estimate for p:
$$E = z\,s_{\hat{p}} \quad \text{where} \quad s_{\hat{p}} = \sqrt{\hat{p}\,\hat{q}/n}$$
- Determining sample size for estimating p: $n = z^2\,p\,q/E$

CHAPTER 9 · HYPOTHESIS TESTS ABOUT THE MEAN AND PROPORTION

- Test statistic z for a test of hypothesis about μ for a large sample:
$$z = \frac{\bar{x} - \mu}{\sigma_{\bar{x}}} \quad \text{if } \sigma \text{ is known, where } \sigma_{\bar{x}} = \frac{\sigma}{\sqrt{n}}$$
or $$z = \frac{\bar{x} - \mu}{s_{\bar{x}}} \quad \text{if } \sigma \text{ is not known, where } s_{\bar{x}} = \frac{s}{\sqrt{n}}$$
- Test statistic for a test of hypothesis about μ for a small sample:
$$t = \frac{\bar{x} - \mu}{s_{\bar{x}}} \quad \text{where} \quad s_{\bar{x}} = \frac{s}{\sqrt{n}}$$
- Test statistic for a test of hypothesis about p for a large sample:
$$z = \frac{\hat{p} - p}{\sigma_{\hat{p}}} \quad \text{where} \quad \sigma_{\hat{p}} = \sqrt{\frac{p\,q}{n}}$$

CHAPTER 10 · ESTIMATION AND HYPOTHESIS TESTING: TWO POPULATIONS

- Mean of the sampling distribution of $\bar{x}_1 - \bar{x}_2$:
$$\mu_{\bar{x}_1 - \bar{x}_2} = \mu_1 - \mu_2$$
- Confidence interval for $\mu_1 - \mu_2$ for two large and independent samples:
$$(\bar{x}_1 - \bar{x}_2) \pm z\,\sigma_{\bar{x}_1 - \bar{x}_2} \quad \text{if } \sigma_1 \text{ and } \sigma_2 \text{ are known}$$
or $$(\bar{x}_1 - \bar{x}_2) \pm z\,s_{\bar{x}_1 - \bar{x}_2} \quad \text{if } \sigma_1 \text{ and } \sigma_2 \text{ are not known}$$

where $\sigma_{\bar{x}_1 - \bar{x}_2} = \sqrt{\dfrac{\sigma_1^2}{n_1} + \dfrac{\sigma_2^2}{n_2}}$ and $s_{\bar{x}_1 - \bar{x}_2} = \sqrt{\dfrac{s_1^2}{n_1} + \dfrac{s_2^2}{n_2}}$

Test statistic for a test of hypothesis about $\mu_1 - \mu_2$ for two large and independent samples:

$$z = \frac{(\bar{x}_1 - \bar{x}_2) - (\mu_1 - \mu_2)}{\sigma_{\bar{x}_1 - \bar{x}_2}}$$

If σ_1 and σ_2 are not known, then replace $\sigma_{\bar{x}_1 - \bar{x}_2}$ by its point estimator $s_{\bar{x}_1 - \bar{x}_2}$.

- For two small and independent samples taken from two populations with equal standard deviations:

  Pooled standard deviation:

  $$s_p = \sqrt{\frac{(n_1 - 1) s_1^2 + (n_2 - 1) s_2^2}{n_1 + n_2 - 2}}$$

  Estimate of the standard deviation of $\bar{x}_1 - \bar{x}_2$:

  $$s_{\bar{x}_1 - \bar{x}_2} = s_p \sqrt{\frac{1}{n_1} + \frac{1}{n_2}}$$

  Confidence interval for $\mu_1 - \mu_2$:

  $$(\bar{x}_1 - \bar{x}_2) \pm t\, s_{\bar{x}_1 - \bar{x}_2}$$

  Test statistic: $\quad t = \dfrac{(\bar{x}_1 - \bar{x}_2) - (\mu_1 - \mu_2)}{s_{\bar{x}_1 - \bar{x}_2}}$

- For two small and independent samples selected from two populations with unequal standard deviations:

  Degrees of freedom: $\quad df = \dfrac{\left(\dfrac{s_1^2}{n_1} + \dfrac{s_2^2}{n_2}\right)^2}{\dfrac{\left(\dfrac{s_1^2}{n_1}\right)^2}{n_1 - 1} + \dfrac{\left(\dfrac{s_2^2}{n_2}\right)^2}{n_2 - 1}}$

  Estimate of the standard deviation of $\bar{x}_1 - \bar{x}_2$:

  $$s_{\bar{x}_1 - \bar{x}_2} = \sqrt{\frac{s_1^2}{n_1} + \frac{s_2^2}{n_2}}$$

  Confidence interval for $\mu_1 - \mu_2$:

  $$(\bar{x}_1 - \bar{x}_2) \pm t\, s_{\bar{x}_1 - \bar{x}_2}$$

  Test statistic: $\quad t = \dfrac{(\bar{x}_1 - \bar{x}_2) - (\mu_1 - \mu_2)}{s_{\bar{x}_1 - \bar{x}_2}}$

- For two paired or matched samples:

  Sample mean for paired differences: $\quad \bar{d} = \dfrac{\Sigma d}{n}$

  Sample standard deviation for paired differences:

  $$s_d = \sqrt{\frac{\Sigma d^2 - \dfrac{(\Sigma d)^2}{n}}{n - 1}}$$

  Mean and standard deviation of the sampling distribution of $\bar{d}$:

  $$\mu_{\bar{d}} = \mu_d \quad \text{and} \quad s_{\bar{d}} = \frac{s_d}{\sqrt{n}}$$

Confidence interval for μ_d:

$$\bar{d} \pm t\, s_{\bar{d}} \quad \text{where} \quad s_{\bar{d}} = \frac{s_d}{\sqrt{n}}$$

Test statistic for a test of hypothesis about μ_d:

$$t = \frac{\bar{d} - \mu_d}{s_{\bar{d}}}$$

- For two large and independent samples, confidence interval for $p_1 - p_2$:

  $$(\hat{p}_1 - \hat{p}_2) \pm z\, s_{\hat{p}_1 - \hat{p}_2}$$

  where $\quad s_{\hat{p}_1 - \hat{p}_2} = \sqrt{\dfrac{\hat{p}_1 \hat{q}_1}{n_1} + \dfrac{\hat{p}_2 \hat{q}_2}{n_2}}$

- For two large and independent samples, for a test of hypothesis about $p_1 - p_2$ with $H_0: p_1 - p_2 = 0$:

  Pooled sample proportion:

  $$\bar{p} = \frac{x_1 + x_2}{n_1 + n_2} \quad \text{or} \quad \frac{n_1 \hat{p}_1 + n_2 \hat{p}_2}{n_1 + n_2}$$

  Estimate of the standard deviation of $\hat{p}_1 - \hat{p}_2$:

  $$s_{\hat{p}_1 - \hat{p}_2} = \sqrt{\bar{p}\,\bar{q} \left(\frac{1}{n_1} + \frac{1}{n_2}\right)}$$

  Test statistic: $\quad z = \dfrac{(\hat{p}_1 - \hat{p}_2) - (p_1 - p_2)}{s_{\hat{p}_1 - \hat{p}_2}}$

CHAPTER 11 · CHI-SQUARE TESTS

- Expected frequency for a category for a goodness-of-fit test:

  $$E = np$$

- Degrees of freedom for a goodness-of-fit test:

  $df = k - 1 \quad$ where k is the number of categories

- Expected frequency for a cell for an independence or homogeneity test:

  $$E = \frac{(\text{Row total}) (\text{Column total})}{n}$$

- Degrees of freedom for a test of independence or homogeneity:

  $$df = (R - 1) (C - 1)$$

  where R and C are the total number of rows and columns, respectively, in the contingency table

- Test statistic for a goodness-of-fit test and a test of independence or homogeneity:

  $$\chi^2 = \Sigma \frac{(O - E)^2}{E}$$

- Confidence interval for the population variance σ^2:

  $$\frac{(n - 1) s^2}{\chi^2_{\alpha/2}} \quad \text{to} \quad \frac{(n - 1) s^2}{\chi^2_{1 - \alpha/2}}$$

- Test statistic for a test of hypothesis about σ^2:

  $$\chi^2 = \frac{(n - 1) s^2}{\sigma^2}$$

CHAPTER 12 · ANALYSIS OF VARIANCE

Let:

k = the number of different samples (or treatments)

n_i = the size of sample i

T_i = the sum of the values in sample i

n = the number of values in all samples

$\quad = n_1 + n_2 + n_3 + \ldots$

Σx = the sum of the values in all samples

$\quad = T_1 + T_2 + T_3 + \ldots$

Σx^2 = the sum of the squares of values in all samples

- For the F distribution:

  Degrees of freedom for the numerator $= k - 1$

  Degrees of freedom for the denominator $= n - k$

- Between-samples sum of squares:

$$SSB = \left(\frac{T_1^2}{n_1} + \frac{T_2^2}{n_2} + \frac{T_3^2}{n_3} + \cdots \right) - \frac{(\Sigma x)^2}{n}$$

- Within-samples sum of squares:

$$SSW = \Sigma x^2 - \left(\frac{T_1^2}{n_1} + \frac{T_2^2}{n_2} + \frac{T_3^2}{n_3} + \cdots \right)$$

- Total sum of squares:

$$SST = SSB + SSW = \Sigma x^2 - \frac{(\Sigma x)^2}{n}$$

- Variance between samples: $\quad MSB = \dfrac{SSB}{k - 1}$

- Variance within samples: $\quad MSW = \dfrac{SSW}{n - k}$

- Test statistic for a one-way ANOVA test: $\quad F = \dfrac{MSB}{MSW}$

CHAPTER 13 · SIMPLE LINEAR REGRESSION

- Simple linear regression model:

$$y = A + Bx + \epsilon$$

- Estimated simple linear regression model:

$$\hat{y} = a + bx$$

- Sum of squares of xy, xx, and yy:

$$SS_{xy} = \Sigma xy - \frac{(\Sigma x)(\Sigma y)}{n},$$

$$SS_{xx} = \Sigma x^2 - \frac{(\Sigma x)^2}{n}, \text{ and } SS_{yy} = \Sigma y^2 - \frac{(\Sigma y)^2}{n}$$

- Least squares estimates of A and B:

$$b = \frac{SS_{xy}}{SS_{xx}} \text{ and } a = \bar{y} - b\bar{x}$$

- Standard deviation of the sample errors:

$$s_e = \sqrt{\frac{SS_{yy} - b\, SS_{xy}}{n - 2}}$$

- Error sum of squares: $\quad SSE = \Sigma e^2 = \Sigma(y - \hat{y})^2$

- Total sum of squares: $\quad SST = \Sigma y^2 - \dfrac{(\Sigma y)^2}{n}$

- Regression sum of squares: $\quad SSR = SST - SSE$

- Coefficient of determination: $\quad r^2 = b\, \dfrac{SS_{xy}}{SS_{yy}}$

- Confidence interval for B:

$$b \pm t\, s_b \quad \text{where} \quad s_b = \frac{s_e}{\sqrt{SS_{xx}}}$$

- Test statistic for a test of hypothesis about B:

$$t = \frac{b - B}{s_b}$$

- Linear correlation coefficient:

$$r = \frac{SS_{xy}}{\sqrt{SS_{xx}\, SS_{yy}}}$$

- Confidence interval for $\mu_{y\,|\,x}$:

$$\hat{y} \pm t\, s_{\hat{y}_m} \quad \text{where} \quad s_{\hat{y}_m} = s_e \sqrt{\frac{1}{n} + \frac{(x_0 - \bar{x})^2}{SS_{xx}}}$$

- Prediction interval for y_p:

$$\hat{y} \pm t\, s_{\hat{y}_p} \quad \text{where} \quad s_{\hat{y}_p} = s_e \sqrt{1 + \frac{1}{n} + \frac{(x_0 - \bar{x})^2}{SS_{xx}}}$$

CHAPTER 2 · ORGANIZING DATA

- Relative frequency of a class = $f/\Sigma f$
- Percentage of a class = (Relative frequency) $\times$ 100
- Class midpoint or mark = (Upper limit + Lower limit)/2
- Class width = Upper boundary − Lower boundary
- Cumulative relative frequency
$$= \frac{\text{Cumulative frequency}}{\text{Total observations in the data set}}$$
- Cumulative percentage
$$= (\text{Cumulative relative frequency}) \times 100$$

CHAPTER 3 · NUMERICAL DESCRIPTIVE MEASURES

- Mean for ungrouped data: $\mu = \Sigma x/N$ and $\bar{x} = \Sigma x/n$
- Mean for grouped data: $\mu = \Sigma mf/N$ and $\bar{x} = \Sigma mf/n$
  where m is the midpoint and f is the frequency of a class
- Median for ungrouped data
$$= \text{Value of the } \left(\frac{n+1}{2}\right) \text{th term in a ranked data set}$$
- Range = Largest value − Smallest value
- Variance for ungrouped data:
$$\sigma^2 = \frac{\Sigma x^2 - \dfrac{(\Sigma x)^2}{N}}{N} \quad \text{and} \quad s^2 = \frac{\Sigma x^2 - \dfrac{(\Sigma x)^2}{n}}{n-1}$$
  where σ^2 is the population variance and s^2 is the sample variance
- Standard deviation for ungrouped data:
$$\sigma = \sqrt{\frac{\Sigma x^2 - \dfrac{(\Sigma x)^2}{N}}{N}} \quad \text{and} \quad s = \sqrt{\frac{\Sigma x^2 - \dfrac{(\Sigma x)^2}{n}}{n-1}}$$
  where σ and s are the population and sample standard deviations, respectively
- Variance for grouped data:
$$\sigma^2 = \frac{\Sigma m^2 f - \dfrac{(\Sigma mf)^2}{N}}{N} \quad \text{and} \quad s^2 = \frac{\Sigma m^2 f - \dfrac{(\Sigma mf)^2}{n}}{n-1}$$
- Standard deviation for grouped data:
$$\sigma = \sqrt{\frac{\Sigma m^2 f - \dfrac{(\Sigma mf)^2}{N}}{N}} \quad \text{and} \quad s = \sqrt{\frac{\Sigma m^2 f - \dfrac{(\Sigma mf)^2}{n}}{n-1}}$$

- Chebyshev's theorem:
  For any number k greater than 1, at least $(1 - 1/k^2)$ of the values for any distribution lie within k standard deviations of the mean.
- Empirical rule:
  For a specific bell-shaped distribution, about 68% of the observations fall in the interval $(\mu - \sigma)$ to $(\mu + \sigma)$, about 95% fall in the interval $(\mu - 2\sigma)$ to $(\mu + 2\sigma)$, and about 99.7% fall in the interval $(\mu - 3\sigma)$ to $(\mu + 3\sigma)$.
- Interquartile range: $IQR = Q_3 - Q_1$
  where Q_3 is the third quartile and Q_1 is the first quartile
- The kth percentile:
$$P_k = \text{Value of the } \left(\frac{kn}{100}\right)\text{th term in a ranked data set}$$
- Percentile rank of x_i
$$= \frac{\text{Number of values less than } x_i}{\text{Total number of values in the data set}} \times 100$$

CHAPTER 4 · PROBABILITY

- Classical probability rule for a simple event:
$$P(E_i) = \frac{1}{\text{Total number of outcomes}}$$
- Classical probability rule for a compound event:
$$P(A) = \frac{\text{Number of outcomes in } A}{\text{Total number of outcomes}}$$
- Relative frequency as an approximation of probability:
$$P(A) = \frac{f}{n}$$
- Conditional probability of an event:
$$P(A \mid B) = \frac{P(A \text{ and } B)}{P(B)} \quad \text{and} \quad P(B \mid A) = \frac{P(A \text{ and } B)}{P(A)}$$
- Condition for independence of events:
$$P(A) = P(A \mid B) \quad \text{and/or} \quad P(B) = P(B \mid A)$$
- For complementary events: $P(A) + P(\bar{A}) = 1$
- Multiplication rule for dependent events:
$$P(A \text{ and } B) = P(A) \, P(B \mid A)$$
- Multiplication rule for independent events:
$$P(A \text{ and } B) = P(A) \, P(B)$$
- Joint probability of two mutually exclusive events:
$$P(A \text{ and } B) = 0$$
- Addition rule for mutually nonexclusive events:
$$P(A \text{ or } B) = P(A) + P(B) - P(A \text{ and } B)$$
- Addition rule for mutually exclusive events:
$$P(A \text{ or } B) = P(A) + P(B)$$

THIRD EDITION
INTRODUCTORY STATISTICS

THIRD EDITION
INTRODUCTORY STATISTICS

PREM S. MANN
EASTERN CONNECTICUT STATE UNIVERSITY

JOHN WILEY & SONS, INC.
NEW YORK • CHICHESTER • WEINHEIM • BRISBANE • SINGAPORE • TORONTO

ACQUISITIONS EDITOR	Ruth Baruth
MARKETING MANAGER	Jay Kirsch
PROJECT MANAGEMENT	J. Carey Publishing Service
DESIGNER	Laura Boucher
DEVELOPMENTAL EDITOR	Joan Carrafiello
COVER PHOTO	©Mark Tomalty / Masterfile
ILLUSTRATION COORDINATOR	Jamie Perea

This book was set in Times Roman by CR Waldman and York Graphic Services and printed and bound by Von Hoffman Press. The cover was printed by The Lehigh Press, Inc.

Library of Congress Cataloging-in-Publication Data
Mann, Prem S.
    Introductory statistics / Prem S. Mann. — 3rd ed.
        p. cm.
    Includes index.
    ISBN 0-471-16546-8 (cloth:alk.paper)
    1. Statistics. I. Title
QA276. 12.M29 1998   97-28302
519.5—dc21   CIP

Printed in the United States of America

10 9 8 7 6 5 4 3 2 1

To my parents

PREFACE

Introductory Statistics is written for a first course in applied statistics. The book is intended for students who do not have a strong background in mathematics. The only prerequisite for this text is a knowledge of elementary algebra.

Today, college students from almost all fields of study are required to take at least one course in statistics. Consequently, the study of statistical methods has taken on a prominent role in the education of students majoring in all fields of study. The goal of this text is to make the subject of statistics both interesting and accessible to such a wide and varied audience. Three major characteristics of this text support this goal: the realistic content of its examples and exercises that draw on a comprehensive range of applications from all facets of life, the clarity and brevity of its presentation, and the soundness of its pedagogical approach. These characteristics are exhibited through the interplay of a variety of significant text features.

The following are two of the many unsolicited comments from students and professors who are using *Introductory Statistics*. These comments serve as evidence of the success of the text in making statistics interesting and accessible—a goal of the author from its very first edition. Such comments have also motivated the author to pursue this same goal through the refinements and updates in this third edition so that *Introductory Statistics* will continue to provide a successful experience in statistics to growing numbers of students and professors.

" . . . I am currently enrolled in an Introductory Stats class as a prerequisite for grad school. We're using your text, and I wanted to compliment you—it is superb. . . . Without a clear, well-organized text, my classmates and I would be in serious trouble! I can only imagine what a poor stats text would be like . . . yikes. Thanks for making my first stats class much easier to bear."

—Jane Tanton
Ann Arbor, MI

"My students found the book very easy to read and the examples clear. They were able to cover course basics on their own as we did other activities in class."

—Gerald Busald
Professor of Mathematics
San Antonio College

MAJOR CHANGES IN THE THIRD EDITION

The following are some of the many significant changes made in the third edition of the text without changing its main features.

- Presentation of concepts has been refined at many places to enhance clarity.

- A large number of examples, case studies, and exercises have been changed and updated to capture interest of students and maintain relevancy.

- A large number of exercises of greater difficulty have been added to the supplementary exercises of all chapters as a response to requests for an extension of these exercises in this direction. These exercises of greater difficulty are indicated by an asterisk next to them.

- The MINITAB sections have been updated. In addition, these sections now include both the MINITAB FOR WINDOWS and the MINITAB COMMAND LANGUAGE instructions.

- For those who wish to have the option of using a graphing calculator in solving statistical exercises, a few exercises from each chapter that are suitable for this purpose have been identified with a calculator icon. However, these exercises can also be solved without using the graphing calculator.

- All data sets given in Appendix B have been updated.

MAIN FEATURES OF THIS TEXT

Style and Pedagogy

Clear and Concise Exposition   The explanation of statistical methods and concepts is clear and concise. Moreover, the style is user-friendly and easy to understand. In chapter introductions and in transitions from section to section, new ideas are related to those discussed earlier.

Abundant Examples

Examples   The text contains a wealth of examples, a total of 196 in 13 chapters and an appendix. The examples are usually given in a format showing a problem and its solution. They are well-sequenced and thorough, displaying all facets of concepts. Furthermore, the examples capture students' interest because they cover a wide variety of relevant topics. They are based on situations practicing statisticians encounter every day. Finally, a large number of examples are based on real data that are taken from sources such as books, government and private data sources and reports, magazines, newspapers, and professional journals.

Realistic Settings

Solutions   A clear, concise solution follows each problem presented in an example. When the solution to an example involves many steps, it is presented in a step-by-step format. For instance, examples related to tests of hypotheses contain five steps that are consistently used to solve such examples in all chapters. Thus, procedures are presented in the concrete settings of applications rather than as isolated abstractions. Frequently, solutions contain highlighted remarks that recall and reinforce ideas critical to the solution of the problem. Such remarks add to the clarity of presentation.

Guideposts

Margin Notes for Examples   A margin note appears beside each example that briefly describes what is being done in that example. Students can use these margin notes to assist them as they read through sections and to quickly locate appropriate model problems as they work through exercises.

Frequent Use of Diagrams   Concepts can often be made more understandable by describing them visually, with the help of diagrams. This text uses diagrams frequently to help students understand concepts and solve problems. For example, tree diagrams are used extensively in Chapters 4 and 5 to assist in explaining probability concepts and in computing probabilities. Similarly, solutions to all examples about tests of hypotheses contain diagrams showing rejection regions, nonrejection regions, and critical values.

Highlighting   Definitions of important terms, formulas, and key concepts are enclosed in color boxes so that students can easily locate them. A similar use of color is found in the *Using MINITAB* sections where a color tint highlights MINITAB commands, their explanations, and their usage along with MINITAB solutions. Important terms appear in the text either in boldface or italic type.

☞ **Cautions** Certain items need special attention. These may deal with potential trouble spots that commonly cause errors. Or they may deal with ideas that students often overlook. Special emphasis is placed on such items through the headings: *Remember, An Observation,* or *Warning.* An icon is used to identify such items.

Realistic Applications **Case Studies** Case studies, which appear in many chapters, provide additional illustrations of the applications of statistics in research and statistical analysis. Most of these case studies are based on articles published in journals, magazines, or newspapers. All case studies are based on real data.

Abundant Exercises **Exercises and Supplementary Exercises** The text contains an abundance of exercises, a total of 1315 in 13 chapters and an appendix (excluding Computer Assignments). Moreover, a large number of these exercises contain several parts. Exercise sets appearing at the end of each section (or sometimes at the end of two or three sections) include problems on the topics of that section. These exercises are divided into two parts: **Concepts and Procedures** that emphasize key ideas and techniques, and **Applications** that use these ideas and techniques in concrete settings. Supplementary exercises appear at the end of each chapter and contain exercises on all sections and topics discussed in that chapter. A large number of these exercises are based on real data taken from varied data sources such as books, government and private data sources and reports, magazines, newspapers, and professional journals. Exercises given in the text do not merely provide practice for students, but the real data contained in exercises provide interesting information and insight into economic, political, social, psychological, and other aspects of life. The exercise sets also contain many problems that demand critical thinking skills. The answers to selected odd-numbered exercises appear in the *Answers Section* at the back of the book. **Optional exercises** are indicated by an asterisk (*).

More Challenging Exercises A set of especially challenging exercises appears at the end of each block of three chapters. In addition, Chapter 13 is followed by its own *More Challenging Exercises* section. These exercises, which are optional, use material presented in the three previous chapters, or any material already covered. These exercise blocks appear after Chapters 3, 6, 9, 12, and 13. Answers to these problems are not given in the *Answers Section* at the end of the book. They appear only in the *Instructor's Solutions Manual.*

Summary and Review **Glossary** Each chapter has a glossary that lists the key terms introduced in that chapter, along with a brief explanation of each term. Almost all the terms that appear in boldface type in the text are in the glossary.

Key Formulas Each chapter contains a list of key formulas used in that chapter. This list appears before the *Supplementary Exercises* for each chapter.

Self-Review Tests Each chapter contains a *Self-Review Test*, which appears immediately after the *Supplementary Exercises*. These problems can help students to test their grasp of the concepts and skills presented in respective chapters and to monitor their understanding of statistical methods. The problems marked by an asterisk (*) in the *Self-Review Tests* are **optional.** The answers to almost all problems of the *Self-Review Tests* appear in the *Answers Section*.

Formula Card A formula card that contains key formulas from all chapters is included at the beginning of the book.

Technology

Computer Usage

Another feature of this text is the detailed instructions on the use of **MINITAB.**[1] A *Using MINITAB* section follows 12 of the 13 chapters and Appendix A. Each of these *Using MINITAB* sections contains a detailed description of the MINITAB commands that are used to perform the statistical analysis presented in that chapter. These sections now include both the MINITAB FOR WINDOWS step-by-step instructions and the MINITAB COMMAND LANGUAGE instructions to perform statistical data analysis. In addition, each of these sections contains several illustrations that demonstrate how MINITAB can be used to solve statistical problems. These MINITAB instructions and illustrations are so complete that students do not need to purchase any other MINITAB supplement. Computer assignments are also given at the end of each MINITAB section so that students can further practice MINITAB. A total of 68 computer assignments are contained in these sections.

Graphing Calculator

More and more instructors are now using graphing calculators, such as the TI-83, to teach statistics. A few exercises from each chapter in the text have been identified and marked with a **calculator icon** as suitable exercises for graphing calculators. The complete solutions to these exercises along with the TI-83 screen reproductions are provided in the *Instructor's Solutions Manual.* Note that these exercises can also be solved without using a graphing calculator.

Calculator Usage

The text contains many footnotes that explain how a calculator can be used to evaluate complex mathematical expressions.

Data Sets

Four data sets appear in Appendix B. These data sets, collected from different sources, contain information on many variables. They can be used to perform statistical analysis with statistical computer software such as MINITAB. **These data sets are available from the publisher on a diskette in MINITAB format and in ASCII format.**

WEBSITE

http://www.wiley.com/college/mann

The Mann Statistics Website provides resources for instructors and students. The graphing calculator programs for running complex statistical tests on the TI-83 can be downloaded from this site. Other resources include complete data sets for the problems in the text and links to interesting statistical sites on the web. Instructors can find information on changes in the new edition and contact Wiley for examination copies and supplements.

OPTIONAL SECTIONS

Because each instructor has different preferences, the text does not indicate optional sections. This decision has been left to the instructor. Instructors may cover the sections or chapters that they think are important. However, a few alternative one-semester syllabi are available on the Website for this text.

Complete Learning System

SUPPLEMENTS

The following supplements are available to accompany this text.

Instructor's Solutions Manual   This manual contains complete solutions to all exercises, the *Self-Review Test* problems, and *More Challenging Exercises.*

---

[1]MINITAB is a registered trademark of Minitab, Inc., 3081 Enterprise Drive, State College, PA 16801. Phone: 814-238-3280; fax: 814-238-4383; telex: 881612. The author would like to take this opportunity to thank Minitab, Inc., for their help.

Instructor's Resource Guide   Sample syllabi, discussion questions, sample exams and suggested homework assignments, and transparency masters are included in this resource. Written for both the experienced and inexperienced instructor, the manual offers a clear overview of the course while outlining major topics and key points of each chapter.

Students' Solutions Manual   This manual contains complete solutions to all of the odd-numbered exercises, a few even-numbered exercises, and to all the *Self-Review Test* problems.

Student Study Guide   This guide contains review material about studying and learning patterns for a first course in statistics. Special attention is given to the critical material of each chapter. Reviews of mathematical notation, formulas, and table reading are also included.

Printed Test Bank   The printed copy of the test bank contains a large number of multiple choice questions, essay questions, and quantitative problems for each chapter.

Computerized Test Bank   All questions that are in the printed *Test Bank* are available on a diskette. This diskette can be obtained from the publisher.

Graphing Calculator Manual   This manual is a basic guide for beginners on the TI-83 and TI-82. The authors guide students through the important facets of using a graphing calculator in statistics by presenting clear examples, pictures of actual calculator screens, and programs specific to each calculator.

Data Diskette   All data sets given in Appendix B are available on a diskette in MINITAB and ASCII formats to the adopters of this text. These data sets can be downloaded from the John Wiley & Sons, Inc. Website at http://www.wiley.com/college/mann.

ACKNOWLEDGMENTS

I thank the following reviewers of this, or previous editions of this book, whose comments and suggestions were invaluable in improving the text.

K. S. Asal	Broward Community College, Coconut Creek, Florida
Louise Audette	Manchester Community College, Manchester, Connecticut
Joan Bookbinder	Johnson & Wales University, Providence, Rhode Island
Dean Burbank	Gulf Coast Community College, Panama City, Florida
Jayanta Chandra	University of Notre Dame, Notre Dame, Indiana
James Curl	Modesto Community College, Modesto, California
Fred H. Dorner	Trinity University, San Antonio, Texas
William D. Ergle	Roanoke College, Salem, Virginia
Ronald Ferguson	San Antonio College, San Antonio, Texas
Larry Griffey	Florida Community College, Jacksonville, Florida
Frank Goulard	Portland Community College, Portland, Oregon
Robert Graham	Jacksonville State University, Jacksonville, Alabama
A. Eugene Hileman	Northeastern State University, Tahlequah, Oklahoma
Gary S. Itzkowitz	Rowan State College, Glassboro, New Jersey
Jean Johnson	Governors State University, University Park, Illinois
Michael Karelius	American River College, Sacramento, California
Dix J. Kelly	Central Connecticut State University, New Britain, Connecticut
Linda Kohl	University of Michigan, Ann Arbor, Michigan
Martin Kotler	Pace University, Pleasantville, New York
Marlene Kovaly	Florida Community College, Jacksonville, Florida

Carlos de la Lama	San Diego City College, San Diego, California
Richard McGowan	University of Scranton, Scranton, Pennsylvania
Daniel S. Miller	Central Connecticut State University, New Britain, Connecticut
Jeffrey Mock	Diablo Valley College, Pleasant Hill, California
Luis Moreno	Broome Community College, Binghamton, New York
Robert A. Nagy	University of Wisconsin, Green Bay, Wisconsin
Paul T. Nkansah	Florida Agricultural and Mechanical University, Tallahassee, Florida
Mary Parker	Austin Community College, Austin, Texas
Roger Peck	University of Rhode Island, Kingston, Rhode Island
Chester Piascik	Bryant College, Smithfield, Rhode Island
Joseph Pigeon	Villanova University, Villanova, Pennsylvania
Gerald Rogers	New Mexico State University, Las Cruces, New Mexico
Phillis Schumacher	Bryant College, Smithfield, Rhode Island
Kathryn Schwartz	Scottsdale Community College, Scottsdale, Arizona
Ronald Schwartz	Wilkes University, Wilkes-Barre, Pennsylvania
Larry Stephens	University of Nebraska, Omaha, Nebraska
Bruce Trumbo	California State University, Hayward, California
Jean Weber	University of Arizona, Tucson, Arizona
Terry Wilson	San Jacinto College, Pasadena, Texas
K. Paul Yoon	Fairleigh Dickinson University, Madison, New Jersey

I express my special thanks to Professor Gerald Geissert of Eastern Connecticut State University and Professor Daniel S. Miller of Central Connecticut State University. Professor Miller wrote the *More Challenging Exercises.* Professor Geissert checked solutions to all examples for mathematical accuracy, prepared answers to selected odd-numbered exercises for the *Answers Section,* read proofs, prepared the index, and helped in writing new problems. I thank Ms. Jane Tanton and Professor Gerald Busald for letting me use their comments about the previous edition of this text in the beginning of this preface. In addition, I thank Eastern Connecticut State University for all the support I received. I thank many of my students, especially Treesa George, Shruti Iyer, Crystal Morin, and Michelle Terreiro, who were of immense help during this revision. I also offer my thanks to my colleagues, friends, and family, whose support was a source of encouragement during the period when I spent long hours working on this project.

It is of utmost importance that a textbook is accompanied by complete and accurate supplements. I take pride in mentioning that the supplements prepared for this text possess these qualities and much more. I appreciatively thank the following authors of these supplements: Professor Jim Curl for writing the Study Guide, Professor Deborah Betthauser Britt for preparing the Instructor's Resource Guide, Professor John Horner and Professor Virginia Deus for preparing the Graphing Calculator Manual, Ms. Georgia Mederer for preparing the Solutions Manuals, and Professor Larry Stephens for writing the Test Bank.

It is my pleasure to thank all the professionals at John Wiley with whom I enjoyed working. Among these are Wayne Anderson, former Publisher; Ruth M. Baruth, Acquisitions Editor; Joan Carrafiello, Freelance Developmental Editor; Charlotte Hyland, Production Manager; Jennifer Carey, Project Management; Jay Kirsch, Marketing Manager; Barbara Bredenko, Assistant Editor; Laura Boucher, Senior Designer; Jamie Perea, Illustrations Coordinator; and many others.

Any suggestions from readers for future revisions would be greatly appreciated. These suggestions can be sent to the author through the Internet at MANN@ECSUC.CTSTATEU.EDU.

Prem S. Mann
Willimantic, CT 06226
June 1997

CONTENTS

CASE STUDIES

1 INTRODUCTION

The study of statistics has become more popular than ever during the past two decades. The increasing availability of computers and statistical software packages has enlarged the role of statistics as a tool for empirical research. As a result, statistics is used for research in almost all professions, from medicine to sports. Today, college students in almost all disciplines are required to take at least one statistics course.

Every field of study has its own terminology. Statistics is no exception. This introductory chapter explains the basic terms of statistics. These terms will bridge our understanding of the concepts and techniques presented in subsequent chapters.

1.1 WHAT IS STATISTICS?

The word **statistics** has two meanings. In the more common usage, statistics refers to numerical facts. The numbers that represent the income of a family, the age of a student, the percentage of passes completed by the quarterback of a football team, and the starting salary of a typical college graduate are examples of statistics in this sense of the word. A 1988 article in the *U.S. News & World Report* declared "Statistics are an American obsession."[1] During the 1988 baseball World Series between the Los Angeles Dodgers and the Oakland A's, NBC commentator Joe Garagiola reported to the viewers numerical facts about the players' performances. In response, fellow commentator Vin Scully said, "I love it when you talk statistics." In these examples, the word *statistics* refers to numbers. The following examples present a few statistics.

1. Automated teller machines are used for 27 million transactions each day in the United States.
2. In 1995, there were 17,274 alcohol-related traffic deaths in the United States.
3. About 7000 new marriages and 3000 divorces take place in the United States each day.
4. According to the National Football League Players' Association, the average length of a professional football player's career is 3.2 years.
5. The U.S. Postal Service handles 495 million pieces of mail per day.
6. According to the Direct Marketing Association and Conservatree, 140 million trees are used to produce the mail-order catalogs sent to U.S. households each year.
7. Every day about 1.9 million people in America visit doctors.
8. In 1996, 74% of the teachers in the United States were women.
9. Handgun violence in the United States costs $20 billion a year.

The second meaning of *statistics* refers to the field or discipline of study. In this sense of the word, statistics is defined as follows.

STATISTICS

Statistics is a group of methods that are used to collect, analyze, present, and interpret data and to make decisions.

---

[1]"The Numbers Racket: How Polls and Statistics Lie," *U.S. News & World Report*, July 11, 1988, pp. 44–47.

Every day we make decisions that may be personal, business related, or of some other kind. Usually these decisions are made under conditions of uncertainty. Many times, the situations or problems we face in the real world have no precise or definite solution. Statistical methods help us to make scientific and intelligent decisions in such situations. Decisions made by using statistical methods are called *educated guesses*. Decisions made without using statistical (or scientific) methods are *pure guesses* and, hence, may prove to be unreliable.

Like almost all fields of study, statistics has two aspects: theoretical and applied. *Theoretical* or *mathematical statistics* deals with the development, derivation, and proof of statistical theorems, formulas, rules, and laws. *Applied statistics* involves the applications of those theorems, formulas, rules, and laws to solve real-world problems. This text is concerned with applied statistics and not with theoretical statistics. By the time you finish studying this book, you will learn how to think statistically and how to make educated guesses.

1.2 TYPES OF STATISTICS

Broadly speaking, applied statistics can be divided into two areas: *descriptive statistics* and *inferential statistics*.

1.2.1 DESCRIPTIVE STATISTICS

Suppose we have information on the test scores of students enrolled in a statistics class. In statistical terminology, the whole set of numbers that represents the scores of students is called a **data set**, the name of each student is called an **element**, and the score of each student is called an **observation**. (These terms are defined in more detail in Section 1.4.)

A data set in its original form is usually very large. Consequently, such a data set is not very helpful in drawing conclusions or making decisions. It is easier to draw conclusions from summary tables and diagrams than from the original version of a data set. So, we reduce data to a manageable size by constructing tables, drawing graphs, or calculating summary measures such as averages. The portion of statistics that helps us to do this type of statistical analysis is called **descriptive statistics**.

> **DESCRIPTIVE STATISTICS**
>
> *Descriptive statistics* consists of methods for organizing, displaying, and describing data by using tables, graphs, and summary measures.

Case Study 1–1 presents an example of descriptive statistics. It shows a chart and a table that are constructed to organize a data set.

Both Chapters 2 and 3 discuss descriptive statistical methods. In Chapter 2, we learn how to construct tables like the one presented in Case Study 1–1 and how to graph data. In Chapter 3, we learn to calculate numerical summary measures such as averages.

CASE STUDY 1-1 UNSCHEDULED EMPLOYEE ABSENCES

USA SNAPSHOTS ®

A look at statistics that shape your finances

I won't be in today

Workplace absenteeism is up 14% since 1992 and costs U.S. companies an average $668 per employee annually.

Reasons most often given for unscheduled absences:

IMPORTANT MESSAGE
Mr. Smith will be out sick

Personal illness **45%**

Family issues **27%**

Personal needs **13%**

Feel entitled **9%**

Stress **6%**

Source: Survey of midsize and large companies by CCH, provider of human resources information and software

By Cindy Hall and Julie Stacey, USA TODAY

Using the information given in the above chart, we can write the table given below. This table lists the percentage of employees who gave different reasons for unscheduled absences from work. For example, 45% of the employees said they were absent from work due to personal illness, 27% because of family issues, and so forth.

Reason for Absence	Percent of Employees
Personal illness	45
Family issues	27
Personal needs	13
Feel entitled	9
Stress	6

Source: *USA Today*, November 3, 1995. Copyright © 1995, *USA Today*. Chart reproduced with permission.

1.2.2 INFERENTIAL STATISTICS

In statistics, the collection of all elements of interest is called a **population**. The selection of a few elements from this population is called a **sample**. (Population and sample are discussed in more detail in Section 1.3.)

A major portion of statistics deals with making decisions, inferences, predictions, and forecasts about populations based on results obtained from samples. For example, we may make some decisions about the political views of all college and university students based on the political views of 1000 students selected from a few colleges and universities. As another example, we may want to find the starting salary of a typical college graduate. To do so, we may select 2000 recent college graduates, find their starting salaries, and make a decision based on this information. The area of statistics that deals with such decision-making procedures is referred to as **inferential statistics**. This branch of statistics is also called *inductive reasoning* or *inductive statistics*.

> **INFERENTIAL STATISTICS**
>
> *Inferential statistics* consists of methods that use sample results to help make decisions or predictions about a population.

Chapters 8 through 13 and parts of Chapter 7 deal with inferential statistics.

Probability, which gives a measurement of the likelihood that a certain outcome will occur, acts as a link between descriptive and inferential statistics. Probability is used to make statements about the occurrence or nonoccurrence of a certain event under uncertain conditions. Probability and probability distributions are discussed in Chapters 4 through 6 and parts of Chapter 7.

EXERCISES

Concepts and Procedures

1.1 Briefly describe the two meanings of the word *statistics*.

1.2 Briefly explain the types of statistics.

1.3 POPULATION VERSUS SAMPLE

We will encounter the terms *population* and *sample* on almost every page of this text.[2] Consequently, understanding the meaning of each of these two terms and the difference between them is crucial.

Suppose a statistician is interested in knowing

1. The percentage of all voters who will vote for a particular candidate in an election
2. The 1996 gross sales of all companies in New York City
3. The prices of all statistics books published in the United States during the past five years

In these examples, the statistician is interested in *all* voters, *all* companies, and *all* statistics books. Each of these groups is called the population for the respective example. In statistics, a population does not necessarily mean a collection of people. It can, in fact, be a collection of people or of any kind of item such as books, television sets, or cars. The population of interest is usually called the **target population**.

---

[2]To learn more about sampling and sampling techniques, refer to Appendix A.

> **POPULATION OR TARGET POPULATION**
>
> A *population* consists of all elements—individuals, items, or objects—whose characteristics are being studied. The population that is being studied is also called the *target population*.

Most of the time, decisions are made based on portions of populations. For example, the various election polls conducted in the United States to estimate the percentage of voters favoring various candidates in any presidential election are based on only a few hundred or a few thousand voters selected from across the country. In this case, the population consists of all registered voters in the United States. The sample is made up of the few hundred or few thousand voters who are included in an opinion poll. Thus, the collection of a few elements selected from a population is called a **sample**.

> **SAMPLE**
>
> A portion of the population selected for study is referred to as a *sample*.

Figure 1.1 illustrates the selection of a sample from a population.

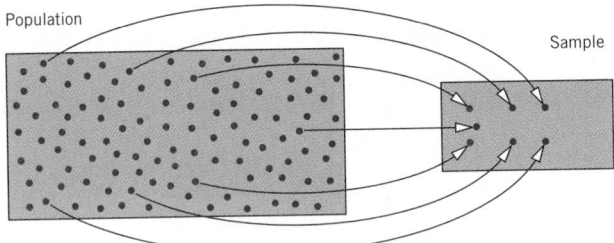

Figure 1.1 Population and sample.

The collection of information from the elements of a population or a sample is called a **survey**. A survey that includes every element of the target population is called a **census**. Often the size of the target population is large. Hence, in practice, a census is rarely taken because it is very expensive and time consuming. In many cases it is even impossible to identify each element of the target population. Usually, t oc onducta s urvey, w es electa sample and collect the required information from the elements included in that sample. We then make decisions based on this sample information. Such a survey conducted on a sample is called a **sample survey**. As an example, if we collect information on the 1996 incomes of all families in Connecticut, it will be referred to as a census. On the other hand, if we collect information on the 1996 incomes of 50 families from Connecticut, it will be called a sample survey.

> **CENSUS AND SAMPLE SURVEY**
>
> A survey that includes every member of the population is called a *census*. The technique of collecting information from a portion of the population is called a *sample survey*.

Case Study 1–2 presents an example of a sample survey.

CASE STUDY 1-2 CAUGHT DIRTY HANDED: MANY FAIL TO WASH WHEN THEY SHOULD

Researchers lurking in public restrooms have discovered a nasty truth: Americans don't always wash up after using the toilet.

A national survey . . . found that while 94% of those surveyed by telephone say they always wash their hands after going to the bathroom, direct observation of 6,333 people in five major cities found only 68% really do.

The dirt on dirty hands:

Of the five cities, New Yorkers are the worst offenders. Only 60% of the 2,129 people watched in Penn Station restrooms washed up.

Chicagoans have the cleanest hands. Of 1,251 observed at Navy Pier, 78% washed their hands after using the bathroom.

In between are people at a casino in New Orleans, where 71% of 1,251 men and women observed washed up; at Golden Gate Park, San Francisco (69% of 888 washed); and at a Braves baseball game in Atlanta (64% of 814 washed).

Overall, women washed their hands (74%) more often than men (61%).

The survey, conducted in August by Wirthlin Worldwide, . . . included phone interviews with 1,004 adults. Respondents say they're most likely to wash their hands after changing a diaper (78%) and before handling or eating food (81%). But washing up after other activities is less frequent: petting an animal (48%), coughing or sneezing (33%), or handling money (22%).

Source: *USA Today*, September 17, 1996. Copyright © 1996, *USA Today*. Excerpted and reproduced with permission.

The purpose of conducting a sample survey is to make decisions about the corresponding population. It is important that the results obtained from a sample survey closely match the results that we would obtain by conducting a census. Otherwise, any decision based on a sample survey will not apply to the corresponding population. As an example, to find the average income of families living in New York City by conducting a sample survey, the sample must contain families who belong to different income groups in almost the same proportion as they exist in the population. Such a sample is called a **representative sample**. Inferences derived from a representative sample will be more reliable.

REPRESENTATIVE SAMPLE

A sample that represents the characteristics of the population as closely as possible is called a *representative sample*.

Case Study 1–3 shows how different studies can produce conflicting results. Whether or not the results of a study are valid depends on how the sample is selected, whether or not the sample is representative of the population, how the study is conducted, and many other factors.

CASE STUDY 1-3   DIFFERENT STUDIES, CONFLICTING RESULTS

The following are the excerpts from the *CBS Evening News* of August 14, 1995.

DAN RATHER, anchor: . . . Americans, in general, are having a hard time figuring out what to do and what to eat to stay healthy. That's because new health studies have a way of turning today's best medical advice into tomorrow's medical no-no. What's going on here? CBS News health correspondent Dr. Bob Arnot has a guide for the perplexed in tonight's Eye on America.

MR. HARRY SMITH (Co-host, *CBS This Morning*): (From October 1992) So you gave up butter because of fat and cholesterol and switched to margarine because it was better for you. Well, wrong!

UNIDENTIFIED MAN #1: So am I supposed to use neither?

DR. BOB ARNOT reporting: Americans are bombarded on a daily basis with conflicting medical advice. Oat bran, coffee, salt, red wine, saccharin, beta carotene—it seems the culinary heroes and villains pop into prominence, change places and fall from favor almost every week.

UNIDENTIFIED WOMAN #1: People don't know what to believe anymore.

UNIDENTIFIED MAN #2: (From September 1986 news broadcast) Lately scientists have proven what Mother knew all along: Fish oil is good for your health.

UNIDENTIFIED WOMAN #2: (From April 1995 news broadcast) But a new study indicates that might not be true after all.

ARNOT: What in the name of good health is going on? Surveys show that Americans are more mixed up than ever before about what foods are good or bad for their health. People are so overloaded and frustrated that they've tuned out, and that has some doctors alarmed.

DR. DONALD LOURIA (Specialist in Nutrition and Preventive Medicine): What we are seeing is a rejection of all advice.

ARNOT: Dr. Donald Louria specializes in nutrition and preventive medicine.

DR. LOURIA: The authors of studies want publicity. The journals in which the studies are produced want publicity. The news media wants stories and they want to make them as dramatic as possible. So these three occurrences inadvertently are conspiring to confuse the public.

ARNOT: That's exactly what happened 20 years ago in the rush to recommend that Americans switch from butter to margarine.

DR. WALTER WILLETT (Harvard University): This was a—a guess, but surprisingly little evidence was there to support that recommendation.

ARNOT: Dr. Walter Willett of Harvard University startled everyone several years ago when he found that margarine may be just as unhealthy as butter. Yet Willett himself had recommended margarine for years, even though he was uncertain of its benefit.

DR. WILLETT: If we had just, first of all, been more careful about collecting good scientific evidence in the first place and, second, about sharing the uncertainty about our pronouncements, then I think we'd have less problem with confusion in the public.

ARNOT: Like thousands of consumers, Marty Gold has gone back to butter, lots of it. Suffering from very high cholesterol, he had taken his doctor's advice to use margarine.

MR. MARTY GOLD (Consumer): My cholesterol was still hovering around 350, so I figured nobody knows what they're talking about.

ARNOT: So how is it possible that two studies on exactly the same subject could reach completely opposite conclusions? Actually, it's easy. Take for example, the case of vitamin E.

A number of studies have shown that 'Vitamin E greatly reduces the risk of heart disease.' Then last year another major study found, 'Vitamin E and other supplements do not guard against diseases.'

But let's examine it more closely. In the studies that showed a benefit, patients took large amounts of vitamin E, but in the study that showed no benefit, patients took 75 percent less vitamin E. And what's more, they were smokers. So how do you know what to believe?

Read beyond the headlines and don't jump to conclusions. Make a checklist. Are the researchers' methods explained? Are the risks spelled out? Who funded the study? And is this a large study conducted over many years that builds on the conclusions of many previous studies?

DR. WILLETT: The public would like a quick and absolute truth. The truth is, though, that except for a very few things, like cigarette smoking and keeping lean and active, most of our recommendations are subject to change.

ARNOT: But that may be a tough pill to swallow for a society hooked on the quick fix and the easy answer. There isn't one. In New York, this is Dr. Bob Arnot for Eye on America.

Source: *CBS Evening News,* August 14, 1995. Copyright © 1995, *CBS Evening News.* Reproduced with permission.

A sample may be random or nonrandom. In a **random sample**, each element of the population has some chance of being included in the sample. However, in a nonrandom sample this may not be the case.

> **RANDOM SAMPLE**
>
> A sample drawn in such a way that each element of the population has some chance of being selected is called a *random sample*. If the chance of being selected is the same for each element of the population, it is called a **simple random sample**.

One way to select a random sample is by lottery or draw. For example, if we are to select five students from a class of 50, we write each of the 50 names on a separate piece of paper. Then we place all 50 slips in a box and mix them thoroughly. Finally, we randomly draw five slips from the box. The five names drawn will give a random sample. On the other hand, if we arrange all 50 names alphabetically and then select the first five names on the list, it would be a nonrandom sample because the students listed sixth to fiftieth have no chance of being included in the sample.

A sample may be selected with or without replacement. In sampling **with replacement**, each time we select an element from the population, we put it back in the population before we select the next element. Thus, in sampling with replacement, the population contains the same number of items each time a selection is made. As a result, we may select the same item more than once in such a sample. Consider a box that contains 25 balls of different colors. Suppose we draw a ball, record its color, and put it back in the box before drawing the next ball. Every time we draw a ball from this box, the box contains 25 balls. This is an example of sampling with replacement.

Sampling **without replacement** occurs when the selected element is not replaced in the population. In this case, each time we select an item, the size of the population is reduced by one element. Thus, we cannot select the same item more than once in this type of sampling. Most of the time, samples taken in statistics are without replacement. Consider an opinion poll based on a certain number of voters selected from the population of all eligible voters. In this case, the same voter would not be selected more than once. Therefore, this is an example of sampling without replacement.

EXERCISES

Concepts and Procedures

1.3 Briefly explain the terms *population*, *sample*, *representative sample*, *random sample*, *sampling with replacement*, and *sampling without replacement*.

1.4 Give one example each of sampling with and sampling without replacement.

1.5 Briefly explain the difference between a census and a sample survey. Why is conducting a sample survey preferable to conducting a census?

Applications

1.6 Explain whether each of the following constitutes a population or a sample.

    a. Ages of all members of a family
    b. Number of days missed by all employees of a company during the past month
    c. Marital status of 50 persons selected from a large city
    d. Number of VCRs owned by all families in Chicago
    e. Weights of 100 packages

1.7 Explain whether each of the following constitutes a population or a sample.

    a. Scores of all students in a statistics class
    b. Yield of potatoes per acre for 10 pieces of land
    c. Weekly salaries of all employees of a company
    d. Cattle owned by 100 farmers in Iowa
    e. Number of computers sold during the past week at all computer stores in Chicago

1.4 BASIC TERMS

It is very important to understand the meaning of some basic terms that will be used frequently in this text. This section explains the meaning of an element (or member), a variable, an observation, and a data set. An element and a data set were briefly defined earlier in Section 1.2. This section defines these terms formally and illustrates them with the help of an example.

Table 1.1 gives information on 1995 profits (in billions of U.S. dollars) of six U.S. companies. We can call this group of companies a sample of six companies. Each company listed in this table is called an **element** or a **member** of the sample. Table 1.1 contains information on six elements. Note that elements are also called *observational units*.

Table 1.1   1995 Profits of Six U.S. Companies

Company	1995 Profits ⟵ Variable (billions of dollars)
General Motors	6.9
IBM	4.2
Intel	3.6
Merck	3.3 ⟵ An observation or measurement
Coca-Cola	3.0
Procter & Gamble	2.8

An element or a member → Merck

Source: Business Week, March 25, 1996.

ELEMENT OR MEMBER

An *element* or *member* of a sample or population is a specific subject or object (for example, a person, firm, item, state, or country) about which the information is collected.

The *1995 profits of companies* in our example is called a **variable**. The *1995 profits* is a characteristic of companies that we are investigating.

VARIABLE

A *variable* is a characteristic under study that assumes different values for different elements. In contrast to a variable, the value of a *constant* is fixed.

A few other examples of variables are the incomes of households, the number of houses built in a city per month during the past year, the makes of cars owned by people, the gross sales of companies, and the number of insurance policies sold by a salesperson per day during the past month.

In general, a variable assumes different values for different elements, as does the 1995 profits of the six companies in Table 1.1. For some elements, however, the value of the variable may be the same. For example, if we collect information on incomes of households, these households are expected to have different incomes, although some of them may have the same income.

A variable is often denoted by x, y, or z. For instance, in Table 1.1, the 1995 profits of companies may be denoted by any one of these letters. Starting with Section 1.8, we will begin to use these letters to denote variables.

Each of the values representing the 1995 profits of the six companies in Table 1.1 is called an **observation** or **measurement**.

> **OBSERVATION OR MEASUREMENT**
>
> The value of a variable for an element is called an *observation* or *measurement*.

According to Table 1.1, the 1995 profits of Coca-Cola were $3 billion. The value $3 billion is an observation or measurement. Table 1.1 contains six observations, one for each of the six companies.

The information given in Table 1.1 on 1995 profits of companies is called the **data** or a **data set**.

> **DATA SET**
>
> A *data set* is a collection of observations on one or more variables.

Another example of a data set would be a list of prices of 25 recently sold homes.

EXERCISES

Concepts and Procedures

1.8   Explain the meaning of an element, a variable, an observation, and a data set.

Applications

1.9   The following table gives the scores of five students on a statistics test.

Student	Score
Bill	83
Susan	91
Allison	78
Jeff	69
Neil	87

Briefly explain the meaning of a member, a variable, a measurement, and a data set with reference to this table.

1.10   The following table lists the 1995 per capita incomes for the 10 highest ranking states (*Source: U.S. Department of Commerce*).

State	Per Capita Income ($)
Connecticut	30,303
New Jersey	28,858
Massachusetts	26,994
New York	26,782
Maryland	25,927
New Hampshire	25,151
Nevada	25,013
Illinois	24,763
Hawaii	24,738
Alaska	24,182

Briefly explain the meaning of a member, a variable, an observation, and a data set with reference to this table.

1.11 Refer to the data set given in Exercise 1.9.

    a. What is the variable for this data set?
    b. How many observations does this data set contain?
    c. How many elements does this data set contain?

1.12 Refer to the data set given in Exercise 1.10.

    a. What is the variable for this data set?
    b. How many observations does this data set contain?
    c. How many elements does this data set contain?

1.5 TYPES OF VARIABLES

In Section 1.4, we learned that a variable is a characteristic under investigation that assumes different values for different elements. The incomes of families, heights of persons, gross sales of companies, prices of college textbooks, makes of cars owned by families, number of accidents, and status (freshman, sophomore, junior, or senior) of students enrolled at a university are a few examples of variables.

A variable may be classified as quantitative or qualitative. These two types of variables are explained next.

1.5.1 QUANTITATIVE VARIABLES

Some variables can be measured numerically whereas others cannot. A variable that can assume numerical values is called a **quantitative variable**.

> **QUANTITATIVE VARIABLE**
>
> A variable that can be measured numerically is called a *quantitative variable*. The data collected on a quantitative variable are called *quantitative data*.

Incomes, heights, gross sales, prices of homes, number of cars owned, and accidents are examples of quantitative variables since each of them can be expressed numerically. For instance, the income of a family may be $41,520.75 per year, the gross sales for a company may be $567 million for the past year, and so forth. Such quantitative variables can be classified as either *discrete variables* or *continuous variables*.

Discrete Variables

The values that a certain quantitative variable can assume may be countable or not. For example, we can count the number of cars owned by a family but we cannot count the income of a family. The variable whose values are countable is called a **discrete variable**. Note that there are no possible intermediate values between consecutive values of a discrete variable.

> **DISCRETE VARIABLE**
>
> A variable whose values are countable is called a *discrete variable*. In other words, a discrete variable can assume only certain values with no intermediate values.

For example, the number of cars sold on any day at a car dealership is a discrete variable because the number of cars sold must be 0, 1, 2, 3, ... The number of cars sold cannot be between 0 and 1, or between 1 and 2. A few other examples of discrete variables are the number of people visiting a bank on any day, the number of cars in a parking lot, the number of cattle owned by a farmer, and the number of students in a class.

Continuous Variables

A **continuous variable** can assume any value over a certain range, and we cannot count these values.

> **CONTINUOUS VARIABLE**
>
> A variable that can assume any numerical value over a certain interval or intervals is called a *continuous variable*.

The time taken to complete an examination is an example of a continuous random variable because it can assume any value, let us say, between 30 and 60 minutes. The time taken may be 42.6 minutes, 42.67 minutes, or 42.674 minutes. (Theoretically, we can measure time as precisely as we want.) Similarly, the height of a person can be measured to the tenth of an inch or to the hundredth of an inch. However, neither time nor height can be counted in a discrete fashion. A few other examples of continuous variables are weights of people, amount of soda in a 12-ounce can (note that a can will not contain exactly 12 ounces of soda), and the yield of potatoes per acre. Note that any variable that involves money is considered a continuous variable.

1.5.2   QUALITATIVE OR CATEGORICAL VARIABLES

Variables that cannot be measured numerically but can be divided into different categories are called **qualitative** or **categorical variables**.

> **QUALITATIVE OR CATEGORICAL VARIABLE**
>
> A variable that cannot assume a numerical value but can be classified into two or more nonnumeric categories is called a *qualitative* or *categorical variable*. The data collected on such a variable are called *qualitative data*.

For example, the status of an undergraduate college student is a qualitative variable because a student can fall into any one of four categories: freshman, sophomore, junior, or senior. A few other examples of qualitative variables are the gender of a person, hair color, and the make of a car. Figure 1.2 illustrates the types of variables.

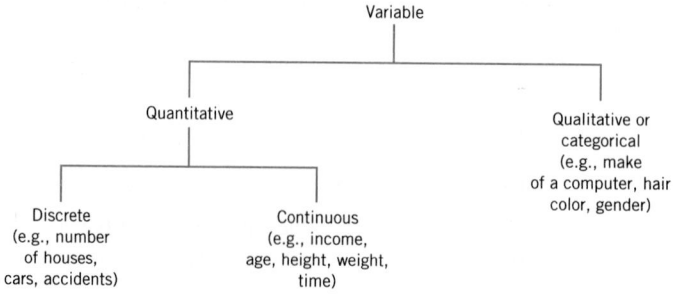

Figure 1.2   Types of variables.

EXERCISES

Concepts and Procedures

1.13 Explain the meaning of the following terms.
   a. Quantitative variable   b. Qualitative variable
   c. Discrete variable   d. Continuous variable
   e. Quantitative data   f. Qualitative data

Applications

1.14 Explain which of the following variables are quantitative and which are qualitative.
   a. Number of persons in a family   b. Color of cars
   c. Marital status of people   d. Length of a frog's jump
   e. Number of students in a class

1.15 Explain which of the following variables are quantitative and which are qualitative.
   a. Number of homes owned   b. Rent paid by tenants
   c. Types of cars owned by families   d. Monthly phone bills
   e. Color of eyes

1.16 Classify the quantitative variables in Exercise 1.14 as discrete or continuous.

1.17 Classify the quantitative variables in Exercise 1.15 as discrete or continuous.

1.6 CROSS-SECTION VERSUS TIME-SERIES DATA

Based on the time over which they are collected, data can be classified as either cross-section or time-series data.

1.6.1 CROSS-SECTION DATA

Cross-section data contain information on different elements of a population or sample for the *same* period of time. The information on incomes of 100 families for the year 1996 is an example of cross-section data. All examples of data already presented in this chapter have been those of cross-section data.

> **CROSS-SECTION DATA**
>
> Data collected on different elements at the same point in time or for the same period of time are called *cross-section data*.

Table 1.2 gives the total population (in millions) of the six New England states for the same period, 1995. This is an example of cross-section data.

Table 1.2 Population of Six States

State	Total population (in millions)
Connecticut	3.28
Maine	1.24
Massachusetts	6.07
New Hampshire	1.15
Rhode Island	.99
Vermont	.59

1.6.2   TIME-SERIES DATA

Time-series data contain information on the same element for *different* periods of time. Information on U.S. exports for the years 1980 to 1996 is an example of time-series data.

TIME-SERIES DATA

Data collected on the same element for the same variable at different points in time or for different periods of time are called *time-series data*.

The data given in Table 1.3 are an example of time-series data. This table gives the number of new life insurance policies purchased in the United States between 1989 and 1994.

Table 1.3   New Life Insurance Policies Purchased Between 1989 and 1994

Year	Number of Life Insurance Policies Purchased
1989	29,960
1990	28,791
1991	29,813
1992	28,382
1993	31,238
1994	31,955

Source: American Council of Life Insurance.

1.7   SOURCES OF DATA

The availability of accurate and appropriate data is essential for deriving reliable results.[3] Data may be obtained from internal sources, external sources, or surveys and experiments.

Many times data come from *internal sources*, such as a company's own personnel files or accounting records. For example, a company that wants to forecast the future sales of its product may use the data of past periods from its own records. However, for most studies, all the data that are needed are not usually available from internal sources. In such cases, one may have to depend on outside sources to obtain data. These sources are called *external sources*. For instance, the *Statistical Abstract of the United States* (published annually), which contains various kinds of data on the United States, is an external source of data.

A large number of government and private publications can be used as external sources of data. The following is a list of some of the government publications.[4]

1. *Statistical Abstract of the United States*
2. *Employment and Earnings*
3. *Handbook of Labor Statistics*
4. *Source Book of Criminal Justice Statistics*
5. *Economic Report of the President*
6. *County & City Data Book*
7. *State & Metropolitan Area Data Book*

---

[3]Sources of data are discussed in more detail in Appendix A.
[4]All of these books, and the catalog of U.S. government publications, can be obtained from the Superintendent of Documents, U.S. Government Printing Office, Washington, D.C. 20402.

8. *Digest of Education Statistics*
9. *Health United States*
10. *Agricultural Statistics*

Besides these government publications, a large number of private publications (e.g., *Standard and Poors' Security Owner's Stock Guide* and *World Almanac and Book of Facts*) and periodicals (e.g., *The Wall Street Journal, USA Today, Fortune, Forbes,* and *Business Week*) can be used as external data sources.

Sometimes the needed data may not be available from either internal or external sources. In such cases, the investigator may have to conduct a survey or experiment to obtain the required data.

EXERCISES

Concepts and Procedures

1.18 Explain the difference between cross-section and time-series data. Give an example of each of these two types of data.

1.19 Briefly describe the internal and external sources of data.

Applications

1.20 Classify the following as cross-section or time-series data.
- a. IBM's gross sales for the period 1985 to 1996
- b. Weights of 20 chickens
- c. Poverty rates in the United States for 1980 to 1996
- d. Auto insurance premiums paid by 100 persons

1.21 Classify the following as cross-section or time-series data.
- a. Average prices of houses in 100 cities
- b. Salaries of 50 employees
- c. Students enrolled at Eastern Connecticut State University during the years 1980 to 1996
- d. Number of employees employed by a company each year from 1985 to 1996

1.8 SUMMATION NOTATION

Sometimes mathematical notation helps to express a mathematical relationship concisely. This section describes the **summation notation** that is used to denote the sum of values.

Suppose a sample consists of five books and the prices of these five books are $25, $60, $37, $53, and $16. The variable *price of a book* can be denoted by x. The prices of the five books can be written as follows.

$$\text{Price of the first book} = x_1 = \$25$$

$\uparrow$
Subscript of x denotes
the number of the book

Similarly,

$$\text{Price of the second book} = x_2 = \$60$$

$$\text{Price of the third book} = x_3 = \$37$$

$$\text{Price of the fourth book} = x_4 = \$53$$

$$\text{Price of the fifth book} = x_5 = \$16$$

In the above notation, x represents the price and the subscript denotes a particular book.

Now, suppose we want to add the prices of all five books. Then,

$$x_1 + x_2 + x_3 + x_4 + x_5 = 25 + 60 + 37 + 53 + 16 = \$191$$

The uppercase Greek letter Σ (pronounced *sigma*) is used to denote the sum of all values. Using Σ notation, we can write the foregoing sum as follows.

$$\Sigma x = x_1 + x_2 + x_3 + x_4 + x_5 = \$191$$

The notation Σx in this expression represents the sum of all the values of x.

Using summation notation: one variable.

EXAMPLE 1–1 Suppose the ages of four persons are 35, 47, 28, and 60 years. Find
(a) Σx (b) $(\Sigma x)^2$ (c) Σx^2

Solution Let x_1, x_2, x_3, and x_4 be the ages (in years) of the first, second, third, and fourth person, respectively. Then,

$$x_1 = 35, \quad x_2 = 47, \quad x_3 = 28, \quad \text{and} \quad x_4 = 60$$

(a) $\Sigma x = x_1 + x_2 + x_3 + x_4 = 35 + 47 + 28 + 60 = \mathbf{170}$
(b) Note that $(\Sigma x)^2$ is the square of the sum of all x values. Thus,

$$(\Sigma x)^2 = (170)^2 = \mathbf{28{,}900}$$

(c) The expression Σx^2 is the sum of the squares of x values. To calculate Σx^2, we first square each of the x values and then add these squared values. Thus,

$$\Sigma x^2 = (35)^2 + (47)^2 + (28)^2 + (60)^2 = 1225 + 2209 + 784 + 3600 = \mathbf{7818}$$

$\blacksquare$

Using summation notation: two variables.

EXAMPLE 1–2 The following table lists four pairs of m and f values:

m	12	15	20	30
f	5	9	10	16

Compute the following:

(a) Σm (b) Σf^2 (c) Σmf (d) $\Sigma m^2 f$

Solution We can write

$$m_1 = 12, \quad m_2 = 15, \quad m_3 = 20, \quad m_4 = 30,$$
$$f_1 = 5, \quad f_2 = 9, \quad f_3 = 10, \quad f_4 = 16$$

(a) $\Sigma m = 12 + 15 + 20 + 30 = \mathbf{77}$
(b) $\Sigma f^2 = (5)^2 + (9)^2 + (10)^2 + (16)^2 = 25 + 81 + 100 + 256 = \mathbf{462}$
(c) To compute Σmf, we multiply the corresponding values of m and f and add the products as follows:

$$\Sigma mf = m_1 f_1 + m_2 f_2 + m_3 f_3 + m_4 f_4$$
$$= 12(5) + 15(9) + 20(10) + 30(16) = \mathbf{875}$$

(d) To calculate $\Sigma m^2 f$, we square each m value, then multiply the corresponding m^2 and f values and add the products.

$$\Sigma m^2 f = (m_1)^2 f_1 + (m_2)^2 f_2 + (m_3)^2 f_3 + (m_4)^2 f_4$$
$$= (12)^2(5) + (15)^2(9) + (20)^2(10) + (30)^2(16) = \mathbf{21{,}145}$$

The calculations done in parts (a) through (d) to find the values of Σm, Σf^2, Σmf, and $\Sigma m^2 f$ can be performed in tabular form, as shown in Table 1.4.

Table 1.4

m	f	f^2	mf	$m^2 f$
12	5	$5 \times 5 = 25$	$12 \times 5 = 60$	$12 \times 12 \times 5 = 720$
15	9	$9 \times 9 = 81$	$15 \times 9 = 135$	$15 \times 15 \times 9 = 2{,}025$
20	10	$10 \times 10 = 100$	$20 \times 10 = 200$	$20 \times 20 \times 10 = 4{,}000$
30	16	$16 \times 16 = 256$	$30 \times 16 = 480$	$30 \times 30 \times 16 = 14{,}400$
$\Sigma m = 77$	$\Sigma f = 40$	$\Sigma f^2 = 462$	$\Sigma mf = 875$	$\Sigma m^2 f = 21{,}145$

The columns of Table 1.4 can be explained as follows:

1. The first column lists the values of m. The sum of these values gives $\Sigma m = 77$.
2. The second column lists the values of f. The sum of this column gives $\Sigma f = 40$.
3. The third column lists the squares of the f values. For example, the first value, 25, is the square of 5. The sum of the values in this column gives $\Sigma f^2 = 462$.
4. The fourth column records products of the corresponding m and f values. For example, the first value, 60, in this column is obtained by multiplying 12 by 5. The sum of the values in this column gives $\Sigma mf = 875$.
5. Next, the m values are squared and multiplied by the corresponding f values. The resulting products, denoted by $m^2 f$, are recorded in the fifth column of the table. For example, the first value, 720, is obtained by squaring 12 and multiplying this result by 5. The sum of the values in this column gives $\Sigma m^2 f = 21{,}145$. ■

EXERCISES

Concepts and Procedures

1.22 The following table lists five pairs of m and f values.

m	5	10	15	20	25
f	12	8	5	16	4

Compute the value of each of the following:

      a. Σm b. Σf^2 c. Σmf d. $\Sigma m^2 f$

1.23 The following table lists six pairs of m and f values.

m	3	6	9	12	15	18
f	16	11	6	8	4	14

Calculate the value of each of the following:

      a. Σf b. Σm^2 c. Σmf d. $\Sigma m^2 f$

1.24   The following table lists five pairs of x and y values.

x	15	20	11	8	5
y	10	7	14	9	18

Compute   a. Σx b. Σy c. Σxy d. Σx^2 e. Σy^2

1.25   The following table lists six pairs of x and y values.

x	4	18	25	9	12	20
y	12	5	14	7	12	8

Compute   a. Σx b. Σy c. Σxy d. Σx^2 e. Σy^2

Applications

1.26   The scores of five students in a statistics class are 75, 80, 97, 91, and 63. Find
   a. Σx b. $(\Sigma x)^2$ c. Σx^2

1.27   The number of cars owned by six families are 3, 2, 1, 4, 1, and 2. Find
   a. Σx b. $(\Sigma x)^2$ c. Σx^2

1.28   The electric heating bills for November 1996 for four families were \$122, 72, 96, and 110. Find
   a. Σx b. $(\Sigma x)^2$ c. Σx^2

1.29   The weights of seven newborn babies are 7, 9, 6, 12, 10, 9, and 8 pounds. Find
   a. Σx b. $(\Sigma x)^2$ c. Σx^2

GLOSSARY

Census   A survey that includes all members of the population.

Continuous variable   A (quantitative) variable that can assume any numerical value over a certain interval or intervals.

Cross-section data   Data collected on different elements at the same point in time or for the same period of time.

Data or **data set**   Collection of observations or measurements on a variable.

Descriptive statistics   Collection of methods that are used for organizing, displaying, and describing data using tables, graphs, and summary measures.

Discrete variable   A (quantitative) variable whose values are countable.

Element or **member**   A specific subject or object included in a sample or population.

Inferential statistics   Collection of methods that help make decisions about a population based on sample results.

Observation or **measurement**   The value of a variable for an element.

Population or **target population**   The collection of all elements whose characteristics are being studied.

Qualitative or **categorical data**   Data generated by a qualitative variable.

Qualitative or **categorical variable**   A variable that cannot assume numerical values but is classified into two or more categories.

Quantitative data   Data generated by a quantitative variable.

Quantitative variable   A variable that can be measured numerically.

Random sample A sample drawn in such a way that each element of the population has some chance of being included in the sample.

Representative sample A sample that contains the characteristics of the corresponding population.

Sample A portion of the population of interest.

Sample survey A survey that includes elements of a sample.

Simple random sample A sample drawn in such a way that each element of the population has the same chance of being selected for the sample.

Statistics Collection of methods that are used to collect, analyze, present, and interpret data and to make decisions.

Survey Collecting data on the elements of a population or sample.

Time-series data Data that give the values of the same variable for the same element at different points in time or for different periods of time.

Variable A characteristic under study or investigation that assumes different values for different elements.

SUPPLEMENTARY EXERCISES

1.30 The following table gives the average hourly earnings (in current dollars) of production or non-supervisory workers on private nonagricultural payrolls in the United States for nine months, from October 1995 through June 1996.

Month and Year	Average Hourly Earnings ($)
October 1995	11.55
November 1995	11.59
December 1995	11.61
January 1996	11.62
February 1996	11.65
March 1996	11.68
April 1996	11.72
May 1996	11.73
June 1996	11.83

Source: U.S. Bureau of Labor Statistics.

Describe the meaning of a variable, a measurement, and a data set with reference to this table.

1.31 The following table gives the total number of marriages during March 1996 for the five Pacific states (*Monthly Vital Statistics Report*, September 26, 1996).

States	Number of Marriages
Alaska	309
California	19,409
Hawaii	1591
Oregon	1716
Washington	4030

Describe the meaning of an element, a variable, a measurement, and a data set with reference to this table.

1.32 Refer to Exercises 1.30 and 1.31. Classify these data sets as either cross-section or time-series.

1.33 Indicate whether each of the following examples refers to a population or to a sample.

a. A group of 25 patients selected to test a new drug
b. Total items produced on a machine during one week
c. Yearly expenditure on clothes for 50 persons
d. Number of houses sold by each of the 10 employees of a real estate agency

1.34 Indicate whether each of the following examples refers to a population or to a sample.

a. Ages of CEOs of all companies in New York City
b. Prices of 100 houses selected from Chicago
c. Number of subscribers to five magazines
d. Salaries of all employees of a bank

1.35 State which of the following is an example of sampling with replacement and which is an example of sampling without replacement.

a. Selecting 10 patients out of 100 to test a new drug
b. Selecting one professor to be a member of the university senate and then selecting one professor from the same group to be a member of the curriculum committee

1.36 State which of the following is an example of sampling with replacement and which is an example of sampling without replacement.

a. Selecting five students to form a committee
b. Selecting one student for a leadership award and then selecting one student from the same group for a scholarship award

1.37 The number of ties owned by six persons are 10, 9, 14, 12, 7, and 4. Find

a. Σx b. $(\Sigma x)^2$ c. Σx^2

1.38 The number of hours worked during the past week by five employees of a company are 43, 39, 44, 31, and 40. Find

a. Σx b. $(\Sigma x)^2$ c. Σx^2

1.39 The following table lists five pairs of m and f values.

m	3	16	11	9	20
f	7	32	17	12	34

Compute the value of each of the following.

a. Σm b. Σf^2 c. Σmf d. $\Sigma m^2 f$ e. Σm^2

1.40 The following table lists six pairs of x and y values.

x	7	11	8	4	14	18
y	5	15	7	10	9	12

Compute the value of each of the following.

a. Σy b. Σx^2 c. Σxy d. $\Sigma x^2 y$ e. Σy^2

SELF-REVIEW TEST

1. A population in statistics means

a. a collection of all men and women
b. a collection of all subjects or objects of interest
c. a collection of all people living in a country

2. A sample in statistics means
    a. a portion of the people selected from the population of a country
    b. a portion of the people selected from the population of an area
    c. a portion of the population of interest

3. Indicate which of the following is an example of a sample with replacement or a sample without replacement.
    a. Ten students are selected from a statistics class in such a way that as soon as a student is selected his or her name is deleted from the list before the next student is selected.
    b. A box contains five balls of different colors. A ball is drawn from this box, its color is recorded, and it is put back in the box before the next ball is drawn. This experiment is repeated 12 times.

4. Indicate which of the following variables are quantitative and which are qualitative. Classify the quantitative variables as discrete or continuous.
    a. Brand of coffee   b. Number of TV sets owned by families
    c. Weekly earnings of employees

5. The following table gives the starting salaries of five recent college graduates.

Name	Salary
Matt	$29,200
Lucia	42,450
Alison	27,920
Warren	32,350
Lori	38,100

Explain the meaning of a member, a variable, a measurement, and a data set with reference to this table.

6. The values (in thousands of dollars) of cars owned by six persons are 13, 9, 3, 28, 7, and 16. Calculate
    a. Σx b. Σx^2 c. $(\Sigma x)^2$

7. The following table lists five pairs of m and f values.

m	3	6	9	12	15
f	15	25	40	20	12

Calculate   a. Σm b. Σf c. Σm^2 d. Σmf e. $\Sigma m^2 f$ f. Σf^2

USING MINITAB
AN INTRODUCTION

In recent years the use of computers has significantly increased in almost every aspect of life. Such usage of computers has reduced the computation time for quantitative analysis to a negligible amount.

In the real world, when doing a statistical analysis we usually deal with hundreds or thousands of observations. For this reason, it is either very time consuming or almost impossible to make all the required calculations manually. The use of computers is of invaluable assistance in such situations. Consequently, learning to use a statistical software package has become an important part of learning statistics.

A large number of statistical software packages, both for mainframe computers and for microcomputers, have been developed in recent years. Most of these software packages are user friendly.

Four of the major statistical software packages are BMDP, MINITAB, SAS, and SPSS.[5] All four of these packages are available for mainframe computers and for microcomputers. Besides these four packages, a large number of software packages have been developed for personal computers. The *Using MINITAB* sections, given at the end of most of the chapters of this text, provide brief instructions on how to use MINITAB to solve statistical problems. Three formal manuals that explain MINITAB commands in detail are

1. *MINITAB User's Guide*, Release 11 for Windows, Windows 95, and Windows NT. MINITAB Inc., June 1996.

2. *MINITAB Reference Manual*, Release 11 for Windows, Windows 95, and Windows NT. MINITAB Inc., June 1996.

3. *MINITAB Reference Manual*, Release 11 for Macintosh. MINITAB Inc., June 1996.

MINITAB is available for both **microcomputers** and **mainframe computers**. The procedures to start a system and to access MINITAB are different for the two systems. If you are using a mainframe computer system, the first step is to *log on* to the system using an account number and a password. Your instructor or an assistant in the computer lab can explain how to use your school's mainframe computer system. Also, remember that you must *log off* the system at the end of each session. You must also ask either your instructor or an assistant at the computer lab how to obtain a *hard (printed) copy* of the MINITAB *output*. After you log on to the mainframe system, the next step is to enter the MINITAB *environment*. Again, your instructor or an assistant at the computer lab can show you how to enter the MINITAB environment.

If you are working on a microcomputer, double-click the blue MINITAB icon to load MINITAB and to enter the MINITAB environment. If you cannot locate the blue MINITAB icon, seek help from your instructor or the computer lab assistant.

If you are using a microcomputer, you have the option of using the **MINITAB FOR WINDOWS pull-down menus**, the **MINITAB COMMAND LANGUAGE**, or both together. However, if you are using the mainframe computer system, you will have to use the MINITAB COMMAND LANGUAGE. In the MINITAB sections of this text, MINITAB FOR WINDOWS refers to using the pull-down menus

---

[5]BMDP is a registered trademark of BMDP Statistical Software, Inc. MINITAB is a registered trademark of Minitab, Inc. SAS is a registered trademark of SAS Institute, Inc. SPSS is a registered trademark of SPSS, Inc.

that you see at the top of the screen. The MINITAB COMMAND LANGUAGE refers to typing the MINITAB commands on the screen instead of using the pull-down menus.

The MINITAB sections in this text that illustrate the use of MINITAB FOR WINDOWS and MINITAB COMMAND LANGUAGE for statistical analysis are based on **MINITAB Release 11 for Windows, Windows 95,** and **Windows NT**. Most of the commands described under MINITAB COMMAND LANGUAGE will work on mainframe computer systems also. Most of the MINITAB applications described in the text will also work on Macintosh computers.

In the following sections we will first describe MINITAB FOR WINDOWS and then explain the MINITAB COMMAND LANGUAGE.

MINITAB FOR WINDOWS

Assuming you know how to use Windows and that you are in the Windows environment, double-click the blue MINITAB icon to load MINITAB and to enter the MINITAB environment. Once you enter the MINITAB environment, you will see the heading ''MINITAB–Untitled Worksheet'' at the top of the screen. Below it you will see the following row of **pull-down menus**.

File   Edit   Manip   Calc   Stat   Graph   Editor   Window   Help

By clicking any of these menus, you can obtain the options included in that menu. For example, if you click the **File** menu, you will see the following options appear on the screen. You can select any of these options by clicking it.

File	
New Worksheet. . .	**Ctrl + N**
Open Worksheet. . .	**Ctrl + O**
Merge Worksheet. . .	
Query Database [ODBC]. . .	
Save Worksheet	**Ctrl + S**
Save Worksheet As. . .	
Worksheet Description. . .	
Open Graph. . .	
Save Window As. . .	
Other Files. . .	▶
Print Window. . .	**Ctrl + P**
Print Setup. . .	
Restart Minitab	
Exit	
1.	
2.	
3.	
4.	

The computer screen below the row of pull-down menus is divided into two parts—**Session** and **Data** windows. The Session window is used to type the MINITAB commands if you are using the MINITAB COMMAND LANGUAGE. Also, the MINITAB output will appear in this part of the screen. You can move the Session window up and down by clicking the bar to the right of it.

Entering Data

You will usually enter data in the Data window unless you are using the MINITAB COMMAND LANGUAGE. The Data window contains columns that are denoted by C1, C2, . . . , and so on. In MINITAB, data are always entered in columns. For example, if you have data on heights of people, you will enter these data in column C1 or any other column. Similarly, if you have information on weights of people, you can enter these data in column C2. You can use any column to enter data on a variable. You can enter data on a qualitative variable (also called an alpha-variable), such as names, in a column in the Data window. You enter only one name or data value in a cell of a column. In the cell immediately below C1 or C2 (etc.) you can enter the name of the variable such as Height or Weight (etc). Note that this cell is not numbered. The first piece of information on a variable will be entered in the cell of a column in the row labeled 1, the second value on the same variable will be entered in the cell of the same column in the row labeled 2, and so on. Each time you enter a data value, either hit the Enter/Return key or press the downward arrow ($\downarrow$) key to move down to the next cell. You can use the various arrow keys ($\downarrow$, $\uparrow$, $\rightarrow$, $\leftarrow$) to move down, up, right, or left in the Data window. You can move the cursor to any cell by clicking that cell. You can move up or down in the Data window by clicking the bar to the right of it.

To enter new data into the MINITAB worksheet, click the cell in the row labeled 1 and the column in which you want to enter data and then start entering data. Once you have finished entering data, you can use the pull-down menus to do any statistical analysis. Remember, *do not use commas while entering data*. For example, to enter 45,763, you must type **45763** and not **45,763**. Illustration M1–1 shows how to enter data in the MINITAB Data window.

ILLUSTRATION M1–1   Suppose you want to enter the following data on the names and scores of eight students into the MINITAB Data window in columns C1 and C2.

Name	Score
Susan	91
Mark	85
Robert	62
Peter	96
Allison	81
Deborah	73
Scott	90
Maureen	88

To enter these data in columns C1 and C2 of the Data window, first click the unnumbered cell below C1 and type **Name** in it. Then move the cursor to the cell below it either by hitting the Enter/Return key or by pressing the downward arrow ($\downarrow$) key. Next, type the first name, Susan. Continue this process until all names have been entered. Then click in the cell below C2 and type **Score**. Note that this is the unnumbered cell in column C2. Next, move the cursor to the next cell down and type the first score, which is 91. Continue this process until all scores have been entered. After you have entered all eight names and scores, the Data window will look as follows.

	C1-T	C2	C3	C4	C5	C6	C7	
	Name	Score						
1	Susan	91						
2	Mark	85						
3	Robert	62						
4	Peter	96						
5	Allison	81						
6	Deborah	73						
7	Scott	90						
8	Maureen	88						
9								

If you make any mistakes entering information, move the cursor to the desired cell by clicking that cell and then enter the correct information. The old information will be replaced by the new information.

Saving a MINITAB Data File

After you have entered data into the columns of the Data window, you can save them as a MINITAB data file for future use. Suppose you want to save the file containing the data on the names and scores of the eight students given in Illustration M1–1 as the SCORES file. To do so, perform the following steps.

Step 1.　Click the **File** pull-down menu at the top of the screen.

Step 2.　Click **Save Worksheet As** from the selections in the File menu.

Step 3.　You will see a dialog box entitled **Save Worksheet As** appear on the screen. Type the file name in the box below **File Name**. Note that this file name usually should have the extension ".MTW," which indicates that it is a MINITAB worksheet file. For example, for the data entered in Illustration M1–1, type **SCORES.MTW** in this box. Next, choose the appropriate drive from the options under **Drives**. For example, you can save this file on drive A, drive C, or on any other drive you may have the option to save to. To see these options, click the arrow next to the box below **Drives**. Then click the appropriate drive. Next, if you have more than one directory, select the directory you want to save the file in under **Directories** and click **OK**. Make sure the box below **Save File As Type** shows the word *MINITAB* indicating that it is a MINITAB data file.

Step 4.　After you have made all the selections in the dialog box of Step 3, click the **OK** button. The file will be saved as a MINITAB worksheet. The file containing the data of Illustration M1–1 will be saved as SCORES.MTW.

Retrieving a Previously Saved MINITAB Data File

You can retrieve a previously saved MINITAB data file by performing the following steps.

Step 1.　Click the **File** pull-down menu at the top of the screen.

Step 2.　Click **Open Worksheet** from the selections available in the **File** menu.

Step 3.　You will see a dialog box entitled **Open Worksheet** appear on the screen. Select the drive that contains the file you want to open from the options under **Drives**. If you select drive A, you will see *a:* appear in the box under **Directories**, and all MINITAB files on the diskette in drive A will appear in the box below **File Name**. Highlight the file you want to retrieve. If you select drive C, you will see *c:* appear in the box under **Directories**, and all directories in drive C will appear in that box. First highlight *c:* and click **OK**. Then highlight the directory that contains the MINITAB file to be opened and click **OK**. All files in this directory will appear in the box below **File Name**. Click the file that you want to retrieve. For example, if you want to retrieve the SCORES.MTW file, click the SCORES.MTW file that you see in this box.

Step 4.　After you have made all the selections in the dialog box of Step 3, click the **OK** button. The file SCORES.MTW will be opened, and you will see the data entered earlier in columns C1 and C2.

Saving a MINITAB Output File

A MINITAB output file contains answers to all data analysis procedures you perform. You can save this output file for future use as a *text* file. Suppose you performed some statistical analysis on the data of Illustration M1–1 and you want to save this output file as SCORES.TXT. To do so, perform the following steps.

Step 1.　Click the **File** pull-down menu at the top of the screen.

Step 2.　Click **Save Window As** from the selections available in the **File** menu.

Step 3.　You will see a dialog box entitled **Save As** appear on the screen. Type the file name in the box below **File Name**. Note that this file name usually should have the extension ".TXT," which indicates that it is a MINITAB *output* file. For example, to save the output file for the data of Illustration M1–1, type **SCORES.TXT** in this box. Next, choose the appropriate drive that you want to save this file to, from the options under **Drives**. For example, you can save this file on drive A, drive C, or on any other drive you may have. Next, under **Directories**, select the directory you want to save the file in.

Step 4.　After you have made all the selections in the dialog box of Step 3, click the **OK** button. The file will be saved as a text file under the name SCORES.TXT.

　　You can open a text (output) file in any word processor, such as Microsoft Word, WordPerfect, and so on, and also either insert it in another document or print it from the word processor.

If you constructed a graph using MINITAB, you can save that graph using the four steps just described. The only difference is that the file name should have an extension ''.MGF.''

You can **copy a graph** from the MINITAB window and paste it in any other document in a word processor. When you make a graph for a data set, the graph appears on the screen in what we can call the *Graph window*. To copy a graph, click anywhere inside the **Graph window**, then click the **Edit** pull-down menu, and finally click **Copy Graph** from the options available in the Edit pull-down menu. Then you can exit MINITAB and paste the graph in any other document, for example, a Microsoft Word or WordPerfect document.

To close this Graph window after you have either copied the graph or printed it, click the top of the left-hand corner of the Graph window. Then click the **Close** option.

You can copy any part of the MINITAB output into another file outside MINITAB. To do so, highlight the required portion of the MINITAB output, click the **Edit** pull-down menu, and click the **Copy** option. Then exit MINITAB and paste this text in any document you want.

Printing a MINITAB Output File

You can print the MINITAB output, including a graph, from within MINITAB. To do so, perform the following steps.

Step 1.   Click the **File** pull-down menu at the top of the screen.

Step 2.   Click **Print Window** from the selections available in the **File** menu.

Step 3.   You will see a dialog box entitled **Print** appear on the screen. If you want to print the whole output, which should usually be the case, check the box next to **All**. Otherwise, enter the pages that you want to print in the boxes next to **From** and **To**.

Step 4.   Click **OK**. The MINITAB output text or graph, whichever appears on the screen, will be printed.

You can also save a MINITAB output file as described earlier, open it in a word processor, and then print it from there.

Selecting a Sample Using MINITAB

MINITAB can be used to select a sample from a population. Suppose the data on names and scores of eight students entered in Illustration M1–1 belong to a population of eight students in a small class. You want to select a sample of three observations from this population and store the sample data in columns C3 and C4. Note that the population data are in columns C1 and C2. You can select this sample by performing the following steps in MINITAB for Windows.

Step 1.   Click the **Calc** pull-down menu at the top of the screen.

Step 2.   Click **Random Data** from the selections available in the **Calc** menu.

Step 3.   Click **Sample From Columns** from the selections available in the **Random Data** menu.

Step 4.   You will see a dialog box entitled **Sample From Columns** appear on the screen. Type **3** in the box next to **Sample** and **C1–C2** in the box below **Samples**. This tells MINITAB to select a sample of **three** rows (observations) from the data entered in columns C1 and C2. Next, type **C3–C4** in the box below **Store samples in**. If you want the sample to be selected with replacement, then check the box next to **Sample with replacement**. Otherwise make sure this box is not checked.

Step 5.   Click the **OK** button at the bottom of this dialog box. A sample of three observations will appear in columns C3 and C4 of the Data window.

Exiting MINITAB

Before you exit MINITAB, make sure you have saved the output and data files if you want to use these at a later date. To exit MINITAB, perform the following steps.

Step 1.   Click the **File** pull-down menu at the top of the screen.

Step 2.   Click **Exit** from the selections available in the **File** menu.

Step 3.   MINITAB will ask you a few questions. Answer *yes* or *no* to these questions as you like. After you answer these questions, the MINITAB session will end.

MINITAB COMMAND LANGUAGE

The alternative to using the MINITAB FOR WINDOWS pull-down menus is to use the MINITAB COMMAND LANGUAGE. Most of the commands described under MINITAB COMMAND LANGUAGE can also be used on mainframe computers. Here, instead of using the pull-down menus, you type the MINITAB commands in the Session window and obtain MINITAB output in response to a single command or a set of commands.

When you enter the MINITAB environment, the cursor is in the Data window, which means MINITAB is ready for you to use MINITAB FOR WINDOWS. To change to the MINITAB COMMAND LANGUAGE, perform the following steps.

Step 1. Click anywhere inside the **Session window**. The cursor will move from the Data window to the Session window.

Step 2. Click the **Editor** pull-down menu at the top of the screen.

Step 3. Click the **Enable Command Language** option from the selections available in the **Editor** menu.

As a result of these steps, your computer terminal will display the following message on the screen.

```
MTB >
```

The message ''MTB >'' is called the **MINITAB prompt**. At this point, the computer is ready to receive MINITAB commands. The MINITAB commands are entered next to the MINITAB prompt, and the Enter/Return key is pressed after typing each command. The MINITAB commands can be entered in uppercase letters, lowercase letters, or in any combination of the two.

To change from MINITAB COMMAND LANGUAGE to MINITAB FOR WINDOWS, first click the **Editor** pull-down menu, and then click the **Disable Command Language** option from the selections available in the **Editor** menu.

Note that you can use the MINITAB FOR WINDOWS pull-down menus and MINITAB COMMAND LANGUAGE at the same time after you enable the MINITAB COMMAND LANGUAGE.

Entering Data

As mentioned earlier, in MINITAB data are entered in columns that are denoted by C1, C2, . . . , and so on. Illustration M1–2 demonstrates how you can enter data by typing MINITAB commands in the MINITAB COMMAND LANGUAGE environment.

ILLUSTRATION M1–2   Refer to Illustration M1–1. Suppose you want to enter the data on the names and scores of eight students into a MINITAB worksheet in columns C1 and C2.

To enter data into MINITAB using MINITAB COMMAND LANGUAGE, you can use the **SET** or **READ** command. The **SET** command is used to enter data on one variable at a time. The **READ** command can be used to enter data on one or more variables. A MINITAB command that begins with **NOTE** is not processed by MINITAB. This command is only for the information of the user. By using the **NOTE** command, you can enter any comments for your own information. To enter data on a variable, first you will create a column and then enter data. You have to use a special set of commands to enter data on an alpha-variable such as ''names''. The set of commands presented in Figure 1.3 shows how to enter data on the names and scores of eight students.

Figure 1.3   Entering data into MINITAB using MINITAB COMMAND LANGUAGE.

```
MTB > NOTE: DATA ON NAMES AND SCORES OF EIGHT STUDENTS
MTB > SET C1;   ←——— This command instructs MINITAB to create column C1. The semicolon at
                      the end of this command instructs MINITAB that there is more information
                      to be entered. As a result of this semicolon, MINITAB will come back
                      with SUBC >, which stands for SUBCOMMAND.
```

Figure 1.3   (*Continued*)

SUBC > FORMAT(A7).   ←—— This command tells MINITAB that the variable to be entered in
                                        column C1 is an alpha-variable and the maximum length of a data
                                        value is seven characters. Note that the longest name in eight
                                        names of students has seven letters. The period at the end of this
                                        subcommand tells MINITAB that all information has been entered.
                                        Note that there is no space between FORMAT and (A7) in this
                                        subcommand.

DATA > SUSAN   ←—— This DATA prompt indicates that MINITAB is ready for you to enter data.
                                Type the first name and hit the return key. Continue this process until all
                                names have been entered. Note that when entering data on an alpha-variable,
                                enter only one data value in a row.

DATA > MARK
DATA > ROBERT
DATA > PETER
DATA > ALLISON
DATA > DEBORAH
DATA > SCOTT
DATA > MAUREEN
DATA > END   ←—— This command indicates the end of data entry in column C1.
MTB > SET C2   ←—— This command instructs MINITAB to create column C2. Note that there is
                                no semicolon at the end of this command because the data on scores
                                involves numbers and not letters.
DATA > 91 85 62 96 81 73 90 88   ←—— When entering data on a quantitative variable,
                                                                you can enter data in as many rows as you
                                                                want. Here we entered all scores in one row.
DATA > END   ←—— This command indicates the end of data entry in column C2.

In the MINITAB display in Figure 1.3, the ''DATA >'' prompt indicates that MINITAB is ready for data entry. When the **SET** command is used to enter data on an alpha-variable, you type a semicolon at the end of the first command. Then in the subcommand you tell MINITAB to format the indicated column to enter data on the alpha-variable and mention the maximum length of a name inside the parentheses next to **A**, which indicates an alpha-variable. In our example, the maximum length of a name is seven letters, as each of the names Allison, Deborah, and Maureen have seven letters. Also, when entering information on an alpha-variable, you must enter only one piece of information (such as a name) in one row. After you have entered all names, type **END** in the next **DATA** command and hit the Enter/Return key.

To enter quantitative data, you can enter all pieces of information in one row, two rows, three rows, and so on. If you are entering data in more than one row, hit the Enter/Return key after entering a few data values and continue entering data. After all data values have been entered, type **END** next to the **DATA** command and hit the Enter/Return key.   ▬▬

You can assign a name to the data entered in a column by using the **NAME** command. The following MINITAB commands will name the data entered in columns C1 and C2 as NAME and SCORE (see Figure 1.4).

Figure 1.4   Assigning names to data entered in columns.

```
MTB > NAME C1 'NAME'
MTB > NAME C2 'SCORE'
```

Note that the name you assign to a column should be enclosed within single quotation marks. Either of the following MINITAB commands will display the data on names and scores on the computer screen (see Figure 1.5).

Figure 1.5   Displaying data on the computer screen.

```
MTB > PRINT C1-C2
MTB > PRINT 'NAME' 'SCORE'
```

Note that when you use the assigned name for a column, such as NAME or SCORE in our example, this name must always be enclosed within single quotation marks.

Saving a MINITAB Data File

If you plan to use the data entered in MINITAB again at some later time, you need to save it as a file before you end the current MINITAB session. To save a data file, you must give it a name. Suppose you want to save the data file of Illustration M1–2 as the file SCORES. Either of the two commands given in Figure 1.6 will save this file. Use only one of these commands, depending on whether you want to save your file on drive C or drive A. Note that the file name is enclosed within single quotation marks.

Figure 1.6   Saving a MINITAB data file.

```
MTB > SAVE 'SCORES'     ←——— This command will save the data file as SCORES.MTW on the
                                   hard disk.
MTB > SAVE 'A:SCORES'   ←——— This command will save the data file as SCORES. MTW on
                                   the floppy disk in drive A.
```

MINITAB will attach the extension ''.MTW'' (which stands for MINITAB Worksheet) at the end of the file name when you save a data file unless you use another extension. The file SCORES will be saved as SCORES.MTW.

The first command given in Figure 1.6 will save the SCORES file on drive C. However, if you want to save this file in a particular directory on drive C, you will have to change this command slightly. For example, suppose you want to save this file in a directory named STAT on drive C; then type the following command.

```
MTB > SAVE 'C:\STAT\SCORES'
```

Note that if you are using a mainframe computer, you will always use the first command given in Figure 1.6 to save a MINITAB data file.

Retrieving a Previously Saved MINITAB Data File

To work on the SCORES file at a later date, use the **RETRIEVE** command to bring it back into the current worksheet. The MINITAB commands listed in Figure 1.7 will retrieve a previously saved file.

Figure 1.7   Retrieving a MINITAB data file.

```
MTB > RETRIEVE 'SCORES'     ←——— This command is used if the saved file is on the hard disk
                                      drive C.
MTB > RETRIEVE 'A:SCORES'   ←——— This command is used if the saved file is on a floppy
                                      disk in drive A.
```

If the file to be retrieved is in a directory on drive C, then you must mention the directory name when using the **RETRIEVE** command. For example, suppose the SCORES file is saved in a directory named STAT. Then the first command in Figure 1.7 will be changed to

```
MTB > RETRIEVE 'C:\STAT\SCORES'
```

MINITAB recognizes a command from its first four letters. Thus, if you use the command **RETR** **'SCORES'** instead of **RETRIEVE 'SCORES'**, the computer will respond with the same answer. This is true of all MINITAB commands. Remember that whenever you use the file name, whether to save, retrieve, or work on it, you must enclose the file name within single quotation marks.

The **INFORMATION** command, shown in Figure 1.8, helps you know what is in a retrieved file.

Figure 1.8

```
MTB > INFO
```

Saving a MINITAB Output File

As mentioned earlier, a MINITAB output file contains answers to all data analysis procedures you perform. You can save this output file for future use.

If you are using the MINITAB COMMAND LANGUAGE, the **OUTFILE** command can be used to save the MINITAB output to print at a later time. For example, suppose you plan to perform some statistical procedures on the data of Illustration M1–2 and you want to save the output file as SCORES. To do so, type **OUTFILE** followed by the name of the file enclosed within single quotation marks, as shown in Figure 1.9, at the very first MINITAB prompt before you perform any statistical analysis.

Figure 1.9

```
MTB > OUTFILE 'SCORES'    ←── This command will save the output file as SCORES.LIS on
                               the hard disk drive. You will use this command if you are
                               using a mainframe computer system.
MTB > OUTFILE 'A:SCORES'  ←── This command will save the output file as SCORES.LIS
                               on the floppy disk in drive A.
```

The first command given in Figure 1.9 will save the SCORES output file on drive C. However, if you want to save this file in a particular directory on drive C, you will have to change this command slightly. For example, suppose you want to save this output file in a directory named STAT on drive C; then type the following command.

```
MTB > OUTFILE 'C:\STAT\SCORES'
```

After you are finished using MINITAB, leave the MINITAB environment and print the SCORES.LIS file or copy it in any other document. Note that in the **OUTFILE** commands of Figure 1.9, you did not use any extension for the file and MINITAB automatically assigned the extension ''.LIS'' to your file. If you want, you can use an extension of your own. For example, you can enter **OUTFILE 'SCORES.TXT'** and **OUTFILE 'A:SCORES.TXT'** in the MINITAB commands in Figure 1.9. In this case, the output file will be saved as SCORES.TXT.

Printing a MINITAB Output File

You cannot print an output file, such as SCORES.LIS or SCORES.TXT from within MINITAB if you are using the MINITAB COMMAND LANGUAGE. To do so, either you use the **File** pull-down menu as explained in the MINITAB FOR WINDOWS section of this chapter or you exit MINITAB and then print the output file. If you are using a mainframe computer system, you will have to exit MINITAB to print the output file. Your instructor or computer lab assistant can help you to print an output file in this case.

Entering Data Using the Read Command

If you are using either the MINITAB COMMAND LANGUAGE or mainframe computers and you have data on two (or more) quantitative variables with the same number of observations for each variable, you can use the **READ** command to enter data. Illustration M1–3 explains how data on two variables can be entered into MINITAB using the **READ** command.

ILLUSTRATION M1-3   Suppose you need to enter the following data on the heights and weights of six persons into MINITAB.

Height (inches)	Weight (pounds)
69	178
67	135
65	121
71	210
68	149
66	142

To enter these data using the **READ** command, enter both the height and weight of each person in one row. The table in Illustration M1–3 contains information on six persons. Therefore, enter these data in six rows, with each row containing information on the height and weight of one person, as shown in Figure 1.10.

Figure 1.10   Entering data using the READ command.

```
MTB > READ C1 C2     ◄──── This command instructs MINITAB that you are to enter data on two
                             variables in two columns, C1 and C2.

DATA > 69 178
DATA > 67 135
DATA > 65 121
DATA > 71 210
DATA > 68 149
DATA > 66 142
DATA > END
```

Selecting a Sample Using MINITAB

Suppose the data entered on the names and scores of eight students in Illustration M1–2 belong to a population of eight students in a small class. You want to select a sample of three observations from this population and store the sample data in columns C3 and C4. Using the procedure described in Figure 1.11, a sample of three observations can be selected from this population.

Figure 1.11   Selecting a sample.

```
MTB > NOTE: SELECTING A SAMPLE OF 3 OBSERVATIONS
MTB > SAMPLE 3 FROM C1-C2 PUT IN C3-C4     ◄──── This command instructs
                                                   MINITAB to select a sample of
                                                   three observations (without
                                                   replacement) from the data of
                                                   columns C1 and C2 and put the
                                                   sample data in columns C3 and
                                                   C4.
```

Figure 1.11   (*Continued*)

```
MTB > PRINT C3-C4   ←—— This command will display the sample data of columns C3 and C4
                        on the computer screen.
```

Exiting MINITAB

If you are using the MINITAB COMMAND LANGUAGE, type **STOP** next to the MINITAB prompt, as shown in Figure 1.12, and hit the Enter/Return key to end the MINITAB session.

Figure 1.12

```
MTB > STOP
```

MINITAB will ask you a few questions. Answer *yes* or *no* to these questions as you like. After you answer these questions, the MINITAB session will end.

Additional MINITAB Commands

Figure 1.13 gives some additional MINITAB commands and their explanations. (Note that these commands will not be used in the sequence in which they are presented here.)

Figure 1.13   Some additional MINITAB commands.

```
MTB > HELP HELP   ←—— This command can be used to seek help about MINITAB commands.
MTB > HELP COMMANDS   ←—— This command also provides help about MINITAB commands.
MTB > HELP OVERVIEW   ←—— This command can be used for an overview of MINITAB.
MTB > COPY C1 to C2   ←—— This command will copy all data values from column C1 to
                          column C2.
MTB > ERASE C2   ←—— This command will delete all data values entered in column C2.
MTB > DELETE ROW 2 C1   ←—— This command will delete the second value entered in column
                            C1.
MTB > DELETE ROW 2 C1-C2   ←—— This command will delete the data values entered in the
                               second row of columns C1 and C2.
MTB > INSERT BETWEEN 2 AND 3 C1-C2   ←—— This command will insert a new row
                                         between the second and third rows for
                                         columns C1 and C2.
MTB > LET C1(4) = 10   ←—— This command will replace the fourth entry in column C1 with
                           10.
MTB > SORT C1 PUT IN C3   ←—— This command will sort the data of column C1 in
                              increasing order and put the results in column C3.
MTB > ADD C1 C2 PUT IN C4   ←—— This command will add the corresponding values of
                                columns C1 and C2 and put the new data in column
                                C4.
MTB > SUBTRACT C2 FROM C1 PUT IN C5   ←—— This command will subtract each
                                          value of column C2 from the
                                          corresponding value of column C1 and
                                          put the new data in column C5.
MTB > MULTIPLY C1 BY C2 PUT IN C6   ←—— This command will multiply the
                                        corresponding values of columns C1 and
                                        C2 and put the new data in column C6.
MTB > DIVIDE C1 BY C2 PUT IN C7   ←—— This command will divide each value of
                                      column C1 by the corresponding value of
                                      column C2 and put the new data in column
                                      C7.
```

Figure 1.13 (*Continued*)

```
MTB > LET C8 = C1*C2      ←── This command will multiply the corresponding values of
                              columns C1 and C2 and put the new data in column C8.
MTB > LET C9 = C1**2      ←── This command will square each value entered in column C1
                              and put the new data in column C9.
MTB > ADD 5 TO C1 PUT IN C10    ←── This command will add 5 to each value of column
                                    C1 and put the new data in column C10.
MTB > SUBTRACT 8 FROM C1 PUT IN C11    ←── This command will subtract 8 from
                                           each value of column C1 and put the
                                           new data in column C11.
MTB > MULTIPLY C1 BY 2 PUT IN C12      ←── This command will multiply each value
                                           of column C1 by 2 and put the new data
                                           in column C12.
MTB > DIVIDE C1 BY 3 PUT IN C13    ←── This command will divide each value of
                                       column C1 by 3 and put the new data in
                                       column C13.
```

The MINITAB **HELP** command followed by a specific command provides information about that command. For example, the command **HELP SET** can be used to find information about the **SET** command and its use.

COMPUTER ASSIGNMENTS

M1.1   Refer to Data Set III of Appendix B on the heights and weights of all NBA players. Enter the data on names, heights, and weights of all players using MINITAB FOR WINDOWS. If you are using the data disk that contains these data, you do not have to enter these data again. Take a sample of 15 players using MINITAB FOR WINDOWS and the MINITAB COMMAND LANGUAGE. Then print the sample data.

M1.2   The following table gives the names, hours worked, and the salary for the past week for five workers.

Name	Hours Worked	Salary (dollars)
John	42	725
Shannon	33	483
Kathy	28	355
David	47	790
Steve	40	820

a.   Enter these data into MINITAB using MINITAB FOR WINDOWS. Save the data file as WORKER.MTW. Exit MINITAB. Then restart MINITAB and retrieve the file WORKER.MTW.

b.   Enter these data into columns C1, C2, and C3 using the MINITAB COMMAND LANGUAGE. Save the data file as WORKER.MTW. Then try the following commands for this assignment and analyze the output for each command.

```
MTB > PRINT C1-C3
MTB > NAME C1 'NAME' C2 'HOURS' C3 'SALARY'
MTB > PRINT 'NAME' 'HOURS' 'SALARY'
MTB > LET K2 = SUM(C2)
MTB > PRINT K2
MTB > LET K3 = SUM(C3)
MTB > PRINT K3
```

```
MTB > PRINT C4
MTB > LET K4 = SUM(C4)
MTB > PRINT K4
MTB > LET C5 = C3*C3
MTB > LET K5 = SUM(C5)
MTB > PRINT C2—C5
MTB > PRINT K2—K5
MTB > SAMPLE 2 FROM C1—C3 PUT IN C6—C8
MTB > PRINT C6—C8
MTB > SAVE 'ASSIGN1'
MTB > DIR
MTB > RETRIEVE 'ASSIGN1'
MTB > INFO
MTB > STOP
```

2 ORGANIZING DATA

I n addition to thousands of private organizations and individuals, a large number of U.S. government agencies such as the Bureau of the Census, the Bureau of Labor Statistics, the National Agricultural Statistics Service, the National Center for Education Statistics, the National Center for Health Statistics, and the Bureau of Justice Statistics conduct hundreds of surveys every year. The data collected from each of these surveys fill hundreds of thousands of pages. In their original form, these data sets may be so large that they do not make sense to most of us. Descriptive statistics, however, supplies the techniques that help to condense large data sets by using tables, graphs, and summary measures. We see such tables, graphs, and summary measures in newspapers and magazines every day. At a glance, these tabular and graphical displays present information on every aspect of life. Consequently, descriptive statistics is of immense importance because it provides efficient and effective methods for summarizing and analyzing information.

This chapter explains how to organize and display data using tables and graphs. We will learn how to prepare frequency distribution tables for qualitative and quantitative data; how to construct bar graphs, pie charts, histograms, and polygons for such data; and how to prepare stem-and-leaf displays.

2.1   RAW DATA

When data are collected, the information obtained from each member of a population or sample is recorded in the sequence in which it becomes available. This sequence of data recording is random and unranked. Such data, before they are grouped or ranked, are called **raw data**.

> **RAW DATA**
>
> Data recorded in the sequence in which they are collected and before they are processed or ranked are called *raw data*.

Suppose we collect information on the ages (in years) of 50 students selected from a university. The data values, in the order they are collected, are recorded in Table 2.1. For instance, the first student's age is 21, the second student's age is 19 (second number in the first row), and so forth. The data given in Table 2.1 are quantitative raw data.

Table 2.1   Ages of 50 Students

21	19	24	25	29	34	26	27	37	33	18	20	19
22	19	19	25	22	25	23	25	19	31	19	23	18
23	19	23	26	22	28	21	20	22	22	21	20	19
21	25	23	18	37	27	23	21	25	21	24		

Suppose we ask the same 50 students about their student status. The responses of the students are recorded in Table 2.2. In this table, F, SO, J, and SE are the abbreviations for

Table 2.2   Status of 50 Students

J	F	SO	SE	J	J	SE	J	J	J	F	F	J
F	F	F	SE	SO	SE	J	J	F	SE	SO	SO	F
J	F	SE	SE	SO	SE	J	SO	SO	J	J	SO	F
SO	SE	SE	F	SE	J	SO	F	J	SO	SO		

freshman, sophomore, junior, and senior, respectively. This is an example of qualitative (or categorical) raw data.

The data presented in Tables 2.1 and 2.2 are also called **ungrouped data**. An ungrouped data set contains information on each member of a sample or population individually.

2.2   ORGANIZING AND GRAPHING QUALITATIVE DATA

This section discusses how to organize and display qualitative (or categorical) data. A data set is organized and displayed by constructing tables and by making graphs, respectively.

2.2.1   FREQUENCY DISTRIBUTIONS

A sample of 100 students enrolled at a university were asked what they intended to do after graduation. Forty-four said they wanted to work for private companies/businesses; 16 said they wanted to work for the federal government; 23 wanted to work for state or local governments; and 17 intended to start their own businesses. Table 2.3 lists the type of employment and the number of students who intend to engage in each type of employment. In this table, the variable is the *type of employment*, which is a qualitative variable. The categories (representing the type of employment) listed in the first column of the table are mutually exclusive. In other words, each of the 100 students belongs to one and only one of these categories. The number of students who belong to a certain category is called the *frequency* of that category. A **frequency distribution** exhibits how the frequencies are distributed over various categories. Table 2.3 is called a *frequency distribution table* or simply a *frequency table*.

Table 2.3   Type of Employment Students Intend to Engage In

Variable ⟶ **Type of Employment**	**Number of Students** ⟵ Frequency column
Private companies/businesses	44
Category ⟶ Federal government	16 ⟵ Frequency
State/local government	23
Own business	17
	Sum = 100

FREQUENCY DISTRIBUTION FOR QUALITATIVE DATA

A *frequency distribution* for qualitative data lists all categories and the number of elements that belong to each of the categories.

Example 2–1 illustrates how a frequency distribution table is constructed for qualitative data.

Constructing frequency distribution table for qualitative data.

EXAMPLE 2-1   A sample was taken of 25 high school seniors who were planning to go to college. Each of the students was asked which of the following majors he or she intended to choose: business, economics, management information systems (MIS), behavioral sciences (BS), other. The responses of these students are as follows:

Economics	MIS	Economics	Business	Business
Business	Business	Other	Other	Other
BS	BS	MIS	Other	MIS
Other	Business	MIS	Business	Other
Economics	MIS	Other	Other	MIS

Construct a frequency distribution table for these data.

Solution   Note that the *major a student intends to choose* is the variable in this example. This variable is classified into five categories: business, economics, MIS, BS, and other. We record these categories in the first column of Table 2.4. Then we read each student's response from the given data and mark a *tally*, denoted by the symbol ''|,'' in the second column of Table 2.4 next to the corresponding category. For example, the first student intends to major in economics. We show this in the frequency distribution table by marking a tally in the second column next to the category *Economics*. Note that the tallies are marked in blocks of five for counting convenience. Finally, we record the total of tallies for each category in the third column of the table. This column is called the *column of frequencies* and is usually denoted by f. The sum of the entries in the frequency column gives the sample size or total frequency. In Table 2.4, this total is 25, which is the sample size.

Table 2.4   Frequency Distribution of Majors

Major	Tally	Frequency (f)
Business	||||		6
Economics	|||	3
MIS	||||		6
BS	||	2
Other	|||| |||	8
		Sum = 25

2.2.2   RELATIVE FREQUENCY AND PERCENTAGE DISTRIBUTIONS

The **relative frequency** of a category is obtained by dividing the frequency of that category by the sum of all frequencies. Thus, the relative frequency shows what fractional part or proportion of the total frequency belongs to the corresponding category. A *relative frequency distribution* lists the relative frequencies for all categories.

> **RELATIVE FREQUENCY OF A CATEGORY**
>
> $$\text{Relative frequency of a category} = \frac{\text{Frequency of that category}}{\text{Sum of all frequencies}}$$

The **percentage** for a category is obtained by multiplying the relative frequency of that category by 100. A *percentage distribution* lists the percentages for all categories.

> **PERCENTAGE**
>
> Percentage = (Relative frequency) · 100

Constructing relative frequency and percentage distributions.

EXAMPLE 2-2 Construct the relative frequency and percentage distributions for Table 2.4.

Solution The relative frequencies and percentages for Table 2.4 are calculated and listed in Table 2.5. Based on this table, we can state that .24 or 24% of the students in the sample said that they intend to major in business. By adding the percentages for the first two categories, we can state that 36% of the students said they intend to major in business or economics. The other numbers in Table 2.5 can be interpreted the same way.

Notice that the sum of the relative frequencies is always 1.00 (or approximately 1.00 if the relative frequencies are rounded), and the sum of the percentages is always 100 (or approximately 100 if the percentages are rounded).

Table 2.5 Relative Frequency and Percentage Distributions Table

Major	Relative Frequency	Percentage
Business	6/25 = .24	.24 (100) = 24
Economics	3/25 = .12	.12 (100) = 12
MIS	6/25 = .24	.24 (100) = 24
BS	2/25 = .08	.08 (100) = 8
Other	8/25 = .32	.32 (100) = 32
	Sum = 1.00	Sum = 100

2.2.3 GRAPHICAL PRESENTATION OF QUALITATIVE DATA

All of us have heard the saying "a picture is worth a thousand words." A graphic display can reveal at a glance the main characteristics of a data set. The *bar graph* and the *pie chart* are two types of graphs used to display qualitative data.

Bar Graphs

To construct a **bar graph** (also called a *bar chart*), we mark the various categories on the horizontal axis as in Figure 2.1. Note that all categories are represented by intervals of the same width. We mark the frequencies on the vertical axis. Then we draw one bar for each category such that the height of the bar represents the frequency of the corresponding category.

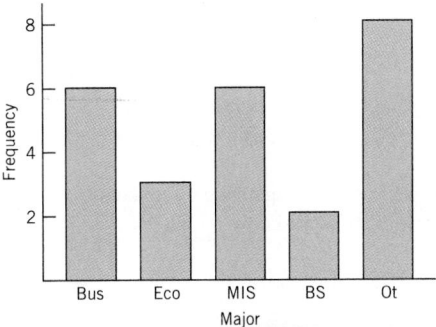

Figure 2.1 Bar graph for the frequency distribution of Table 2.4.

We leave a small gap between adjacent bars. Figure 2.1 gives the bar graph for the frequency distribution of Table 2.4.

BAR GRAPH

A graph made of bars whose heights represent the frequencies of respective categories is called a *bar graph*.

Case Study 2–1 presents a bar graph, reproduced from *USA Today*, that is based on the results of a survey seeking the opinions of people on environmental concerns.

CASE STUDY 2-1   PASSIVELY GREEN

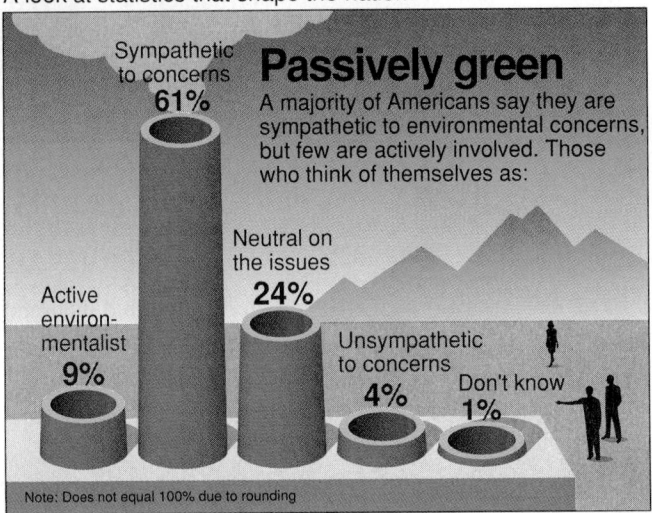

The following table, which is constructed using the above chart, presents the percentages of people in the sample who held different opinions on environmental issues.

Opinion	Percent of People
Active environmentalist	9
Sympathetic to concerns	61
Neutral on the issues	24
Unsympathetic to concerns	4
Don't know	1

Source: *USA Today*, January 30, 1997. Copyright © 1997, *USA Today*. Chart reproduced with permission.

The bar graphs for relative frequency and percentage distributions can be drawn simply by marking the relative frequencies or percentages, instead of the class frequencies, on the vertical axis.

Sometimes in a bar graph the categories are marked on the vertical axis and the frequencies on the horizontal axis.

Pie Charts

A **pie chart** is more commonly used to display percentages, although it can be used to display frequencies or relative frequencies. The whole pie (or circle) represents the total sample or population. The pie is divided into different portions that represent the percentages of the population or sample belonging to different categories.

PIE CHART

A circle divided into portions that represent the relative frequencies or percentages of a population or a sample belonging to different categories is called a *pie chart*.

As we know, a circle contains 360 degrees. To construct a pie chart, we multiply 360 by the relative frequency for each category to obtain the degree measure or size of the angle for the corresponding category. Table 2.6 shows the calculation of angle sizes for the various categories of Table 2.5.

Table 2.6 Calculating Angle Sizes for the Pie Chart

Major	Relative Frequency	Angle Size
Business	.24	360 (.24) = 86.4
Economics	.12	360 (.12) = 43.2
MIS	.24	360 (.24) = 86.4
BS	.08	360 (.08) = 28.8
Other	.32	360 (.32) = 115.2
	Sum = 1.00	Sum = 360

Figure 2.2 shows the pie chart for the percentage distribution of Table 2.5, which uses the angle sizes calculated in Table 2.6.

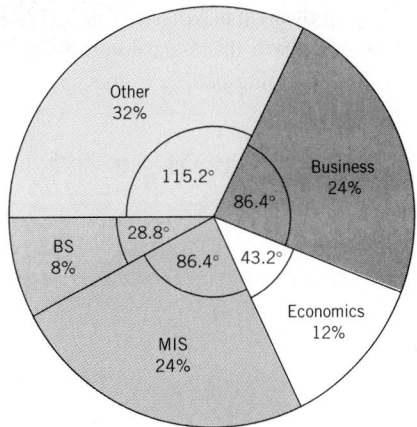

Figure 2.2 Pie chart for the percentage distribution of majors.

44

Case Studies 2–2 and 2–3 both show pie charts that are reproduced from *USA Today*. Case Study 2–2 shows preferences of American adults for different kinds of pizzas. Case Study 2–3 presents the results of a survey indicating the times or events that cause us the most stress. We can all relate to these case studies. Most of us love pizza and many of us feel stress.

CASE STUDY 2–2   PEPPERONI ON TOP

USA SNAPSHOTS®

A look at statistics that shape our lives

Pepperoni on top
American adults' favorite pizza toppings:

4% Onions

43% Pepperoni

7% Others

13% Vegetables

14% Mushrooms

19% Sausage

Source: Market Facts for Bolla wines   By Patti Stang and Nick Galifianakis, USA TODAY

The above pie chart shows what American adults like as toppings on their pizzas. As can be observed from this pie chart, the most favorite pizza topping is pepperoni, favored by 43% of the adults. Of the remaining people in the survey, 19% said they like sausages, 14% favor mushrooms, and so on.

Source: *USA Today*, November 7, 1996. Copyright © 1996, *USA Today*. Chart reproduced with permission.

CASE STUDY 2-3 MOST STRESSFUL TIMES

USA SNAPSHOTS®
A look at statistics that shape our lives

Most stressful times of the year or seasonal events

Family gatherings 25%

Income tax season 20%

Clothes shopping 10%

Vacations 6%

Birthdays 5%

Don't know 2%

Holidays 32%

Source: Survey for Devonsheer Melba Toast

By Scott Boeck and Genevieve Lynn, USA TODAY

The above pie chart shows the most stressful times of the year or seasonal events. According to this survey, holidays are the most stressful time, mentioned by 32% of the people surveyed. Family gatherings are the next event that causes stress, expressed by 25% of the people. Income tax season, clothes shopping, vacations, and birthdays are also mentioned as stressful events.

Source: USA Today, May 30, 1995. Copyright © 1995, *USA Today*. Chart reproduced with permission.

EXERCISES

Concepts and Procedures

2.1   Why do we need to group data in the form of a frequency table? Explain briefly.

2.2   How are the relative frequencies and percentages of categories obtained from the frequencies of categories? Illustrate with the help of an example.

2.3   The following data give the results of a sample survey. The letters A, B, and C represent the three categories.

A	B	B	A	C	B	C	C	C	A	C	B	C	C	C
C	B	C	C	A	A	B	C	C	B	C	B	B	C	A

    a.   Prepare a frequency distribution table.
    b.   Calculate the relative frequencies and percentages for all categories.

   c.   What percentage of the elements in this sample belong to category B?

   d.   What percentage of the elements in this sample belong to categories A or C?

   e.   Draw a bar graph for the frequency distribution.

2.4   The following data give the results of a sample survey. The letters Y, N, and D represent the three categories.

D	N	N	Y	Y	Y	N	Y	D	N	Y	Y	Y	Y
N	Y	Y	N	N	Y	N	Y	Y	N	D	N	Y	N
Y	Y	Y	Y	N	N	Y	Y	N	N	D	Y		

   a.   Prepare a frequency distribution table.

   b.   Calculate the relative frequencies and percentages for all categories.

   c.   What percentage of the elements in this sample belong to category Y?

   d.   What percentage of the elements in this sample belong to categories N or D?

   e.   Draw a pie chart for the percentage distribution.

Applications

2.5   Data on the status of 50 students given in Table 2.2 of Section 2.1 are reproduced below.

J	F	SO	SE	J	J	SE	J	J	J
F	F	J	F	F	F	SE	SO	SE	J
J	F	SE	SO	SO	F	J	F	SE	SE
SO	SE	J	SO	SO	J	J	SO	F	SO
SE	SE	F	SE	J	SO	F	J	SO	SO

   a.   Prepare a frequency distribution table.

   b.   Calculate the relative frequencies and percentages for all categories.

   c.   What percentage of these students are juniors or seniors?

   d.   Draw a bar graph for the frequency distribution.

2.6   The following are the responses of 20 students from a statistics class who were asked to evaluate their instructor. The students were asked to choose one of five answers: Excellent (E), Above average (AA), Average (A), Below average (B), and Poor (P).

AA	B	A	A	E	AA	P
E	AA	B	E	AA	E	B
E	A	B	P	AA	E	

   a.   Construct a frequency distribution table.

   b.   Calculate the relative frequencies and percentages for all categories.

   c.   What percentage of these students ranked this instructor as excellent or above average?

   d.   Draw a bar graph for the relative frequency distribution.

2.7   Fifteen professors at a university were asked whether or not community service should be required of all students as a condition for graduation. The responses of these professors are listed below. (F, A, and N indicate that a professor is in favor, against, or has no opinion, respectively.)

F	A	A	F	N	N	F	F	F	A	A	N	F	F	A

   a.   Prepare a frequency distribution table.

   b.   Compute the relative frequencies and percentages for all categories.

   c.   What percentage of the professors in this sample are in favor of this issue?

   d.   Draw a pie chart for the percentage distribution.

2.8   Twelve persons were asked to taste two types of soft drinks, *A* and *B*, and indicate if the taste of *A* was superior (S), the same (M), or inferior (I) to that of *B*. Their responses are listed on the following page.

S   I   I   M   M   S   M   M   S   I   S   S

a. Construct a frequency distribution table.
b. Calculate the relative frequencies and percentages for all categories.
c. Draw a pie chart for the percentage distribution.

2.9 A poll conducted by Bruskin/Goldring Research for TurboTax asked a sample of taxpayers about the most difficult part of federal income tax return preparation (*USA Today*, February 22, 1995). The responses obtained from these taxpayers are summarized and presented in the following table, which shows the percentages of taxpayers in the poll who mentioned each of the listed items to be the most difficult.

Problem Area	Percentage
Understanding IRS jargon	43
Knowing deductions	29
Getting the right forms	10
Calculations	8
Do not know	10

Draw a pie chart for this percentage distribution.

2.10 According to 1994 data from the Tax Foundation, 2% of federal tax dollars were spent on transportation, 3% on education, 3% on veterans' benefits, 14% on income security, 14% on interest on debt, 17% on health, 19% on defense, 21% on social security, and 7% on other items (*Business Week*, May 1, 1995). Draw a bar graph to display these data.

2.3 ORGANIZING AND GRAPHING QUANTITATIVE DATA

In the previous section we learned how to group and display qualitative data. This section explains how to group and display quantitative data.

2.3.1 FREQUENCY DISTRIBUTIONS

Table 2.7 gives the weekly earnings of 100 employees of a large company. The first column of this table lists the *classes*, which represent the (quantitative) variable *the weekly earnings*. For quantitative data, an interval that includes all the values that fall within two numbers, the lower and upper limits, is called a **class**. Note that the classes always represent a variable. As

Table 2.7 Weekly Earnings of 100 Employees of a Company

Variable → **Weekly Earnings (dollars)**	**Number of Employees** *f* ← Frequency column
301 to 400	9
401 to 500	16
Third class → 501 to 600	33 ← Frequency of the third class
601 to 700	20
701 to 800	14
→ 801 to 900 ←	8

Lower limit of the sixth class   Upper limit of the sixth class we can observe, the classes are nonoverlapping; that is, each value on earnings belongs to one and only one class. The second column in the table lists the number of employees who have earnings within each class. For example, nine employees of this company earn $301 to $400 per week. The numbers listed in the second column of this table are called the **frequencies**, which give the number of values that belong to different classes. The frequencies are denoted by f.

For quantitative data, the frequency of a class represents the number of values in the data set that fall in that class. Table 2.7 contains six classes. Each class has a *lower limit* and an *upper limit*. The values 301, 401, 501, 601, 701, and 801 give the lower limits and the values 400, 500, 600, 700, 800, and 900 give the upper limits of the six classes, respectively. The data presented in Table 2.7 are an illustration of a **frequency distribution table** for quantitative data. Data presented in the form of a frequency distribution table are called **grouped data**.

FREQUENCY DISTRIBUTION FOR QUANTITATIVE DATA

A *frequency distribution* for quantitative data lists all the classes and the number of values that belong to each class. Data presented in the form of a frequency distribution are called *grouped data*.

To find the midpoint of the upper limit of the first class and the lower limit of the second class in Table 2.7, we divide the sum of these two limits by 2. Thus, this midpoint is

$$\frac{400 \,+\, 401}{2} = 400.5$$

The value 400.5 is called the *upper boundary* of the first class and the *lower boundary* of the second class. By using this technique, we can convert the class limits of Table 2.7 to **class boundaries**, which are also called *real class limits*. The second column of Table 2.8 lists the boundaries for Table 2.7.

CLASS BOUNDARY

The *class boundary* is given by the midpoint of the upper limit of one class and the lower limit of the next class.

Table 2.8   Class Boundaries, Class Widths, and Class Midpoints for Table 2.7

Class Limits	Class Boundaries	Class Width	Class Midpoint
301 to 400	300.5 to less than 400.5	100	350.5
401 to 500	400.5 to less than 500.5	100	450.5
501 to 600	500.5 to less than 600.5	100	550.5
601 to 700	600.5 to less than 700.5	100	650.5
701 to 800	700.5 to less than 800.5	100	750.5
801 to 900	800.5 to less than 900.5	100	850.5

The difference between the two boundaries of a class gives the **class width**. The class width is also called the **class size**.

CLASS WIDTH

$$\text{Class width} = \text{Upper boundary} - \text{Lower boundary}$$

Thus, in Table 2.8,

$$\text{Width of the first class} = 400.5 - 300.5 = 100$$

The class widths for the frequency distribution of Table 2.7 are listed in the third column of Table 2.8. Each class in Table 2.8 (and Table 2.7) has the same width of 100.

The **class midpoint** or **mark** is obtained by dividing the sum of the two limits (or the two boundaries) of a class by 2.

CLASS MIDPOINT OR MARK

$$\text{Class midpoint or mark} = \frac{\text{Lower limit} + \text{Upper limit}}{2}$$

Thus, the midpoint of the first class in Table 2.7 or Table 2.8 is calculated as follows.

$$\text{Midpoint of the first class} = \frac{301 + 400}{2} = 350.5$$

The class midpoints for the frequency distribution of Table 2.7 are listed in the fourth column of Table 2.8.

Note that in Table 2.8, when we write classes using class boundaries, we write *to less than* in order to ensure that each value belongs to one and only one class. As we can see, the upper boundary of the preceding class and the lower boundary of the succeeding class are the same.

2.3.2 CONSTRUCTING FREQUENCY DISTRIBUTION TABLES

While constructing a frequency distribution table, the following three major decisions need to be made.

Number of Classes

Usually the number of classes for a frequency distribution table varies from 5 to 20, depending mainly on the number of observations in the data set.[1] It is preferable to have more classes as the size of a data set increases. The decision about the number of classes is arbitrarily made by the data organizer.

Class Width

Although it is not uncommon to have classes of different sizes, most of the time it is preferable to have the same width for all classes. To determine the class width when all classes are of

---

[1] One rule to help decide about the number of classes is the Sturge's formula, which is
$$c = 1 + 3.3 \log n$$
where c is the number of classes and n is the number of observations in the data set. The value of $\log n$ can be obtained by entering the value of n on the calculator and pressing the *log* key.

the same size, first find the difference between the largest and the smallest values in the data. Then, the approximate width of a class is obtained by dividing this difference by the number of desired classes.

CALCULATION OF CLASS WIDTH

$$\text{Approximate class width} = \frac{\text{Largest value} - \text{Smallest value}}{\text{Number of classes}}$$

Usually this approximate class width is rounded to a convenient number, which is then used as the class width. Note that rounding this number may slightly change the number of classes initially intended.

Lower Limit of the First Class or the Starting Point

Any convenient number, which is equal to or less than the smallest value in the data set, can be used as the lower limit of the first class.

Example 2–3 illustrates the procedure for constructing a frequency distribution table for quantitative data.

Constructing frequency distribution table for quantitative data.

EXAMPLE 2-3   Refer to Data Set III (NBA Data) given in Appendix B of the text. The following data give the heights (in inches) of a random sample of 30 National Basketball Association players selected from that data set.

81	84	79	76	73	74	77	82	75	81
76	76	80	82	78	72	80	83	80	77
78	78	79	84	73	86	83	79	83	79

Construct a frequency distribution table.

Solution   In these data, the minimum value is 72 and the maximum value is 86. Suppose we decide to group these data using five classes of equal width. Then,

$$\text{Approximate width of a class} = \frac{86 - 72}{5} = 2.8$$

Suppose we round this approximate width to a convenient number, say, 3. The lower limit of the first class can be taken as 72 or any number less than 72. Suppose we take 72 as the lower limit of the first class. Then, our classes will be

$$72–74, \quad 75–77, \quad 78–80, \quad 81–83, \quad \text{and} \quad 84–86$$

We record these five classes in the first column of Table 2.9.

Now we read each value from the given data and make a tally mark in the second column of Table 2.9 next to the corresponding class. The first value in our original data is 81, which belongs to the 81–83 class. To record it, we make a tally mark in the second column of Table 2.9 next to the 81–83 class. We continue this process until all the data values have been read and entered in the tally column. Note that tallies are marked in blocks of five for counting convenience. After the tally column is completed, we count the tally marks for each class and write those numbers in the third column. This gives the column of frequencies. These frequencies represent the number of players that belong to each of the five different classes. For example, 4 of the 30 players have a height of 72 to 74 inches, and so forth.

Table 2.9 Frequency Distribution of Heights of NBA Players

Height (in inches)	Tally	f
72–74	||||	4
75–77	⫿⫿⫿⫿⫿ |	6
78–80	⫿⫿⫿⫿⫿ ⫿⫿⫿⫿⫿	10
81–83	⫿⫿⫿⫿⫿ ||	7
84–86	|||	3
		$\Sigma f = 30$

In Table 2.9, we can denote the frequencies of the five classes by f_1, f_2, f_3, f_4, and f_5, respectively. Therefore,

$$f_1 = \text{frequency of the first class} = 4$$

Similarly,

$$f_2 = 6, \qquad f_3 = 10, \qquad f_4 = 7, \qquad \text{and} \qquad f_5 = 3$$

Using the Σ notation (see Section 1.8 of Chapter 1), we can denote the sum of the frequencies of all classes by Σf. Hence,

$$\Sigma f = f_1 + f_2 + f_3 + f_4 + f_5 = 4 + 6 + 10 + 7 + 3 = 30$$

The number of observations in a sample is usually denoted by n. Thus, for the sample data, Σf is equal to n. The number of observations in a population is denoted by N. Consequently, Σf is equal to N for population data. Because the data set on heights in Table 2.9 is for only 30 NBA players, it represents a sample. Therefore, in Table 2.9 we can denote the sum of frequencies by n instead of Σf. ▬

Note that when we present the data in the form of a frequency distribution table, as in Table 2.9, we lose the information on individual observations. We cannot know the exact height of any particular player from Table 2.9. All we know is that three players are 72 to 74 inches tall, and so forth.

2.3.3 RELATIVE FREQUENCY AND PERCENTAGE DISTRIBUTIONS

Using Table 2.9, we can compute the relative frequency and percentage columns the same way we did for qualitative data in Section 2.2.2. The relative frequencies and percentages for a quantitative data set are obtained as follows.

RELATIVE FREQUENCY AND PERCENTAGE

$$\text{Relative frequency of a class} = \frac{\text{Frequency of that class}}{\text{Sum of all frequencies}} = \frac{f}{\Sigma f}$$

$$\text{Percentage} = (\text{Relative frequency}) \cdot 100$$

Example 2–4 illustrates how to construct relative frequency and percentage distributions.

Constructing relative frequency and percentage distributions.

EXAMPLE 2–4   Calculate the relative frequencies and percentages for Table 2.9.

Solution   The relative frequencies and percentages for Table 2.9 are calculated and listed in the third and fourth columns, respectively, of Table 2.10. Note that the class boundaries are listed in the second column of Table 2.10.

Table 2.10   Relative Frequency and Percentage Distributions for Table 2.9

Height	Class Boundaries	Relative Frequency	Percentage
72–74	71.5 to less than 74.5	4/30 = .133	13.3
75–77	74.5 to less than 77.5	6/30 = .200	20.0
78–80	77.5 to less than 80.5	10/30 = .333	33.3
81–83	80.5 to less than 83.5	7/30 = .233	23.3
84–86	83.5 to less than 86.5	3/30 = .100	10.0
		Sum = .999	Sum = 99.9%

From Table 2.10, we can make statements about the percentage of players with heights within a certain interval. For example, about 13.3% of the players in this sample are 72 to 74 inches tall. By adding the percentages for the first two classes, we can state that about 33.3% of the players are 72 to 77 inches tall. Similarly, by adding the percentages of the last two classes, we can state that about 33.3% of the players are 81 to 86 inches tall.

Note that in Table 2.10 the sum of the relative frequency and percentage columns do not add up to 1.00 and 100%, respectively, due to rounding off.   ▬

2.3.4   GRAPHING GROUPED DATA

Grouped (quantitative) data can be displayed by using a *histogram* or a *polygon*. This section describes how to construct such graphs. We can also draw a pie chart to display the percentage distribution for a quantitative data set. The procedure to construct a pie chart is similar to the one for qualitative data explained in Section 2.2.3; it will not be repeated in this section.

Histograms

A **histogram** is a certain kind of graph that can be drawn for a frequency distribution, a relative frequency distribution, or a percentage distribution. To draw a histogram, we first mark classes on the horizontal axis and frequencies (or relative frequencies or percentages) on the vertical axis. Next, we draw a bar for each class so that its height represents the frequency of that class. The bars in a histogram are drawn adjacent to each other without leaving any gap between them. A histogram is called a **frequency histogram**, a **relative frequency histogram**, or a **percentage histogram** depending on whether the frequencies, relative frequencies, or percentages are marked on the vertical axis.

HISTOGRAM

A *histogram* is a graph in which classes are marked on the horizontal axis and either the frequencies, relative frequencies, or percentages are marked on the vertical axis. The frequencies, relative frequencies, or percentages are represented by the heights of the bars. In a histogram, the bars are drawn adjacent to each other.

Figures 2.3 and 2.4 show the frequency and the relative frequency histograms, respectively, for the data of Tables 2.9 and 2.10 of Sections 2.3.2 and 2.3.3. The two histograms look alike because they represent the same data. A percentage histogram can be drawn for the percentage distribution of Table 2.10 by marking the percentages on the vertical axis.

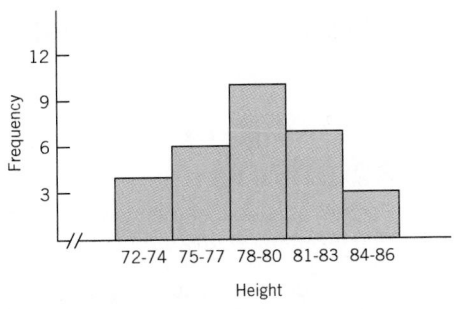

Figure 2.3 Frequency histogram for Table 2.9.

The symbol ''–//–'' used in the horizontal axes of Figures 2.3 and 2.4 represents a break, called the **truncation**, in the horizontal axis. It indicates that the entire horizontal axis is not shown in these figures. Notice that the 0 to 71 portion of the horizontal axis has been omitted in each figure.

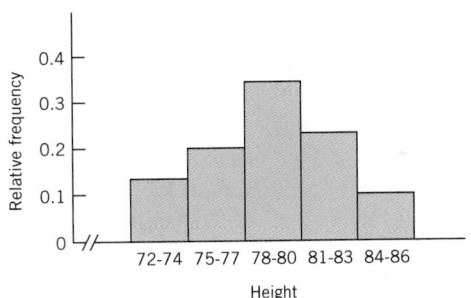

Figure 2.4 Relative frequency histogram for Table 2.10.

In Figures 2.3 and 2.4, we have used class limits to mark classes on the horizontal axis. However, we can show the intervals on the horizontal axis by using the class boundaries instead of the class limits.

Polygons

A **polygon** is another device that can be used to present quantitative data in graphic form. To draw a **frequency polygon**, we first mark a dot above the midpoint of each class at a height equal to the frequency of that class. This is the same as marking the midpoint at the top of each bar in a histogram. Next, we mark two more classes, one at each end, and mark their midpoints. Note that these two classes have zero frequencies. In the last step, we join the adjacent dots with straight lines. The resulting line graph is called a frequency polygon or simply a polygon.

CASE STUDY 2-4 DRINKING, DRIVING, AND DYING

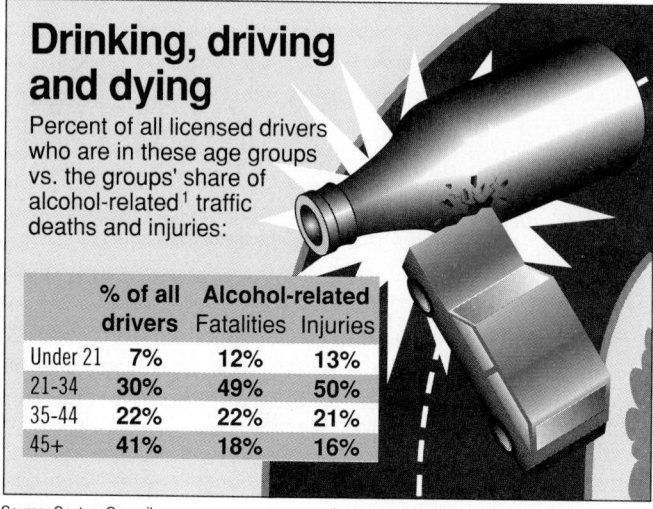

USA SNAPSHOTS ®

A look at statistics that shape the nation

Drinking, driving and dying

Percent of all licensed drivers who are in these age groups vs. the groups' share of alcohol-related[1] traffic deaths and injuries:

	% of all drivers	Alcohol-related Fatalities	Injuries
Under 21	7%	12%	13%
21-34	30%	49%	50%
35-44	22%	22%	21%
45+	41%	18%	16%

Source: Century Council

By Anne R. Carey and Gary Visgaitis, USA TODAY

The above chart, reproduced from *USA Today*, gives three percentage distributions—the percentage of all drivers who belong to different age groups, the percentage of alcohol-related fatalities, and the percentage of alcohol-related injuries caused by drivers in different age groups. For example, of all drivers in the United States, 7% are under the age of 21; they cause 12% of all the alcohol-related fatalities and 13% of all the alcohol-related injuries. Forty-one percent of all drivers are 45 years of age or older, but only 18% of the fatalities and 16% of the injuries that occur due to drunk driving are caused by these drivers. Notice that there is no lower limit for the first class and no upper limit for the fourth class in this distribution. Such classes are called open-ended classes.

Source: *USA Today*, December 30, 1996. Copyright © 1996, *USA Today*. Chart reproduced with permission.

A polygon with relative frequencies marked on the vertical axis is called a *relative frequency polygon*. Similarly, a polygon with percentages marked on the vertical axis is called a *percentage polygon*.

POLYGON

A graph formed by joining the midpoints of the tops of successive bars in a histogram with straight lines is called a *polygon*.

Figure 2.5 shows the frequency polygon for the frequency distribution of Table 2.9.

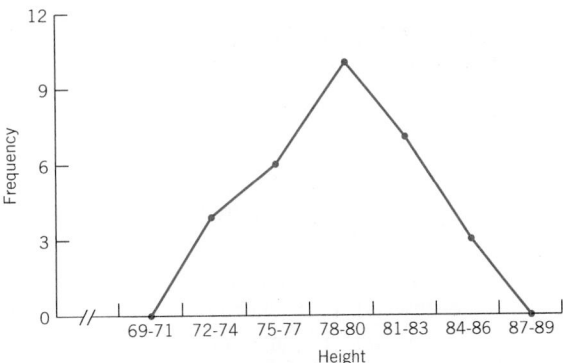

Figure 2.5 Frequency polygon for Table 2.9.

For a very large data set, as the number of classes is increased (and the width of classes is decreased), the frequency polygon eventually becomes a smooth curve. Such a curve is called a *frequency distribution curve* or simply a *frequency curve*. Figure 2.6 shows the frequency curve for a large data set with a large number of classes.

Figure 2.6 Frequency distribution curve.

2.3.5 MORE ON CLASSES AND FREQUENCY DISTRIBUTIONS

This section presents two alternative methods for writing classes to construct a frequency distribution for quantitative data.

Less Than Method for Writing Classes

The classes in the frequency distribution given in Table 2.9 for the data on heights of 30 players were written as 72–74, 75–77, and so on. Alternatively, we can write the classes in a frequency distribution table using the *less than* method. The technique for writing classes shown in Table 2.9 is more commonly used for data sets that do not contain fractional values. The *less than* method is more appropriate when a data set contains fractional values. Example 2–5 illustrates the *less than* method.

Constructing frequency distribution using less than *method.*

EXAMPLE 2–5 The following data give the gasoline tax rates (in cents per gallon) for 1995 for 50 states.[2] (*Source*: U.S. Federal Highway Administration.)

18.00	8.00	18.00	18.70	18.00	22.00	34.00	23.00	12.30	7.50
16.00	21.00	19.00	15.00	20.00	18.00	16.40	20.00	19.00	23.50
21.00	15.00	20.00	18.40	15.00	27.00	25.40	24.00	18.70	10.50
18.00	21.92	21.60	18.00	22.00	17.00	24.00	22.35	29.00	16.00
18.00	20.00	20.00	19.00	16.00	17.50	23.00	25.35	23.40	9.00

[2]The data for the 50 states are entered (by row) in the following order: Alabama, Alaska, Arizona, Arkansas, California, Colorado, Connecticut, Delaware, Florida, Georgia, Hawaii, Idaho, Illinois, Indiana, Iowa, Kansas, Kentucky, Louisiana, Maine, Maryland, Massachusetts, Michigan, Minnesota, Mississippi, Missouri, Montana, Nebraska, Nevada, New Hampshire, New Jersey, New Mexico, New York, North Carolina, North Dakota, Ohio, Oklahoma, Oregon, Pennsylvania, Rhode Island, South Carolina, South Dakota, Tennessee, Texas, Utah, Vermont, Virginia, Washington, West Virginia, Wisconsin, Wyoming.

Construct a frequency distribution table. Calculate the relative frequencies and percentages for all classes.

Solution   The minimum value in this data set is 7.50 and the maximum value is 34.00. Suppose we decide to group these data using six classes of equal width. Then,

$$\text{Approximate width of a class} = \frac{34.00 - 7.50}{6} = 4.42$$

We round this number to a more convenient number, say, 5. Then, we take 5 as the width of each class. If we start the first class at 5, the classes will be written as 5 *to less than* 10, 10 *to less than* 15, and so on. The six classes, which cover all the data, are recorded in the first column of Table 2.11. The second column of that table lists the frequencies of classes. A value in the data set that is 5 or larger but less than 10 belongs to the first class, and a value that is 10 or larger but less than 15 falls in the second class, and so on. The relative frequencies and percentages for classes are recorded in the third and fourth columns, respectively, of Table 2.11. Note that Table 2.11 does not contain a column of tallies.

Table 2.11   Gasoline Tax Rates for 50 States

Gasoline Tax Rate (cents per gallon)	f	Relative Frequency	Percentage
5 to less than 10	3	.06	6
10 to less than 15	2	.04	4
15 to less than 20	22	.44	44
20 to less than 25	18	.36	36
25 to less than 30	4	.08	8
30 to less than 35	1	.02	2
	$\Sigma f = 50$	Sum = 1.00	Sum = 100%

A histogram and a polygon for the data of Table 2.11 can be drawn in the same way as for the data of Tables 2.9 and 2.10.

Single-Valued Classes

If the observations in a data set assume only a few distinct values, it may be appropriate to prepare a frequency distribution table using *single-valued classes*, that is, classes that are made of single values and not of intervals. This technique is especially useful in cases of discrete data with only a few possible values. Example 2–6 exhibits such a situation.

Constructing frequency distribution using single-valued classes.

EXAMPLE 2–6   The administration in a large city wanted to know the distribution of vehicles owned by households in that city. A sample of 40 randomly selected households from this city produced the following data on the number of vehicles owned.

5	1	1	2	0	1	1	2
1	1	1	3	3	0	2	5
1	2	3	4	2	1	2	2
1	2	2	1	1	1	4	2
1	1	2	1	1	4	1	3

Construct a frequency distribution table for these data using single-valued classes.

Solution   The observations in this data set assume only six distinct values: 0, 1, 2, 3, 4, and 5. Each of these six values is used as a class in the frequency distribution Table 2.12, and these six classes are listed in the first column of that table. To obtain the frequencies of these classes, the observations in the data that belong to each class are counted, and the results are recorded in the second column of Table 2.12. Thus, in these data, 2 households own no vehicle, 18 own one vehicle each, 11 own two vehicles each, and so on.

Table 2.12   Frequency Distribution of Vehicles Owned

Vehicles Owned	Number of Households (f)
0	2
1	18
2	11
3	4
4	3
5	2
	$\Sigma f = 40$

The data of Table 2.12 can also be displayed by drawing a bar graph, as shown in Figure 2.7. To construct a bar graph, we mark the classes, as intervals, on the horizontal axis with a little gap between consecutive intervals. The bars represent the frequencies of respective classes.

The frequencies of Table 2.12 can be converted to relative frequencies and percentages the same way as in Table 2.10. Then, a bar graph can be constructed to display the relative frequency or percentage distribution by marking the relative frequencies or percentages, respectively, on the vertical axis.

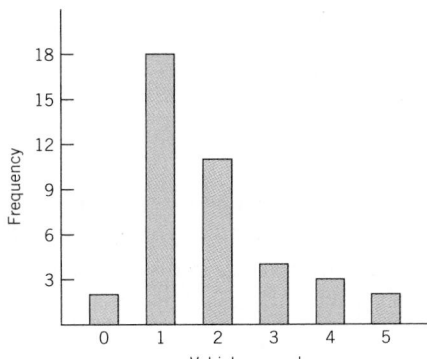

Figure 2.7   Bar graph for vehicles owned.

2.4   SHAPES OF HISTOGRAMS

A histogram can assume any one of a large number of shapes. The most common of these shapes are

1. Symmetric
2. Skewed
3. Uniform or rectangular

A **symmetric histogram** is identical on both sides of its central point. The histograms shown in Figure 2.8 are symmetric around the dashed lines that represent their central points.

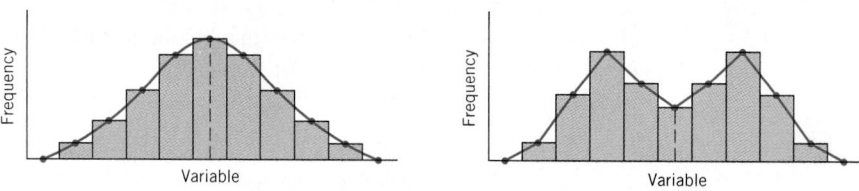

Figure 2.8   Symmetric histograms.

A **skewed histogram** is nonsymmetric. For a skewed histogram, the tail on one side is longer than the tail on the other side. A **skewed-to-the-right histogram** has a longer tail on the right side (see Figure 2.9a). A **skewed-to-the-left histogram** has a longer tail on the left side (see Figure 2.9b).

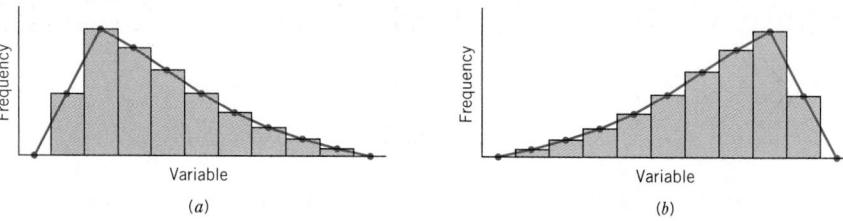

Figure 2.9   (a) A histogram skewed to the right. (b) A histogram skewed to the left.

A **uniform** or **rectangular histogram** has the same frequency for each class. Figure 2.10 is an illustration of such a case.

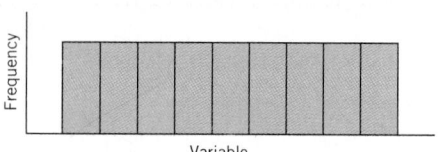

Figure 2.10   A histogram with uniform distribution.

Figures 2.11a and 2.11b display symmetric frequency curves. Figures 2.11c and 2.11d show frequency curves skewed to the right and to the left, respectively.

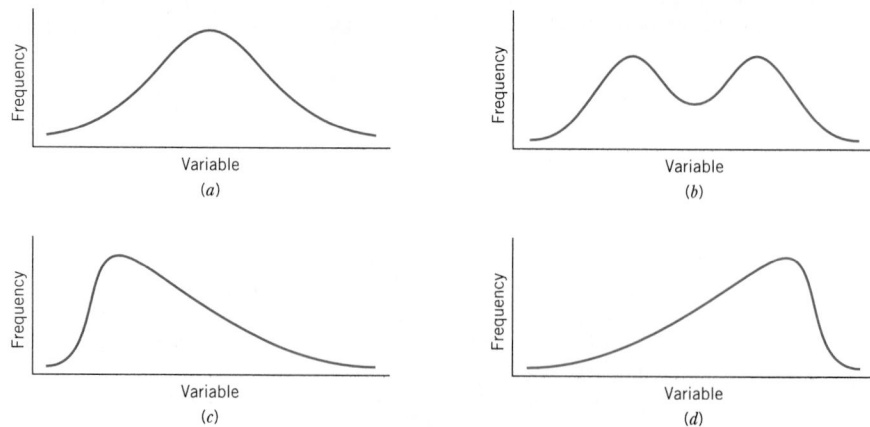

Figure 2.11   (a) and (b) Symmetric frequency curves. (c) Frequency curve skewed to the right. (d) Frequency curve skewed to the left.

☞ **WARNING**

Describing data using graphs helps to give us insights into the main characteristics of the data. But graphs, unfortunately, can also be used, intentionally or unintentionally, to distort the facts and to deceive the reader. The following are two ways to manipulate graphs to convey a particular opinion or impression.

1. *Changing the scale* either on one or on both axes, that is, shortening or stretching one or both of the axes
2. *Truncating the frequency axis*, that is, starting the frequency axis at a number greater than zero

When interpreting a graph, we should be very cautious. We should observe carefully whether the frequency axis has been truncated or whether any axis has been unnecessarily shortened or stretched.

CASE STUDY 2-5 USING TRUNCATED AXES

Consider the following table, which is based on a survey of service and retail businesses with less than 20 employees conducted by Padgett Business Services (*Source*: *USA Today*, February 20, 1996). These business owners were asked what qualities they want most in their new employees. This table lists the qualities that new employees should possess and the percent of owners favoring those qualities.

Quality	Percent of Business Owners
Dependability (D)	35
Honesty (H)	27
Competence (C)	19
Good attitude (G)	19

The following two bar graphs are constructed for the percentage distribution given in the above table. As you would observe, the figure on the left side shows the complete vertical axis and the one on the right side shows the truncated vertical axis. As you can observe, the two graphs give completely different impressions of the results of the survey and the differences in the percentages for different categories.

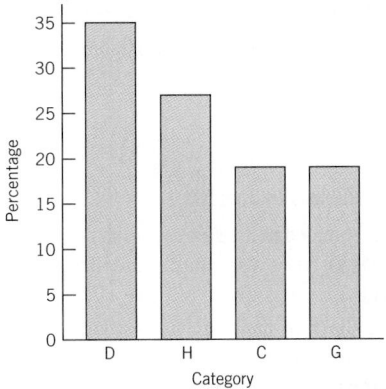 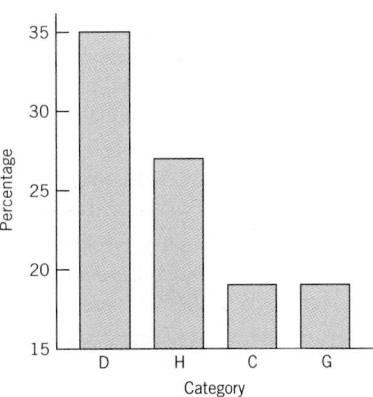

EXERCISES

Concepts and Procedures

2.11 Briefly explain the three decisions that have to be made to group a data set in the form of a frequency distribution table.

2.12 How are the relative frequencies and percentages of classes obtained from the frequencies of classes? Illustrate with the help of an example.

2.13 Three methods—writing classes using limits, using the *less than* method, and grouping data using single-valued classes—were discussed to group quantitative data into classes. Explain these three methods and give one example of each.

Applications

2.14 The following table gives the frequency distribution of weekly earnings for a sample of 100 employees selected from a large company.

Weekly Earnings (dollars)	Number of Employees
200 to 349	12
350 to 499	23
500 to 649	31
650 to 799	19
800 to 949	15

    a. Find the class boundaries and class midpoints.
    b. Do all classes have the same width? If yes, what is that width?
    c. Prepare the relative frequency and percentage distribution columns.
    d. What percentage of these employees earn $650 or more per week?

2.15 The following table gives the frequency distribution of ages for all 50 employees of a company.

Age	Number of Employees
18 to 30	12
31 to 43	17
44 to 56	14
57 to 69	7

    a. Find the class boundaries and class midpoints.
    b. Do all classes have the same width? If yes, what is that width?
    c. Prepare the relative frequency and percentage distribution columns.
    d. What percentage of the employees of this company are 43 years old or younger?

2.16 A data set on the political contributions (rounded to the nearest dollar) made during the past year by 200 households has a lowest value of $1 and a highest value of $683. Suppose we want to group these data into seven classes of equal widths.

    a. Assuming we take the lower limit of the first class as $1 and the width of each class equal to $100, write the class limits for all seven classes.
    b. Write the class boundaries and class midpoints.

2.17 A data set on weekly expenditures (rounded to the nearest dollar) on bakery products for a sample of 500 households has a minimum value of $1 and a maximum value of $18. Suppose we want to group these data into five classes of equal widths.

    a. Assuming we take the lower limit of the first class as $1 and the upper limit of the fifth class as $20, write the class limits for all five classes.
    b. Determine the class boundaries and class widths.
    c. Find the class midpoints.

2.18 The following data give the market value (in millions of U.S. dollars) as of May 31, 1996, of 25 Canadian companies included in the Business Week Global 1000 (*Business Week*, July 8, 1996).

13,900	12,918	12,540	11,318	9729	8115	
7493	7386	7045	7017	6806	6224	
6008	5631	5473	5431	4864	4720	
4693	4615	3979	3773	3704	3487	3194

a. Construct a frequency distribution table. Take the classes as 3000–5999, 6000–8999, 9000–11,999, and 12,000–14,999.
b. Calculate the relative frequencies and percentages for all classes.
c. Based on the frequency distribution, can you say whether the data are symmetric or skewed?
d. What percentage of these companies have a market value of $9 billion or higher?

2.19 Nixon Corporation manufactures computer terminals. The following data give the number of computer terminals produced at the company for a sample of 30 days.

24	32	27	23	33	33	29	25	23	28
21	26	31	22	27	33	27	23	28	29
31	35	34	22	26	28	23	35	31	27

a. Construct a frequency distribution table using the classes 21–23, 24–26, 27–29, 30–32, and 33–35.
b. Calculate the relative frequencies and percentages for all classes.
c. Construct a histogram and a polygon for the percentage distribution.
d. For what percentage of the days is the number of computer terminals produced in the interval 27 to 29?

2.20 The following data give the number of computer keyboards assembled at the Twentieth Century Electronics Company for a sample of 25 days.

45	52	48	41	56	46	44	42	48	53	51	53	51
48	46	43	52	50	54	47	44	47	50	49	52	

The following is the frequency distribution table for these data.

Classes	Frequency
41–44	5
45–48	8
49–52	8
53–56	4

a. Construct the relative frequencies for all classes.
b. Construct a histogram for the relative frequency distribution.
c. Construct a polygon for the relative frequency distribution.

2.21 The following data give the number of people unemployed (rounded to the nearest thousand) as of December 1995 for 20 northeastern and midwestern states (*Source:* U.S. Bureau of Labor Statistics).

90	328	132	51	51	33	167	231	90	101
24	20	270	10	287	347	31	11	13	107

a. Construct a frequency distribution table. Take the classes as 1–70, 71–140, and so on.
b. Prepare the relative frequency and percentage columns for the table of part a.

Exercises 2.22 through 2.26 are based on the following data.

The following table, based on the American Chamber of Commerce Researchers Association Survey for the second quarter of 1996, gives the prices of five items (in dollars) in 25 urban areas across the United States. (See Data Set I of Appendix B.)

City	Apartment Rent	Price of House	Phone Bill	Cost of Hospital Room	Price of Wine
Huntsville (AL)	468	113,920	23.07	352.50	6.16
Anchorage (AK)	750	186,061	14.43	684.00	5.99
Phoenix (AZ)	625	126,566	19.03	472.14	4.43
Little Rock (AR)	430	93,875	23.63	221.80	4.93
San Diego (CA)	821	204,951	15.64	614.36	4.39
Denver (CO)	699	154,868	20.51	484.24	4.47
Orlando (FL)	525	123,966	19.65	469.10	4.99
Bloomington (IN)	607	123,426	15.31	493.00	4.11
Des Moines (IA)	510	119,840	19.76	393.00	5.83
New Orleans (LA)	535	105,335	23.23	366.00	4.66
Minneapolis (MN)	600	124,463	22.28	536.80	5.21
Lincoln (NE)	460	99,438	16.58	355.00	4.60
Reno-Sparks (NV)	716	170,462	13.58	546.33	4.41
Manchester (NH)	610	143,950	20.09	456.00	4.99
Albuquerque (NM)	645	132,000	22.20	358.25	5.54
Albany (NY)	660	145,000	30.25	334.00	5.43
Charlotte (NC)	486	131,600	17.51	349.33	4.74
Cincinnati (OH)	562	128,839	20.51	375.00	4.99
Salem (OR)	513	153,072	19.82	380.00	4.72
Sioux Falls (SD)	595	114,950	22.99	387.00	5.31
Memphis (TN)	646	119,216	20.37	269.80	5.31
Houston (TX)	624	108,863	17.52	395.81	5.38
Salt Lake City (UT)	517	124,840	18.77	542.00	5.45
Charleston (WV)	517	144,560	29.93	275.40	5.19
Green Bay (WI)	547	129,500	15.94	318.00	4.63

Explanation of variables
Apartment rent Monthly rent of an unfurnished two-bedroom apartment (excluding all utilities except water), 1½ or 2 baths, approximately 950 square feet
Price of house Purchase price of a new house with 1800 square feet of living area, on an 8000-square-foot lot in an urban area with all utilities
Phone bill Monthly telephone charges for a private residential line (customer owns instruments)
Cost of a hospital room Average cost per day of a semiprivate room in a hospital
Price of wine Price of Paul Masson Chablis, 1.5-liter bottle

2.22   **a.**   Prepare a frequency distribution table for apartment rents using five classes of equal widths.
       b.   Construct the relative frequency and percentage distribution columns.
       c.   Write the class midpoints.

2.23   **a.**   Prepare a frequency distribution table for the purchase prices of houses using five classes of equal widths.
       b.   Calculate the relative frequencies and percentages for all classes.
       c.   Write the class midpoints.

2.24   **a.**   Prepare a frequency distribution table for phone bills.
       b.   Construct the relative frequency and percentage distribution columns.
       c.   Draw a histogram and a polygon for the relative frequency distribution.

2.25   **a.**   Prepare a frequency distribution table for the average cost per day of rooms in hospitals.
       b.   Calculate the relative frequencies and percentages for all classes.
       c.   Draw a histogram and a polygon for percentage distribution.

2.26 **a.** Prepare a frequency distribution table for the price of wine. Take $4 as the lower boundary of the first class and $.75 as the width of each class.

  b. Construct the relative frequency and percentage distribution columns.

2.27 The following data give the 1995 total annual compensation (in thousands of dollars) for 29 chief executive officers of U.S. banks and bank holding companies (*Business Week*, April 22, 1996).

2395	1578	4731	4542	849	2307	1175	4300
1131	1829	751	2586	967	2892	1275	
2780	1159	2407	909	3500	812	1567	
963	721	1206	1028	1384	1012	2813	

  a. Construct a frequency distribution table. Take $1 thousand as the lower limit of the first class and $1000 thousand as the width of each class.

  b. Prepare the relative frequency and percentage columns for the table of part a.

2.28 The following data give the number of women holding statewide elective executive offices in each of the 50 states in January 1995 (*Statistical Abstract of the United States*, 1995). The data, entered in row-wise, are for states arranged in alphabetical order.

3	1	3	2	2	3	2	4	1	1
1	2	2	4	1	4	1	2	0	1
0	2	3	0	2	1	1	1	0	1
2	1	0	5	2	3	1	2	1	1
3	1	2	2	1	0	4	0	0	2

  a. Prepare a frequency distribution table for these data using single-valued classes.

  b. Calculate the relative frequencies and percentages for all classes.

  c. How many states had two or three women in statewide elective executive offices in 1995?

  d. Draw a bar graph for the frequency distribution of part a.

2.29 The following data give the number of children less than 18 years of age for 30 randomly selected families.

2	1	2	0	3	1	1	2	2	0
1	2	0	1	0	2	1	2	0	0
1	0	0	2	1	2	3	2	0	0

  a. Prepare a frequency distribution table for these data using single-valued classes.

  b. Calculate the relative frequencies and percentages for all classes.

  c. How many families in this sample have two or three children under 18 years of age?

  d. Draw a bar graph for the frequency distribution.

2.30 The following table lists the frequency distribution for the number of members for 200 households.

Members	Households
1	30
2	40
3	65
4	37
5	28

Draw two bar graphs for these data, one without truncating the frequency axis and the second by truncating the frequency axis. In the second case, mark the frequencies on the vertical axis starting with 25. Briefly comment on the two bar graphs.

2.31   The following table lists the frequency distribution for scores of students in a class.

Score	Frequency
61– 70	11
71– 80	13
81– 90	20
91–100	15

Draw two histograms for these data, one without truncating the frequency axis and the second by truncating the frequency axis. In the second case, mark the frequencies on the vertical axis starting with 10. Briefly comment on the two histograms.

2.5   CUMULATIVE FREQUENCY DISTRIBUTIONS

Consider again Example 2–3 of Section 2.3.2 about the heights of 30 NBA players. Suppose we want to know how many players are 80 inches tall or shorter. Such a question can be answered using a **cumulative frequency distribution**. Each class in a cumulative frequency distribution table gives the total number of values that fall below a certain value. A cumulative frequency distribution is constructed for quantitative data only.

CUMULATIVE FREQUENCY DISTRIBUTION

A *cumulative frequency distribution* gives the total number of values that fall below the upper boundary of each class.

In a *less than* cumulative frequency distribution table, each class has the same lower limit but a different upper limit. Example 2–7 illustrates the procedure to prepare a cumulative frequency distribution.

Constructing cumulative frequency distribution table.

EXAMPLE 2–7   Using the frequency distribution of Table 2.9, reproduced below, prepare a cumulative frequency distribution for the heights of 30 NBA players.

Height (in inches)	f
72–74	4
75–77	6
78–80	10
81–83	7
84–86	3

Solution   Table 2.13 gives the cumulative frequency distribution for the heights of 30 NBA players. As we can observe, 72 (which is the lower limit of the first class in Table 2.9) is taken as the lower limit of each class in Table 2.13. The upper limits of all classes in Table 2.13 are the same as those in Table 2.9. To obtain the cumulative frequency of a class, we have added the frequency of that class in Table 2.9 to the frequencies of all preceding classes. The cumulative frequencies are recorded in the third column of Table 2.13. The second column of this table lists the class boundaries.

Table 2.13 Cumulative Frequency Distribution of Heights of 30 NBA Players

Class Limits	Class Boundaries	Cumulative Frequency
72–74	71.5 to less than 74.5	4
72–77	71.5 to less than 77.5	4 + 6 = 10
72–80	71.5 to less than 80.5	4 + 6 + 10 = 20
72–83	71.5 to less than 83.5	4 + 6 + 10 + 7 = 27
72–86	71.5 to less than 86.5	4 + 6 + 10 + 7 + 3 = 30

From Table 2.13, we can determine the number of observations that fall below the upper boundary of a class. For example, from Table 2.13, 20 players in this sample are 80 inches tall or shorter. ▬

The **cumulative relative frequencies** are obtained by dividing the cumulative frequencies by the total number of observations in the data set. The **cumulative percentages** are obtained by multiplying the cumulative relative frequencies by 100.

CUMULATIVE RELATIVE FREQUENCY AND CUMULATIVE PERCENTAGE

$$\text{Cumulative relative frequency} = \frac{\text{Cumulative frequency}}{\text{Total observations in the data set}}$$

$$\text{Cumulative percentage} = (\text{Cumulative relative frequency}) \cdot 100$$

Table 2.14 contains both the cumulative relative frequencies and the cumulative percentages for Table 2.13. We can observe from this table, for example, that about 66.7% of the players in this sample are 80 inches tall or shorter.

Table 2.14 Cumulative Relative Frequency and Cumulative Percentage Distributions of Heights of 30 NBA Players

Class Limits	Cumulative Relative Frequency	Cumulative Percentage
72–74	4/30 = .133	13.3
72–77	10/30 = .333	33.3
72–80	20/30 = .667	66.7
72–83	27/30 = .900	90.0
72–86	30/30 = 1.000	100.0

Ogives

When plotted on a diagram, the cumulative frequencies give a curve that is called an **ogive** (pronounced *o-jive*). Figure 2.12 gives an ogive for the cumulative frequency distribution of Table 2.13. To draw the ogive in Figure 2.12, the variable, height, is marked on the horizontal axis and the cumulative frequencies on the vertical axis. Then, the dots are marked above the upper boundaries of various classes at the heights equal to the corresponding cumulative frequencies. The ogive is obtained by joining consecutive points with straight lines. Note that the ogive starts at the lower boundary of the first class and ends at the upper boundary of the last class.

> **OGIVE**
>
> An *ogive* is a curve drawn for the cumulative frequency distribution by joining with straight lines the dots marked above the upper boundaries of classes at heights equal to the cumulative frequencies of respective classes.

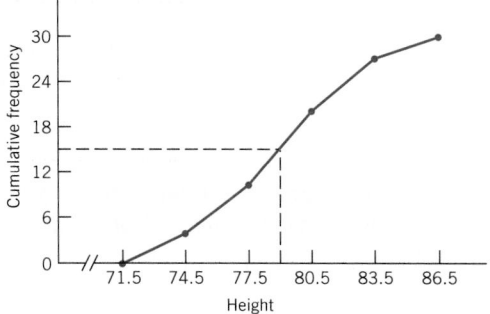

Figure 2.12   Ogive for the cumulative frequency of Table 2.13.

One advantage of an ogive is that it can be used to approximate the cumulative frequency for any interval. For example, to find the number of players with a height of 79 inches or less, first draw a vertical line from 79 on the horizontal axis up to the ogive. Then draw a horizontal line from the point where this line intersects the ogive to the vertical axis. This point gives the cumulative frequency of the class 72 to 79. In Figure 2.12, this cumulative frequency is 15. Therefore, 15 players in this sample are 79 inches tall or shorter.

We can draw an ogive for cumulative relative frequency and cumulative percentage distributions in the same way as we did for the cumulative frequency distribution.

EXERCISES

Concepts and Procedures

2.32   Briefly explain the concept of cumulative frequency distribution. How are the cumulative relative frequencies and cumulative percentages calculated?

2.33   Explain for what kind of frequency distribution an ogive is drawn. Can you think of any use of an ogive? Explain.

Applications

2.34   The following table, reproduced from Exercise 2.14, gives the frequency distribution of weekly earnings for a sample of 100 employees selected from a company.

Weekly Earnings (dollars)	Number of Employees
200 to 349	12
350 to 499	23
500 to 649	31
650 to 799	19
800 to 949	15

a. Prepare a cumulative frequency distribution table.
b. Calculate the cumulative relative frequencies and cumulative percentages for all classes.
c. What percentage of the workers earned $649 or less per week?
d. Draw an ogive for the cumulative percentage distribution.
e. Using the ogive, find the percentage of workers who earned $600 or less per week.

2.35 The following table, reproduced from Exercise 2.15, gives the frequency distribution of ages for all 50 employees of a company.

Age	Number of Employees
18 to 30	12
31 to 43	17
44 to 56	14
57 to 69	7

a. Prepare a cumulative frequency distribution table.
b. Calculate the cumulative relative frequencies and cumulative percentages for all classes.
c. What percentage of the employees of this company are 44 years of age or older?
d. Draw an ogive for the cumulative percentage distribution.
e. Using the ogive, find the percentage of employees who are 40 years old or younger.

2.36 Using the frequency distribution table constructed in Exercise 2.18, prepare the cumulative frequency, cumulative relative frequency, and cumulative percentage distributions.

2.37 Using the frequency distribution table constructed in Exercise 2.19, prepare the cumulative frequency, cumulative relative frequency, and cumulative percentage distributions.

2.38 Using the frequency distribution table constructed in Exercise 2.20, prepare the cumulative frequency, cumulative relative frequency, and cumulative percentage distributions.

2.39 Prepare the cumulative frequency, cumulative relative frequency, and cumulative percentage distributions using the frequency distribution constructed in Exercise 2.23.

2.40 Using the frequency distribution table constructed for the data of Exercise 2.25, prepare the cumulative frequency, cumulative relative frequency, and cumulative percentage distributions.

2.41 Refer to the frequency distribution table constructed in Exercise 2.26. Prepare the cumulative frequency, cumulative relative frequency, and cumulative percentage distributions by using that table.

2.42 Using the frequency distribution table constructed for the data of Exercise 2.21, prepare the cumulative frequency, cumulative relative frequency, and cumulative percentage distributions. Draw an ogive for the cumulative frequency distribution. Using the ogive, find the (approximate) number of northeastern and midwestern states for which the number of unemployed people is 180 thousand or less.

2.43 Refer to the frequency distribution table constructed in Exercise 2.27. Prepare the cumulative frequency, cumulative relative frequency, and cumulative percentage distributions by using that table. Draw an ogive for the cumulative percentage distribution. Using the ogive, find the (approximate) percentage of CEOs in this sample who had a total annual compensation of $2500 thousand or less.

2.6 STEM-AND-LEAF DISPLAYS

Another technique that is used to present quantitative data in condensed form is the **stem-and-leaf display**. An advantage of a stem-and-leaf display over a frequency distribution is that by preparing a stem-and-leaf display we do not lose information on individual observations. A stem-and-leaf display is constructed only for quantitative data.

STEM-AND-LEAF DISPLAY

In a *stem-and-leaf display* of quantitative data, each value is divided into two portions—a stem and a leaf. Then the leaves for each stem are shown separately in a display.

Example 2–8 describes the procedure for constructing a stem-and-leaf display.

Constructing stem-and-leaf display for two-digit numbers.

EXAMPLE 2–8   The following are the scores of 30 college students on a statistics test.

75	52	80	96	65	79	71	87	93	95
69	72	81	61	76	86	79	68	50	92
83	84	77	64	71	87	72	92	57	98

Construct a stem-and-leaf display.

Solution   To construct a stem-and-leaf display for these scores, we split each score into two parts. The first part contains the first digit, which is called the *stem*. The second part contains the second digit, which is called the *leaf*. Thus, for the score of the first student, which is 75, 7 is the stem and 5 is the leaf. For the score of the second student, which is 52, the stem is 5 and the leaf is 2. We observe from the above data that stems for all scores are 5, 6, 7, 8, and 9 as all the scores lie in the range 52 to 98. To obtain a stem-and-leaf display, we draw a vertical line and write the stems on the left side of it, arranged in increasing order, as shown in Figure 2.13.

After we have listed the stems, we read the leaves for all scores and record them next to the corresponding stems on the right side of the vertical line. For example, for the first score we write the leaf 5 next to the stem 7; for the second score we write the leaf 2 next to the stem 5. The recording of these two scores in a stem-and-leaf display is shown in Figure 2.13.

Stems

```
  5 │ 2 ⟵── Leaf for 52
  6 │
  7 │ 5 ⟵── Leaf for 75
  8 │
  9 │
```

Figure 2.13   Stem-and-leaf display.

Now we read all the scores and write the leaves on the right side of the vertical line in the rows of corresponding stems. The complete stem-and-leaf display for scores is shown in Figure 2.14.

```
  5 │ 2 0 7
  6 │ 5 9 1 8 4
  7 │ 5 9 1 2 6 9 7 1 2
  8 │ 0 7 1 6 3 4 7
  9 │ 6 3 5 2 2 8
```

Figure 2.14   Stem-and-leaf display of test scores.

By looking at the stem-and-leaf display of Figure 2.14, we can observe how the data values are distributed. For example, the stem 7 has the highest frequency, followed by stems 8, 9, 6, and 5.

The leaves of the stem-and-leaf display of Figure 2.14 are *ranked* (in increasing order) and presented in Figure 2.15.

```
5 │ 0 2 7
6 │ 1 4 5 8 9
7 │ 1 1 2 2 5 6 7 9 9
8 │ 0 1 3 4 6 7 7
9 │ 2 2 3 5 6 8
```

Figure 2.15   Ranked stem-and-leaf display of test scores.

As mentioned earlier, one advantage of a stem-and-leaf display is that we do not lose information on individual observations. We can rewrite the individual scores of the 30 college students from the stem-and-leaf display of Figure 2.14 or 2.15. By contrast, the information on individual observations is lost when data are grouped into a frequency table.

Constructing stem-and-leaf display for three- and four-digit numbers.

EXAMPLE 2-9   The following data give the monthly rents paid by a sample of 30 households selected from a city.

429	585	732	675	550	989	1020	620	750	660
540	578	956	1030	1070	930	871	765	880	975
650	1020	950	840	780	870	900	800	750	820

Construct a stem-and-leaf display for these data.

Solution   Each of the values in the given data set contains either three or four digits. We will take the first digit for three-digit numbers and the first two digits for four-digit numbers as stems. Then, we will use the last two digits of each number as a leaf. Thus, for the first value, which is 429, the stem is 4 and the leaf is 29. The stems for the entire data set are 4, 5, 6, 7, 8, 9, and 10. They are recorded on the left side of the vertical line in Figure 2.16. The leaves for various numbers are recorded on the right side.

```
 4 │ 29
 5 │ 85 50 40 78
 6 │ 75 20 60 50
 7 │ 32 50 65 80 50
 8 │ 71 80 40 70 00 20
 9 │ 89 56 30 75 50 00
10 │ 20 30 70 20
```

Figure 2.16   Stem-and-leaf display of rents.

Sometimes a data set may contain too many stems, with each stem containing only a few leaves. In such cases, we may want to condense the stem-and-leaf display by *grouping the stems*. The following example describes this procedure.

Preparing a grouped stem-and-leaf display.

EXAMPLE 2-10   The following is the stem-and-leaf display prepared for the stock prices of 21 companies.

```
1 | 3 5
2 | 2 5 6
3 | 0 1
4 | 2 3 6
5 | 0
6 | 5 6
7 | 0 3 9
8 | 1 5 7
9 | 2 6
```

Prepare a new stem-and-leaf display by grouping the stems.

Solution   To condense the given stem-and-leaf display, we can combine the first three rows, the middle three rows, and the last three rows, thus getting the stems 1–3, 4–6, and 7–9. The leaves for each stem of a group are separated by an asterisk (∗), as shown in Figure 2.17. Thus, the leaves 3 and 5 in the first row of Figure 2.17 correspond to stem 1; the leaves 2, 5, and 6 correspond to stem 2; and leaves 0 and 1 belong to stem 3.

```
1–3 | 3 5 ∗ 2 5 6 ∗ 0 1
4–6 | 2 3 6 ∗ 0 ∗ 5 6
7–9 | 0 3 9 ∗ 1 5 7 ∗ 2 6
```

Figure 2.17   Grouped stem-and-leaf display.

If a stem does not contain a leaf, this can be indicated in the grouped stem-and-leaf display by two consecutive asterisks. For example, in the following stem-and-leaf display there is no leaf for 3, that is, there is no number in the 30s. The numbers in this display are 21, 25, 43, 48, and 50.

$$2–5 \mid 1\ 5 \ast \ast 3\ 8 \ast 0$$

EXERCISES

Concepts and Procedures

2.44   Briefly explain how a stem-and-leaf display for a data set is prepared. You may use an example to illustrate.

2.45   What advantage does preparing a stem-and-leaf display have over grouping a data set using a frequency distribution? Give one example.

2.46   Consider the following stem-and-leaf display.

```
4 | 3 6
5 | 0 1 4 5 9
6 | 3 4 6 7 7 7 8 9
7 | 2 2 3 5 6 6
8 | 0 7 8 9
```

Write the data set that is represented by this stem-and-leaf display.

2.47   Consider the following stem-and-leaf display.

```
2–3 | 18 45 56 ∗ 29 67 83 97
4–5 | 04 27 33 71 ∗ 23 37 51 63 81 92
6–8 | 22 36 47 55 78 89 ∗ ∗ 10 41
```

Write the data set that is represented by this stem-and-leaf display.

Applications

2.48 The following data give the time (in minutes) that each of 20 students took to complete a statistics test.

55	49	53	59	38	56	39	58	47	53
58	42	37	43	47	44	55	51	46	45

Construct a stem-and-leaf display for these data. Arrange the leaves for each stem in increasing order.

2.49 Following are the SAT scores (out of a maximum possible score of 1600) of 12 students who took this test recently.

785	890	996	1169	881	1042	995	1083	773	980	1128	1066

Prepare a stem-and-leaf display. Arrange the leaves for each stem in increasing order.

2.50 Reconsider the data on the number of computer terminals produced at the Nixon Corporation for a sample of 30 days given in Exercise 2.19. Prepare a stem-and-leaf display for those data. Arrange the leaves for each stem in increasing order.

2.51 Reconsider the data on the number of computer keyboards assembled at the Twentieth Century Electronics Company given in Exercise 2.20. Prepare a stem-and-leaf display for those data. Arrange the leaves for each stem in increasing order.

2.52 Refer to Exercise 2.25. Rewrite the data on the daily average cost of rooms in hospitals by rounding each observation to the nearest dollar. Prepare a stem-and-leaf display for those data. Arrange the leaves for each stem in increasing order.

2.53 The following data give the time (in minutes) taken to commute from home to work for 20 workers.

10	50	65	33	48	5	11	23	37	26
26	32	17	7	13	19	29	43	21	22

Construct a stem-and-leaf display for these data. Arrange the leaves for each stem in increasing order. (*Note:* To prepare a stem-and-leaf display, each number in this data set can be written as a two-digit number. For example, 5 can be written as 05 for which the stem is 0 and the leaf is 5.)

2.54 The following data give the time served (in months) by 35 prison inmates who were released recently.

37	6	20	5	25	30	24	10	12	20
24	8	26	15	13	22	72	80	96	30
84	80	70	40	92	36	25	90	36	32
72	45	38	18	9					

    a. Prepare a stem-and-leaf display for these data.
    b. Condense the stem-and-leaf display by grouping the stems as 0–2, 3–5, and 6–9.

2.55 The following data give the money (in dollars) spent on textbooks by 35 college students during the 1996–97 academic year.

475	418	180	110	155	288	110	175	250
295	420	610	380	98	230	415	357	357
409	611	455	318	395	612	468	610	380
450	280	490	490	626	350	188	388	

    a. Prepare a stem-and-leaf display for these data using the last two digits as leaves.
    b. Condense the stem-and-leaf display by grouping the stems as 0–1, 2–3, and 4–6.

GLOSSARY

Bar graph   A graph made of bars whose heights represent the frequencies of respective categories.

Class   An interval that includes all the values in a (quantitative) data set that fall within two numbers, the lower and upper limits of the class.

Class boundary   The midpoint of the upper limit of one class and the lower limit of the next class.

Class frequency   The number of values in a data set that belong to a certain class.

Class midpoint or **mark**   The class midpoint or mark is obtained by dividing the sum of the lower and upper limits (or boundaries) of a class by 2.

Class width or **size**   The difference between the two boundaries of a class.

Cumulative frequency   The frequency of a class that includes all values in a data set that fall below the upper boundary of that class.

Cumulative frequency distribution   A table that lists the total number of values that fall below the upper boundary of each class.

Cumulative percentage   The cumulative relative frequency multiplied by 100.

Cumulative relative frequency   The cumulative frequency of a class divided by the total number of observations.

Frequency distribution   A table that lists all the categories or classes and the number of values that belong to each of these categories or classes.

Grouped data   A data set presented in the form of a frequency distribution.

Histogram   A graph in which classes are marked on the horizontal axis and either frequencies, relative frequencies, or percentages are marked on the vertical axis. The frequencies, relative frequencies, or percentages of various classes are represented by bars that are drawn adjacent to each other.

Ogive   A curve drawn for a cumulative frequency distribution.

Percentage   The percentage for a class or category is obtained by multiplying the relative frequency of that class or category by 100.

Pie chart   A circle divided into portions that represent the relative frequencies or percentages of different categories or classes.

Polygon   A graph formed by joining the midpoints of the tops of successive bars in a histogram by straight lines.

Raw data   Data recorded in the sequence in which they are collected and before they are processed.

Relative frequency   The frequency of a class or category divided by the sum of all frequencies.

Skewed-to-the-left histogram   A histogram with a longer tail on the left side.

Skewed-to-the-right histogram   A histogram with a longer tail on the right side.

Stem-and-leaf display   A display of data in which each value is divided into two portions, a stem and a leaf.

Symmetric histogram   A histogram that is identical on both sides of its central point.

Uniform or **rectangular histogram**   A histogram with the same frequency for all classes.

KEY FORMULAS

1. **Relative frequency of a class**

$$\text{Relative frequency of a class} = \frac{\text{Frequency of that class}}{\text{Sum of all frequencies}} = \frac{f}{\Sigma f}$$

2. **Percentage of a class**

$$\text{Percentage} = (\text{Relative frequency}) \times 100$$

3. **Class midpoint or mark**

$$\text{Class midpoint} = \frac{\text{Upper limit} + \text{Lower limit}}{2}$$

4. **Class width or size**

$$\text{Class width} = \text{Upper boundary} - \text{Lower boundary}$$

5. **Cumulative relative frequency**

$$\text{Cumulative relative frequency} = \frac{\text{Cumulative frequency}}{\text{Total observations in the data set}}$$

6. **Cumulative percentage**

$$\text{Cumulative percentage} = (\text{Cumulative relative frequency}) \times 100$$

SUPPLEMENTARY EXERCISES

2.56 The following data give the political party of each of the first 30 U.S. presidents. In the data, D stands for Democrat, DR for Democratic Republican, F for Federalist, R for Republican, and W for Whig.

F	F	DR	DR	DR	DR	D	D	W
W	D	W	W	D	D	R	D	R
R	R	R	D	R	D	R	R	R
D	R	R						

a. Prepare a frequency distribution table for these data.
b. Calculate the relative frequency and percentage distributions.
c. Draw a bar graph for the relative frequency distribution and a pie chart for the percentage distribution.
d. What percentage of these presidents were Whigs?

2.57 The following data indicate the country of origin of each of the top 40 corporations in *Fortune* magazine's Global 500 with the highest 1995 revenues (*Fortune*, August 5, 1996).

Japan	Japan	Japan	U.S.	Japan
Japan	U.S.	Japan	U.S.	Brit./Neth.
Japan	U.S.	Japan	Japan	Japan
U.S.	Germany	U.S.	Japan	U.S.
Japan	U.S.	Japan	Germany	Germany
Japan	Britain	Switzerland	U.S.	U.S.
U.S.	Japan	Japan	S. Korea	Japan
Japan	Japan	Brit./Neth.	Switzerland	Japan

a. Prepare a frequency distribution table for these data.
b. Calculate the relative frequency and percentage distributions.
c. Draw a bar graph for the frequency distribution and a pie chart for the percentage distribution.
d. What percentage of these corporations are from the United States?

2.58 The following data give the number of television sets owned by 40 randomly selected households.

1	1	2	3	2	4	1	3	2	1
3	0	2	1	2	3	2	3	2	2
1	2	1	1	1	3	1	1	1	2
2	4	2	3	1	3	1	2	2	4

a.   Prepare a frequency distribution table for these data using single-valued classes.
b.   Compute the relative frequency and percentage distributions.
c.   Draw a bar graph for the frequency distribution.
d.   What percentage of the households own two or more television sets?

2.59   The following data give the number of persons in each of 30 groups who made reservations at a restaurant on a recent Friday night.

2	3	2	2	1	2	4	4	2	1
5	2	3	2	2	3	1	4	2	4
4	2	5	4	1	2	2	4	3	2

a.   Prepare a frequency distribution table for these data using single-valued classes.
b.   Compute the relative frequency and percentage distributions.
c.   What percentage of the groups in this sample have two or three persons?
d.   Draw a bar graph for the relative frequency distribution.

2.60   The following data give the amount spent on video rentals (in dollars) during 1996 by 30 households randomly selected from those households who rented videos in 1996.

595	24	6	100	100	40	622	405	90
55	135	760	405	90	205	70	180	88
808	100	40	127	83	310	350	130	22
111	70	15						

a.   Construct a frequency distribution table. Take $1 as the lower limit of the first class and $200 as the width of each class.
b.   Calculate the relative frequencies and percentages for all classes.
c.   What percentage of the households in this sample spent more than $400 on video rentals in 1996?

2.61   The following data give the number of orders received for a sample of 30 hours at the Timesaver Mail Order Company.

34	44	31	52	41	47	38	35	32	39
28	24	46	41	49	53	57	33	27	37
30	27	45	38	34	46	36	30	47	50

a.   Construct a frequency distribution table. Take 23 as the lower limit of the first class and 7 as the width of each class.
b.   Calculate the relative frequencies and percentages for all classes.
c.   For what percentage of the hours in this sample was the number of orders more than 36?

2.62   The following data give the weekly expenditures (in dollars) on fruit and vegetables for 30 households randomly selected from the households who incurred such expenses.

4.57	3.95	6.95	3.80	1.50	2.99	7.84	5.05
8.00	14.75	9.33	1.05	5.08	7.00	9.60	18.99
9.15	11.32	4.75	9.95	3.63	1.99	1.39	13.09
19.31	11.15	7.73	12.00	7.58	16.35		

a.   Construct a frequency distribution table using the *less than* method to write classes. Take $0 as the lower boundary of the first class and $4 as the width of each class.
b.   Calculate the relative frequencies and percentages for all classes.
c.   Draw a histogram for the frequency distribution.

2.63 The following data give the repair costs (in dollars) for 30 cars randomly selected from a list of cars that were involved in collisions.

2300	750	2500	410	555	1576
2460	1795	2108	897	989	1866
2105	335	1344	1159	1236	1395
6108	4995	5891	2309	3950	3950
6655	4900	1320	2901	1925	6896

    a. Construct a frequency distribution table. Take $1 as the lower limit of the first class and $1400 as the width of each class.

    b. Compute the relative frequencies and percentages for all classes.

    c. Draw a histogram and a polygon for the relative frequency distribution.

    d. What are the class boundaries and the width of the fourth class?

2.64 Refer to Exercise 2.60. Prepare the cumulative frequency, cumulative relative frequency, and cumulative percentage distributions by using the frequency distribution table of that exercise.

2.65 Refer to Exercise 2.61. Prepare the cumulative frequency, cumulative relative frequency, and cumulative percentage distributions using the frequency distribution table constructed for the data of that exercise.

2.66 Refer to Exercise 2.62. Prepare the cumulative frequency, cumulative relative frequency, and cumulative percentage distributions using the frequency distribution table constructed for the data of that exercise.

2.67 Construct the cumulative frequency, cumulative relative frequency, and cumulative percentage distributions by using the frequency distribution constructed for the data of Exercise 2.63.

2.68 Refer to Exercise 2.60. Prepare a stem-and-leaf display for the data of that exercise.

2.69 Construct a stem-and-leaf display for the data given in Exercise 2.61.

2.70 The following table gives the divorce rate (as the number of divorces per hundred marriages) for the United States, Sweden, England and Wales, and Canada based on data from the Population Council (*U.S. News & World Report*, June 12, 1995).

Country	Divorce Rate
United States	54.8
Sweden	44.1
England & Wales	41.7
Canada	38.3

Draw two bar graphs for these data, one without truncating the axis on which the divorce rate is marked and the second by truncating this axis. In the second case, mark the divorce rate on the vertical axis starting with 35. Briefly comment on the two bar graphs.

2.71 The following table gives the average number of days in the school year for five countries (*U.S. News & World Report*, April 1, 1996).

Country	Average Number of School Days
Japan	240
Korea	222
Israel	215
Canada	188
United States	178

Draw two bar graphs for these data, one without truncating the axis on which school days are marked and the second by truncating this axis. In the second case, mark the number of school days on the vertical axis starting with 175. Briefly comment on the two bar graphs.

*2.72   Consider the graphs shown below that give information on suspicious fires in the state of Connecticut. (*Source: The Hartford Courant*, March 17, 1996. Reproduced with permission.)

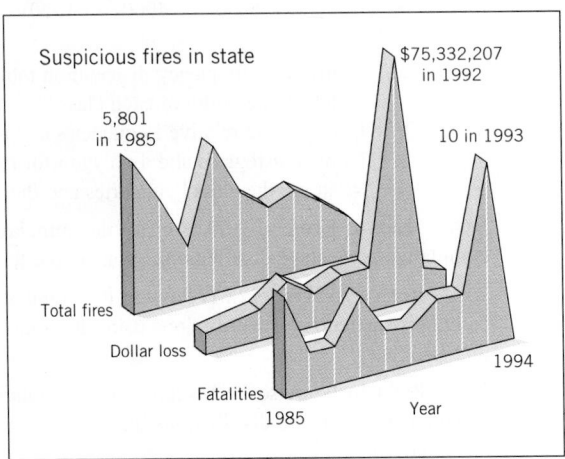

a.   Can you obtain even a rough estimate for total fires, dollar loss, or fatalities in years other than those for which such information is given? Why or why not? Explain.
b.   Is it clear at what value the vertical axis for each graph starts? In other words, can you tell by looking at these graphs if the vertical axes are truncated?
c.   What improvements in the presentation of these graphs would you suggest?

*2.73   The following frequency distribution table gives the age distribution of drivers who were at fault in auto accidents that occurred during a one-week period in a city.

Age	f
18 to less than 20	7
20 to less than 25	12
25 to less than 30	18
30 to less than 40	14
40 to less than 50	15
50 to less than 60	16
60 and over	35

a.   Draw a relative frequency histogram for this table.
b.   In what way(s) is this histogram misleading?
c.   How could you change the frequency distribution so that the resulting histogram would give a clearer picture?

*2.74   Suppose a data set contains the ages of 135 auto workers ranging from 20 to 53.

a.   Using Sturge's formula given in footnote 1 at the beginning of section 2.3.2, find an appropriate number of classes for a frequency distribution for this data set.
b.   Find an appropriate class width based on the number of classes in part a.

*2.75   The following table gives unemployment rates (as percentage of the civilian labor force) for several occupational categories as of March 1996 (*U.S. Bureau of Labor Statistics*, April 5, 1996).

Category	Unemployment Rate
Mining	6.8
Construction	10.0
Manufacturing	5.3
Transportation and public utilities	4.2
Wholesale and retail trade	6.9
Finance, insurance, and real estate	2.5
Services	5.4
Government workers	2.8
Agricultural wage and salary workers	10.7

a. Construct a bar graph for this data set.

b. Would a pie chart be appropriate for this data set? Why or why not?

c. If you combine construction and manufacturing groups into a single category, could you determine the unemployment rate for this new category with the information given in the table? If so, find this unemployment rate. If not, indicate what additional information you would need to find this rate.

*2.76 In some applications, it may be desirable to use open-ended classes or classes of different widths in order to accommodate the data. Consider the following data on the ages of 35 students enrolled in an introductory statistics class.

17	18	18	18	18	19	19	19	19	19	19
19	19	19	20	20	20	20	20	20	20	21
21	21	22	22	23	26	28	30	38	43	51
62	64									

a. Construct a stem-and-leaf display for these data.

b. Do you think a stem-and-leaf display is appropriate for this data set? Explain.

c. Construct a frequency distribution for these data using equal class widths. What problems do you encounter?

d. Construct a frequency distribution using the first four classes with a width of 2 each, the next two classes having a width of 10 each, and the last class being an open-ended class.

e. Draw a histogram based on the frequency distribution table in part d.

SELF-REVIEW TEST

1. Briefly explain the difference between ungrouped and grouped data and give one example of each type.

2. The following table gives the frequency distribution of the duration (in minutes) of 100 long-distance phone calls made by persons using TVI long-distance service.

Duration (minutes)	Frequency
0 to 4	8
5 to 9	22
10 to 14	35
15 to 19	20
20 to 24	15

Circle the correct answer for each of the following questions, which are based on this table.

   a. The number of classes in the table is   5, 100, 80

   b. The class width is   4, 5, 10

   c. The midpoint of the third class is   11.5, 12, 12.5

   d. The lower boundary of the second class is   4.5, 5, 5.5

   e. The upper limit of the second class is   8.5, 9, 9.5

   f. The sample size is   5, 100, 50

   g. The relative frequency of the first class is   .04, .08, .16

3. Briefly explain and illustrate with the help of graphs a symmetric histogram, a histogram skewed to the right, and a histogram skewed to the left.

4. Twenty elementary school children were asked if they live with both parents (B), father only (F), mother only (M), or someone else (S). The responses of the children are as follows.

M	B	B	M	F	S	B	M	F	B
B	F	B	M	M	B	B	F	B	M

   a. Construct a frequency distribution table.

   b. Write the relative frequencies and percentages for all categories.

   c. What percentage of the children in this sample live with their mothers only?

   d. Draw a bar graph for the frequency distribution and a pie chart for the percentages.

5. The following data set gives the number of years for which 24 workers have been with their current employers.

15	12	9	10	5	12	3	7	16	13	11	14
11	8	7	14	11	8	4	13	2	18	6	19

   a. Construct a frequency distribution table. Take 1 as the lower limit of the first class and 4 as the width of each class.

   b. Calculate the relative frequencies and percentages for all classes.

   c. What percentage of the employees have been with their current employers for 8 or fewer years?

   d. Draw the frequency histogram and polygon.

6. Refer to the frequency distribution prepared in Problem 5. Prepare the cumulative percentage distribution using that table. Draw an ogive for the cumulative percentage distribution. Using the ogive, find the percentage of employees in the sample who have been with their current employers for 11 or fewer years.

7. Construct a stem-and-leaf display for the following data, which give the time (in minutes) taken by an accountant to prepare 20 income tax returns.

34	21	67	53	18	38	45	56	62	48
75	58	43	69	56	38	71	50	42	36

8. Consider the following stem-and-leaf display.

```
3 | 0 3 7
4 | 2 4 6 7 9
5 | 1 3 3 6
6 | 0 7 7
7 | 1 9
```

Write the data set that was used to construct this stem-and-leaf display.

USING MINITAB

This section illustrates how to use MINITAB to obtain a bar graph, a pie chart, a histogram, and a stem-and-leaf display. We will illustrate the required steps using examples.

BAR GRAPH AND PIE CHART

Illustration M2–1 shows how to make a bar graph and a pie chart for the frequency distribution of a qualitative data set. The procedure is the same to construct a bar graph and a pie chart for quantitative data.

ILLUSTRATION M2-1   Refer to the frequency distribution of student majors recorded in Table 2.4 of Example 2–1. Using MINITAB, prepare a bar graph and a pie chart for that frequency distribution.

Solution   If you are using MINITAB FOR WINDOWS, perform the following steps to construct a *bar graph* for the frequency distribution of Table 2.4.

Step 1.   Enter the data containing categories and frequencies (from the first and third columns of Table 2.4) into columns C1 and C2, respectively, of the Data window.
Step 2.   Click the **Graph** pull-down menu at the top of the screen.
Step 3.   Click the **Chart** from the selections available in the **Graph** menu.
Step 4.   You will see a dialog box entitled **Chart** appear on the screen. Type **C2** in the box below **Y** and **C1** in the box below **X**.
Step 5.   Click the **OK** button at the bottom of the dialog box. The bar graph will appear on the screen.

If you are using the MINITAB COMMAND LANGUAGE, enter the given data on categories and frequencies (from the first and third columns of Table 2.4) into columns C1 and C2, respectively, using the **SET** command. Note that you will have to use the **FORMAT** command to enter the categories. Then, type the following MINITAB command to obtain the bar graph.

```
MTB > CHART C2*C1
```

Whether you use MINITAB FOR WINDOWS or the MINITAB COMMAND LANGUAGE, you will obtain the bar graph given in Figure 2.18 for the data of Table 2.4.

Figure 2.18   Bar graph for the frequency distribution of Table 2.4.

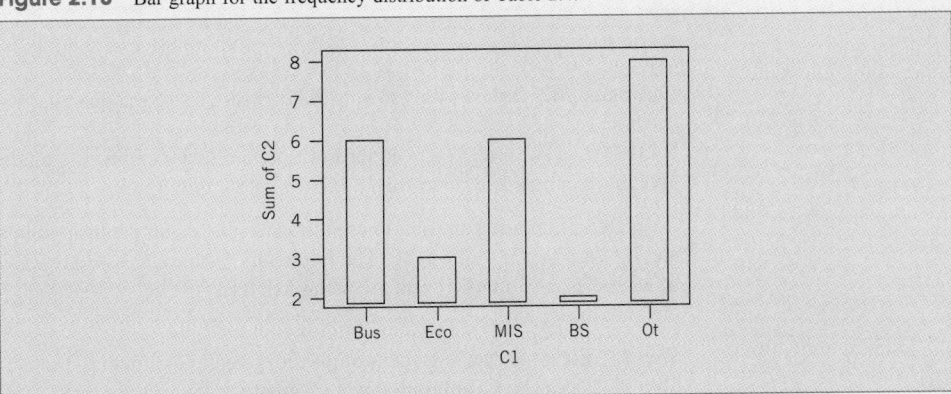

To obtain a *pie chart* for the frequency distribution of Table 2.4 of Example 2-1, perform the following steps if you are using MINITAB FOR WINDOWS.

Step 1. Enter the data containing categories and frequencies (from the first and third columns of Table 2.4) into columns C1 and C2, respectively, of the Data window.

Step 2. Click the **Graph** pull-down menu at the top of the screen.

Step 3. Click the **Pie chart** from the selections available in the **Graph** menu.

Step 4. You will see a dialog box entitled **Pie Chart** appear on the screen. Click inside the circle next to the **Chart table**. Then, type **C1** in the box next to **Categories in** and **C2** in the box next to **Frequencies in**.

Step 5. Click the **OK** button at the bottom of this dialog box. The pie chart will appear on the screen.

If you are using the MINITAB COMMAND LANGUAGE, enter the given data on categories and frequencies (from the first and third columns of Table 2.4) into columns C1 and C2 using the **SET** command. Then, type the following MINITAB commands to obtain the pie chart.

```
MTB  > %PIE C1;
SUBC > COUNTS C2.
```

Here, the first command tells MINITAB to make a pie chart for the categories listed in column C1. The subcommand tells it to use the frequencies listed in column C2.

Whether you use MINITAB FOR WINDOWS or the MINITAB COMMAND LANGUAGE, you will obtain the pie chart given in Figure 2.19 for the data of Table 2.4. In the pie chart, MINITAB lists frequencies and percentages within parentheses next to the categories.

Figure 2.19 Pie chart for the frequency distribution of Table 2.4.

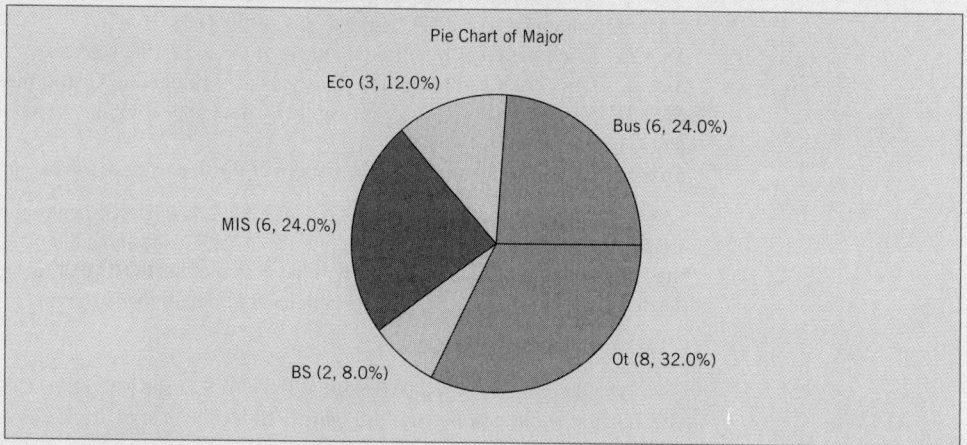

HISTOGRAM

Illustration M2–2 shows how to construct a histogram for a quantitative data set.

ILLUSTRATION M2–2   Refer to the data on the heights of a random sample of 30 NBA players given in Example 2–3. Construct a histogram for that data set.

Solution   Suppose you want to use the same classes for the histogram as used in Example 2-3, which are 72–74, 75–77, . . . , 84–86. The midpoints of these classes are 73, 76, 79, 82, and 85. The width of each class is 3 units. If you are using MINITAB FOR WINDOWS, perform the following steps to obtain a histogram.

Step 1. Enter the data on heights of players given in Example 2–3 into column C1 of the Data window.

Step 2. Click the **Graph** pull-down menu at the top of the screen.

Step 3. Click the **Histogram** from the selections available in the **Graph** menu.

Step 4. You will see a dialog box entitled **Histogram** appear on the screen. Type **C1** in the first box below **X**.

Step 5. Click the **Options** button that appears at the bottom of this dialog box.

Step 6. Click inside the circle next to **MidPoint** below **Type of Intervals** and then click inside the circle next to **Midpoint/cutpoint positions** below **Definition of Intervals**. Next, type **73:85/3** in the box next to **Midpoint/cutpoint positions**. Note that, here, 73 is the midpoint of the first class, 85 is the midpoint of the last class, and 3 is the width of a class.

Step 7. Click the **OK** button at the bottom of this dialog box. Click **OK** again. The histogram will appear on the screen.

If you are using the MINITAB COMMAND LANGUAGE, enter the given data on heights into column C1 using the **SET** command. Then, type the following MINITAB commands to obtain the histogram.

```
MTB > HISTOGRAM C1;
SUBC > MIDPOINTS 73:85/3.
```

Observe these MINITAB commands carefully for semicolons and periods. Note that when we put a semicolon at the end of a MINITAB command, it instructs MINITAB that a subcommand **(SUBC)** is to follow with some additional information. A semicolon at the end of a subcommand would indicate that another subcommand is to follow with some more information. A period at the end of a subcommand instructs MINITAB that all MINITAB commands and subcommands have been entered.

Whether you use MINITAB FOR WINDOWS or the MINITAB COMMAND LANGUAGE, you will obtain the histogram given in Figure 2.20 for the data on heights of NBA players from Example 2–3.

Figure 2.20 Histogram for heights of NBA players.

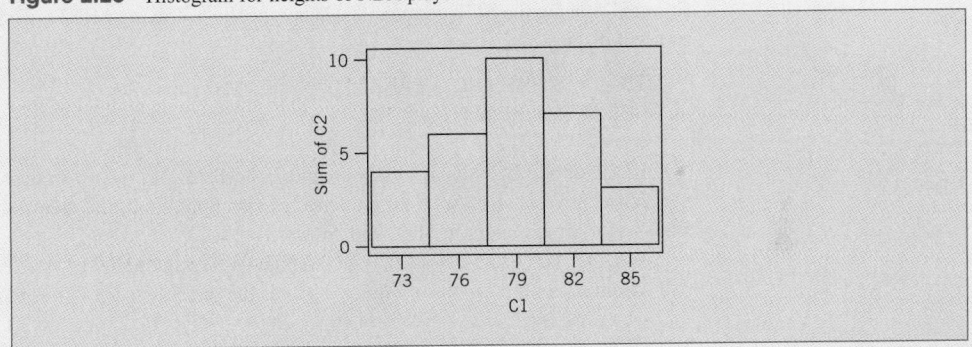

In the above MINITAB commands, we selected our own classes. If you want MINITAB to choose classes and widths, skip Steps 5 and 6 of the MINITAB FOR WINDOWS steps. In the MINITAB COMMAND LANGUAGE, just type **HISTOGRAM C1** next to *MTB* > and skip the subcommand.

You can also use the cutpoints (which are the same as the boundaries of classes) instead of the midpoints to label the horizontal axis of a histogram. To do so, in MINITAB FOR WINDOWS, click the **Cutpoints** below **Type of Intervals** in Step 6. Then type **71.5:86.5/3** in the box next to **Midpoint/ cutpoint positions**. Then follow Step 7. In the MINITAB COMMAND LANGUAGE, type **CUT-POINTS 71.5:86.5/3** instead of **MIDPOINTS 73:85/3** in the subcommand. Note, here, 71.5 is the lower boundary of the first class, 86.5 is the upper boundary of the last class, and 3 is the width of each class.

If you want to construct a **frequency polygon** using MINITAB FOR WINDOWS, then click the **Graph** pull-down menu, select **Histogram**, type **C1** in the box below **X**, click the arrow next to **Display**, click **Connect**, click the **Options** button at the bottom of the dialog box, click **Midpoint/cutpoint positions**, type **73:85/3** or **71.5:86.5/3** in the box, and finally, click **OK** twice. MINITAB will produce the frequency polygon. If you are using the MINITAB COMMAND LANGUAGE, then use the following commands.

```
MTB  > HISTOGRAM C1;
SUBC > MIDPOINTS 73:85/3;
SUBC > CONNECT.
```

You can use **CUTPOINTS 71.5:86.5/3** instead of **MIDPOINTS 73:85/3** in the first subcommand above.

STEM-AND-LEAF DISPLAY

Illustration M2–3 shows how to construct a stem-and-leaf display for a data set.

ILLUSTRATION M2-3   Refer to the data on test scores for 30 students given in Example 2–8. Prepare a stem-and-leaf display for that data set.

Solution   If you are using MINITAB FOR WINDOWS, perform the following steps to obtain a stem-and-leaf display.

Step 1.   Enter the data on scores given in Example 2–8 into column C1 of the Data window.
Step 2.   Click the **Graph** pull-down menu at the top of the screen.
Step 3.   Click **Character Graphs** from the selections available in the **Graph** menu.
Step 4.   Click **Stem-and-Leaf** from the selections available in the **Character Graphs** menu.
Step 5.   You will see a dialog box entitled **Stem-and-Leaf** appear on the screen. Type **C1** in the box below **Variables**, and **10** in the box next to **Increment**.
Step 6.   Click the **OK** button at the bottom of this dialog box. The stem-and-leaf display will appear on the screen.

If you are using the MINITAB COMMAND LANGUAGE, enter the given data on scores into column C1 using the **SET** command. Then, type the following MINITAB commands to obtain the stem-and-leaf display.

```
MTB  > STEM-AND-LEAF C1;
SUBC > INCREMENT = 10.
```

In the above MINITAB subcommand, INCREMENT = 10 indicates the distance between any two consecutive stems. As a consequence of this command, the stems will be 5 (for the numbers in 50s), 6 (for the numbers in 60s), and so forth.

Whether you use MINITAB FOR WINDOWS or the MINITAB COMMAND LANGUAGE, you will obtain the output given in Figure 2.21 for the data on scores from Example 2–8.

Figure 2.21   Stem-and-leaf display for Illustration M2–3.

```
Character Stem-and-Leaf Display

Stem-and-leaf of C1        N = 30
Leaf Unit = 1.0


     3     5 027
     8     6 14589
    (9)    7 112256799
    13     8 0134677
     6     9 223568
```

In the MINITAB printout of Figure 2.21, N = 30 is the number of observations in the data. LEAF UNIT = 1.0 means that the decimal point is after one leaf digit in the printout. Thus, the first number is 50, the second is 52, and so on. If the leaf unit is .10, the numbers in the above stem-and-leaf display will be 5.0, 5.2, 5.7, and so on. (See Computer Assignment M2.9 as an example of this case.) On the other hand, if the leaf unit is 10, then the numbers will be 500, 520, 570, and so on.

The numbers in the first column of the stem-and-leaf display of Figure 2.21 are called depths, which give the cumulative frequencies from above and below. The depth, which appears in parentheses (9 in this MINITAB output), gives the number of leaves in the row that contains the median value. The depths before this row give the total number of leaves in the corresponding row and the row or rows before it. Thus, the first depth (which is 3) gives the number of leaves belonging to stem 5. The second number (which is 8) gives the cumulative number of leaves belonging to the first two stems. After the stem that contains the median data value, the depths are cumulative from the bottom of the stem-and-leaf display. For example, the depth of 13 for the stem of 8 indicates the total number of leaves belonging to stems 8 and 9. The last depth, which is 6, indicates the number of leaves for stem 9. ▄▄

DOTPLOT

We can also display data using a **dotplot**. A dotplot shows each observation of the data set by a dot on the graph. The following illustration shows how to make a dotplot.

ILLUSTRATION M2–4   Refer to the data on heights of a random sample of 30 NBA players given in Example 2–3. Construct a dotplot for that data set.

Solution   The following steps will be used to prepare a dotplot if you are using MINITAB FOR WINDOWS.

Step 1.   Enter the given data on heights of players into column C1 of the Data window.
Step 2.   Click the **Graph** pull-down menu at the top of the screen.
Step 3.   Click the **Character graphs** from the selections available in the **Graph** menu.
Step 4.   Click the **Dotplot** from the selections available in the **Character graphs** menu.
Step 5.   You will see a dialog box entitled **Dotplot** appear on the screen. Type **C1** in the box below **Variables**.
Step 6.   Click the **OK** button at the bottom of this dialog box. The dotplot will appear on the screen.

If you are using the MINITAB COMMAND LANGUAGE, enter the given data on heights of players into column C1 using the **SET** command. Then, type the following MINITAB command to obtain the dotplot.

```
MTB > DOTPLOT C1
```

By using either MINITAB FOR WINDOWS or MINITAB COMMAND LANGUAGE, you will obtain the dotplot shown in Figure 2.22.

Figure 2.22   Dotplot for Illustration M2–4.

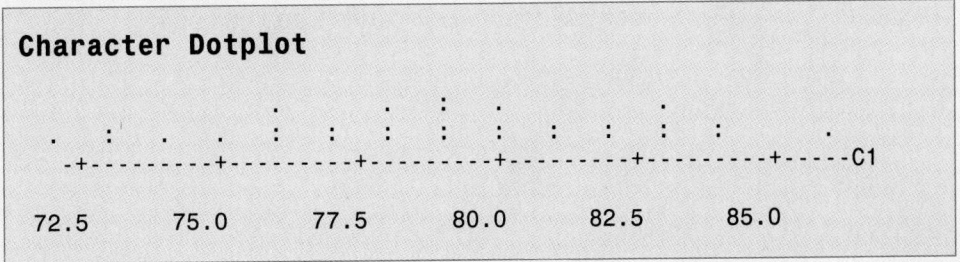

COMPUTER ASSIGNMENTS

M2.1   Using MINITAB, construct a bar graph and a pie chart for the frequency distribution prepared in Exercise 2.5.

M2.2 Using MINITAB, construct a bar graph and a pie chart for the frequency distribution prepared in Exercise 2.6.

M2.3 Refer to Data Set IV of Appendix B on the time taken to run the Manchester Road Race for a sample of 500 participants. From that data set, select the 6th value and then select every 10th value after that (i.e., select the 6th, 16th, 26th, 36th, . . . values). This subsample will give you 50 measurements. (Such a sample selected from a population is called a *systematic random sample*.) Using MINITAB, construct a histogram for these data. Let MINITAB decide on classes and class limits.

M2.4 Refer to Data Set I of Appendix B on the prices of various products in different cities across the country. Using MINITAB, select a subsample of 60 from column C7 (telephone charges) and then construct a histogram and a polygon for these data.

M2.5 Using MINITAB, construct a histogram for the data on the number of computer keyboards assembled that were given in Exercise 2.20. Use the classes mentioned in that exercise. Use the midpoints to mark the horizontal axis in the histogram.

M2.6 Using MINITAB, construct a polygon for the data on the number of computer keyboards assembled that were given in Exercise 2.20. Use the classes mentioned in that exercise. Use the cutpoints to mark the horizontal axis in the polygon.

M2.7 Using MINITAB, prepare a stem-and-leaf display for the data given in Exercise 2.48.

M2.8 Using MINITAB, prepare a stem-and-leaf display for the data of Exercise 2.53.

M2.9 The following data give the weights (in pounds) of 20 of the parcels mailed from a post office during the past week.

1.8	7.5	8.2	3.4	5.1	9.3	1.9	2.5	7.3	5.8
6.2	8.6	2.0	6.3	8.5	0.7	3.8	7.3	5.2	3.7

Using MINITAB, prepare a stem-and-leaf display for these data. Use INCREMENT = 1. Observe the value of the leaf unit in the output.

M2.10 Using MINITAB, prepare a dotplot for the data obtained in Computer Assignment M2.3.

M2.11 Using MINITAB, prepare a dotplot for the data obtained in Computer Assignment M2.4.

M2.12 Using MINITAB, prepare a bar graph for the frequency distribution obtained in Exercise 2.28.

M2.13 Using MINITAB, prepare a bar graph for the frequency distribution obtained in Exercise 2.29.

M2.14 Using MINITAB, make a pie chart for the frequency distribution obtained in Exercise 2.19.

M2.15 Using MINITAB, make a pie chart for the frequency distribution obtained in Exercise 2.29.

3

NUMERICAL DESCRIPTIVE MEASURES

In Chapter 2 we discussed how to organize and display large data sets. The techniques presented in that chapter, however, are not helpful when we need to describe verbally the main characteristics of a data set. The numerical summary measures, such as the ones that give the center and spread of a distribution, provide us with the main features of a data set. For example, the techniques learned in Chapter 2 can help us to graph the data on family incomes. However, we may want to know the income of a "typical" family (given by the center of the distribution), the spread of the distribution of incomes, or the location of a family with a specific income. Figure 3.1 shows these three concepts. Such questions can

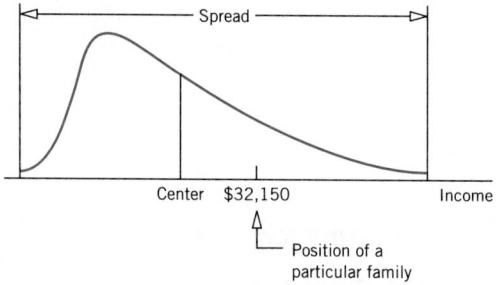

Figure 3.1

be answered using the numerical summary measures discussed in this chapter. Included among these are (1) measures of central tendency, (2) measures of dispersion, and (3) measures of position.

3.1 MEASURES OF CENTRAL TENDENCY FOR UNGROUPED DATA

We often represent a data set by numerical summary measures, usually called the *typical values*. A measure of central tendency gives the center of a histogram or a frequency distribution curve. This section discusses three different measures of central tendency: the mean, the median, and the mode. We will learn how to calculate each of these measures for ungrouped data. Recall from Chapter 2 that data containing information on each member of the population or sample individually are called *ungrouped data*, whereas *grouped data* refer to the data presented in the form of a frequency distribution table.

3.1.1 MEAN

The **mean**, also called the *arithmetic mean*, is the most frequently used measure of central tendency. This book will use the words *mean* and *average* synonymously. For ungrouped data, the mean is obtained by dividing the sum of all values by the number of values in the data set.

$$\text{Mean} = \frac{\text{Sum of all values}}{\text{Number of values}}$$

The mean calculated for sample data is denoted by $\bar{x}$ (read as "x bar"), and the mean calculated for population data is denoted by μ (Greek letter *mu*). We know from the discussion in Chapter 2 that the number of values in a data set is denoted by n for a sample and by N for a population. In Chapter 1 we learned that a variable is denoted by x and the sum of all values of x is denoted by Σx. Using these notations, we can write the following formulas for the mean.

MEAN FOR UNGROUPED DATA

The *mean for ungrouped data* is obtained by dividing the sum of all values by the number of values in the data set. Thus,

Mean for population data: $\quad \mu = \dfrac{\Sigma x}{N}$

Mean for sample data: $\quad \bar{x} = \dfrac{\Sigma x}{n}$

where Σx is the sum of all values, N is the population size, n is the sample size, μ is the population mean, and $\bar{x}$ is the sample mean.

Calculating sample mean for ungrouped data.

EXAMPLE 3–1   The following data give the prices (rounded to thousands of dollars) of five homes sold recently in Seattle.

$$158 \quad 189 \quad 265 \quad 127 \quad 191$$

Find the mean sale price for these homes.

Solution   The variable in this example is the *sale price of homes*. Let us denote it by x. Then the five values of x are

$$x_1 = 158, \quad x_2 = 189, \quad x_3 = 265, \quad x_4 = 127, \quad \text{and} \quad x_5 = 191$$

where x_1 represents the sale price of the first home, x_2 denotes the sale price of the second home, and so on. The sum of the sale prices of these five homes is

$$\Sigma x = x_1 + x_2 + x_3 + x_4 + x_5$$
$$= 158 + 189 + 265 + 127 + 191 = \$930 \text{ thousand}$$

Note that the given information is on only five homes. Hence, it represents a sample. Because the data set contains five values, $n = 5$. Substituting the values of Σx and n in the sample formula, the mean sale price of these five homes is

$$\bar{x} = \frac{\Sigma x}{n} = \frac{930}{5} = \textbf{\$186 thousand}$$

Thus, these five homes were sold for an average price of \$186,000.   ▬

Physically, the mean is the point that balances a histogram. If we consider the mean as a fulcrum, the histogram will balance on the fulcrum. This is shown in Figure 3.2 for the data of Example 3–1.

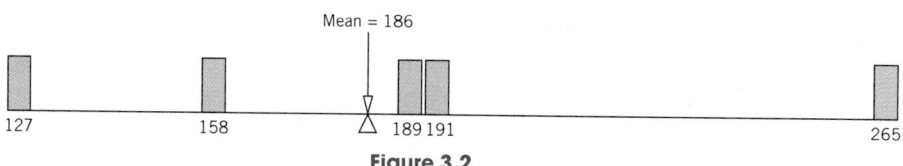

Figure 3.2

Calculating population mean for ungrouped data.

EXAMPLE 3–2 The following are the weekly earnings of *all* eight employees of a small company.

$450   530   825   370   615   480   910   560

Find the mean weekly earnings of these employees.

Solution Since the given data set includes *all* eight employees of the company, it represents the population. Hence, $N = 8$.

$$\Sigma x = 450 + 530 + 825 + 370 + 615 + 480 + 910 + 560 = 4740$$

The population mean is

$$\mu = \frac{\Sigma x}{N} = \frac{4740}{8} = \mathbf{\$592.50}$$

Thus, the mean weekly earnings of the employees of this company are $592.50. ■

Reconsider Example 3–2. Now if we take a sample of three employees from this company and calculate the mean weekly earnings of those three employees, this mean will be denoted by $\bar{x}$. Suppose the three values included in the sample are $530, $480, and $910. Then, the mean weekly earnings for this sample are

$$\bar{x} = (530 + 480 + 910)/3 = \mathbf{\$640}$$

If we take a second sample of three employees of this company, the value of $\bar{x}$ will (most likely) be different. Suppose the second sample includes the values $450, $825, and $480. Then, the mean weekly earnings for this sample are

$$\bar{x} = (450 + 825 + 480)/3 = \mathbf{\$585}$$

Consequently, we can state that the value of the population mean μ is constant. However, the value of the sample mean $\bar{x}$ varies from sample to sample. The value of $\bar{x}$ for a particular sample will depend on what values of the population are included in that sample.

Sometimes a data set may contain a few very small or a few very large values. Such values are called **outliers** or **extreme values**.

OUTLIERS OR EXTREME VALUES

Values that are very small or very large relative to the majority of the values in a data set are called *outliers* or *extreme values*.

A major shortcoming of the mean as a measure of central tendency is that it is very sensitive to outliers. Example 3–3 illustrates this point.

Illustrating the effect of an outlier on the mean.

EXAMPLE 3-3 Table 3.1 lists the 1995 population (in thousands) for the five Pacific states.

Table 3.1

State	Population (thousands)
Washington	5,431
Oregon	3,141
Alaska	604
Hawaii	1,187
California	31,589 ⟵ An outlier

Notice that the population of California is very large compared to the populations of the other four states. Hence, it is an outlier. Show how the inclusion of this outlier affects the value of the mean.

Solution If we do not include the population of California (the outlier), the mean population of the remaining four states (Washington, Oregon, Alaska, and Hawaii) is

$$\text{Mean} = \frac{5431 + 3141 + 604 + 1187}{4} = \textbf{2590.75 thousand}$$

Now, to see the impact of the outlier on the value of the mean, we include the population of California and find the mean population of all five Pacific states. This mean is

$$\text{Mean} = \frac{5431 + 3141 + 604 + 1187 + 31,589}{5} = \textbf{8390.4 thousand}$$

Thus, including California causes more than a threefold increase in the value of the mean, as it changes from 2590.75 thousand to 8390.4 thousand. ▬

The above example should encourage us to be cautious. We should remember that the mean is not always the best measure of central tendency because it is heavily influenced by outliers. Sometimes other measures of central tendency give a more accurate impression of a data set.

Case Study 3–1 shows how averages may be used to summarize very large data sets. You may also notice that outliers in the original data might have influenced the two averages—the average income before taxes of U.S. households and the average number of vehicles owned by U.S. households—mentioned in the case study.

CASE STUDY 3-1   AN AVERAGE U.S. HOUSEHOLD

USA SNAPSHOTS®
A look at statistics that shape your finances

Portrait of average U.S. household

Income before taxes	$33,854
Number of vehicles	1.9
Own home	61%

Source: Bureau of Labor Statistics   By Cindy Hall and Sam Ward, USA TODAY

The above chart shows the main characteristics of a *typical* U.S. household. For example, according to the information given on U.S. households in this chart, their average income before taxes was $33,854, they owned an average of 1.9 vehicles, and 61% of them owned homes at the time of the survey. These summary measures are based on the data collected by the U.S. Bureau of Labor Statistics. Usually the surveys conducted by the Bureau of Labor Statistics are based on many thousands of members. Thus, the summary measures given in the chart are based on a data set that contained information on thousands of U.S. households. The data obtained on each variable for households included in the survey are represented by one number in the chart.

Source: USA Today, September 30, 1994. Copyright © 1994, *USA Today*. Chart reproduced with permission.

3.1.2 MEDIAN

Another important measure of central tendency is the **median**. It is defined as follows.

MEDIAN

The *median* is the value of the middle term in a data set that has been ranked in increasing order.

As is obvious from the definition of the median, it divides a ranked data set into two equal parts. The calculation of the median consists of the following two steps.

1. Rank the given data set in increasing order.
2. Find the middle term. The value of this term is the median.[1]

The position of the middle term in a data set with n values is obtained as follows.

$$\text{Position of the middle term} = \frac{n + 1}{2}$$

Thus, we can redefine the median as follows.

MEDIAN FOR UNGROUPED DATA

$$\text{Median} = \text{the value of the } \left(\frac{n + 1}{2}\right)\text{th term in a ranked data set}$$

If the given data set represents a population, replace n by N.

If the number of observations in a data set is *odd*, then the median is given by the value of the middle term in the ranked data. If the number of observations is *even*, then the median is given by the average of the values of the two middle terms.

Calculating median for ungrouped data: odd number of data values.

EXAMPLE 3–4 The following data give the weight lost (in pounds) by a sample of five members of a health club at the end of two months of membership.

$$10 \quad 5 \quad 19 \quad 8 \quad 3$$

Find the median.

Solution   First, we rank the given data in increasing order as follows.

$$3 \quad 5 \quad 8 \quad 10 \quad 19$$

There are five observations in the data set. Consequently, $n = 5$ and

$$\text{Position of the middle term} = \frac{n + 1}{2} = \frac{5 + 1}{2} = 3$$

Therefore, the median is the value of the third term in the ranked data.

$$3 \quad 5 \quad \boxed{8} \quad 10 \quad 19$$
$$\uparrow$$
$$\text{Median}$$

Thus, the median weight loss for this sample of five members of this health club is 8 pounds. ■

---

[1]The value of the middle term in a data set ranked in decreasing order will also give the value of the median.

CASE STUDY 3-2   MEDIAN INCOME OF MEN AGED 30 OR OLDER

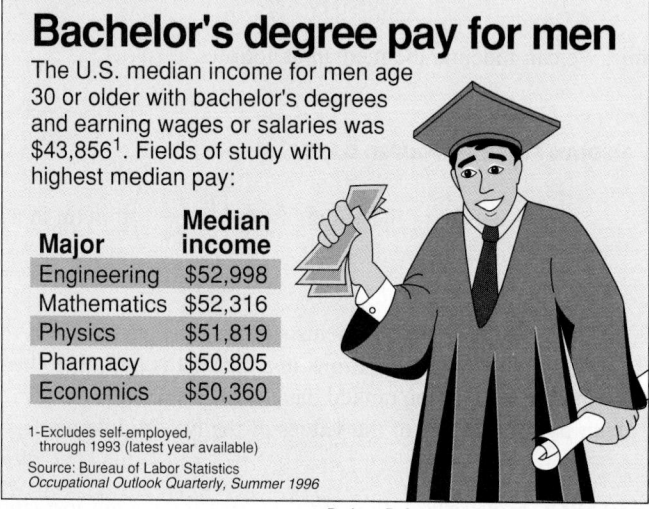

USA SNAPSHOTS ®

A look at statistics that shape the nation

Bachelor's degree pay for men

The U.S. median income for men age 30 or older with bachelor's degrees and earning wages or salaries was $43,856[1]. Fields of study with highest median pay:

Major	Median income
Engineering	$52,998
Mathematics	$52,316
Physics	$51,819
Pharmacy	$50,805
Economics	$50,360

1-Excludes self-employed, through 1993 (latest year available)

Source: Bureau of Labor Statistics
Occupational Outlook Quarterly, Summer 1996

By Anne R. Carey and Grant Jerding, USA TODAY

The above chart shows that the median income of U.S. men aged 30 and older who possess a bachelor's degree is $43,856. It also shows that such men with degrees in engineering, mathematics, physics, pharmacy, and economics earn the highest median income. These results are based on a survey conducted by the U.S. Bureau of Labor Statistics.

Source: USA Today, January 15, 1997. Copyright © 1997, *USA Today*. Chart reproduced with permission.

Calculating median for ungrouped data: even number of data values.

EXAMPLE 3-5   The following table lists the 1995 total sales of the 12 U.S. companies with the highest sales for 1995. (*Source: Business Week*, March 25, 1996.)

Company	1995 Sales (billions of dollars)
AT&T	79.6
Chevron	37.1
Chrysler	53.2
DuPont	42.2
Exxon	109.6
Ford Motor	137.1
General Electric	70.0
General Motors	168.8
IBM	71.9
Mobil	74.9
Philip Morris	53.1
Wal-Mart Stores	93.6

Find the 1995 median sales for these companies.

Solution   First, we rank the data in increasing order as follows.

37.1   42.2   53.1   53.2   70.0   71.9   74.9   79.6   93.6   109.6   137.1   168.8

There are 12 values in the data set. Hence, $n = 12$ and

$$\text{Position of the middle term} = \frac{n + 1}{2} = \frac{12 + 1}{2} = 6.5$$

Therefore, the median is given by the mean of the sixth and seventh values in the ranked data.

37.1   42.2   53.1   53.2   70.0   71.9   74.9   79.6   93.6   109.6   137.1   168.8

$$\text{Median} = \frac{71.9 + 74.9}{2} = \textbf{73.40}$$

Thus, the median 1995 sales of these 12 companies were $73.40 billion.

The median gives the center of a histogram, with half of the data values to the left of the median and half to the right of the median. The advantage of using the median as a measure of central tendency is that it is not influenced by outliers. Consequently, the median is preferred over the mean as a measure of central tendency for data sets that contain outliers. Case Study 3–3 is an example of such a controversial case. This case study is reproduced from the *U.S. News & World Report*. Note that the last paragraph in the case study is added by the author.

CASE STUDY 3-3   WHAT'S THE TAX BITE?

In political campaigns, there are lies, damn lies, and economic statistics. One statistic that keeps showing up in speeches is this: The typical family pays 38.2 percent of its income in taxes. That estimate comes from the Tax Foundation, a Washington-based group that favors lower taxes. (Recently) . . . , a far different number emerged from the Washington-based Center on Budget and Policy Priorities, which focuses on issues involving low- and middle-income Americans. The typical family, it reported, pays 27.3 percent of income in federal, state, and local taxes.

Richard Kogan, the report's author, claims that the Tax Foundation incorrectly counted certain government revenues, like receipts from Medicare premiums, as taxes, artificially raising the percentage. And, he says, the 38.2 percent figure was computed using average income, meaning the result was distorted by the incomes of the exceptionally wealthy. To describe the typical family, Kogan chose to use median income, which is lower than average income.

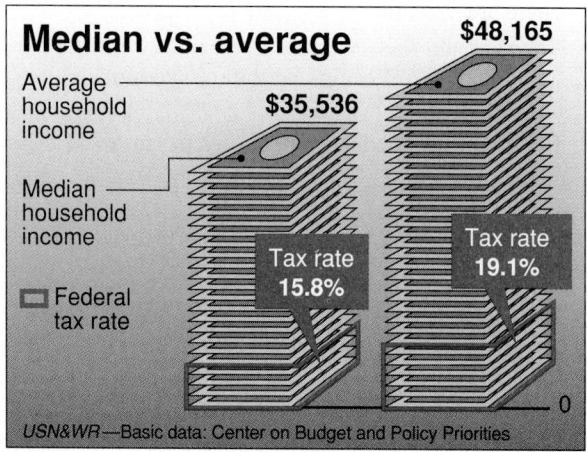

Richard Gage—*USN&WR*

The Tax Foundation's Patrick Fleenor stands by his estimate. Fleenor was looking at the typical family—dual-income with two kids—which has a higher tax bite than the average American. But Kogan says that even the richest 1 percent of Americans, who have the highest federal tax rate, don't pay more than 35 percent of their income in taxes.

One thing remains clear: In Washington, few statistics go unchallenged.

Note that the chart shows the percentages of the median and average household incomes (15.8% and 19.1% respectively) paid in federal taxes alone. These percentages do not include the state and local taxes, which are included in the 27.3% and 38.2% mentioned by Mr. Richard Kogan and Mr. Patrick Fleenor, respectively.

Source: Kevin Whitelaw, "What is the tax bite?" *U.S. News & World Report*, September 16, 1996. Copyright © 1996, U.S. News & World Report, Inc. Reproduced with permission.

3.1.3   MODE

Mode is a French word that means *fashion*—an item that is most popular or common. In statistics, the mode represents the most common value in a data set.

> **MODE**
>
> The *mode* is the value that occurs with the highest frequency in a data set.

Calculating mode for ungrouped data.

EXAMPLE 3-6 The following data give the speeds (in miles per hour) of eight cars that were stopped on I-95 for speeding violations.

$$77 \quad 69 \quad 74 \quad 81 \quad 71 \quad 68 \quad 74 \quad 73$$

Find the mode.

Solution In this data set, 74 occurs twice and each of the remaining values occurs only once. Because 74 occurs with the highest frequency, it is the mode. Therefore,

$$\text{Mode} = \textbf{74 miles per hour} \quad \blacksquare$$

A major shortcoming of the mode is that a data set may have none or may have more than one mode, whereas it will have only one mean and only one median. For instance, a data set with each value occurring only once has no mode. A data set with only one value occurring with highest frequency has only one mode. The data set in this case is called **unimodal**. A data set with two values occurring with the same (highest) frequency has two modes. The distribution, in this case, is said to be **bimodal**. If more than two values in a data set occur with the same (highest) frequency, then the data set contains more than two modes and it is said to be **multimodal**.

Data set with no mode.

EXAMPLE 3-7 Last year's incomes of five randomly selected families were $26,150, $65,750, $34,985, $47,490, and $13,740. Find the mode.

Solution As each value in this data set occurs only once, this data set contains no mode. $\quad \blacksquare$

Data set with two modes.

EXAMPLE 3-8 The prices of the same brand of television set at eight stores are found to be $495, $486, $503, $495, $470, $505, $470, and $499. Find the mode.

Solution In this data set, each of the two values $495 and $470 occurs twice and each of the remaining values occurs only once. Therefore, this data set has two modes: **$495** and **$470**. $\quad \blacksquare$

Data set with three modes.

EXAMPLE 3-9 The ages of 10 randomly selected students from a class are 21, 19, 27, 22, 29, 19, 25, 21, 22, and 30. Find the mode.

Solution This data set has three modes: **19**, **21**, and **22**. Each of these three values occurs with a (highest) frequency of 2. $\quad \blacksquare$

One advantage of the mode is that it can be calculated for both kinds of data, quantitative and qualitative, whereas the mean and median can be calculated only for quantitative data.

Finding mode for qualitative data.

EXAMPLE 3-10 The status of five students, who are members of the student senate at a college, are senior, sophomore, senior, junior, senior. Find the mode.

Solution As *senior* occurs more frequently than the other categories, it is the mode for this data set. However, we cannot calculate the mean and median for this data set. $\quad \blacksquare$

To sum up, we cannot conclude which of the three measures of central tendency is a better measure overall. Each of them may be better under different situations. Probably the mean is the most used measure of central tendency followed by the median. The mean has the advantage that its calculation includes each value of the data set. The median is a better measure when a data set includes outliers. The mode is simple to locate, but it is not of much use in practical applications.

3.1.4   RELATIONSHIP BETWEEN THE MEAN, MEDIAN, AND MODE

As discussed in Chapter 2, two of the many shapes that a histogram or a frequency distribution curve can assume are symmetric and skewed. This section describes the relationship between the mean, median, and mode for three such histograms and frequency curves. Knowing the values of the mean, median, and mode can give us some idea about the shape of a frequency curve.

1.   For a symmetric histogram and frequency curve with one peak (see Figure 3.3), the values of the mean, median, and mode are identical, and they lie at the center of the distribution.

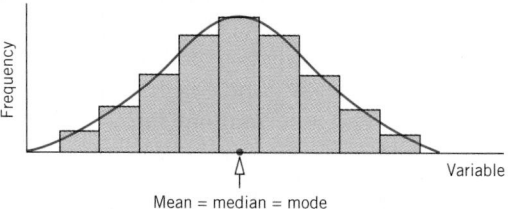

Figure 3.3   Mean, median, and mode for a symmetric histogram and a frequency curve.

2.   For a histogram and a frequency curve skewed to the right (see Figure 3.4), the value of the mean is the largest, that of the mode is the smallest, and the value of the median lies between these two. (Notice that the mode always occurs at the peak point.) The value of the mean is the largest in this case because it is sensitive to outliers that occur in the right tail. These outliers pull the mean to the right.

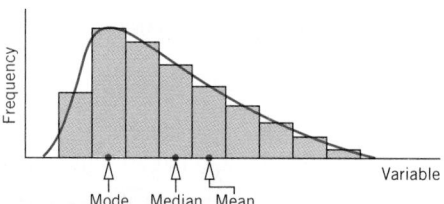

Figure 3.4   Mean, median, and mode for a histogram and a frequency curve skewed to the right.

3.   If a histogram and a distribution curve are skewed to the left (see Figure 3.5), the value of the mean is the smallest and that of the mode is the largest, with the value of the median lying between these two. In this case, the outliers in the left tail pull the mean to the left.

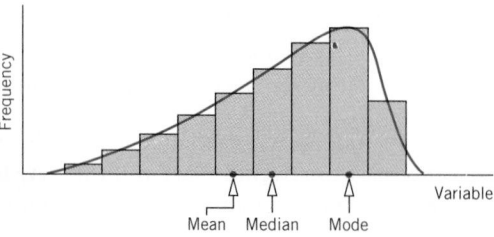

Figure 3.5   Mean, median, and mode for a histogram and a frequency curve skewed to the left.

EXERCISES

Concepts and Procedures

3.1 Explain how the value of the median is determined for a data set that contains an odd number of observations and for a data set that contains an even number of observations.

3.2 Briefly explain the meaning of an outlier. Is the mean or the median a better measure of central tendency for a data set that contains an outlier? Illustrate with the help of an example.

3.3 Using an example, show how an outlier can affect the value of the mean.

3.4 Which of the three measures of central tendency (the mean, the median, and the mode) can be calculated for quantitative data only, and which ones can be calculated for both quantitative and qualitative data? Illustrate with examples.

3.5 Which of the three measures of central tendency (the mean, the median, and the mode) can assume more than one value for a data set? Give an example of a data set for which this summary measure assumes more than one value.

3.6 Is it possible for a (quantitative) data set to have no mean, no median, or no mode? Give an example of a data set for which this summary measure does not exist.

3.7 Explain the relationship between the mean, median, and mode for symmetric and skewed histograms. Illustrate these relationships with graphs.

3.8 Prices of cars have a distribution that is skewed to the right with outliers in the right tail. Which of the measures of central tendency is the best to summarize this data set? Explain.

3.9 The following data set belongs to a population.

$$5 \quad -7 \quad 2 \quad 0 \quad -9 \quad 12 \quad 10 \quad 7$$

Calculate the mean, median, and mode.

3.10 The following data set belongs to a sample.

$$14 \quad 11 \quad -10 \quad 8 \quad 8 \quad -16$$

Calculate the mean, median, and mode.

Applications

Exercises 3.11 through 3.15 are based on the following data.

The following table, based on the American Chamber of Commerce Researchers Association Survey, gives the prices of five items in 12 urban areas across the United States. (See Data Set I of Appendix B.)

City	Apartment Rent	Price of House	Phone Bill	Cost of a Hospital Room	Price of Wine
Huntsville (AL)	468	113,920	23.07	352.50	6.16
Phoenix (AZ)	625	126,566	19.03	472.14	4.43
San Diego (CA)	821	204,951	15.64	614.36	4.39
Denver (CO)	699	154,868	20.51	484.24	4.47
Bloomington (IN)	607	123,426	15.31	493.00	4.11
New Orleans (LA)	535	105,335	23.23	366.00	4.66
Manchester (NH)	610	143,950	20.09	456.00	4.99
Albuquerque (NM)	645	132,000	22.20	358.25	5.54
Albany (NY)	660	145,000	30.25	334.00	5.43
Charlotte (NC)	486	131,600	17.51	349.33	4.74
Salem (OR)	513	153,072	19.82	380.00	4.72
Charleston (WV)	517	144,560	29.93	275.40	5.19

Explanation of variables

Apartment rent   Monthly rent of an unfurnished two-bedroom apartment (excluding all utilities except water), 1½ or 2 baths, approximately 950 square feet

Price of house   Purchase price of a new house with 1800 square feet of living area and on an 8000-square-foot lot in an urban area with all utilities

Phone bill   Monthly telephone charges for a private residential line (customer owns instruments)

Cost of a hospital room   Average cost per day of a semiprivate room in a hospital

Price of wine   Price of Paul Masson Chablis, 1.5-liter bottle

3.11   Calculate the mean and median for data on apartment rents.

3.12   Find the mean and median for data on prices of houses.

3.13   Calculate the mean and median for data on phone bills. Do these data have a mode?

3.14   Calculate the mean and median for data on cost of hospital rooms. Do these data have a mode?

3.15   Find the mean and median for data on price of wine.

3.16   The following data give the number of car thefts that occurred in a city during the past 12 days.

> 6   3   7   11   5   3   8   7   2   6   9   13

Find the mean, median, and mode.

3.17   The following data give the 1994 total cash receipts from the sale of farm produce (in billions of Canadian dollars) for the six largest (based on the population size) provinces of Canada *(1996 Canadian Global Almanac)*. These values are for the provinces of Quebec, Ontario, Manitoba, Saskatchewan, Alberta, and British Columbia, respectively, entered in that order.

> 3.7   5.9   2.1   4.6   5.1   1.5

Compute the mean and median. Do these data have a mode?

3.18   Following are the temperatures (in Fahrenheit) observed during eight wintry days in a Midwestern city.

> 23   14   6   −7   −2   9   16   19

Compute the mean and median. Do these data have a mode?

3.19   The following data give the number of hours spent partying by 10 randomly selected college students during the past week.

> 7   14   5   0   2   7   10   4   0   8

Compute the mean, median, and mode.

3.20   The following data give the 1994 corn production (in millions of bushels) for 13 states *(Statistical Abstract of the United States, 1995)*. The data, entered by row, are for the states of Iowa, Illinois, Nebraska, Minnesota, Indiana, Ohio, Wisconsin, South Dakota, Kansas, Missouri, Michigan, Texas, and Kentucky, respectively.

> 1930   1786   1154   916   858   487   437
> 367   305   274   261   239   156

Calculate the mean and median. Do these data have a mode? Why or why not? Explain.

3.21   Nixon Corporation manufactures computer terminals. The following data give the number of computer terminals produced at the company for a sample of 10 days.

> 24   32   27   23   35   33   29   21   23   28

Calculate the mean, median, and mode for these data.

3.22 The following data give the number of computer keyboards assembled at the Twentieth Century Electronics Company for a sample of 12 days.

| 45 | 52 | 48 | 41 | 56 | 46 | 44 | 42 | 48 | 53 | 43 | 50 |

Calculate the mean, median, and mode for these data.

3.23 The following data give the 1994 total area of farmland (in millions of acres) for 10 states (*Statistical Abstract of the United States*, 1995). The data, entered in that order, are for the states of Colorado, Iowa, Kansas, Minnesota, Missouri, Nebraska, North Dakota, Oklahoma, South Dakota, and Texas, respectively.

| 33 | 33 | 48 | 30 | 30 | 47 | 40 | 34 | 44 | 129 |

a. Calculate the mean and median for these data.
b. Do these data contain an outlier? If yes, drop this value and recalculate the mean and median. Which of the two summary measures changes by a larger amount when you drop the outlier?
c. Is the mean or the median a better summary measure for these data? Explain.

3.24 The following data give the 1995 population of "market-areas" (in millions) for nine major league baseball teams (*American Demographics*, April 1996). These data, entered in that order, are for the Colorado Rockies, Cleveland Indians, Toronto Blue Jays, Milwaukee Brewers, Texas Rangers, Boston Red Sox, New York Yankees, Chicago White Sox, and California Angels, respectively.

| 1.83 | 2.91 | 4.31 | 1.78 | 4.16 | 6.52 | 7.04 | 6.91 | 15.44 |

a. Calculate the mean and median for these data.
b. Do these data contain an outlier? If yes, drop this value and recalculate the mean and median. Which of the two summary measures changes by a larger amount when you drop the outlier?
c. Is the mean or the median a better summary measure for these data? Explain.

*3.25 One property of the mean is that if we know the means and sample sizes of two (or more) data sets, we can calculate the **combined mean** of both (or all) data sets. The combined mean for two data sets is calculated by using the formula

$$\text{Combined mean} = \bar{x} = \frac{n_1 \bar{x}_1 + n_2 \bar{x}_2}{n_1 + n_2}$$

where n_1 and n_2 are the sample sizes of the two data sets and $\bar{x}_1$ and $\bar{x}_2$ are the means of the two data sets, respectively. Suppose a sample of 10 statistics books gave a mean price of $41 and a sample of 8 mathematics books gave a mean price of $43. Find the combined mean. (*Hint:* For this example $n_1 = 10$, $n_2 = 8$, $\bar{x}_1 = \$41$, $\bar{x}_2 = \$43$.)

*3.26 The mean score of 20 female students on a statistics test is 77 and the mean score of 15 male students on the same test is 74. Find the combined mean score.

*3.27 For any data, the sum of all values is equal to the product of the sample size and mean, that is, $\Sigma x = n\bar{x}$. Suppose the average amount of money spent on shopping by 10 persons during a given week is $85.50. Find the total amount of money spent on shopping by these 10 persons.

*3.28 The mean 1996 income for five families was $39,520. What was the total 1996 income of these five families?

*3.29 The mean age of six persons is 46 years. The ages of five of these six persons are 57, 39, 44, 51, and 37 years. Find the age of the sixth person.

*3.30 The mean score on a statistics test for eight students is 76. The scores of seven of these eight students are 81, 65, 93, 66, 71, 84, and 80. Find the score of the eighth student.

∗3.31   Consider the following two data sets.

Data Set I:	12	25	37	8	41
Data Set II:	19	32	44	15	48

Notice that each value of the second data set is obtained by adding 7 to the corresponding value of the first data set. Calculate the mean for each of these two data sets. Comment on the relationship between the two means.

∗3.32   Consider the following two data sets.

Data Set I:	4	8	15	9	11
Data Set II:	8	16	30	18	22

Notice that each value of the second data set is obtained by multiplying the corresponding value of the first data set by 2. Calculate the mean for each of these two data sets. Comment on the relationship between the two means.

∗3.33   The **trimmed mean** is calculated by dropping a certain percentage of values from each end of a ranked data set. The trimmed mean is especially useful as a measure of central tendency when a data set contains a few outliers at each end. Suppose the following data give the ages of 10 employees of a company.

$$47 \quad 53 \quad 38 \quad 26 \quad 39 \quad 49 \quad 19 \quad 67 \quad 31 \quad 23$$

To calculate the 10% trimmed mean, first rank these data values in increasing order, then drop 10% of the smallest values and 10% of the largest values. The mean of the remaining 80% of the values will give the 10% trimmed mean. As this data set contains 10 values, 10% of 10 is 1. Hence, drop the smallest value and the largest value from this data set. The mean of the remaining 8 values will be the 10% trimmed mean. Calculate the 10% trimmed mean for this data set.

∗3.34   The following data give the prices (in thousands of dollars) of 20 houses sold recently in a city.

184	197	145	209	245	187	169	238	161	290
223	278	310	179	107	271	357	195	259	199

Find the 20% trimmed mean for this data set.

3.2   MEASURES OF DISPERSION FOR UNGROUPED DATA

The measures of central tendency, such as the mean, median, and mode, do not reveal the whole picture of the distribution of a data set. Two data sets with the same mean may have completely different spreads. The variation among values of observations for one data set may be much larger or smaller than for the other data set. (Note that the words *dispersion*, *spread*, and *variation* have the same meaning.) Consider the following two data sets on the ages of all workers for each of two small companies.

Company 1:	47	38	35	40	36	45	39
Company 2:		70	33	18	52	27	

The mean age of workers of each of these two companies is the same, 40 years. If we do not know the ages of individual workers for these two companies and are told only that the mean age of the workers for both companies is the same, we may deduce that the workers of these two companies have a similar age distribution. But, as we can observe, the variation in the workers' ages for each of these two companies is very different. As illustrated in the diagram, the ages of the workers of the second company have a much larger variation than the ages of the workers of the first company.

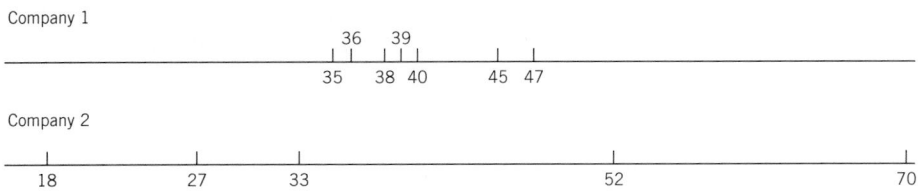

Thus, the mean, median, or mode is usually not by itself a sufficient measure to reveal the shape of the distribution of a data set. We also need a measure that can provide some information about the variation among data values. The measures that help us to know about the spread of a data set are called the **measures of dispersion**. The measures of central tendency and dispersion taken together give a better picture of a data set than the measures of central tendency alone. This section discusses three measures of dispersion: range, variance, and standard deviation.

3.2.1 RANGE

The **range** is the simplest measure of dispersion to calculate. It is obtained by taking the difference between the largest and the smallest values in a data set.

RANGE FOR UNGROUPED DATA

$$\text{Range} = \text{Largest value} - \text{Smallest value}$$

Calculating range for ungrouped data.

EXAMPLE 3–11 The following data give the total area in square miles of the four western South-Central states of the United States.

State	Total Area (in square miles)
Arkansas	53,182
Louisiana	49,651
Oklahoma	69,903
Texas	267,277

Find the range for this data set.

Solution The maximum total area for a state in this data set is 267,277 square miles, and the smallest area is 49,651 square miles. Therefore,

Range = Largest value − Smallest value = 267,277 − 49,651 = **217,626 square miles**

Thus, the total areas of these four states are spread over a range of 217,626 square miles.

The range, like the mean, has the disadvantage of being influenced by outliers. In Example 3–11, if the state of Texas with a total area of 267,277 square miles is dropped, the range decreases from 217,626 square miles to 20,252 square miles. Consequently, the range is not a good measure of dispersion to use for a data set that contains outliers.

Another disadvantage of using the range as a measure of dispersion is that its calculation is based on two values only: the largest and the smallest. All other values in a data set are ignored while calculating the range. Thus, the range is not a very satisfactory measure of dispersion.

3.2.2 VARIANCE AND STANDARD DEVIATION

The **standard deviation** is the most used measure of dispersion. The value of the standard deviation tells how closely the values of a data set are clustered around the mean. In general, a lower value of the standard deviation for a data set indicates that the values of that data set are spread over a relatively smaller range around the mean. On the other hand, a larger value of the standard deviation for a data set indicates that the values of that data set are spread over a relatively larger range around the mean.

The standard deviation is obtained by taking the positive square root of the **variance**. The variance calculated for population data is denoted by σ^2 (read as *sigma squared*),[2] and the variance calculated for sample data is denoted by s^2. Consequently, the standard deviation calculated for population data is denoted by σ, and the standard deviation calculated for sample data is denoted by s. Following are the *basic formulas* that are used to calculate the variance.[3]

$$\sigma^2 = \frac{\Sigma(x - \mu)^2}{N} \quad \text{and} \quad s^2 = \frac{\Sigma(x - \bar{x})^2}{n - 1}$$

where σ^2 is the population variance and s^2 is the sample variance.

The quantity $x - \mu$ or $x - \bar{x}$ in the above formulas is called the *deviation* of x value from the mean.

The sum of the deviations of x values from the mean is always zero. That is, $\Sigma(x - \mu) = 0$ and $\Sigma(x - \bar{x}) = 0$. For example, suppose the midterm scores of a sample of four students are 82, 95, 67, and 92. Then, the mean score for these four students is

$$\bar{x} = (82 + 95 + 67 + 92)/4 = 84$$

The deviations of the four scores from the mean are calculated in Table 3.2.

Table 3.2

x	$x - \bar{x}$
82	$82 - 84 = -2$
95	$95 - 84 = +11$
67	$67 - 84 = -17$
92	$92 - 84 = +8$
	$\Sigma(x - \bar{x}) = 0$

[2]Note that Σ is uppercase sigma and σ is lowercase sigma of the Greek alphabet.

[3]From the formula for σ^2, it can be stated that the population variance is the mean of the squared deviations of x values from the mean. However, this is not true for the variance calculated for a sample data set.

As we can observe from Table 3.2, the sum of the deviations of x values from the mean is zero, that is, $\Sigma(x - \bar{x}) = 0$. For this reason we square the deviations to calculate the variance and standard deviation.

From the computational point of view, it is easier and more efficient to use *short-cut formulas* to calculate the variance and standard deviation. By using the short-cut formula, we reduce the computation time and round off errors. Use of the basic formulas for ungrouped data is illustrated in Section A3.1.1 of Appendix 3.1 of this chapter. The short-cut formulas for calculating the variance and standard deviation are as follows.

SHORT-CUT FORMULAS FOR THE VARIANCE AND STANDARD DEVIATION FOR UNGROUPED DATA

$$\sigma^2 = \frac{\Sigma x^2 - \dfrac{(\Sigma x)^2}{N}}{N} \quad \text{and} \quad s^2 = \frac{\Sigma x^2 - \dfrac{(\Sigma x)^2}{n}}{n - 1}$$

where σ^2 is the population variance and s^2 is the sample variance.

The standard deviation is obtained by taking the positive square root of the variance.

The population standard deviation: $\quad \sigma = \sqrt{\sigma^2}$

The sample standard deviation: $\quad s = \sqrt{s^2}$

Note that the denominator in the formula for population variance is N but that in the formula for sample variance it is $n - 1$.[4]

Calculating variance and standard deviation for ungrouped data.

EXAMPLE 3–12 The following table lists the annual energy cost (in dollars) of five models of refrigerators. (*Source: Consumer Reports*, May 1996, p. 36. Copyright © 1996 by Consumers Union of United States, Inc., Yonkers, NY. Adapted and reproduced with permission. Results and conclusions not endorsed by Consumers Union.)

Refrigerator Model	Annual Energy Cost (dollars)
General Electric TBX21ZAX	72
Hotpoint CTH14CYX	49
Kenmore 65271	79
Amana TR22S4	55
Whirlpool ET21DKXD	57

Find the variance and standard deviation for these data.

Solution Let x denote the annual energy cost of a model of refrigerator. The values of Σx and Σx^2 are calculated in Table 3.3.

---

[4]The reason that the denominator in the sample formula is $n - 1$ and not n is the following. The sample variance underestimates the population variance when the denominator in the sample formula for variance is n. However, the sample variance does not underestimate the population variance if the denominator in the sample formula for variance is $n - 1$. In Chapter 8 we will learn that $n - 1$ is called the degrees of freedom.

Table 3.3

x	x^2
72	5184
49	2401
79	6241
55	3025
57	3249
$\Sigma x = 312$	$\Sigma x^2 = 20{,}100$

The calculation of variance involves the following steps.

Step 1.   *Calculate Σx*

The sum of the entries in the first column of Table 3.3 gives the value of Σx, which is 312.

Step 2.   *Find Σx^2*

The value of Σx^2 is obtained by squaring each value of x and then adding the squared values. The results of this step are shown in the second column of Table 3.3. Notice that $\Sigma x^2 = 20{,}100$.

Step 3.   *Determine the variance*

Substitute all the values in the variance formula and simplify. Because the given data belong to a sample of five refrigerators, we use the formula for the sample variance.

$$s^2 = \frac{\Sigma x^2 - \dfrac{(\Sigma x)^2}{n}}{n-1} = \frac{20{,}100 - \dfrac{(312)^2}{5}}{5-1} = \frac{20{,}100 - 19{,}468.8}{4} = \mathbf{157.8}$$

Step 4.   *Obtain the standard deviation*

The standard deviation is obtained by taking the positive square root of the variance.

$$s = \sqrt{157.8} = \mathbf{\$12.56}$$

Thus, the standard deviation of the annual energy costs of these five refrigerators is $12.56.

☞   **TWO OBSERVATIONS**

1.   **The values of variance and standard deviation are never negative.** That is, the numerator in the formula for variance should never produce a negative value. Usually the values of the variance and standard deviation are positive, but if a data set has no variation, then the variance and standard deviation are both zero. For example, if four persons in a group are of the same age, say, 35 years, then the four values in the data set are

$$35 \quad 35 \quad 35 \quad 35$$

If we calculate the variance and standard deviation for these data, their values will be zero. This will be so because there is no variation in the values of this data set.

2.   **The measurement units of variance are always the square of the measurement units of the original data.** This is so because the original values are squared to calculate the variance. In Example 3–12, the measurement units of the original data are dollars. However, the measurement units of the variance are squared dollars, which, of course, does not make any sense. Thus, the variance of the annual energy costs of five refrigerators

in Example 3–12 is 157.8 squared dollars. But the measurement units of the standard deviation are the same as the measurement units of the original data because the standard deviation is obtained by taking the square root of the variance.

Calculating variance and standard deviation for ungrouped data.

EXAMPLE 3–13   Following are the 1996 earnings (in thousands of dollars) before taxes for *all* six employees of a small company.

$$29.50 \qquad 16.20 \qquad 35.45 \qquad 21.35 \qquad 49.70 \qquad 24.60$$

Calculate the variance and standard deviation for these data.

Solution   Let x denote the 1996 earnings before taxes of employees of this company. The values of Σx and Σx^2 are calculated in Table 3.4.

Table 3.4

x	x^2
29.50	870.2500
16.20	262.4400
35.45	1256.7025
21.35	455.8225
49.70	2470.0900
24.60	605.1600
$\Sigma x = 176.80$	$\Sigma x^2 = 5920.4650$

Since the data on earnings are for *all* employees of this company, we will use the population formula to compute the variance. Thus, the variance is

$$\sigma^2 = \frac{\Sigma x^2 - \dfrac{(\Sigma x)^2}{N}}{N} = \frac{5920.4650 - \dfrac{(176.80)^2}{6}}{6} = \mathbf{118.4597}$$

The standard deviation is obtained by taking the (positive) square root of the variance.

$$\sigma = \sqrt{118.4597} = \mathbf{\$10.884 \text{ thousand}} = \mathbf{\$10{,}884}$$

Thus, the standard deviation of the 1996 earnings of all six employees of this company is $10,884. ▬

☞ **WARNING**

Note that Σx^2 is not the same as $(\Sigma x)^2$. The value of Σx^2 is obtained by squaring the x values and adding them. The value of $(\Sigma x)^2$ is obtained by squaring the value of Σx.

The uses of the standard deviation are discussed in Section 3.4. Later chapters will explain how the mean and the standard deviation taken together can help in making inferences about the population.

3.2.3   POPULATION PARAMETERS AND SAMPLE STATISTICS

A numerical measure such as the mean, median, mode, range, variance, or standard deviation calculated for a population data set is called a *population parameter*, or simply a **parameter**. A summary measure calculated for a sample data set is called a *sample statistic*, or simply a **statistic**. Thus, μ and σ are population parameters and $\bar{x}$ and s are sample statistics. As an illustration, $\bar{x} = \$186$ thousand in Example 3–1 is a sample statistic and $\mu = \$592.50$ in Example 3–2 is a population parameter. Similarly, $s = \$12.56$ in Example 3–12 is a sample statistic whereas $\sigma = \$10,884$ in Example 3–13 is a population parameter.

EXERCISES

Concepts and Procedures

3.35   The range, as a measure of spread, has the disadvantage of being influenced by outliers. Illustrate this with an example.

3.36   Can the standard deviation have a negative value? Explain.

3.37   When is the value of the standard deviation for a data set zero? Give one example. Calculate the standard deviation for this example and show that its value is zero.

3.38   Briefly explain the difference between a population parameter and a sample statistic. Give one example of each of these.

3.39   The following data set belongs to a population.

$$5 \quad -7 \quad 2 \quad 0 \quad -9 \quad 12 \quad 10 \quad 7$$

Calculate the range, variance, and standard deviation.

3.40   The following data set belongs to a sample.

$$14 \quad 11 \quad -10 \quad 8 \quad 8 \quad -16$$

Calculate the range, variance, and standard deviation.

Applications

3.41   The following data give the weekly food expenditures for a sample of five families.

$$\$82 \quad 116 \quad 65 \quad 170 \quad 92$$

    a.   Find the mean for these data. Calculate the deviations of the data values from the mean. Is the sum of these deviations zero?
    b.   Calculate the range, variance, and standard deviation.

3.42   A sample of seven statistics books produced the following data on their prices.

$$\$56 \quad 75 \quad 68 \quad 61 \quad 71 \quad 66 \quad 78$$

    a.   Find the mean for these data. Calculate the deviations of the data values from the mean. Is the sum of these deviations zero?
    b.   Calculate the range, variance, and standard deviation.

3.43   The following data give the number of car thefts that occurred in a city during the past 12 days.

$$6 \quad 3 \quad 7 \quad 11 \quad 5 \quad 3 \quad 8 \quad 7 \quad 2 \quad 6 \quad 9 \quad 13$$

Calculate the range, variance, and standard deviation.

3.44   The following table gives the 1995 revenues (rounded to billions of dollars) of the top 10 companies in *Fortune* magazine's Global 500 (*Fortune*, August 5, 1996).

Company	1995 Revenue (in billions of U.S. dollars)
Mitsubishi (Japan)	184
Mitsui (Japan)	182
Itochu (Japan)	169
General Motors (U.S.)	169
Sumitomo (Japan)	168
Marubeni (Japan)	161
Ford Motor (U.S.)	137
Toyota Motor (Japan)	111
Exxon (U.S.)	110
Royal Dutch/Shell Group (Brit/Neth)	110

Find the range, variance, and standard deviation for these data.

3.45 The following data give the number of cars that stopped at a service station during each of the 10 hours observed.

29 35 42 31 24 18 16 27 39 34

Find the range, variance, and standard deviation.

3.46 The following data give the number of new cars sold at a dealership during a 12-day period.

13 5 9 6 8 11 9 15 4 11 7 5

Find the range, variance, and standard deviation.

3.47 The following data give the weight (in pounds) lost by 15 new members of a health club at the end of their first two months of membership.

5 10 8 7 25 12 5 14
11 10 21 9 8 11 18

Compute the range, variance, and standard deviation.

3.48 The following data give the speeds (in miles per hour), as measured by radar, of 13 cars traveling on interstate highway I-84.

67 72 63 66 76 69 71
76 65 79 68 67 71

Calculate the range, variance, and standard deviation.

3.49 Following are the temperatures (in degrees Fahrenheit) observed during eight wintry days in a Midwestern city.

23 14 6 −7 −2 9 16 19

Compute the range, variance, and standard deviation.

3.50 The following data give the number of hours spent partying by 10 randomly selected college students during the past week.

7 14 5 0 2 7 10 4 0 8

Compute the range, variance, and standard deviation.

3.51 The following data give the market value (in billions of U.S. dollars) for the 8 largest Canadian companies as of May 31, 1996 (*Business Week*, July 8, 1996). The data, entered in that order, are for the companies Northern Telecom, Seagram, BCE, Barrick Gold, Thomson, Imperial Oil, Royal Bank of Canada, and Alcan Aluminum, respectively.

$$13.9 \quad 12.9 \quad 12.5 \quad 11.3 \quad 9.7 \quad 8.1 \quad 7.5 \quad 7.4$$

Find the range, variance, and standard deviation.

3.52   The following data give the typical charges (in thousands of dollars) for a radical prostatectomy for eight western states for the year 1994 (*Statistical Bulletin*, July-September 1996). These data are for Idaho, Colorado, New Mexico, Arizona, Utah, Washington, Oregon, and California, respectively.

$$11.2 \quad 16.8 \quad 13.6 \quad 19.4 \quad 11.9 \quad 12.8 \quad 14.0 \quad 23.9$$

Find the range, variance, and standard deviation.

3.53   The following data give the hourly wage rate of eight employees of a company.

$$\$12 \quad 12 \quad 12 \quad 12 \quad 12 \quad 12 \quad 12 \quad 12$$

Calculate the standard deviation. Is its value zero? If yes, why?

3.54   The following data give the ages (in years) of six students.

$$19 \quad 19 \quad 19 \quad 19 \quad 19 \quad 19$$

Calculate the standard deviation. Is its value zero? If yes, why?

∗3.55   One disadvantage of the standard deviation as a measure of dispersion is that it is a measure of absolute variability and not of relative variability. Sometimes we may need to compare the variability for two different data sets that have different units of measurement. The **coefficient of variation** is one such measure. The coefficient of variation, denoted by CV, expresses standard deviation as a percentage of the mean and is computed as follows.

$$\text{For population data:} \qquad \text{CV} = \frac{\sigma}{\mu} \times 100\%$$

$$\text{For sample data:} \qquad \text{CV} = \frac{s}{\bar{x}} \times 100\%$$

The yearly salaries of all employees working for a company have a mean of $42,350 and a standard deviation of $3,820. The years of schooling for the same employees have a mean of 15 years and a standard deviation of 2 years. Is the relative variation in the salaries higher or lower than that in years of schooling for these employees?

∗3.56   The SAT scores of 100 students have a mean of 915 and a standard deviation of 105. The GPAs of the same 100 students have a mean of 3.06 and a standard deviation of .22. Is the relative variation in SAT scores higher or lower than that in GPAs?

∗3.57   Consider the following two data sets.

Data Set I:	12	25	37	8	41
Data Set II:	19	32	44	15	48

Note that each value of the second data set is obtained by adding 7 to the corresponding value of the first data set. Calculate the standard deviation for each of these two data sets using the formula for sample data. Comment on the relationship between the two standard deviations.

∗3.58   Consider the following two data sets.

Data Set I:	4	8	15	9	11
Data Set II:	8	16	30	18	22

Note that each value of the second data set is obtained by multiplying the corresponding value of the first data set by 2. Calculate the standard deviation for each of these two data sets using the formula for population data. Comment on the relationship between the two standard deviations.

3.3 MEAN, VARIANCE, AND STANDARD DEVIATION FOR GROUPED DATA

In Sections 3.1.1 and 3.2.2, we learned how to calculate the mean, variance, and standard deviation for ungrouped data. In this section we will learn how to calculate the mean, variance, and standard deviation for grouped data.

3.3.1 MEAN FOR GROUPED DATA

We learned in Section 3.1.1 that the mean is obtained by dividing the sum of all values by the number of values in a data set. However, if the data are given in the form of a frequency table, we will no longer know the values of individual observations. Consequently, in such cases, we cannot obtain the sum of individual values. We find an approximation for the sum of these values using the procedure explained in the next paragraph and example. The formulas used to calculate the mean for grouped data are as follows.

MEAN FOR GROUPED DATA

Mean for population data: $\quad \mu = \dfrac{\Sigma mf}{N}$

Mean for sample data: $\quad \bar{x} = \dfrac{\Sigma mf}{n}$

where m is the midpoint and f is the frequency of a class.

To calculate the mean for grouped data, first find the midpoint of each class and then multiply the midpoints by the frequencies of the corresponding classes. The sum of these products, denoted by Σmf, gives an approximation for the sum of all values. To find the value of the mean, divide this sum by the total number of observations in the data.

Calculating population mean for grouped data.

EXAMPLE 3–14 The following table gives the frequency distribution of the daily commuting time (in minutes) from home to work for *all* 25 employees of a company.

Daily Commuting Time (minutes)	Number of Employees
0 to less than 10	4
10 to less than 20	9
20 to less than 30	6
30 to less than 40	4
40 to less than 50	2

Calculate the mean of the daily commuting times.

Solution   Note that because the data set includes *all* 25 employees of the company, it represents the population. Table 3.5 shows the calculation of Σmf. Note that in Table 3.5, *m* denotes the midpoints of the classes.

Table 3.5

Daily Commuting Time (minutes)	*f*	*m*	*mf*
0 to less than 10	4	5	20
10 to less than 20	9	15	135
20 to less than 30	6	25	150
30 to less than 40	4	35	140
40 to less than 50	2	45	90
	$N = 25$		$\Sigma mf = 535$

To calculate the mean, we first find the midpoint of each class. The class midpoints are recorded in the third column of Table 3.5. The products of the midpoints and the corresponding frequencies are listed in the fourth column of that table. The sum of the fourth column, denoted by Σmf, gives the approximate total daily commuting time (in minutes) for all 25 employees. The mean is obtained by dividing this sum by the total frequency. Therefore,

$$\mu = \frac{\Sigma mf}{N} = \frac{535}{25} = \textbf{21.40 minutes}$$

Thus, the employees of this company spend an average of 21.40 minutes a day commuting from home to work. ▬

What do the numbers 20, 135, 150, 140, and 90 in the column labeled *mf* in Table 3.5 represent? We know from this table that 4 employees spend 0 to less than 10 minutes commuting per day. Assuming that the time spent commuting by these 4 employees is evenly spread in the interval 0 to less than 10, the midpoint of this class (which is 5) gives the mean time spent commuting by these 4 employees. Hence, $4 \times 5 = 20$ is the approximate total time (in minutes) spent commuting per day by these 4 employees. Similarly, 9 employees spend 10 to less than 20 minutes commuting per day, and the total time spent commuting by these 9 employees is approximately 135 minutes a day. The other numbers in this column can be interpreted in the same way. Note that these numbers give the approximate commuting times for these employees based on the assumption of an even spread within classes. The total commuting time for all 25 employees is approximately 535 minutes. Consequently, 21.40 minutes is an approximate and not the exact value of the mean. We can find the exact value of the mean only if we know the exact commuting time for each of the 25 employees of the company.

Calculating sample mean for grouped data.

EXAMPLE 3–15   The following table gives the frequency distribution of the number of orders received each day during the past 50 days at the office of a mail-order company.

Number of Orders	Number of Days
10–12	4
13–15	12
16–18	20
19–21	14

Calculate the mean.

Solution   Because the data set includes only 50 days, it represents a sample. The value of Σmf is calculated in Table 3.6.

Table 3.6

Number of Orders	f	m	mf
10–12	4	11	44
13–15	12	14	168
16–18	20	17	340
19–21	14	20	280
	$n = 50$		$\Sigma mf = 832$

The value of the sample mean is

$$\bar{x} = \frac{\Sigma mf}{n} = \frac{832}{50} = \textbf{16.64 orders}$$

Thus, this mail-order company received an average of 16.64 orders per day during these 50 days. ■

3.3.2   VARIANCE AND STANDARD DEVIATION FOR GROUPED DATA

Following are the *basic formulas* used to calculate the population and sample variances for grouped data.

$$\sigma^2 = \frac{\Sigma f(m - \mu)^2}{N} \quad \text{and} \quad s^2 = \frac{\Sigma f(m - \bar{x})^2}{n - 1}$$

where σ^2 is the population variance, s^2 is the sample variance, and m is the midpoint of a class.

In either case, the standard deviation is obtained by taking the positive square root of the variance.

Again, the *short-cut formulas* are more efficient for calculating the variance and standard deviation. Section A3.1.2 of Appendix 3.1 at the end of this chapter shows how to use the basic formulas to calculate the variance and standard deviation for grouped data.

SHORT-CUT FORMULAS FOR THE VARIANCE AND STANDARD DEVIATION FOR GROUPED DATA

$$\sigma^2 = \frac{\Sigma m^2 f - \dfrac{(\Sigma mf)^2}{N}}{N} \quad \text{and} \quad s^2 = \frac{\Sigma m^2 f - \dfrac{(\Sigma mf)^2}{n}}{n - 1}$$

where σ^2 is the population variance, s^2 is the sample variance, and m is the midpoint of a class.

The standard deviation is obtained by taking the positive square root of the variance.

The population standard deviation: $\sigma = \sqrt{\sigma^2}$

The sample standard deviation: $s = \sqrt{s^2}$

---

*Calculating population
variance and standard
deviation for grouped
data.*

EXAMPLE 3–16   The following table, reproduced from Example 3–14, gives the frequency distribution of the daily commuting time (in minutes) from home to work for all 25 employees of a company.

Daily Commuting Time (minutes)	Number of Employees
0 to less than 10	4
10 to less than 20	9
20 to less than 30	6
30 to less than 40	4
40 to less than 50	2

Calculate the variance and standard deviation.

Solution   All four steps needed to calculate the variance and standard deviation for grouped data are shown after Table 3.7.

Table 3.7

Daily Commuting Time (minutes)	f	m	mf	m^2f
0 to less than 10	4	5	20	100
10 to less than 20	9	15	135	2,025
20 to less than 30	6	25	150	3,750
30 to less than 40	4	35	140	4,900
40 to less than 50	2	45	90	4,050
	$N = 25$		$\Sigma mf = 535$	$\Sigma m^2f = 14{,}825$

Step 1.   *Calculate the value of Σmf*
   To calculate the value of Σmf, first find the midpoint m of each class (see the third column in Table 3.7) and then multiply the corresponding class midpoints and class frequencies (see the fourth column in Table 3.7). The value of Σmf is obtained by adding these products. Thus,

$$\Sigma mf = 535$$

Step 2.   *Find the value of Σm^2f*
   To find the value of Σm^2f, square each m value and multiply this squared value of m by the corresponding frequency (see the fifth column in Table 3.7). The sum of these products (that is, the sum of the fifth column in Table 3.7) gives Σm^2f. Hence,

$$\Sigma m^2f = 14{,}825$$

Step 3.   *Calculate the variance*
   Because the data set includes all 25 employees of the company, it represents the population. Therefore, we will use the formula for the population variance. Thus,

$$\sigma^2 = \frac{\Sigma m^2f - \dfrac{(\Sigma mf)^2}{N}}{N} = \frac{14{,}825 - \dfrac{(535)^2}{25}}{25} = \frac{3376}{25} = \mathbf{135.04}$$

Step 4. *Calculate the standard deviation*

To obtain the standard deviation, take the (positive) square root of the variance.

$$\sigma = \sqrt{\sigma^2} = \sqrt{135.04} = \textbf{11.62 minutes}$$

Thus, the standard deviation of the daily commuting times for these employees is 11.62 minutes. ■

Note that the values of the variance and standard deviation calculated in Example 3–16 for the grouped data are approximations. The exact values of the variance and standard deviation can be obtained only by using the ungrouped data on the daily commuting times of the 25 employees.

Calculating sample variance and standard deviation for grouped data.

EXAMPLE 3–17   The following table, reproduced from Example 3–15, gives the frequency distribution of the number of orders received each day during the past 50 days at the office of a mail-order company.

Number of Orders	f
10–12	4
13–15	12
16–18	20
19–21	14

Calculate the variance and standard deviation.

Solution   All the information required for the calculation of the variance and standard deviation appears in Table 3.8.

Table 3.8

Number of Orders	f	m	mf	m^2f
10–12	4	11	44	484
13–15	12	14	168	2,352
16–18	20	17	340	5,780
19–21	14	20	280	5,600
	$n = 50$		$\Sigma mf = 832$	$\Sigma m^2f = 14{,}216$

Because the data set includes only 50 days, it represents a sample. Hence, we will use the sample formulas to calculate the variance and standard deviation. By substituting the values in the formula for the sample variance, we obtain

$$s^2 = \frac{\Sigma m^2 f - \dfrac{(\Sigma mf)^2}{n}}{n-1} = \frac{14{,}216 - \dfrac{(832)^2}{50}}{50 - 1} = \textbf{7.5820}$$

Hence, the standard deviation is

$$s = \sqrt{s^2} = \sqrt{7.5820} = \textbf{2.75 orders}$$

Thus, the standard deviation of the number of orders received at the office of this mail-order company during the past 50 days is 2.75. ■

EXERCISES

Concepts and Procedures

3.59 Are the values of the mean and standard deviation that are calculated using grouped data exact or approximate values of the mean and standard deviation, respectively? Explain.

3.60 Using the population formulas, calculate the mean, variance, and standard deviation for the following grouped data.

x	2–4	5–7	8–10	11–13	14–16
f	5	9	14	7	5

3.61 Using the population formulas, find the mean, variance, and standard deviation for the grouped data recorded in the following table.

x	0–3	4–7	8–11	12–15	16–19
f	7	4	19	12	8

3.62 Using the sample formulas, calculate the mean, variance, and standard deviation for the following data.

x	f
1 to less than 5	16
5 to less than 9	27
9 to less than 13	38
13 to less than 17	14
17 to less than 21	5

3.63 Using the sample formulas, find the mean, variance, and standard deviation for the grouped data displayed in the following table.

x	f
0 to less than 4	17
4 to less than 8	23
8 to less than 12	15
12 to less than 16	11
16 to less than 20	8
20 to less than 24	6

Applications

3.64 The following table gives the frequency distribution of the amounts of telephone bills for the month of October 1996 for a sample of 50 families.

Amount of Telephone Bill (dollars)	Number of Families
20 to less than 40	8
40 to less than 60	13
60 to less than 80	17
80 to less than 100	9
100 to less than 120	3

Calculate the mean, variance, and standard deviation.

3.65 The following table gives the frequency distribution of entertainment expenditures (in dollars) incurred by 50 families during the past week.

Entertainment Expenditure (dollars)	Number of Families
0 to less than 10	5
10 to less than 20	10
20 to less than 30	15
30 to less than 40	12
40 to less than 50	5
50 to less than 60	3

Find the mean, variance, and standard deviation.

3.66 The following table gives the frequency distribution of total hours spent studying statistics during the semester for all 40 students enrolled in an introductory statistics course during Spring 1997.

Hours of Study	Number of Students
24 to less than 40	3
40 to less than 56	5
56 to less than 72	10
72 to less than 88	12
88 to less than 104	5
104 to less than 120	5

Find the mean, variance, and standard deviation.

3.67 The following table gives the grouped data on the weights of all 100 babies born at a hospital in 1996.

Weight (pounds)	Number of Babies
3 to less than 5	5
5 to less than 7	30
7 to less than 9	40
9 to less than 11	20
11 to less than 13	5

Find the mean, variance, and standard deviation.

3.68 The following table gives the frequency distribution of the amount of snowfall for January 1997 for 40 cities.

Snowfall (inches)	Number of Cities
0 to less than 4	5
4 to less than 8	6
8 to less than 12	8
12 to less than 16	10
16 to less than 20	8
20 to less than 24	3

Find the mean, variance, and standard deviation. Give a brief interpretation of the values in the column labeled mf in your table of calculations. What does Σmf represent?

3.69   The following table gives the frequency distribution of the number of personal computers sold during the past month at 40 computer stores located in New York City.

Computers Sold	Number of Stores
4 to 12	6
13 to 21	9
22 to 30	14
31 to 39	7
40 to 48	4

Calculate the mean, variance, and standard deviation. Give a brief interpretation of the values in the column labeled mf in your table of calculations. What does Σmf represent?

3.70   The following table gives information on the amounts (in dollars) of electric bills for August 1996 for a sample of 50 families.

Amount of Electric Bill (dollars)	Number of Families
0 to less than 20	5
20 to less than 40	16
40 to less than 60	11
60 to less than 80	10
80 to less than 100	8

Find the mean, variance, and standard deviation. Give a brief interpretation of the values in the column labeled mf in your table of calculations. What does Σmf represent?

3.71   For 50 airplanes that arrived late at an airport during a week, the time by which they were late was observed. In the following table, x denotes the time (in minutes) by which an airplane was late and f denotes the number of airplanes.

x	f
0 to less than 20	14
20 to less than 40	18
40 to less than 60	9
60 to less than 80	5
80 to less than 100	4

Find the mean, variance, and standard deviation.

3.72   The following table gives the frequency distribution of the number of cars owned by 100 households.

Number of Cars Owned	Number of Households
0	12
1	40
2	30
3	15
4	3

Calculate the mean, variance, and standard deviation. (*Hint:* The classes in this example are single-valued. These values of classes [the number of cars owned] will be used as values of m in the formula for the mean, variance, and standard deviation.)

3.73 The following table gives the number of television sets owned by 80 households.

Number of Television Sets Owned	Number of Households
0	4
1	33
2	28
3	10
4	5

Find the mean, variance, and standard deviation. (*Hint:* The classes in this example are single-valued. These values of classes [the number of television sets owned] will be used as values of m in the formula for the mean, variance, and standard deviation.)

3.4 USE OF STANDARD DEVIATION

By using the mean and standard deviation, we can find the proportion or percentage[5] of the total observations that fall within a given interval about the mean. This section briefly discusses Chebyshev's theorem and the empirical rule, both of which demonstrate this use of the standard deviation.

3.4.1 CHEBYSHEV'S THEOREM

Chebyshev's theorem gives a lower bound for the area under a curve between two points that are on opposite sides of the mean and at the same distance from the mean.

> **CHEBYSHEV'S THEOREM**
>
> For any number k greater than 1, at least $(1 - 1/k^2)$ of the data values lie within k standard deviations of the mean.

Figure 3.6 illustrates Chebyshev's theorem.

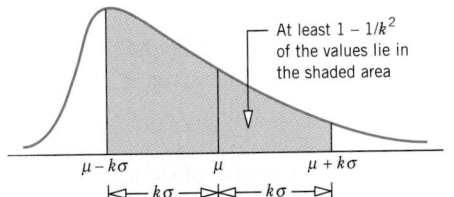

Figure 3.6 Chebyshev's theorem.

[5]Proportion refers to the relative frequency of a category or class and is expressed as a decimal. Percentage is the relative frequency multiplied by 100.

Thus, if $k = 2$, then,

$$1 - \frac{1}{k^2} = 1 - \frac{1}{(2)^2} = 1 - \frac{1}{4} = 1 - .25 = .75 \text{ or } 75\%$$

Therefore, according to Chebyshev's theorem, at least .75 or 75% of the values of a data set lie within two standard deviations of the mean. This is shown in Figure 3.7.

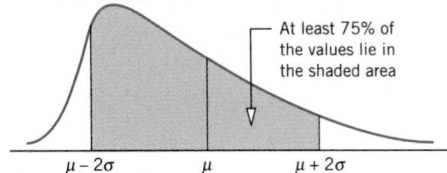

Figure 3.7   Percentage of values within two standard deviations of the mean for Chebyshev's theorem.

If $k = 3$, then,

$$1 - \frac{1}{k^2} = 1 - \frac{1}{(3)^2} = 1 - \frac{1}{9} = 1 - .11 = .89 \text{ or } 89\% \text{ approximately}$$

Therefore, according to Chebyshev's theorem, at least .89 or 89% of the values fall within three standard deviations of the mean. This is shown in Figure 3.8.

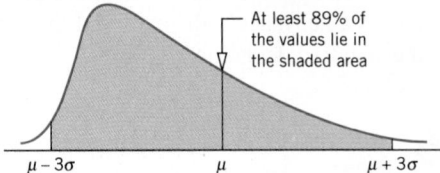

Figure 3.8   Percentage of values within three standard deviations of the mean for Chebyshev's theorem.

Although in Figures 3.6 through 3.8 we have used the population notation for the mean and standard deviation, the theorem applies to both sample and population data. Note that Chebyshev's theorem is applicable to a distribution of any shape. However, Chebyshev's theorem can be used only for $k > 1$. This is so because when $k = 1$, the value of $1 - 1/k^2$ is zero, and when $k < 1$ the value of $1 - 1/k^2$ is negative.

Applying Chebyshev's theorem.

EXAMPLE 3–18   For a statistics class, the mean for the midterm scores is 75 and the standard deviation is 8. Using Chebyshev's theorem, find the percentage of students who scored between 59 and 91.

Solution   Let μ and σ be the mean and the standard deviation, respectively, of the midterm scores. Then from the given information,

$$\mu = 75 \quad \text{and} \quad \sigma = 8$$

To find the percentage of students who scored between 59 and 91, the first step is to determine k. As shown below, each of the two points, 59 and 91, is 16 units away from the mean.

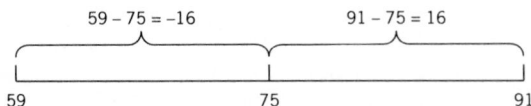

The value of k is obtained by dividing the distance between the mean and each point by the standard deviation. Thus,

$$k = 16/8 = 2$$

$$1 - \frac{1}{k^2} = 1 - \frac{1}{(2)^2} = 1 - \frac{1}{4} = 1 - .25 = \textbf{.75 or 75\%}$$

Hence, according to Chebyshev's theorem, at least 75% of the students scored between 59 and 91. This percentage is shown in Figure 3.9.

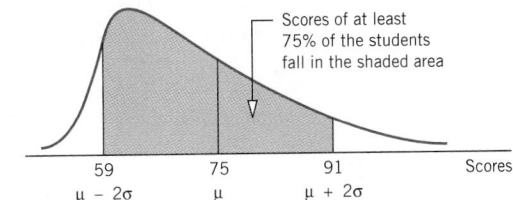

Figure 3.9 Percentage of students who scored between 59 and 91.

3.4.2 EMPIRICAL RULE

Whereas Chebyshev's theorem is applicable to any kind of distribution, the **empirical rule** applies only to a specific type of distribution called a *bell-shaped distribution*, as shown in Figure 3.10. More will be said about such a distribution in Chapter 6, where it will be called a *normal curve*. In this section, only the following three rules for such a curve are given.

EMPIRICAL RULE

For a bell-shaped distribution, approximately

1. 68% of the observations lie within one standard deviation of the mean
2. 95% of the observations lie within two standard deviations of the mean
3. 99.7% of the observations lie within three standard deviations of the mean

---

Figure 3.10 illustrates the empirical rule. Again, the empirical rule applies to population data as well as sample data.

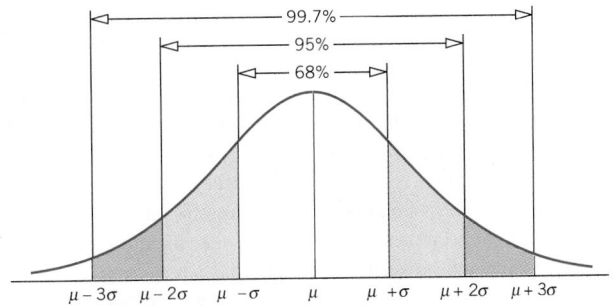

Figure 3.10 Illustration of the empirical rule.

Applying the empirical rule.

EXAMPLE 3-19 The age distribution of a sample of 5000 persons is bell-shaped with a mean of 40 years and a standard deviation of 12 years. Determine the approximate percentage of people who are 16 to 64 years old.

Solution We will use the empirical rule to find the required percentage because the distribution of ages follows a bell-shaped curve. From the given information, for this distribution,

$$\bar{x} = 40 \text{ years} \quad \text{and} \quad s = 12 \text{ years}$$

Each of the two points, 16 and 64, is 24 units away from the mean. Dividing 24 by 12, we convert the distance between each of the two points and the mean in terms of standard deviation. Thus, the distance between 16 and 40 and between 40 and 64 is each equal to $2s$. Consequently, as shown in Figure 3.11, the area from 16 to 64 is the area from $\bar{x} - 2s$ to $\bar{x} + 2s$.

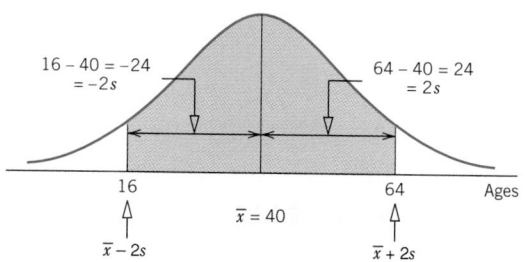

Figure 3.11 Percentage of people who are 16 to 64 years old.

Because the area within two standard deviations of the mean is approximately 95% for a bell-shaped curve, approximately 95% of the people in the sample are 16 to 64 years old.

The following case study, reproduced from the *Fortune* magazine, describes the application of the standard deviation to annual returns on mutual funds, assuming these returns are bell shaped. The readers should read this case study again after finishing Chapter 6.

CASE STUDY 3-4 HERE COMES THE SD

When your servant first became a *Fortune* writer several decades ago, it was hard doctrine that "several" meant three to eight, also that writers must not refer to "gross national product" without pausing to define this arcane term. GNP was in fact a relatively new concept at the time, having been introduced to the country only several years previously—in Roosevelt's 1944 budget message—so the presumption that readers had to be told repeatedly it was the "value of all goods and services produced by the economy" seemed entirely reasonable to this young writer, who personally had to look up the definition every time.

Numeracy lurches on. Nowadays the big question for editors is whether an average college-educated bloke needs a handhold when confronted with the term "standard deviation." The SD is suddenly onstage because the Securities and Exchange Commission is wondering aloud whether investment companies should be required to tell investors the standard deviation of their mutual funds' total returns over various past periods. Barry Barbash, SEC director of investment management, favors the requirement but confessed to the Washington *Post* that he worries about investors who will think a standard deviation is the dividing line on a highway or something.

The view around here is that the SEC is performing a noble service, but only partly because the requirement would enhance folk's insights into mutual funds. The commission's underlying idea is to give investors a better and more objective measure than is now available of the risk associated with different kinds of portfolios. The SD is a measure of variability, and funds with unusually variable returns—sometimes very high, sometimes very low—are presumed to be more risky.

What one really likes about the proposal, however, is the prospect that it will incentivize millions of greedy Americans to learn a little elementary statistics. One already has a list of issues that could be discussed much more thrillingly if only your average liberal arts graduate had a glimmer about the SD and the normal curve. The bell-shaped normal curve, or rather, the area underneath the curve, shows you how Providence arranged for things to be distributed in our world—with people's heights, or incomes, or IQs, or investment returns bunched around middling outcomes, and fewer and fewer cases as you move down and out toward the extremes. A line down the center of the curve represents the mean outcome, and deviations from the mean are measured by the SD.

An amazing property of the SD is that exactly 68.26% of all normally distributed data are within one SD of the mean. We once asked a professor of statistics a question that seemed to us quite profound, to wit, why that particular figure? The Prof answered dismissively that God had decided on 68.26% for exactly the same reason He had landed on 3.14 as the ratio between circumferences and diameters—because He just felt like it. The Almighty has also proclaimed that 95.44% of all data are within two SDs of the mean, and 99.73% within three SDs. When you know the mean and SD of some outcome, you can instantly establish the percentage probability of its occurrence. White men's heights in the U.S. average 69.2 inches, with an SD of 2.8 inches (according to the National Center for Health Statistics), which means that a 6-foot-5 chap is in the 99th percentile. In 1994, scores on the verbal portion of the Scholastic Assessment Test had a mean of 423 and an SD of 113, so if you scored 649—two SDs above the mean—you were in the 95th percentile.

As the SEC is heavily hinting, average outcomes are interesting but for many purposes inadequate; one also yearns to know the variability around that average. From 1926 through 1994, the S&P 500 had an average annual return of just about 10%. The SD accompanying that figure was just about 20%. Since returns will be within 1 SD some 68% of the time, they will be more than 1 SD from the mean 32% of the time. And since half these swings will be on the downside, we expect fund owners to lose more than 10% of their money about one year out of six and to lose more than 30% (two SDs below the mean) about one year out of 20. If your time horizon is short and you can't take losses like that, you arguably don't belong in stocks. If you think SDs are highway dividers, you arguably don't belong in cars.

Source: Daniel Seligman, ''Here comes the SD,'' *Fortune*, May 15, 1995. Copyright © 1995, The Time Inc. Reproduced with permission. All rights reserved.

EXERCISES

Concepts and Procedures

3.74 Briefly explain Chebyshev's theorem and its applications.

3.75 Briefly explain the empirical rule. To what kind of distribution is it applied?

3.76 A sample of 2000 observations has a mean of 74 and a standard deviation of 12. Using Chebyshev's theorem, find at least what percentage of the observations fall in the intervals $\bar{x} \pm 2s$, $\bar{x} \pm 2.5s$, and $\bar{x} \pm 3s$, respectively.

3.77 A large population has a mean of 230 and a standard deviation of 41. Using Chebyshev's theorem, find at least what percentage of the observations fall in the intervals $\mu \pm 2\sigma$, $\mu \pm 2.5\sigma$, and $\mu \pm 3\sigma$, respectively.

3.78 A large population has a mean of 310 and a standard deviation of 37. Using the empirical rule, find what percentage of the observations fall in the intervals $\mu \pm 1\sigma$, $\mu \pm 2\sigma$, and $\mu \pm 3\sigma$, respectively.

3.79 A sample of 3000 observations has a mean of 82 and a standard deviation of 16. Using the empirical rule, find what percentage of the observations fall in the intervals $\bar{x} \pm 1s$, $\bar{x} \pm 2s$, and $\bar{x} \pm 3s$, respectively.

Applications

3.80 The mean time taken by all participants to run a road race was found to be 220 minutes with a standard deviation of 20 minutes. Using Chebyshev's theorem, find the percentage of runners who ran this road race in

    a. 180 to 260 minutes   **b.** 160 to 280 minutes   **c.** 170 to 270 minutes

3.81 The 1996 gross sales of all firms in a large city have a mean of $2.3 million and a standard deviation of $.6 million. Using Chebyshev's theorem, find at least what percentage of firms in this city had 1996 gross sales of

    a. $1.1 to $3.5 million   **b.** $.8 to $3.8 million   **c.** $.5 to $4.1 million

3.82 According to data from A. C. Nielsen, Americans spent an average of $36.32 per pair of sneakers between October 1994 and September 1995 (*American Demographics*, March 1996). Assume that the purchase prices of all sneakers bought by Americans have a mean of $36.32 and a standard deviation of $8.00.

    a. Using Chebyshev's theorem, find at least what percentage of all sneaker purchase prices are between
        i. $16.32 and $56.32   **ii.** $12.32 and $60.32
    *b. Using Chebyshev's theorem, find the interval that contains at least 75% of all purchase prices for sneakers.

3.83 The mean monthly mortgage paid by all home owners in a city is $1365 with a standard deviation of $240.

    a. Using Chebyshev's theorem, find at least what percentage of all home owners in this city pay a monthly mortgage of
        i. $885 to $1845   **ii.** $645 to $2085
    *b. Using Chebyshev's theorem, find the interval that contains monthly mortgage payments of at least 84% of all home owners.

3.84 The mean life of a certain brand of auto batteries is 44 months with a standard deviation of 3 months. Assume that the lives of all auto batteries of this brand have a bell-shaped distribution. Using the empirical rule, find the percentage of auto batteries of this brand that have a life of

    a. 41 to 47 months   **b.** 38 to 50 months   **c.** 35 to 53 months

3.85 According to the National Education Association, the mean salary of public school teachers was $35,819 in 1994. Assume that the 1994 salaries of all public school teachers had a bell-shaped distribution with a mean of $35,819 and a standard deviation of $3500. Using the empirical rule, find the percentage of public school teachers whose 1994 salaries were between

    a. $28,819 and $42,819   **b.** $32,319 and $39,319   **c.** $25,319 and $46,319

3.86 According to Metropolitan Life Insurance company's claims data for 1995, the average hospital's and physician's charges for total hip replacement revision surgery were $27,760 (*Statistical Bulletin*, October-December 1996). Assume that these charges for all such surgeries done in 1995 have a bell-shaped distribution with a mean of $27,760 and a standard deviation of $2700.

a. Using the empirical rule, find the percentage of 1995 total hip replacement revision surgeries for which such charges were between
  i. $22,360 and $33,160   ii. $25,060 and $30,460
*b. Using the empirical rule, find the interval that contains such charges for 99.7% of all total hip replacement revision surgeries for 1995.

3.87 The ages of cars owned by all employees of a large company have a bell-shaped distribution with a mean of 7 years and a standard deviation of 2 years.

a. Using the empirical rule, find the percentage of cars owned by these employees that are
  i. 5 to 9 years old   ii. 1 to 13 years old
*b. Using the empirical rule, find the interval that contains the ages of the cars owned by 95% of the employees of this company.

3.5 MEASURES OF POSITION

A **measure of position** determines the position of a single value in relation to other values in a sample or a population data set. There are many measures of position. However, only quartiles, percentiles, and percentile rank are discussed in this section.

3.5.1 QUARTILES AND INTERQUARTILE RANGE

Quartiles are the summary measures that divide a ranked data set into four equal parts. Three measures will divide any data set into four equal parts. These three measures are the **first quartile** (denoted by Q_1), the **second quartile** (denoted by Q_2), and the **third quartile** (denoted by Q_3). The data should be ranked in increasing order before the quartiles are determined. The quartiles are defined as follows.

QUARTILES

Quartiles are three summary measures that divide a ranked data set into four equal parts. The second quartile is the same as the median of a data set. The first quartile is the value of the middle term among the observations that are less than the median, and the third quartile is the value of the middle term among the observations that are greater than the median.

Figure 3.12 describes the positions of the three quartiles.

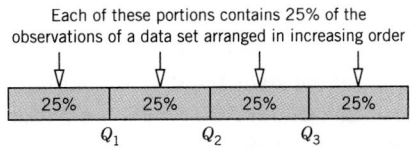

Each of these portions contains 25% of the observations of a data set arranged in increasing order

Figure 3.12 Quartiles.

Approximately 25% of the values in a ranked data set are less than Q_1 and about 75% are greater than Q_1. The second quartile, Q_2, divides a ranked data set into two equal parts;

hence, the second quartile and the median are the same. Approximately 75% of the data values are less than Q_3 and about 25% are greater than Q_3.

The difference between the third quartile and the first quartile for a data set is called the **interquartile range (IQR)**.

INTERQUARTILE RANGE

The difference between the third and the first quartiles gives the *interquartile range*. That is,

$$IQR = \text{Interquartile range} = Q_3 - Q_1$$

Examples 3–20 and 3–21 show the calculation of the quartiles and the interquartile range.

Finding quartiles and interquartile range.

EXAMPLE 3–20   The following are the scores of 12 students in a mathematics class.

75	80	68	53	99	58	76	73	85	88	91	79

(a)   Find the values of the three quartiles. Where does the score of 88 lie in relation to these quartiles?

(b)   Find the interquartile range.

Solution

Finding quartiles for even number of data values.

(a)   First, we rank the given scores in increasing order. Then we calculate the three quartiles as follows.

Values less than the median   Values greater than the median

53   58   68   73   75   76   79   80   85   88   91   99

$$Q_1 = \frac{68 + 73}{2} \qquad Q_2 = \frac{76 + 79}{2} \qquad Q_3 = \frac{85 + 88}{2}$$

$$= 70.5 \qquad\qquad = 77.5 \qquad\qquad = 86.5$$

Also the median

The value of Q_2, which is also the median, is given by the value of the middle term in the ranked data set. For the data of this example, this value is given by the average of the sixth and seventh terms. Consequently, Q_2 is 77.5. The value of Q_1 is given by the value of the middle term of the six values that fall below the median (or Q_2). Thus, it is obtained by taking the average of the third and fourth terms. So, Q_1 is 70.5. The value of Q_3 is given by the value of the middle term of the six values that fall above the median. For the data of this example, Q_3 is obtained by taking the average of the ninth and tenth terms, and it is 86.5.

The value of $Q_1 = 70.5$ indicates that the scores of (approximately) 25% of the students in this sample are less than 70.5 and those of (approximately) 75% of the students are greater than this value. Similarly, we can state that the scores of about half the students are less than 77.5 (which is Q_2) and those of the other half are greater than this value. The value of $Q_3 = 86.5$ indicates that the scores of (approximately) 75% of the students in this sample are less than 86.5 and those of (approximately) 25% of the students are greater than this value.

By looking at the position of 88, we can state that the score of 88 lies in the top 25% of the scores.

Finding interquartile range.

(b) The interquartile range is given by the difference between the values of the third and the first quartiles. Thus,

$$IQR = \text{Interquartile range} = Q_3 - Q_1 = 86.5 - 70.5 = \mathbf{16} \quad \blacksquare$$

Finding quartiles and interquartile range.

EXAMPLE 3–21 The following are the ages of nine employees of an insurance company.

$$47 \quad 28 \quad 39 \quad 51 \quad 33 \quad 37 \quad 59 \quad 24 \quad 33$$

(a) Find the values of the three quartiles. Where does the age of 28 fall in relation to the ages of these employees?

(b) Find the interquartile range.

Solution

Finding quartiles for odd number of data values.

(a) First we rank the given data in increasing order. Then we calculate the three quartiles as follows.

Values less than the median   Values greater than the median

$$24 \quad 28 \quad 33 \quad 33 \quad 37 \quad 39 \quad 47 \quad 51 \quad 59$$

$$Q_1 = \frac{28 + 33}{2} \qquad Q_2 = 37 \qquad Q_3 = \frac{47 + 51}{2}$$

$$= 30.5 \qquad\qquad\qquad\qquad = 49$$

Also the median

Thus, the values of the three quartiles are

$$Q_1 = \mathbf{30.5}, \qquad Q_2 = \mathbf{37}, \qquad \text{and} \qquad Q_3 = \mathbf{49}$$

The age of 28 falls in the lowest 25% of the ages.

Finding interquartile range.

(b) The interquartile range is

$$IQR = \text{Interquartile range} = Q_3 - Q_1 = 49 - 30.5 = \mathbf{18.5} \quad \blacksquare$$

3.5.2 PERCENTILES AND PERCENTILE RANK

Percentiles are the summary measures that divide a ranked data set into 100 equal parts. Each (ranked) data set has 99 percentiles that divide it into 100 equal parts. The data should be ranked in increasing order to compute percentiles. The kth percentile is denoted by P_k, where k is an integer in the range 1 to 99. For instance, the 25th percentile is denoted by P_{25}. Figure 3.13 shows the positions of the 99 percentiles.

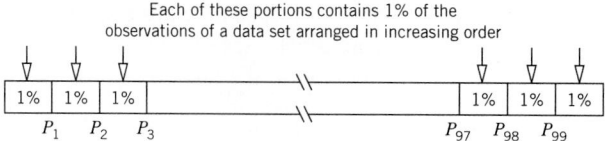

Figure 3.13 Percentiles.

Thus, the kth percentile, P_k, can be defined as a value in a data set such that about k% of the measurements are smaller than the value of P_k and about $(100 - k)$% of the measurements are greater than the value of P_k.

The approximate value of the kth percentile is determined as explained next.

PERCENTILES

The value of the kth *percentile*, denoted by P_k, is

$$P_k = \text{the value of the } \left(\frac{kn}{100}\right)\text{th term in a ranked data set}$$

where k denotes the number of the percentile and n represents the sample size.

Example 3–22 describes the procedure to calculate the percentiles.

Finding percentile for a data set.

EXAMPLE 3–22 The following are the test scores of 12 students in a mathematics class.

| 75 | 80 | 68 | 53 | 99 | 58 | 76 | 73 | 85 | 88 | 91 | 79 |

Find the value of the 62nd percentile. Give a brief interpretation of the 62nd percentile.

Solution First, we arrange the given scores in increasing order as follows.

| 53 | 58 | 68 | 73 | 75 | 76 | 79 | 80 | 85 | 88 | 91 | 99 |

The position of the 62nd percentile is

$$\frac{k\,n}{100} = \frac{62\,(12)}{100} = 7.44\text{th term}$$

The value of 7.44th term can be approximated by the average of the seventh and eighth terms in the ranked data. Therefore,

$$P_{62} = 62\text{nd percentile} = \frac{79 + 80}{2} = \mathbf{79.5}$$

Thus, approximately 62% of the scores are less than 79.5 and 38% are greater than 79.5 in the given data.

Note that if a data set contains only a few observations, then the number of values less than the 62nd percentile may not be exactly 62% and the number of values greater than the

62nd percentile may not be exactly 38%. For example, in our data on 12 scores, 7 scores (which is approximately 58% of 12) are less than 79.5 and 5 scores (which is approximately 42% of 12) are greater than 79.5. However, these percentages will be more accurate in a larger data set.   ▬

We can also calculate the **percentile rank** for a particular value x_i of a data set by using the following formula. The percentile rank of x_i gives the percentage of values in the data set that are smaller than x_i.

PERCENTILE RANK OF A VALUE

$$\text{Percentile rank of } x_i = \frac{\text{Number of values less than } x_i}{\text{Total number of values in the data set}} \times 100$$

Example 3–23 shows how the percentile rank is calculated for a data value.

Finding percentile rank for a data value.

EXAMPLE 3–23   Refer to the math scores of the 12 students given in Example 3–22, which are reproduced below.

| 75 | 80 | 68 | 53 | 99 | 58 | 76 | 73 | 85 | 88 | 91 | 79 |

Find the percentile rank for the score 85. Give a brief interpretation of this percentile rank.

Solution   First, we arrange the scores in increasing order as follows.

| 53 | 58 | 68 | 73 | 75 | 76 | 79 | 80 | 85 | 88 | 91 | 99 |

In this data set, 8 of the 12 scores are less than 85. Hence,

$$\text{Percentile rank of } 85 = \frac{8}{12} \times 100 = \mathbf{66.67\%}$$

Rounding this answer to the nearest integral value, we can state that about 67% of the scores in this sample are less than 85. In other words, about 67% of the students scored lower than 85.   ▬

EXERCISES

Concepts and Procedures

3.88   Briefly describe how the three quartiles are calculated for a data set. Illustrate by calculating the three quartiles for two examples, one with an odd number of observations and the second with an even number of observations.

3.89   Explain how the interquartile range is calculated. Give one example.

3.90   Briefly describe how the percentiles are calculated for a data set.

3.91   Explain the concept of the percentile rank for an observation of a data set.

Applications

3.92 The following data give the weights lost by 15 members of a health club at the end of two months after joining the club.

5	10	8	7	25	12	5	14
11	10	21	9	8	11	18	

   a. Compute the values of the three quartiles and the interquartile range.
   b. Calculate the (approximate) value of the 82nd percentile.
   c. Find the percentile rank of 10.

3.93 The following data give the speeds of 13 cars, measured by radar, traveling on interstate highway I-84.

67	72	63	66	76	69	71
76	65	79	68	67	71	

   a. Find the values of the three quartiles and the interquartile range.
   b. Calculate the (approximate) value of the 35th percentile.
   c. Compute the percentile rank of 71.

3.94 The following data give the number of computer keyboards assembled at the Twentieth Century Electronics Company for a sample of 25 days.

45	52	48	41	56	46	44	42	48	53
51	53	51	48	46	43	52	50	54	47
44	47	50	49	52					

   a. Calculate the values of the three quartiles and the interquartile range.
   b. Determine the (approximate) value of the 53rd percentile.
   c. Find the percentile rank of 50.

3.95 The following data give the hours worked last week by 30 employees of a company.

42	45	40	38	35	47	40	27	39	43
40	53	23	51	42	48	40	36	51	40
48	34	21	40	31	34	16	39	41	36

   a. Calculate the values of the three quartiles and the interquartile range.
   b. Find the (approximate) value of the 79th percentile.
   c. Calculate the percentile rank of 39.

3.96 The following data give the number of car thefts that occurred in a city during the past 12 days.

6	3	7	11	5	3	8	7	2	6	9	13

   a. Determine the values of the three quartiles and the interquartile range. Where does the value of 9 lie in relation to these quartiles?
   b. Calculate the (approximate) value of the 55th percentile.
   c. Find the percentile rank of 8.

3.97 Nixon Corporation manufactures computer terminals. The following data give the number of computer terminals produced at the company for a sample of 30 days.

24	32	27	23	33	33	29	25	23	28
21	26	31	20	27	33	27	23	28	29
31	35	34	22	26	28	23	35	31	27

a. Calculate the values of the three quartiles and the interquartile range. Where does the value of 31 lie in relation to these quartiles?

b. Find the (approximate) value of the 65th percentile. Give a brief interpretation of this percentile.

c. For what percentage of the days was the number of computer terminals produced 32 or higher? Answer by finding the percentile rank of 32.

3.98 The following data give the number of new cars sold at a dealership during a 20-day period.

8	5	12	3	9	10	6	3	8	8
4	6	10	11	7	7	3	5	9	11

a. Calculate the values of the three quartiles and the interquartile range. Where does the value of 4 lie in relation to these quartiles?

b. Find the (approximate) value of the 25th percentile. Give a brief interpretation of this percentile.

c. Find the percentile rank of 10. Give a brief interpretation of this percentile rank.

3.99 The following data give the scores of 19 students in a statistics class.

84	92	63	75	81	97	73	69	46	58
94	84	78	43	77	82	69	98	84	

a. Calculate the values of the three quartiles and the interquartile range. Where does the value of 81 lie in relation to these quartiles?

b. Compute the (approximate) value of the 93rd percentile. Give a brief interpretation of this percentile.

c. Find the percentile rank of 82. Give a brief interpretation of this percentile rank.

3.6 BOX-AND-WHISKER PLOT

A **box-and-whisker plot** gives a graphic presentation of data using five measures: the median, the first quartile, the third quartile, and the smallest and the largest values in the data set between the lower and the upper inner fences. A box-and-whisker plot can help us visualize the center, the spread, and the skewness of a data set. It also helps detect outliers. We can compare the different distributions by making box-and-whisker plots for each of them.

> **BOX-AND-WHISKER PLOT**
>
> A plot that shows the center, spread, and skewness of a data set. It is constructed by drawing a box and two whiskers that use the median, the first quartile, the third quartile, and the smallest and the largest values in the data set between the lower and the upper inner fences.

Example 3–24 explains all the steps needed to make a box-and-whisker plot.

Constructing a box-and-whisker plot.

EXAMPLE 3–24 The following data give the incomes (in thousands of dollars) for a sample of 12 households.

23	17	32	60	22	52	29	38	42	92	27	46

Construct a box-and-whisker plot for these data.

Solution   The following five steps are performed to construct a box-and-whisker plot.

Step 1.   First, rank the data in increasing order and calculate the values of the median, the first quartile, the third quartile, and the interquartile range. The ranked data are

17   22   23   27   29   32   38   42   46   52   60   92

For these data,

$$\text{Median} = (32 + 38)/2 = 35$$

$$Q_1 = (23 + 27)/2 = 25$$

$$Q_3 = (46 + 52)/2 = 49$$

$$\text{IQR} = Q_3 - Q_1 = 49 - 25 = 24$$

Step 2.   Find the points that are $1.5 \times \text{IQR}$ below Q_1 and $1.5 \times \text{IQR}$ above Q_3. These two points are called the **lower** and the **upper inner fences**, respectively.

$$1.5 \times \text{IQR} = 1.5 \times 24 = 36$$

$$\text{Lower inner fence} = Q_1 - 36 = 25 - 36 = -11$$

$$\text{Upper inner fence} = Q_3 + 36 = 49 + 36 = 85$$

Step 3.   Determine the smallest and the largest values in the given data set within the two inner fences. These two values for our example are as follows.

$$\text{Smallest value within the two inner fences} = 17$$

$$\text{Largest value within the two inner fences} = 60$$

Step 4.   Draw a horizontal line and mark the income levels on it such that all the values in the given data set are covered. Above the horizontal line, draw a box with its left side at the position of the first quartile and the right side at the position of the third quartile. Inside the box, draw a vertical line at the position of the median. The result of this step is shown in Figure 3.14.

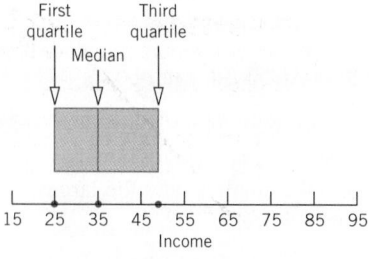

Figure 3.14

Step 5.   By drawing two lines, join the points of the smallest and the largest values within the two inner fences to the box. These values are 17 and 60 in this example as listed in Step 3. The two lines that join the box to these two values are called **whiskers**. A value that falls outside the two inner fences is shown by marking an asterisk and is called an outlier. This completes the box-and-whisker plot, as shown in Figure 3.15.

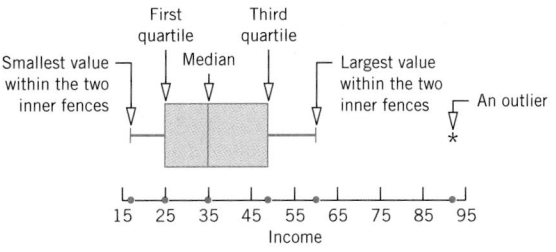

Figure 3.15

In Figure 3.15, about 50% of the data values fall within the box, about 25% of the values fall on the left side of the box, and about 25% fall on the right side of the box. Also, 50% of the values fall on the left side of the median and 50% lie on the right side of the median. The data of this example are skewed to the right because the lower 50% of the values are spread over a smaller range than the upper 50% of the values.

The observations that fall outside the two inner fences are called outliers. These outliers can be classified into two kinds of outliers—mild and extreme outliers. To do so, we define two outer fences—a **lower outer fence** at 3.0 × IQR below the first quartile and an **upper outer fence** at 3.0 × IQR above the third quartile. If an observation is outside either of the two inner fences but within either of the two outer fences, it is called a *mild outlier*. However, an observation that is outside either of the two outer fences is called an *extreme outlier*. For this example, the outer fences occur at −47 and 121. Since 92 is outside the upper inner fence but inside the upper outer fence, it is a mild outlier. ▬

For a symmetric data set, the line representing the median will be in the middle of the box and the spread of the values will be over almost the same range on both sides of the box.

EXERCISES

Concepts and Procedures

3.100 Briefly explain what summary measures are used to construct a box-and-whisker plot.

3.101 Prepare a box-and-whisker plot for the following data.

36	43	28	52	41	59	47	61
24	55	63	73	32	25	35	49
31	22	61	42	58	65	98	34

Does this data set contain any outliers?

3.102 Prepare a box-and-whisker plot for the following data.

11	8	26	31	62	19	7	3	14	75
33	30	42	15	18	23	29	13	16	6

Does this data set contain any outliers?

Applications

3.103 The following data give the time (in minutes) taken to complete a statistics test by 14 students.

93	87	96	77	73	91	82	71	98	74	95	89	79	88

Prepare a box-and-whisker plot. Comment on the skewness of these data.

 3.104 The following data give the scores of 19 students in a statistics class.

84	92	63	75	81	97	73	69	46	58
94	84	78	43	77	82	69	98	84	

Make a box-and-whisker plot. Are the data skewed in any direction?

3.105 The following data give the weights (in pounds) lost by 15 members of a health club at the end of two months after joining the club.

5	10	8	7	25	12	5	14
11	10	21	9	8	11	18	

Construct a box-and-whisker plot. Are the data symmetric or skewed?

 3.106 The following data give the number of computer keyboards assembled at the Twentieth Century Electronics Company for a sample of 25 days.

45	52	48	41	56	46	44	42	48	53
51	53	51	48	46	43	52	50	54	47
44	47	50	49	52					

Prepare a box-and-whisker plot. Comment on the skewness of these data.

 3.107 The following data give the hours worked last week by 30 employees of a company.

42	45	40	38	35	47	40	27	39	43
40	53	23	51	42	48	40	36	51	40
48	34	21	40	31	34	16	39	41	36

Make a box-and-whisker plot. Comment on the skewness of these data.

3.108 The following data give the number of car thefts that occurred in a city during the past 12 days.

6	3	7	11	5	3	8	7	2	6	9	13

Make a box-and-whisker plot. Comment on the skewness of these data.

 3.109 Nixon Corporation manufactures computer terminals. The following data give the number of computer terminals produced at the company for a sample of 30 days.

24	32	27	23	33	33	29	25	23	28
21	26	31	20	27	33	27	23	28	29
31	35	34	22	26	28	23	35	31	27

Prepare a box-and-whisker plot. Comment on the skewness of these data.

3.110 The following data give the number of new cars sold at a dealership during a 20-day period.

8	5	12	3	9	10	6	3	8	8
4	6	10	11	7	7	3	5	9	11

Make a box-and-whisker plot. Comment on the skewness of these data.

GLOSSARY

Bimodal distribution A distribution that has two modes.

Box-and-whisker plot A plot that shows the center, spread, and skewness of a data set by drawing a box and two whiskers using the median, the first quartile, the third quartile, and the smallest and the largest values in the data set between the lower and the upper inner fences.

Chebyshev's theorem For any number k greater than 1, at least $(1 - 1/k^2)$ of the values for any distribution lie within k standard deviations of the mean.

Coefficient of variation A measure of relative variability that expresses standard deviation as a percentage of the mean.

Empirical rule For a specific bell-shaped distribution, about 68% of the observations fall in the interval $(\mu - \sigma)$ to $(\mu + \sigma)$, about 95% fall in the interval $(\mu - 2\sigma)$ to $(\mu + 2\sigma)$, and about 99.7% fall in the interval $(\mu - 3\sigma)$ to $(\mu + 3\sigma)$.

First quartile The value in a ranked data set such that about 25% of the measurements are smaller than this value and about 75% are larger. It is the median of the values that are smaller than the median of the whole data set.

Interquartile range The difference between the third and the first quartiles.

Lower inner fence The value in a data set that is $1.5 \times$ IQR below the first quartile.

Lower outer fence The value in a data set that is $3.0 \times$ IQR below the first quartile.

Mean A measure of central tendency calculated by dividing the sum of all values by the number of values in the data set.

Measures of central tendency Measures that describe the center of a distribution. The mean, median, and mode are three of the measures of central tendency.

Measures of dispersion Measures that give the spread of a distribution. The range, variance, and standard deviation are three such measures.

Measures of position Measures that determine the position of a single value in relation to other values in a data set. Quartiles, percentiles, and percentile rank are examples of measures of position.

Median The value of the middle term in a ranked data set. The median divides a ranked data set into two equal parts.

Mode A value (or values) that occurs with highest frequency in a data set.

Multimodal distribution A distribution that has more than two modes.

Outliers or **extreme values** Values that are very small or very large relative to the majority of the values in a data set.

Parameter A summary measure calculated for population data.

Percentile rank The percentile rank of a value gives the percentage of values in the data set that are smaller than this value.

Percentiles Ninety-nine values that divide a ranked data set into 100 equal parts.

Quartiles Three summary measures that divide a ranked data set into four equal parts.

Range A measure of spread obtained by taking the difference between the largest and the smallest values in a data set.

Second quartile Middle or second of the three quartiles that divide a ranked data set into four equal parts. About 50% of the values in the data set are smaller and about 50% are larger than the second quartile. The second quartile is the same as the median.

Standard deviation A measure of spread that is given by the positive square root of the variance.

Statistic A summary measure calculated for sample data.

Third quartile Third of the three quartiles that divide a ranked data set into four equal parts. About

75% of the values in a data set are smaller than the value of the third quartile and about 25% are larger. It is the median of the values that are greater than the median of the whole data set.

Trimmed mean The k% trimmed mean is obtained by dropping k% of the smallest values and k% of the largest values from the given data and then calculating the mean of the remaining $(100 - 2k)$% of the values.

Unimodal distribution A distribution that has only one mode.

Upper inner fence The value in a data set that is $1.5 \times$ IQR above the third quartile.

Upper outer fence The value in a data set that is $3.0 \times$ IQR above the third quartile.

Variance A measure of spread.

KEY FORMULAS

1. **Mean for ungrouped data**

   For population data: $\mu = \dfrac{\Sigma x}{N}$

   For sample data: $\bar{x} = \dfrac{\Sigma x}{n}$

2. **Mean for grouped data**

   For population data: $\mu = \dfrac{\Sigma mf}{N}$

   For sample data: $\bar{x} = \dfrac{\Sigma mf}{n}$

   where m is the midpoint and f is the frequency of a class.

3. **Median for ungrouped data**

   $$\text{Median} = \text{Value of the } \left(\frac{n+1}{2}\right)\text{th term in a ranked data set}$$

4. **Range**

   $$\text{Range} = \text{Largest value} - \text{Smallest value}$$

5. **Variance for ungrouped data**

   For population data: $\sigma^2 = \dfrac{\Sigma x^2 - \dfrac{(\Sigma x)^2}{N}}{N}$

   For sample data: $s^2 = \dfrac{\Sigma x^2 - \dfrac{(\Sigma x)^2}{n}}{n-1}$

6. **Variance for grouped data**

   For population data: $\sigma^2 = \dfrac{\Sigma m^2 f - \dfrac{(\Sigma mf)^2}{N}}{N}$

   For sample data: $s^2 = \dfrac{\Sigma m^2 f - \dfrac{(\Sigma mf)^2}{n}}{n-1}$

   where m is the midpoint and f is the frequency of a class.

7. **Standard deviation**

For population data: $\quad \sigma = \sqrt{\sigma^2}$

For sample data: $\quad s = \sqrt{s^2}$

8. **Coefficient of variation**

For population data: $\quad CV = \dfrac{\sigma}{\mu} \times 100\%$

For sample data: $\quad CV = \dfrac{s}{x} \times 100\%$

9. **Interquartile range**

$$IQR = \text{Interquartile range} = Q_3 - Q_1$$

10. **Percentiles**

The kth percentile is given by:

$$P_k = \text{Value of the } \left(\dfrac{kn}{100}\right)\text{th term in a ranked data set}$$

11. **Percentile rank of a value**

The percentile rank for a particular value x_i of a data set is calculated as follows.

$$\text{Percentile rank of } x_i = \dfrac{\text{Number of values less than } x_i}{\text{Total number of values in the data set}} \times 100$$

SUPPLEMENTARY EXERCISES

3.111 The following data give the 1995 profits (in millions of dollars) for the 7 largest chemical companies in *Fortune* magazine's Global 500 (*Fortune*, August 5, 1996). The data, entered in that order, are for the companies DuPont, Hoechst, BASF, Bayer, Dow Chemical, CIBA-GEIGY, and Mitsubishi Chemical, respectively.

3293	1193	1724	1671	2078	1824	233

a. Calculate the mean and median.
b. Does this data set contain any outlier(s)? If yes, drop the outlier(s) and recalculate the mean and median. Which of these measures changes by a larger amount when you drop the outlier(s)?
c. Is the mean or the median a better summary measure for these data? Explain.

 3.112 The following data give the 1995 population (in millions) of the world's 10 most populous nations (*Statistical Bulletin*, October-December 1996). The data, entered in that order, are for Bangladesh, Japan, India, Pakistan, China, Nigeria, Indonesia, the United States, Brazil, and the Russian Federation, respectively.

128	126	937	132	1203	101	204	263	161	149

a. Find the mean and median for these data. Do these data have a mode?
b. Find the range, variance, and standard deviation.

3.113 The following data give the 1995 total pay (in millions of dollars) of the 10 highest-paid CEOs in the United States (*Business Week*, April 22, 1996). The data are for the CEOs of Green Tree Financial, Travelers Group, General Electric, AMGEN, DSC Communications, U.S. Robotics, Andrew Corporation, Forest Laboratories, Goodyear Tire & Rubber, and Sears, Roebuck, respectively.

66   50   22   22   19   19   18   17   17   16

   a. Calculate the mean and median. Do these data have a mode?
   b. Find the range, variance, and standard deviation.

3.114 The following data give the total number of driving citations previously received by 12 drivers.

4   8   0   3   11   7   4   14   8   13   7   9

   a. Find the mean, median, and mode for these data.
   b. Calculate the range, variance, and standard deviation.

3.115 The following table gives the distribution of the amount of rainfall (in inches) for July 1996 for 50 cities.

Rainfall	Number of Cities
0 to less than 2	6
2 to less than 4	10
4 to less than 6	18
6 to less than 8	9
8 to less than 10	4
10 to less than 12	3

Find the mean, variance, and standard deviation.

3.116 The following table gives the distribution of the number of days for which *all* 40 employees of a company were absent during the last year.

Number of Days Absent	Number of Employees
0 to 2	13
3 to 5	14
6 to 8	6
9 to 11	4
12 to 14	3

Calculate the mean, variance, and standard deviation. Are the values of these summary measures population parameters or sample statistics?

3.117 The mean time taken to learn the basics of a word processor by all students is 200 minutes with a standard deviation of 20 minutes.

   a. Using Chebyshev's theorem, find at least what percentage of students will learn the basics of this word processor in
       i.   160 to 240 minutes   ii.   140 to 260 minutes
  *b. Using Chebyshev's theorem, find the interval that contains the time taken by at least 75% of all students to learn this word processor.

3.118 According to the American Association of University Professors, the mean 1995–96 salary of university and college professors was $50,980 (*Academe*, March-April 1996). Assume that the standard deviation of these salaries is $6300.

   a. Using Chebyshev's theorem, find at least what percentage of university and college professors have their salaries between
       i. $38,380 and $63,580   ii. $32,080 and $69,880

*b. Using Chebyshev's theorem, find the interval that contains the salaries of at least 84% of all university and college professors.

3.119 Refer to Exercise 3.117. Suppose the time taken to learn the basics of this word processor by all students has a bell-shaped distribution with a mean of 200 minutes and a standard deviation of 20 minutes.

    a. Using the empirical rule, find the percentage of students who will learn the basics of this word processor in

        i. 180 to 220 minutes   **ii.** 160 to 240 minutes

    *b. Using the empirical rule, find the interval that contains the time taken by 99.7% of all students to learn this word processor.

3.120 Refer to Exercise 3.118. Assume that the salaries of university and college professors have a bell-shaped distribution with a mean of $50,980 and a standard deviation of $6300.

    a. Using the empirical rule, find the percentage of university and college professors whose salaries are between

        i. $38,380 and $63,580   **ii.** $32,080 and $69,880

    *b. Using the empirical rule, find the interval that contains the salaries of 68% of all university and college professors.

3.121 Refer to the data of Exercise 3.113 on the 1995 total pay (in millions of dollars) of the 10 highest-paid CEOs in the United States.

    a. Determine the values of the three quartiles and the interquartile range. Where does the value of 18 lie in relation to these quartiles?

    b. Calculate the (approximate) value of the 70th percentile. Give a brief interpretation of this percentile.

    c. Find the percentile rank of 18. Give a brief interpretation of this percentile rank.

3.122 Refer to the data of Exercise 3.114 on the total number of driving citations received by 12 drivers.

    a. Determine the values of the three quartiles and the interquartile range. Where does the value of 4 lie in relation to these quartiles?

    b. Calculate the (approximate) value of the 70th percentile. Give a brief interpretation of this percentile.

    c. Find the percentile rank of 3. Give a brief interpretation of this percentile rank.

3.123 The following data give the ages of 15 employees of a company.

36	47	23	55	42	31	27	19
38	65	52	47	39	25	44	

Prepare a box-and-whisker plot. Is this data set skewed in any direction? If yes, is it skewed to the right or to the left? Does this data set contain any outliers?

3.124 The following data give the prices (in thousands of dollars) of 16 recently sold houses in an area.

141	163	127	104	197	203	113	179
256	228	183	119	133	199	871	191

Make a box-and-whisker plot. Comment on the skewness of this data set. Does this data set contain any outliers?

*3.125 Melissa's grade in her math class is determined by three 100-point tests and a 200-point final exam. To determine the grade of a student in this class, the instructor will add the four scores together and divide this sum by 5 to obtain a percentage. This percentage must be at least 80 for a grade of B. If Melissa's three test scores are 75, 69, and 87, what is the minimum score she needs on the final exam to obtain a B grade?

*3.126 Jeffrey is serving on a six-person jury for a personal-injury lawsuit. All six jurors want to award damages to the plaintiff but cannot agree on the amount of the award. The jurors have decided that each of them will suggest an amount that he or she thinks should be awarded; then they will use the mean of these six numbers to recommend the award to the plaintiff.

    a. Jeffrey thinks the plaintiff should receive $20,000, but he thinks the mean of the other five jurors' recommendations will be about $12,000. He decides to suggest an inflated amount so that the mean for all six jurors is $20,000. What amount would Jeffrey have to suggest?

    b. How might this jury revise its procedure to prevent a juror like Jeffrey from having an undue influence on the amount of damages to be awarded to the plaintiff?

*3.127 The heights of five starting players on a basketball team have a mean of 76 inches, a median of 78 inches, and a range of 11 inches.

    a. If the tallest of these five players is replaced by a substitute who is 2 inches taller, find the new mean, median, and range.

    b. If the tallest player is replaced by a substitute who is 4 inches shorter, which of the new values (mean, median, range) could you determine, and what would their new values be?

*3.128 The ZT500 car gives 12 miles per gallon in city driving and 23 miles per gallon on highways. Rick owns a ZT500 car. He does 70% of his driving in the city and 30% on highways. What mean mileage per gallon should he expect?

*3.129 A small country bought oil from three different sources in one week, as shown in the following table.

Source	Barrels Purchased	Price per Barrel ($)
Mexico	1000	20
Venezuela	200	25
Spot Market	100	40

Find the mean price per barrel for all 1300 barrels of oil purchased in that week.

*3.130 An investor bought 47 ounces of gold bullion during 1996 as shown in the following table.

Month	Ounces Bought	Price per Ounce ($)
January	10	400
March	5	410
July	12	390
October	20	380

The investor tells you that the mean price he paid for the gold is $(400 + 410 + 390 + 380)/4 = \$395$ per ounce. Do you agree? If not, explain to the investor why this method of calculating the mean is wrong in this case. Then find the correct value of the mean.

*3.131 In the Olympic Games, when events require that a subjective judgment of an athlete's performance be made, the highest and lowest of the judges' scores may be dropped. Consider a gymnast whose performance is judged by seven judges and the highest and the lowest of the seven scores are dropped.

    a. Gymnast A's scores in this event are 9.4, 9.7, 9.5, 9.5, 9.4, 9.6, and 9.5. Find this gymnast's mean score after dropping the highest and the lowest scores.

    b. The answer to part a is an example of what percentage of trimmed mean?

    c. Write another set of scores for a gymnast B so that gymnast A has a higher mean score than gymnast B based on the trimmed mean, but gymnast B would win if all seven scores were counted. Do not use any scores lower than 9.0.

*3.132 A study of young people's buying habits was conducted by the International Mass Retail Association in 1995. Among the results were that buyers aged 8-17 make an average of 12.7 shopping trips per month, the mean amount spent per trip by buyers aged 8–12 is $18.50, and the mean amount spent per trip by buyers aged 13–17 is $31.20.

    a. Suppose you want to find the mean amount spent per trip by all buyers aged 8–17 included in this study. Why can you not just average $18.50 and $31.20? What additional information would you need, and how would you calculate the mean for the entire group?

    b. Suppose you want to determine the mean amount spent per month by buyers aged 13–17 included in this study. What additional information would you need and how would you find this mean?

*3.133 Records of surgeries on 1000 patients were obtained to compare the safety of two anesthetics, A and B. These 1000 patients were classified as low- or high-risk based on their age, state of health, and the nature of the surgery. The following table gives the number of patients in each risk group who were administered each type of anesthetic and the number of deaths for each of the two anesthetics and risk groups.

	Low-Risk Patients		High-Risk Patients	
	A	**B**	**A**	**B**
Number of patients	100	500	300	100
Number of deaths	1	10	15	8

    a. Compare the death rates for the two anesthetics for the low-risk surgeries.
    b. Compare the death rates for the two anesthetics in the high-risk surgeries.
    c. Compare the overall death rates for the two anesthetics.
    d. Explain how the anesthetic with the higher death rates for both levels of risk can have a lower overall death rate. This phenomenon is called Simpson's Paradox.

*3.134 The U.S. Bureau of Justice compiles information on sentences imposed and time actually served in state prisons for crimes of violence. The following table gives such information for 1994 for the four regions of the United States. The cases where life or death sentences were imposed are not included in this information. The percentages in the table give the average percentage of the sentence time served. For example, 53% means that on average 53% of 102 months, which is equal to 54.06 months, is actually served by the inmates in the Northeast region. Also, the maximum sentence is the time the prisoner would serve if he/she were given no early release.

Region	Mean Maximum Sentence Imposed (in months)	Percentage of the Sentence Time Served
Northeast	102	53
Midwest	120	31
South	105	43
West	51	71

A politician from the Midwest claims that his region is very strict with violent offenders since the mean maximum sentence of 120 months for the Midwest is the longest of the four regions. A politician from the West argues that her region is stricter because violent criminals there serve an average of 71% of their sentences, which is the highest of the four regions. Use the information given in the table to show which of the four regions is actually strictest with violent offenders.

*3.135 The test scores for a large statistics class have an unknown distribution with a mean of 70 and a standard deviation of 10.

    a. Find k so that at least 50% of the scores are within k standard deviations of the mean.
    b. Find k so that at most 10% of the scores are greater than k standard deviations above the mean.

*3.136   The test scores for a very large statistics class have a normal distribution with a mean of 70 points.

    a.   If 16% of all students in the class scored above 85, what is the standard deviation of the scores?

    b.   If 95% of the scores are between 60 and 80, what would the standard deviation be?

APPENDIX 3.1

A3.1.1 BASIC FORMULAS FOR THE VARIANCE AND STANDARD DEVIATION FOR UNGROUPED DATA

Example 3–25 illustrates how to use the basic formulas to calculate the variance and standard deviation for ungrouped data. From Section 3.2.2, the basic formulas for variance for ungrouped data are

$$\sigma^2 = \frac{\Sigma(x - \mu)^2}{N} \quad \text{and} \quad s^2 = \frac{\Sigma(x - \bar{x})^2}{n - 1}$$

where σ^2 is the population variance and s^2 is the sample variance.

    In either case, the standard deviation is obtained by taking the square root of the variance.

Calculating variance and standard deviation for ungrouped data using basic formulas.

EXAMPLE 3–25   Refer to Example 3–12. In that example, we used the short-cut formula to compute the variance and standard deviation for the data on annual energy costs of five models of refrigerators. Calculate the variance and standard deviation for those data using the basic formula.

Solution   All the required calculations to find the variance and standard deviation are made in Table 3.9.

Table 3.9

x	$(x - \bar{x})$		$(x - \bar{x})^2$
72	$72 - 62.40 =$	9.60	92.16
49	$49 - 62.40 =$	-13.40	179.56
79	$79 - 62.40 =$	16.60	275.56
55	$55 - 62.40 =$	-7.40	54.76
57	$57 - 62.40 =$	-5.40	29.16
$\Sigma x = 312$			$\Sigma(x - \bar{x})^2 = 631.20$

The following steps are performed to compute the variance and standard deviation.

Step 1.   Find the mean as follows.

$$\bar{x} = \Sigma x / n = 312/5 = \$62.40$$

Step 2.   Calculate $x - \bar{x}$, the deviation of each value of x from the mean. The results are shown in the second column of Table 3.9.

Step 3.   Square each of the deviations of x from $\bar{x}$, that is, calculate each of the $(x - \bar{x})^2$ values. These are called the *squared deviations*, and they are recorded in the third column of Table 3.9.

Step 4.   Add all the squared deviations to obtain $\Sigma(x - \bar{x})^2$, that is, sum all the values given in the third column of Table 3.9. This gives

$$\Sigma(x - \bar{x})^2 = 631.20$$

Step 5.   Obtain the sample variance by dividing the sum of the squared deviations by $n - 1$. Thus,

$$s^2 = \frac{\Sigma(x - \bar{x})^2}{n - 1} = \frac{631.20}{5 - 1} = \mathbf{157.8}$$

Step 6. Obtain the sample standard deviation by taking the positive square root of the variance. Hence,

$$s = \sqrt{s^2} = \sqrt{157.8} = \mathbf{\$12.56} \qquad \blacksquare$$

A3.1.2 BASIC FORMULAS FOR THE VARIANCE AND STANDARD DEVIATION FOR GROUPED DATA

Example 3–26 demonstrates how to use the basic formulas to calculate the variance and standard deviation for grouped data. The basic formulas for these calculations are

$$\sigma^2 = \frac{\Sigma f(m - \mu)^2}{N} \qquad \text{and} \qquad s^2 = \frac{\Sigma f(m - \bar{x})^2}{n - 1}$$

where σ^2 is the population variance, s^2 is the sample variance, m is the midpoint of a class, and f is the frequency of a class.

In either case, the standard deviation is obtained by taking the square root of the variance.

Calculating variance and standard deviation for grouped data using basic formulas.

EXAMPLE 3–26 In Example 3–17, we used the short-cut formula to compute the variance and standard deviation for the data on the number of orders received each day during the past 50 days at the office of a mail-order company. Calculate the variance and standard deviation for those data using the basic formula.

Solution All the required calculations to find the variance and standard deviation appear in Table 3.10.

Table 3.10

Number of Orders	f	m	mf	$m - \bar{x}$	$(m - \bar{x})^2$	$f(m - \bar{x})^2$
10 − 12	4	11	44	− 5.64	31.8096	127.2384
13 − 15	12	14	168	− 2.64	6.9696	83.6352
16 − 18	20	17	340	.36	.1296	2.5920
19 − 21	14	20	280	3.36	11.2896	158.0544
	$n = 50$		$\Sigma mf = 832$			$\Sigma f(m - \bar{x})^2 = 371.5200$

The following steps are performed to compute the variance and standard deviation using the basic formula.

Step 1. Find the midpoint of each class. Multiply the corresponding values of m and f. Find Σmf. From Table 3.10, $\Sigma mf = 832$.

Step 2. Find the mean as follows.

$$\bar{x} = \Sigma mf/n = 832/50 = 16.64$$

Step 3. Calculate $m - \bar{x}$, the deviation of each value of m from the mean. These calculations are done in the fifth column of Table 3.10.

Step 4. Square each of the deviations $m - \bar{x}$, that is, calculate each of the $(m - \bar{x})^2$ values. These are called *squared deviations*, and they are recorded in the sixth column of Table 3.10.

Step 5. Multiply the squared deviations by the corresponding frequencies (see the seventh column of Table 3.10). Adding the values of the seventh column, we obtain

$$\Sigma f(m - \bar{x})^2 = 371.5200$$

Step 6. Obtain the sample variance by dividing $\Sigma f(m - \bar{x})^2$ by $n - 1$. Thus,

$$s^2 = \frac{\Sigma f(m - \bar{x})^2}{n - 1} = \frac{371.5200}{50 - 1} = \textbf{7.5820}$$

Step 7. Obtain the standard deviation by taking the positive square root of the variance.

$$s = \sqrt{s^2} = \sqrt{7.5820} = \textbf{2.75 orders}$$

SELF-REVIEW TEST

1. The value of the middle term in a ranked data set is called the
    a. mean   **b.** median   **c.** mode

2. Which of the following summary measures is/are influenced by extreme values?
    a. mean   **b.** median   **c.** mode   **d.** range

3. Which of the following summary measures can be calculated for qualitative data?
    a. mean   **b.** median   **c.** mode

4. Which of the following can have more than one value?
    a. mean   **b.** median   **c.** mode

5. Which of the following is obtained by taking the difference between the largest and the smallest values of a data set?
    a. variance   **b.** range   **c.** mean

6. Which of the following is the mean of the squared deviations of x values from the mean?
    a. standard deviation   **b.** population variance   **c.** sample variance

7. The values of the variance and standard deviation are
    a. never negative   **b.** always positive   **c.** never zero

8. A summary measure calculated for the population data is called
    a. a population parameter   **b.** a sample statistic   **c.** an outlier

9. A summary measure calculated for the sample data is called
    a. a population parameter   **b.** a sample statistic   **c.** a boxplot

10. Chebyshev's theorem can be applied to
    a. any distribution   **b.** bell-shaped distributions only
    c. skewed distributions only

11. The empirical rule can be applied to
    a. any distribution   **b.** bell-shaped distributions only
    c. skewed distributions only

12. The first quartile is a value in a ranked data set such that
    a. about 75% of the values are smaller and about 25% are larger than this value
    b. about 50% of the values are smaller and about 50% are larger than this value
    c. about 25% of the values are smaller and about 75% are larger than this value

13. The third quartile is a value in a ranked data set such that
    a. about 75% of the values are smaller and about 25% are larger than this value
    b. about 50% of the values are smaller and about 50% are larger than this value
    c. about 25% of the values are smaller and about 75% are larger than this value

14. The 75th percentile is a value in a ranked data set such that
   a. about 75% of the values are smaller and about 25% are larger than this value
   b. about 25% of the values are smaller and about 75% are larger than this value

15. The following data give the number of times 10 persons used their credit cards during the past three months.

9   6   22   14   2   18   7   3   11   6

Calculate the mean, median, mode, range, variance, and standard deviation.

16. The mean, as a measure of central tendency, has the disadvantage of being influenced by extreme values. Illustrate this point with an example.

17. The range, as a measure of spread, has the disadvantage of being influenced by extreme values. Illustrate this point with an example.

18. When is the value of the standard deviation for a data set zero? Give one example of such a data set. Calculate the standard deviation for that data set to show that it is zero.

19. The following table gives the frequency distribution of the number of computers sold during the past 25 weeks at a computer store.

Computers Sold	Frequency
4 to 9	2
10 to 15	5
16 to 21	10
22 to 27	5
28 to 33	3

   a. What does the frequency column in the table represent?
   b. Calculate the mean, variance, and standard deviation.

20. The cars owned by all people living in a city are, on average, 7.3 years old with a standard deviation of 2.2 years.
   a. Using Chebyshev's theorem, find at least what percentage of the cars in this city are
       i. 1.8 to 12.8 years old   ii. .7 to 13.9 years old
   b. Using Chebyshev's theorem, find the interval that contains the ages of at least 75% of the cars owned by all people in this city.

21. The ages of cars owned by all people living in a city have a bell-shaped distribution with a mean of 7.3 years and a standard deviation of 2.2 years.
   a. Using the empirical rule, find the percentage of cars in this city that are
       i. 5.1 to 9.5 years old   ii. .7 to 13.9 years old
   b. Using the empirical rule, find the interval that contains the ages of 95% of the cars owned by all people in this city.

22. The following data give the number of books purchased by 16 adults during the past year.

8   12   20   16   0   11   18   4
10   6   17   24   15   9   2   6

   a. Calculate the three quartiles and the interquartile range. Where does the value of 4 lie in relation to these quartiles?
   b. Find the (approximate) value of the 68th percentile. Give a brief interpretation of this value.
   c. Calculate the percentile rank of 16. Give a brief interpretation of this value.

23. Make a box-and-whisker plot for the data on the number of books purchased by 16 adults during the past year given in Problem 22. Comment on the skewness of this data set.

*24.   The mean weekly wages of a sample of 15 employees of a company are $435. The mean weekly wages of a sample of 20 employees of another company are $490. Find the combined mean for these 35 employees.

*25.   The mean GPA of five students is 3.21. The GPAs of four of these five students are 3.85, 2.67, 3.45, and 2.91. Find the GPA of the fifth student.

*26.   The following are the prices (in thousands of dollars) of 10 houses sold recently in a city.

179   166   58   207   287   149   193   2534   163   238

Calculate the 10% trimmed mean for this data set. Do you think the 10% trimmed mean is a better summary measure for these data than the simple mean (i.e., the mean of all 10 values)? Briefly explain why, or why not.

*27.   Consider the following two data sets.

Data Set I: 8   16   20   35
Data Set II: 5   13   17   32

Note that each value of the second data set is obtained by subtracting three from the corresponding value of the first data set.

   a.   Calculate the mean for each of these two data sets. Comment on the relationship between the two means.
   b.   Calculate the standard deviation for each of these two data sets. Comment on the relationship between the two standard deviations.

USING MINITAB

This section illustrates how to use MINITAB to calculate some of the summary measures for a data set and to prepare a box-and-whisker plot. We will illustrate the required steps for these data analyses using examples.

CALCULATING SUMMARY MEASURES

Illustration M3–1 shows how to calculate the mean, median, range, standard deviation, and some of the other summary measures.

ILLUSTRATION M3–1   Refer to the data on the 1995 total sales of the top 12 U.S. companies given in Example 3–5. Using MINITAB, find the mean, median, range, and standard deviation for those data.

Solution   You can use either the **Stat** or **Calc** pull-down menu to find the mean, median, range, and standard deviation for any data set. The **Stat** menu gives the values of many summary measures at the same time, whereas the **Calc** menu helps to find the value of one summary measure at a time. If you are using MINITAB FOR WINDOWS, perform the following steps to find these summary measures using the **Stat** pull-down menu.

Step 1.   Enter the data on sales of companies into column C1 of the Data window.
Step 2.   Click the **Stat** pull-down menu at the top of the screen.
Step 3.   Click **Basic Statistics** from the selections available in the **Stat** menu.
Step 4.   Click **Descriptive Statistics** from the selections available in the **Basic Statistics** menu.
Step 5.   You will see a dialog box entitled **Descriptive Statistics** appear on the screen. Type **C1** in the box below **Variables**.
Step 6.   Click the **OK** button at the bottom of this dialog box. The MINITAB output will appear on the screen.

If you are using the MINITAB COMMAND LANGUAGE, enter the given data on sales of companies into column C1 using the **SET** command. Then, type the following MINITAB command to obtain the various summary measures.

```
MTB > DESCRIBE C1
```

In response to the above MINITAB commands, you will obtain the MINITAB output shown in Figure 3.16.

Most of the entries in the MINITAB output of Figure 3.16 are self-explanatory. For example, from this output, the mean is 82.6, the median is 73.4, the standard deviation is 39.3, the minimum value in the data set is 37.1, the maximum value in the data set is 168.8, the first quartile (Q1) is 53.1, and the third quartile (Q3) is 105.6. The fourth entry, labeled *Tr Mean* (called the trimmed mean and briefly

145

Figure 3.16   MINITAB output in response to **DESCRIPTIVE STATISTICS** or **DESCRIBE** command.

```
Descriptive Statistics

Variable      N      Mean    Median   Tr Mean    StDev    SE Mean
sales        12      82.6     73.4      78.5      39.3      11.4

Variable     Min     Max       Q1        Q3
sales       37.1    168.8     53.1      105.6
```

explained in Exercise 3.33), is the mean of the data when approximately 5% of the values at each end of the ranked data are dropped. In this example, it is calculated by dropping the smallest and the largest values of the data set. The sixth entry, *SE Mean*, gives the standard error (also called the standard deviation) of the sample mean, which will be discussed in Chapter 7.

As mentioned earlier, you can also use the **Calc** pull-down menu to find the various summary measures for a data set. Note that you will find the value of one summary measure at a time if you use the **Calc** menu. If you are using MINITAB FOR WINDOWS, perform the following steps to find these summary measures using the **Calc** pull-down menu.

Step 1.   Enter the data on sales of companies into column C1 of the Data window.
Step 2.   Click the **Calc** pull-down menu at the top of the screen.
Step 3.   Click **Column Statistics** from the selections available in the **Calc** menu.
Step 4.   You will see a dialog box entitled **Column Statistics** appear on the screen. Click next to the summary measure you want to find. For example, click inside the circle next to **Mean** to find the value of the mean.
Step 5.   Type **C1** in the box next to **Input Variable**.
Step 6.   If you want to store the answer permanently as a constant, type K1 in the box next to **Store Result in**. Remember, if you have used K1 earlier to store another answer, then type K2 in this box, and so on. However, if you do not want to store the answer, skip this step.
Step 7.   Click the **OK** button at the bottom of this dialog box. The MINITAB output will appear on the screen.

If you are using the MINITAB COMMAND LANGUAGE, enter the given data on sales of companies into column C1 using the **SET** command. Then, type the following MINITAB command to obtain the mean.

```
MTB > MEAN C1
```

In response to the above MINITAB commands, you will obtain the MINITAB output shown in Figure 3.17.

Figure 3.17   Mean for the 1995 sales of 12 U.S. companies.

```
Column Mean

Mean of C1 = 82.592
```

Similarly, you can calculate the median, range, or standard deviation by clicking next to the required summary measure in Step 4 of the MINITAB FOR WINDOWS steps described above. In the case of MINITAB COMMAND LANGUAGE, you type the command **MEDIAN C1**, **RANGE C1**, or **STANDARD DEVIATION C1** next to the MTB prompt to find the required measure.

BOX-AND-WHISKER PLOT

Illustration M3–2 shows how to prepare a box-and-whisker plot using MINITAB.

ILLUSTRATION M3–2   Refer to the data on incomes (in thousands of dollars) for a sample of 12 households given in Example 3–24. Using MINITAB, construct a box-and-whisker plot for those data.

Solution To obtain a box-and-whisker plot for the data on incomes given in Example 3–24, perform the following steps if you are using MINITAB FOR WINDOWS.

Step 1. Enter the data on incomes into column C1 of the Data window.
Step 2. Click the **Graph** pull-down menu at the top of the screen.
Step 3. Click **Boxplot** from the selections available in the **Graph** menu.
Step 4. You will see a dialog box entitled **Boxplot** appear on the screen. Type **C1** in the box below **Y**.
Step 5. Click the **OK** button at the bottom of this dialog box. The boxplot will appear on the screen.

If you are using the MINITAB COMMAND LANGUAGE, enter the given data on incomes into column C1 using the **SET** command. Then, type the following MINITAB command to obtain the box-and-whisker plot.

```
MTB > BOXPLOT C1
```

Whether you use MINITAB FOR WINDOWS or the MINITAB COMMAND LANGUAGE, you will obtain the box-and-whisker plot given in Figure 3.18 for the data on incomes. The asterisk in this graph indicates an outlier. Note that the box-and-whisker plot here appears vertically.

Figure 3.18 Box-and-whisker plot for incomes.

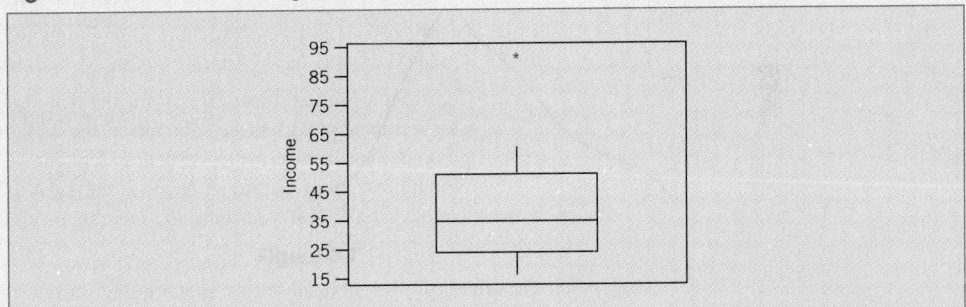

COMPUTER ASSIGNMENTS

M3.1 Refer to the subsample taken in the MINITAB Computer Assignment M2.3 of Chapter 2 from the sample data on the time taken to run the Manchester Road Race given in Appendix B. Using MINITAB, find the mean, median, range, and standard deviation for those data.

M3.2 Refer to the data on phone charges given in column C7 of Data Set I of Appendix B. From that data set, select the 4th value and then select every 10th value after that (i.e., select the 4th, 14th, 24th, 34th, . . . values). This subsample will give you 20 measurements. (Such a sample taken from a population is called a *systematic random sample*.) Using MINITAB, find the mean, median, standard deviation, first quartile, and third quartile for the phone charges for these 20 observations.

M3.3 Refer to Data Set I of Appendix B on prices of various products in different cities across the country. Using MINITAB, select a subsample of the price of regular unleaded gas (column C8) for 40 cities. Using MINITAB, find the mean, median, and standard deviation for the data of this subsample.

M3.4 Refer to the data on the 1995 gasoline tax rates (in cents per gallon) for 50 states given in Example 2–5 of Chapter 2. Using MINITAB, compute the mean, median, and standard deviation for those data.

M3.5 Refer to Data Set II of Appendix B, which contains data on different variables for 50 states. Using MINITAB, make a box-and-whisker plot for the data on the 1995 per capita incomes for 50 states given in column C5.

M3.6 Refer to Data Set I of Appendix B on prices of various products in different cities across the country. Using MINITAB, make a box-and-whisker plot for the data on the monthly telephone charges given in column C7.

M3.7 Refer to the data on the number of computer keyboards assembled at the Twentieth Century Electronics Company for a sample of 25 days given in Exercise 3.106. Using MINITAB, prepare a box-and-whisker plot for those data.

MORE CHALLENGING EXERCISES (OPTIONAL)
CHAPTERS 1 TO 3

1. Statisticians often need to know the shape of a population in order to make inferences. Suppose you are asked to specify the shape of the population of weights of all college students.
   a. Sketch a graph of what you think the weights of all college students would look like.
   b. The following data give the weights (in pounds) of a random sample of 44 college students. (In these data, F and M indicate whether the student is a female or a male.)

123 F	195 M	138 M	115 F	179 M	119 F	148 F	147 F	180 M
146 F	179 M	189 M	175 M	108 F	193 M	114 F	179 M	147 M
108 F	128 F	164 F	174 M	128 F	159 M	193 M	204 M	125 F
133 F	115 F	168 M	123 F	183 M	116 F	182 M	174 M	102 F
123 F	99 F	161 M	162 M	155 F	202 M	110 F	132 M	

   i. Construct a stem-and-leaf display for the data on weights.
   ii. Can you explain why these data appear the way they do?
   iii. Compute the three measures of central tendency for the data on weights.
   iv. Which measure of central tendency is the most informative?
   c. Now sketch a picture of what you think the weights of all college students look like. Is this similar to your sketch in part a?

2. How much does the typical American family spend going away on vacation each year? Twenty-five randomly selected households reported the following vacation expenditures during the past year (rounded to the nearest hundred dollars).

$2500	500	800	0	100
0	200	2200	0	200
0	1000	900	321,500	400
500	100	0	8200	900
0	1700	1100	600	3400

   a. Using both graphical and numerical methods, organize and interpret these data.
   b. What measure of central tendency would you say best answers the original question?

3. Actuaries at an insurance company must determine a premium for a new type of insurance. A random sample of 40 potential purchasers of this type of insurance were found to have suffered the following losses during the past year. These losses would have been covered by the insurance if it were available.

$100	32	0	0	470	50	0	14,589	212	93
0	0	1127	421	0	87	135	420	0	250
12	0	309	0	177	295	501	0	143	0
167	398	54	0	141	0	3709	122	0	0

   a. Use both the graphical and numerical methods you have learned to organize and interpret these data. Write a report summarizing the loss characteristics for the insurance company.
   b. Which measures of central tendency and variation would you suggest the actuaries use when determining the premium for this insurance?

4.   A local golf club has men's and women's summer leagues. The following data give the scores for a round of 18 holes of golf for 17 men and 15 women randomly selected from their respective leagues.

Men	87	68	92	79	83	67	71	92	112
	75	77	102	79	78	85	75	72	
Women	101	100	87	95	98	81	117	107	103
	97	90	100	99	94	94			

   a.   Make a box plot of each of the data sets and use them to discuss the similarities and differences between the scores of the men and women golfers.
   b.   Compute the various descriptive statistics you have learned for each sample. How do they compare?

5.   The final exam scores of the students in an elementary statistics course are summarized in the following frequency distribution table.

Score	Frequency
81–90	6
91–100	24
101–110	33
111–120	39
121–130	30
131–140	9
141–150	9

Construct a cumulative percentage ogive and use it to approximate

   a.   the 80th percentile
   b.   the percentile rank of a student who scored 95 on the exam
   c.   the interquartile range
   d.   the median

6.   A state police officer clocks the speeds of randomly selected drivers on an interstate highway. The following is a cumulative frequency distribution of the speeds (in miles per hour).

Speed	Cumulative Frequency
51–55	1
51–60	7
51–65	19
51–70	36
51–75	45
51–80	47
51–85	50

   a.   Make a cumulative frequency ogive of the speeds.
   b.   Based on the shape of the cumulative frequency ogive of the speeds, what can you tell about the observations?
   c.   Explain the meaning of a cumulative frequency ogive that is concave up or concave down.
   d.   What proportion of these drivers exceed 70 miles per hour?
   e.   What proportion of these drivers drive between 61 and 75 miles per hour?

7.   The following is a stem-and-leaf diagram of the distances (in thousands of miles) driven during the past year by a sample of drivers in the United States.

```
0 | 3 6 9
1 | 2 8 5 1 0 5
2 | 5 1 6
3 | 8
4 | 1
5 |
6 | 2
```

a. Compute the sample mean, median, and mode of the distances.
b. Compute the range, variance, and standard deviation for these data.
c. Compute the first and third quartiles.
d. An often used measure of variation called the interquartile range is computed as the difference $Q_3 - Q_1$.
   i.   Compute the interquartile range of the distances driven.
   ii. Describe what properties the interquartile range has. When would it be preferable to using the standard deviation when measuring variation?

8. Find the missing entries in the following frequency distribution table. Assume that all classes have the same width.

Class Limits	Frequency	Relative Frequency	Cumulative Frequency	Cumulative Percentage
8 to —	—	—	—	25
— to —	—	.05	—	—
— to —	—	—	9	—
— to —	—	.30	15	—
— to 32	—	—	—	—
	—	—		

9. In each of the following situations, find the missing quantity.
   a. The total weight of all pieces of luggage loaded onto an airplane is 12,372 pounds, which works out to be an average of 51.55 pounds per piece. How many pieces of luggage are on the plane?
   b. A group of seven friends, having just gotten back a chemistry exam, discuss their scores. Six of the students reveal that they got 81, 75, 93, 88, 82, and 85, but the seventh student is reluctant to say what she got. After some calculation she announces the group averaged 81 on the exam. What is her score?
   c. A statistician observes the daily cost of parking in five cities and computes the sample standard deviation to be $4. He later realizes he needs the individual costs but can only remember four: $4, $10, $0, and $8. However, he recalls the missing cost was less than the average. What was the daily cost of parking in the fifth city?

10. The time needed to complete an examination is bell-shaped with a mean of 50 minutes and a variance of 25 square minutes.
   a. What interval contains approximately 99.7% of the completion times?
   b. What proportion of those taking the exam need longer than an hour?
   c. What proportion of those taking the exam need between 45 minutes and 50 minutes?
   d. By what time will approximately 2.5% of those taking the exam finish?
   e. What proportion of those taking the exam need between 55 minutes and 60 minutes to finish?

11. The distribution of the lengths of fish in a certain lake is not known, but it is definitely *not* bell-shaped. It is estimated that the mean length is 6 inches with a standard deviation of 2 inches.
   a. At least what proportion of fish in the lake are between 3 inches and 9 inches?
   b. What is the smallest interval that will contain the lengths of at least 84% of the fish?
   c. Find an interval so that fewer than 36% of the fish have lengths outside this interval.

4 | PROBABILITY

We often make statements about probability. For example, a weather forecaster may predict that there is an 80% chance of rain tomorrow. A health-news reporter may state that a smoker has a much greater chance of getting cancer than a nonsmoker. A college student may ask an instructor about the chances of passing a course or getting an A if he or she did not do well on the midterm examination.

Probability, which measures the likelihood that an event will occur, is an important part of statistics. It is the basis of inferential statistics, which will be introduced in later chapters. In inferential statistics, we make decisions under conditions of uncertainty. Probability theory is used to evaluate the uncertainty involved in those decisions. For example, estimating next year's sales for a company is based on many assumptions, some of which may happen to be true and others may not. Probability theory will help us make decisions under such conditions of imperfect information and uncertainty. Combining probability and probability distributions (which are discussed in Chapters 5 through 7) with descriptive statistics will help us make decisions about populations based on information obtained from samples. This chapter presents the basic concepts of probability and the rules for computing probability.

4.1 EXPERIMENT, OUTCOMES, AND SAMPLE SPACE

Quality control inspector Jack Cook of Tennis Products Company picks up a tennis ball from the production line to check whether it is good or defective. Jack Cook's act of inspecting a tennis ball is an example of a statistical **experiment**. The result of his inspection will be that the ball is either ''good'' or ''defective.'' Each of these two observations is called an **outcome** (also called a *basic* or *final outcome*) of the experiment, and these outcomes taken together constitute the **sample space** for this experiment.

> **EXPERIMENT, OUTCOMES, AND SAMPLE SPACE**
>
> An *experiment* is a process that, when performed, results in one and only one of many observations. These observations are called the *outcomes* of the experiment. The collection of all outcomes for an experiment is called a *sample space*.

A sample space is denoted by S. The sample space for the example of inspecting a tennis ball is written as

$$S = \{\text{good, defective}\}$$

The elements of a sample space are called **sample points**.

Table 4.1 lists a few examples of experiments, their outcomes, and their sample spaces.

Table 4.1 Examples of Experiments, Outcomes, and Sample Spaces

Experiment	Outcomes	Sample Space
Toss a coin once	Head, Tail	$S = \{\text{Head, Tail}\}$
Roll a die once	1, 2, 3, 4, 5, 6	$S = \{1, 2, 3, 4, 5, 6\}$
Toss a coin twice	*HH, HT, TH, TT*	$S = \{HH, HT, TH, TT\}$
Birth of a baby	Boy, Girl	$S = \{\text{Boy, Girl}\}$
Take a test	Pass, Fail	$S = \{\text{Pass, Fail}\}$
Select a student	Male, Female	$S = \{\text{Male, Female}\}$

The sample space for an experiment can also be described by drawing either a Venn diagram or a tree diagram. A **Venn diagram** is a picture (a closed geometric shape such as a rectangle, a square, or a circle) that depicts all the possible outcomes for an experiment. In a **tree diagram**, each outcome is represented by a branch of the tree. Venn and tree diagrams help us understand probability concepts by presenting them visually. Examples 4–1 through 4–3 describe how to draw these diagrams for statistical experiments.

Venn and tree diagrams: one toss of a coin.

EXAMPLE 4–1   Draw the Venn and tree diagrams for the experiment of tossing a coin once.

Solution   This experiment has two possible outcomes: head and tail. Consequently, the sample space is given by

$$S = \{H, T\} \qquad \text{where } H = \text{Head}, \qquad T = \text{Tail}$$

To draw a Venn diagram for this example, we draw a rectangle and mark two points inside this rectangle that represent the two outcomes, head and tail. The rectangle is labeled *S* because it represents the sample space (see Figure 4.1*a*). To draw a tree diagram, we draw two branches starting at the same point, one representing the head and the second representing the tail. The two final outcomes are listed at the end of the branches (see Figure 4.1*b*).

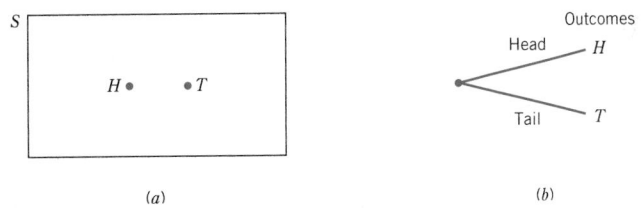

Figure 4.1   (*a*) Venn diagram and (*b*) tree diagram for one toss of a coin.

Venn and tree diagrams: two tosses of a coin.

EXAMPLE 4–2   Draw the Venn and tree diagrams for the experiment of tossing a coin twice.

Solution   This experiment can be split into two parts: the first toss and the second toss. Suppose the first time the coin is tossed we obtain a head. Then, on the second toss, we can still obtain a head or a tail. This gives us the two outcomes: *HH* (head on both tosses) and *HT* (head on the first toss and tail on the second toss). Now suppose we observe a tail on the first toss. Again, either a head or a tail can occur on the second toss, giving the remaining two outcomes: *TH* (tail on the first toss and head on the second toss) and *TT* (tail on both tosses). Thus, the sample space for two tosses of a coin is

$$S = \{HH, HT, TH, TT\}$$

The Venn and tree diagrams are given in Figure 4.2. Both these diagrams show the sample space for this experiment.

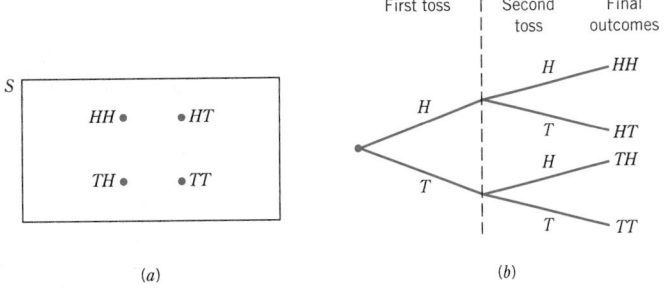

Figure 4.2   (*a*) Venn diagram and (*b*) tree diagram for two tosses of a coin.

Venn and tree diagrams: two selections.

EXAMPLE 4-3 Suppose we randomly select two persons from the members of a club and observe whether the person selected each time is a man or a woman. Write all the outcomes for this experiment. Draw the Venn and tree diagrams for this experiment.

Solution Let us denote the selection of a man by M and that of a woman by W. We can compare the selection of two persons to two tosses of a coin. Just as each toss of a coin can result in one of two outcomes, head or tail, each selection from the members of this club can result in one of two outcomes, man or woman. As we can see from the Venn and tree diagrams of Figure 4.3, there are four final outcomes: MM, MW, WM, WW. Hence, the sample space is written as

$$S = \{MM, MW, WM, WW\}$$

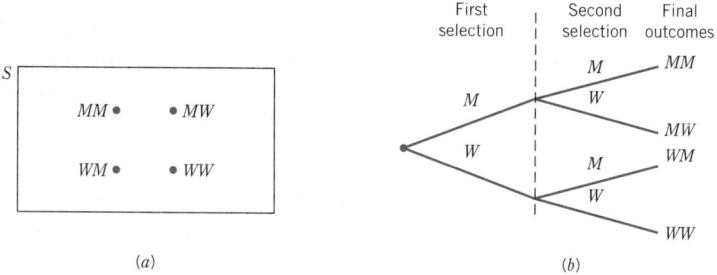

Figure 4.3 (*a*) Venn diagram and (*b*) tree diagram for selecting two persons.

4.1.1 SIMPLE AND COMPOUND EVENTS

An **event** consists of one or more of the outcomes of an experiment.

> **EVENT**
>
> An *event* is a collection of one or more of the outcomes of an experiment.

An event may be a *simple event* or a *compound event*. A simple event is also called an *elementary event*, and a compound event is also called a *composite event*.

Simple Event

Each of the final outcomes for an experiment is called a **simple event**. In other words, a simple event includes one and only one outcome. Usually, a simple event is denoted by E_1, E_2, E_3, and so forth. However, we can denote it by any of the other letters too, that is, by A, B, C, and so forth.

> **SIMPLE EVENT**
>
> An event that includes one and only one of the (final) outcomes for an experiment is called a *simple event* and is usually denoted by E_i.

Example 4–4 describes simple events.

Illustrating simple events.

EXAMPLE 4–4 Reconsider Example 4–3 about selecting two persons from the members of a club and observing whether the person selected each time is a man or a woman. Each of the final four outcomes (*MM*, *MW*, *WM*, and *WW*) for this experiment is a simple event. These four events can be denoted by E_1, E_2, E_3, and E_4, respectively. Thus,

$$E_1 = (MM), \quad E_2 = (MW), \quad E_3 = (WM), \quad \text{and} \quad E_4 = (WW) \quad ▬$$

Compound Event

A **compound event** consists of more than one outcome.

> **COMPOUND EVENT**
>
> A *compound event* is a collection of more than one outcome for an experiment.

Compound events are denoted by $A, B, C, D, \ldots$, or by $A_1, A_2, A_3, \ldots, B_1, B_2, B_3, \ldots$, and so forth. Examples 4–5 and 4–6 describe compound events.

Illustrating compound event: two selections.

EXAMPLE 4–5 Reconsider Example 4–3 about selecting two persons from the members of a club and observing whether the person selected each time is a man or a woman. Let *A* be the event that at most one man is selected. Event *A* will occur if either no man or one man is selected. Hence, the event *A* is given by

$$A = \{MW, WM, WW\}$$

Since event *A* contains more than one outcome, it is a compound event. The Venn diagram of Figure 4.4 gives a graphic presentation of compound event *A*.

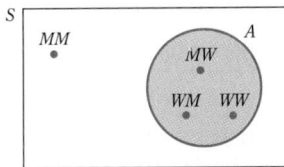

Figure 4.4 Venn diagram for event *A*. ▬

Illustrating simple and compound events: two selections.

EXAMPLE 4–6 In a group of people, some are in favor of genetic engineering and others are against it. Two persons are selected at random from this group and asked whether they are in favor of or against genetic engineering. How many distinct outcomes are possible? Draw a Venn diagram and a tree diagram for this experiment. List all the outcomes included in each of the following events and mention whether they are simple or compound events.

(a) Both persons are in favor of genetic engineering.

(b) At most one person is against genetic engineering.

(c) Exactly one person is in favor of genetic engineering.

Solution Let

$$F = \text{a person is in favor of genetic engineering}$$

$$A = \text{a person is against genetic engineering}$$

This experiment has the following four outcomes.

FF = both persons are in favor of genetic engineering

FA = the first person is in favor and the second is against

AF = the first person is against and the second is in favor

AA = both persons are against genetic engineering

The Venn and tree diagrams in Figure 4.5 show these four outcomes.

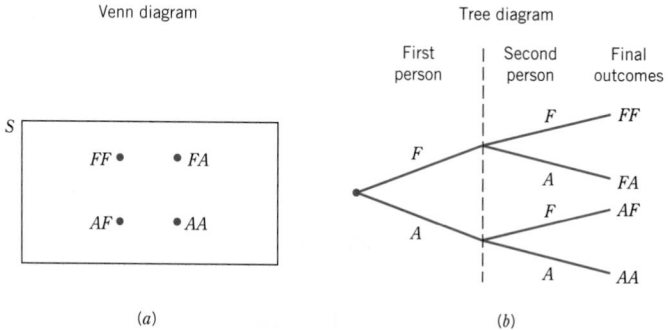

Figure 4.5 Venn and tree diagrams.

(a) The event "both persons are in favor of genetic engineering" will occur if FF is obtained. Thus,

Both persons are in favor of genetic engineering = $\{FF\}$

Because this event includes only one of the final four outcomes, it is a simple event.

(b) The event "at most one person is against genetic engineering" will occur if either none or one of the persons selected is against genetic engineering. Consequently,

At most one person is against genetic engineering = $\{FF, FA, AF\}$

Because this event includes more than one outcome, it is a compound event.

(c) The event "exactly one person is in favor of genetic engineering" will occur if one of the two persons selected is in favor and the other is against genetic engineering. Hence, it includes the following two outcomes.

Exactly one person is in favor of genetic engineering = $\{FA, AF\}$

Because this event includes more than one outcome, it is a compound event. ▄▄

EXERCISES

Concepts and Procedures

4.1 Define the following terms: experiment, outcome, sample space, simple event, compound event.

4.2 List the simple events for each of the following statistical experiments in a sample space S.
  a. One roll of a die.
  b. Three tosses of a coin.
  c. One toss of a coin and one roll of a die.

4.3 A box contains three items that are labeled A, B, and C. Two items are selected at random (without replacement) from this box. List all the possible outcomes for this experiment. Write the sample space *S*.

Applications

4.4 Two students are randomly selected from a statistics class, and it is observed whether or not they suffer from math anxiety. How many total outcomes are possible? Draw a tree diagram for this experiment. Draw a Venn diagram.

4.5 A hat contains a few red and a few green marbles. If two marbles are randomly drawn and the colors of these marbles are observed, how many total outcomes are possible? Draw a tree diagram for this experiment. Show all the outcomes in a Venn diagram.

4.6 A test contains two multiple-choice questions. If a student makes a random guess to answer each question, how many outcomes are possible? Depict all these outcomes in a Venn diagram. Also draw a tree diagram for this experiment. (*Hint:* Consider two outcomes for each question—either the answer is correct or it is wrong.)

4.7 A box contains a certain number of computer parts, a few of which are defective. Two parts are selected at random from this box and inspected to determine if they are good or defective. How many total outcomes are possible? Draw a tree diagram for this experiment.

4.8 In a group of people, some are in favor of a tax increase on rich people to reduce the federal deficit and others are against it. (Assume that there is no other outcome such as ''no opinion'' and ''do not know.'') Three persons are selected at random from this group and their opinions in favor or against raising such taxes are noted. How many total outcomes are possible? Write these outcomes in a sample space *S*. Draw a tree diagram for this experiment.

4.9 Draw a tree diagram for three tosses of a coin. List all outcomes for this experiment in a sample space *S*.

4.10 Refer to Exercise 4.4. List all the outcomes included in each of the following events. Indicate which are simple and which are compound events.

    a.   Both students suffer from math anxiety.
    b.   Exactly one student suffers from math anxiety.
    c.   The first student does not suffer and the second suffers from math anxiety.
    d.   None of the students suffers from math anxiety.

4.11 Refer to Exercise 4.5. List all the outcomes included in each of the following events. Indicate which are simple and which are compound events.

    a.   Both marbles are of different colors.
    b.   At least one marble is red.
    c.   Not more than one marble is green.
    d.   The first marble is green and the second is red.

4.12 Refer to Exercise 4.6. List all the outcomes included in each of the following events and mention which are simple and which are compound events.

    a.   Both answers are correct.
    b.   At most one answer is wrong.
    c.   The first answer is correct and the second is wrong.
    d.   Exactly one answer is wrong.

4.13 Refer to Exercise 4.7. List all the outcomes included in each of the following events. Indicate which are simple and which are compound events.

    a.   At least one part is good.
    b.   Exactly one part is defective.
    c.   The first part is good and the second is defective.
    d.   At most one part is good.

4.14 Refer to Exercise 4.8. List all the outcomes included in each of the following events and mention which are simple and which are compound events.

- **a.** At most one person is against a tax increase on rich people.
- **b.** Exactly two persons are in favor of a tax increase on rich people.
- **c.** At least one person is against a tax increase on rich people.
- **d.** More than one person is against a tax increase on rich people.

4.2 CALCULATING PROBABILITY

Probability, which gives the likelihood of occurrence of an event, is denoted by P. The probability that a simple event E_i will occur is denoted by $P(E_i)$, and the probability that a compound event A will occur is denoted by $P(A)$.

PROBABILITY

Probability is a numerical measure of the likelihood that a specific event will occur.

☞ **TWO PROPERTIES OF PROBABILITY**

The following are two important properties of probability.

1. The probability of an event always lies in the range 0 to 1.

Whether it is a simple or a compound event, the probability of an event is never less than 0 or greater than 1. Using mathematical notation, we can write this property as follows.

$$0 \leq P(E_i) \leq 1$$
$$0 \leq P(A) \leq 1$$

An event that cannot occur has zero probability; such an event is called an **impossible event**. An event that is certain to occur has a probability equal to 1 and is called a **sure event**. That is,

For an impossible event M: $P(M) = 0$
For a sure event C: $P(C) = 1$

2. The sum of the probabilities of all simple events (or final outcomes) for an experiment, denoted by $\Sigma P(E_i)$, is always 1.

Thus, for an experiment

$$\Sigma P(E_i) = P(E_1) + P(E_2) + P(E_3) + \ldots = 1$$

From this property, for the experiment of one toss of a coin

$$P(H) + P(T) = 1$$

For the experiment of two tosses of a coin

$$P(HH) + P(HT) + P(TH) + P(TT) = 1$$

For one game of football by a National Football League team

$$P(\text{Win}) + P(\text{Loss}) + P(\text{Tie}) = 1$$

4.2.1 THREE CONCEPTUAL APPROACHES TO PROBABILITY

There are three conceptual approaches to probability: (1) classical probability, (2) the relative frequency concept of probability, and (3) the subjective probability concept. These three concepts of probability are explained next.

Classical Probability

Outcomes that have the same probability of occurrence are called **equally likely outcomes**. The classical probability rule is applied to compute the probabilities of events for an experiment all of whose outcomes are equally likely.

EQUALLY LIKELY OUTCOMES

Two or more outcomes (or events) that have the same probability of occurrence are said to be *equally likely outcomes* (or events).

According to the **classical probability rule**, the probability of a simple event is equal to 1 divided by the total number of outcomes for the experiment. This is obvious, as the sum of the probabilities of all final outcomes for an experiment is 1, and all the final outcomes are equally likely. On the other hand, the probability of a compound event A is equal to the number of outcomes favorable to event A divided by the total number of outcomes for the experiment.

CLASSICAL PROBABILITY RULE

$$P(E_i) = \frac{1}{\text{Total number of outcomes for the experiment}}$$

$$P(A) = \frac{\text{Number of outcomes favorable to } A}{\text{Total number of outcomes for the experiment}}$$

Examples 4–7 through 4–9 illustrate how probabilities of events are calculated using the classical probability rule.

Calculating probability of a simple event.

EXAMPLE 4–7 Find the probability of obtaining a head and the probability of obtaining a tail for one toss of a coin.

Solution The two outcomes, head and tail, are equally likely outcomes. Therefore,[1]

$$P(\text{head}) = \frac{1}{\text{Total number of outcomes}} = \frac{1}{2} = .50$$

Similarly,

$$P(\text{tail}) = \frac{1}{2} = .50$$

Calculating probability of a compound event: one roll of a die.

EXAMPLE 4-8 Find the probability of obtaining an even number in one roll of a die.

Solution This experiment has a total of six outcomes: 1, 2, 3, 4, 5, and 6. All these outcomes are equally likely. Let A be an event that an even number is observed on the die. Event A includes three outcomes: 2, 4, and 6, that is,

$$A = \{2, 4, 6\}$$

If any one of these three numbers is obtained, event A is said to occur. Hence,

$$P(A) = \frac{\text{Number of outcomes included in } A}{\text{Total number of outcomes}} = \frac{3}{6} = .50$$

Calculating probability of a compound event.

EXAMPLE 4-9 A club has 100 members, of whom 60 are men and 40 are women. Suppose one of these members is randomly selected to be the president of the club. What is the probability that a woman is selected?

Solution Because the selection is to be made randomly, each of the 100 members of the club has the same probability of being selected. Consequently, this experiment has a total of 100 equally likely outcomes. Forty of these 100 outcomes are included in the event that "a woman is selected." Hence,

$$P(\text{a woman is selected}) = \frac{40}{100} = .40$$

Relative Frequency Concept of Probability

Suppose we want to calculate the following probabilities.

1. The probability that the next car that comes out of an auto factory is a "lemon"
2. The probability that a randomly selected family owns two cars
3. The probability that the next baby born at a hospital is a girl
4. The probability that an 80-year-old person will live for at least one more year
5. The probability that the tossing of an unbalanced coin will result in a head
6. The probability that we will observe a 1-spot if we roll a loaded die

These probabilities cannot be computed using the classical probability rule because the various outcomes for the corresponding experiments are not equally likely. For example, the next car manufactured at an auto factory may or may not be a lemon. The two outcomes, "it is a lemon" and "it is not a lemon," are not equally likely. If they were, then (approximately) half the cars manufactured by this company would be lemons, and this might prove disastrous to the survival of the firm.

Although the various outcomes for each of these experiments are not equally likely, each of these experiments can be performed again and again to generate data. In such cases, to calculate probabilities, we either use past data or generate new data by performing the exper-

---

[1]If the final answer for the probability of an event does not terminate within three decimal places, usually it will be rounded to four decimal places.

iment a large number of times. The relative frequency of an event is used as an approximation for the probability of that event. This method of assigning a probability to an event is called the **relative frequency concept of probability**. Because relative frequencies are determined by performing an experiment, the probabilities calculated using relative frequencies may change almost each time an experiment is repeated. For example, every time a new sample of 500 cars is selected from the production line of an auto factory, the number of lemons in those 500 cars is expected to be different. However, the variation in the percentage of lemons will be small if the sample size is large. Note that if we are considering the complete population, the relative frequency will give an exact probability.

RELATIVE FREQUENCY AS AN APPROXIMATION OF PROBABILITY

If an experiment is repeated n times and an event A is observed f times, then, according to the relative frequency concept of probability:

$$P(A) = \frac{f}{n}$$

Examples 4–10 and 4–11 illustrate how the probabilities of events are approximated using the relative frequencies.

Approximating probability by relative frequency: sample data.

EXAMPLE 4–10   Ten of the 500 randomly selected cars manufactured at a certain auto factory are found to be lemons. Assuming that the lemons are manufactured randomly, what is the probability that the next car manufactured at this auto factory is a lemon?

Solution   Let n denote the total number of cars in the sample and f the number of lemons in n. Then,

$$n = 500 \quad \text{and} \quad f = 10$$

Using the relative frequency concept of probability, we obtain:

$$P(\text{next car is a lemon}) = \frac{f}{n} = \frac{10}{500} = .02$$

This probability is actually the relative frequency of lemons in 500 cars. Table 4.2 lists the frequency and relative frequency distributions for this example.

Table 4.2   Frequency and Relative Frequency Distributions for the Sample of Cars

Car	f	Relative Frequency
Good	490	$490/500 = .98$
Lemon	10	$10/500 = .02$
	$n = 500$	Sum $= 1.00$

The column of relative frequencies in Table 4.2 is used as the column of approximate probabilities. Thus, from the relative frequency column:

$$P(\text{next car is a lemon}) = .02$$

and
$$P(\text{next car is a good car}) = .98$$

Note that relative frequencies are not probabilities but approximate probabilities. However, if the experiment is repeated again and again, this approximate probability of an outcome obtained from the relative frequency will approach the actual probability of that outcome. This is called the **Law of Large Numbers**.

LAW OF LARGE NUMBERS

If an experiment is repeated again and again, the probability of an event obtained from the relative frequency approaches the actual or theoretical probability.

Approximating probability by relative frequency.

EXAMPLE 4-11 Allison wants to determine the probability that a randomly selected family from New York State owns a home. How would she determine this probability?

Solution There are two outcomes for a randomly selected family from New York State: "This family owns a home" or "this family does not own a home." These two events are not equally likely. (Note that these two outcomes will be equally likely if exactly half of the families in New York State own homes and exactly half do not own homes.) Hence, the classical probability rule cannot be applied. However, we can repeat this experiment again and again. In other words, we can select a sample of families from New York State and observe whether or not each of them owns a home. Hence, we will use the relative frequency approach to probability.

Suppose Allison selects a random sample of 1000 families from New York State and observes that 670 of them own homes and 330 do not own homes. Then,

$$n = \text{sample size} = 1000$$

and

$$f = \text{number of families who own homes} = 670$$

Consequently,

$$P(\text{a randomly selected family owns a home}) = \frac{f}{n} = \frac{670}{1000} = \textbf{.670}$$

Again, note that .670 is just an approximation of the probability that a randomly selected family from New York State owns a home. Every time Allison repeats this experiment she may obtain a different probability for this event. However, because the sample size ($n = 1000$) in this example is large, the variation is expected to be very small. ■

Subjective Probability

Many times we face experiments that neither have equally likely outcomes nor can be repeated to generate data. In such cases, we cannot compute the probabilities of events using the classical probability rule or the relative frequency concept. For example, consider the following probabilities of events.

1. The probability that Carol, who is taking statistics, will earn an A in this course
2. The probability that the Dow Jones industrial average will be higher at the end of the next trading day
3. The probability that the New York Giants will win the Super Bowl next season
4. The probability that Joe will lose the lawsuit that he has filed against his landlord

Neither the classical probability rule nor the relative frequency concept of probability can be applied to calculate probabilities for these examples. All these examples belong to experiments that have neither equally likely outcomes nor the potential of being repeated. For example, Carol, who is taking statistics, will take the test (or tests) only once and based on that she will either earn an A or not. The two events ''she will earn an A'' and ''she will not earn an A'' are not equally likely. The probability assigned to an event in such cases is called **subjective probability**. It is based on the individual's own judgment, experience, information, and belief. Carol may assign a high probability to the event that she will earn an A in statistics, whereas her instructor may assign a low probability to the same event.

SUBJECTIVE PROBABILITY

Subjective probability is the probability assigned to an event based on subjective judgment, experience, information, and belief.

Subjective probability is assigned arbitrarily. It is usually influenced by the biases, preferences, and experience of the person assigning the probability.

4.2.2 ODDS

Another concept related to probability is that of odds. The odds are obtained by finding the ratio of the probability that an event will occur to the probability that this event will not occur. Case Study 4–1 is an example of the application of odds.

CASE STUDY 4-1   PROBABILITY AND ODDS

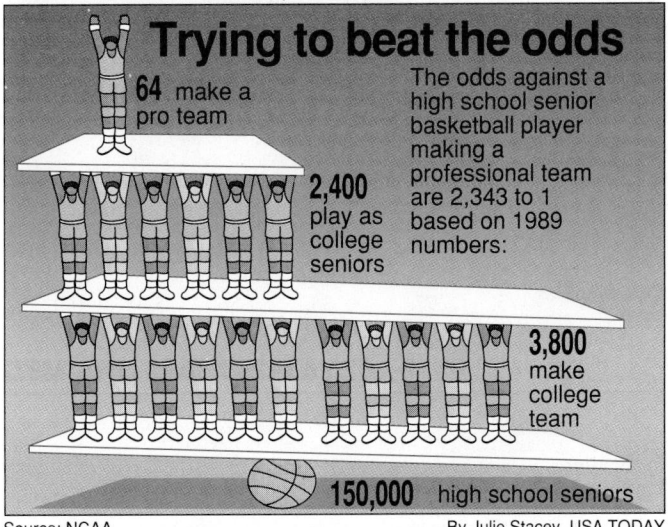

USA SNAPSHOTS®

A look at statistics that shape the sports world

Trying to beat the odds

64 make a pro team

2,400 play as college seniors

3,800 make college team

150,000 high school seniors

The odds against a high school senior basketball player making a professional team are 2,343 to 1 based on 1989 numbers:

Source: NCAA

By Julie Stacey, USA TODAY

According to the information given in the chart, the odds against a high school senior basketball player making a professional team, based on 1989 data, are 2343 to 1. In other words, out of every 2344 high school senior basketball players, 2343 will not be able to make a professional team and one will make it. The probabilities of the two events, that a high school senior basketball player will not make a professional team and that he will make it, are calculated by dividing the respective odds by the sum of the two odds. Let

N = a high school senior basketball player will not make a pro team

M = a high school senior basketball player will make a pro team

Then,

$$P(N) = \frac{2343}{2343 + 1} = .9995734$$

$$P(M) = \frac{1}{2343 + 1} = .0004266$$

If we take the ratio of $P(N)$ and $P(M)$, we obtain the odds.

How are these odds calculated from the data given in the chart? As we can observe, out of 150,000 high school senior basketball players, 64 made a pro team and 149,936 (which is 150,000 minus 64) did not. Hence, the odds against a high school senior basketball player making a pro team are 149,936 to 64. Dividing these two numbers by 64, the odds become approximately 2343 to 1. Note that we can also state that the odds in favor of a high school senior basketball player making a pro team are 1 to 2343.

Source: Chart reproduced with permission from *USA Today*, March 21, 1990. Copyright © 1990, *USA Today*.

EXERCISES

Concepts and Procedures

4.15 Briefly explain the two properties of probability.

4.16 Briefly describe an impossible and a sure event. What is the probability of the occurrence of each of these two events?

4.17 Briefly explain the three approaches to probability. Give one example of each of these three approaches.

4.18 Briefly explain for what kind of experiments we use the classical approach to calculate probabilities of events and for what kind of experiments we use the relative frequency approach.

4.19 Which of the following values cannot be probabilities of events and why?

$$1/5, \quad .97, \quad -.35, \quad 1.56, \quad 5/3, \quad 0.0, \quad -2/7, \quad 1.0$$

4.20 Which of the following values cannot be probabilities of events and why?

$$.46, \quad 2/3, \quad -.09, \quad 1.42, \quad .56, \quad 9/4, \quad -1/4, \quad .02$$

Applications

4.21 Suppose a family is selected at random from New York City. Consider the following two outcomes: This family's yearly income is less than $50,000; this family's yearly income is $50,000 or higher. Are these two outcomes equally likely? Explain why or why not. If you are to find the probabilities of these two outcomes, would you use the classical approach or the relative frequency approach? Explain why.

4.22 A class has a total of 35 students. Of them, 15 are business majors and 20 are nonbusiness majors. Suppose one student is selected at random from this class. Consider the following two events: This student is a business major; this student is a nonbusiness major. If you are to find the probabilities of these two events, would you use the classical approach or the relative frequency approach? Explain why.

4.23 The president of a company has a hunch that there is a .80 probability that the company will be successful in marketing a new brand of ice cream. Is this a case of classical, relative frequency, or subjective probability? Explain why.

4.24 The coach of a college football team thinks there is a .75 probability that the team will win the national championship this year. Is this a case of classical, relative frequency, or subjective probability? Explain why.

4.25 A hat contains 40 marbles. Of them, 16 are red and 24 are green. If one marble is randomly selected out of this hat, what is the probability that this marble is

     a. red   **b.** green?

4.26 A die is rolled once. What is the probability that

     a. a number less than 3 is obtained

     b. a number 3 to 6 is obtained?

4.27 A random sample of 800 college students showed that 240 of them are politically liberal. What is the (approximate) probability that a randomly selected student is a liberal?

4.28 In a statistics class of 45 students, 12 have a strong interest in statistics. Find the probability that a randomly selected student from this class has a strong interest in statistics.

4.29 In a group of 50 executives, 27 have a type A personality. If one executive is selected at random from this group, what is the probability that this executive has a type A personality?

4.30 Out of the 3000 families living in an apartment complex in New York City, 600 paid no income tax last year. What is the probability that a randomly selected family from these 3000 families paid income tax last year?

4.31 A multiple-choice question in a test contains five answers. If Dianne chooses one answer based on ''pure guess,'' what is the probability that her answer is

     a. correct   **b.** wrong?

Do these two probabilities add up to 1.0? If yes, why?

4.32 A university has a total of 320 professors and 64 of them are female. What is the probability that a randomly selected professor from this university is a

     a. female   **b.** male?

Do these two probabilities add up to 1.0? If yes, why?

4.33 A company that plans to hire one new employee has prepared a final list of six candidates, all of whom are equally qualified. Two of these six candidates are women. If the company decides to select at random one person out of these six candidates, what is the probability that this person will be a woman? What is the probability that this person will be a man? Do these two probabilities add up to 1.0? If yes, why?

4.34 A sample of 500 large companies showed that 80 of them offer free psychiatric help to their employees who suffer from psychological problems. If one company is selected at random from this sample, what is the probability that this company offers free psychiatric help to its employees who suffer from psychological problems? What is the probability that this company does not offer free psychiatric help to its employees who suffer from psychological problems? Do these two probabilities add up to 1.0? If yes, why?

4.35 A sample of 400 large companies showed that 120 of them offer free health fitness centers to their employees within the company premises. If one company is selected at random from this sample, what is the probability that this company offers a free health fitness center to its employees within the company premises? What is the probability that this company does not offer a free health fitness center to its employees within the company premises? Do these two probabilities add up to 1.0? If yes, why?

4.36 According to the U.S. Bureau of Labor Statistics, there were 6210 fatal work injuries in the United States in 1995. Of these, 2560 involved transportation incidents, 1262 were from assaults and violent acts, 915 were caused by contact with objects and equipment, 643 resulted from falls, and 830 were due to other causes. If one of these fatalities is selected at random, find the probability that it was due to

    a. a transportation incident   **b.** an assault or violent act

    c. contact with objects or equipment   **d.** a fall   **e.** another cause

Do these five probabilities add up to 1.0? If so, why?

4.37 A sample of 1000 families showed that 34 of them own no cars, 208 own one car each, 376 own two cars each, 265 own three cars each, and 117 own four or more cars each. Write the frequency distribution table for this problem. Calculate the relative frequencies for all categories. Suppose one family is randomly selected from these 1000 families. Find the probability that this family owns

    a. two cars   **b.** four or more cars

4.38 In a sample of 500 families, 95 have a yearly income of less than $20,000, 272 have a yearly income of $20,000 to $50,000, and the remaining families have a yearly income of more than $50,000. Write the frequency distribution table for this problem. Calculate the relative frequencies for all classes. Suppose one family is randomly selected from these 500 families. Find the probability that this family has a yearly income of

    a. less than $20,000   **b.** more than $50,000

4.39 Suppose you want to find the (approximate) probability that a randomly selected family from Los Angeles earns more than $75,000 a year. How would you find this probability? What procedure would you use? Explain briefly.

4.40 Suppose you have a loaded die and you want to find the (approximate) probabilities of different outcomes for this die. How would you find these probabilities? What procedure would you use? Explain briefly.

4.3 COUNTING RULE

The experiments dealt with so far in this chapter have had only a few outcomes, which were easy to list. However, for experiments with a large number of outcomes, it may not be easy to list all outcomes. In such cases, we may use the **counting rule** to find the total number of outcomes.

COUNTING RULE

If an experiment consists of three steps and if the first step can result in m outcomes, the second step in n outcomes, and the third step in k outcomes, then,

$$\text{Total outcomes for the experiment} = m \cdot n \cdot k$$

The counting rule can easily be extended to apply to an experiment with less or more than three steps.

Applying the counting rule: 3 steps.

EXAMPLE 4–12 Suppose we toss a coin three times. This experiment has three steps: the first toss, the second toss, and the third toss. Each step has two outcomes: a head and a tail. Thus,

$$\text{Total outcomes for three tosses of a coin} = 2 \times 2 \times 2 = \mathbf{8}$$

The eight outcomes for this experiment are *HHH, HHT, HTH, HTT, THH, THT, TTH*, and *TTT*. ∎

Applying the counting rule: 2 steps.

EXAMPLE 4–13 A prospective car buyer can choose between a fixed or a variable interest rate and can also choose a payment period of 36 months, 48 months, or 60 months. How many total outcomes are possible?

Solution This experiment is made up of two steps: choosing an interest rate and selecting a loan payment period. There are two outcomes (a fixed or a variable interest rate) for the first step and three outcomes (a payment period of 36 months, 48 months, or 60 months) for the second step. Hence,

$$\text{Total outcomes} = 2 \times 3 = \mathbf{6}$$ ∎

Applying the counting rule: 16 steps.

EXAMPLE 4–14 A National Football League team will play 16 games during a regular season. Each game can result in one of three outcomes: a win, a loss, or a tie. The total possible outcomes for 16 games are calculated as follows.

$$\text{Total outcomes} = 3 \cdot 3 \cdot 3 \cdot 3 \cdot 3 \cdot 3 \cdot 3 \cdot 3 \cdot 3 \cdot 3 \cdot 3 \cdot 3 \cdot 3 \cdot 3 \cdot 3 \cdot 3$$

$$= 3^{16} = \mathbf{43,046,721}$$

One of the 43,046,721 possible outcomes is all 16 wins.[2] ∎

4.4 MARGINAL AND CONDITIONAL PROBABILITIES

Suppose all 100 employees of a company were asked whether they are in favor of or against paying high salaries to CEOs of U.S. companies. Table 4.3 gives a two-way classification of the responses of these 100 employees.

Table 4.3 Two-way Classification of the Responses of Employees

	In Favor	Against
Male	15	45
Female	4	36

Table 4.3 gives the distribution of 100 employees based on two variables or characteristics: gender (male or female) and opinion (in favor or against). Such a table is called a *contingency table*. In Table 4.3, each box that contains a number is called a *cell*. Notice that there are four cells in Table 4.3. Each cell gives the frequency for two characteristics. For example, 15 employees in this group possess two characteristics: They are "male" and "in favor of paying high salaries to CEOs." We can interpret the numbers in other cells the same way.

By adding the row of totals and the column of totals to Table 4.3, we write Table 4.4, which is given on the next page.

Suppose one employee is selected at random from these 100 employees. This employee may be classified either on the basis of gender alone or on the basis of opinion. If only one

---

[2]Using a calculator to evaluate 3^{16}: If your calculator contains a y^x or an x^y key, you can use that key to simplify 3^{16} as follows: First enter 3 on the calculator, then press the y^x key, next enter 16, and finally press the " = " key. The screen of the calculator will display 43046721 as the answer.

Table 4.4 Two-way Classification of Employees

	In Favor	Against	Total
Male	15	45	60
Female	4	36	40
Total	19	81	100

characteristic is considered at a time, the employee selected can be a male, a female, in favor, or against. The probability of each of these four characteristics or events is called **marginal probability** or *simple probability*. These probabilities are called marginal probabilities because they are calculated by dividing the corresponding row margins (totals for the rows) or column margins (totals for the columns) by the grand total.

MARGINAL PROBABILITY

Marginal probability is the probability of a single event without consideration of any other event. Marginal probability is also called *simple probability*.

For Table 4.4, the four marginal probabilities are calculated as follows.

$$P(\text{male}) = \frac{\text{Number of males}}{\text{Total number of employees}} = \frac{60}{100} = .60$$

As we can observe, the probability that a male will be selected is obtained by dividing the total of the row labeled "Male" (60) by the grand total (100).

Similarly,
$$P(\text{female}) = 40/100 = .40$$

$$P(\text{in favor}) = 19/100 = .19$$

and
$$P(\text{against}) = 81/100 = .81$$

These four marginal probabilities are shown along the right side and along the bottom of Table 4.5.

Table 4.5 Listing the Marginal Probabilities

	In Favor (*A*)	Against (*B*)	Total	
Male (*M*)	15	45	60	$P(M) = 60/100 = .60$
Female (*F*)	4	36	40	$P(F) = 40/100 = .40$
Total	19	81	100	
	$P(A) = 19/100$ $= .19$	$P(B) = 81/100$ $= .81$		

Now suppose that one employee is selected at random from these 100 employees. Furthermore, assume that it is known that this (selected) employee is a male. In other words, the event that the employee selected is a male has already occurred. What is the probability that the employee selected is in favor of paying high salaries to CEOs? This probability is written as follows.

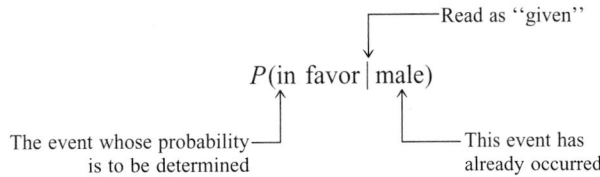

This probability, $P(\text{in favor} \mid \text{male})$, is called the **conditional probability** of "in favor" and it is read as "the probability that the employee selected is in favor given that this employee is a male."

CONDITIONAL PROBABILITY

Conditional probability is the probability that an event will occur given that another event has already occurred. If A and B are two events, then the conditional probability of A given B is written as

$$P(A \mid B)$$

and read as "the probability of A given that B has already occurred."

Calculating conditional probability: two-way table.

EXAMPLE 4-15 Compute the conditional probability $P(\text{in favor} \mid \text{male})$ for the data on 100 employees given in Table 4.4.

Solution The probability $P(\text{in favor} \mid \text{male})$ is the conditional probability that a randomly selected employee is in favor given that this employee is a male. It is known that the event "male" has already occurred. Based on the information that the employee selected is a male, we can infer that the employee selected must be one of the 60 males and, hence, must belong to the first row of Table 4.4. Therefore, we are concerned only with the first row of that table.

	In Favor	**Against**	**Total**
Male	15	45	60

Males who are in favor ↑

Total number of males ↑

The required conditional probability is calculated as follows.

$$P(\text{in favor} \mid \text{male}) = \frac{\text{Number of males who are in favor}}{\text{Total number of males}} = \frac{15}{60} = .25$$

As we can observe from this computation of conditional probability, the total number of males (the event that has already occurred) is written in the denominator and the number of males who are in favor (the event whose probability we are to find) is written in the numerator. Note that we are considering the row of the event that has already occurred. The tree diagram in Figure 4.6 illustrates Example 4–15.

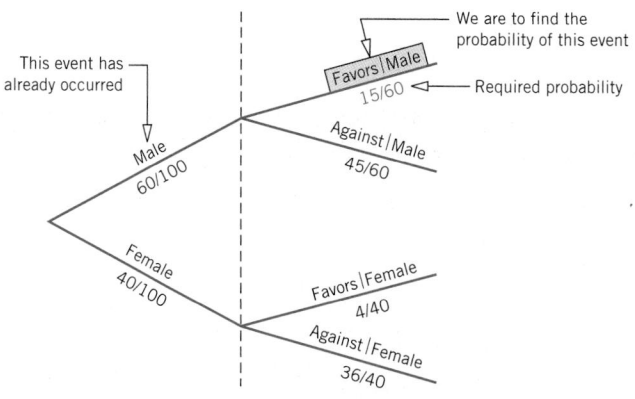

Figure 4.6 Tree diagram.

EXAMPLE 4–16 For the data of Table 4.4, calculate the conditional probability that a randomly selected employee is a female given that this employee is in favor of paying high salaries to CEOs.

Solution We are to compute the probability, $P(\text{female} \mid \text{in favor})$.

Since it is known that the employee selected is in favor of paying high salaries to CEOs, this employee must belong to the first column (the column labeled ''in favor'') and must be one of the 19 employees who are in favor.

In Favor

15

4 ⟵——— Females who are in favor

19 ⟵——— Total number of employees who are in favor

Hence, the required probability is

$$P(\text{female} \mid \text{in favor}) = \frac{\text{Number of females who are in favor}}{\text{Total number of employees who are in favor}}$$

$$= \frac{4}{19} = \mathbf{.2105}$$

The tree diagram in Figure 4.7 illustrates this example.

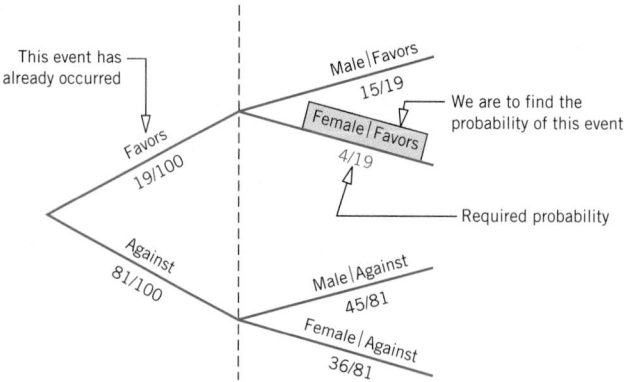

Figure 4.7 Tree diagram.

Case Study 4–2 illustrates the conditional probability of listening to oldies radio stations given the age group of people.

CASE STUDY 4-2   PROBABILITIES OF LISTENING TO OLDIES RADIO STATIONS

USA SNAPSHOTS®

A look at statistics that shape our lives

Oldies fans in the middle

Age of adults who listen to the about 700 Oldies radio stations in the USA:

Note: Exceeds 100% due to rounding

| 18-24 | 25-34 | 35-44 | 45-54 | 55 and older |

12% 22% 32% 23% 12%

Source: Interep Research   By Anne R. Carey and Marcy E. Mullins, USA TODAY

The above chart shows age distribution of U.S. adults who listen to 700 Oldies radio stations. For example, of such adults, 12% are 18 to 24 years of age, 22% are 25 to 34 years of age, and so on. By converting these percentages to relative frequencies, the percentages given in the chart can be used to write the conditional probabilities. Suppose one adult is selected at random from the population of all U.S. adults irrespective of whether or not they listen to such radio stations. Then, the fact that 12% of adults who listen to such radio stations are aged 18 to 24 means that given that this selected adult listens to Oldies radio stations, the probability that he/she is 18 to 24 years of age is approximately .12. That is,

P(the selected adult is aged 18–24 | he/she listens to Oldies radio stations) = .12

Similarly, we can write the following conditional probabilities using the above chart.

P(the selected adult is aged 25–34 | he/she listens to Oldies radio stations) = .22
P(the selected adult is aged 35–44 | he/she listens to Oldies radio stations) = .32
P(the selected adult is aged 45–54 | he/she listens to Oldies radio stations) = .23
P(the selected adult is aged 55 or older | he/she listens to Oldies radio stations) = .12

Note that these are approximate probabilities due to the fact that the given age distribution is based on a sample survey. Also, the five probabilities do not add up to exactly 1.00 due to rounding.

Source: USA Today, March 7, 1997. Copyright © 1997, *USA Today*. Chart reproduced with permission.

4.5 MUTUALLY EXCLUSIVE EVENTS

Events that cannot occur together are called **mutually exclusive events**. Such events do not have any common outcomes. If two or more events are mutually exclusive, then at most one of them will occur every time we repeat the experiment. Thus, the occurrence of one event excludes the occurrence of the other event or events.

MUTUALLY EXCLUSIVE EVENTS

Events that cannot occur together are said to be *mutually exclusive events*.

For any experiment, the final outcomes are always mutually exclusive because one and only one of these outcomes is expected to occur in one repetition of the experiment. For example, consider tossing a coin twice. This experiment has four outcomes: *HH*, *HT*, *TH*, and *TT*. These outcomes are mutually exclusive because one and only one of them will occur when we toss this coin twice.

Illustrating mutually exclusive and mutually nonexclusive events.

EXAMPLE 4–17   Consider the following events for one roll of a die.

$$A = \text{an even number is observed} = \{2, 4, 6\}$$

$$B = \text{an odd number is observed} = \{1, 3, 5\}$$

$$C = \text{a number less than 5 is observed} = \{1, 2, 3, 4\}$$

Are events *A* and *B* mutually exclusive? Are events *A* and *C* mutually exclusive?

Solution   Figures 4.8 and 4.9 show the diagrams of events *A* and *B* and events *A* and *C*, respectively.

As we can observe from the definitions of events *A* and *B* and from Figure 4.8, events *A* and *B* have no common element. For one roll of a die, only one of the two events, *A* and *B*, can happen. Hence, these are two mutually exclusive events. On the other hand, we can observe from the definitions of events *A* and *C* and from Figure 4.9 that events *A* and *C* have two common outcomes: 2-spot and 4-spot. Thus, if we roll a die and obtain either a 2-spot or a 4-spot, then *A* and *C* happen at the same time. Hence, events *A* and *C* are not mutually exclusive.

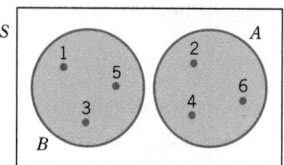

Figure 4.8   Mutually exclusive events *A* and *B*.

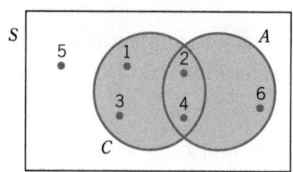

Figure 4.9   Mutually nonexclusive events *A* and *C*.

*Illustrating mutually
exclusive events.*

EXAMPLE 4–18 Suppose an employee is selected at random from a large company. Consider the following two events.

$$D = \text{the employee selected holds a college degree}$$

$$N = \text{the employee selected does not hold a college degree}$$

Are events D and N mutually exclusive?

Solution Event D consists of all employees at this company who possess a college degree and event N includes all employees who do not hold a college degree. These two events are shown in Figure 4.10.

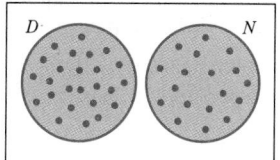

Figure 4.10 Mutually exclusive events D and N.

As we can see from the definitions of events D and N and from Figure 4.10, events D and N have no common outcome. They represent two distinct sets of employees: the ones who hold a college degree and the ones who do not possess a college degree. Hence, these two events are mutually exclusive events. ■

4.6 INDEPENDENT VERSUS DEPENDENT EVENTS

In the case of two **independent events**, the occurrence of one event does not change the probability of the occurrence of the other event.

INDEPENDENT EVENTS

Two events are said to be *independent* if the occurrence of one does not affect the probability of the occurrence of the other. In other words, A and B are *independent events* if

$$\text{either} \quad P(A \mid B) = P(A) \quad \text{or} \quad P(B \mid A) = P(B)$$

It can be shown that if one of these two conditions is true, then the second will also be true, and if one is not true then the second will also not be true.

If the occurrence of one event affects the probability of the occurrence of the other event, then the two events are said to be **dependent events**. Using probability notation, the two events will be dependent if either $P(A \mid B) \neq P(A)$ or $P(B \mid A) \neq P(B)$.

*Illustrating two
dependent events:
two-way table.*

EXAMPLE 4–19 Refer to the information on 100 employees given in Table 4.4 in Section 4.4. Are events "female (F)" and "in favor (A)" independent?

Solution Events F and A will be independent if

$$P(F) = P(F \mid A)$$

Otherwise they will be dependent.

Using the information given in Table 4.4, we compute the following two probabilities.

$$P(F) = 40/100 = \textbf{.40} \quad \text{and} \quad P(F\,|\,A) = 4/19 = \textbf{.2105}$$

Because these two probabilities are not equal, the two events are dependent. Here, dependence of events means that the percentages of males who are in favor of and against paying high salaries to CEOs are different from the percentages of females who are in favor and against.

In this example, the dependence of A and F can also be proved by showing that the probabilities $P(A)$ and $P(A\,|\,F)$ are not equal. ◼

Illustrating two independent events.

EXAMPLE 4–20   A box contains a total of 100 cassettes that were manufactured on two machines. Of them, 60 were manufactured on Machine I. Of the total cassettes, 15 are defective. Of the 60 cassettes that were manufactured on Machine I, 9 are defective. Let D be the event that a randomly selected cassette is defective and A be the event that a randomly selected cassette was manufactured on Machine I. Are events D and A independent?

Solution   From the given information,

$$P(D) = 15/100 = .15 \quad \text{and} \quad P(D\,|\,A) = 9/60 = .15$$

Hence, $$P(D) = P(D\,|\,A)$$

Consequently, the two events, D and A, are independent.

Independence, in this example, means that the probability for any cassette to be defective is the same, .15, irrespective of the machine on which it is manufactured. In other words, the two machines are producing the same percentage of defective items. For example, 9 of the 60 cassettes manufactured on Machine I are defective and 6 of the 40 cassettes manufactured on Machine II are defective. Thus, for each of the two machines, 15% of the cassettes produced are defective.

Actually, using the given information, we can prepare Table 4.6 for Example 4–20. The numbers in the shaded cells are given to us. The remaining numbers are calculated by doing some arithmetic manipulations.

Table 4.6   Two-way Classification Table

	Defective (D)	Good (G)	Total
Machine I (A)	9	51	60
Machine II (B)	6	34	40
Total	15	85	100

Using this table, we can find the following probabilities.

$$P(D) = 15/100 = .15$$

and $$P(D\,|\,A) = 9/60 = .15$$

Since these two probabilities are the same, the two events are independent. ◼

☞ **TWO IMPORTANT OBSERVATIONS**

The following are two important observations about mutually exclusive, independent, and dependent events.

1. Two events are either mutually exclusive or independent.[3] In other words,
   a. mutually exclusive events are always dependent
   b. independent events are never mutually exclusive
2. Dependent events may or may not be mutually exclusive.

4.7 COMPLEMENTARY EVENTS

Two mutually exclusive events that taken together include all the outcomes for an experiment are called **complementary events**. Note that two complementary events are always mutually exclusive.

COMPLEMENTARY EVENTS

The complement of event A, denoted by $\overline{A}$ and read as "A bar" or "A complement," is the event that includes all the outcomes for an experiment that are not in A.

Events A and $\overline{A}$ are complements of each other. The Venn diagram in Figure 4.11 shows the complementary events A and $\overline{A}$.

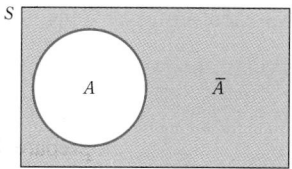

Figure 4.11 Venn diagram of two complementary events.

Because two complementary events, taken together, include all the outcomes for an experiment and because the sum of the probabilities of all outcomes is 1, it is obvious that

$$P(A) + P(\overline{A}) = 1$$

From this equation we can deduce that

$$P(A) = 1 - P(\overline{A}) \quad \text{and} \quad P(\overline{A}) = 1 - P(A)$$

Thus, if we know the probability of an event, we can find the probability of its complementary event by subtracting the given probability from 1.0.

Calculating probabilities of complementary events.

EXAMPLE 4–21 In a lot of five washing machines, two are defective. If one machine is randomly selected, what are the two complementary events for this experiment and what are their probabilities?

Solution The two complementary events for this experiment are

$$A = \text{the machine selected is defective}$$

$$\overline{A} = \text{the machine selected is not defective}$$

---

[3]The exception to this rule occurs when at least one of the two events has a zero probability.

Since there are two defective and three nondefective machines, the probabilities of events A and $\overline{A}$ are

$$P(A) = 2/5 = \mathbf{.40} \quad \text{and} \quad P(\overline{A}) = 3/5 = \mathbf{.60}$$

As we can observe, the sum of these two probabilities is 1. Figure 4.12 shows a Venn diagram for this example.

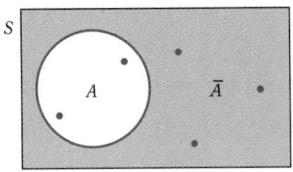

Figure 4.12 Venn diagram.

Calculating probability of the complement of an event.

EXAMPLE 4–22 There are a total of 120 professors at a college and 90 of them possess a Ph.D. degree. If one professor is selected at random from this college, what are the two complementary events and their probabilities?

Solution The two complementary events are

A = the randomly selected professor possesses a Ph.D. degree

$\overline{A}$ = the randomly selected professor does not possess a Ph.D. degree

The probabilities of these two events are

$$P(A) = 90/120 = \mathbf{.75} \quad \text{and} \quad P(\overline{A}) = 1 - .75 = \mathbf{.25}$$

Figure 4.13 shows a Venn diagram for this example.

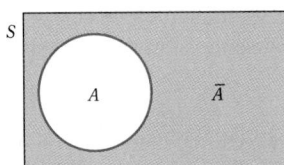

Figure 4.13 Venn diagram.

EXERCISES

Concepts and Procedures

4.41 Briefly explain the difference between the marginal and conditional probabilities of events. Give one example of each.

4.42 What is meant by two mutually exclusive events? Give one example of two mutually exclusive events and another example of two mutually nonexclusive events.

4.43 Briefly explain the meaning of independent and dependent events. Suppose A and B are two events. What formula would you use to prove whether A and B are independent or dependent?

4.44 What is the complement of an event? What is the sum of the probabilities of two complementary events?

4.45   How many different outcomes are possible for four rolls of a die?

4.46   How many different outcomes are possible for 10 tosses of a coin?

4.47   A statistical experiment has eight equally likely outcomes that are denoted by 1, 2, 3, 4, 5, 6, 7, and 8. Let event $A = \{2, 5, 7\}$ and event $B = \{2, 4, 8\}$.

   a.   Are events A and B mutually exclusive events?
   b.   Are events A and B independent events?
   c.   What are the complements of events A and B, respectively, and their probabilities?

4.48   A statistical experiment has 10 equally likely outcomes that are denoted by 1, 2, 3, 4, 5, 6, 7, 8, 9, and 10. Let event $A = \{3, 4, 6, 9\}$ and event $B = \{1, 2, 5\}$.

   a.   Are events A and B mutually exclusive events?
   b.   Are events A and B independent events?
   c.   What are the complements of events A and B, respectively, and their probabilities?

Applications

4.49   A specific model of a car comes in five exterior colors and three interior colors. All exterior colors can be combined with any of the interior colors. How many different selections of one exterior and one interior color are possible?

4.50   A man just bought four suits, eight shirts, and nine ties. All of these suits, shirts, and ties coordinate with each other. If he is to randomly select one suit, one shirt, and one tie to wear on a certain day, how many different outcomes (selections) are possible?

4.51   A restaurant menu has four kinds of soups, eight kinds of main courses, five kinds of desserts, and six kinds of drinks. If a customer randomly selects one item from each of these four categories, how many different outcomes are possible?

4.52   A student is to select three courses for next semester. If this student decides to randomly select one course from each of eight economics courses, six mathematics courses, and five computer courses, how many different outcomes are possible?

4.53   All 420 employees of a company were asked if they are smokers or nonsmokers and whether or not they are college graduates. Based on this information, the following two-way classification table was prepared.

	College Graduate	Not a College Graduate
Smoker	35	80
Nonsmoker	130	175

   a.   If one employee is selected at random from this company, find the probability that this employee is a
       i.   college graduate
       ii.   nonsmoker
       iii. smoker given the employee is not a college graduate
       iv.   college graduate given the employee is a nonsmoker
   b.   Are the events "smoker" and "college graduate" mutually exclusive? What about the events "smoker" and "nonsmoker"? Why or why not?
   c.   Are the events "smoker" and "not a college graduate" independent? Why or why not?

4.54   The following table gives a two-way classification of all high school dropouts (in thousands) in the U.S. civilian labor force in 1994–95 (*U.S. Bureau of Labor Statistics*, April 25, 1996).

	Employed	Unemployed
Male	179	72
Female	109	49

a. If one person is selected at random from these high school dropouts, find the probability that this person is
    i. unemployed
    ii. a male
    iii. employed given the person is a female
    iv. a male given the person is employed

b. Are the events "employed" and "unemployed" mutually exclusive? What about the events "unemployed" and "male"? Why or why not?

c. Are the events "female" and "unemployed" independent? Why or why not?

4.55 A group of 2000 randomly selected adults were asked if they are in favor of or against abortion. The following table gives the results of this survey.

	In Favor	Against
Male	495	405
Female	620	480

a. If one person is selected at random from these 2000 adults, find the probability that this person is
    i. in favor of abortion
    ii. against abortion
    iii. in favor of abortion given the person is a female
    iv. a male given the person is against abortion

b. Are the events "male" and "in favor" mutually exclusive? What about the events "in favor" and "against"? Why or why not?

c. Are the events "female" and "in favor" independent? Why or why not?

4.56 Five hundred employees were selected from a city's large private companies, and they were asked whether or not they have any retirement benefits provided by their companies. Based on this information, the following two-way classification table was prepared.

	Have Retirement Benefits	
	Yes	No
Men	225	75
Women	150	50

a. If one employee is selected at random from these 500 employees, find the probability that this employee
    i. is a woman
    ii. has retirement benefits
    iii. has retirement benefits given the employee is a man
    iv. is a woman given that she does not have retirement benefits

b. Are the events "man" and "yes" mutually exclusive? What about the events "yes" and "no"? Why or why not?

c. Are the events "woman" and "yes" independent? Why or why not?

4.57 The following table gives the two-way classification of 2000 randomly selected employees from a city based on gender and commuting time from home to work.

	Commuting Time from Home to Work		
	Less Than 30 Minutes	30 Minutes to One Hour	More Than One Hour
Men	524	455	221
Women	413	263	124

    a. If one employee is selected at random from these 2000 employees, find the probability that this employee

        i. commutes for more than one hour

        ii. commutes for less than 30 minutes

        iii. is a man given that he commutes for 30 minutes to one hour

        iv. commutes for more than one hour given the employee is a woman

    b. Are the events "man" and "commutes for more than one hour" mutually exclusive? What about the events "less than 30 minutes" and "more than one hour"? Why or why not?

    c. Are the events "woman" and "commutes for 30 minutes to one hour" independent? Why or why not?

4.58 Two thousand randomly selected adults were asked if they think they are financially better off than their parents. The following table gives the two-way classification of the responses based on the education levels of the persons included in the survey and whether they are financially better off, the same, or worse off than their parents.

	Education Level		
	Less Than High School	High School	More Than High School
Better Off	140	450	420
Same	60	250	110
Worse Off	200	300	70

    a. If one adult is selected at random from these 2000 adults, find the probability that this adult is

        i. financially better off than his/her parents

        ii. financially better off than his/her parents given he/she has less than high school education

        iii. financially worse off than his/her parents given he/she has high school education

        iv. financially the same as his/her parents given he/she has more than high school education

    b. Are the events "better off" and "high school education" mutually exclusive? What about the events "less than high school" and "more than high school"? Why or why not?

    c. Are the events "worse off" and "more than high school" independent? Why or why not?

4.59 There are a total of 160 practicing physicians in a city. Of them, 55 are female and 25 are pediatricians. Of the 55 females, eight are pediatricians. Are the events "female" and "pediatrician" independent? Are they mutually exclusive? Explain why or why not.

4.60 Of a total of 100 diskettes manufactured on two machines, 20 are defective. Sixty of the total diskettes were manufactured on Machine I and 10 of these 60 are defective. Are the events "machine type" and "defective diskettes" independent? (*Note:* Compare this exercise with Example 4–20.)

4.61 A company hired 30 new graduates last week. Of these, 16 are female and 11 are business majors. Of the 16 females, 7 are business majors. Are the events "female" and "business major" independent? Are they mutually exclusive? Explain why or why not.

4.62 Define the following two events for two tosses of a coin.

$$A = \text{at least one head is obtained}$$

$$B = \text{both tails are obtained}$$

    a. Are A and B mutually exclusive events? Are they independent? Explain why or why not.

    b. Are A and B complementary events? If yes, first calculate the probability of B and then calculate the probability of A using the complementary event rule.

4.63 Let A be the event that a number less than 3 is obtained if we roll a die once. What is the probability of A? What is the complementary event of A, and what is its probability?

4.64 According to the U.S. Bureau of Justice Statistics, there were 1,127,132 prisoners under the jurisdiction of state or federal correctional authorities in the United States at the end of 1995, and 1,058,588 of them were male. If one person is selected at random from these 1,127,132 prisoners, what are the two complementary events and their probabilities?

4.65 The probability that an adult reads a newspaper every day is .65. What is its complementary event? What is the probability of this complementary event?

4.8 INTERSECTION OF EVENTS AND THE MULTIPLICATION RULE

This section discusses the intersection of two events and the application of the multiplication rule to compute the probability of the intersection of events.

4.8.1 INTERSECTION OF EVENTS

The **intersection of two events** is given by the outcomes that are common to both events.

INTERSECTION OF EVENTS

Let A and B be two events defined in a sample space. The *intersection* of A and B represents the collection of all outcomes that are common to both A and B and is denoted by

$$A \text{ and } B$$

The intersection of events A and B is also denoted by either $A \cap B$ or AB. Let

$$A = \text{ the event that a family owns a VCR}$$

$$B = \text{ the event that a family owns a telephone answering machine}$$

Figure 4.14 illustrates the intersection of events A and B. The shaded area in this figure gives the intersection of events A and B, and it includes all the families who own both a VCR and a telephone answering machine.

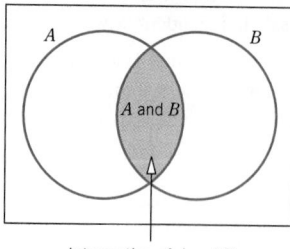

Intersection of A and B

Figure 4.14 Intersection of events A and B.

4.8.2 MULTIPLICATION RULE

The probability that events A and B happen together is called the **joint probability** of A and B and is written as $P(A \text{ and } B)$.

JOINT PROBABILITY

The probability of the intersection of two events is called their *joint probability*. It is written as

$$P(A \text{ and } B)$$

The probability of the intersection of two events is obtained by multiplying the marginal probability of one event by the conditional probability of the second event. This rule is called the **multiplication rule**.

MULTIPLICATION RULE

The probability of the intersection of two events A and B is

$$P(A \text{ and } B) = P(A) \, P(B \,|\, A)$$

The joint probability of events A and B can also be denoted by $P(A \cap B)$ or $P(AB)$.

Calculating joint probability of two events: two-way table.

EXAMPLE 4–23 The following table gives the classification of all employees of a company by gender and college degree.

	College Graduate (G)	Not a College Graduate (N)	Total
Male (M)	7	20	27
Female (F)	4	9	13
Total	11	29	40

If one of these employees is selected at random for membership on the employee-management committee, what is the probability that this employee is a female and a college graduate?

Solution We are to calculate the probability of the intersection of events "female" (denoted by F) and "college graduate" (denoted by G). This probability will be computed using the formula

$$P(F \text{ and } G) = P(F) \, P(G \,|\, F)$$

The shaded area in Figure 4.15 gives the intersection of events "female" and "college graduate."

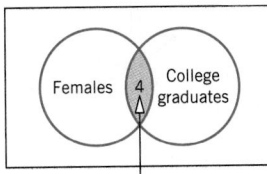

Females and college graduates   **Figure 4.15**

Notice that there are 13 females among 40 employees. Hence, the probability that a female is selected is

$$P(F) = 13/40$$

To calculate the probability $P(G|F)$, we know that F has already occurred. Consequently, the employee selected is one of the 13 females. In the table, there are 4 college graduates among 13 female employees. Hence, the conditional probability of G given F is

$$P(G|F) = 4/13$$

The joint probability of F and G is

$$P(F \text{ and } G) = P(F)\, P(G|F) = (13/40)\,(4/13) = \mathbf{.100}$$

Thus, the probability is .100 that a randomly selected employee is a female and a college graduate.

The probability in this example can also be calculated without using the multiplication rule. As we can notice from Figure 4.15 and from the table, there are four employees out of a total of 40 who are female and college graduates. Hence, if any of these four employees is selected, events "female" and "college graduate" both happen. Hence, the required probability is

$$P(F \text{ and } G) = 4/40 = \mathbf{.100}$$

Similarly, we can compute three other joint probabilities for the table as follows.

$$P(M \text{ and } G) = P(M)\, P(G|M) = (27/40)\,(7/27) = \mathbf{.175}$$

$$P(M \text{ and } N) = P(\text{M})\, P(N|M) = (27/40)\,(20/27) = \mathbf{.500}$$

$$P(F \text{ and } N) = P(F)\, P(N|F) = (13/40)\,(9/13) = \mathbf{.225}$$

The tree diagram in Figure 4.16 shows all four joint probabilities for this example. The joint probability of F and G is highlighted in the tree diagram.

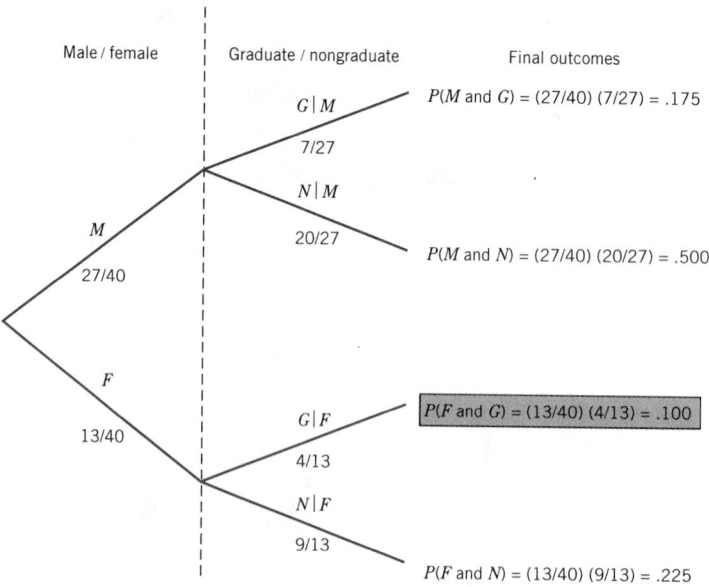

Figure 4.16 Tree diagram.

Calculating joint probability of two events.

EXAMPLE 4-24 A box contains 20 cassettes, 4 of which are defective. If 2 cassettes are selected at random (without replacement) from this box, what is the probability that both are defective?

Solution Let us define the following events for this experiment.

G_1 = event that the first cassette selected is good

D_1 = event that the first cassette selected is defective

G_2 = event that the second cassette selected is good

D_2 = event that the second cassette selected is defective

We are to calculate the joint probability of D_1 and D_2, which is given by

$$P(D_1 \text{ and } D_2) = P(D_1)\, P(D_2 | D_1)$$

As we know, there are 4 defective cassettes in 20. Consequently, the probability of selecting a defective cassette at the first selection is

$$P(D_1) = 4/20$$

To calculate the probability $P(D_2 | D_1)$, we know that the first cassette selected is defective because D_1 has already occurred. Because the selections are made without replacement, there are 19 total cassettes, and 3 of them are defective at the time of the second selection. Therefore,

$$P(D_2 | D_1) = 3/19$$

Hence, the required probability is

$$P(D_1 \text{ and } D_2) = P(D_1)\, P(D_2 | D_1) = (4/20)\,(3/19) = \mathbf{.0316}$$

The tree diagram in Figure 4.17 shows the selection procedure and the final four outcomes for this experiment along with their probabilities. The joint probability of D_1 and D_2 is highlighted in the tree diagram.

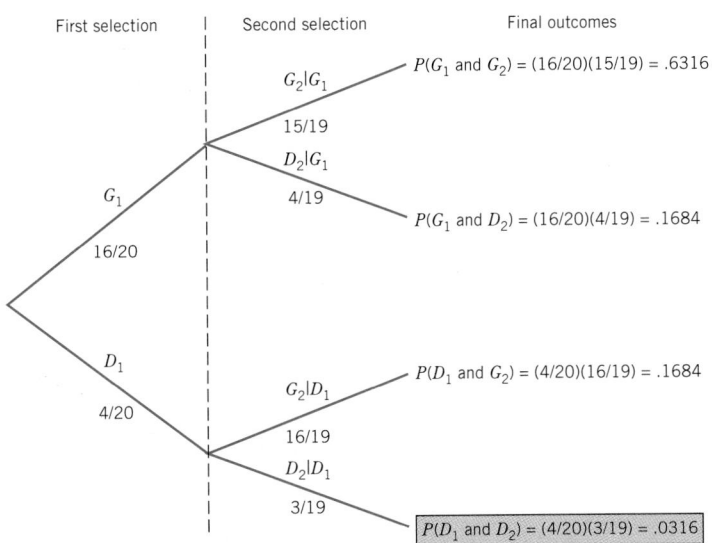

Figure 4.17 Selecting two cassettes.

Conditional probability was discussed in Section 4.4. It is obvious from the formula for joint probability that if we know the probability of an event A and the joint probability of events A and B, then we can calculate the conditional probability of B given A.

CONDITIONAL PROBABILITY

If A and B are two events, then,

$$P(B \mid A) = \frac{P(A \text{ and } B)}{P(A)} \quad \text{and} \quad P(A \mid B) = \frac{P(A \text{ and } B)}{P(B)}$$

given that $P(A) \neq 0$ and $P(B) \neq 0$.

Calculating conditional probability of an event.

EXAMPLE 4–25   The probability that a randomly selected student from a college is a senior is .20, and the joint probability that the student is a computer science major and a senior is .03. Find the conditional probability that a student selected at random is a computer science major given that he/she is a senior.

Solution   Let us define the following two events.

$$A = \text{the student selected is a senior}$$

$$B = \text{the student selected is a computer science major}$$

From the given information,

$$P(A) = .20 \quad \text{and} \quad P(A \text{ and } B) = .03$$

Hence,

$$P(B \mid A) = \frac{P(A \text{ and } B)}{P(A)} = \frac{.03}{.20} = \mathbf{.15}$$

Thus, the (conditional) probability is .15 that a student selected at random is a computer science major given that he or she is a senior.   ■

MULTIPLICATION RULE FOR INDEPENDENT EVENTS

The foregoing discussion of the multiplication rule was based on the assumption that the two events are dependent. Now suppose that events A and B are independent. Then,

$$P(A) = P(A \mid B) \quad \text{and} \quad P(B) = P(B \mid A)$$

By substituting $P(B)$ for $P(B \mid A)$ into the formula for the joint probability of A and B, we obtain

$$P(A \text{ and } B) = P(A)\, P(B)$$

MULTIPLICATION RULE FOR INDEPENDENT EVENTS

The probability of the intersection of two independent events A and B is

$$P(A \text{ and } B) = P(A)\, P(B)$$

Calculating joint probability of two independent events.

EXAMPLE 4-26 An office building has two fire detectors. The probability is .02 that any fire detector of this type will fail to go off during a fire. Find the probability that both of these fire detectors will fail to go off in case of a fire.

Solution In this example, the two fire detectors are independent. This is so because whether or not one fire detector goes off during a fire has no effect on the second fire detector. Define the following two events.

$$A = \text{the first fire detector fails to go off during a fire}$$

$$B = \text{the second fire detector fails to go off during a fire}$$

Then the joint probability of A and B is

$$P(A \text{ and } B) = P(A) P(B) = (.02)(.02) = \mathbf{.0004} \qquad \blacksquare$$

The multiplication rule can be extended to calculate the joint probability of more than two events. Example 4–27 illustrates such a case for independent events.

Calculating joint probability of three events.

EXAMPLE 4-27 The probability that a patient is allergic to penicillin is .20. Suppose this drug is administered to three patients.

(a) Find the probability that all three of them are allergic to it.
(b) Find the probability that at least one of them is not allergic to it.

Solution

(a) Let A, B, and C denote the events that the first, second, and third patients, respectively, are allergic to penicillin. We are to find the joint probability of A, B, and C. All three events are independent because whether or not one patient is allergic does not depend on whether or not any of the other patients is allergic. Hence,

$$P(A \text{ and } B \text{ and } C) = P(A) P(B) P(C) = (.20)(.20)(.20) = \mathbf{.008}$$

The tree diagram in Figure 4.18, given on the next page, shows all the outcomes for this experiment. Events $\overline{A}$, $\overline{B}$, and $\overline{C}$ are the complementary events of A, B, and C, respectively. They represent the events that the respective patients are not allergic to penicillin. Note that the intersection of events A, B, and C is written as ABC in the tree diagram.

(b) Let us define the following events.

$$G = \text{all three patients are allergic}$$

$$H = \text{at least one patient is not allergic}$$

Events G and H are two complementary events. Event G consists of the intersection of events A, B, and C. Hence, from part (a),

$$P(G) = P(A \text{ and } B \text{ and } C) = .008$$

Therefore, using the complementary event rule, we obtain:

$$P(H) = 1 - P(G) = 1 - .008 = \mathbf{.992}$$

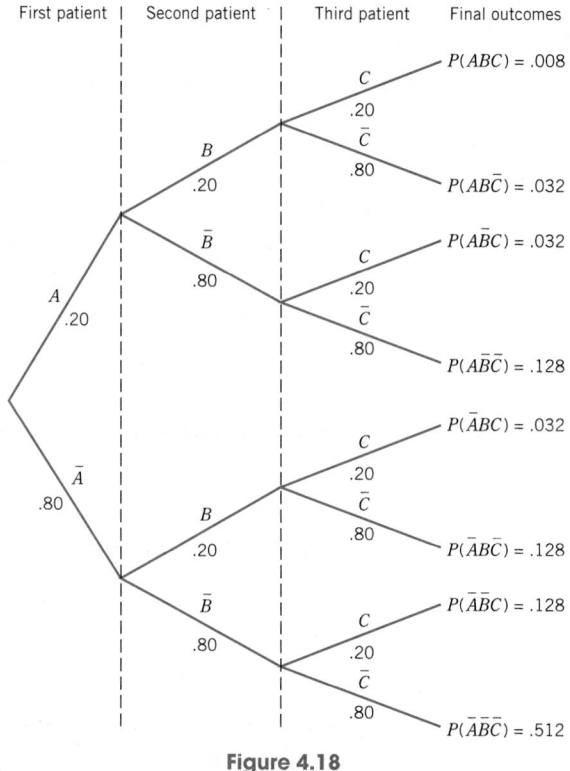

First patient | Second patient | Third patient | Final outcomes

$P(ABC) = .008$

$P(AB\bar{C}) = .032$

$P(A\bar{B}C) = .032$

$P(A\bar{B}\bar{C}) = .128$

$P(\bar{A}BC) = .032$

$P(\bar{A}B\bar{C}) = .128$

$P(\bar{A}\bar{B}C) = .128$

$P(\bar{A}\bar{B}\bar{C}) = .512$

Figure 4.18

Case Study 4–3 calculates the probability of a hitless streak in baseball by using the multiplication rule.

CASE STUDY 4-3 BASEBALL PLAYERS HAVE "SLUMPS" AND "STREAKS"

Going '0 for July,' as former infielder Bob Aspromonte once put it, is enough to make a baseball player toss out his lucky bat or start seriously searching for flaws in his hitting technique. But the culprit is usually just simple mathematics.

Statistician Harry Roberts of the University of Chicago's Graduate School of Business studied the records of major-league baseball players and found that a batter is no more likely to hit worse when he is in a slump than when he is in a hot streak. The occurrences of hits followed the same pattern as purely random events such as pulling marbles out of a hat. If there were one white marble and three black ones in the hat, for example, then a white marble would come out about one quarter of the time—a .250 average. In the same way, a player who hits .250 will in the long run get a hit every four times at bat.

But that doesn't mean the player will hit the ball exactly every fourth time he comes to the plate—just as it's unlikely that the white marble will come out exactly every fourth time.

Even a batter who goes hitless 10 times in a row might safely be able to pin the blame on statistical fluctuations. The odds of pulling a black marble out of a hat 10 times in a row are about 6 percent—not a frequent occurrence, but not impossible, either. Only in the long run do these statistical fluctuations even out. . . .

If we assume a player hits .250 in the long run, the probability that this player does not hit during a specific trip to the plate is .75. Hence, we can calculate the probability that he goes hitless 10 times in a row as follows.

$$P(\text{hitless 10 times in a row}) = (.75)(.75)\ldots(.75) \text{ ten times}$$
$$= (.75)^{10} = .0563$$

Note that each trip to the plate is independent, and the probability that a player goes hitless 10 times in a row is given by the intersection of 10 hitless trips. This probability has been rounded off to "about 6%" in this illustration.

Source: *U.S. News & World Report*, July 11, 1988, p. 46. Copyright © 1988, by U.S. News & World Report, Inc. Excerpts reprinted with permission.

JOINT PROBABILITY OF MUTUALLY EXCLUSIVE EVENTS

We know from an earlier discussion that two mutually exclusive events cannot happen together. Consequently, their joint probability is 0.

> **JOINT PROBABILITY OF MUTUALLY EXCLUSIVE EVENTS**
>
> The joint probability of two mutually exclusive events is always 0. If A and B are two mutually exclusive events, then,
>
> $$P(A \text{ and } B) = 0$$

Illustrating probability of two mutually exclusive events.

EXAMPLE 4–28 Consider the following two events for an application filed by a person to obtain a car loan.

$$A = \text{event that the loan application is approved}$$

$$R = \text{event that the loan application is rejected}$$

What is the joint probability of A and R?

Solution The two events A and R are mutually exclusive. Either the loan application will be approved or it will be rejected. Hence,

$$P(A \text{ and } R) = \mathbf{0}$$

EXERCISES

Concepts and Procedures

4.66 Explain the meaning of the intersection of two events. Give one example.

4.67 What is meant by the joint probability of two or more events? Give one example.

4.68 How is the multiplication rule of probability for two dependent events different from the one for two independent events?

4.69 What is the joint probability of two mutually exclusive events? Give one example.

4.70 Find the joint probability of A and B for the following.

    a. $P(A) = .40$ and $P(B \mid A) = .32$
    b. $P(B) = .65$ and $P(A \mid B) = .36$

4.71 Find the joint probability of A and B for the following.

    a. $P(B) = .59$ and $P(A \mid B) = .77$
    b. $P(A) = .28$ and $P(B \mid A) = .15$

4.72 Given that A and B are two independent events, find their joint probability for the following.

    a. $P(A) = .61$ and $P(B) = .27$
    b. $P(A) = .39$ and $P(B) = .73$

4.73 Given that A and B are two independent events, find their joint probability for the following.

    a. $P(A) = .20$ and $P(B) = .86$
    b. $P(A) = .57$ and $P(B) = .32$

4.74 Given that A, B, and C are three independent events, find their joint probability for the following.

    a. $P(A) = .20$, $P(B) = .46$, and $P(C) = .15$
    b. $P(A) = .44$, $P(B) = .27$, and $P(C) = .33$

4.75 Given that A, B, and C are three independent events, find their joint probability for the following.

    a. $P(A) = .39$, $P(B) = .67$, and $P(C) = .75$
    b. $P(A) = .71$, $P(B) = .34$, and $P(C) = .41$

4.76 Given that $P(A) = .30$ and $P(A \text{ and } B) = .24$, find $P(B \mid A)$.

4.77 Given that $P(B) = .65$ and $P(A \text{ and } B) = .45$, find $P(A \mid B)$.

4.78 Given that $P(A \mid B) = .40$ and $P(A \text{ and } B) = .36$, find $P(B)$.

4.79 Given that $P(B \mid A) = .80$ and $P(A \text{ and } B) = .58$, find $P(A)$.

Applications

4.80 The following table gives a two-way classification of the U.S. population (in millions) based on gender and health insurance coverage.

	Covered by Health Insurance	Not Covered by Health Insurance
Male	105.3	21.7
Female	114.8	12.1

Source: U.S. Bureau of the Census.

    a.   If one person is randomly selected from this population, find the following probabilities.
        i. P(male and not covered)   ii. P(female and covered)
    b.   Find P(male and female). Is this probability 0? Explain why or why not.

4.81 The following table gives a two-way classification of all faculty members of a university based on gender and tenure.

	Tenured	Nontenured
Male	74	28
Female	29	12

    a.   If one of these faculty members is selected at random, find the following probabilities.
        i. P(male and nontenured)   ii P(tenured and female)
    b.   Find P(tenured and nontenured). Is this probability 0? If yes, why?

4.82 Five hundred employees were selected from a city's large private companies and asked whether or not they have any retirement benefits provided by their companies. Based on this information, the following two-way classification table was prepared.

	Have Retirement Benefits	
	Yes	**No**
Men	225	75
Women	150	50

a. Suppose one employee is selected at random from these 500 employees. Find the following probabilities.
   i. Probability of the intersection of events ''woman'' and ''has retirement benefits''
   ii. Probability of the intersection of events ''does not have retirement benefits'' and ''man''

b. Mention what other joint probabilities you can calculate for this table and then find them. You may draw a tree diagram to find these probabilities.

4.83 All 420 employees of a company were asked whether they are smokers or nonsmokers and whether or not they are college graduates. Based on this information, the following two-way classification table was prepared.

	College Graduate	**Not a College Graduate**
Smoker	35	80
Nonsmoker	130	175

a. Suppose one employee is selected at random from this company. Find the following probabilities.
   i. P(college graduate and nonsmoker)
   ii. P(smoker and not a college graduate)

b. Mention what other joint probabilities you can calculate for this table and then find them. You may draw a tree diagram to find these probabilities.

4.84 The following table gives the two-way classification of 2000 randomly selected employees from a city based on gender and commuting time from home to work.

	Commuting Time		
	Less Than 30 Minutes	**30 Minutes to One Hour**	**More Than One Hour**
Men	524	455	221
Women	413	263	124

a. Suppose one employee is selected at random from these 2000 employees. Find the following probabilities.
   i. P(commutes for more than one hour and man)
   ii. P(woman and commutes for less than 30 minutes)

b. Find the joint probability of events ''commutes for 30 minutes to one hour'' and ''commutes for more than one hour.'' Is this probability zero? Explain why or why not.

4.85 Two thousand randomly selected adults were asked if they think they are financially better off than their parents. The following table gives the two-way classification of the responses based on the education levels of the persons included in the survey and whether they are financially better off, the same, or worse off than their parents.

	Education Level		
	Less Than High School	High School	More Than High School
Better Off	140	450	420
Same	60	250	110
Worse Off	200	300	70

    a. Suppose one adult is selected at random from these 2000 adults. Find the following probabilities.

       i. *P*(better off and high school)

      ii. *P*(more than high school and worse off)

    b. Find the joint probability of events ''worse off'' and ''better off.'' Is this probability zero? Explain why or why not.

4.86   In a statistics class of 45 students, 12 have a strong interest in statistics. If two students are selected at random from this class, what is the probability that both of them have a strong interest in statistics? Draw a tree diagram for this problem.

4.87   In a group of 15 students, 5 have liberal views. If 2 students are randomly selected from this group, what is the probability that the first of them has liberal views and the second does not? Draw a tree diagram for this problem.

4.88   A company is to hire two new employees. They have prepared a final list of eight candidates, all of whom are equally qualified. Of these eight candidates, five are women. If the company decides to select two persons randomly from these eight candidates, what is the probability that both of them are women? Draw a tree diagram for this problem.

4.89   In a group of 10 persons, 4 have a type A personality and 6 have a type B personality. If 2 persons are selected at random from this group, what is the probability that the first of them has a type A personality and the second has a type B personality? Draw a tree diagram for this problem.

4.90   The probability is .35 that an adult has never flown on an airplane. If two adults are selected at random, what is the probability that the first of them has never flown on an airplane and the second has? Draw a tree diagram for this problem.

4.91   The probability is .76 that a family owns a house. If two families are randomly selected, what is the probability that neither of them owns a house?

4.92   A contractor has submitted bids for two state construction projects. The probability that he will win any contract is .30, and it is the same for each of the two contracts.

    a. What is the probability that he will win both contracts?

    b. What is the probability that he will win neither contract?

Draw a tree diagram for this problem.

4.93   Five percent of all items sold by a mail-order company are returned by customers for a refund. Find the probability that of two items sold during a given hour by this company

    a. both will be returned for a refund

    b. neither will be returned for a refund

Draw a tree diagram for this problem.

4.94   The probability that any given person is allergic to a certain drug is .03. What is the probability that none of three randomly selected persons is allergic to this drug? Assume that all three persons are independent.

4.95   The probability that a farmer is in debt is .75. What is the probability that three randomly selected farmers are all in debt? Assume independence of events.

4.96   The probability that a household owns a house is .76. The probability that a household owns a house and is a married couple is .69. Find the conditional probability that a randomly selected household is a married couple given that it owns a house.

4.97 The probability that an employee at a company is a female is .36. The probability that an employee is a female and married is .19. Find the conditional probability that a randomly selected employee from this company is married given that she is a female.

4.98 According to the U.S. National Center for Health Statistics, there were 92,814 thousand visits to emergency rooms in U.S. hospitals in 1993. Of these visits, 44,426 thousand were made by males, and 20,848 thousand were urgent visits by males. Find the probability that a randomly selected visit was urgent given that the patient was male.

4.99 According to the estimates of the U.S. National Center for Education Statistics, the probability that a randomly selected student from the population of all students enrolled at U.S. institutions of higher education is a female is .55, and the probability that this student is a female and a part-time student is .26. What is the probability that this student is part-time given that she is a female?

4.9 UNION OF EVENTS AND THE ADDITION RULE

This section discusses the union of events and the addition rule that is applied to compute the probability of the union of events.

4.9.1 UNION OF EVENTS

The **union of two events** A and B includes all outcomes that are either in A or in B or in both A and B.

UNION OF EVENTS

Let A and B be two events defined in a sample space. The *union of events A and B* is the collection of all outcomes that belong either to A or to B or to both A and B and is denoted by

$$A \text{ or } B$$

The union of events A and B is also denoted by "$A \cup B$." Example 4–29 illustrates the union of events A and B.

Illustrating the union of two events.

EXAMPLE 4–29   According to the U.S. Department of Education, there are 14,210 thousand students enrolled at institutions of higher education in the United States. Of them, 7829 thousand are female, 8064 thousand are full-time students, and 4161 thousand are female and full-time students. Describe the union of events "female" and "full-time."

Solution   The union of events "female" and "full-time" for students enrolled at institutions of higher education includes all students who are either female or full-time or both. The number of such students is

$$7829 + 8064 - 4161 = 11,732 \text{ thousand}$$

Thus, there are a total of 11,732 thousand enrolled at institutions of higher education who are either female or full-time students or both.

Why did we subtract 4161 thousand from the sum of 7829 thousand and 8064 thousand? The reason is that 4161 thousand students (which represent the intersection of events ''female'' and ''full-time'') are common to both events ''female'' and ''full-time'' and, hence, are counted twice. To avoid double counting, we subtracted 4161 thousand from the sum of the other two numbers. We can observe this double counting from Table 4.7, which is constructed using the given information. The sum of the numbers in the three shaded cells gives the students who are either female or full-time or both. However, if we add the totals of the row labeled ''Female'' and the column labeled ''Full-time,'' we count 4161 twice. Note that the numbers in the table are in thousands.

Table 4.7

	Full-time	**Part-time**	**Total**
Male	3,903	2,478	6,381
Female	4,161 ←	3,668	7,829
Total	8,064	6,146	14,210

└— Counted twice

Figure 4.19 shows the diagram for the union of events ''female'' and ''full-time.''

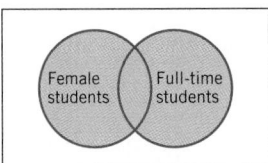

Shaded area gives the union of two events
and includes 11,732 thousand students

Figure 4.19 Union of events ''female'' and ''full-time.''

4.9.2 ADDITION RULE

The method used to calculate the probability of the union of events is called the **addition rule**. It is defined as follows.

ADDITION RULE

The probability of the union of two events A and B is

$$P(A \text{ or } B) = P(A) + P(B) - P(A \text{ and } B)$$

Thus, to calculate the probability of the union of two events A and B, we add their marginal probabilities and subtract their joint probability from this sum. We must subtract the joint probability of A and B from the sum of their marginal probabilities to avoid double counting due to common outcomes in A and B.

Calculating probability of union of two events: two-way table.

EXAMPLE 4–30 A university president has proposed that all students must take a course in ethics as a requirement for graduation. Three hundred faculty members and students from this university were asked about their opinion on this issue. The following table gives a two-way classification of the responses of these faculty members and students.

	Opinion			
	Favor	**Oppose**	**Neutral**	**Total**
Faculty	45	15	10	70
Student	90	110	30	230
Total	135	125	40	300

Find the probability that one person selected at random from these 300 persons is a faculty member or is in favor of this proposal.

Solution   Let us define the following events.

$$A = \text{the person selected is a faculty member}$$

$$B = \text{the person selected is in favor of the proposal}$$

From the information given in the table,

$$P(A) = 70/300 = .2333$$

$$P(B) = 135/300 = .4500$$

and $\qquad P(A \text{ and } B) = P(A)\, P(B \mid A) = (70/300)\,(45/70) = .1500$

Using the addition rule

$$P(A \text{ or } B) = P(A) + P(B) - P(A \text{ and } B) = .2333 + .4500 - .1500 = \mathbf{.5333}$$

Thus, the probability that a randomly selected person from these 300 persons is a faculty member or is in favor of this proposal is .5333.

The probability in this example can also be calculated without using the addition rule. The total number of persons in the given table who are either faculty members or are in favor of this proposal is

$$45 + 15 + 10 + 90 = 160$$

Hence, the required probability is

$$P(A \text{ or } B) = 160/300 = \mathbf{.5333}$$

Calculating probability of union of two events.

EXAMPLE 4–31   There are a total of 7693 thousand persons with multiple jobs in the United States. Of them, 4139 thousand are male, 3395 thousand are single, and 1456 thousand are male and single. (*Source:* U.S. Bureau of Labor Statistics.) What is the probability that a randomly selected person with multiple jobs is a male or single? Note that here single persons include never-married single, widowed, divorced, and separated.

Solution   Let us define the following two events.

$$M = \text{the randomly selected person is a male}$$

$$A = \text{the randomly selected person is single}$$

From the given information,

$$P(M) = 4139/7693 = .5380$$

$$P(A) = 3395/7693 = .4413$$

and $\qquad P(M \text{ and } A) = 1456/7693 = .1893$

Hence,

$$P(M \text{ or } A) = P(M) + P(A) - P(M \text{ and } A) = .5380 + .4413 - .1893 = \mathbf{.7900}$$

Actually, using the given information, we can prepare Table 4.8 for this example. The numbers in the shaded cells are given to us. The remaining numbers are calculated by doing some arithmetic manipulations.

Table 4.8   Two-way Classification Table

	Single (A)	Married (B)	Total
Male (M)	1456	2683	4139
Female (F)	1939	1615	3554
Total	3395	4298	7693

Now, using this table, we can find the required probabilities as follows.

$$P(M) = 4139/7693 = .5380$$

$$P(A) = 3395/7693 = .4413$$

$$P(M \text{ and } A) = 1456/7693 = .1893$$

$$P(M \text{ or } A) = P(M) + P(A) - P(M \text{ and } A) = .5380 + .4413 - .1893 = \mathbf{.7900} \ \blacksquare$$

ADDITION RULE FOR MUTUALLY EXCLUSIVE EVENTS

We know from an earlier discussion that the joint probability of two mutually exclusive events is zero. When A and B are mutually exclusive events, the term $P(A \text{ and } B)$ in the addition rule becomes zero and is dropped from the formula. Thus, the probability of the union of two mutually exclusive events is given by the sum of their marginal probabilities.

ADDITION RULE FOR MUTUALLY EXCLUSIVE EVENTS

The probability of the union of two mutually exclusive events A and B is

$$P(A \text{ or } B) = P(A) + P(B)$$

Calculating probability of union of two mutually exclusive events: two-way table.

EXAMPLE 4-32   A university president has proposed that all students must take a course in ethics as a requirement for graduation. Three hundred faculty members and students from this university were asked about their opinion on this issue. The following table, reproduced from Example 4–30, gives a two-way classification of the responses of these faculty members and students.

	Opinion			
	Favor	Oppose	Neutral	Total
Faculty	45	15	10	70
Student	90	110	30	230
Total	135	125	40	300

What is the probability that a randomly selected person from these 300 faculty members and students is in favor of the proposal or is neutral?

Solution   Let us define the following events.

$$F = \text{the person selected is in favor of the proposal}$$

$$N = \text{the person selected is neutral}$$

As shown in Figure 4.20, events F and N are mutually exclusive because a person selected can be either in favor or neutral but not both.

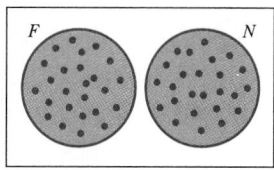

Figure 4.20

From the given information,

$$P(F) = 135/300 = .4500$$

and

$$P(N) = 40/300 = .1333$$

Hence,

$$P(F \text{ or } N) = P(F) + P(N) = .4500 + .1333 = \textbf{.5833}$$ ▄

The addition rule formula can easily be extended to apply to more than two events. The following example illustrates this.

Calculating joint probability of three mutually exclusive events.

EXAMPLE 4–33   Consider the experiment of rolling a die twice. Find the probability that the sum of the numbers obtained on two rolls is 5, 7, or 10.

Solution   The experiment of rolling a die twice has a total of 36 outcomes, which are listed in Table 4.9. Assuming that the die is balanced, these 36 outcomes are equally likely.

Table 4.9   Two Rolls of a Die

		Second Roll of the Die					
		1	**2**	**3**	**4**	**5**	**6**
	1	(1,1)	(1,2)	(1,3)	(1,4)	(1,5)	(1,6)
	2	(2,1)	(2,2)	(2,3)	(2,4)	(2,5)	(2,6)
First Roll of the Die	**3**	(3,1)	(3,2)	(3,3)	(3,4)	(3,5)	(3,6)
	4	(4,1)	(4,2)	(4,3)	(4,4)	(4,5)	(4,6)
	5	(5,1)	(5,2)	(5,3)	(5,4)	(5,5)	(5,6)
	6	(6,1)	(6,2)	(6,3)	(6,4)	(6,5)	(6,6)

The events that give the sum of two numbers equal to 5 or 7 or 10 are marked in the table. As we can observe, the three events "the sum is 5," "the sum is 7," and "the sum is 10" are mutually exclusive. Four outcomes give a sum of 5, six give a sum of 7, and three outcomes give a sum of 10. Thus,

$$P(\text{sum is 5 or 7 or 10}) = P(\text{sum is 5}) + P(\text{sum is 7}) + P(\text{sum is 10})$$

$$= 4/36 + 6/36 + 3/36 = 13/36 = \textbf{.3611} \quad \blacksquare$$

Calculating joint probability of three mutually exclusive events.

EXAMPLE 4–34   The probability that a person is in favor of genetic engineering is .55 and that a person is against it is .45. Two persons are randomly selected, and it is observed whether they favor or oppose genetic engineering.

(a)   Draw a tree diagram for this experiment.

(b)   Find the probability that at least one of the two persons favors genetic engineering.

Solution

(a)   Let

$$F = \text{a person is in favor of genetic engineering}$$

$$A = \text{a person is against genetic engineering}$$

This experiment has four outcomes: both persons are in favor (*FF*), the first person is in favor and the second is against (*FA*), the first person is against and the second is in favor (*AF*), and both persons are against genetic engineering (*AA*). The tree diagram in Figure 4.21 shows these four outcomes and their probabilities.

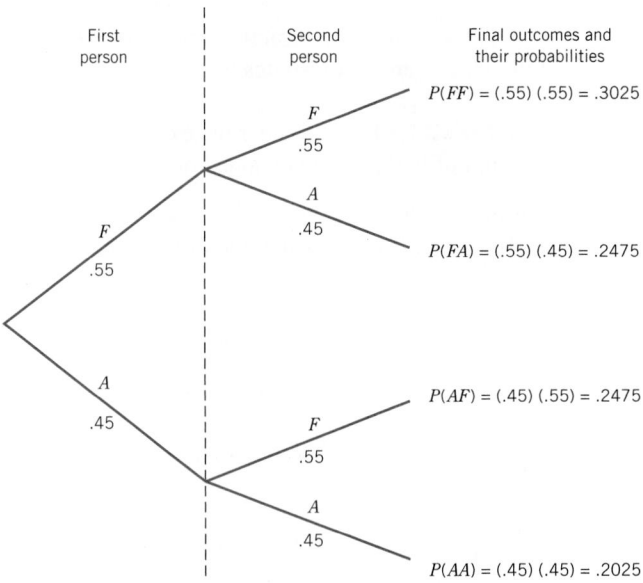

Figure 4.21   Tree diagram.

(b)   The probability that at least one person favors genetic engineering is given by the union of events *FF*, *FA*, and *AF*. These three outcomes are mutually exclusive. Hence,

$$P(\text{at least one person favors}) = P(FF \text{ or } FA \text{ or } AF)$$

$$= P(FF) + P(FA) + P(AF)$$

$$= .3025 + .2475 + .2475 = \textbf{.7975} \quad \blacksquare$$

EXERCISES

Concepts and Procedures

4.100 Explain the meaning of the union of two events. Give one example.

4.101 How is the addition rule of probability for two mutually exclusive events different from the one for two mutually nonexclusive events?

4.102 Consider the following addition rule to find the probability of the union of two events A and B.

$$P(A \text{ or } B) = P(A) + P(B) - P(A \text{ and } B)$$

When and why is the term $P(A \text{ and } B)$ subtracted from the sum of $P(A)$ and $P(B)$? Give one example where you might use this formula.

4.103 When is the following addition rule used to find the probability of the union of two events A and B?

$$P(A \text{ or } B) = P(A) + P(B)$$

Give one example where you might use this formula.

4.104 Find $P(A \text{ or } B)$ for the following.
a. $P(A) = .58$, $P(B) = .66$, and $P(A \text{ and } B) = .47$
b. $P(A) = .72$, $P(B) = .42$, and $P(A \text{ and } B) = .33$

4.105 Find $P(A \text{ or } B)$ for the following.
a. $P(A) = .18$, $P(B) = .49$, and $P(A \text{ and } B) = .13$
b. $P(A) = .83$, $P(B) = .71$, and $P(A \text{ and } B) = .68$

4.106 Given that A and B are two mutually exclusive events, find $P(A \text{ or } B)$ for the following.
a. $P(A) = .57$ and $P(B) = .32$
b. $P(A) = .16$ and $P(B) = .49$

4.107 Given that A and B are two mutually exclusive events, find $P(A \text{ or } B)$ for the following.
a. $P(A) = .25$ and $P(B) = .17$
b. $P(A) = .38$ and $P(B) = .09$

Applications

4.108 The following table gives a two-way classification of the U.S. population (in millions) based on gender and health insurance coverage.

	Covered by Health Insurance	Not Covered by Health Insurance
Male	105.3	21.7
Female	114.8	18.1

Source: U.S. Bureau of the Census.

If one person is randomly selected from this population, find the following probabilities.
a. $P(\text{female or not covered})$ b. $P(\text{male or covered})$

4.109 The following table gives a two-way classification of all faculty members of a university based on gender and tenure.

	Tenured	Nontenured
Male	74	28
Female	29	12

If one of these faculty members is selected at random, find the following probabilities.
a. $P(\text{female or nontenured})$ b. $P(\text{tenured or male})$

4.110 Five hundred employees were selected from a city's large private companies and they were asked whether or not they have any retirement benefits provided by their companies. Based on this information, the following two-way classification table was prepared.

	Have Retirement Benefits	
	Yes	**No**
Men	225	75
Women	150	50

Suppose one employee is selected at random from these 500 employees. Find the following probabilities.

    a. The probability of the union of events "woman" and "has retirement benefits"
    b. The probability of the union of events "does not have retirement benefits" and "man"

4.111 All 420 employees of a company were asked whether they are smokers or nonsmokers and whether or not they are college graduates. Based on this information, the following two-way classification table was prepared.

	College Graduate	Not a College Graduate
Smoker	35	80
Nonsmoker	130	175

Suppose one employee is selected at random from this company. Find the following probabilities.

    a. P(college graduate or smoker)
    b. P(smoker or not a college graduate)
    c. P(smoker or nonsmoker)

4.112 The following table gives the two-way classification of 2000 randomly selected employees from a city based on gender and commuting time from home to work.

	Commuting Time		
	Less Than 30 Minutes	**30 Minutes to One Hour**	**More Than One Hour**
Men	524	455	221
Women	413	263	124

If one employee is selected at random from these 2000 employees, find the following probabilities.

    a. P(commutes for more than one hour or man)
    b. P(woman or commutes for less than 30 minutes)
    c. P(man or woman)

4.113 Two thousand randomly selected adults were asked if they think they are financially better off than their parents. The following table gives the two-way classification of the responses based on the education levels of the persons included in the survey and whether they are financially better off, the same, or worse off than their parents.

	Education Level		
	Less Than High School	**High School**	**More Than High School**
Better Off	140	450	420
Same	60	250	110
Worse Off	200	300	70

Suppose one adult is selected at random from these 2000 adults. Find the following probabilities.

    a. *P*(better off or high school)

    b. *P*(more than high school or worse off)

    c. *P*(better off or worse off)

4.114 The probability that a football player weighs more than 230 pounds is .69, that he is at least 75 inches tall is .55, and that he weighs more than 230 pounds and is at least 75 inches tall is .43. Find the probability that a randomly selected football player weighs more than 230 pounds or is at least 75 inches tall.

4.115 The probability that a family owns a washing machine is .78, that it owns a VCR is .71, and that it owns both a washing machine and a VCR is .58. What is the probability that a randomly selected family owns a washing machine or a VCR?

4.116 The probability that a person has a checking account is .74, a savings account is .31, and both accounts is .22. Find the probability that a randomly selected person has a checking or a savings account.

4.117 The probability that a randomly selected elementary or secondary school teacher from a city is a female is .68, holds a second job is .42, and is a female and holds a second job is .29. Find the probability that an elementary or secondary school teacher selected at random from this city is a female or holds a second job.

4.118 According to an estimate by the U.S. Bureau of the Census, there are a total of 69,305 thousand families living in the United States. Of them, 29,176 thousand have two members per family, and 14,624 thousand have four members per family. If one family is selected at random from all these families, what is the probability that it has two or four members? Explain why this probability is not equal to 1.0.

4.119 According to an estimate by the U.S. Bureau of the Census, there are a total of 189,987 thousand persons aged 18 and over living in the United States. Of them, 44,185 thousand are single and 115,141 thousand are married. If one person is randomly selected from this population, what is the probability that this person is single or married? Explain why this probability is not equal to 1.0.

4.120 The probability of a student getting an A grade in an economics class is .24 and that of getting a B grade is .28. What is the probability that a randomly selected student from this class will get an A or a B in this class? Explain why this probability is not equal to 1.0.

4.121 Seventy-two percent of a town's voters favor the recycling issue, 12% oppose it, and 16% are indifferent. What is the probability that a randomly selected voter from this town will either favor recycling or be indifferent? Explain why this probability is not equal to 1.0.

4.122 The probability that a corporation makes charitable contributions is .68. Two corporations are selected at random, and it is noted whether or not they make charitable contributions.

    a. Draw a tree diagram for this experiment.

    b. Find the probability that at most one corporation makes charitable contributions.

4.123 The probability that an open-heart operation is successful is .72. What is the probability that in two randomly selected open-heart operations at least one will be successful? Draw a tree diagram for this experiment.

GLOSSARY

Classical probability rule The method of assigning probabilities to outcomes or events of an experiment with equally likely outcomes.

Complementary events Two events that taken together include all the outcomes for an experiment but do not contain any common outcome.

Compound event An event that contains more than one outcome of an experiment. It is also called a *composite event*.

Conditional probability   The probability of an event subject to the condition that another event has already occurred.

Dependent events   Two events for which the occurrence of one changes the probability of the occurrence of the other.

Equally likely outcomes   Two (or more) outcomes or events that have the same probability of occurrence.

Event   A collection of one or more outcomes of an experiment.

Experiment   A process with well-defined outcomes that, when performed, results in one and only one of the outcomes per repetition.

Impossible event   An event that cannot occur.

Independent events   Two events for which the occurrence of one does not change the probability of occurrence of the other.

Intersection of events   Intersection of events is given by the outcomes that are common to two (or more) events.

Joint probability   The probability that two (or more) events occur together.

Law of Large Numbers   If an experiment is repeated again and again, the probability of an event obtained from the relative frequency approaches the actual or theoretical probability.

Marginal probability   The probability of one event or characteristic without consideration of any other event.

Mutually exclusive events   Two or more events that do not contain any common outcome and, hence, cannot occur together.

Outcome   The result of the performance of an experiment.

Probability   A numerical measure of the likelihood that a specific event will occur.

Relative frequency as an approximation of probability   Probability assigned to an event based on the results of an experiment or based on historical data.

Sample point   An outcome of an experiment.

Sample space   The collection of all sample points or outcomes of an experiment.

Simple event   An event that contains one and only one outcome of an experiment. It is also called an *elementary event*.

Subjective probability   The probability assigned to an event based on the information and judgment of a person.

Sure event   An event that is certain to occur.

Tree diagram   A diagram in which each outcome of an experiment is represented by a branch of a tree.

Union of two events   All outcomes that belong either to one or to both events.

Venn diagram   A picture that represents a sample space or certain events.

KEY FORMULAS

1. **Classical probability rule**

For a simple event E_i:

$$P(E_i) = \frac{1}{\text{Total number of outcomes for the experiment}}$$

For a compound event A:

$$P(A) = \frac{\text{Number of outcomes favorable to } A}{\text{Total number of outcomes for the experiment}}$$

2. **Relative frequency as an approximation of probability**

$$P(\text{an event}) = \frac{\text{Frequency of that event}}{\text{Sample size}} = \frac{f}{n}$$

3. **Counting rule**

If an experiment consists of three steps and if the first step can result in m outcomes, the second step in n outcomes, and the third step in k outcomes, then

$$\text{Total outcomes for the experiment} = m \cdot n \cdot k$$

4. **Conditional probability of an event**

$$P(B \mid A) = \frac{P(A \text{ and } B)}{P(A)} \quad \text{and} \quad P(A \mid B) = \frac{P(A \text{ and } B)}{P(B)}$$

5. **Independent events**

Two events A and B are independent if

$$P(A) = P(A \mid B) \quad \text{and/or} \quad P(B) = P(B \mid A)$$

6. **Complementary events**

For two complementary events A and $\overline{A}$:

$$P(A) + P(\overline{A}) = 1, \quad P(A) = 1 - P(\overline{A}), \quad \text{and} \quad P(\overline{A}) = 1 - P(A)$$

7. **Multiplication rule for joint probability of events**

If A and B are dependent events, then

$$P(A \text{ and } B) = P(A) \, P(B \mid A)$$

If A and B are independent events, then

$$P(A \text{ and } B) = P(A) \, P(B)$$

8. **Joint probability of two mutually exclusive events**

For two mutually exclusive events A and B:

$$P(A \text{ and } B) = 0$$

9. **Addition rule for the probability of union of events**

If A and B are mutually nonexclusive events, then

$$P(A \text{ or } B) = P(A) + P(B) - P(A \text{ and } B)$$

If A and B are mutually exclusive events, then

$$P(A \text{ or } B) = P(A) + P(B)$$

SUPPLEMENTARY EXERCISES

4.124 A lawyers' association has 80 members. Of them, 12 are corporate lawyers. One lawyer is selected at random. Find the probability that this lawyer is

    a. a corporate lawyer   **b.** not a corporate lawyer

4.125 In a class of 35 students, 13 are seniors, 9 are juniors, 8 are sophomores, and 5 are freshmen. If one student is selected at random from this class, what is the probability that this student is

    a. a junior   **b.** a freshman

4.126 A group of 150 randomly selected CEOs was tested for personality type. The following table gives the results of this survey.

	Type A	**Type B**
Men	78	42
Women	19	11

a.   If one CEO is selected at random from this group, find the probability that this CEO
   i.   has a type A personality
   ii. is a woman
   iii. is a man given that he has a type A personality
   iv. has a type B personality given that she is a woman
   v.   has a type A personality and is a woman
   vi. is a man or has a type B personality
b.   Are the events ''woman'' and ''type A personality'' mutually exclusive? What about the events ''type A personality'' and ''type B personality''? Why or why not?
c.   Are the events ''type A personality'' and ''man'' independent? Why or why not?

4.127   A random sample of 250 adults was taken, and they were asked whether they prefer watching sports or opera on television. The following table gives the two-way classification of these 250 adults.

	Prefers Watching Sports	Prefers Watching Opera
Male	96	24
Female	45	85

a.   If one adult is selected at random from this group, find the probability that this adult
   i.   prefers watching opera
   ii. is a male
   iii. prefers watching sports given that the adult is a female
   iv. is a male given that he prefers watching sports
   v.   is a female and prefers watching opera
   vi. prefers watching sports or is a male
b.   Are the events ''female'' and ''prefers watching sports'' independent? Are they mutually exclusive? Explain why or why not.

4.128   A random sample of 80 lawyers was taken, and they were asked if they are in favor of or against capital punishment. The following table gives the two-way classification of these 80 lawyers.

	Favors Capital Punishment	Opposes Capital Punishment
Male	32	26
Female	13	9

a.   If one lawyer is randomly selected from this group, find the probability that this lawyer
   i.   favors capital punishment
   ii. is a female
   iii. opposes capital punishment given that the lawyer is a female
   iv. is a male given that he favors capital punishment
   v.   is a female and favors capital punishment
   vi. opposes capital punishment or is a male
b.   Are the events ''female'' and ''opposes capital punishment'' independent? Are they mutually exclusive? Explain why or why not.

4.129   A random sample of 400 college students was asked if college athletes should be paid. The following table gives a two-way classification of the responses.

	Should Be Paid	Should Not Be Paid
Student Athlete	90	10
Student Nonathlete	210	90

a. If one student is randomly selected from these 400 students, find the probability that this student
   i. is in favor of paying college athletes
   ii. favors paying college athletes given that the student selected is a nonathlete
   iii. is an athlete and favors paying student athletes
   iv. is a nonathlete or is against paying student athletes
b. Are the events ''student athlete'' and ''should be paid'' independent? Are they mutually exclusive? Explain why or why not.

4.130 A survey conducted about job satisfaction showed that 20% of workers are not happy with their current jobs. Assume that this result is true for the population of all workers. Two workers are selected at random, and it is observed whether or not they are happy with their current jobs. Draw a tree diagram. Find the probability that in this sample of two workers
   a. both are not happy with their current jobs
   b. at least one of them is happy with the current job

4.131 According to the Centers for Disease Control and Prevention, 34.8% of ninth-through-twelfth-grade students smoked in 1995 (Millicent Lawton, ''More Students Report Smoking in '95 Than Two Years Earlier,'' *Education Week*, June 5, 1996). Assume that this result is true for the current population of ninth-through-twelfth-grade students. Two such students are selected at random and asked whether or not they smoke. Draw a tree diagram for this problem. Find the probability that in this sample of two students
   a. both are smokers   b. at most one smokes

4.132 Refer to Exercise 4.124. Two lawyers are selected at random from this group of 80 lawyers. Find the probability that both of these lawyers are corporate lawyers.

4.133 Refer to Exercise 4.125. Two students are selected at random from this class of 35 students. Find the probability that the first student selected is a junior and the second is a sophomore.

4.134 A company has installed a generator to back up the power in case there is a power failure. The probability that there will be a power failure during a snowstorm is .30. The probability that the generator will stop working during a snowstorm is .09. What is the probability that during a snowstorm the company will lose both sources of power?

4.135 Terry & Sons Inc. makes bearings for autos. The production system involves two independent processing machines so that each bearing passes through these two processes. The probability that the first processing machine is not working properly at any time is .08, and the probability that the second machine is not working properly at any time is .06. Find the probability that both machines will not be working properly at any given time.

*4.136 During the National Basketball Association (NBA) finals in June 1995, Nick Anderson of the Orlando Magic missed four consecutive free throws in the closing seconds of regulation time in game one (*Sports Illustrated*, June 19, 1995). Anderson had made 70% of his free throw attempts during the regular season. One of the TV announcers commented that there was less than a 1% chance that a player with a 70% success rate would miss four consecutive free throws. Do you agree with the announcer's comment? Support your answer by calculating the appropriate probability. Assume that each time the player attempts a free throw he has a 70% chance of success and that all free throws are independent of each other.

*4.137 A certain state's auto license plate has three letters of the alphabet followed by a three-digit number.
   a. How many different license plates are possible if all three-letter sequences are permitted and any number from 000 to 999 is allowed?
   b. Arnold witnessed a hit-and-run accident. He knows that the first two letters on the license plate were ''BX'' and that the last number was a 7. How many license plates would fit this description?

*4.138 The median life of Brand LT5 batteries is 100 hours. What is the probability that in a set of three such batteries, exactly two will last longer than 100 hours?

*4.139   A friend of yours prepares six envelopes for you; two of these envelopes contain a $100 bill each, and the remaining four are empty. Your friend asks you to randomly select two envelopes from these six. (Adapted from a question in the *Ask Marilyn* column of *Parade* magazine, April 30, 1995).

  a.   Which of the two events "at least one of the selected envelopes contains a $100 bill" or "both envelopes selected are empty" is more likely to occur? Answer without doing the calculations.

  b.   Find the probability that both envelopes selected are empty.

  c.   Find the probability that at least one of the selected envelopes contains a $100 bill.

  d.   Do your answers to parts b and c support your guess in part a?

*4.140   A trimotor plane has three engines—a central engine, and an engine on each wing. The plane will crash only if the central engine fails *and* at least one of the two wing engines fails. The probability of failure during any given flight is .005 for the central engine and .008 for each of the wing engines. Assuming that the three engines operate independently, what is the probability that the plane will crash during a flight?

*4.141   A box contains 10 red marbles and 10 green marbles.

  a.   Sampling at random from the box five times with replacement, you have drawn a red marble all five times. What is the probability of drawing a red marble the sixth time?

  b.   Sampling at random from the box five times without replacement, you have drawn a red marble all five times. Without replacing any of the marbles, what is the probability of drawing a red marble the sixth time?

  c.   You have tossed a fair coin five times and have obtained heads all five times. A friend argues that, according to the law of averages, a tail is due to occur and, hence, the probability of obtaining a head on the sixth toss is less than .50. Is he right? Is coin tossing mathematically equivalent to the procedure mentioned in part a or in part b? Explain.

*4.142   A gambler has four cards, two of which are diamonds and two are clubs. The gambler proposes the following game to you: You will leave the room and the gambler will put the cards face down on a table. When you return to the room, you pick two cards at random. If both cards are either clubs or diamonds, you win $10. Otherwise you lose $10. Assuming there is no cheating, should you accept this proposition? Support your answer by calculating your probability of winning $10.

*4.143   A thief has stolen Roger's automatic teller card. The card has a four-digit personal identification number (PIN). The thief knows the first two digits are 3 and 5, but he does not know the last two digits. Thus the PIN could be any number from 3500 to 3599. To protect the customer, the automatic teller machine will not allow more than three unsuccessful attempts to enter the PIN. After a wrong PIN is entered the third time, the machine keeps the card and allows no further attempts. What is the probability that the thief will find the correct PIN within three tries?

SELF-REVIEW TEST

 1.   The collection of all outcomes for an experiment is called

  a.   a sample space **b.**   intersection of events **c.**   joint probability

 2.   A final outcome of an experiment is called

  a.   a compound event **b.**   a simple event **c.**   a complementary event

 3.   A compound event includes

  a.   all final outcomes **b.**   exactly two outcomes

  c.   more than one outcome for an experiment

4. Two equally likely events

    a. have the same probability of occurrence

    b. cannot occur together

    c. have no effect on the occurrence of each other

5. Which of the following probability approaches can be applied only to experiments with equally likely outcomes?

    a. Classical probability   **b.** Empirical probability   **c.** Subjective probability

6. Two mutually exclusive events

    a. have the same probability   **b.** cannot occur together

    c. have no effect on the occurrence of each other

7. Two independent events

    a. have the same probability   **b.** cannot occur together

    c. have no effect on the occurrence of each other

8. The probability of an event is always

    a. less than 0   **b.** in the range 0 to 1.0   **c.** greater than 1.0

9. The sum of the probabilities of all final outcomes of an experiment is always

    a. 100   **b.** 1.0   **c.** 0

10. The joint probability of two mutually exclusive events is always

    a. 1.0   **b.** between 0 and 1   **c.** 0

11. Two independent events are

    a. always mutually exclusive   **b.** never mutually exclusive

    c. always complementary

12. A magazine editor is refurbishing her office. She can choose one of three kinds of paint, one of four kinds of furniture, and one of three kinds of telephones. If she randomly selects one paint, one kind of furniture, and one telephone, how many different outcomes are possible?

13. Lucia graduated this year with an accounting degree from Eastern Connecticut State University. She has received job offers from an accounting firm, an insurance company, and an airline. She cannot decide which of the three job offers she should accept. Suppose she decides to randomly select one of these three job offers. Find the probability that the job offer selected is

    a. from the insurance company

    b. not from the accounting firm

14. A company has 500 employees. Of them, 300 are male and 280 are married. Of the 300 males, 190 are married.

    a. Are the events "male" and "married" independent? Are they mutually exclusive? Explain why or why not.

    b. If one employee of this company is selected at random, what is the probability that this employee is

        i. a female   ii. a male given that he is married

15. Reconsider Problem 14. If one employee is selected at random from this company, what is the probability that this employee is a male or married?

16. Reconsider Problem 14. If two employees are selected at random from this company, what is the probability that both of them are female?

17. The probability that an American adult has ever experienced a migraine headache is .35. If two American adults are randomly selected, what is the probability that neither of them has ever experienced a migraine headache?

18. A hat contains five green, eight red, and seven blue marbles. Let A be the event that a red marble is drawn if we randomly select one marble out of this hat. What is the probability of A? What is the complementary event of A, and what is its probability?

19.   The probability that a student is a male is .48 and that a student likes statistics is .35. If these two events are independent, what is the probability that a student selected at random is

    a.   a male and likes statistics

    b.   a female or likes statistics

20.   Five hundred married men and women were asked whether or not they would marry their current spouses if they were given a chance to do it over again. Their responses are recorded in the following table.

	Would Marry the Current Spouse	
	Yes (Y)	No (N)
Male (M)	125	175
Female (F)	55	145

    a.   If one person is selected at random from these 500 persons, find the following probabilities

        i.   Yes

        ii.   Yes given female

        iii.   Male and no

        iv.   Yes or female

    b.   Are the events "male" and "yes" independent? Are they mutually exclusive? Explain why or why not.

5

DISCRETE RANDOM VARIABLES AND THEIR PROBABILITY DISTRIBUTIONS

C hapter 4 discussed the concepts and rules of probability. This chapter extends the concept of probability to explain probability distributions. As was seen in Chapter 4, any given statistical experiment has more than one outcome. It is impossible to predict which of the many possible outcomes will occur if an experiment is performed. Consequently, decisions are made under uncertain conditions. For example, a lottery player does not know in advance whether or not he is going to win that lottery. If he knows that he is not going to win, he will definitely not play. It is the uncertainty about winning (some positive probability of winning) that makes him play. This chapter shows that if the outcomes and their probabilities for a statistical experiment are known, we can find out what will happen, on average, if that experiment is performed many times. For the lottery example, we can find out what a lottery player can expect to win (or lose), on average, if he continues playing this lottery again and again.

In this chapter, first, random variables and types of random variables are explained. Then, the concept of a probability distribution and its mean and standard deviation are discussed. Finally, two special probability distributions for a discrete random variable—the binomial probability distribution and the Poisson probability distribution—are developed.

5.1 RANDOM VARIABLES

Suppose Table 5.1 gives the frequency and relative frequency distributions of the number of vehicles owned by all 2000 families living in a small town.

Table 5.1 Frequency and Relative Frequency Distributions of the Number of Vehicles Owned by Families

Number of Vehicles Owned	Frequency	Relative Frequency
0	30	30/2000 = .015
1	470	470/2000 = .235
2	850	850/2000 = .425
3	490	490/2000 = .245
4	160	160/2000 = .080
	$N = 2000$	Sum = 1.000

Suppose one family is randomly selected from this population. The act of randomly selecting a family is called a *random* or *chance experiment*. Let x denote the number of vehicles owned by the selected family. Then x can assume any of the five possible values (0, 1, 2, 3, and 4) listed in the first column of Table 5.1. The value assumed by x depends on which family is selected. Thus, this value depends on the outcome of a random experiment. Consequently, x is called a **random variable** or a **chance variable**. In general, a random variable is denoted by x or y.

RANDOM VARIABLE

A *random variable* is a variable whose value is determined by the outcome of a random experiment.

As explained next, a random variable can be discrete or continuous.

5.1.1 DISCRETE RANDOM VARIABLE

A **discrete random variable** assumes values that can be counted. In other words, the consecutive values of a discrete random variable are separated by a certain gap.

> **DISCRETE RANDOM VARIABLE**
>
> A *random variable* that assumes countable values is called a *discrete random variable*.

In Table 5.1, *the number of vehicles owned by a family* is an example of a discrete random variable because the values of the random variable x are countable: 0, 1, 2, 3, and 4. Some other examples of discrete random variables are

1. The number of cars sold at a dealership during a given month
2. The number of people coming to a theater on a certain day
3. The number of shoe pairs a person owns
4. The number of complaints received at the office of an airline on a given day
5. The number of customers visiting a bank during any given hour
6. The number of heads obtained in three tosses of a coin

5.1.2 CONTINUOUS RANDOM VARIABLE

A random variable whose values are not countable is called a **continuous random variable**. A continuous random variable can assume any value over an interval or intervals.

> **CONTINUOUS RANDOM VARIABLE**
>
> A random variable that can assume any value contained in one or more intervals is called a *continuous random variable*.

Because the number of values contained in any interval is infinite, the possible number of values that a continuous random variable can assume is also infinite. Moreover, we cannot count these values. Consider the life of a battery. We can measure it as precisely as we want. For instance, the life of this battery may be 40 hours, or 40.25 hours, or 40.247 hours. Assume that the maximum life of such a battery is 200 hours. Let x denote the life of a randomly selected battery of this kind. Then, x can assume any value in the interval 0 to 200. Consequently, x is a continuous random variable. As shown below, every point on the line representing the interval 0 to 200 gives a possible value of x.

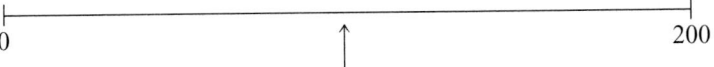

Every point on this line represents a possible value of x that denotes the life of a battery. There are an infinite number of points on this line. The values represented by points on this line are uncountable.

The following are a few examples of continuous random variables.

1. The height of a person
2. The time taken to complete an examination
3. The amount of milk in a gallon (note that we do not expect a gallon to contain exactly one gallon of milk but either slightly more or slightly less than a gallon)
4. The weight of a baby
5. The price of a house

This chapter is limited to the discussion of discrete random variables and their probability distributions. Continuous random variables will be discussed in Chapter 6.

EXERCISES

Concepts and Procedures

5.1   Explain the meaning of a random variable, a discrete random variable, and a continuous random variable. Give one example each of a discrete random variable and a continuous random variable.

5.2   Classify the following random variables as discrete or continuous.
   a.   The number of students in a class
   b.   The amount of soda in a 12-oz can
   c.   The number of cattle owned by a farmer
   d.   The age of a house
   e.   The number of pages in a book that contain at least one error
   f.   The time spent by a physician examining a patient

5.3   Indicate which of the following random variables are discrete and which are continuous.
   a.   The number of new accounts opened at a bank during a certain month
   b.   The time taken to run a marathon
   c.   The price of a concert ticket
   d.   The number of rotten eggs in a randomly selected box
   e.   The points scored in a football game
   f.   The weight of a randomly selected package

Applications

5.4   A household can watch news on any of the three networks—ABC, CBS, or NBC. On a certain day, five households randomly and independently decide which channel to watch. Let x be the number of households among these five who decide to watch news on ABC. Is x a discrete or a continuous random variable? Explain.

5.5   One of the four gas stations located at an intersection of two major roads is an Exxon station. Suppose the next six cars that stop at any of these four gas stations make the selections randomly and independently. Let x be the number of cars in these six that stop at the Exxon station. Is x a discrete or a continuous random variable? Explain.

5.2   PROBABILITY DISTRIBUTION OF A DISCRETE RANDOM VARIABLE

Let x be a discrete random variable. The **probability distribution** of x describes how the probabilities are distributed over all the possible values of x.

> **PROBABILITY DISTRIBUTION OF A DISCRETE RANDOM VARIABLE**
>
> The *probability distribution of a discrete random variable* lists all the possible values that the random variable can assume and their corresponding probabilities.

Example 5–1 illustrates the concept of the probability distribution of a discrete random variable.

Probability distribution of a discrete random variable.

EXAMPLE 5-1   Recall the frequency and relative frequency distributions of the number of vehicles owned by families given in Table 5.1. That table is reproduced below as Table 5.2. Let x be the number of vehicles owned by a randomly selected family. Write the probability distribution of x.

Table 5.2   Frequency and Relative Frequency Distributions of the Number of Vehicles Owned by Families

Number of Vehicles Owned	Frequency	Relative Frequency
0	30	.015
1	470	.235
2	850	.425
3	490	.245
4	160	.080
	$N = 2000$	Sum $= 1.000$

Solution   In Chapter 4 we learned that the relative frequencies obtained from an experiment or a sample can be used as approximate probabilities. However, when the relative frequencies are known for the population as in Table 5.2, they give the actual (theoretical) probabilities of outcomes. Using the relative frequencies of Table 5.2, we can write the *probability distribution* of the discrete random variable x in Table 5.3.

Table 5.3   Probability Distribution of the Number of Vehicles Owned by Families

Number of Vehicles Owned x	Probability $P(x)$
0	.015
1	.235
2	.425
3	.245
4	.080
	$\Sigma P(x) = 1.000$

The probability distribution of a discrete random variable possesses the following *two characteristics*.

1. The probability assigned to each value of a random variable x lies in the range 0 to 1, that is, $0 \leq P(x) \leq 1$ for each x.

2. The sum of the probabilities assigned to all possible values of x is equal to 1.0, that is,

$\Sigma P(x) = 1$. (Remember, if the probabilities are rounded, the sum may not be exactly 1.0.)

TWO CHARACTERISTICS OF A PROBABILITY DISTRIBUTION

The probability distribution of a discrete random variable possesses the following two characteristics.

1. $0 \le P(x) \le 1$ for each value of x
2. $\Sigma P(x) = 1$

These two characteristics are also called the *two conditions* that a probability distribution must satisfy. Notice that in Table 5.3, each probability listed in the column labeled $P(x)$ is between 0 and 1. Also, $\Sigma P(x) = 1.0$. Because both conditions are satisfied, Table 5.3 represents the probability distribution of x.

From Table 5.3, the probability for any value of x can be read. For example, the probability that a randomly selected family from this town owns two vehicles is .425. This probability is written as

$$P(x = 2) = .425$$

The probability that the selected family owns more than two vehicles is given by the sum of the probabilities of three and four vehicles, respectively. This probability is .245 + .080 = .325, which can be written as

$$P(x > 2) = P(x = 3) + P(x = 4) = .245 + .080 = .325$$

The probability distribution of a discrete random variable can be presented in the form of a *mathematical formula, a table, or a graph*. Table 5.3 presented the probability distribution in tabular form. Figure 5.1 shows the graphical presentation of the probability distribution of Table 5.3. In this figure, each value of x is marked on the horizontal axis. The probability for each value of x is exhibited by the height of the corresponding line. Such a graph is called the **line graph**. This section does not discuss the presentation of a probability distribution using a mathematical formula.

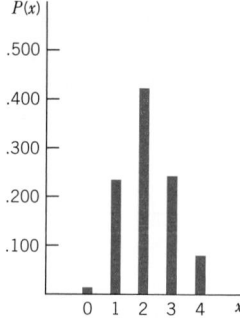

Figure 5.1 Graphical presentation of the probability distribution of Table 5.3.

Verifying conditions of a probability distribution.

EXAMPLE 5–2 Each of the following tables lists certain values of x and their probabilities. Determine whether or not each table represents a valid probability distribution.

(a)	x	$P(x)$
	0	.08
	1	.11
	2	.39
	3	.27

(b)	x	$P(x)$
	2	.25
	3	.34
	4	.28
	5	.13

(c)	x	$P(x)$
	7	.70
	8	.50
	9	−.20

Solution

(a) Because each probability listed in this table is in the range 0 to 1, it satisfies the first condition of a probability distribution. However, the sum of all probabilities is not equal to 1.0 because $\Sigma P(x) = .08 + .11 + .39 + .27 = .85$. Therefore, the second condition is not satisfied. Consequently, this table does not represent a valid probability distribution.

(b) Each probability listed in this table is in the range 0 to 1. Also, $\Sigma P(x) = .25 + .34 + .28 + .13 = 1.0$. Consequently, this table represents a valid probability distribution.

(c) Although the sum of all probabilities listed in this table is equal to 1.0, one of the probabilities is negative. This violates the first condition of a probability distribution. Therefore, this table does not represent a valid probability distribution. ▬

EXAMPLE 5–3 The following table lists the probability distribution of the number of breakdowns per week for a machine based on past data.

Breakdowns per week	0	1	2	3
Probability	.15	.20	.35	.30

(a) Present this probability distribution graphically.

(b) Find the probability that the number of breakdowns for this machine during a given week is

    (i) exactly 2   (ii) 0 to 2

    (iii) more than 1   (iv) at most 1

Solution Let x denote the number of breakdowns for this machine during a given week. Table 5.4 lists the probability distribution of x.

Table 5.4 Probability Distribution of Breakdowns

x	$P(x)$
0	.15
1	.20
2	.35
3	.30
	$\Sigma P(x) = 1.00$

Graph of a probability distribution.

(a) Figure 5.2 shows the line graph of the probability distribution of Table 5.4.

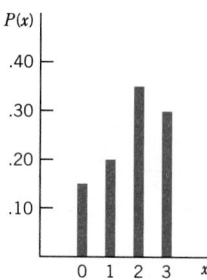

Figure 5.2 Graphical presentation of the probability distribution of Table 5.4.

Probabilities of events for a discrete random variable.

(b) Using Table 5.4, we can calculate the required probabilities as follows.

    (i)   The probability of exactly two breakdowns is

$$P(\text{exactly 2 breakdowns}) = P(x = 2) = \textbf{.35}$$

   (ii)   The probability of zero to two breakdowns is given by the sum of the probabilities of 0, 1, and 2 breakdowns.

$$P(\text{0 to 2 breakdowns}) = P(0 \leq x \leq 2)$$
$$= P(x = 0) + P(x = 1) + P(x = 2)$$
$$= .15 + .20 + .35 = \textbf{.70}$$

  (iii)   The probability of more than one breakdown is obtained by adding the probabilities of 2 and 3 breakdowns.

$$P(\text{more than 1 breakdown}) = P(x > 1)$$
$$= P(x = 2) + P(x = 3)$$
$$= .35 + .30 = \textbf{.65}$$

  (iv)   The probability of at most one breakdown is given by the sum of the probabilities of 0 and 1 breakdown.

$$P(\text{at most 1 breakdown}) = P(x \leq 1)$$
$$= P(x = 0) + P(x = 1)$$
$$= .15 + .20 = \textbf{.35}$$ ▬

Constructing a probability distribution.

EXAMPLE 5–4 According to a survey, 60% of all students at a large university suffer from math anxiety. Two students are randomly selected from this university. Let x denote the number of students in this sample who suffer from math anxiety. Develop the probability distribution of x.

Solution Let us define the following two events.

$$N = \text{the student selected does not suffer from math anxiety}$$
$$M = \text{the student selected suffers from math anxiety}$$

As we can observe from the tree diagram of Figure 5.3, there are four possible outcomes for this experiment: NN (neither of the students suffers from math anxiety), NM (the first student does not suffer from math anxiety and the second does), MN (the first student suffers from math anxiety and the second does not), and MM (both students suffer from math anxiety).

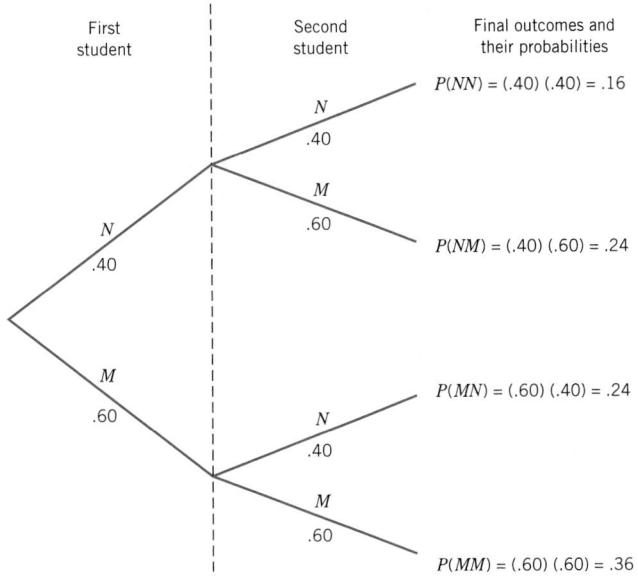

Figure 5.3 Tree diagram.

The probabilities of these four outcomes are listed in the tree diagram. Because 60% of the students suffer from math anxiety and 40% do not, the probability is .60 that any student selected suffers from math anxiety and .40 that he or she does not.

In a sample of two students, the number who suffer from math anxiety can be 0 (*NN*), 1 (*NM* or *MN*), or 2 (*MM*). Thus, *x* can assume any of three possible values: 0, 1, or 2. The probabilities of these three outcomes are calculated as follows.

$$P(x = 0) = P(NN) = .16$$

$$P(x = 1) = P(NM \text{ or } MN) = P(NM) + P(MN) = .24 + .24 = .48$$

$$P(x = 2) = P(MM) = .36$$

Using these probabilities, we can write the probability distribution of *x* as in Table 5.5.

Table 5.5 Probability Distribution of the Number of Students with Math Anxiety in a Sample of Two Students

x	*P(x)*
0	.16
1	.48
2	.36
	$\Sigma P(x) = 1.00$

EXERCISES

Concepts and Procedures

5.6 Explain the meaning of the probability distribution of a discrete random variable. Give one example of such a probability distribution. What are the three ways to present the probability distribution of a discrete random variable?

5.7 Briefly explain the two characteristics (conditions) of the probability distribution of a discrete random variable.

5.8 Each of the following tables lists certain values of *x* and their probabilities. Verify if each of them represents a valid probability distribution.

a.	*x*	*P(x)*		b.	*x*	*P(x)*		c.	*x*	*P(x)*
	0	.10			2	.35			7	− .25
	1	.05			3	.28			8	.85
	2	.45			4	.23			9	.40
	3	.32			5	.14				

5.9 Each of the following tables lists certain values of *x* and their probabilities. Determine if each one satisfies the two conditions required for a valid probability distribution.

a.	*x*	*P(x)*		b.	*x*	*P(x)*		c.	*x*	*P(x)*
	5	− .36			1	.16			0	.15
	6	.48			2	.24			1	.00
	7	.62			3	.49			2	.35
	8	.26							3	.50

5.10 The following table gives the probability distribution of a discrete random variable *x*.

x	0	1	2	3	4	5	6
P(x)	.11	.19	.28	.15	.12	.09	.06

Find the following probabilities.

  a. $P(x = 3)$ b. $P(x \le 2)$ c. $P(x \ge 4)$ d. $P(1 \le x \le 4)$
  e. Probability that *x* assumes a value less than 4
  f. Probability that *x* assumes a value greater than 2
  g. Probability that *x* assumes a value in the interval 2 to 5

5.11 The following table gives the probability distribution of a discrete random variable *x*.

x	0	1	2	3	4	5
P(x)	.03	.13	.22	.31	.19	.12

Find the following probabilities.

  a. $P(x = 1)$ b. $P(x \le 1)$ c. $P(x \ge 3)$ d. $P(0 \le x \le 2)$
  e. Probability that *x* assumes a value less than 3
  f. Probability that *x* assumes a value greater than 3
  g. Probability that *x* assumes a value in the interval 2 to 4

Applications

5.12 Elmo's Sporting Goods sells exercise machines as well as other sporting goods. On different days, it sells different numbers of these machines. The following table, constructed using past data, lists the probability distribution of the number of exercise machines sold per day at Elmo's.

Machines sold per day	4	5	6	7	8	9	10
Probability	.08	.11	.14	.19	.23	.16	.09

a. Graph the probability distribution.
b. Determine the probability that the number of exercise machines sold at Elmo's on a given day is
    i. exactly 6   ii. more than 8   iii. 5 to 8   iv. at most 6

 5.13 Let x denote the number of auto accidents that occur in a city during a week. The following table lists the probability distribution of x.

x	0	1	2	3	4	5	6
$P(x)$	.12	.16	.22	.18	.14	.12	.06

a. Draw a line graph for this probability distribution.
b. Determine the probability that the number of auto accidents that will occur during a given week in this city is
    i. exactly 4   ii. at least 3   iii. less than 3   iv. 3 to 5

5.14 A consumer agency surveyed all 2500 families living in a small town to collect data on the number of television sets owned by them. The following table lists the frequency distribution of the data collected by this agency.

Number of TV sets owned	0	1	2	3	4
Number of families	120	970	730	410	270

a. Construct a probability distribution table for the number of television sets owned by these families. Draw a line graph of the probability distribution.
b. Are the probabilities listed in the table of part a exact or approximate probabilities of various outcomes? Explain.
c. Let x denote the number of television sets owned by a randomly selected family from this town. Find the following probabilities.
    i. $P(x = 1)$ ii. $P(x > 2)$
    iii. $P(x \le 1)$ iv. $P(1 \le x \le 3)$

5.15 The Webster Mail Order Company sells expensive stereos by mail. The following table lists the frequency distribution of the number of orders received per day by this company during the past 100 days.

Number of orders received per day	2	3	4	5	6
Number of days	12	21	34	19	14

a. Construct a probability distribution table for the number of orders received per day. Draw a graph of the probability distribution.
b. Are the probabilities listed in the table of part a exact or approximate probabilities of various outcomes? Explain.
c. Let x denote the number of orders received on any given day. Find the following probabilities.
    i. $P(x = 3)$ ii. $P(x \ge 3)$
    iii. $P(2 \le x \le 4)$ iv. $P(x < 4)$

5.16 Five percent of all cars manufactured at a large auto company are lemons. Suppose two cars are selected at random from the production line of this company. Let x denote the number of lemons in this sample. Write the probability distribution of x. Draw a tree diagram for this problem.

5.17 According to a survey by the National Center for Education Information, 74% of U.S. teachers in 1996 were women. Assume that this result holds true for the current population of all U.S. teachers. Suppose two persons are randomly selected from the population of all U.S. teachers. Let x denote the number of women in this sample. Construct the probability distribution table of x. Draw a tree diagram for this problem.

5.18 According to a survey, 30% of adults are against using animals for research. Assume that this result holds true for the current population of all adults. Let x be the number of adults who are against using animals for research in a random sample of two adults. Obtain the probability distribution of x. Draw a tree diagram for this problem.

5.19 According to the Third International Mathematics and Science Study, 38% of U.S. eighth-graders spend 3 or more hours watching television or videos after school each day (*USA Today*, November 21, 1996). Assume that this result holds true for the current population of all U.S. eighth-graders. Suppose two students are randomly selected from the population of all U.S. eighth-graders and x denotes the number of eighth-graders in this sample who spend 3 or more hours watching television or videos after school each day. Construct the probability distribution table of x. Draw a tree diagram for this problem.

*5.20 A box contains 10 parts, 3 of which are known to be defective. Suppose 2 parts are randomly selected from this box. Let x denote the number of good parts in this sample. Write the probability distribution of x. You may draw a tree diagram and use it to write the probability distribution. (*Hint*: Note that the draws are made without replacement from a small population. Hence, the probabilities of outcomes do not remain constant for each draw. Also, refer to Example 4–24.)

*5.21 A statistics class has 20 students; 12 of them are female. Suppose 2 students are randomly selected from this class. Let x denote the number of females in this sample. Find the probability distribution of x. You may draw a tree diagram and use that to write the probability distribution. (*Hint*: Note that the draws are made without replacement from a small population. Hence, the probabilities of outcomes do not remain constant for each draw. Also, refer to Example 4–24.)

5.3 MEAN OF A DISCRETE RANDOM VARIABLE

The **mean of a discrete random variable**, denoted by μ, is actually the mean of its probability distribution. The mean of a discrete random variable x is also called its *expected value* and is denoted by $E(x)$. The mean (or expected value) of a discrete random variable is the value that we expect to observe per repetition, on average, if we perform an experiment a large number of times. For example, we may expect a car salesperson to sell, on average, 2.4 cars per week. This does not mean that every week this salesperson will sell exactly 2.4 cars. (Obviously she cannot sell exactly 2.4 cars.) This simply means that if we observe for many weeks, this salesperson will sell a different number of cars during different weeks. However, the average for all these weeks will be 2.4 cars.

To calculate the mean of a discrete random variable x, we multiply each value of x by the corresponding probability and sum the resulting products. This sum gives the mean (or expected value) of the discrete random variable x.

MEAN OF A DISCRETE RANDOM VARIABLE

The *mean of a discrete random variable* x is the value that is expected to occur per repetition, on average, if an experiment is repeated a large number of times. It is denoted by μ and calculated as

$$\mu = \Sigma x P(x)$$

The mean of a discrete random variable x is also called its expected value and denoted by $E(x)$, that is,

$$E(x) = \Sigma x P(x)$$

Example 5–5 illustrates the calculation of the mean of a discrete random variable.

Calculating and interpreting the mean of a discrete random variable.

EXAMPLE 5–5   Recall Example 5–3 of Section 5.2. The probability distribution Table 5.4 from that example is reproduced below. In this table, x represents the number of breakdowns for a machine during a given week, and $P(x)$ is the probability of the corresponding value of x.

x	$P(x)$
0	.15
1	.20
2	.35
3	.30

Find the mean number of breakdowns per week for this machine.

Solution   To find the mean number of breakdowns per week for this machine, we multiply each value of x by its probability and add these products. This sum gives the mean of the probability distribution of x. The products $xP(x)$ are listed in the third column of Table 5.6. The sum of these products gives $\Sigma xP(x)$, which is the mean of x.

Table 5.6   Calculating the Mean for the Probability Distribution of Breakdowns

x	$P(x)$	$xP(x)$
0	.15	$0(.15) = .00$
1	.20	$1(.20) = .20$
2	.35	$2(.35) = .70$
3	.30	$3(.30) = .90$
		$\Sigma xP(x) = 1.80$

The mean is

$$\mu = \Sigma xP(x) = \mathbf{1.80}$$

Thus, on average, this machine is expected to break down 1.80 times per week over a period of time. In other words, if this machine is used for many weeks, then for certain weeks we will observe no breakdowns; for some other weeks we will observe one breakdown per week; and for still other weeks we will observe two or three breakdowns per week. The mean number of breakdowns is expected to be 1.80 per week for the entire period.

Note that $\mu = 1.80$ is also the expected value of x. It can also be written as

$$E(x) = 1.80$$

Case Study 5–1 illustrates the calculation of the mean amount that an instant lottery player is expected to win.

CASE STUDY 5–1 BANK ROLL INSTANT LOTTERY

EXAMPLE: Ticket as Printed

EXAMPLE: Scratched-off Ticket

Currently the state of Connecticut has in circulation an instant lottery game called *Bank Roll*. The cost of each ticket for this lottery is $1. A player can instantly win $2500, $100, $50, $18, $9, $6, $3, $2, or $1. Each ticket has nine erasable spots, each of which contains one of the many symbols. A player can win a prize by getting three matching symbols in three of these nine spots. The prize won by the player is revealed by scratching off the *Prize* spot.

The following table lists the number of tickets with different prizes in a total of 4,080,000 tickets printed in the first batch for this lottery. Note that the lottery commission could print more tickets but the proportion of tickets with these prizes remains the same in each batch. As is obvious from this table, out of a total of 4,080,000 tickets, 3,099,355 are nonwinning tickets (the ones with a prize of $0 in this table). Of the remaining tickets, 432,650 have a prize of $1 each, 273,700 have a prize of $2 each, and so on.

Prize (in dollars)	Number of Tickets
0	3,099,355
1	432,650
2	273,700
3	163,030
6	58,990
9	32,640
18	16,320
50	2618
100	680
2500	17
	Total = 4,080,000

The net gain to a player for each of the instant winning tickets is equal to the amount of the prize minus $1, which is the cost of the ticket. Thus, the net gain for each of the non-winning tickets is $-\$1$, which is the cost of the ticket. Let

$$x = \text{the net amount a player wins by playing this lottery}$$

The following table shows the probability distribution of x and all the calculations required to compute the mean of x for that probability distribution. The probability of an outcome (net winnings) is calculated by dividing the number of tickets with that outcome by the total number of tickets.

x (in dollars)	$P(x)$	$xP(x)$
-1	3,099,355/4,080,000 = .75964583	$-.759646$
0	432,650/4,080,000 = .10604167	.000000
1	273,700/4,080,000 = .06708333	.067083
2	163,030/4,080,000 = .03995833	.079917
5	58,990/4,080,000 = .01445833	.072292
8	32,640/4,080,000 = .00800000	.064000
17	16,320/4,080,000 = .00400000	.068000
49	2618/4,080,000 = .00064167	.031442
99	680/4,080,000 = .00016667	.016500
2499	17/4,080,000 = .000004167	.010413
		$\Sigma xP(x) = -.349999$

Hence, the mean of x is

$$\mu = \Sigma xP(x) = -\$.349999 \approx -\$.35$$

This mean gives the expected value of the random variable x, that is,

$$E(x) = \Sigma xP(x) = -\$.35$$

Thus, the mean of net winnings for this lottery is $-\$.35$. In other words, all players taken together will lose an average of $.35 (or 35 cents) per ticket. This can also be interpreted as follows: Only 65% ($= 100 - 35$) of the total money spent by all players on buying lottery tickets for this lottery will be returned to them in the form of prizes and 35% will not be returned. (The money that will not be returned to players will cover the costs of operating the lottery, the commission paid to agents, and revenue to the state of Connecticut.)

Source: The State of Connecticut Lottery Commission publications. Lottery ticket reproduced with permission.

5.4   STANDARD DEVIATION OF A DISCRETE RANDOM VARIABLE

The **standard deviation of a discrete random variable**, denoted by σ, measures the spread of its probability distribution. A higher value for the standard deviation of a discrete random variable indicates that x can assume values over a larger range about the mean. On the other hand, a smaller value for the standard deviation indicates that most of the values that x can assume are clustered closely about the mean. The basic formula to compute the standard deviation of a discrete random variable is

$$\sigma = \sqrt{\Sigma[(x - \mu)^2 \cdot P(x)]}$$

However, it is more convenient to use the following short-cut formula to compute the standard deviation of a discrete random variable.

STANDARD DEVIATION OF A DISCRETE RANDOM VARIABLE

The *standard deviation of a discrete random variable x* measures the spread of its probability distribution and is computed as

$$\sigma = \sqrt{\Sigma x^2 P(x) - \mu^2}$$

Note that the variance σ^2 of a discrete random variable is obtained by squaring its standard deviation.

Example 5–6 illustrates how to use the short-cut formula to compute the standard deviation of a discrete random variable.

Calculating standard deviation of a discrete random variable.

EXAMPLE 5–6   Baier's Electronics manufactures computer parts that are supplied to many computer companies. Despite the fact that two quality control inspectors at Baier's Electronics check every part for defects before it is shipped to another company, a few defective parts do pass through these inspections undetected. Let x denote the number of defective computer parts in a shipment of 400. The following table gives the probability distribution of x.

x	0	1	2	3	4	5
$P(x)$	.02	.20	.30	.30	.10	.08

Compute the standard deviation of x.

Solution   Table 5.7 shows all the calculations required for the computation of the standard deviation of x.

Table 5.7   Computations to Find the Standard Deviation

x	$P(x)$	$xP(x)$	x^2	$x^2P(x)$
0	.02	.00	0	.00
1	.20	.20	1	.20
2	.30	.60	4	1.20
3	.30	.90	9	2.70
4	.10	.40	16	1.60
5	.08	.40	25	2.00
		$\Sigma xP(x) = 2.50$		$\Sigma x^2P(x) = 7.70$

We perform the following steps to compute the standard deviation of x.

Step 1. Compute the mean of the discrete random variable.

The sum of the products $xP(x)$, recorded in the third column of Table 5.7, gives the mean of x.

$$\mu = \Sigma xP(x) = 2.50 \text{ defective computer parts in } 400$$

Step 2. Compute the value of $\Sigma x^2 P(x)$.

First we square each value of x and record it in the fourth column of Table 5.7. Then we multiply these values of x^2 by the corresponding values of $P(x)$. The resulting values of $x^2 P(x)$ are recorded in the fifth column of Table 5.7. The sum of this column gives

$$\Sigma x^2 P(x) = 7.70$$

Step 3. Substitute the values of μ and $\Sigma x^2 P(x)$ in the formula for the standard deviation of x and simplify.

By performing this step, we obtain:

$$\sigma = \sqrt{\Sigma x^2 P(x) - \mu^2} = \sqrt{7.70 - (2.50)^2} = \sqrt{1.45}$$

$$= \mathbf{1.20} \text{ defective computer parts}$$

Thus, a given shipment of 400 computer parts is expected to contain an average of 2.50 defective parts with a standard deviation of 1.20. ■

☞   *Remember*   Because the standard deviation of a discrete random variable is obtained by taking the positive square root, its value is never negative.

EXAMPLE 5-7   Loraine Corporation is planning to market a new makeup product. According to the analysis made by the financial department of the company, it will earn an annual profit of $4.5 million if this product has high sales, an annual profit of $1.2 million if the sales are mediocre, and it will lose $2.3 million a year if the sales are low. The probabilities of these three scenarios are .32, .51, and .17, respectively.

(a)   Let x be the profits (in millions of dollars) earned per annum by the company from this product. Write the probability distribution of x.

(b)   Calculate the mean and standard deviation of x.

Solution

Probability distribution of a discrete random variable.

(a)   The following table lists the probability distribution of x. Note that since x denotes profits earned by the company, the loss is written as *negative profits* in the table.

x	$P(x)$
4.5	.32
1.2	.51
−2.3	.17

Calculating mean and standard deviation of a discrete random variable.

(b)   Table 5.8 shows all the calculations needed for the computation of the mean and standard deviation of x.

The mean of x is

$$\mu = \Sigma xP(x) = \mathbf{\$1.661 \text{ million}}$$

Table 5.8   Computations to Find the Mean and Standard Deviation

x	$P(x)$	$xP(x)$	x^2	$x^2P(x)$
4.5	.32	1.440	20.25	6.4800
1.2	.51	.612	1.44	.7344
−2.3	.17	−.391	5.29	.8993
		$\Sigma xP(x) = 1.661$		$\Sigma x^2P(x) = 8.1137$

The standard deviation of x is

$$\sigma = \sqrt{\Sigma x^2 P(x) - \mu^2} = \sqrt{8.1137 - (1.661)^2} = \mathbf{\$2.314 \ million}$$

Thus, it is expected that Loraine Corporation will earn an average of $1.661 million in profit per year from the new makeup product with a standard deviation of $2.314 million.   ▬

☞   **INTERPRETATION OF THE STANDARD DEVIATION**

The standard deviation of a discrete random variable can be interpreted or used the same way as the standard deviation of a data set in Section 3.4 of Chapter 3. In that section, we learned that according to Chebyshev's theorem, at least $[1 - (1/k^2)] \times 100\%$ of the total area under a curve lies within k standard deviations of the mean where k is any number greater than 1. Thus, if $k = 2$, then 75% of the area under a curve lies between $\mu - 2\sigma$ and $\mu + 2\sigma$. In Example 5–6,

$$\mu = 2.50 \quad \text{and} \quad \sigma = 1.20$$

Hence,

$$\mu - 2\sigma = 2.50 - 2 \ (1.20) = .10$$

and

$$\mu + 2\sigma = 2.50 + 2 \ (1.20) = 4.90$$

Using Chebyshev's theorem, we can state that at least 75% of the shipments (each containing 400 computer parts) are expected to contain .10 to 4.90 defective computer parts each.

EXERCISES

Concepts and Procedures

5.22   Briefly explain the concept of the mean and standard deviation of a discrete random variable.

5.23   Find the mean and standard deviation for each of the following probability distributions.

a.	x	$P(x)$	b.	x	$P(x)$
	0	.12		6	.36
	1	.27		7	.26
	2	.43		8	.21
	3	.18		9	.17

5.24   Find the mean and standard deviation for each of the following probability distributions.

a.

x	P(x)
3	.09
4	.21
5	.34
6	.23
7	.13

b.

x	P(x)
0	.43
1	.31
2	.17
3	.09

Applications

5.25 Let x be the number of errors contained on a randomly selected page of a book. The following table lists the probability distribution of x.

x	0	1	2	3	4
P(x)	.73	.16	.06	.04	.01

Find the mean and standard deviation of x.

5.26 Let x be the number of newborn pigs per year on a pig farm. The following table lists the probability distribution of x.

x	4	5	6	7	8	9	10
P(x)	.17	.22	.19	.13	.11	.10	.08

Find the mean and standard deviation of x.

 5.27 The following table gives the probability distribution of camcorders sold on a given day at an electronics store.

Camcorders sold	0	1	2	3	4	5	6
Probability	.05	.12	.23	.30	.16	.10	.04

Calculate the mean and standard deviation for this probability distribution. Give a brief interpretation of the value of the mean.

 5.28 The following table, reproduced from Exercise 5.12, lists the probability distribution of the number of exercise machines sold per day at Elmo's Sporting Goods store.

Machines sold per day	4	5	6	7	8	9	10
Probability	.08	.11	.14	.19	.23	.16	.09

Calculate the mean and standard deviation for this probability distribution. Give a brief interpretation of the value of the mean.

5.29 Let x be the number of heads obtained in two tosses of a coin. The following table lists the probability distribution of x.

x	0	1	2
P(x)	.25	.50	.25

Calculate the mean and standard deviation of x. Give a brief interpretation of the value of the mean.

5.30 Let x be a random variable that represents the number of students who are absent on a given day from a class of 25. The following table lists the probability distribution of x.

x	0	1	2	3	4	5
$P(x)$	.08	.18	.32	.22	.14	.06

Calculate the mean and standard deviation for this probability distribution and give a brief interpretation of the value of the mean.

5.31 Refer to Exercise 5.14. Find the mean and standard deviation for the probability distribution you developed for the number of television sets owned by all 2500 families living in a town. Give a brief interpretation of the values of the mean and standard deviation.

5.32 Refer to Exercise 5.15. Find the mean and standard deviation for the probability distribution you developed for the number of orders received per day for the past 100 days at the Webster Mail Order Company. Give a brief interpretation of the values of the mean and standard deviation.

5.33 Refer to the probability distribution developed in Exercise 5.16 for the number of lemons in two selected cars. Calculate the mean and standard deviation of x for that probability distribution.

5.34 Refer to the probability distribution developed in Exercise 5.17 for the number of women in a sample of two teachers. Compute the mean and standard deviation of x for that probability distribution.

5.35 A farmer will earn a profit of $30 thousand next year in case of heavy rain, $60 thousand in case of moderate rain, and $15 thousand in case of little rain. A meteorologist forecasts that the probability is .35 for heavy rain, .40 for moderate rain, and .25 for little rain next year. Let x be the random variable that represents next year's profits in thousands of dollars for this farmer. Write the probability distribution of x. Find the mean and standard deviation of x. Give a brief interpretation of the values of the mean and standard deviation.

5.36 An instant lottery ticket costs $2. Out of a total of 10,000 tickets printed for this lottery, 1000 tickets contain a prize of $5 each, 100 tickets have a prize of $10 each, 5 tickets have a prize of $1000 each, and 1 ticket has a prize of $5000. Let x be the random variable that denotes the net amount a player wins by playing this lottery. Write the probability distribution of x. Determine the mean and standard deviation of x. How will you interpret the values of the mean and standard deviation of x?

***5.37** Refer to the probability distribution developed in Exercise 5.20 for the number of good parts in a random sample of two parts selected from a box. Calculate the mean and standard deviation of x for that distribution.

***5.38** Refer to the probability distribution developed in Exercise 5.21 for the number of females in a random sample of two students selected from a class. Calculate the mean and standard deviation of x for that distribution.

5.5 FACTORIALS AND COMBINATIONS

This section introduces factorials and combinations, which will be used in the binomial formula discussed in Section 5.6.

5.5.1 FACTORIALS

The symbol ''!'' (read as *factorial*) is used to denote **factorials**. The value of the factorial of a number is obtained by multiplying all integers from that number to 1. For example, ''7!'' is read as *seven factorial* and is evaluated by multiplying all integers from 7 to 1.

> **FACTORIALS**
>
> The symbol $n!$, read as "n factorial," represents the product of all integers from n to 1. In other words,
>
> $$n! = n\,(n-1)\,(n-2)\,(n-3)\ldots 3 \cdot 2 \cdot 1$$
>
> By definition,
>
> $$0! = 1$$

Evaluating factorial.

EXAMPLE 5-8 Evaluate 7!.

Solution To evaluate 7!, we multiply all integers from 7 to 1.

$$7! = 7 \cdot 6 \cdot 5 \cdot 4 \cdot 3 \cdot 2 \cdot 1 = \mathbf{5040}$$

Thus, the value of 7! is 5040.[1]

Evaluating factorial.

EXAMPLE 5-9 Evaluate 10!.

Solution The value of 10! is given by the product of all integers from 10 to 1. Thus,

$$10! = 10 \cdot 9 \cdot 8 \cdot 7 \cdot 6 \cdot 5 \cdot 4 \cdot 3 \cdot 2 \cdot 1 = \mathbf{3{,}628{,}800}$$

Factorial of difference between two numbers.

EXAMPLE 5-10 Evaluate $(12 - 4)!$.

Solution The value of $(12 - 4)!$ is

$$(12 - 4)! = 8! = 8 \cdot 7 \cdot 6 \cdot 5 \cdot 4 \cdot 3 \cdot 2 \cdot 1 = \mathbf{40{,}320}$$

Factorial of zero.

EXAMPLE 5-11 Evaluate $(5 - 5)!$.

Solution As shown below, the value of $(5 - 5)!$ is 1.

$$(5 - 5)! = 0! = \mathbf{1}$$

Note that 0! is always equal to 1.

We can read the value of $n!$ for $n = 1$ through $n = 25$ from Table II of Appendix C. Example 5–12 illustrates how to read that table.

Using the table of factorials.

EXAMPLE 5-12 Find the value of 15! by using Table II of Appendix C.

Solution To find the value of 15! from Table II, we locate 15 in the column labeled n. Then we read the value in the column for $n!$ entered next to 15. Thus,

$$15! = \mathbf{1{,}307{,}674{,}368{,}000}$$

---

[1]**Using a calculator to evaluate $n!$:** Most calculators have an $n!$ or $x!$ function key. If your calculator has such a key, you can use it to evaluate 7! as follows: First enter 7 and then press the $n!$ or $x!$ key. The calculator screen will display 5040 as the answer.

5.5.2   COMBINATIONS

Quite often we face the problem of selecting a few elements from a large number of distinct elements. For example, a student may be required to attempt any two questions out of four in an examination. As another example, the faculty in a department may need to select 3 professors from 20 to form a committee. Or a lottery player may have to pick 6 numbers from 49. The question arises: In how many ways can we make the selections in each of these examples? For instance, how many possible selections exist for the student who is to choose any two questions out of four? The answer is six. Let the four questions be denoted by the numbers 1, 2, 3, and 4. Then the six selections are

$$(1 \text{ and } 2) \quad (1 \text{ and } 3) \quad (1 \text{ and } 4) \quad (2 \text{ and } 3) \quad (2 \text{ and } 4) \quad (3 \text{ and } 4)$$

The student can choose questions 1 and 2, or 1 and 3, or 1 and 4, and so on.

Each of the possible selections in this list is called a **combination**. All six combinations are distinct; that is, each combination contains a different set of questions. It is important to remember that the order in which the selections are made is not significant in the case of combinations. Thus, whether we write (1 and 2) or (2 and 1), both these arrangements represent only one combination.

COMBINATIONS NOTATION

Combinations give the number of ways x elements can be selected from n elements. The notation used to denote the total number of combinations is

$$\binom{n}{x}$$

which is read as "the number of combinations of n elements selected x at a time."

Suppose there are a total of n elements from which we want to select x elements. Then,

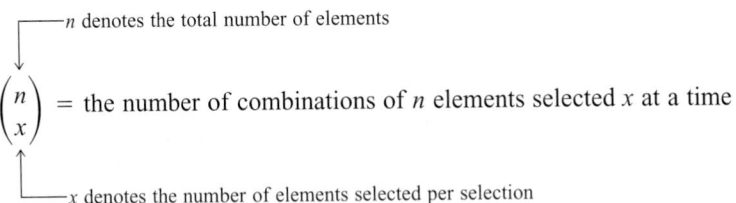

$\binom{n}{x}$ = the number of combinations of n elements selected x at a time

---

NUMBER OF COMBINATIONS

The *number of combinations* for selecting x from n distinct elements is given by the formula

$$\binom{n}{x} = \frac{n!}{x! \, (n - x)!}$$

where $n!$, $x!$, and $(n - x)!$ are read as "n factorial," "x factorial," and "n minus x factorial," respectively.

In the combinations formula,

$$n! = n(n-1)(n-2)(n-3)\ldots 3 \cdot 2 \cdot 1$$

$$x! = x(x-1)(x-2)\ldots 3 \cdot 2 \cdot 1$$

and

$$(n-x)! = (n-x)(n-x-1)(n-x-2)\ldots 3 \cdot 2 \cdot 1$$

Note that in combinations, n is always greater than or equal to x. If n is smaller than x, then we cannot select x distinct elements from n.

Finding the number of combinations by using formula.

EXAMPLE 5-13 Reconsider the example of a student who is to select two questions from four. Using the combinations formula, find the number of ways this student can select two questions from four.

Solution   For this example,

$$n = \text{total number of questions} = 4$$

$$x = \text{questions to be selected} = 2$$

Therefore, the number of ways this student can select two questions from four is

$$\binom{4}{2} = \frac{4!}{2!\,(4-2)!} = \frac{4!}{2!\,2!} = \frac{4 \cdot 3 \cdot 2 \cdot 1}{2 \cdot 1 \cdot 2 \cdot 1} = 6$$

We listed these six combinations earlier in this section.[2]

Finding the number of combinations and listing them.

EXAMPLE 5-14 Three members of a jury will be randomly selected from five persons. How many different combinations are possible?

Solution   There are a total of five persons, and we are to select three of them. Hence,

$$n = 5 \quad \text{and} \quad x = 3$$

Applying the combinations formula,

$$\binom{5}{3} = \frac{5!}{3!\,(5-3)!} = \frac{5!}{3!\,2!} = \frac{120}{6 \cdot 2} = 10$$

If we assume that the five persons are A, B, C, D, and E, then the 10 possible combinations for the selection of three members of the jury are

ABC, ABD, ABE, ACD, ACE, ADE, BCD, BCE, BDE, CDE

Case Study 5-2 describes the number of ways a lottery player can select 6 numbers from 49 in a lotto game.

---

[2]**Using a calculator to evaluate** $\binom{n}{x}$: Most calculators have a Cn,r or Cn,x or $\binom{n}{r}$ or $\binom{n}{x}$ function key. (Note that all these notations are used to denote combinations.) If your calculator has such a key, read the manual that accompanies the calculator to find out how this key functions and then evaluate $\binom{4}{2}$ using that key.

CASE STUDY 5-2   PLAYING LOTTO

During the past few years, many states have initiated the popular lottery game called lotto. To play lotto, a player picks any 6 numbers from a list of numbers usually starting with 1, for example from 1 through 49. At the end of the lottery period, the state lottery commission randomly selects 6 numbers from the same list. If all 6 numbers picked by a player are the same as the ones randomly selected by the lottery commission, the player wins.

USA SNAPSHOTS®

A look at statistics that shape the nation

Playing to win
Number of $1 tickets someone would have to buy to cover every 6-number combination in selected lotto games. Tickets in millions:

Calif. (51 numbers) 18.0
Fla., Mass. (49) 13.9
Ill., N.Y., Lotto America (54)[1] 12.9
Mich., Ohio (47) 10.7
N.J. (46) 9.4
Va.,Conn.,La. (44) 7.0

1-Players get two plays for $1

Source: USA TODAY research   By Ron Coddington, USA TODAY

The above chart shows the number of combinations (in millions) for picking 6 numbers for lotto games played in a few states. For example, in California a player has to pick 6 numbers from 1 through 51. As shown in the chart, there are approximately 18 million ways (combinations) to select 6 numbers from 1 through 51. In Florida and Massachusetts, a player has to pick 6 numbers from 1 through 49. For this lotto, there are approximately 13.9 million combinations.

Let us find the probability that a player who picks 6 numbers from 49 wins this game. The total combinations of selecting 6 numbers from 49 numbers are obtained as follows.

$$\binom{49}{6} = \frac{49!}{6!\,(49-6)!} = 13{,}983{,}816$$

Thus, there are a total of 13,983,816 different ways to select 6 numbers from 49 numbers. Hence, the probability that a player (who plays this lottery once) wins is

$$P(\text{player wins}) = 1/13{,}983{,}816 = .0000000715$$

Source: Chart reprinted with permission from *USA Today*, February 27, 1992. Copyright © 1992, *USA Today*.

FOOTNOTE TO THE CASE STUDY

The state of Texas currently has a lotto CASH 5 lottery. To play this game, a player selects five numbers from 1 through 39. Ten percent of the ticket sales for this game is awarded to the players who match all five of their numbers to the five numbers (randomly) selected by the Texas Lottery Commission; 15% is shared by the players who match 4 of 5 numbers (this share is increased to 25% if there is no 5 of 5 match); and 25% of the ticket sales is awarded to the players who match 3 of 5 numbers. The lottery commission's advertisement cards displayed at retailer shops made the following claim.

Match 5 numbers and win $75,000
Match 4 numbers and win $500
Match 3 numbers and win $25
All prizes are an average. You could win more
or less depending on the number of winners
and total sales for each drawing.

Professor Gerald Busald, who teaches at San Antonio College and uses this textbook, had his Spring 1997 statistics class analyze the CASH 5 lottery. Surprisingly, Professor Busald and his students found out that the commission's advertised estimated winnings for this game were incorrect. They found out that the expected winnings for a 5 of 5 match should be $57,575.70 instead of the advertised average of $75,000.

As shown below, using the combinations formula, we can find out that there are a total of 575,757 possible combinations of selecting 5 numbers from 39.

$$\binom{39}{5} = \frac{39!}{5! \, (39 - 5)!} = 575{,}757$$

Hence,

Probability of a 5 of 5 match $= 1/575{,}757$

Since 10% of the ticket sales are shared by 5 of 5 match winners, the expected winnings of these players are:

$$\frac{.10}{1/575{,}757} = \$57{,}575.70$$

Thus, the lottery commission's advertised claim of average winnings of $75,000 was incorrect.

Then Professor Busald and his students used the actual data for the first 154 drawings (from the inception of the game to January 27, 1997) of CASH 5 lottery to calculate the actual average winnings of all 5 of 5 match winners. These calculations showed that all 5 of 5 match winners in the first 154 drawings collected an average of $59,384.91, which is very close to the theoretical average of $57,575.70. The Texas Lottery Commission admitted that their advertised claim was incorrect and they have removed all advertisement cards claiming that the average winnings for a 5 of 5 match are $75,000.

5.5.3 USING THE TABLE OF COMBINATIONS

Table III in Appendix C lists the number of combinations of n elements selected x at a time. The following example illustrates how to read that table to find combinations.

Using the table of combinations.

EXAMPLE 5–15 Marv & Sons advertised to hire a financial analyst. The company has received applications from 10 candidates, who seem to be equally qualified. The company manager has decided to call only 3 of these candidates for an interview. If she randomly selects 3 candidates from the 10, how many total selections are possible?

Solution The total number of ways to select 3 applicants from 10 is given by $\binom{10}{3}$. To find the value of $\binom{10}{3}$ from Table III, we locate 10 in the column labeled n and 3 in the row labeled x. The relevant part of that table is reproduced here as Table 5.9.

Table 5.9 Determining the Value of $\binom{10}{3}$

n	x	0	1	2	3	...	20
1		1	1				
2		1	2	1			
3		1	3	3	1		
⋮		⋮	⋮	⋮	⋮		
$n = 10 \rightarrow$ 10		1	10	45	120	...	...
⋮		⋮	⋮	⋮	⋮	...	

$x = 3$

The value of $\binom{10}{3}$

The number at the intersection of the row for $n = 10$ and the column for $x = 3$ gives the value of $\binom{10}{3}$, which is

$$\binom{10}{3} = 120$$

Thus, the company manager can select 3 applicants from 10 in 120 ways. ∎

☞ *Remember* If the total number of elements and the number of elements to be selected are the same, then there is only one combination. In other words,

$$\binom{n}{n} = 1$$

Also, the number of combinations for selecting zero items from n is 1. That is,

$$\binom{n}{0} = 1$$

For example,

$$\binom{5}{5} = \frac{5!}{5!\,(5-5)!} = \frac{5!}{5!\,0!} = \frac{120}{(120)\,(1)} = 1$$

and

$$\binom{8}{0} = \frac{8!}{0!\,(8-0)!} = \frac{8!}{0!\,8!} = \frac{40{,}320}{(1)\,(40{,}320)} = 1$$

EXERCISES

Concepts and Procedures

5.39   Determine the value of each of the following using the appropriate formula.

3!   (7 − 3)!   9!   (14 − 12)! $\binom{5}{3}$ $\binom{7}{4}$ $\binom{9}{3}$ $\binom{6}{0}$ $\binom{3}{3}$

Verify the calculated values by using Tables II and III of Appendix C.

5.40   Find the value of each of the following using the appropriate formula.

6!   11!   (7 − 2)!   (13 − 5)! $\binom{8}{2}$ $\binom{4}{0}$ $\binom{5}{5}$ $\binom{6}{4}$ $\binom{11}{7}$

Verify the calculated values by using Tables II and III of Appendix C.

Applications

5.41   An English department at a university has 15 faculty members. Two of the faculty members will be randomly selected to represent the department on a committee. In how many ways can the department select 2 faculty members from 15? Use the appropriate formula.

5.42   A person will randomly select 3 varieties of cookies from 10 varieties for a party. In how many ways can this person select 3 varieties from 10? Use the appropriate formula.

5.43   A superintendent of schools has 12 schools under her jurisdiction. She plans to visit 3 of these schools during the next week. If she randomly selects 3 schools from these 12, how many total selections are possible? Use the appropriate formula. Verify your answer by using Table III of Appendix C.

5.44   An environmental agency will randomly select 4 houses from a block containing 20 houses for a radon check. How many total selections are possible? Use the appropriate formula. Verify your answer by using Table III of Appendix C.

5.45   An investor will randomly select 5 stocks from 20 for an investment. How many total combinations are possible? Use the appropriate formula. Verify your answer by using Table III of Appendix C.

5.46   A company employs a total of 16 workers. The management has asked these employees to select 2 workers who will negotiate a new contract with management. The employees have decided to select the 2 workers randomly. How many total selections are possible? Use the appropriate formula. Verify your answer by using Table III of Appendix C.

5.47   In how many ways can a sample (without replacement) of 8 items be selected from a population of 20 items?

5.48   In how many ways can a sample (without replacement) of 4 items be selected from a population of 15 items?

5.6   THE BINOMIAL PROBABILITY DISTRIBUTION

The **binomial probability distribution** is one of the most widely used discrete probability distributions. It is applied to find the probability that an outcome will occur x times in n performances of an experiment. For example, given that the probability is .05 that a VCR manufactured at a firm is defective, we may be interested in finding the probability that in a random sample of three VCRs manufactured at this firm, exactly one will be defective. As a second example, we may be interested in finding the probability that a baseball player, with a batting average of .250, will have no hits in 10 trips to the plate.

To apply the binomial probability distribution, the random variable x must be a discrete dichotomous random variable. In other words, the variable must be a discrete random variable and each repetition of the experiment must result in one of two possible outcomes. The binomial distribution is applied to experiments that satisfy the four conditions of a *binomial experiment*. (These conditions are described in Section 5.6.1.) Each repetition of a binomial experiment is called a **trial** or a **Bernoulli trial** (after Jacob Bernoulli). For example, if an experiment is defined as one toss of a coin and this experiment is repeated 10 times, then each repetition (toss) is called a trial. Consequently, there are 10 total trials for this experiment.

5.6.1   THE BINOMIAL EXPERIMENT

An experiment that satisfies the following four conditions is called a **binomial experiment**.

1. There are n identical trials. In other words, the given experiment is repeated n times. All these repetitions are performed under identical conditions.
2. Each trial has two and only two outcomes. These outcomes are usually called a *success* and a *failure*.
3. The probability of success is denoted by p and that of failure by q, and $p + q = 1$. The probabilities p and q remain constant for each trial.
4. The trials are independent. In other words, the outcome of one trial does not affect the outcome of another trial.

CONDITIONS OF A BINOMIAL EXPERIMENT

A binomial experiment must satisfy the following four conditions.

1. There are n identical trials.
2. Each trial has only two possible outcomes.
3. The probabilities of the two outcomes remain constant.
4. The trials are independent.

Note that one of the two outcomes of a trial is called a *success* and the other a *failure*. Notice that a success does not mean that the corresponding outcome is considered favorable or desirable. Similarly, a failure does not necessarily refer to an unfavorable or undesirable outcome. Success and failure are simply the names used to denote the two possible outcomes of a trial. The outcome to which the question refers is usually called a success; the outcome to which it does not refer is called a failure.

Verifying conditions of a binomial experiment.

EXAMPLE 5–16   Consider the experiment consisting of 10 tosses of a coin. Determine if it is a binomial experiment.

Solution   As described below, the experiment consisting of 10 tosses of a coin satisfies all four conditions of a binomial experiment.

1. There are a total of 10 trials (tosses), and they are all identical. All 10 tosses are performed under identical conditions.
2. Each trial (toss) has only two possible outcomes: a head and a tail. Let a head be called a success and a tail be called a failure.
3. The probability of obtaining a head (a success) is $1/2$ and that of a tail (a failure) is $1/2$ for any toss. That is,

$$p = P(H) = 1/2 \quad \text{and} \quad q = P(T) = 1/2$$

The sum of these two probabilities is 1.0. Also, these probabilities remain the same for each toss.

4. The trials (tosses) are independent. The result of any preceding toss has no bearing on the result of any succeeding toss.

   Consequently, the experiment consisting of 10 tosses is a binomial experiment.

Verifying conditions of a binomial experiment.

EXAMPLE 5–17   Five percent of all VCRs manufactured by a large electronics company are defective. Three VCRs are randomly selected from the production line of this company. The selected VCRs are inspected to determine if each of them is defective or good. Is this experiment a binomial experiment?

Solution

1. This example consists of three identical trials. A trial represents the selection of a VCR.
2. Each trial has two outcomes: A VCR is defective or a VCR is good. Let a defective VCR be called a success and a good VCR be called a failure.
3. Five percent of all VCRs are defective. So, the probability p that a VCR is defective is .05. As a result, the probability q that a VCR is good is .95. These two probabilities add up to 1.
4. Each trial (VCR) is independent. In other words, if one VCR is defective it does not affect the outcome of another VCR being defective or good. This is so because the size of the population is very large as compared to the sample size.

Since all four conditions of a binomial experiment are satisfied, this is an example of a binomial experiment.

5.6.2   THE BINOMIAL PROBABILITY DISTRIBUTION AND BINOMIAL FORMULA

The random variable x that represents the number of successes in n trials for a binomial experiment is called a *binomial random variable*. The probability distribution of x in such experiments is called the **binomial probability distribution** or simply *binomial distribution*. Thus, the binomial probability distribution is applied to find the probability of x successes in n trials for a binomial experiment. The number of successes x in such an experiment is a discrete random variable. Consider Example 5–17. Let x be the number of defective VCRs in a sample of three. Since we can obtain any number of defective VCRs from zero to three in a sample of three, x can assume any of the values 0, 1, 2, and 3. Since the values of x are countable, it is a discrete random variable.

BINOMIAL FORMULA

For a binomial experiment, the probability of exactly x successes in n trials is given by the binomial formula

$$P(x) = \binom{n}{x} p^x \, q^{n-x}$$

where

$$n = \text{total number of trials}$$
$$p = \text{probability of success}$$
$$q = 1 - p = \text{probability of failure}$$
$$x = \text{number of successes in } n \text{ trials}$$
$$n - x = \text{number of failures in } n \text{ trials}$$

In the binomial formula, n is the total number of trials and x is the number of successes. The difference between the total number of trials and the total number of successes, $n - x$, gives the total number of failures in n trials. The value of $\binom{n}{x}$ gives the number of ways to obtain x successes in n trials. As mentioned earlier, p and q are the probabilities of success and failure, respectively. Again, although it does not matter which of the two outcomes is called a success and which one a failure, usually the outcome to which the question refers is called a success.

To solve a binomial problem, we determine the values of n, x, $n - x$, p, and q and then substitute these values in the binomial formula. To find the value of $\binom{n}{x}$, we can use either the combinations formula from Section 5.5.2 or the table of combinations (Table III of Appendix C).

To find the probability of x successes in n trials for a binomial experiment, the only values needed are those of n and p. These are called the *parameters of the binomial probability distribution* or simply the **binomial parameters**. The value of q is obtained by subtracting the value of p from 1.0. Thus, $q = 1 - p$.

Next we solve a binomial problem, first without using the binomial formula and then by using the binomial formula.

Calculating probability: using a tree diagram and the binomial formula.

EXAMPLE 5–18   Five percent of all VCRs manufactured by a large electronics company are defective. A quality control inspector randomly selects three VCRs from the production line. What is the probability that exactly one of these three VCRs is defective?

Solution   Let

$$D = \text{a selected VCR is defective}$$

and

$$G = \text{a selected VCR is good}$$

As the tree diagram in Figure 5.4 shows, there are a total of eight outcomes, and three of them contain exactly one defective VCR. These three outcomes are

$$DGG, \quad GDG, \quad \text{and} \quad GGD$$

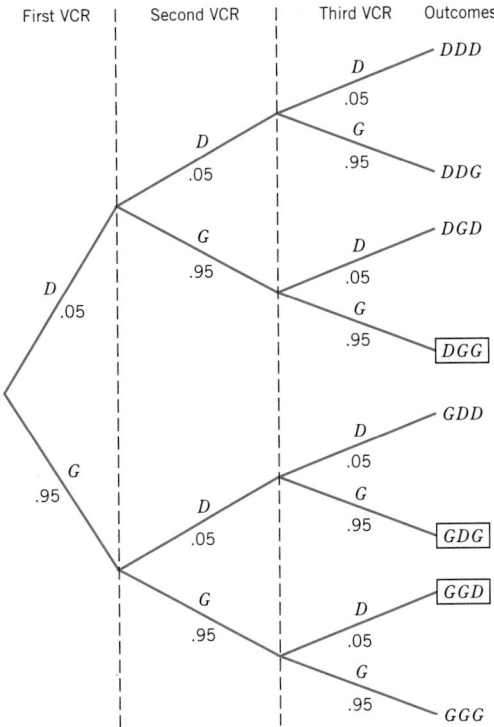

Figure 5.4 Tree diagram for selecting three VCRs.

We know that 5% of all VCRs manufactured at this company are defective. As a result, 95% of all VCRs are good. So the probability that a randomly selected VCR is defective is .05 and the probability that it is good is .95.

$$P(D) = .05 \quad \text{and} \quad P(G) = .95$$

Because the size of the population is large (note that it is a large company), the selections can be considered to be independent. The probability of each of the three outcomes, which give exactly one defective VCR, is calculated as follows.

$$P(DGG) = P(D) \cdot P(G) \cdot P(G) = (.05)(.95)(.95) = .0451$$
$$P(GDG) = P(G) \cdot P(D) \cdot P(G) = (.95)(.05)(.95) = .0451$$
$$P(GGD) = P(G) \cdot P(G) \cdot P(D) = (.95)(.95)(.05) = .0451$$

Note that DGG is simply the intersection of the three events D, G, and G. In other words, $P(DGG)$ is the joint probability of three events: the first VCR selected is defective, the second is good, and the third is good. To calculate this probability, we use the multiplication rule for independent events learned in Chapter 4. The same is true about the probabilities of the other two outcomes: GDG and GGD.

Exactly one defective VCR will be selected if either DGG or GDG or GGD occurs. These are three mutually exclusive outcomes. Therefore, applying the addition rule of Chapter 4, the probability of the union of these three outcomes is simply the sum of their individual probabilities.

$$P(1 \text{ VCR is defective in } 3) = P(DGG \text{ or } GDG \text{ or } GGD)$$

$$= P(DGG) + P(GDG) + P(GGD)$$

$$= .0451 + .0451 + .0451 = \mathbf{.1353}$$

Now let us use the binomial formula to compute this probability. Let us call the selection of a defective VCR a *success* and the selection of a good VCR a *failure*. The reason we have called a defective VCR a *success* is that the question refers to selecting exactly one defective VCR. Then,

$$n = \text{total number of trials} = 3 \text{ VCRs}$$

$$x = \text{number of successes} = \text{number of defective VCRs} = 1$$

$$n - x = \text{number of failures} = \text{number of good VCRs} = 3 - 1 = 2$$

$$p = P(\text{success}) = .05$$

$$q = P(\text{failure}) = 1 - p = .95$$

The probability of 1 success is denoted by $P(x = 1)$ or simply by $P(1)$. By substituting all the values in the binomial formula, we obtain:

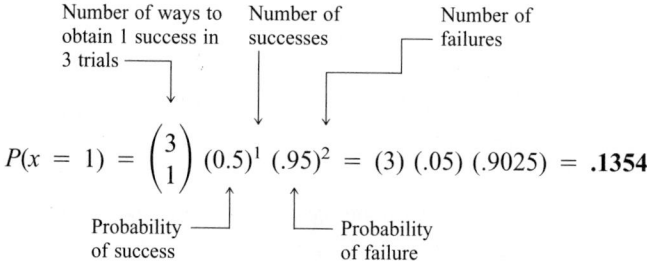

$$P(x = 1) = \binom{3}{1} (0.5)^1 (.95)^2 = (3) (.05) (.9025) = \mathbf{.1354}$$

Note that the value of $\binom{3}{1}$ in the above formula can either be read from Table III of Appendix C or it can be computed as follows.

$$\binom{3}{1} = \frac{3!}{1! \, (3 - 1)!} = \frac{3 \cdot 2 \cdot 1}{1 \cdot 2 \cdot 1} = 3$$

In the above computation, $\binom{3}{1}$ gives the three ways to select one defective VCR in three selections. As listed earlier, these three ways to select one defective VCR are *DGG*, *GDG*, and *GGD*. The probability .1354 is slightly different from the earlier calculation (.1353) because of rounding.

Calculating probability by using the binomial formula.

EXAMPLE 5–19   At the Express House Delivery Service, providing high-quality service to its customers is the top priority of the management. The company guarantees a refund of all charges if a package it is delivering does not arrive at its destination by the specified time. It is known from past data that despite all efforts, 2% of the packages mailed through this company do not arrive at their destinations within the specified time. A corporation mailed 10 packages through Express House Delivery Service on Monday.

(a) Find the probability that exactly one of these 10 packages will not arrive at its destination within the specified time.

(b) Find the probability that at most one of these 10 packages will not arrive at its destination within the specified time.

Solution Let us call it a success if a package does not arrive at its destination within the specified time and a failure if it does arrive within the specified time. Then,

$$n = \text{total number of packages mailed} = 10$$

$$p = P(\text{success}) = .02$$

and

$$q = P(\text{failure}) = 1 - .02 = .98$$

(a) For this part,

$$x = \text{number of successes} = 1$$

$$n - x = \text{number of failures} = 10 - 1 = 9$$

Substituting all values in the binomial formula, we obtain:

$$P(x = 1) = \binom{10}{1} (.02)^1 (.98)^9 = \frac{10!}{1! \, (10 - 1)!} (.02)^1 (.98)^9$$

$$= (10) (.02) (.83374776) = \mathbf{.1667}$$

Thus, there is a .1667 probability that exactly one of the 10 packages mailed will not arrive at its destination within the specified time.[3]

(b) The probability that at most one of the 10 packages will not arrive at its destination within the specified time is given by the sum of the probabilities of $x = 0$ and $x = 1$. Thus,

$$P(x \leq 1) = P(x = 0) + P(x = 1)$$

$$= \binom{10}{0} (.02)^0 (.98)^{10} + \binom{10}{1} (.02)^1 (.98)^9$$

$$= (1) (1) (.81707281) + (10) (.02) (.83374776)$$

$$= .8171 + .1667 = \mathbf{.9838}$$

Thus, the probability that at most one of the 10 packages will not arrive at its destination within the specified time is .9838. ◼

Constructing a binomial probability distribution and its graph.

EXAMPLE 5–20 According to a *Time*/CNN poll of U.S. adults conducted by Yankelovich Partners, Inc., 60% said that "compared with their parents they are better off economically" (*Time*, January 16, 1995). Assume that this result holds true for the current population of all U.S. adults. Let *x* denote the number in a random sample of three U.S. adults who hold this view. Write the probability distribution of *x* and draw a line graph for this probability distribution.

---

[3]**Using a calculator to simplify the expression for $P(x = 1)$:** As explained in Chapter 4, you can evaluate $(.98)^9$ by using the y^x key on your calculator. To evaluate $(.98)^9$, first enter .98, then press y^x key, then enter 9, and finally press the "=" key. The calculator screen will display the answer.

Solution Let x denote the number of adults in a sample of three who hold the given view. Then $n - x$ is the number of adults in the sample who do not hold this view. From the given information,

$$n = \text{total number of adults in the sample} = 3$$

$$p = P(\text{an adult holds the given view}) = .60$$

and $\quad q = P(\text{an adult does not hold the given view}) = 1 - .60 = .40$

The possible values that x can assume are 0, 1, 2, and 3. In other words, the number of adults in a sample of three who hold the given view can be 0, 1, 2, or 3. The probability of each of these four outcomes is calculated as follows.

If $x = 0$, then $n - x = 3$. Using the binomial formula, the probability of $x = 0$ is

$$P(x = 0) = \binom{3}{0} (.60)^0 (.40)^3 = (1)\,(1)\,(.064) = \mathbf{.0640}$$

Note that $\binom{3}{0}$ is equal to 1 by definition and $(.60)^0$ is equal to 1 because any number raised to the power zero is always 1.

If $x = 1$, then $n - x = 2$. The probability $P(x = 1)$ is

$$P(x = 1) = \binom{3}{1} (.60)^1 (.40)^2 = (3)\,(.60)\,(.16) = \mathbf{.2880}$$

If $x = 2$, then $n - x = 1$. The probability $P(x = 2)$ is

$$P(x = 2) = \binom{3}{2} (.60)^2 (.40)^1 = (3)\,(.36)\,(.40) = \mathbf{.4320}$$

If $x = 3$, then $n - x = 0$. The probability $P(x = 3)$ is

$$P(x = 3) = \binom{3}{3} (.60)^3 (.40)^0 = (1)\,(.216)\,(1) = \mathbf{.2160}$$

These probabilities are written in tabular form in Table 5.10. Figure 5.5 shows the line graph for the probability distribution of Table 5.10.

Table 5.10 Probability Distribution of x

x	$P(x)$
0	.0640
1	.2880
2	.4320
3	.2160

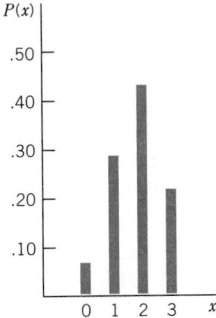

Figure 5.5 Line graph of the probability distribution of x.

CASE STUDY 5-3   MISSING WOMEN

[This case study is based on] the 1968 trial of the pediatrician-author Dr. Benjamin Spock and others in the U.S. District Court in Boston for conspiracy to violate the Selective Service Act by encouraging resistance to the war in Vietnam. In that trial, the defense challenged the legality of the jury-selection method. Although more than half of all eligible jurors in Boston were women, there were no women on Dr. Spock's jury. Yet he, more than any defendant, would have wanted some because so many mothers have raised their children "according to Dr. Spock''; moreover, the opinion polls showed women in general to be more opposed to the Vietnam war than men.

The question was whether this total absence of women jurors was an accident of this particular jury or whether it had resulted from systematic discrimination. Statistical reasoning was to provide the answer.

In the Boston District Court, jurors are selected in three stages. The City Directory is used for the first stage; from it, the Clerk of the Court is supposed to select 300 names at random, that is, by a lotterylike method, and put a slip with each of these names into a box. The City Directory is renewed annually by censuslike household visits of the police, and it lists all adult individuals in the Boston area. The Directory lists slightly more women than men. The second selection stage occurs when a trial is about to begin. From the 300 names in the box, the names of 30 or more potential jurors are drawn. These people are ordered to appear in court on the morning of the trial. The subgroup of 30 or more is called a venire. In the third stage, the one that most of us think of as jury selection, 12 actual jurors are selected after interrogation by both the prosecutor and the defense counsel.

The average proportion of women drawn by the six judicial colleagues of the Spock trial judge was 29%, and furthermore, the averages of these six judges bunched closely around the group average. This suggests that the proportion of women among the names in the 300-name panels in the jury box was somewhere close to that 29% mark. But . . . the Spock judge's venires had consistently lower percentages of women, with an overall average of only 14.6% women, almost exactly half of that of his colleagues.

It is possible, of course, that the selection method used by the trial judge was the same as that of his colleagues. But what is the probability that a difference as large (or larger) as that between 14.6 and 29% could arise by chance? Statistical computation revealed the probability to be 1 in 1,000,000,000,000,000,000 that the "luck of the draw" would yield the distribution of women jurors obtained by the trial judge or a more extreme one. The conclusion, therefore, was virtually inescapable: the venires for the trial judge must have been drawn from the central jury lists in a fashion that somehow systematically reduced the proportion of women jurors.

Thus the proportion of women among the potential jurors twice suffered an improper reduction—first when the court clerk reduced their share from a majority in the City Directory to 29% in the jury lists and, second, when the judge managed to lower the 29% to his private average of 14.6%. In the Spock trial, only one potential woman juror came before the court, and she was easily eliminated in stage 3 by the prosecutor under his quota of peremptory challenges (for which he need not give any reasons).

For further discussion, see H. Zeisel, ''Dr. Spock and the Case of the Vanishing Women Jurors,'' *University of Chicago Law Review*, 37, 1969, 1–18.

Source: From Judith M. Tanur et al., *Statistics: A Guide to the Unknown*, 2d ed. Copyright © 1985 by Wadsworth, Inc. Reprinted by permission of Wadsworth & Brooks/Cole Advanced Books & Software, Pacific Grove, CA 93950.

Quiz

Fifty percent of the adult population in a large city are women. A court is to randomly select a jury of 12 adults from the population of all adults of this city. Using the binomial formula, find the probability that none of the 12 jury members is a woman.

5.6.3 USING THE TABLE OF BINOMIAL PROBABILITIES

The probabilities for a binomial experiment can also be read from Table IV, the table of binomial probabilities, given in Appendix C. That table lists the probabilities of x for $n = 1$ to $n = 25$ and for selected values of p. Example 5–21 illustrates how to read Table IV.

Using the binomial table to find probabilities and to construct the probability distribution and graph.

EXAMPLE 5–21 In a poll of U.S. adults conducted by Peter Hart and Robert Teeter for *The Wall Street Journal*/NBC News, about 20% of adults rated their own morals and values as 96 to 100 on a scale from 1 to 100 where 1 is the lowest and 100 is the highest rating (*The Wall Street Journal*, December 13, 1996). Suppose that 20% of all current U.S. adults would rate their own morals and values as 96 to 100 on a scale from 1 to 100. A sample of six U.S. adults is selected. Using Table IV of Appendix C, answer the following.

(a) Find the probability that exactly three adults in this sample rate their own morals and values as 96 to 100.

(b) Find the probability that at most two adults in this sample rate their own morals and values as 96 to 100.

(c) Find the probability that at least three adults in this sample rate their own morals and values as 96 to 100.

(d) Find the probability that one to three adults in this sample rate their own morals and values as 96 to 100.

(e) Let x be the number of adults in this sample who rate their own morals and values as 96 to 100. Write the probability distribution of x and draw a line graph for this probability distribution.

Solution

(a) To read the required probability from Table IV of Appendix C, we first determine the values of n, x, and p. These values are

n = number of adults in the sample = 6

x = number of adults in 6 who rate their own morals and values as 96 to 100 = 3

p = P(that an adult rates his/her own morals and values as 96 to 100) = .20

Then we locate $n = 6$ in the column labeled n in Table IV. The relevant portion of Table IV with $n = 6$ is reproduced here as Table 5.11. Next, we locate 3 in the column for x in the portion of the table for $n = 6$ and locate $p = .20$ in the row for p at the top of the table. The entry at the intersection of the row for $x = 3$ and the column for $p = .20$ gives the probability of 3 successes in 6 trials when the probability of success is .20. From Table IV or Table 5.11,

$$P(x = 3) = \mathbf{.0819}$$

Table 5.11   Determining $P(x = 3)$ for $n = 6$ and $p = .20$

$p = .20$

n	x	.05	.10	.20	...	.95
$n = 6$ → 6	0	.7351	.5314	.2621	...	.0000
	1	.2321	.3543	.3932	...	.0000
	2	.0305	.0984	.2458	...	.0001
$x = 3$ → 3		.0021	.0146	.0819 ←	...	.0021
	4	.0001	.0012	.0154	...	.0305
	5	.0000	.0001	.0015	...	.2321
	6	.0000	.0000	.0001	...	.7351

$P(x = 3) = .0819$

Using Table IV or Table 5.11, we write Table 5.12, which can be used to answer the remaining parts of this example.

Table 5.12   Portion of Table IV for $n = 6$ and $p = .20$

n	x	p .20
6	0	.2621
	1	.3932
	2	.2458
	3	.0819
	4	.0154
	5	.0015
	6	.0001

(b)   The event that at most two adults in a sample of six would rate their own morals and values as 96 to 100 will occur if x is equal to 0, 1, or 2. Using Table IV of Appendix C or Table 5.12, the required probability is

$$P(\text{at most 2}) = P(0 \text{ or } 1 \text{ or } 2) = P(x = 0) + P(x = 1) + P(x = 2)$$

$$= .2621 + .3932 + .2458 = \mathbf{.9011}$$

(c)   The probability that at least three adults in a sample of six would rate their own morals and values as 96 to 100 is given by the sum of the probabilities of 3, 4, 5, or 6 adults. Using Table IV of Appendix C or Table 5.12,

$$P(\text{at least 3}) = P(3 \text{ or } 4 \text{ or } 5 \text{ or } 6)$$

$$= P(x = 3) + P(x = 4) + P(x = 5) + P(x = 6)$$

$$= .0819 + .0154 + .0015 + .0001 = \mathbf{.0989}$$

(d) The probability that one to three adults in a sample of six would rate their own morals and values as 96 to 100 is given by the sum of the probabilities $x = 1$, 2, or 3. Using Table IV of Appendix C or Table 5.12,

$$P(1 \text{ to } 3) = P(x = 1) + P(x = 2) + P(x = 3)$$

$$= .3932 + .2458 + .0819 = \mathbf{.7209}$$

(e) Using Table IV of Appendix C or Table 5.12, we list the probability distribution of x for $n = 6$ and $p = .20$ in Table 5.13. Figure 5.6 shows the line graph of the probability distribution of x.

Table 5.13   Probability Distribution of x for $n = 6$ and $p = .20$

x	$P(x)$
0	.2621
1	.3932
2	.2458
3	.0819
4	.0154
5	.0015
6	.0001

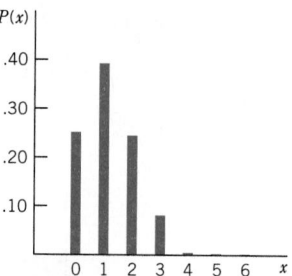

Figure 5.6 Line graph for the probability distribution of x.

5.6.4 PROBABILITY OF SUCCESS AND THE SHAPE OF THE BINOMIAL DISTRIBUTION

For any number of trials n:

1. The binomial probability distribution is symmetric if $p = .50$.
2. The binomial probability distribution is skewed to the right if p is less than .50.
3. The binomial probability distribution is skewed to the left if p is greater than .50.

These three cases are illustrated next with examples and graphs.

1. Let $n = 4$ and $p = .50$. Using Table IV of Appendix C, we have written the probability distribution of x in Table 5.14 and plotted it in Figure 5.7. As we can observe from Table 5.14 and Figure 5.7, the probability distribution of x is symmetric.

Table 5.14 Probability
Distribution of x for
$n = 4$ and $p = .50$

x	$P(x)$
0	.0625
1	.2500
2	.3750
3	.2500
4	.0625

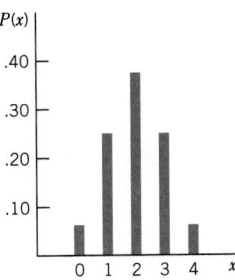

Figure 5.7 Graph of the probability
distribution of Table 5.14.

2. Let $n = 4$ and $p = .30$ (which is less than .50). Table 5.15, which is written by using Table IV of Appendix C, and the graph of the probability distribution in Figure 5.8 show that the probability distribution of x for $n = 4$ and $p = .30$ is skewed to the right.

Table 5.15 Probability
Distribution of x for
$n = 4$ and $p = .30$

x	$P(x)$
0	.2401
1	.4116
2	.2646
3	.0756
4	.0081

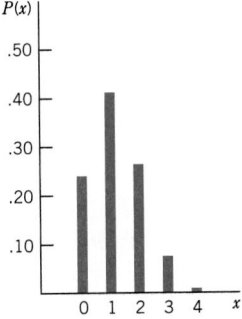

Figure 5.8 Graph of the probability
distribution of Table 5.15.

3. Let $n = 4$ and $p = .80$ (which is greater than .50). Table 5.16, which is written by using Table IV of Appendix C, and the graph of the probability distribution in Figure 5.9 show that the probability distribution of x for $n = 4$ and $p = .80$ is skewed to the left.

Table 5.16   Probability
Distribution of x for
$n = 4$ and $p = .80$

x	$P(x)$
0	.0016
1	.0256
2	.1536
3	.4096
4	.4096

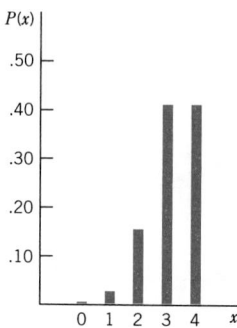

Figure 5.9   Graph of the probability distribution of Table 5.16.

5.6.5   MEAN AND STANDARD DEVIATION OF THE BINOMIAL DISTRIBUTION

Sections 5.3 and 5.4 explained how to compute the mean and standard deviation, respectively, for a probability distribution of a discrete random variable. When a discrete random variable has a binomial distribution, the formulas learned in Sections 5.3 and 5.4 could still be used to compute its mean and standard deviation. However, it is simpler and more convenient to use the following formulas to find the mean and standard deviation in such cases.

MEAN AND STANDARD DEVIATION OF A BINOMIAL DISTRIBUTION

The *mean and standard deviation for a binomial distribution* are

$$\mu = np \quad \text{and} \quad \sigma = \sqrt{npq}$$

where n is the total number of trials, p is the probability of success, and q is the probability of failure.

Example 5–22 describes the calculation of the mean and standard deviation for a binomial distribution.

Mean and standard deviation of a binomial random variable.

EXAMPLE 5–22   In a poll conducted by the Center for Media and Public Affairs, 52% of adults said that the press abuses its freedom in the United States (*USA Today*, December 18, 1996). Assume that this result holds true for the current population of all U.S. adults. A sample of 25 U.S. adults is selected. Let x denote the number of adults in this sample who hold this view. Find the mean and standard deviation of the probability distribution of x.

Solution   This is a binomial experiment with 25 total trials. Each trial has two outcomes: (1) the selected adult holds the view that the press abuses its freedom in the United States, (2) the selected adult does not hold this view. (Note that the second outcome includes the possibility of an adult saying that the press does not abuse its freedom in the United States or having no opinion.) The probabilities p and q for these two outcomes are .52 and .48, respectively. Thus,

$$n = 25, \quad p = .52, \quad \text{and} \quad q = .48$$

Using the formulas for the mean and standard deviation of the binomial distribution, we obtain:

$$\mu = np = 25 \, (.52) = \mathbf{13}$$
$$\sigma = \sqrt{npq} = \sqrt{(25) \, (.52) \, (.48)} = \mathbf{2.498}$$

Thus, the mean of the probability distribution of x is 13 and the standard deviation is 2.498. The value of the mean is what we expect to obtain, on average, per repetition of the experiment. In this example, if we take many samples of 25 U.S. adults, we expect that each sample will contain an average of 13 adults, with a standard deviation of 2.498, who would say that the press abuses its freedom in the United States. ▄▄

EXERCISES

Concepts and Procedures

5.49 Briefly explain the following.
   a. A binomial experiment   b. A trial   c. A binomial random variable

5.50 What are the parameters of the binomial probability distribution and what do they mean?

5.51 Which of the following are binomial experiments? Explain why.
   a. Rolling a die many times and observing the number of spots
   b. Rolling a die many times and observing whether the number obtained is even or odd
   c. Selecting a few voters from a very large population of voters and observing whether or not each of them favors a certain proposition in an election when 54% of all voters are known to be in favor of this proposition

5.52 Which of the following are binomial experiments? Explain why.
   a. Drawing 3 balls with replacement from a box that contains 10 balls, 6 of which are red and 4 blue, and observing the colors of the drawn balls
   b. Drawing 3 balls without replacement from a box that contains 10 balls, 6 of which are red and 4 blue, and observing the colors of the drawn balls
   c. Selecting a few households from New York City and observing whether or not they own stocks when it is known that 28% of all households in New York City own stocks

5.53 Let x be a discrete random variable that possesses a binomial distribution. Using the binomial formula, find the following probabilities.
   a. $P(x = 5)$ for $n = 8$ and $p = .60$
   b. $P(x = 3)$ for $n = 4$ and $p = .30$
   c. $P(x = 2)$ for $n = 6$ and $p = .20$
Verify your answer by using Table IV of Appendix C.

5.54 Let x be a discrete random variable that possesses a binomial distribution. Using the binomial formula, find the following probabilities.
   a. $P(x = 0)$ for $n = 5$ and $p = .10$
   b. $P(x = 4)$ for $n = 7$ and $p = .80$
   c. $P(x = 7)$ for $n = 10$ and $p = .40$
Verify your answer by using Table IV of Appendix C.

5.55 Let x be a discrete random variable that possesses a binomial distribution.

a. Using Table IV of Appendix C, write the probability distribution of x for $n = 7$ and $p = .30$ and graph it.

b. What are the mean and standard deviation of the probability distribution developed in part a?

5.56 Let x be a discrete random variable that possesses a binomial distribution.

a. Using Table IV of Appendix C, write the probability distribution of x for $n = 5$ and $p = .80$ and graph it.

b. What are the mean and standard deviation of the probability distribution developed in part a?

5.57 The binomial probability distribution is symmetric for $p = .50$, skewed to the right for $p < .50$, and skewed to the left for $p > .50$. Illustrate each of these three cases by writing a probability distribution table and by drawing a graph. Choose any values of n and p and use the table of binomial probabilities (Table IV of Appendix C) to write the probability distribution tables.

Applications

5.58 Louis Harris and Associates interviewed 408 senior executives at major U.S. corporations for *Business Week* magazine in May 1995 (*Business Week*, June 5, 1995). Eighty-six percent of these executives felt that the federal budget will not be balanced by the year 2002. Suppose this result is true of all senior executives at U.S. corporations.

a. Let x be a binomial random variable that denotes the number of senior executives in a random sample of 10 who hold this opinion. What are the possible values that x can assume?

b. Find the probability that exactly 7 senior executives in a random sample of 10 will hold this opinion.

5.59 The Metropolitan Life Survey of the American Teacher estimated that 54% of the nation's teachers were "very satisfied" with their jobs in 1995 (Ann Bradley, "Nation's Teachers Feeling Better About Jobs, Salaries, Survey Finds," *Education Week*, December 6, 1995). Assume that this percentage is true for the current population of all U.S. teachers.

a. Let x be a binomial random variable that denotes the number of teachers in a sample of 12 who are "very satisfied" with their jobs. What are the possible values that x can assume?

b. Find the probability that in a random sample of 12 teachers, exactly 9 will say that they are "very satisfied" with their jobs.

5.60 According to a *U.S. News*/CNN poll conducted by the Gallup Organization, 50% of Americans recognize their postal carrier by sight (*U.S. News & World Report*, March 13, 1995). Assume that this result is true for the current population of Americans. Using the binomial probabilities table (Table IV of Appendix C), find the probability that the number of Americans in a random sample of 12 who recognize their postal carrier by sight is

a. at most 6   b. at least 10   c. 5 to 8

5.61 In August 1994, Clark Martire and Bartolomeo conducted a survey for *Fortune* magazine to examine CEOs' attitudes toward office romances between unmarried employees. Seventy percent of the CEOs interviewed felt that office romances were none of the company's business (*Fortune*, October 3, 1994). Assume that this result is true for the current population of CEOs in the United States. Using the binomial probabilities table (Table IV of Appendix C), find the probability that the number of CEOs in a random sample of 15 who hold this view is

a. at least 10   b. 9 to 12   c. at most 7

5.62 In a survey conducted in January 1995 by Louis Harris & Associates for *Business Week*, 55% of adults felt that U.S. federal agencies and departments were "very efficient" or "somewhat efficient" (*Business Week*, January 23, 1995). Find the probability that in a random sample of 8 adults the number who hold this view is

a. exactly 6   b. none   c. exactly 8

5.63 In a telephone poll of adult Americans conducted for *Time*/CNN in June 1995 by Yankelovich Partners, Inc., 52% of these adults were "very concerned" about the amount of violence in movies, television, and popular music (*Time*, June 12, 1995). Assuming that this result holds true for the current population of all adult Americans, find the probability that in a random sample of 10 adults the number who share this concern is

  a. exactly 4   **b.** none   **c.** exactly 8

5.64 According to Case Study 4–3 in Chapter 4, the probability that a baseball player will have no hits in 10 trips to the plate is .056, given that this player has a hitting percentage of 25%. Using the binomial formula, show that this probability is indeed .056.

5.65 A professional basketball player makes 85% of the free throws he tries. Assuming this percentage will hold true for future attempts, find the probability that in the next eight tries the number of free throws he will make is

  a. exactly 8   **b.** exactly 5

5.66 Yankelovich Partners, Inc., conducted a survey in May 1995 for *Time*/CNN to examine opinions of American adults on reducing the federal budget deficit (*Time*, May 22, 1995). Twenty percent of the adults polled were in favor of eliminating the U.S. Department of Education. Assume this result is true for the current population of all American adults.

  a. Using the binomial formula, find the probability that in a random sample of 12 American adults the number who would favor eliminating the Department of Education is

      i. exactly 6   ii. none

  b. Using the binomial probabilities table, find the probability that in a random sample of 12 American adults the number who would favor eliminating the Department of Education is

      i. at least 6   ii. at most 3   iii. 2 to 5

5.67 A poll on college athletics conducted by Institute for Social Inquiry in June 1996 found that 60% of men felt that athletics at larger colleges are overemphasized (*The Hartford Courant*, July 6, 1996). Assume this result holds true for the current population of all American men.

  a. Using the binomial formula, find the probability that in a random sample of 10 men the number who would say that athletics are overemphasized at larger colleges is

      i. exactly 4   ii. exactly 10

  b. Using the binomial probabilities table, find the probability that in a random sample of 10 men the number who would say that athletics are overemphasized at larger colleges is

      i. less than 6   ii. more than 5   iii. 4 to 7

5.68 Johnson Electronics makes calculators. Consumer satisfaction is one of the top priorities of the company's management. The company guarantees a refund or a replacement for any calculator that malfunctions within 2 years from the date of purchase. It is known from past data that despite all efforts, 5% of the calculators manufactured by the company malfunction within a two-year period. The company mailed a package of 10 randomly selected calculators to a store.

  a. Let *x* denote the number of calculators in this package of 10 that will be returned for refund or replacement within a 2-year period. Using the binomial probabilities table, obtain the probability distribution of *x* and draw a graph of the probability distribution. Determine the mean and standard deviation of *x*.

  b. Using the probability distribution of part a, find the probability that exactly 2 of the 10 calculators will be returned for refund or replacement within a 2-year period.

5.69 A fast food chain store conducted a taste survey before marketing a new hamburger. The results of the survey showed that 70% of the people who tried this hamburger liked it. Encouraged by this result, the company decided to market the new hamburger. Assume that 70% of all people like this hamburger. On a certain day, eight customers bought it.

  a. Let *x* denote the number of customers in this sample of eight who will like this hamburger. Using the binomial probabilities table, obtain the probability distribution of *x* and draw a graph of the probability distribution. Determine the mean and standard deviation of *x*.

  b. Using the probability distribution of part a, find the probability that exactly three of the eight customers will like this hamburger.

5.7 THE POISSON PROBABILITY DISTRIBUTION

The **Poisson probability distribution**, named after the French mathematician Simeon D. Poisson, is another important probability distribution of a discrete random variable that has a large number of applications. Suppose a washing machine in a laundromat breaks down an average of three times a month. We may want to find the probability of exactly two breakdowns during the next month. This is an example of a Poisson probability distribution problem. Each breakdown is called an *occurrence* in Poisson probability distribution terminology. The Poisson probability distribution is applied to experiments with random and independent occurrences. The occurrences are random in the sense that they do not follow any pattern and, hence, they are unpredictable. Independence of occurrences means that one occurrence (or nonoccurrence) of an event does not influence the successive occurrences or nonoccurrences of that event. The occurrences are always considered with respect to an interval. In the example of the washing machine, the interval represents one month. The interval may be a time interval, a space interval, or a volume interval. The actual number of occurrences within an interval is random and independent. If the average number of occurrences for a given interval is known, then by using the Poisson probability distribution we can compute the probability of a certain number of occurrences, x, in that interval. Note that the number of actual occurrences in an interval is denoted by x.

CONDITIONS TO APPLY POISSON PROBABILITY DISTRIBUTION

The following three conditions must be satisfied to apply the Poisson probability distribution.

1. x is a discrete random variable.
2. The occurrences are random.
3. The occurrences are independent.

The following are a few examples of discrete random variables for which the occurrences are random and independent. Hence, these are examples to which the Poisson probability distribution can be applied.

1. Consider the number of patients arriving at the emergency ward of a hospital during a one-hour interval. In this example, an occurrence is the arrival of a patient at the emergency ward, the interval is one hour (an interval of time), and the occurrences are random. The total number of patients who may arrive at this emergency ward during a one-hour interval may be 0, 1, 2, 3, 4, ... The independence of occurrences in this example means that patients arrive individually and the arrival of any two (or more) patients is not related.

2. Consider the number of defective items in the next 100 items manufactured on a machine. In this case, the interval is a volume interval (100 items). The occurrences (number of defective items) are random because there may be 0, 1, 2, 3, . . . , 100 defective items in 100 items. We can assume the occurrence of defective items to be independent of one another.

3. Consider the number of defects in a 5-foot-long iron rod. The interval, in this example, is a space interval (5 feet). The occurrences (defects) are random because there may be any number of defects in a 5-foot iron rod. We can assume that these defects are independent of one another.

The following examples also qualify for the application of the Poisson probability distribution.

1. The number of accidents that occur on a given highway during a one-week period.
2. The number of customers coming to a grocery store during a one-hour interval.
3. The number of television sets sold at a department store during a given week.

On the other hand, consider the arrival of patients at a physician's office. These arrivals will be nonrandom if the patients have to make appointments to see the doctor. The arrival of commercial airplanes at an airport is nonrandom because all planes are scheduled to arrive at certain times, and airport authorities know the exact number of arrivals for any period (although this number may change slightly because of late or early arrivals and cancellations). The Poisson probability distribution cannot be applied to these examples.

In the Poisson probability distribution terminology, the average number of occurrences in an interval is denoted by λ (Greek letter *lambda*). The actual number of occurrences in that interval is denoted by x. Then, using the Poisson probability distribution, we find the probability of x occurrences during an interval given that the mean occurrences during that interval are λ.

POISSON PROBABILITY DISTRIBUTION FORMULA

According to the *Poisson probability distribution*, the probability of x occurrences in an interval is

$$P(x) = \frac{\lambda^x e^{-\lambda}}{x!}$$

where λ (pronounced *lambda*) is the mean number of occurrences in that interval and the value of e is approximately 2.71828.

The mean number of occurrences in an interval, denoted by λ, is called the *parameter of the Poisson probability distribution* or the **Poisson parameter**. As is obvious from the Poisson probability distribution formula, we need to know only the value of λ to compute the probability of any given value of x. We can read the value of $e^{-\lambda}$ for a given λ from Table V of Appendix C. Examples 5–23 through 5–25 illustrate the use of the Poisson probability distribution formula.

Using Poisson formula: x equals a specific value.

EXAMPLE 5–23 According to a survey by the International Mass Retail Association, buyers aged 8 to 17 make an average of 12.7 shopping trips per month. Assume that this result is true for the current population of all buyers aged 8 to 17 and that all the conditions of the Poisson distribution are satisfied. Find the probability that a randomly selected buyer aged 8 to 17 would make exactly nine shopping trips next month.

Solution Let λ be the mean number of shopping trips per month by all buyers aged 8 to 17. Then, $\lambda = 12.7$. Let x be the number of shopping trips next month by the selected buyer aged 8 to 17. We are to find the probability of $x = 9$. Substituting all the values in the Poisson probability distribution formula, we obtain:

$$P(x = 9) = \frac{\lambda^x e^{-\lambda}}{x!} = \frac{(12.7)^9 e^{-12.7}}{9!} = \frac{(8594754749)(.00000305)}{362880} = .0723$$

In these calculations, we can find the value of 9! from Table II of Appendix C and the value of $e^{-12.7}$ by using the e^x key on a calculator.[4] ■

Calculating probabilities by using the Poisson formula.

EXAMPLE 5–24 A washing machine in a laundromat breaks down an average of three times per month. Using the Poisson probability distribution formula, find the probability that during the next month this machine will have

(a) exactly two breakdowns   (b) at most one breakdown

Solution Let λ be the mean number of breakdowns per month and x be the actual number of breakdowns observed during the next month for this machine. Then,

$$\lambda = 3$$

(a) The probability that exactly two breakdowns will be observed during the next month is

$$P(x = 2) = \frac{\lambda^x e^{-\lambda}}{x!} = \frac{(3)^2 e^{-3}}{2!} = \frac{(9)(.04978707)}{2} = .2240$$

(b) The probability that at most one breakdown will be observed during the next month is given by the sum of the probabilities of zero and one breakdown. Thus,

$$P(\text{at most 1 breakdown}) = P(0 \text{ or } 1 \text{ breakdown}) = P(x = 0) + P(x = 1)$$

$$= \frac{(3)^0 e^{-3}}{0!} + \frac{(3)^1 e^{-3}}{1!}$$

$$= \frac{(1)(.04978707)}{1} + \frac{(3)(.04978707)}{1}$$

$$= .0498 + .1494 = .1992$$ ■

 Remember One important point to remember about the Poisson probability distribution is that *the intervals for λ and x must be equal*. If they are not, the mean λ should be redefined to make them equal. Example 5–25 illustrates this point.

Calculating probability by using the Poisson formula.

EXAMPLE 5–25 Cynthia's Mail Order Company provides free examination of its products for seven days. If not completely satisfied, a customer can return the product within that period and get a full refund. According to past records of the company, an average of 2 of every 10 products sold by this company are returned for a refund. Using the Poisson probability distribution formula, find the probability that exactly 6 of the 40 products sold by this company on a given day will be returned for a refund.

Solution Let x denote the number of products in 40 that will be returned for a refund. We are to find $P(x = 6)$. The given mean is defined per 10 products, but x is defined for 40 products. As a result, we should first find the mean for 40 products. Because, on average, 2 out of 10 products are returned, the mean number of products returned out of 40 will be 8. Thus, $\lambda = 8$. Substituting $x = 6$ and $\lambda = 8$ in the Poisson probability distribution formula, we obtain:

$$P(x = 6) = \frac{\lambda^x e^{-\lambda}}{x!} = \frac{(8)^6 e^{-8}}{6!} = \frac{(262144)(.00033546)}{720} = .1221$$

Thus, the probability is .1221 that exactly 6 products out of 40 sold on a given day will be returned. ■

[4]**Using a calculator to simplify the expression for $P(x = 9)$:** In the expression for $P(x = 9)$, you can evaluate $(12.7)^9$ and 9! by using the y^x and $n!$ keys, respectively. If your calculator has the e^x key, you can evaluate $e^{-12.7}$ as follows. First enter 12.7, then press the $+/-$ key. Finally press the e^x key.

Note that Example 5–25 is actually a binomial problem with $p = 2/10 = .20$, $n = 40$, and $x = 6$. In other words, the probability of success (that is, the probability that a product is returned) is .20 and the number of trials (products sold) is 40. We are to find the probability of 6 successes (returns). We have used the Poisson distribution to solve this problem. This is referred to as *using the Poisson distribution as an approximation to the binomial distribution*. We can also use the binomial distribution to find this probability as follows.

$$P(x = 6) = \binom{40}{6} (.20)^6 (.80)^{34} = \frac{40!}{6! \, (40 - 6)!} (.20)^6 (.80)^{34}$$

$$= (3838380) (.000064) (.00050706) = \textbf{.1246}$$

Thus, the probability $P(x = 6)$ is .1246 when we use the binomial distribution.

As we can observe, simplifying the above calculations for the binomial formula is quite complicated when n is large. It is much easier to solve this problem using the Poisson distribution. As a general rule, if it is a binomial problem with $n > 25$ but $\mu \leq 25$, then we can use the Poisson distribution as an approximation to the binomial distribution. However, if $n > 25$ and $\mu > 25$, we prefer to use the normal distribution as an approximation to the binomial. The latter case will be discussed in Chapter 6.

Case Study 5–4 presents applications of the binomial and Poisson probability distributions.

CASE STUDY 5–4 ASK MR. STATISTICS

Fortune magazine frequently publishes a column titled *Ask Mr. Statistics*, which contains questions and answers to statistical problems. The following excerpts are reprinted from one such column.

"Dear Oddgiver: I am in the seafood distribution business and find myself endlessly wrangling with supermarkets about appropriate order sizes, especially with high-end tidbit products like our matjes herring in superspiced wine, which we let them have for $4.25, and still they take only a half-dozen jars, thereby running the risk of getting sold out early in the week and causing the better class of customers to storm out empty-handed. How do I get them to realize that lowballing on inventories is usually bad business, also to at least try a few jars of our pickled crappie balls?

—HEADED FOR A BREAKDOWN

Dear Picklehead: The science of statistics has much to offer people puzzled by seafood inventory problems. Your salvation lies in the Poisson distribution, "poisson" being French for fish and, of arguably greater relevance, the surname of a 19th-century French probabilist.

Simeon Poisson's contribution was to develop a method for calculating the likelihood that a specified number of successes will occur given that (a) the probability of success on any one trial is very low but (b) the number of trials is very high. A real world example often mentioned in the literature concerns the distribution of Prussian cavalry deaths from getting kicked by horses in the period 1875–94.

As you would expect of Teutons, the Prussian military kept meticulous records on horse-kick deaths in each of its army corps, and the data are neatly summarized in a 1963 book called *Lady Luck*, by the late Warren Weaver. There were a total of 196 kicking deaths—these being the, er, "successes." The "trials" were each army corps' observations on the number of kicking deaths sustained in the year. So with 14 army corps and data for 20 years, there were 280 trials. We shall not detain you with the Poisson formula, but it predicts, for example, that there will be 34.1 instances of a corps' having exactly two deaths in a year. In fact, there were 32 such cases. Pretty good, eh?

Back to seafood. The Poisson calculation is appropriate to your case, since the likelihood of any one customer's buying your overspiced herring is extremely small, but the number of trials—i.e., customers in the store during a typical week—is very large. Let us say that one customer in 1,000 deigns to buy the herring, and 6,000 customers visit the store in a week. So six jars are sold in an average week.

But the store manager doesn't care about average weeks. What he's worried about is having too much or not enough. He needs to know the probabilities assigned to different sales levels. Our Poisson distribution shows the following morning line: The chance of fewer than three sales—only 6.2%. Of four to six sales: 45.5%. Chances of losing some sales if the store elects to start the week with six jars because that happens to be the average: 39.4%. If the store wants to be 90% sure of not losing sales, it needs to start with nine jars.

There is no known solution to the problem of pickled crappie balls.''

Source: Daniel Seligman, ''Ask Mr. Statistics,'' *Fortune*, March 7, 1994. Copyright © 1994, Time Inc. Reprinted with permission. All rights reserved.

Quiz:

Using the Poisson probability distribution, calculate the probabilities mentioned in the last paragraph of this case study.

5.7.1   USING THE TABLE OF POISSON PROBABILITIES

The probabilities for a Poisson distribution can also be read from Table VI, the table of Poisson probabilities, given in Appendix C. The following example describes how to read that table.

Using the table of Poisson probabilities.

EXAMPLE 5–26   On average, two new accounts are opened per day at an Imperial Savings Bank branch. Using Table VI of Appendix C, find the probability that on a given day the number of new accounts opened at this bank will be

(a)   exactly 6   (b)   at most 3   (c)   at least 7

Solution   Let

λ = the mean number of new accounts opened per day at this bank

x = the number of new accounts opened at this bank on a given day

(a)   The values of λ and x are

$$\lambda = 2 \quad \text{and} \quad x = 6$$

In Table VI of Appendix C, we first locate the column that corresponds to $\lambda = 2$. In this column, we then read the value for $x = 6$. The relevant portion of that table is shown here as Table 5.17. The probability that exactly 6 new accounts will be opened on a given day is .0120. Therefore,

$$P(x = 6) = \mathbf{.0120}$$

Table 5.17   Portion of Table VI for $\lambda = 2.0$

x	1.1	1.2	...	2.0	$\longleftarrow \lambda = 2.0$
0				.1353	
1				.2707	
2				.2707	
3				.1804	
4				.0902	
5				.0361	
$x = 6 \longrightarrow$ 6				.0120	$\longleftarrow P(x = 6)$
7				.0034	
8				.0009	
9				.0002	

Actually, Table 5.17 gives the probability distribution of x for $\lambda = 2.0$. Note that the sum of the 10 probabilities given in Table 5.17 is .9999 and not 1.0. This is so for two reasons. First, these probabilities are rounded to four decimal places. Second, on a given day more than 9 new accounts might be opened at this bank. However, the probabilities of 10, 11, 12, ... new accounts are very small and they are not listed in the table.

(b)   The probability that at most three new accounts are opened on a given day is obtained by adding the probabilities of 0, 1, 2, and 3 new accounts. Thus, using Table VI of Appendix C or Table 5.17, we obtain:

$$P(\text{at most } 3) = P(x = 0) + P(x = 1) + P(x = 2) + P(x = 3)$$
$$= .1353 + .2707 + .2707 + .1804 = \textbf{.8571}$$

(c)   The probability that at least 7 new accounts are opened on a given day is obtained by adding the probabilities of 7, 8, or 9 new accounts. Note that 9 is the last value of x for $\lambda = 2.0$ in Table VI of Appendix C or Table 5.17. Hence, 9 is the last value of x whose probability is included in the sum. However, this does not mean that on a given day more than 9 new accounts cannot be opened. It simply means that the probability of 10 or more accounts is close to zero. Thus,

$$P(\text{at least } 7) = P(x = 7) + P(x = 8) + P(x = 9)$$
$$= .0034 + .0009 + .0002 = \textbf{.0045}$$

Constructing a Poisson probability distribution and graphing it.

EXAMPLE 5-27   An auto salesperson sells an average of .9 cars per day. Let x be the number of cars sold by this salesperson on any given day. Using the Poisson probability distribution table, write the probability distribution of x. Draw a graph of the probability distribution.

Solution   Let λ be the mean number of cars sold per day by this salesperson. Hence, $\lambda = .9$. Using the portion of Table VI corresponding to $\lambda = .9$, we write the probability distribution of x in Table 5.18. Figure 5.10 shows the line graph for the probability distribution of Table 5.18.

Table 5.18   Probability
Distribution of x for $\lambda = .9$

x	$P(x)$
0	.4066
1	.3659
2	.1647
3	.0494
4	.0111
5	.0020
6	.0003

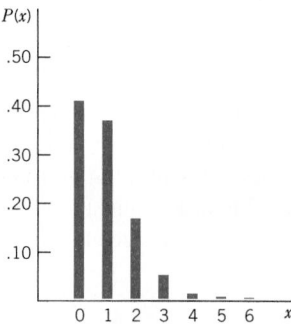

Figure 5.10   Graph of the probability
distribution of Table 5.18.

Note that 6 is the largest value of x for $\lambda = .9$ listed in Table VI for which the probability is greater than zero. However, this does not mean that this salesperson cannot sell more than six cars on a given day. What this means is that the probability of selling seven or more cars is very small. Actually, the probability of $x = 7$ for $\lambda = .9$ calculated by using the Poisson formula is .000039. When rounded to four decimal places, this probability is .0000, as listed in Table VI.   ■

5.7.2   MEAN AND STANDARD DEVIATION OF THE POISSON PROBABILITY DISTRIBUTION

For the Poisson probability distribution, the mean and variance both are equal to λ, and the standard deviation is equal to $\sqrt{\lambda}$. That is, for the Poisson probability distribution

$$\mu = \lambda$$
$$\sigma^2 = \lambda$$
$$\sigma = \sqrt{\lambda}$$

For Example 5–27, $\lambda = .9$. Therefore, for the probability distribution of x listed in Table 5.18, the mean, variance, and standard deviation are

$$\mu = \lambda = .9 \text{ cars}$$
$$\sigma^2 = \lambda = .9$$
$$\sigma = \sqrt{\lambda} = \sqrt{.9} = .949 \text{ cars}$$

EXERCISES

Concepts and Procedures

5.70 What are the conditions that must be satisfied to apply the Poisson probability distribution?

5.71 What is the parameter of the Poisson probability distribution, and what does it mean?

5.72 Using the Poisson formula, find the following probabilities.

    a. $P(x \leq 1)$ for $\lambda = 4$ **b.** $P(x = 2)$ for $\lambda = 2.2$

Verify these probabilities using Table VI of Appendix C.

5.73 Using the Poisson formula, find the following probabilities.

    a. $P(x < 2)$ for $\lambda = 3$ **b.** $P(x = 8)$ for $\lambda = 5.3$

Verify these probabilities using Table VI of Appendix C.

5.74 Let x be a Poisson random variable. Using the Poisson probabilities table, write the probability distribution of x for each of the following. Find the mean, variance, and standard deviation for each of these probability distributions. Draw a graph for each of these probability distributions.

    a. $\lambda = 1.3$ **b.** $\lambda = 2.1$

5.75 Let x be a Poisson random variable. Using the Poisson probabilities table, write the probability distribution of x for each of the following. Find the mean, variance, and standard deviation for each of these probability distributions. Draw a graph for each of these probability distributions.

    a. $\lambda = .6$ **b.** $\lambda = 1.8$

Applications

5.76 A mail-order company receives an average of 7.4 orders per day. Find the probability that it will receive exactly 10 orders on a certain day. Use the Poisson formula.

5.77 A commuter airline receives an average of 9.7 complaints per day from its passengers. Using the Poisson formula, find the probability that on a certain day this airline will receive exactly seven complaints.

 5.78 An average of 8.2 crimes are reported to police per day in a city. Find the probability that exactly three crimes will be reported to police on a certain day in this city. Use the Poisson formula.

5.79 On average, 12.5 rooms stay vacant per day at a large hotel in a city. Find the probability that on a given day exactly three rooms will be vacant. Use the Poisson formula.

 5.80 An average of 2.8 employees of a telephone company are absent per day.

    a. Find the probability that at most one employee will be absent on a given day at this company.

    b. Find the probability that on a given day the number of employees who will be absent at this company is

        i. 1 to 5   **ii.** at least 7   **iii.** at most 3

5.81 A large proportion of small businesses in the United States fail during the first few years of operation. On average, 1.3 businesses file for bankruptcy per day in a large city.

    a. Using the Poisson formula, find the probability that exactly three businesses will file for bankruptcy on a given day in this city.

    b. Using the Poisson probabilities table, find the probability that the number of businesses that will file for bankruptcy on a given day in this city is

        i. 2 to 3   **ii.** more than 3   **iii.** less than 3

5.82 Despite all efforts by the quality control department, the fabric made at Benton Corporation always contains a few defects. A certain type of fabric made at this corporation contains an average of .4 defects per 500 yards.

    a. Using the Poisson formula, find the probability that a given piece of 500 yards of this fabric will contain exactly one defect.

    b. Using the Poisson probabilities table, find the probability that the number of defects in a given 500-yard piece of this fabric will be
          i. 2 to 4   ii. more than 3   iii. less than 2

5.83 The reception office at Tom's Building Corporation receives an average of 4.9 phone calls per half hour.

    a. Using the Poisson formula, find the probability that exactly six phone calls will be received at this office during a certain hour.

    b. Using the Poisson probabilities table, find the probability that the number of phone calls received at this office during a certain hour will be
          i. less than 8   ii. more than 12   iii. 5 to 8

5.84 An average of 4.5 customers come to Columbia Savings and Loan every half hour.

    a. Find the probability that exactly two customers will come to this savings and loan during a given hour.

    b. Find the probability that during a given hour, the number of customers who will come to this savings and loan is
          i. 2 or less   ii. 10 or more

5.85 A certain newspaper contains an average of 1.1 typographical errors per page.

    a. Using the Poisson formula, find the probability that a randomly selected page of this newspaper will contain exactly 4 typographical errors.

    b. Using the Poisson probabilities table, find the probability that the number of typographical errors on a randomly selected page will be
          i. more than 3   ii. less than 4

5.86 An insurance salesperson sells an average of 1.2 insurance policies per day.

    a. Using the Poisson formula, find the probability that this salesperson will sell no insurance policy on a certain day.

    b. Let x denote the number of insurance policies that this salesperson will sell on a given day. Using the Poisson probabilities table, write the probability distribution of x.

    c. Find the mean, variance, and standard deviation of the probability distribution developed in part b.

5.87 An average of .6 accidents occur per day in a large city.

    a. Find the probability that no accident will occur in this city on a given day.

    b. Let x denote the number of accidents that will occur in this city on a given day. Write the probability distribution of x.

    c. Find the mean, variance, and standard deviation of the probability distribution developed in part b.

5.88 On average, 20 households in 50 own answering machines.

    a. Using the Poisson formula, find the probability that in a random sample of 50 households, exactly 25 will own answering machines.

    b. Using the Poisson probabilities table, find the probability that the number of households in 50 who own answering machines is
          i. at most 12   ii. 13 to 17   iii. at least 30

***5.89** Fifteen percent of the students who take a standardized test fail.

    a. Using the Poisson formula, find the probability that in a random sample of 100 students who took this test exactly 20 will fail.

    b. Using the Poisson probabilities table, find the probability that the number of students who fail this test in a randomly selected 100 examinees is
          i. at most 9   ii. 10 to 16   iii. at least 20

GLOSSARY

Bernoulli trial   One repetition of a binomial experiment. Also called a *trial*.

Binomial experiment   An experiment that contains n identical trials such that each of these n trials has only two possible outcomes, the probabilities of these two outcomes remain constant for each trial, and the trials are independent.

Binomial parameters   The total trials n and the probability of success p for the binomial probability distribution.

Binomial probability distribution   The probability distribution that gives the probability of x successes in n trials when the probability of success is p for each trial of a binomial experiment.

Combinations   The number of ways x elements can be selected from n elements.

Continuous random variable   A random variable that can assume any value in one or more intervals.

Discrete random variable   A random variable whose values are countable.

Factorial   Denoted by the symbol "!." The product of all integers from a given number to 1. For example, "$n!$" (read as *n factorial*) represents the product of all integers from n to 1.

Mean of a discrete random variable   The mean of a discrete random variable x is the value that is expected to occur per repetition, on average, if an experiment is performed a large number of times. The mean of a discrete random variable is also called its *expected value*.

Poisson parameter   The average occurrences, denoted by λ, during an interval for a Poisson probability distribution.

Poisson probability distribution   The probability distribution that gives the probability of x occurrences in an interval when the average occurrences in that interval are λ.

Probability distribution of a discrete random variable   A list of all the possible values that a discrete random variable can assume and their corresponding probabilities.

Random variable   A variable, denoted by x, whose value is determined by the outcome of a random experiment. Also called a *chance variable*.

Standard deviation of a discrete random variable   A measure of spread for the probability distribution of a discrete random variable.

KEY FORMULAS

1.   **Mean of a discrete random variable x**

$$\mu = \Sigma x P(x)$$

The mean of a discrete random variable x is also called its expected value and is denoted by $E(x)$.

2.   **Standard deviation of a discrete random variable x**

$$\sigma = \sqrt{\Sigma x^2 P(x) - \mu^2}$$

3.   **Factorials**

$$n! = n\,(n - 1)\,(n - 2) \ldots 3 \cdot 2 \cdot 1$$

4.   **Number of combinations of n items selected x at a time**

$$\binom{n}{x} = \frac{n!}{x!\,(n - x)!}$$

5.   **Binomial probability formula**

$$P(x) = \binom{n}{x} p^x\, q^{n-x}$$

6. **Mean of the binomial probability distribution**

$$\mu = np$$

7. **Standard deviation of the binomial probability distribution**

$$\sigma = \sqrt{npq}$$

8. **Poisson probability formula**

$$P(x) = \frac{\lambda^x e^{-\lambda}}{x!}$$

9. **Mean of the Poisson probability distribution**

$$\mu = \lambda$$

10. **Variance of the Poisson probability distribution**

$$\sigma^2 = \lambda$$

11. **Standard deviation of the Poisson probability distribution**

$$\sigma = \sqrt{\lambda}$$

SUPPLEMENTARY EXERCISES

5.90   Let x be the number of cars that a randomly selected auto mechanic repairs on a given day. The following table lists the probability distribution of x.

x	2	3	4	5	6
$P(x)$	.05	.22	.35	.28	.10

Find the mean and standard deviation of x. Give a brief interpretation of the value of the mean.

5.91   Let x be the number of shopping trips made during a given week by a randomly selected family from a city. The following table lists the probability distribution of x.

x	0	1	2	3	4	5
$P(x)$	.08	.24	.39	.18	.07	.04

Calculate the mean and standard deviation of x. Give a brief interpretation of the value of the mean.

5.92   Based on its analysis of the future demand for its products, the financial department at Tipper Corporation has determined that there is a .17 probability that the company will lose $1.2 million during the next year, a .21 probability that it will lose $.7 million, a .37 probability that it will make a profit of $.9 million, and a .25 probability that it will make a profit of $2.3 million.

    a.   Let x be a random variable that denotes the profit earned by this corporation during the next year. Write the probability distribution of x.

    b.   Find the mean and standard deviation of the probability distribution of part a. Give a brief interpretation of the value of the mean.

5.93   GESCO Insurance Company charges a $350 premium per annum for a $100,000 life insurance policy for a 40-year-old female. The probability that a 40-year-old female will die within one year is .002.

    a.   Let x be a random variable that denotes the gain of the company for next year from a $100,000 life insurance policy sold to a 40-year-old female. Write the probability distribution of x.

b. Find the mean and standard deviation of the probability distribution of part a. Give a brief interpretation of the value of the mean.

5.94 Spoke Weaving Corporation has eight weaving machines of the same kind and of the same age. The probability is .04 that any weaving machine will break down at any time. Find the probability that at any given time

a. all eight weaving machines will be broken down
b. exactly two weaving machines will be broken down
c. none of the weaving machines will be broken down

5.95 At the Bank of California, past data show that 7% of all credit card holders default at some time in their lives. On one recent day, this bank issued 12 credit cards to new customers. Find the probability that of these 12 customers, eventually

a. exactly 3 will default
b. exactly 1 will default
c. none will default

5.96 Maine Corporation buys motors for electric fans from another company that guarantees that at most 5% of its motors are defective and that it will replace all defective motors at no cost to Maine Corporation. The motors are received in large shipments. The quality control department at Maine Corporation randomly selects 20 motors from each shipment and inspects them for being good or defective. If this sample contains more than 2 defective motors, the entire shipment is rejected.

a. Using the appropriate probabilities table from Appendix C, find the probability that a given shipment of motors received by Maine Corporation will be accepted. Assume that 5% of all motors received by Maine Corporation are defective.
b. Using the appropriate probabilities table from Appendix C, find the probability that a given shipment of motors received by Maine Corporation will be rejected.

5.97 One of the toys made by Dillon Corporation is called Speaking Joe, which is sold only by mail. Consumer satisfaction is one of the top priorities of the company's management. The company guarantees a refund or a replacement for any Speaking Joe toy if the chip that is installed inside becomes defective within a year from the date of purchase. It is known from past data that 10% of these chips become defective within a one-year period. The company sold 15 Speaking Joes on a given day.

a. Let x denote the number of Speaking Joes in these 15 that will be returned for a refund or a replacement within a one-year period. Using the appropriate probabilities table from Appendix C, obtain the probability distribution of x and draw a graph of the probability distribution. Determine the mean and standard deviation of x.
b. Using the probability distribution constructed in part a, find the probability that exactly 5 of the 15 Speaking Joes will be returned for a refund or a replacement within a one-year period.

5.98 An average of eight videos are rented per day at a video-rental store.

a. Using the appropriate formula, find the probability that on a given day exactly five videos will be rented at this store.
b. Using the appropriate probabilities table from Appendix C, find the probability that on a given day the number of videos that will be rented at this store is
    i. at least 10   ii. at most 4   iii. 6 to 9

5.99 An average of 5.3 robberies occur per day in a large city.

a. Using the Poisson formula, find the probability that on a given day exactly three robberies will occur in this city.
b. Using the appropriate probabilities table from Appendix C, find the probability that on a given day the number of robberies that will occur in this city is
    i. at least 12   ii. at most 3   iii. 2 to 6

5.100 An average of 1.2 private airplanes arrive per hour at an airport.

a. Find the probability that during a given hour no private airplane will arrive at this airport.
b. Let x denote the number of private airplanes that will arrive at this airport during a given hour. Write the probability distribution of x.

5.101 A machine produces an average of .8 items per minute.

    a. Using the appropriate formula, find the probability that during a certain minute this machine will produce exactly three items.

    b. Let x denote the number of items that this machine will produce during a given minute. Using the appropriate probabilities table from Appendix C, write the probability distribution of x.

*5.102 Scott offers you the following game: You will roll two fair dice. If the sum of the two numbers obtained is 2, 3, 4, 9, 10, 11, or 12, Scott will pay you $20. However, if the sum of the two numbers is 5, 6, 7, or 8, you will pay Scott $20. Scott points out that you have seven winning numbers and only four losing numbers. Is this game fair to you? Should you accept this offer? Support your conclusion with appropriate calculations.

*5.103 Suppose the owner of a salvage company is considering raising a sunken ship. If successful, the venture will yield a net profit of $10 million. Otherwise the owner will lose $4 million. Let p denote the probability of success for this venture. Assume the owner is willing to take the risk to go ahead with this project provided the expected net profit is at least $500,000. What is the smallest value of p for which the owner will risk undertaking this project?

*5.104 Two teams, A and B, will play a best-of-seven series, which will end as soon as one of the teams wins four games. Thus, the series may end in four, five, six, or seven games. Assume that each team has an equal chance of winning each game and that all games are independent of each other. Find the following probabilities.

    a. Team A wins the series in four games.

    b. Team A wins the series in five games.

    c. Seven games are required for a team to win the series.

*5.105 York Steel Corporation produces a special bearing that must meet rigid specifications. When the production process is running properly, 10% of the bearings fail to meet the required specifications. Sometimes problems develop with the production process that cause the rejection rate to exceed 10%. To guard against this higher rejection rate, samples of 15 bearings are taken periodically and carefully inspected. If more than two bearings in a sample of 15 fail to meet the required specifications, the production is suspended for necessary adjustments.

    a. If the true rate of rejection is 10% (that is, the production process is working properly), what is the probability that the production will be suspended based on a sample of 15 bearings?

    b. What assumptions did you make in part a?

*5.106 Residents in an inner city area are concerned about drug dealers entering their neighborhood. Over the past fourteen nights, they have taken turns watching the street from a darkened apartment. Drug deals seem to take place randomly at various times and locations on the street and average about three per night. The residents of this street contacted the local police who informed them that they do not have sufficient resources to set up surveillance. The police suggested videotaping the activity on the street, and if the residents are able to capture five or more drug deals on tape, the police will take action. Unfortunately, none of the residents in this street owns a video camera and, hence, they would have to rent the equipment. Inquiries at the local dealers indicated that the best available rate for renting a video camera is $75 for the first night and $40 for each additional night. To obtain this rate, the residents must sign up in advance for a specified number of nights. The residents hold a neighborhood meeting and invite you to help them decide on the length of the rental period. Since it is difficult for them to pay the rental fees, they want to know the probability of taping at least five drug deals for a given number of nights of videotaping.

    a. Which of the probability distributions you have studied might be helpful here?

    b. What assumption(s) would you have to make?

    c. If the residents tape for two nights, what is the probability of filming at least five drug deals?

    d. For how many nights must the camera be rented so that there is at least .90 probability that five or more drug deals will be taped?

*5.107 A high school history teacher gives a 50-question multiple-choice examination in which each question has four choices. The scoring includes a penalty for guessing. Each correct answer is worth 1 point, and each wrong answer costs 1/2 point. For example, if a student answers 35 questions correctly, 8 questions incorrectly, and does not answer 7 questions, the total score for this student will be $35 - (1/2)(8) = 31$.

a. What is the expected score of a student who answers 38 questions correctly and guesses on the other 12 questions? Assume the student randomly chooses one of the four answers for each of the 12 guessed questions.

b. Does a student increase his expected score by guessing a question if he has no idea what the correct answer is? Explain.

c. Does a student increase her expected score by guessing on a question for which she can eliminate one of the wrong answers? Explain.

*5.108   A baker, who makes fresh cheesecakes daily, sells an average of five such cakes per day. How many cheesecakes should he make each day so that the probability of running out and losing one or more sales is less than .10? Assume that the number of cheesecakes sold each day follows a Poisson distribution. You may use the Poisson probabilities table from Appendix C.

SELF-REVIEW TEST

1.   Briefly explain the meaning of a random variable, a discrete random variable, and a continuous random variable. Give one example each of a discrete and a continuous random variable.

2.   What name is given to a table that lists all the values that a discrete random variable x can assume and their corresponding probabilities?

3.   For the probability distribution of a discrete random variable, the probability of any single value of x is always

   a.   in the range zero to 1   b.   1.0   c.   less than zero

4.   For the probability distribution of a discrete random variable, the sum of the probabilities of all possible values of x is always

   a.   greater than 1   b.   1.0   c.   less than 1.0

5.   The number of combinations of 10 items selected 7 at a time is

   a.   120   b.   200   c.   80

6.   State the four conditions of a binomial experiment. Give one example of such an experiment.

7.   The parameters of the binomial probability distribution are

   a. $n, p,$ and q b. n and p c. $n, p,$ and x

8.   The mean and standard deviation of a binomial probability distribution with $n = 25$ and $p = .20$ are

   a.   5 and 2   b.   8 and 4   c.   4 and 3

9.   The binomial probability distribution is symmetric if

   a. $p < .5$ b. $p = .5$ c. $p > .5$

10.   The binomial probability distribution is skewed to the right if

   a. $p < .5$ b. $p = .5$ c. $p > .5$

11.   The binomial probability distribution is skewed to the left if

   a. $p < .5$ b. $p = .5$ c. $p > .5$

12. The parameter/parameters of the Poisson probability distribution is/are

     a. λ **b.** λ and x **c.** λ and e

13. Describe the three conditions that must be satisfied to apply the Poisson probability distribution.

14. Let x be the number of homes sold per week by all four real estate agents working at a Century 21 office. The following table lists the probability distribution of x.

x	0	1	2	3	4	5
$P(x)$	.15	.24	.31	.14	.10	.06

Calculate the mean and standard deviation of x. Give a brief interpretation of the value of the mean.

15. According to a survey, 70% of adults believe that every college student should be required to take at least one course in ethics. Assume that this percentage is true for the current population of all adults.

     a. Find the probability that the number of adults in a random sample of 12 who hold this view is

         i.   exactly 10 (use the appropriate formula)

         ii.   at least 7 (use the appropriate table from Appendix C)

         iii.   less than 4 (use the appropriate table from Appendix C)

     b. Let x be the number of adults in a random sample of 12 who believe that every college student should be required to take at least one course in ethics. Using the appropriate table from Appendix C, write the probability distribution of x. Find the mean and standard deviation of this probability distribution.

16. A department store sells an average of two electric appliances per day.

     a. Find the probability that on a given day this store will sell

         i.   exactly five electric appliances (use the appropriate formula)

         ii.   at most four electric appliances (use the appropriate table from Appendix C)

         iii.   five to nine electric applicances (use the appropriate table from Appendix C)

     b. Let x be the number of electric appliances sold at this store on a given day. Write the probability distribution of x. Use the appropriate table from Appendix C.

17. The binomial probability distribution is symmetric when $p = .50$; it is skewed to the right when $p < .50$; and it is skewed to the left when $p > .50$. Illustrate these three cases by writing three probability distributions and graphing them. Choose any values of n and p and use the table of binomial probabilities (Table IV of Appendix C).

USING MINITAB

Using MINITAB, we can find the probability of a single outcome for the binomial and Poisson probability distributions or we can list the probabilities for all outcomes of a binomial or a Poisson experiment.

THE BINOMIAL PROBABILITY DISTRIBUTION

Illustration M5–1 shows how to find the probability of a single outcome for the binomial probability distribution, Illustration M5–2 explains how to find the cumulative probability for the binomial probability distribution, and Illustration M5–3 describes the procedure for obtaining the probability distribution of x for a binomial distribution.

ILLUSTRATION M5–1   Reconsider Example 5–18. Five percent of all VCRs manufactured by an electronics company are defective. Find the probability that if three VCRs are randomly selected from the production line of this company, exactly one of them will be defective.

Solution   From the given information,

$$n = \text{total number of VCRs in the sample} = 3$$

$$x = \text{defective VCRs in the sample} = 1$$

$$p = P(\text{a VCR is defective}) = .05$$

If you are using MINITAB FOR WINDOWS, perform the following steps to find the probability of $x = 1$ when $n = 3$ and $p = .05$.

Step 1.   Click the **Calc** pull-down menu at the top of the screen.
Step 2.   Click **Probability Distributions** from the selections available in the **Calc** menu.
Step 3.   Click **Binomial** from the selections available in the **Probability Distributions** menu.
Step 4.   You will see a dialog box entitled **Binomial Distribution** appear on the screen. Click inside the circle next to **Probability**. Type **3** (the value of n) in the box next to **Number of trials** and **.05** (the value of p) in the box next to **Probability of success.** Then click inside the circle next to **Input constant** and type **1** (the value of x) in the box next to it.
Step 5.   Click the **OK** button at the bottom of this dialog box. The probability of $x = 1$ will appear on the screen.

If you are using the MINITAB COMMAND LANGUAGE, type the following MINITAB commands to obtain the probability of $x = 1$.

```
MTB > PDF x=1;
SUBC > BINOMIAL n=3 p=.05.
```

Whether you use MINITAB FOR WINDOWS or the MINITAB COMMAND LANGUAGE, you will obtain the output given in Figure 5.11. From this output, the required probability is .1354.

Figure 5.11   Binomial probability of $x = 1$ for $n = 3$ and $p = .05$.

```
Probability Density Function

Binomial with n = 3 and p = 0.0500000

         x        P( X = x)
      1.00          0.1354
```

ILLUSTRATION M5–2   Refer to Illustration M5–1. Find $P(x \leq 2)$.

Solution   Here, you want to find the cumulative probability, $P(x \leq 2)$, for the binomial distribution problem of Illustration M5–1. In other words, you need to find the probability that at most two VCRs are defective in a sample of three when the probability of success is .05. To find this probability using MINITAB FOR WINDOWS, click inside the circle next to **Cumulative probability** instead of **Probability** in Step 4 of the steps described in Illustration M5–1. Then, type **3** (the value of n) in the box next to **Number of trials** and **.05** (the value of p) in the box next to **Probability of success.** Then click inside the circle next to **Input constant** and type **2** (the value of x) in the box next to it. The remaining steps are the same as in Illustration M5–1.

If you are using the MINITAB COMMAND LANGUAGE, replace the first MINITAB command with

```
MTB > CDF x=2;
```

The subcommand does not change.

Either of these procedures will produce the output shown in Figure 5.12, which gives $P(x \leq 2) = .9999$.

Figure 5.12   Cumulative binomial probability of $P(x \leq 2)$ for $n = 3$ and $p = .05$.

```
Cumulative Distribution Function

Binomial with n = 3 and p = 0.0500000

         x        P( X <= x)
      2.00          0.9999
```

ILLUSTRATION M5–3   Reconsider Illustration M5–1. Let x denote the number of defective VCRs in a random sample of three. Using MINITAB, list the probability distribution of x.

Solution   As we know,

$$n = 3 \quad \text{and} \quad p = .05$$

If you are using MINITAB FOR WINDOWS, perform the following steps to obtain the probability distribution of x for $n = 3$ and $p = .05$.

Step 1.   Enter the values 0, 1, 2, and 3 in column C1 of the Data window. Note that these are the possible values that x can assume in this example.

Step 2.   Click the **Calc** pull-down menu at the top of the screen.

Step 3.   Click **Probability Distributions** from the selections available in the **Calc** menu.

Step 4.   Click **Binomial** from the selections available in the **Probability Distributions** menu.

Step 5. You will see a dialog box entitled **Binomial Distribution** appear on the screen. Click inside the circle next to **Probability.** Type **3** (the value of *n*) in the box next to **Number of trials** and **.05** (the value of *p*) in the box next to **Probability of success.** Then click inside the circle next to **Input column** and type **C1** inside the box next to it.

Step 6. Click the **OK** button at the bottom of this dialog box. The probability distribution of *x* will appear on the screen.

If you are using the MINITAB COMMAND LANGUAGE, type the following MINITAB commands to obtain the probability distribution of *x*.

```
MTB > PDF;
SUBC > BINOMIAL n=3 p=.05.
```

Whether you use MINITAB FOR WINDOWS or the MINITAB COMMAND LANGUAGE, you will obtain the output given in Figure 5.13. In this figure, the column labeled *x* lists the possible values of *x* and the column labeled $P(X = x)$ gives their corresponding probabilities.

Figure 5.13 Binomial probability distribution of *x*.

```
Probability Density Function

Binomial with n = 3 and p = 0.0500000

         x        P( X = x)
      0.00          0.8574
      1.00          0.1354
      2.00          0.0071
      3.00          0.0001
```

Using the MINITAB output of Figure 5.13, we can write the probability of any value of *x* for this example. For instance,

$$P(1 \leq x \leq 3) = P(1) + P(2) + P(3) = .1354 + .0071 + .0001 = .1426$$

$$P(\text{at least 2 defective VCRs}) = P(2) + P(3) = .0071 + .0001 = .0072$$

$$P(\text{at most 1 defective VCR}) = P(0) + P(1) = .8574 + .1354 = .9928$$

THE POISSON PROBABILITY DISTRIBUTION

Illustration M5–4 shows how to find the probability of a single outcome for the Poisson probability distribution, Illustration M5–5 explains how to find the cumulative probability for the Poisson probability distribution, and Illustration M5–6 describes the procedure for obtaining the probability distribution of *x* for a Poisson probability distribution.

ILLUSTRATION M5–4 A washing machine in a laundromat breaks down an average of 1.2 times per month. Find the probability that during a given month it will have exactly three breakdowns.

Solution From the given information,

$$\lambda = \text{mean number of breakdowns per month} = 1.2$$

$$x = \text{number of breakdowns during a given month} = 3$$

If you are using MINITAB FOR WINDOWS, perform the following steps to find the probability of $x = 3$ when $\lambda = 1.2$.

Step 1. Click the **Calc** pull-down menu at the top of the screen.

Step 2. Click **Probability Distributions** from the selections available in the **Calc** menu.

Step 3. Click **Poisson** from the selections available in the **Probability Distributions** menu.

Step 4. You will see a dialog box entitled **Poisson Distribution** appear on the screen. Click inside the circle next to **Probability.** Type **1.2** (the value of λ) in the box next to **Mean.** Then click inside the circle next to **Input constant** and type **3** (the value of x) in the box next to it.

Step 5. Click the **OK** button at the bottom of this dialog box. The probability of $x = 3$ will appear on the screen.

If you are using the MINITAB COMMAND LANGUAGE, type the following MINITAB commands to obtain the probability of $x = 3$.

```
MTB > PDF x=3;
SUBC > POISSON MEAN = 1.2.
```

Whether you use MINITAB FOR WINDOWS or the MINITAB COMMAND LANGUAGE, you will obtain the output given in Figure 5.14. From this output, the required probability is .0867. Note that in the MINITAB output, the statement "Poisson with mu = 1.20000" refers to $\lambda = 1.2$.

Figure 5.14 Poisson probability of $x = 3$ when $\lambda = 1.2$.

```
Probability Density Function

Poisson with mu = 1.20000

        x        P( X = x)
     3.00          0.0867
```

ILLUSTRATION M5–5   Refer to Illustration M5–4. Find $P(x \leq 2)$.

Solution   Here, you want to find the cumulative probability, $P(x \leq 2)$, for the Poisson distribution problem mentioned in Illustration M5–4. In other words, you need to find the probability that there will be at most two breakdowns during a given month when $\lambda = 1.2$. To find this probability using MINITAB FOR WINDOWS, click inside the circle next to **Cumulative probability** instead of **Probability** in Step 4 of the steps described in Illustration M5–4. Then, type **1.2** (the value of λ) in the box next to **Mean.** Then click inside the circle next to **Input constant** and type **2** (the value of x) in the box next to it. The remaining steps are the same as in Illustration M5–4.

If you are using the MINITAB COMMAND LANGUAGE, replace the first MINITAB command mentioned in Illustration M5–4 with

```
MTB > CDF x=2;
```

The subcommand remains unchanged.

Either of these procedures will produce the MINITAB output shown in Figure 5.15, which gives $P(x \leq 2) = .8795$.

Figure 5.15 Cumulative Poisson probability $P(x \leq 2)$ when $\lambda = 1.2$.

```
Cumulative Distribution Function

Poisson with mu = 1.20000

        x       P(X <= x)
     2.00         0.8795
```

ILLUSTRATION M5–6   Reconsider Illustration M5–4. Let x denote the number of breakdowns observed during a given month. Using MINITAB, list the probability distribution of x.

Solution   From the given information, $\lambda = 1.2$.

If you are using MINITAB FOR WINDOWS, perform the following steps to obtain the probability distribution of x for $\lambda = 1.2$.

Step 1.   Enter the values 0, 1, 2, . . . in column C1 of the Data window. Note that these are the possible values that x can assume in this example. As it is difficult to know the largest value that x can assume in a Poisson probability distribution, you may enter as many values as you want.
Step 2.   Click the **Calc** pull-down menu at the top of the screen.
Step 3.   Click **Probability Distributions** from the selections available in the **Calc** menu.
Step 4.   Click **Poisson** from the selections available in the **Probability Distributions** menu.
Step 5.   You will see a dialog box entitled **Poisson Distribution** appear on the screen. Click inside the circle next to **Probability**. Type **1.2** (the value of λ) in the box next to **Mean**. Then click inside the circle next to **Input column** and type **C1** in the box next to it.
Step 6.   Click the **OK** button at the bottom of this dialog box. The probability distribution of x will appear on the screen.

If you are using the MINITAB COMMAND LANGUAGE, type the following MINITAB commands to obtain the probability distribution of x.

```
MTB > PDF;
SUBC > POISSON MEAN = 1.2.
```

Whether you use MINITAB FOR WINDOWS or the MINITAB COMMAND LANGUAGE, you will obtain the output given in Figure 5.16. In this figure, the column labeled x lists the possible values of x and the column labeled $P(X = x)$ gives their corresponding probabilities.

Figure 5.16   Poisson probability distribution of x when $\lambda = 1.2$.

```
Probability Density Function

Poisson with mu = 1.20000

       x          P(X = x)
    0.00           0.3012
    1.00           0.3614
    2.00           0.2169
    3.00           0.0867
    4.00           0.0260
    5.00           0.0062
    6.00           0.0012
    7.00           0.0002
    8.00           0.0000
```

The problem with using MINITAB FOR WINDOWS here will be that you may not know how many values of x should be entered in column C1. In the previous example (Illustration M5–6), if you enter only up to $x = 6$ in column C1, then the probability distribution will list the probabilities only up to $x = 6$. Hence, in a case where you need to list the probability distribution of x, it is easier to use the MINITAB COMMAND LANGUAGE.

Using the output given in Figure 5.16, we can determine the probability of any value of x for this Poisson probability distribution problem. For example, the probability that we will observe less than three breakdowns during a given month is

$$P(x < 3) = P(0) + P(1) + P(2) = .3012 + .3614 + .2169 = .8795$$

Similarly, the probability that we will observe at least four breakdowns during a given month is

$$P(x \geq 4) = P(4) + P(5) + P(6) + P(7) = .0260 + .0062 + .0012 + .0002 = .0336$$

COMPUTER ASSIGNMENTS

M5.1   Forty-five percent of the adult population in a large city are women. A court is to randomly select a jury of five adults from the population of all adults of this city.

    a.   Using MINITAB, find the probability that none of the five jurors is a woman.

    b.   Using the MINITAB CDF command, find the probability that at most two of the five jurors are women.

    c.   Let x denote the number of women in five adults selected for this jury. Using MINITAB, obtain the probability distribution of x.

    d.   Using the probability distribution obtained in part c, find the following probabilities.

        i. $P(x > 2)$ **ii.** $P(x \leq 1)$ **iii.** $P(2 \leq x \leq 4)$

M5.2   The Metropolitan Life Survey of the American Teacher estimated that 54% of the nation's teachers were ''very satisfied'' with their jobs in 1995 (Ann Bradley, ''Nation's Teachers Feeling Better About Jobs, Salaries, Survey Finds,'' *Education Week*, December 6, 1995). Assume that this percentage is true for the current population of all U.S. teachers.

    a.   Using MINITAB, find the probability that the number of teachers in a sample of 12 who are ''very satisfied'' with their jobs is 8.

    b.   Using the MINITAB CDF command, find the probability that at most 5 teachers in a sample of 12 are ''very satisfied'' with their jobs.

    c.   Let x denote the number of teachers in a sample of 12 who are ''very satisfied'' with their jobs. Using MINITAB, obtain the probability distribution of x.

    d.   Using the probability distribution obtained in part c, find the following probabilities.

        i. $P(x \geq 7)$ **ii.** $P(x < 6)$ **iii.** $P(3 < x < 9)$

M5.3   A mail-order company receives an average of 7.4 orders per day.

    a.   Using MINITAB, find the probability that it will receive exactly 10 orders on a certain day.

    b.   Using the MINITAB CDF command, find the probability that it will receive at most six orders on a certain day.

    c.   Let x denote the number of orders received by this company on a given day. Using MINITAB, obtain the probability distribution of x.

    d.   Using the probability distribution obtained in part c, find the following probabilities.

        i. $P(x \geq 5)$ **ii.** $P(x < 9)$ **iii.** $P(5 < x < 10)$

M5.4   A commuter airline receives an average of 3.7 complaints per day from its passengers. Let x denote the number of complaints received by this airline on a given day.

    a.   Using MINITAB, find $P(x = 0)$.

    b.   Using the MINITAB CDF command, find $P(x \leq 7)$.

    c.   Using MINITAB, obtain the probability distribution of x.

    d.   Using the probability distribution obtained in part c, find the following probabilities.

        i. $P(x > 4)$ **ii.** $P(x \leq 5)$ **iii.** $P(4 \leq x \leq 7)$

6

CONTINUOUS RANDOM VARIABLES AND THE NORMAL DISTRIBUTION

Discrete random variables and their probability distributions were presented in Chapter 5. Section 5.1 defined a continuous random variable as a variable that can assume any value in one or more intervals.

The possible values that a continuous random variable can assume are infinite and uncountable. For example, the variable representing the time taken by a worker to commute from home to work is a continuous random variable. Suppose 5 minutes is the minimum time and 130 minutes is the maximum time taken by all workers to commute from home to work. Let x be a continuous random variable that denotes the time taken to commute from home to work by a randomly selected worker. Then x can assume any value in the interval 5 to 130 minutes. This interval contains an infinite number of values that are uncountable.

A continuous random variable can possess one of many probability distributions. In this chapter we discuss the normal probability distribution and the normal distribution as an approximation to the binomial distribution.

6.1 CONTINUOUS PROBABILITY DISTRIBUTION

In Chapter 5 we defined a **continuous random variable** as a random variable whose values are not countable. A continuous random variable can assume any value over an interval or intervals. Because the number of values contained in any interval is infinite, the possible number of values that a continuous random variable can assume is also infinite. Moreover, we cannot count these values. In Chapter 5, it was stated that the life of a battery, heights of persons, time taken to complete an examination, amount of milk in a gallon, weights of babies, and prices of houses are all examples of continuous random variables. Note that although money can be counted, usually all variables involving money are considered to be continuous random variables. This is so because a variable involving money often has a very large number of outcomes.

Table 6.1 Frequency and Relative Frequency Distributions of Heights of Female Students

Height of a Female Student (in inches) x	f	Relative Frequency
60 to less than 61	90	.018
61 to less than 62	170	.034
62 to less than 63	460	.092
63 to less than 64	750	.150
64 to less than 65	970	.194
65 to less than 66	760	.152
66 to less than 67	640	.128
67 to less than 68	440	.088
68 to less than 69	320	.064
69 to less than 70	220	.044
70 to less than 71	180	.036
	$N = 5000$	Sum $= 1.0$

Suppose there are 5000 female students enrolled at a university, and x is the continuous random variable that represents heights of these female students. Table 6.1 lists the frequency and relative frequency distributions of x.

The relative frequencies listed in Table 6.1 can be used as probabilities of respective classes. Note that these are exact probabilities because we are considering the population of all female students.

Figure 6.1 displays the histogram and polygon for the relative frequency distribution of Table 6.1. Figure 6.2 shows the smoothed polygon for the data of Table 6.1. The smoothed polygon is an approximation of the *probability distribution curve* of the continuous random variable x. Note that each class in Table 6.1 has a width equal to 1 inch. If the width of classes is more than 1 unit, we first obtain the *relative frequency densities* and then graph these relative frequency densities to obtain the distribution curve. The relative frequency density of a class is obtained by dividing the relative frequency of that class by the class width. The relative frequency densities are calculated to make the sum of the areas of all rectangles in the histogram equal to 1.0. Case Study 6–1, which appears later in this section, illustrates this procedure. The probability distribution curve of a continuous random variable is also called its *probability density function*.

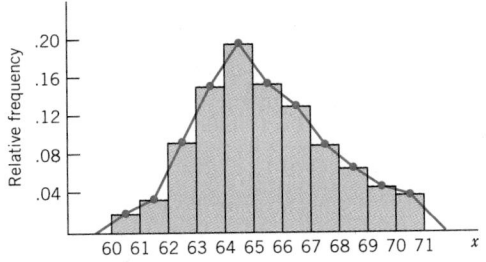

Figure 6.1 Histogram and polygon for Table 6.1.

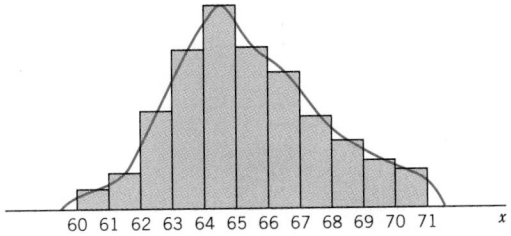

Figure 6.2 Probability distribution curve for heights.

The probability distribution of a continuous random variable possesses the following *two characteristics*.

1. The probability that x assumes a value in any interval lies in the range 0 to 1.

2. The total probability of all the (mutually exclusive) intervals within which x can assume a value is 1.0.

The first characteristic states that the area under the probability distribution curve of a continuous random variable between any two points is between 0 and 1, as shown in Figure 6.3. The second characteristic indicates that the total area under the probability distribution curve of a continuous random variable is always 1.0 or 100%, as shown in Figure 6.4.

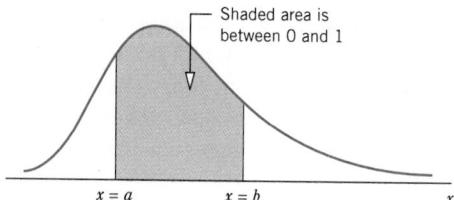

Figure 6.3 Area under a curve between two points.

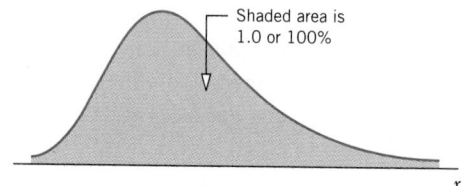

Figure 6.4 Total area under a probability distribution curve.

The probability that a continuous random variable x assumes a value within a certain interval is given by the area under the curve between two limits of the interval, as shown in Figure 6.5. The shaded area under the curve from a to b in this figure gives the probability that x falls in the interval a to b. That is,

$$P(a \leq x \leq b) = \text{Area under the curve from } a \text{ to } b$$

Note that the interval $a \leq x \leq b$ states that x is greater than or equal to a but less than or equal to b.

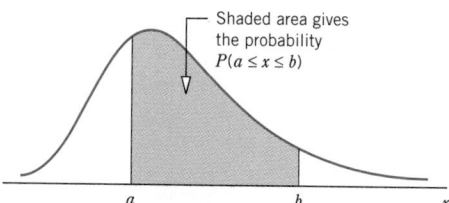

Figure 6.5 Area under the curve as probability.

Reconsider the example on the heights of all female students at a university. The probability that the height of a randomly selected female student from this university lies in the interval 65 to 68 inches is given by the area under the distribution curve of the heights of all female students from $x = 65$ to $x = 68$, as shown in Figure 6.6. This probability is written as

$$P(65 \leq x \leq 68)$$

which states that x is greater than or equal to 65 but less than or equal to 68.

For a continuous probability distribution, the probability is always calculated for an interval. For example, in Figure 6.6, the interval representing the shaded area is from 65 to 68. Consequently, the shaded area in that figure gives the probability for the interval $65 \leq x \leq 68$.

The probability that a continuous random variable x assumes a single value is always zero. This is so because the area of a line, which represents a single point, is zero. For

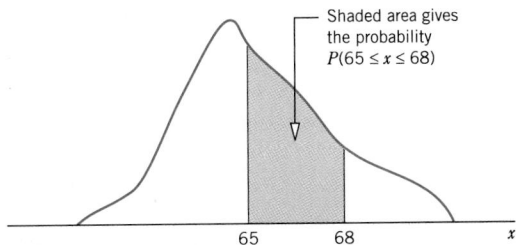

Figure 6.6 Probability that x lies in the interval 65 to 68.

example, if x is the height of a randomly selected female student from that university, then the probability that this student is exactly 67 inches tall is zero. That is,

$$P(x = 67) = 0$$

This probability is shown in Figure 6.7. Similarly, the probability for x to assume any other single value is zero.

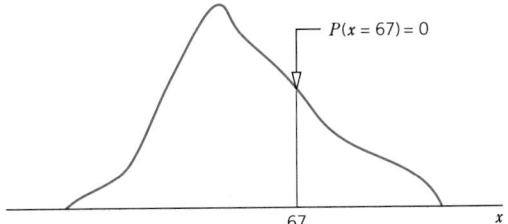

Figure 6.7 Probability of a single value of x is zero.

In general, if a and b are two of the values that x can assume, then,

$$P(a) = 0 \quad \text{and} \quad P(b) = 0$$

From this we can deduce that for a continuous random variable

$$P(a \leq x \leq b) = P(a < x < b)$$

In other words, the probability that x assumes a value in the interval a to b is the same whether or not the values a and b are included in the interval. For the example on the heights of female students, the probability that a randomly selected female student is between 65 and 68 inches tall is the same as the probability that this female is 65 to 68 inches tall. This is shown in Figure 6.8.

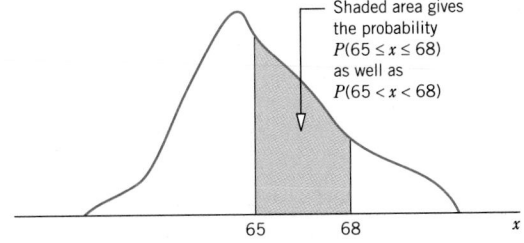

Figure 6.8 Probability "from 65 to 68" and "between 65 and 68."

Note that the interval "between 65 and 68" represents "$65 < x < 68$," and it does not include 65 and 68. On the other hand, the interval "from 65 to 68" represents "$65 \leq x \leq 68$" and it does include 65 and 68. However, as mentioned previously, in the case of a continuous random variable both of these intervals contain the same probability or area under the curve.

Case Study 6–1 describes how we obtain the probability distribution curve of a continuous random variable.

CASE STUDY 6-1   DISTRIBUTION OF TIME TAKEN TO RUN A ROAD RACE

Table 6.2 gives the frequency and relative frequency distributions for the time (in minutes) taken to complete the Manchester Road Race (held on November 28, 1996) for a total of 9071 participants who finished that race. This road race event is held every year on Thanksgiving Day in Manchester, Connecticut. The total distance of the race course is 4.748 miles. The relative frequencies of Table 6.2 are used to construct the histogram and polygon in Figure 6.9.

Table 6.2   Frequency and Relative Frequency Distributions for the Road Race data

Class	Frequency	Relative Frequency
20 to less than 25	37	.0041
25 to less than 30	319	.0352
30 to less than 35	739	.0815
35 to less than 40	1301	.1434
40 to less than 45	1620	.1786
45 to less than 50	1804	.1989
50 to less than 55	1355	.1494
55 to less than 60	582	.0642
60 to less than 65	356	.0392
65 to less than 70	176	.0194
70 to less than 75	215	.0237
75 to less than 80	229	.0252
80 to less than 85	166	.0183
85 to less than 90	99	.0109
90 to less than 95	58	.0064
95 to less than 100	15	.0017
	$\Sigma f = 9071$	Sum = 1.0001

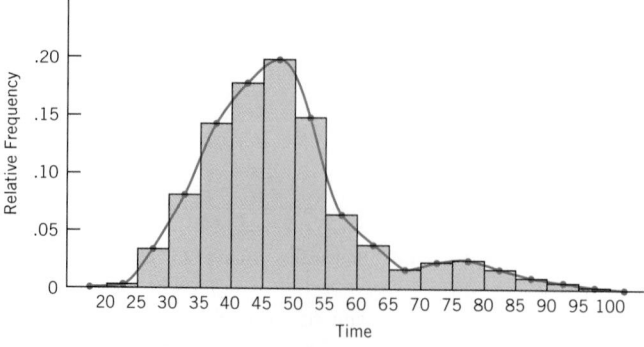

Figure 6.9 Histogram and polygon for the Road Race data.

To derive the probability distribution curve for these data, first we calculate the relative frequency densities that are obtained by dividing the relative frequencies by the class widths. The width of each class in Table 6.2 is five. By dividing the relative frequencies of Table 6.2 by 5, we obtain the relative frequency densities, which are recorded in Table 6.3. Using the relative frequency densities, we draw a histogram and smoothed polygon, as shown in Figure 6.10. The curve in this figure gives the probability distribution curve for the road race data.

Note that the areas of rectangles in Figure 6.9 do not give probabilities (which are approximated by relative frequencies). Rather, it is the heights of these rectangles that give the probabilities. This is so because the base of each rectangle is five in this histogram. Consequently, the area of any rectangle is given by its height multiplied by five. Thus, the total area of all rectangles in Figure 6.9 is 5.0 and not 1.0. However, in Figure 6.10 it is the areas and not the heights of rectangles that give the probabilities of respective classes. Thus, if we add the areas of all rectangles in Figure 6.10, we obtain the sum of all probabilities equal to 1.0. Consequently, the total area under the curve is equal to 1.0.

Table 6.3   Relative Frequency Densities

Class	Relative Frequency Density
20 to less than 25	.00082
25 to less than 30	.00704
30 to less than 35	.01630
35 to less than 40	.02868
40 to less than 45	.03572
45 to less than 50	.03978
50 to less than 55	.02988
55 to less than 60	.01284
60 to less than 65	.00784
65 to less than 70	.00388
70 to less than 75	.00474
75 to less than 80	.00504
80 to less than 85	.00366
85 to less than 90	.00218
90 to less than 95	.00128
95 to less than 100	.00034

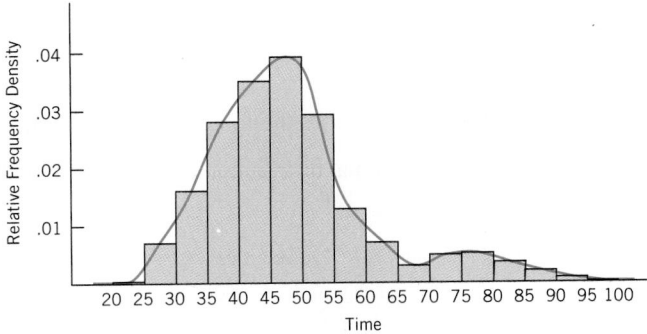

Figure 6.10 Probability distribution curve for Road Race data.

The probability distribution of a continuous random variable has a mean and a standard deviation, which are denoted by μ and σ, respectively. The mean and standard deviation of the probability distribution curve of Figure 6.10 are 48.118 and 13.122 minutes, respectively. These values of μ and σ are calculated by using the raw data on 9071 participants.

Source: This case study is based on data published in *The Hartford Courant*, December 2, 1996.

6.2   THE NORMAL DISTRIBUTION

The normal distribution is one of the many probability distributions that a continuous random variable can possess. The normal distribution is the most important and most widely used of all probability distributions. A large number of phenomena in the real world are normally distributed either exactly or approximately. The continuous random variables representing the heights and weights of people, scores on an examination, weights of packages (e.g., cereal boxes, boxes of cookies), amount of milk in a gallon, life of an item (such as a light bulb or a television set), and the time taken to complete a certain job have all been observed to have a (approximate) normal distribution.

The **normal probability distribution** or the *normal curve* is given by a bell-shaped (symmetric) curve. Such a curve is shown in Figure 6.11. Its mean is denoted by μ and standard deviation by σ. A continuous random variable x that has a normal distribution is called a *normal random variable*. Note that not all bell-shaped curves represent a normal distribution curve. Only a specific kind of bell-shaped curve represents a normal curve.

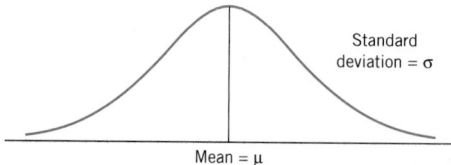

Figure 6.11   Normal distribution with mean μ and standard deviation σ.

NORMAL PROBABILITY DISTRIBUTION

A *normal probability distribution*, when plotted, gives a bell-shaped curve such that

1. The total area under the curve is 1.0.
2. The curve is symmetric about the mean.
3. The two tails of the curve extend indefinitely.

A normal distribution possesses the following three characteristics.

1. The total area under a normal distribution curve is 1.0 or 100%, as shown in Figure 6.12.

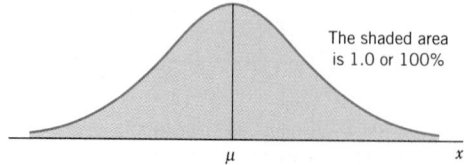

Figure 6.12   Total area under a normal curve.

2. A normal distribution curve is symmetric about the mean, as shown in Figure 6.13. Consequently, 1/2 of the total area under a normal distribution curve lies on the left side of the mean, and 1/2 lies on the right side of the mean.

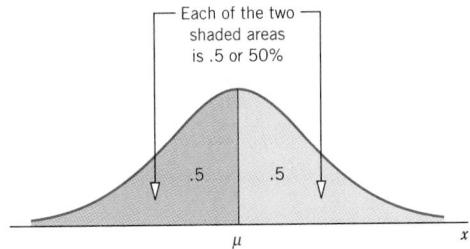

Figure 6.13 A normal curve is symmetric about the mean.

3. The tails of a normal distribution curve extend indefinitely in both directions without touching or crossing the horizontal axis. Although a normal distribution curve never meets the horizontal axis, beyond the points represented by $\mu - 3\sigma$ and $\mu + 3\sigma$ it becomes so close to this axis that the area under the curve beyond these points in both directions can be taken as virtually zero. These areas are shown in Figure 6.14.

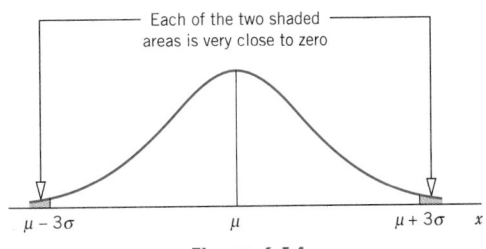

Figure 6.14

The mean, μ, and the standard deviation, σ, are the *parameters* of the normal distribution. Given the values of these two parameters, we can find the area under a normal distribution curve for any interval. Remember, there is not just one normal distribution curve but rather a *family* of normal distribution curves. Each different set of values of μ and σ gives a different normal distribution. The value of μ determines the center of a normal distribution curve on the horizontal axis, and the value of σ gives the spread of the normal distribution curve. The three normal distribution curves drawn in Figure 6.15 have the same mean but different standard deviations. By contrast, the three normal distribution curves in Figure 6.16 have different means but the same standard deviation.

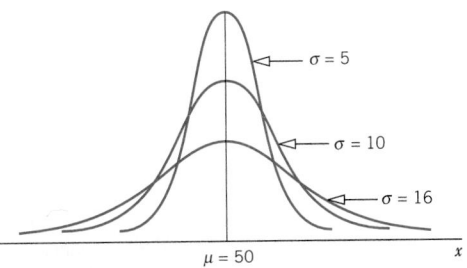

Figure 6.15 Three normal distribution curves with the same mean but different standard deviations.

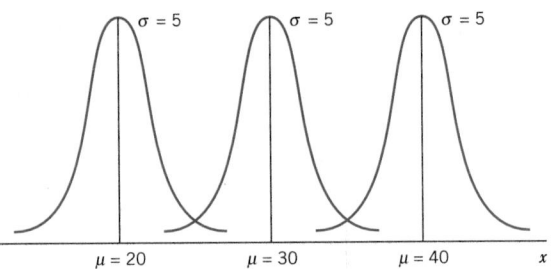

Figure 6.16 Three normal distribution curves with different means but the same standard deviation.

Like the binomial and Poisson probability distributions discussed in Chapter 5, the normal probability distribution can also be expressed by a mathematical equation.[1] However, we will not use this equation to find the area under a normal distribution curve. Instead, we will use Table VII of Appendix C.

6.3   THE STANDARD NORMAL DISTRIBUTION

The **standard normal distribution** is a special case of the normal distribution. For the standard normal distribution, the value of the mean is equal to zero, and the value of the standard deviation is equal to 1.

STANDARD NORMAL DISTRIBUTION

The normal distribution with $\mu = 0$ and $\sigma = 1$ is called the *standard normal distribution*.

Figure 6.17 displays the standard normal distribution curve. The random variable that possesses the standard normal distribution is denoted by z. In other words, the units for the standard normal distribution curve are denoted by z and are called the *z* **values** or *z* **scores**. They are also called *standard units* or *standard scores*.

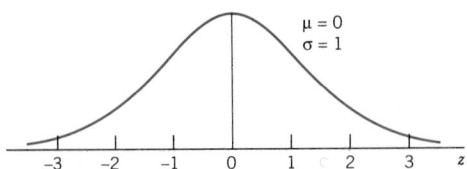

Figure 6.17 The standard normal distribution curve.

---

[1]The equation of the normal distribution is

$$f(x) = \frac{1}{\sigma\sqrt{2\pi}}\, e^{-(1/2)[(x-\mu)/\sigma]^2}$$

where $e = 2.71828$ and $\pi = 3.14159$ approximately; $f(x)$, called the probability density function, gives the vertical distance between the horizontal axis and the curve at point x. For the information of those who are familiar with integral calculus, the definite integral of this equation from a to b gives the probability that x assumes a value between a and b.

> **z VALUES OR z SCORES**
>
> The units marked on the horizontal axis of the standard normal curve are denoted by z and are called the *z values* or *z scores*. A specific value of z gives the distance between the mean and the point represented by z in terms of the standard deviation.

In Figure 6.17, the horizontal axis is labeled z. The z values on the right side of the mean are positive and those on the left side are negative. *The z value for a point on the horizontal axis gives the distance between the mean and that point in terms of the standard deviation.* For example, a point with a value of $z = 2$ is two standard deviations to the right of the mean. Similarly, a point with a value of $z = -2$ is two standard deviations to the left of the mean.

The standard normal distribution table, Table VII of Appendix C, lists the areas under the standard normal curve between $z = 0$ and the values of z from 0.00 to 3.09. To read the standard normal distribution table, we always start at $z = 0$, which represents the mean of the standard normal distribution. We learned earlier that the total area under a normal distribution curve is 1.0. We also learned that, because of symmetry, the area on either side of the mean is .5. This is shown in Figure 6.18.

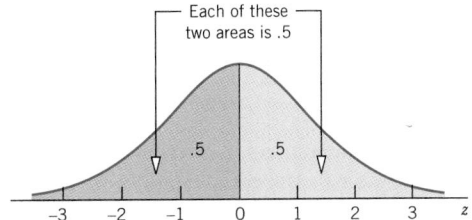

Figure 6.18 Area under the standard normal curve.

☞   *Remember*   Although the values of z on the left side of the mean are negative, the area under the curve is always positive.

The area under the standard normal curve between any two points can be interpreted as the probability that z assumes a value within that interval. Examples 6–1 through 6–4 describe how to read Table VII of Appendix C to find areas under the standard normal curve.

Finding area between $z = 0$ and a positive z.

EXAMPLE 6–1   Find the area under the standard normal curve between $z = 0$ and $z = 1.95$.

Solution   We divide the number 1.95 into two portions: 1.9 (the digit before the decimal and one digit after the decimal) and .05 (the second digit after the decimal). (Note that $1.9 + .05 = 1.95$.) To find the required area under the standard normal curve, we locate 1.9 in the column for z on the left side of Table VII and .05 in the row for z at the top of Table VII. The entry where the row for 1.9 and the column for .05 intersect gives the area under the standard normal curve between $z = 0$ and $z = 1.95$. The relevant portion of Table VII is reproduced here as Table 6.4. From Table VII or Table 6.4, the entry where the row for 1.9 and the column for .05 cross is .4744. Consequently, the area under the standard normal curve between $z = 0$ and $z = 1.95$ is .4744. This area is shown in Figure 6.19. (It is always helpful to sketch the curve and mark the area we are determining.)

Table 6.4   Area Under the Standard Normal Curve
Between $z = 0$ and $z = 1.95$

z	.00	.01	...	.05	...	.09
0.0	.0000	.0040	...	.0199	...	.0359
0.1	.0398	.0438	...	.0596	...	.0753
0.2	.0793	.0832	...	.0987	...	.1141
.	.	.	...	.	...	.
.	.	.	...	.	...	.
1.9	.4713	.4719	...	.4744 ←	...	.4767
.	.	.	...	.	...	.
3.0	.4987	.4987	...	.4989	...	.4990

Required area

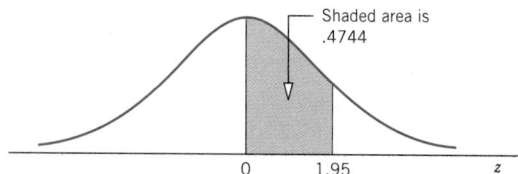

Shaded area is
.4744

0   1.95 z

Figure 6.19   Area between $z = 0$ and $z = 1.95$.

The area between $z = 0$ and $z = 1.95$ can be interpreted as the probability that z assumes a value between 0 and 1.95. That is,

$$\text{Area between 0 and 1.95} = P(0 < z < 1.95) = \textbf{.4744}$$

As mentioned in Section 6.1, the probability that a continuous random variable assumes a single value is zero. Therefore,

$$P(z = 0) = 0 \quad \text{and} \quad P(z = 1.95) = 0$$

Hence,

$$P(0 < z < 1.95) = P(0 \leq z \leq 1.95) = .4744 \quad \blacksquare$$

Finding area between a negative z and z = 0.

EXAMPLE 6-2   Find the area under the standard normal curve from $z = -2.17$ to $z = 0$.

Solution   Because the normal distribution is symmetric about the mean, the area from $z = -2.17$ to $z = 0$ is the same as the area from $z = 0$ to $z = 2.17$, as shown in Figure 6.20.

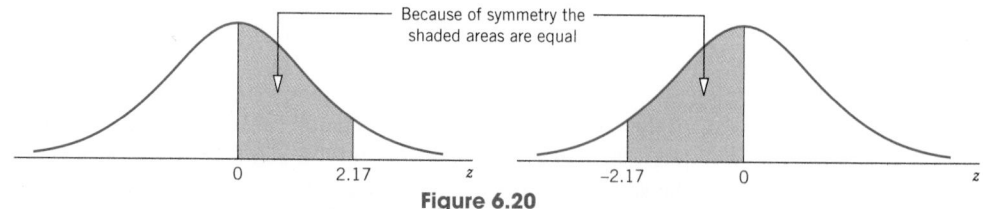

Because of symmetry the
shaded areas are equal

0   2.17 z -2.17 0 z

Figure 6.20

To find the area from $z = -2.17$ to $z = 0$, we look for the area from $z = 0$ to $z = 2.17$ in the standard normal distribution table (Table VII of Appendix C). To do so, first we locate 2.1 in the column for z and .07 in the row for z in that table. Then, we read the number at the intersection of the row for 2.1 and the column for .07. The relevant portion of Table VII is reproduced below as Table 6.5. As shown in Table 6.5 and Figure 6.21, this number is .4850.

Table 6.5 Finding Area Under the Standard Normal Curve from $z = 0$ to $z = 2.17$

z	.00	.01	...	.07	...	.09
0.0	.0000	.0040	...	.0279	...	.0359
0.1	.0398	.0438	...	.0675	...	.0753
0.2	.0793	.0832	...	.1064	...	.1141
.	.	.	...	.	...	.
.	.	.	...	.	...	.
2.1	.4821	.4826	...	.4850	...	.4857
.	.	.	...	.	...	.
.	.	.	...	.	...	.
3.0	.4987	.4987	...	.4989	...	.4990

Required area

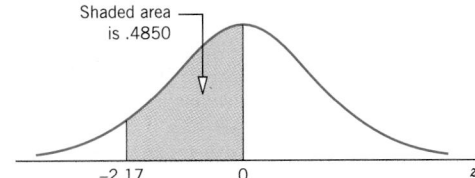

Shaded area is .4850

Figure 6.21 Area from $z = -2.17$ to $z = 0$.

The area from $z = -2.17$ to $z = 0$ gives the probability that z lies in the interval -2.17 to 0. That is,

$$\text{Area from } -2.17 \text{ to } 0 = P(-2.17 \le z \le 0) = \textbf{.4850}$$

■

Finding areas in the right and left tails.

EXAMPLE 6-3 Find the following areas under the standard normal curve.

(a) Area to the right of $z = 2.32$
(b) Area to the left of $z = -1.54$

Solution

(a) As mentioned earlier, to read the normal distribution table we must start with $z = 0$. To find the area to the right of $z = 2.32$, first we find the area between $z = 0$ and $z = 2.32$. Then we subtract this area from .5, which is the total area to the right of $z = 0$. From Table VII, the area between $z = 0$ and $z = 2.32$ is .4898. Consequently, the required area is $.5 - .4898 = .0102$, as shown in Figure 6.22.

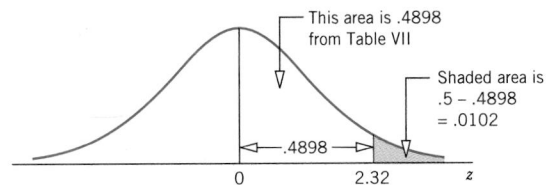

This area is .4898 from Table VII

Shaded area is .5 − .4898 = .0102

.4898

Figure 6.22 Area to the right of $z = 2.32$.

The area to the right of $z = 2.32$ gives the probability that z is greater than 2.32. Thus,

$$\text{Area to the right of } 2.32 = P(z > 2.32) = .5 - .4898 = \textbf{.0102}$$

(b) To find the area under the standard normal curve to the left of $z = -1.54$, first we find the area between $z = -1.54$ and $z = 0$ and then we subtract this area from .5, which is the total area to the left of $z = 0$. From Table VII, the area between $z = -1.54$ and $z = 0$ is .4382. Hence, the required area is $.5 - .4382 = .0618$. This area is shown in Figure 6.23.

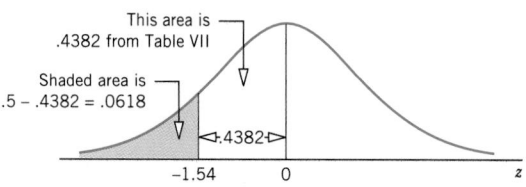

Figure 6.23   Area to the left of $z = -1.54$.

The area to the left of $z = -1.54$ gives the probability that z is less than -1.54. Thus,

$$\text{Area to the left of } -1.54 = P(z < -1.54) = .5 - .4382 = \textbf{.0618} \quad \blacksquare$$

EXAMPLE 6–4   Find the following probabilities for the standard normal curve.

(a) $P(1.19 < z < 2.12)$ (b) $P(-1.56 < z < 2.31)$ (c) $P(z > -.75)$

Solution

Finding area between two positive values of z.

(a) The probability $P(1.19 < z < 2.12)$ is given by the area under the standard normal curve between $z = 1.19$ and $z = 2.12$, which is the shaded area in Figure 6.24.

Both of the points, $z = 1.19$ and $z = 2.12$, are on the same (right) side of $z = 0$. To find the area between $z = 1.19$ and $z = 2.12$, first we find the areas between $z = 0$ and $z = 1.19$ and between $z = 0$ and $z = 2.12$. Then, we subtract the smaller area (the area between $z = 0$ and $z = 1.19$) from the larger area (the area between $z = 0$ and $z = 2.12$).

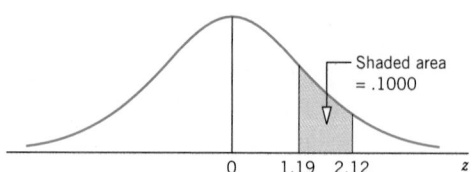

Figure 6.24   Finding $P(1.19 < z < 2.12)$.

From Table VII, for the standard normal distribution,

$$\text{Area between 0 and } 1.19 = .3830$$

and

$$\text{Area between 0 and } 2.12 = .4830$$

Then, the required probability is

$$P(1.19 < z < 2.12) = \text{Area between } 1.19 \text{ and } 2.12$$

$$= .4830 - .3830 = \textbf{.1000}$$

☞ *Remember* As a general rule, when the two points are on the same side of the mean, first find the areas between the mean and each of the two points. Then, subtract the smaller area from the larger area.

Area between a positive and a negative value of z.

(b) The probability $P(-1.56 < z < 2.31)$ is given by the area under the standard normal curve between $z = -1.56$ and $z = 2.31$, which is the shaded area in Figure 6.25.

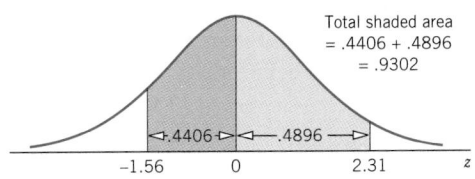

Total shaded area
= .4406 + .4896
= .9302

.4406   .4896

−1.56   0   2.31   z

Figure 6.25   Finding $P(-1.56 < z < 2.31)$.

The two points, $z = -1.56$ and $z = 2.31$, are on different sides of $z = 0$. The area between $z = -1.56$ and $z = 2.31$ is obtained by adding the areas between $z = -1.56$ and $z = 0$ and between $z = 0$ and $z = 2.31$, respectively.

From Table VII, for the standard normal distribution,

$$\text{Area between } -1.56 \text{ and } 0 = .4406$$

and
$$\text{Area between } 0 \text{ and } 2.31 = .4896$$

The required probability is

$$P(-1.56 < z < 2.31) = \text{Area between } -1.56 \text{ and } 2.31$$

$$= .4406 + .4896 = \textbf{.9302}$$

☞ *Remember* As a general rule, when the two points are on different sides of the mean, first find the areas between the mean and each of the two points. Then add these two areas.

Area to the right of a negative value of z.

(c) The probability $P(z > -.75)$ is given by the area under the standard normal curve to the right of $z = -.75$, which is the shaded area in Figure 6.26.

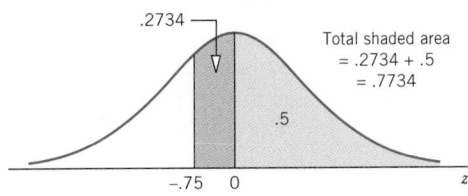

.2734

Total shaded area
= .2734 + .5
= .7734

.5

−.75   0   z

Figure 6.26   Finding $P(z > -.75)$.

The area to the right of $z = -.75$ is obtained by adding the area between $z = -.75$ and $z = 0$ and the area to the right of $z = 0$.

$$\text{Area to the right of } 0 = P(z > 0) = .5$$

From Table VII, for the standard normal distribution,

$$\text{Area between } -.75 \text{ and } 0 = .2734$$

The required probability is

$$P(z > -.75) = \text{Area to the right of } -.75 = .2734 + .5 = \textbf{.7734} \quad ■$$

In the discussion in Section 3.4 of Chapter 3 on the use of the standard deviation, we also discussed the empirical rule for a bell-shaped curve. That empirical rule is based on the standard normal distribution table. By using the normal distribution table, we can now verify the empirical rule as follows.

1. The total area within one standard deviation of the mean is 68.26%. This area is given by the sum of the areas between $z = -1.0$ and $z = 0$ and between $z = 0$ and $z = 1.0$. As shown in Figure 6.27, each of these two areas is .3413 or 34.13%. Consequently, the total area between $z = -1.0$ and $z = 1.0$ is 68.26%.

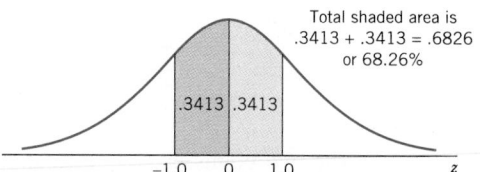

Figure 6.27 Area within one standard deviation of the mean.

2. The total area within two standard deviations of the mean is 95.44%. This area is given by the sum of the areas between $z = -2.0$ and $z = 0$ and between $z = 0$ and $z = 2.0$. As shown in Figure 6.28, each of these two areas is .4772 or 47.72%. Hence, the total area between $z = -2.0$ and $z = 2.0$ is 95.44%.

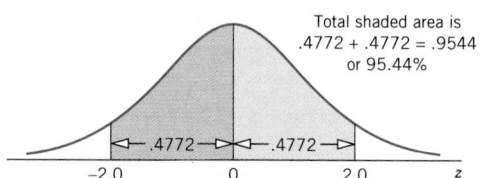

Figure 6.28 Area within two standard deviations of the mean.

3. The total area within three standard deviations of the mean is 99.74%. This area is given by the sum of the areas between $z = -3.0$ and $z = 0$ and between $z = 0$ and $z = 3.0$. As shown in Figure 6.29, each of these two areas is .4987 or 49.87%. Therefore, the total area between $z = -3.0$ and $z = 3.0$ is 99.74%.

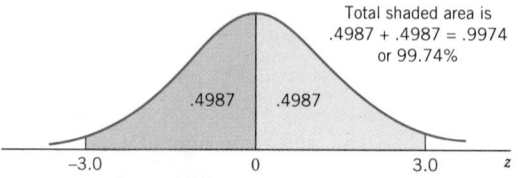

Figure 6.29 Area within three standard deviations of the mean.

Again, note that only a specific bell-shaped curve represents the normal distribution. Now we can state that a bell-shaped curve that contains (about) 68.26% of the total area within one standard deviation of the mean, (about) 95.44% of the total area within two standard deviations of the mean, and (about) 99.74% of the total area within three standard deviations of the mean represents a normal distribution curve.

The standard normal distribution table, Table VII of Appendix C, only goes up to $z = 3.09$. In other words, that table can be read only for $z = 0$ to $z = 3.09$ (or to $z = -3.09$). Consequently, if we need to find the area between $z = 0$ and a z value greater than 3.09 (or between a z value less than -3.09 and $z = 0$) under the standard normal curve, we cannot obtain it from the normal distribution table because it does not contain a z value greater than 3.09. In such cases, the area under the normal distribution curve between $z = 0$ and any z value greater than 3.09 (or less than -3.09) is approximated by .5. From the normal distribution table, the area between $z = 0$ and $z = 3.09$ is .4990. Hence, the area between $z = 0$ and any value of z greater than 3.09 is larger than .4990 and can be approximated by .5. Example 6–5 illustrates this procedure.

EXAMPLE 6–5 Find the following probabilities for the standard normal curve.

(a) $P(0 < z < 5.67)$ (b) $P(z < -5.35)$

Solution

Finding area between $z = 0$ and a value of z greater than 3.09.

(a) The probability $P(0 < z < 5.67)$ is given by the area under the standard normal curve between $z = 0$ and $z = 5.67$. Because $z = 5.67$ is greater than 3.09 and is not in Table VII, the area under the standard normal curve between $z = 0$ and $z = 5.67$ can be approximated by .5. This area is shown in Figure 6.30.

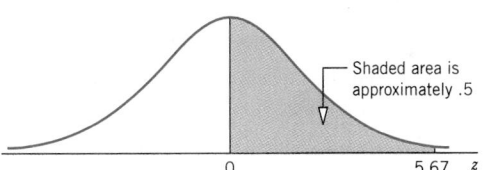

Figure 6.30 Area between $z = 0$ and $z = 5.67$.

The required probability is

$$P(0 < z < 5.67) = \text{Area between 0 and 5.67} = \mathbf{.5} \text{ approximately}$$

Note that the area between $z = 0$ and $z = 5.67$ is not exactly .5 but very close to .5.

Area to the left of a z that is less than -3.09.

(b) The probability $P(z < -5.35)$ represents the area under the standard normal curve to the left of $z = -5.35$. The area between $z = -5.35$ and $z = 0$ is approximately .5. Consequently, the area under the standard normal curve to the left of $z = -5.35$ is approximately zero, as shown in Figure 6.31.

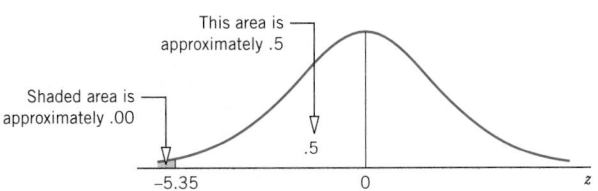

Figure 6.31 Area to the left of $z = -5.35$.

The required probability is

$$P(z < -5.35) = \text{Area to the left of} -5.35$$

$$= .5 - .5 = \textbf{.00 approximately}$$

Again, note that the area to the left of $z = -5.35$ is not exactly .00 but very close to .00. ■

EXERCISES

Concepts and Procedures

6.1 What is the difference between the probability distribution of a discrete random variable and that of a continuous random variable? Explain.

6.2 Let x be a continuous random variable. What is the probability that x assumes a single value, such as a?

6.3 For a continuous probability distribution, why is $P(a < x < b)$ equal to $P(a \leq x \leq b)$?

6.4 Briefly explain the main characteristics of a normal distribution. Illustrate with the help of graphs.

6.5 Briefly describe the standard normal distribution curve.

6.6 What are the parameters of the normal distribution?

6.7 How do the width and height of a normal distribution change when its mean remains the same but its standard deviation decreases?

6.8 Do the width and/or height of a normal distribution change when its standard deviation remains the same but its mean increases?

6.9 For a standard normal distribution, what does z represent?

6.10 For a standard normal distribution, find the area within one standard deviation of the mean, that is, the area between $\mu - \sigma$ and $\mu + \sigma$.

6.11 For a standard normal distribution, find the area within 1.5 standard deviations of the mean, that is, the area between $\mu - 1.5\sigma$ and $\mu + 1.5\sigma$.

6.12 For a standard normal distribution, what is the area within two standard deviations of the mean?

6.13 For a standard normal distribution, what is the area within 2.5 standard deviations of the mean?

6.14 For a standard normal distribution, what is the area within three standard deviations of the mean?

6.15 Find the area under the standard normal curve
a.   between $z = 0$ and $z = 1.90$ b.   between $z = 0$ and $z = -1.75$
c.   between $z = 1.25$ and $z = 2.37$ d.   from $z = -1.53$ to $z = -2.78$
e.   from $z = -1.67$ to $z = 2.34$

6.16 Find the area under the standard normal curve
a.   from $z = 0$ to $z = 2.34$ b.   between $z = 0$ and $z = -2.78$
c.   from $z = .84$ to $z = 1.95$ d.   between $z = -.57$ and $z = -2.39$
e.   between $z = -2.15$ and $z = 1.67$

6.17 Find the area under the standard normal curve
a.   to the right of $z = 1.56$ b.   to the left of $z = -1.97$
c.   to the right of $z = -2.05$ d.   to the left of $z = 1.86$

6.18 Obtain the area under the standard normal curve
a.   to the right of $z = 1.83$ b.   to the left of $z = -1.65$
c.   to the right of $z = -.55$ d.   to the left of $z = .79$

6.19 Find the area under the standard normal curve

a. between $z = 0$ and $z = 4.28$ b. from $z = 0$ to $z = -3.75$
c. to the right of $z = 7.43$ d. to the left of $z = -4.49$

6.20 Find the area under the standard normal curve

a. from $z = 0$ to $z = 3.94$ b. between $z = 0$ and $z = -5.16$
c. to the right of $z = 5.42$ d. to the left of $z = -3.68$

6.21 Determine the following probabilities for the standard normal distribution.

a. $P(-1.83 \leq z \leq 2.67)$ b. $P(0 \leq z \leq 2.12)$
c. $P(-1.89 \leq z \leq 0)$ d. $P(z \geq 1.38)$

6.22 Determine the following probabilities for the standard normal distribution.

a. $P(-2.46 \leq z \leq 1.68)$ b. $P(0 \leq z \leq 1.86)$
c. $P(-2.58 \leq z \leq 0)$ d. $P(z \geq .83)$

6.23 Find the following probabilities for the standard normal distribution.

a. $P(z < -2.04)$ b. $P(.67 \leq z \leq 2.39)$
c. $P(-2.07 \leq z \leq -.83)$ d. $P(z < 1.71)$

6.24 Find the following probabilities for the standard normal distribution.

a. $P(z < -1.21)$ b. $P(1.03 \leq z \leq 2.79)$
c. $P(-2.34 \leq z \leq -1.09)$ d. $P(z < 2.02)$

6.25 Obtain the following probabilities for the standard normal distribution.

a. $P(z > -.78)$ b. $P(-2.47 \leq z \leq 1.09)$
c. $P(0 \leq z \leq 4.25)$ d. $P(-5.36 \leq z \leq 0)$
e. $P(z > 6.07)$ f. $P(z < -5.27)$

6.26 Obtain the following probabilities for the standard normal distribution.

a. $P(z > -1.26)$ b. $P(-.68 \leq z \leq 1.74)$
c. $P(0 \leq z \leq 3.85)$ d. $P(-4.34 \leq z \leq 0)$
e. $P(z > 4.82)$ f. $P(z < -6.12)$

6.4 STANDARDIZING A NORMAL DISTRIBUTION

As was shown in the previous section, Table VII of Appendix C can be used to find areas under the standard normal curve. However, in real-world applications, a (continuous) random variable may have a normal distribution with values of the mean and standard deviation that are different from 0 and 1, respectively. The first step, in such a case, is to convert the given normal distribution to the standard normal distribution. This procedure is called *standardizing a normal distribution*. The units of a normal distribution (which is not the standard normal distribution) are denoted by x. We know from Section 6.3 that units of the standard normal distribution are denoted by z.

CONVERTING AN x VALUE TO A z VALUE

For a normal random variable x, a particular value of x can be converted to its corresponding z value by using the formula

$$z = \frac{x - \mu}{\sigma}$$

where μ and σ are the mean and standard deviation of the normal distribution of x.

Thus, to find the z value for an x value, we calculate the difference between the given x value and the mean, μ, and divide this difference by the standard deviation, σ. If the value of x is equal to μ then its z value is equal to zero. Note that we will always round z values to two decimal places.

☞   *Remember*   The z value for the mean of a normal distribution is always zero.

Examples 6–6 through 6–10 describe how to convert x values to the corresponding z values and how to find areas under a normal distribution curve.

Converting x values to the corresponding z values.

EXAMPLE 6–6   Let x be a continuous random variable that has a normal distribution with a mean of 50 and a standard deviation of 10. Convert the following x values to z values.

(a) $x = 55$ (b) $x = 35$

Solution   For the given normal distribution: $\mu = 50$ and $\sigma = 10$.

(a)   The z value for $x = 55$ is computed as follows.

$$z = \frac{x - \mu}{\sigma} = \frac{55 - 50}{10} = .50$$

Thus, the z value for $x = 55$ is .50. The z values for $\mu = 50$ and $x = 55$ are shown in Figure 6.32. Note that the z value for $\mu = 50$ is zero. The value $z = .50$ for $x = 55$ indicates that the distance between the mean, $\mu = 50$, of the given normal distribution and the point given by $x = 55$ is 1/2 of the standard deviation, $\sigma = 10$. Consequently, we can state that the z value represents the distance between μ and x in terms of the standard deviation. Because $x = 55$ is greater than $\mu = 50$, its z value is positive.

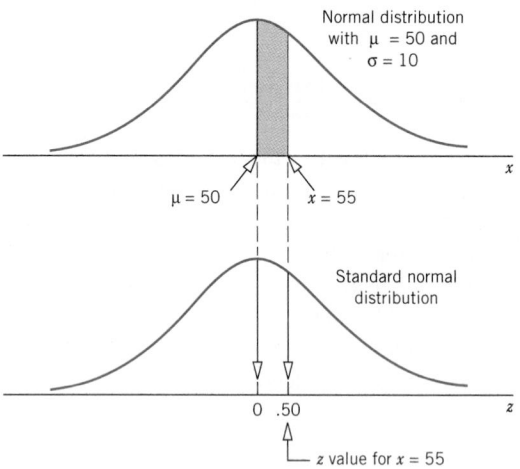

Figure 6.32 z value for $x = 55$.

From this point on, we will usually show only the z axis below the x axis and not the standard normal curve itself.

(b)   The z value for $x = 35$ is computed as follows and is shown in Figure 6.33.

$$z = \frac{x - \mu}{\sigma} = \frac{35 - 50}{10} = -1.50$$

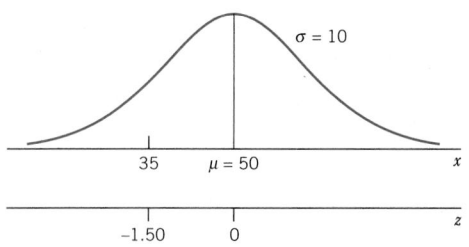

Figure 6.33 z value for x = 35.

Because $x = 35$ is on the left side of the mean (i.e., 35 is less than $\mu = 50$), its z value is negative. As a general rule, whenever an x value is less than the value of μ, its z value is negative. ▪

☞ *Remember* The z value for an x value that is greater than μ is positive; the z value for an x value that is equal to μ is zero, and the z value for an x value that is less than μ is negative.

To find the area between two values of x for a normal distribution, we first convert both values of x to their respective z values. Then we find the area under the standard normal curve between those two z values. The area between the two z values gives the area between the corresponding x values.

EXAMPLE 6–7 Let x be a continuous random variable that is normally distributed with a mean of 25 and a standard deviation of 4. Find the area

(a) between $x = 25$ and $x = 32$ (b) between $x = 18$ and $x = 34$

Solution For the given normal distribution: $\mu = 25$ and $\sigma = 4$.

Area between the mean and a point to its right.

(a) The first step in finding the required area is to standardize the given normal distribution by converting $x = 25$ and $x = 32$ to respective z values using the formula

$$z = \frac{x - \mu}{\sigma}$$

The z value for $x = 25$ is zero because it is the mean of the normal distribution. The z value for $x = 32$ is

$$z = \frac{32 - 25}{4} = 1.75$$

As shown in Figure 6.34, the area between $x = 25$ and $x = 32$ under the given normal distribution curve is equivalent to the area between $z = 0$ and $z = 1.75$ under the standard normal curve. This area from Table VII is .4599. The area between $x = 25$ and $x = 32$ under the normal curve gives the probability that x assumes a value between 25 and 32. This probability can be written as

$$P(25 < x < 32) = P(0 < z < 1.75) = \mathbf{.4599}$$

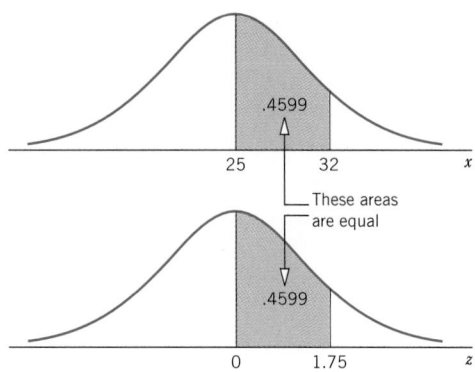

Figure 6.34   Area between $x = 25$ and $x = 32$.

Area between two points that are on different sides of the mean.

(b)   First we calculate the z values for $x = 18$ and $x = 34$ as follows.

$$\text{For } x = 18: \quad z = \frac{18 - 25}{4} = -1.75$$

$$\text{For } x = 34: \quad z = \frac{34 - 25}{4} = 2.25$$

The area under the given normal distribution curve between $x = 18$ and $x = 34$ is given by the area under the standard normal curve between $z = -1.75$ and $z = 2.25$. This area is shown in Figure 6.35. The two values of z are on different sides of $z = 0$. Consequently, the total area is obtained by adding the areas between $z = -1.75$ and $z = 0$ and between $z = 0$ and $z = 2.25$. From Table VII, the area between $z = -1.75$ and $z = 0$ is .4599 and the area between $z = 0$ and $z = 2.25$ is .4878. Hence,

$$P(18 < x < 34) = P(-1.75 < z < 2.25) = .4599 + .4878 = \mathbf{.9477}$$

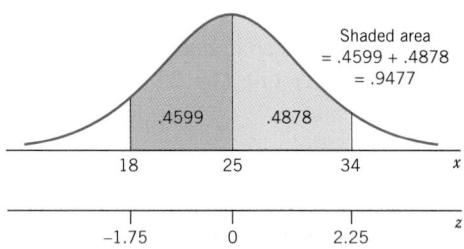

Figure 6.35   Area between $x = 18$ and $x = 34$.

EXAMPLE 6–8   Let x be a normal random variable with its mean equal to 40 and standard deviation equal to 5. Find the following probabilities for this normal distribution.

(a) $P(x > 55)$ (b) $P(x < 49)$

Solution   For the given normal distribution: $\mu = 40$ and $\sigma = 5$.

Probability of x falling in the right tail.

(a)   The probability that x assumes a value greater than 55 is given by the area under the normal distribution curve to the right of $x = 55$, as shown in Figure 6.36. This area is calculated by subtracting the area between $\mu = 40$ and $x = 55$ from .5, which is the total area to the right of the mean.

$$\text{For } x = 55: \quad z = \frac{55 - 40}{5} = 3.00$$

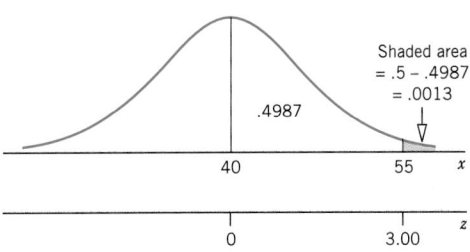

Figure 6.36 Finding $P(x > 55)$.

The required probability is given by the area to the right of $z = 3.00$. To find this area, first we find the area between $z = 0$ and $z = 3.00$, which is .4987. Then we subtract this area from .5. Thus,

$$P(x > 55) = P(z > 3.00) = .5 - .4987 = \textbf{.0013}$$

Probability that x is less than a value that is to the right of the mean.

(b)   The probability that x will assume a value less than 49 is given by the area under the normal distribution curve to the left of 49, which is the shaded area in Figure 6.37. This area is given by the sum of the area to the left of $\mu = 40$ and the area between $\mu = 40$ and $x = 49$.

$$\text{For } x = 49: \quad z = \frac{49 - 40}{5} = 1.80$$

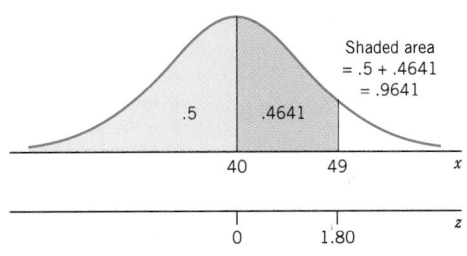

Figure 6.37 Finding $P(x < 49)$.

The required probability is given by the area to the left of $z = 1.80$. This area is obtained by adding the area to the left of $z = 0$ and the area between $z = 0$ and $z = 1.80$. The area to the left of $z = 0$ is .5. From Table VII, the area between $z = 0$ and $z = 1.80$ is .4641. Therefore, the required probability is

$$P(x < 49) = P(z < 1.80) = .5 + .4641 = \textbf{.9641}$$

Area between two x values that are less than the mean.

EXAMPLE 6–9   Let x be a continuous random variable that has a normal distribution with $\mu = 50$ and $\sigma = 8$. Find the probability $P(30 \le x \le 39)$.

Solution   For the given normal distribution: $\mu = 50$ and $\sigma = 8$. The probability $P(30 \le x \le 39)$ is given by the area from $x = 30$ to $x = 39$ under the normal distribution curve. As shown in Figure 6.38, this area is given by the difference between the area from $x = 30$ to $\mu = 50$ and the area from $x = 39$ to $\mu = 50$.

$$\text{For } x = 30: \qquad z = \frac{30 - 50}{8} = -2.50$$

$$\text{For } x = 39: \qquad z = \frac{39 - 50}{8} = -1.38$$

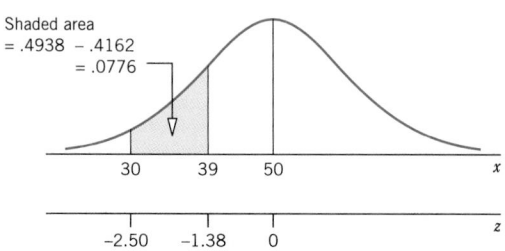

Figure 6.38   Finding $P(30 \leq x \leq 39)$.

To find the required area, we first find the areas from $z = -2.50$ to $z = 0$ and from $z = -1.38$ to $z = 0$ and then take the difference between these two areas. From Table VII, the area from $z = -2.50$ to $z = 0$ is .4938 and the area from $z = -1.38$ to $z = 0$ is .4162. Thus, the required probability is

$$P(30 \leq x \leq 39) = P(-2.50 \leq z \leq -1.38) = .4938 - .4162 = \mathbf{.0776} \quad \blacksquare$$

EXAMPLE 6–10   Let x be a continuous random variable that has a normal distribution with a mean of 80 and a standard deviation of 12. Find the area under the normal distribution curve

(a)   from $x = 70$ to $x = 135$ (b)   to the left of 27

Solution   For the given normal distribution: $\mu = 80$ and $\sigma = 12$.

Area between two values of x that are on different sides of the mean.

(a)   The area from $x = 70$ to $x = 135$ is obtained by adding the areas from $x = 70$ to $\mu = 80$ and from $\mu = 80$ to $x = 135$. This total area is given by the sum of the two shaded areas shown in Figure 6.39.

$$\text{For } x = 70: \qquad z = \frac{70 - 80}{12} = -.83$$

$$\text{For } x = 135: \qquad z = \frac{135 - 80}{12} = 4.58$$

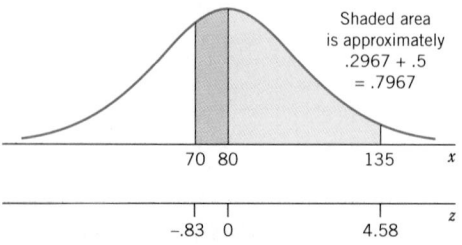

Figure 6.39   Area between $x = 70$ and $x = 135$.

Thus, the required area is obtained by adding the areas from $z = -.83$ to $z = 0$ and from $z = 0$ to $z = 4.58$ under the standard normal curve. From Table VII, the area from $z = -.83$ to $z = 0$ is .2967 and the area from $z = 0$ to $z = 4.58$ is approximately .5. Hence,

$$P(70 \leq x \leq 135) = P(-.83 \leq z \leq 4.58)$$

$$= .2967 + .5 = \textbf{.7967} \text{ approximately}$$

Finding an area in the left tail.

(b) The area to the left of $x = 27$ is obtained by subtracting the area from $x = 27$ to $\mu = 80$ from .5, which is the total area to the left of the mean. This area is calculated as follows.

$$\text{For } x = 27: \quad z = \frac{27 - 80}{12} = -4.42$$

As shown in Figure 6.40, the required area is given by the area under the standard normal distribution curve to the left of $z = -4.42$. This area is

$$P(x < 27) = P(z < -4.42) = .5 - P(-4.42 < z < 0)$$

$$= .5 - .5 = \textbf{.00} \text{ approximately}$$

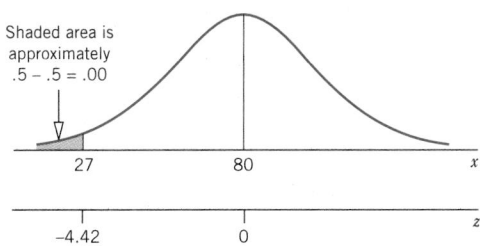

Shaded area is approximately
.5 − .5 = .00

Figure 6.40 Area to the left of $x = 27$.

EXERCISES

Concepts and Procedures

6.27 Find the z value for each of the following x values for a normal distribution with $\mu = 30$ and $\sigma = 5$.

    a. $x = 37$ **b.** $x = 19$ **c.** $x = 23$ **d.** $x = 44$

6.28 Determine the z value for each of the following x values for a normal distribution with $\mu = 16$ and $\sigma = 3$.

    a. $x = 11$ **b.** $x = 22$ **c.** $x = 18$ **d.** $x = 14$

6.29 Find the following areas under a normal distribution curve with $\mu = 20$ and $\sigma = 4$.

    a.   Area between $x = 20$ and $x = 27$
    b.   Area from $x = 23$ to $x = 25$
    c.   Area between $x = 9.5$ and $x = 17$

6.30 Find the following areas under a normal distribution curve with $\mu = 12$ and $\sigma = 2$.

    a.   Area between $x = 7.76$ and $x = 12$
    b.   Area between $x = 14.48$ and $x = 16.34$
    c.   Area from $x = 8.22$ to $x = 11.06$

6.31 Determine the area under a normal distribution curve with $\mu = 55$ and $\sigma = 7$

  a. to the right of $x = 58$
  b. to the right of $x = 43$
  c. to the left of $x = 67$
  d. to the left of $x = 24$

6.32 Find the area under a normal distribution curve with $\mu = 37$ and $\sigma = 3$

  a. to the left of $x = 29$
  b. to the right of $x = 53$
  c. to the left of $x = 42$
  d. to the right of $x = 35$

6.33 Let x be a continuous random variable that is normally distributed with a mean of 25 and a standard deviation of 6. Find the probability that x assumes a value

  a. between 29 and 36   b. between 22 and 33

6.34 Let x be a continuous random variable that has a normal distribution with a mean of 40 and a standard deviation of 4. Find the probability that x assumes a value

  a. between 29 and 35   b. from 34 to 51

6.35 Let x be a continuous random variable that is normally distributed with a mean of 80 and a standard deviation of 12. Find the probability that x assumes a value

  a. greater than 70   b. less than 75
  c. greater than 100   d. less than 89

6.36 Let x be a continuous random variable that is normally distributed with a mean of 65 and a standard deviation of 15. Find the probability that x assumes a value

  a. less than 43   b. greater than 74
  c. greater than 56   d. less than 71

6.5 APPLICATIONS OF THE NORMAL DISTRIBUTION

Sections 6.2 through 6.4 discussed the normal distribution, how to convert a normal distribution to the standard normal distribution, and how to find area under a normal distribution curve. This section presents examples that illustrate the applications of the normal distribution.

Application of the normal distribution: area between two points that are on different sides of the mean.

EXAMPLE 6–11 According to estimates published in *Fortune* magazine, neurosurgeons in the United States earn an average of $263,300 per year (Justin Martin, "How Does Your Pay Really Stack Up?" *Fortune*, June 26, 1995). Assume that the current annual earnings of all neurosurgeons in the United States have a normal distribution with a mean of $263,300 and a standard deviation of $40,000. Find the probability that the annual earnings of a randomly selected U.S. neurosurgeon would be between $210,000 and $285,000.

Solution Let x denote the annual earnings of a randomly selected U.S. neurosurgeon. Then, x is normally distributed with

$$\mu = \$263,300 \quad \text{and} \quad \sigma = \$40,000$$

The probability that the annual earnings of a randomly selected neurosurgeon are between $210,000 and $285,000 is given by the area under the normal curve of x between $x = \$210,000$ and $x = \$285,000$. Because these two points are on different sides of the mean, the required probability is obtained by adding the two shaded areas shown in Figure 6.41.

$$\text{For } x = \$210{,}000: \quad z = \frac{210{,}000 - 263{,}300}{40{,}000} = -1.33$$

$$\text{For } x = \$285{,}000: \quad z = \frac{285{,}000 - 263{,}300}{40{,}000} = .54$$

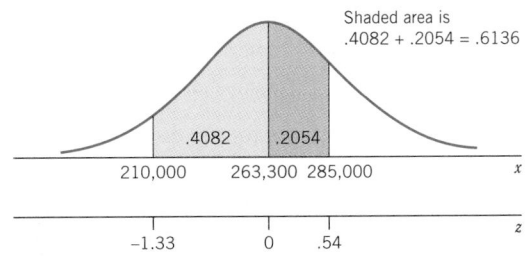

Figure 6.41   Area between $x = \$210{,}000$ and $x = \$285{,}000$.

The required probability is given by the area under the standard normal curve between $z = -1.33$ and $z = .54$, which is obtained by adding the area between $z = -1.33$ and $z = 0$ and the area between $z = 0$ and $z = .54$. From Table VII of Appendix C, the area between $z = -1.33$ and $z = 0$ is .4082 and the area between $z = 0$ and $z = .54$ is .2054. Hence, the required probability is

$$P(210{,}000 < x < 285{,}000) = P(-1.33 < z < .54) = .4082 + .2054 = \mathbf{.6136}$$

Thus, the probability is .6136 that the annual earnings of a randomly selected neurosurgeon are between \$210,000 and \$285,000. Converting this probability to a percentage, we can also state that the annual earnings of about 61.36% of the U.S. neurosurgeons are between \$210,000 and \$285,000. ▬

Application of the normal distribution: probability x is less than a value that is to the right of the mean.

EXAMPLE 6-12 A racing car is one of the many toys manufactured by Mack Corporation. The assembly time for this toy follows a normal distribution with a mean of 55 minutes and a standard deviation of 4 minutes. The company closes at 5 P.M. every day. If one worker starts assembling a racing car at 4 P.M., what is the probability that she will finish this job before the company closes for the day?

Solution   Let x denote the time taken by this worker to assemble a racing car. Then, x is normally distributed with

$$\mu = 55 \text{ minutes} \quad \text{and} \quad \sigma = 4 \text{ minutes}$$

We are to find the probability that this worker can assemble this car in 60 minutes or less (between 4 and 5 P.M.). This probability is given by the area under the normal curve to the left of $x = 60$. This area is obtained by adding the two shaded areas shown in Figure 6.42.

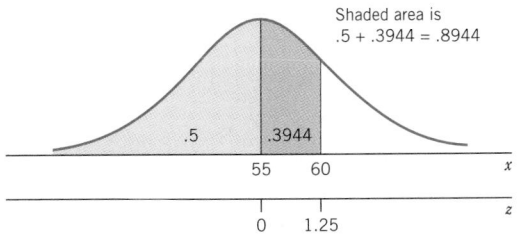

Figure 6.42   Area to the left of $x = 60$.

$$\text{For } x = 60: \qquad z = \frac{60 - 55}{4} = 1.25$$

The required probability is given by the area under the standard normal curve to the left of $z = 1.25$, which is obtained by adding .5 and the area between $z = 0$ and $z = 1.25$. From Table VII of Appendix C, the area between $z = 0$ and $z = 1.25$ is .3944. Therefore, the required probability is

$$P(x \leq 60) = P(z \leq 1.25) = .5 + .3944 = \mathbf{.8944}$$

Thus, the probability is .8944 that this worker will finish assembling this racing car before the company closes for the day.   ▬

Applications of the normal distribution.

EXAMPLE 6–13   Hupper Corporation produces many types of soft drinks, including Orange Cola. The filling machines are adjusted to pour 12 ounces of soda in each 12-ounce can of Orange Cola. However, the actual amount of soda poured into each can is not exactly 12 ounces; it varies from can to can. It has been observed that the net amount of soda in such a can has a normal distribution with a mean of 12 ounces and a standard deviation of .015 ounces.

(a)   What is the probability that a randomly selected can of Orange Cola contains 11.97 to 11.99 ounces of soda?

(b)   What percentage of the Orange Cola cans contain 12.02 to 12.07 ounces of soda?

Solution   Let x be the net amount of soda in a can of Orange Cola. Then, x has a normal distribution with $\mu = 12$ ounces and $\sigma = .015$ ounces.

Probability x is between two points that are to the left of the mean.

(a)   The probability that a randomly selected can contains 11.97 to 11.99 ounces of soda is given by the area under the normal distribution curve from $x = 11.97$ to $x = 11.99$. This area is shown in Figure 6.43.

$$\text{For } x = 11.97: \qquad z = \frac{11.97 - 12}{.015} = -2.00$$

$$\text{For } x = 11.99: \qquad z = \frac{11.99 - 12}{.015} = -.67$$

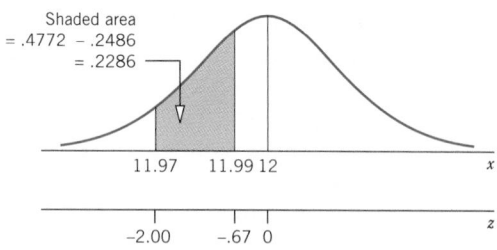

Figure 6.43   Area between $x = 11.97$ and $x = 11.99$.

The required probability is given by the area under the standard normal curve between $z = -2.00$ and $z = -.67$, which is obtained by subtracting the area from $z = -.67$ to $z = 0$ from the area from $z = -2.00$ to $z = 0$. From Table VII of Appendix C, the area from $z = -.67$ to $z = 0$ is .2486 and the area from $z = -2.00$ to $z = 0$ is .4772. Hence, the required probability is

$$P(11.97 \leq x \leq 11.99) = P(-2.00 \leq z \leq -.67) = .4772 - .2486 = \mathbf{.2286}$$

Thus, the probability is .2286 that any randomly selected can of Orange Cola will contain 11.97 to 11.99 ounces of soda. We can also state that about 22.86% of Orange Cola cans contain 11.97 to 11.99 ounces of soda.

Probability x is between two points that are to the right of the mean.

(b) The percentage of Orange Cola cans that contain 12.02 to 12.07 ounces of soda is given by the area under the normal distribution curve from $x = 12.02$ to $x = 12.07$, as shown in Figure 6.44.

$$\text{For } x = 12.02: \quad z = \frac{12.02 - 12}{.015} = 1.33$$

$$\text{For } x = 12.07: \quad z = \frac{12.07 - 12}{.015} = 4.67$$

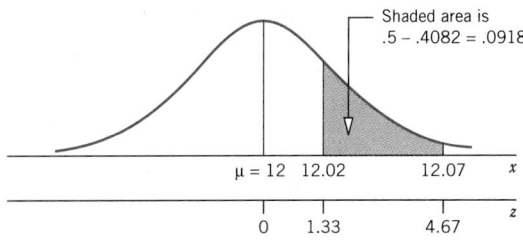

Figure 6.44 Area from $x = 12.02$ to $x = 12.07$.

The required probability is given by the area under the standard normal curve between $z = 1.33$ and $z = 4.67$, which is obtained by subtracting the area from $z = 0$ to $z = 1.33$ from the area from $z = 0$ to $z = 4.67$. From Table VII of Appendix C, the area from $z = 0$ to $z = 1.33$ is .4082. The area from $z = 0$ to $z = 4.67$ is approximately .5. Hence, the required probability is

$$P(12.02 \leq x \leq 12.07) = P(1.33 \leq z \leq 4.67) = .5 - .4082 = \mathbf{.0918}$$

Converting this probability to a percentage, we can state that approximately 9.18% of all Orange Cola cans are expected to contain 12.02 to 12.07 ounces of soda. ■

Area to the left of x that is less than the mean.

EXAMPLE 6-14 The life span of a calculator manufactured by Intal Corporation has a normal distribution with a mean of 54 months and a standard deviation of 8 months. The company guarantees that any calculator that starts malfunctioning within 36 months of the purchase will be replaced by a new one. About what percentage of such calculators made by this company are expected to be replaced?

Solution   Let x be the life span of such a calculator. Then x has a normal distribution with $\mu = 54$ and $\sigma = 8$ months. The probability that a randomly selected calculator will start malfunctioning within 36 months is given by the area under the normal distribution curve to the left of $x = 36$, as shown in Figure 6.45.

$$\text{For } x = 36: \quad z = \frac{36 - 54}{8} = -2.25$$

The required percentage is given by the area under the standard normal curve to the left of $z = -2.25$, which is obtained by subtracting the area between $z = -2.25$ and $z = 0$ from .5. From Table VII of Appendix C, the area between $z = -2.25$ and $z = 0$ is .4878. Hence,

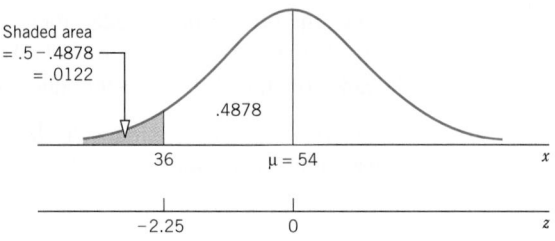

Figure 6.45   Area to the left of $x = 36$.

the required probability is

$$P(x < 36) = P(z < -2.25) = .5 - .4878 = \mathbf{.0122}$$

The probability that any randomly selected calculator manufactured by Intal Corporation will start malfunctioning within 36 months is .0122. Converting this probability to a percentage, we can state that approximately 1.22% of all such calculators manufactured by this company are expected to start malfunctioning within 36 months. Hence, 1.22% of the calculators are expected to be replaced.

EXERCISES

Applications

6.37   Let x denote the time taken to run a road race. Suppose x is approximately normally distributed with a mean of 195 minutes and a standard deviation of 21 minutes. If one runner is selected at random, what is the probability that this runner will complete this road race

  a.   in less than 150 minutes   **b.**   in 205 to 245 minutes

6.38   The U.S. Bureau of Labor Statistics conducts periodic surveys to collect information on the labor market. According to the bureau, workers in the private sector earned an average of $11.66 per hour in February, 1996. Assume that the current hourly wages for all workers in the private sector have a normal distribution with a mean of $11.66 and a standard deviation of $2. Find the percentage of these workers whose current hourly wages are

  a.   between $14 and $16   **b.**   between $10 and $13

 6.39   According to the American Association of University Professors, the mean 1995–96 salary of full-time instructional faculty at two-year colleges was $41,640. Suppose that these salaries have a normal distribution with a mean of $41,640 and a standard deviation of $4000. Find the probability that the 1995–96 salary of a randomly selected faculty member from such colleges was

  a.   more than $45,000   **b.**   between $38,000 and $44,000

6.40   The stress scores (on a scale 1 to 10) of students before a statistics test are found to be approximately normally distributed with a mean of 7.08 and a standard deviation of .63. Find the probability that the stress score before a statistics test for a randomly selected student will be

  a.   more than 6.25   **b.**   between 7.40 and 8.90

 6.41   The speeds of cars traveling on Interstate Highway I-15 are normally distributed with a mean of 69 miles per hour and a standard deviation of 3.5 miles per hour. Find what percentage of the cars traveling on this highway have a speed of

  a.   61 to 66 miles per hour   **b.**   65 to 74 miles per hour

6.42   The Bank of Connecticut issues Visa and Mastercard credit cards. It is estimated that the balances on all Visa credit cards issued by the Bank of Connecticut have a mean of $845 and a standard deviation of $270. Assume that the balances on all these Visa cards follow a normal distribution.

a. What is the probability that a randomly selected Visa card issued by this bank has a balance between $1000 and $1400?

b. What percentage of the Visa cards issued by this bank have a balance of $750 or more?

6.43 According to the U.S. Bureau of the Census, Americans aged 26 to 44 spent an average of 22 minutes traveling to work in 1995 (*USA Today*, November 20, 1995). Assume that such travel times for workers aged 26 to 44 are normally distributed with a mean of 22 minutes and a standard deviation of 5 minutes.

a. Find the probability that a randomly selected 26-to-44-year-old U.S. worker spent more than 30 minutes traveling to work in 1995.

b. What percentage of such workers spent between 10 and 18 minutes traveling to work in 1995?

6.44 The transmission on a model of a specific car has a warranty for 40,000 miles. It is known that the life of such a transmission has a normal distribution with a mean of 72,000 miles and a standard deviation of 12,000 miles.

a. What percentage of the transmissions will fail before the end of the warranty period?

b. What percentage of the transmissions will be good for more than 100,000 miles?

6.45 According to the records of an electric company serving the Boston area, the mean electric consumption for all households during winter is 1650 kilowatt hours per month. Assume that the monthly electric consumption during winter by all households in this area has a normal distribution with a mean of 1650 kilowatt hours and a standard deviation of 320 kilowatt hours.

a. Find the probability that the monthly electric consumption during winter by a randomly selected household from this area is less than 1800 kilowatt hours.

b. What percentage of the households in this area have a monthly electric consumption of 900 to 1300 kilowatt hours?

6.46 The management of a supermarket wants to adopt a new promotional policy of giving a free gift to every customer who spends more than a certain amount per visit at this supermarket. The expectation of the management is that after this promotional policy is advertised, the expenditures for all customers at this supermarket will be normally distributed with a mean of $95 and a standard deviation of $21. If the management decides to give free gifts to all those customers who spend more than $130 at this supermarket during a visit, what percentage of the customers are expected to get free gifts?

6.47 According to the U.S. Bureau of the Census, students require an average of 6.29 years beyond high school to obtain a bachelor's degree. Assume that the current time to obtain a bachelor's degree has a normal distribution with a mean of 6.29 years and a standard deviation of 1.1 years.

a. Find the probability that a randomly selected student will require more than 7 years to obtain a bachelor's degree.

b. What percentage of students take between 5 and 6 years to obtain a bachelor's degree?

6.48 The management at a large insurance company believes that the workers are more productive if they are happy with their jobs. To keep track of workers' satisfaction, the company regularly conducts surveys. According to a recent such survey, the mean job satisfaction score for all workers at this company was 13.10 (on a scale of 1 to 20) and the standard deviation was 1.95. Assume that the job satisfaction scores of workers are normally distributed.

a. Find the probability that the job satisfaction score for a randomly selected worker from this company is less than 11.25.

b. What percentage of the workers have a job satisfaction score between 14.50 and 18.70?

c. A worker with a score of 8 or less is considered to be very unhappy with his/her job. What percentage of the workers are very unhappy with their jobs?

6.49 Almost all high school students who intend to go to college take the SAT test. Suppose the SAT scores of all students who took the test this year have a normal distribution with a mean of 904 and a standard deviation of 153.

a. Sue Hopern scored 1074 on this test. What percentage of the examinees scored higher than Sue?

b. Joe Merck scored 972 on this test. What percentage of the examinees scored lower than Joe?

c. A particular college requires that a student's SAT score must be 1020 or higher for consideration for admission. What percentage of the examinees are eligible for consideration for admission to this college?

6.50 Fast Auto Service guarantees that the maximum waiting time for its customers is 20 minutes for oil and lube service on their cars. It also guarantees that any customer who has to wait for more than 20 minutes for this service will receive a 50% discount on the charges. It is estimated that the mean time taken for oil and lube service at this garage is 15 minutes per car and the standard deviation is 2.4 minutes. Suppose the time taken for oil and lube service on a car follows a normal distribution.

a. What percentage of the customers will receive the 50% discount on their charges?

b. Is it possible that a car may take more than 25 minutes for oil and lube service? Explain.

6.51 The lengths of 3-inch nails manufactured on a machine are normally distributed with a mean of 3.0 inches and a standard deviation of .009 inches. The nails that are either less than 2.98 inches long or more than 3.02 inches long are unusable. What percentage of all the nails produced by this machine are unusable?

6.52 Lewis Corporation manufactures bicycles. The wheels of one type of bicycle are supposed to have a diameter of 30 inches. The processing system that makes these wheels does not produce every wheel with exactly a 30-inch diameter. The diameters of wheels made by this system have a normal distribution with a mean of 30 inches and a standard deviation of .10 inches. The wheels that have a diameter of either less than 29.75 inches or more than 30.25 inches are discarded. What percentage of all the wheels produced by this processing system are discarded?

6.6 DETERMINING THE z AND x VALUES WHEN AN AREA UNDER THE NORMAL DISTRIBUTION CURVE IS KNOWN

So far in this chapter we have discussed how to find the area under a normal distribution curve for an interval of z or x. Now we reverse this procedure and learn how to find the corresponding value of z or x when an area under a normal distribution curve is known. Examples 6–15 through 6–17 describe this procedure for finding the z value.

Finding z when area between the mean and z is known.

EXAMPLE 6–15 Find a point z such that the area under the standard normal curve between 0 and z is .4251 and the value of z is positive.

Solution As shown in Figure 6.46, we are to find the z value such that the area between 0 and z is .4251.

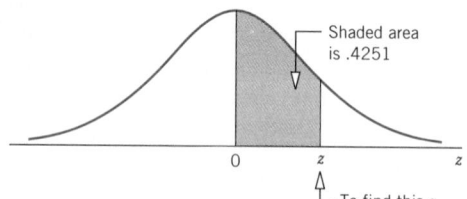

Figure 6.46 Finding the z value.

To find the required value of z, we locate .4251 in the body of the normal distribution table, Table VII of Appendix C. The relevant portion of that table is reproduced below as

Table 6.6. Next we read the numbers, in the column and row for z, which correspond to .4251. As shown in Table 6.6, these numbers are 1.4 and .04, respectively. Combining these two numbers, we obtain the required value of $z = \mathbf{1.44}$.

Table 6.6 Finding the z Value When Area Is Known

z	.00	.01	...	.04	...	.09
0.0	.0000	.0040	...		...	.0359
0.1	.0398	.0438	...		...	.0753
0.2	.0793	.0832	...		...	.1141
.	.	.	...	.	...	.
.	.	.	...	.	...	.
1.4 ←				.4251 ←	...	...
.	.	.	...	.	...	.
.	.	.	...	.	...	.
3.0	.4987	.4987	...	.4988	...	.4990

We locate this
value in Table VII of
Appendix C

EXAMPLE 6-16 Find the value of z such that the area under the standard normal curve in the right tail is .0050.

Solution To find the required value of z, first find the area between 0 and z. The total area to the right of $z = 0$ is .5. Hence,

$$\text{Area between 0 and } z = .5 - .0050 = .4950$$

This area is shown in Figure 6.47.

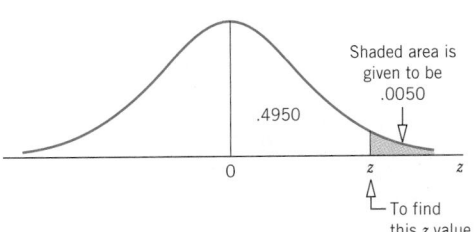

Figure 6.47 Finding the z value.

Now we look for .4950 in the body of the normal distribution table. Table VII does not contain .4950. So we find the value closest to .4950, which is either .4949 or .4951. We can use either of these two values. If we choose .4951, the corresponding z value is 2.58. Hence, the required value of z is **2.58**, and the area to the right of $z = 2.58$ is approximately .0050. Note that there is no apparent reason to choose .4951 and not to choose .4949. We can use either of the two values. If we choose .4949, the corresponding z value will be 2.57. ▬

EXAMPLE 6-17 Find the value of z such that the area under the standard normal curve in the left tail is .05.

Solution Because .05 is smaller than .5 and it is the area in the left tail, the value of z is negative. To find the required value of z, first we find the area between 0 and z. The total area to the left of $z = 0$ is .5. Hence,

$$\text{Area between 0 and } z = .5 - .05 = .4500$$

This area is shown in Figure 6.48.

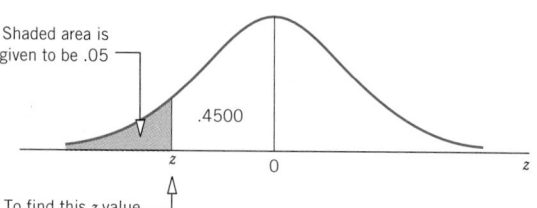

Figure 6.48   Finding the z value.

Next, we look for .4500 in the body of the normal distribution table. The value closest to .4500 in the normal distribution table is either .4495 or .4505. Suppose we use the value .4505. The corresponding z value is 1.65. Because the value of z is negative in our example (see Figure 6.48), the required value of z is **−1.65** and the area to the left of $z = -1.65$ is approximately .05.

To find an x value when an area under a normal distribution curve is given, first we find the z value corresponding to that x value from the normal distribution table. Then, to find the x value, we substitute the values of μ, σ, and z in the following formula, which is obtained from $z = (x - \mu)/\sigma$ by doing some algebraic manipulations. Also, if we know the values of x, z, and σ, we can find μ using this same formula. Exercises 6.63 and 6.64 present such cases.

FINDING AN x VALUE FOR A NORMAL DISTRIBUTION

For a normal curve, with known values of μ and σ and for a given area under the curve between the mean and x, the x value is calculated as

$$x = \mu + z\sigma$$

Examples 6–18 and 6–19 illustrate how to find an x value when an area under a normal distribution curve is known.

Finding x when area in the left tail is known.

EXAMPLE 6-18   Recall Example 6–14. It is known that the life of a calculator manufactured by Intal Corporation has a normal distribution with a mean of 54 months and a standard deviation of 8 months. What should the warranty period be to replace a malfunctioning calculator if the company does not want to replace more than 1% of all the calculators sold?

Solution   Let x be the life of a calculator. Then, x follows a normal distribution with $\mu = 54$ months and $\sigma = 8$ months.

The calculators that would be replaced are the ones that start malfunctioning during the warranty period. The company's objective is to replace at most 1% of all the calculators sold. The shaded area in Figure 6.49 gives the proportion of calculators that are replaced. We are to find the value of x so that the area to the left of x under the normal curve is 1% or .01.

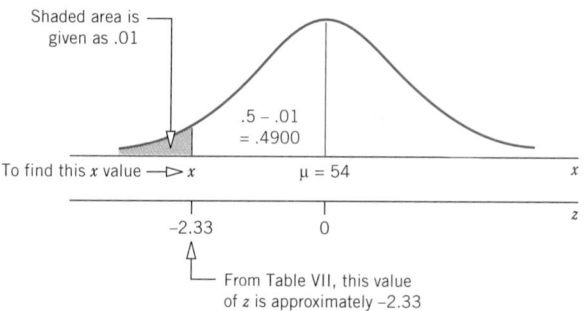

Figure 6.49 Finding an *x* value.

In the first step, we find the *z* value that corresponds to the required *x* value. To find this *z* value, we need to know the area between the mean and the *x* value. This area is calculated as follows.

$$\text{Area between the mean and the } x \text{ value} = .5 - .01 = .4900$$

Next, we find the *z* value from the normal distribution table for .4900. Table VII of Appendix C does not contain a value that is exactly .4900. The value closest to .4900 in the table is .4901 and the *z* value for .4901 is 2.33. Since *z* is on the left side of the mean (see Figure 6.49), the value of *z* is negative. Hence,

$$z = -2.33$$

Substituting the values of μ, σ, and *z* in the formula $x = \mu + z\sigma$, we obtain:

$$x = \mu + z\sigma = 54 + (-2.33)(8) = 54 - 18.64 = \mathbf{35.36}$$

Thus, the company should replace all the calculators that start malfunctioning within 35.36 months (which can be rounded to 35 months) of the date of purchase so that they will not have to replace more than 1% of the calculators. ■

Finding x when area in the right tail is known.

EXAMPLE 6–19 Almost all high school students who intend to go to college take the SAT test. In recent years, the mean SAT score (in verbal and mathematics) of all students has been around 900. Debbie Sears is planning to take this test soon. Suppose the SAT scores of all students who take this test with Debbie will have a normal distribution with a mean of 904 and a standard deviation of 153. What should her score be on this test so that only 10% of all examinees score higher than she does?

Solution: Let *x* represent the SAT scores of examinees. Then, *x* follows a normal distribution with $\mu = 904$ and $\sigma = 153$. We are to find the value of *x* such that the area under the normal distribution curve to the right of *x* is 10%, as shown in Figure 6.50.

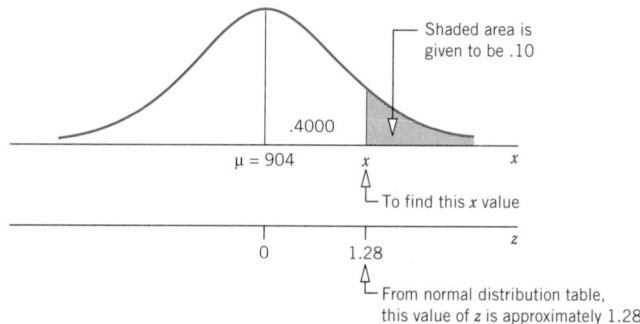

Figure 6.50 Finding an *x* value.

First we find the area under the normal distribution curve between the mean and the x value.

$$\text{Area between } \mu \text{ and the } x \text{ value} = .5 - .10 = .4000$$

To find the z value that corresponds to the required x value, we look for .4000 in the body of the normal distribution table. The value closest to .4000 in Table VII is .3997 and the corresponding z value is 1.28. Hence, the value of x is computed as

$$x = \mu + z\,\sigma = 904 + 1.28\,(153) = 904 + 195.84 = 1099.84 \approx \mathbf{1100}$$

Thus, if Debbie Sears scores 1100 on the SAT, only about 10% of the examinees are expected to score higher than she does.   ▬

EXERCISES

Concepts and Procedures

6.53   Find the value of z so that the area under the standard normal curve
  a.   from 0 to z is .4772 and z is positive
  b.   between 0 and z is (approximately) .4785 and z is negative
  c.   in the left tail is (approximately) .3565
  d.   in the right tail is (approximately) .1530

6.54   Find the value of z so that the area under the standard normal curve
  a.   from 0 to z is (approximately) .1965 and z is positive
  b.   between 0 and z is (approximately) .2740 and z is negative
  c.   in the left tail is (approximately) .2050
  d.   in the right tail is (approximately) .1053

6.55   Determine the value of z so that the area under the standard normal curve
  a.   in the right tail is .0500
  b.   in the left tail is .0250
  c.   in the left tail is .0100
  d.   in the right tail is .0050

6.56   Determine the value of z so that the area under the standard normal curve
  a.   in the right tail is .0250
  b.   in the left tail is .0500
  c.   in the left tail is .0010
  d.   in the right tail is .0100

6.57   Let x be a continuous random variable that follows a normal distribution with a mean of 200 and a standard deviation of 25.
  a.   Find the value of x so that the area under the normal curve to the left of x is approximately .6330.
  b.   Find the value of x so that the area under the normal curve to the right of x is approximately .05.
  c.   Find the value of x so that the area under the normal curve to the right of x is .8051.
  d.   Find the value of x so that the area under the normal curve to the left of x is .0150.
  e.   Find the value of x so that the area under the normal curve between μ and x is .4525 and the value of x is smaller than μ.
  f.   Find the value of x so that the area under the normal curve between μ and x is approximately .4800 and the value of x is greater than μ.

6.58   Let x be a continuous random variable that follows a normal distribution with a mean of 550 and a standard deviation of 75.

a. Find the value of x so that the area under the normal curve to the left of x is .0250.

b. Find the value of x so that the area under the normal curve to the right of x is .9345.

c. Find the value of x so that the area under the normal curve to the right of x is approximately .0275.

d. Find the value of x so that the area under the normal curve to the left of x is approximately .9600.

e. Find the value of x so that the area under the normal curve between μ and x is approximately .4700 and the value of x is smaller than μ.

f. Find the value of x so that the area under the normal curve between μ and x is approximately .4100 and the value of x is greater than μ.

Applications

6.59 Fast Auto Service provides oil and lube service for cars. It is known that the mean time taken for oil and lube service at this garage is 15 minutes per car and the standard deviation is 2.4 minutes. The management wants to promote the business by guaranteeing a maximum waiting time for its customers. If a customer's car is not serviced within that period, the customer will receive a 50% discount on the charges. The company wants to limit this discount to at most 5% of the customers. What should the maximum guaranteed waiting time be? Assume that the times taken for oil and lube service for all cars have a normal distribution.

6.60 The management of a supermarket wants to adopt a new promotional policy of giving a free gift to every customer who spends more than a certain amount per visit at this supermarket. The expectation of the management is that after this promotional policy is advertised, the expenditures for all customers at this supermarket will be normally distributed with a mean of $95 and a standard deviation of $21. If the management wants to give free gifts to at most 8% of the customers, what should the amount be above which a customer would receive a free gift?

6.61 According to the records of an electric company serving the Boston area, the mean electric consumption during winter for all households is 1650 kilowatt hours per month. Assume that the monthly electric consumption during winter by all households in this area has a normal distribution with a mean of 1650 kilowatt hours and a standard deviation of 320 kilowatt hours. The company sent a notice to Bill Johnson informing him that about 90% of the households use less electricity per month than he does. What is Bill Johnson's monthly electric consumption?

6.62 Rockingham Corporation makes electric shavers. The life (period before which a shaver does not need a major repair) of Model J795 of an electric shaver manufactured by this corporation has a normal distribution with a mean of 65 months and a standard deviation of 6 months. The company is to determine the warranty period for this shaver. Any shaver that will need a major repair during this warranty period will be replaced free by the company.

a. What should the warranty period be if the company does not want to replace more than 1% of the shavers?

b. What should the warranty period be if the company does not want to replace more than 5% of the shavers?

*6.63 A study has shown that 20% of all college textbooks have a price of $70 or higher. It is known that the standard deviation of the prices of all college textbooks is $9.50. Suppose the prices of all college textbooks have a normal distribution. What is the mean price of all college textbooks?

*6.64 A machine at Keats Corporation fills 64-ounce detergent jugs. The machine can be adjusted to pour, on average, any amount of detergent into these jugs. However, the machine does not pour exactly the same amount of detergent in each jug; it varies from jug to jug. It is known that the net amount of detergent poured into each jug has a normal distribution with a standard deviation of .4 ounces. The quality control inspector wants to adjust the machine such that at least 95% of the jugs have more than 64 ounces of detergent. What should the mean amount of detergent poured by this machine into these jugs be?

6.7   THE NORMAL APPROXIMATION TO THE BINOMIAL DISTRIBUTION

Recall from Chapter 5 that

1. The binomial distribution is applied to a discrete random variable.
2. Each repetition, called a trial, of a binomial experiment results in one of two possible outcomes, either a success or a failure.
3. The probabilities of the two (possible) outcomes remain the same for each repetition of the experiment.
4. The trials are independent.

The binomial formula, which gives the probability of x successes in n trials, is

$$P(x) = \binom{n}{x} p^x q^{n-x}$$

However, the use of the binomial formula becomes very tedious when n is large. In such cases, the normal distribution can be used to approximate the binomial probability. Note that for a binomial problem, the exact probability is obtained by using the binomial formula. If we apply the normal distribution to solve a binomial problem, the probability that we obtain is an approximation to the exact probability. The approximation obtained by using the normal distribution is very close to the exact probability when n is large and p is very close to .50. However, this does not mean that we should not use the normal approximation when p is not close to .50. The reason for the approximation being closer to the exact probability when p is close to .50 is that the binomial distribution is symmetric when $p = .50$. The normal distribution is always symmetric. Hence, the two distributions are very close to each other when n is large and p is close to .50. However, this does not mean that whenever $p = .50$ the binomial distribution is the same as the normal distribution because not every symmetric bell-shaped curve is a normal distribution curve.

NORMAL DISTRIBUTION AS AN APPROXIMATION TO BINOMIAL DISTRIBUTION

Usually, the normal distribution is used as an approximation to the binomial distribution when np and nq are both greater than 5, that is, when

$$np > 5 \quad \text{and} \quad nq > 5$$

Table 6.7 gives the binomial probability distribution of x for $n = 12$ and $p = .50$. This table is constructed using Table IV of Appendix C. Figure 6.51 shows the histogram and the smoothed polygon for the probability distribution of Table 6.7. As we can observe, the histogram in Figure 6.51 is symmetric and the curve obtained by joining the upper midpoints of the rectangles is approximately bell-shaped.

Examples 6–20 through 6–22 illustrate the application of the normal distribution as an approximation to the binomial distribution.

Table 6.7 The Binomial Probability Distribution for $n = 12$ and $p = .50$

x	$P(x)$
0	.0002
1	.0029
2	.0161
3	.0537
4	.1208
5	.1934
6	.2256
7	.1934
8	.1208
9	.0537
10	.0161
11	.0029
12	.0002

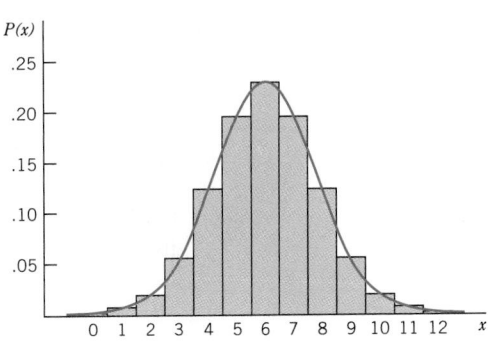

Figure 6.51 Histogram for the probability distribution of Table 6.7.

Normal approximation to binomial: x equals a specific value.

EXAMPLE 6-20 According to an estimate, 50% of the people in America have at least one credit card. If a random sample of 30 persons is selected, what is the probability that 19 of them will have at least one credit card?

Solution Let n be the total number of persons in the sample, x be the number of persons in the sample who have at least one credit card, and p be the probability that a person has at least one credit card. Then, this is a binomial problem with

$$n = 30, \quad p = .50, \quad q = 1 - p = .50,$$
$$x = 19, \quad \text{and} \quad n - x = 30 - 19 = 11$$

Using the binomial formula, the exact probability that 19 persons in a sample of 30 have at least one credit card is

$$P(19) = \binom{30}{19} (.50)^{19} (.50)^{11} = \mathbf{.0509}$$

Now let us solve this problem using the normal distribution as an approximation to the binomial distribution. For this example,

$$np = 30 (.50) = 15 \quad \text{and} \quad nq = 30 (.50) = 15$$

Since np and nq are both greater than 5, we can use the normal distribution as an approximation to solve this binomial problem.

Using the normal distribution as an approximation to the binomial involves the following three steps.

Step 1. *Compute μ and σ for the binomial distribution*

To use the normal distribution, we need to know the mean and standard deviation of the distribution. Hence, the first step in using the normal approximation to the binomial distribution is to compute the mean and standard deviation of the binomial distribution. As we know from Chapter 5, the mean and standard deviation of a binomial distribution are given by np and $\sqrt{npq}$, respectively. Using these formulas, we obtain:

$$\mu = np = 30 (.50) = 15$$
$$\sigma = \sqrt{npq} = \sqrt{30 (.50) (.50)} = 2.73861279$$

Step 2.   *Convert the discrete random variable to a continuous random variable*

The normal distribution applies to a continuous random variable, whereas the binomial distribution applies to a discrete random variable. The second step in applying the normal approximation to the binomial distribution is to convert the discrete random variable to a continuous random variable by making the **correction for continuity**.

CONTINUITY CORRECTION FACTOR

The addition of .5 and/or subtraction of .5 from the value(s) of x when the normal distribution is used as an approximation to the binomial distribution, where x is the number of successes in n trials is called the *continuity correction factor*.

As shown in Figure 6.52, the probability of 19 successes in 30 trials is given by the area of the rectangle for $x = 19$. To make the correction for continuity, we use the interval 18.5 to 19.5 for 19 persons. This interval is actually given by the two boundaries of the rectangle for $x = 19$, which are obtained by subtracting .5 from 19 and by adding .5 to 19. Thus, $P(x = 19)$ for the binomial problem will be approximately equal to $P(18.5 \leq x \leq 19.5)$ for the normal distribution.

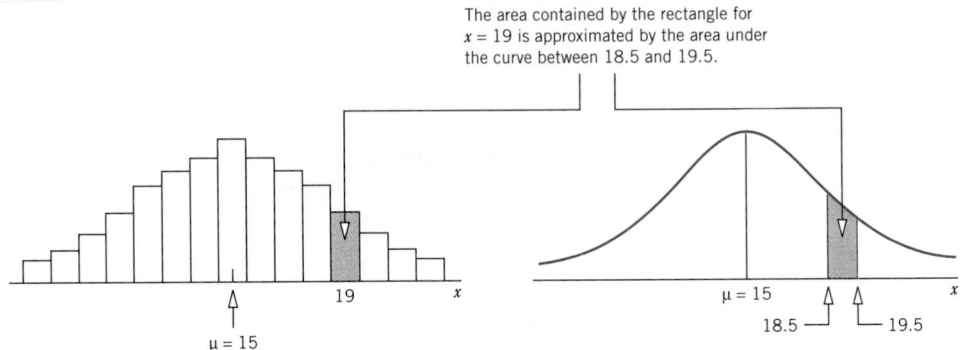

The area contained by the rectangle for $x = 19$ is approximated by the area under the curve between 18.5 and 19.5.

Figure 6.52

Step 3.   *Compute the required probability using the normal distribution*

As shown in Figure 6.53, the area under the normal distribution curve between $x = 18.5$ and $x = 19.5$ will give us the (approximate) probability that 19 persons possess at least one credit card. We calculate this probability as follows.

$$\text{For } x = 18.5: \quad z = \frac{18.5 - 15}{2.73861279} = 1.28$$

$$\text{For } x = 19.5: \quad z = \frac{19.5 - 15}{2.73861279} = 1.64$$

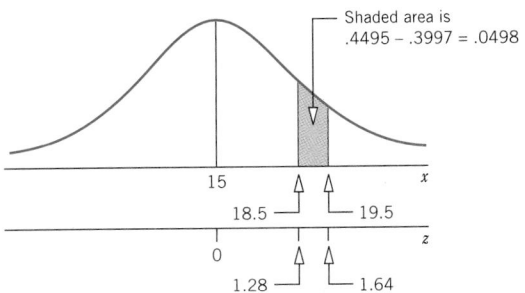

Figure 6.53 Area between $x = 18.5$ and $x = 19.5$.

The required probability is given by the area under the standard normal curve between $z = 1.28$ and $z = 1.64$. This area is obtained by subtracting the area between $z = 0$ and $z = 1.28$ from the area between $z = 0$ and $z = 1.64$. From Table VII of Appendix C, the area between $z = 0$ and $z = 1.28$ is .3997 and the area between $z = 0$ and $z = 1.64$ is .4495. Hence, the required probability is

$$P(18.5 \leq x \leq 19.5) = P(1.28 \leq z \leq 1.64) = .4495 - .3997 = \mathbf{.0498}$$

Thus, based on the normal approximation, the probability that 19 persons in a sample of 30 will possess at least one credit card is approximately .0498. Earlier, using the binomial formula, we obtained the exact probability .0509. The error due to using the normal approximation is .0509 − .0498 = .0011. Thus, the exact probability is underestimated by .0011 if the normal approximation is used. ■

☞ *Remember* When applying the normal distribution as an approximation to the binomial distribution, always make a *correction for continuity*. The continuity correction is made by subtracting .5 from the lower limit of the interval and/or by adding .5 to the upper limit of the interval. For example, the binomial probability $P(7 \leq x \leq 12)$ will be approximated by the probability $P(6.5 \leq x \leq 12.5)$ for the normal distribution; the binomial probability $P(x \geq 9)$ will be approximated by the probability $P(x \geq 8.5)$ for the normal distribution; and the binomial probability $P(x \leq 10)$ will be approximated by the probability $P(x \leq 10.5)$ for the normal distribution. Note that the probability $P(x \geq 9)$ has only the lower limit of 9 and no upper limit, and the probability $P(x \leq 10)$ has only the upper limit of 10 and no lower limit.

Normal approximation to binomial: x assumes a value in an interval.

EXAMPLE 6–21 In a *Time*/CNN poll, 66% of U.S. adults said they approve of more restrictions being placed on what is shown on television to improve the moral climate in the United States (*Time*, June 12, 1995). Suppose this result is true for the current population of all U.S. adults. What is the probability that in a random sample of 100 U.S. adults, 58 to 63 would hold this view?

Solution Let n be the total number of adults in the sample, x the number of adults in the sample who approve placing more restrictions on what is shown on television to improve the moral climate in the United States, and p the probability that an adult approves of the placement of more restrictions on what is shown on television to improve the moral climate in the United States. Then, this is a binomial problem with

$$n = 100, \quad p = .66, \quad \text{and} \quad q = 1 - .66 = .34$$

We are to find the probability of 58 to 63 successes in 100 trials. Because n is large, it is easier to apply the normal approximation than to use the binomial formula. We can check that np and nq are both greater than 5. The mean and standard deviation of the binomial distribution are

$$\mu = np = 100 \, (.66) = 66$$

$$\sigma = \sqrt{npq} = \sqrt{100 \, (.66) \, (.34)} = 4.73708771$$

To make the continuity correction, we subtract .5 from 58 and add .5 to 63 to obtain the interval 57.5 to 63.5. Thus, the probability that 58 to 63 adults hold the given opinion is approximated by the area under the normal distribution curve from $x = 57.5$ to $x = 63.5$. This area is shown in Figure 6.54.

$$\text{For } x = 57.5: \qquad z = \frac{57.5 - 66}{4.73708771} = -1.79$$

$$\text{For } x = 63.5: \qquad z = \frac{63.5 - 66}{4.73708771} = -.53$$

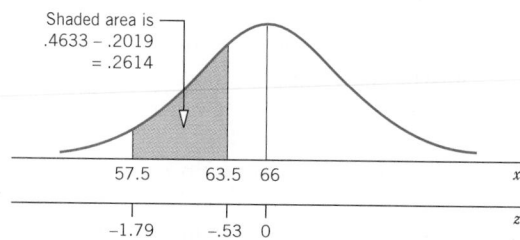

Figure 6.54   Area from $x = 57.5$ to $x = 63.5$.

The required probability is given by the area under the standard normal curve between $z = -1.79$ and $z = -.53$. This area is obtained by subtracting the area between $z = -.53$ and $z = 0$ from the area between $z = -1.79$ and $z = 0$. From Table VII of Appendix C, the area between $z = -.53$ and $z = 0$ is .2019 and the area between $z = -1.79$ and $z = 0$ is .4633. Hence,

$$P(57.5 < x < 63.5) = P(-1.79 < z < -.53) = .4633 - .2019 = \textbf{.2614}$$

Thus, the probability that 58 to 63 adults in a sample of 100 would hold the given view is approximately .2614.   ▪

Normal approximation to binomial: x greater than or equal to a value.

EXAMPLE 6-22   According to a survey of U.S. adults conducted by Louis Harris and Associates for *Business Week*, 67% of adults said that it has become harder to achieve the American Dream of equal opportunity, personal freedom, and social mobility in the past 10 years (*Business Week*, March 13, 1995). Assume that this result is true for the current population of all U.S. adults. What is the probability that in a random sample of 200 U.S. adults, 142 or more would hold this view?

Solution   Let n be the total number of adults in the sample, x the number of adults in the sample who hold the given view, and p the probability that a randomly selected adult holds the given view. Then, this is a binomial problem with

$$n = 200, \qquad p = .67, \qquad \text{and} \qquad q = 1 - .67 = .33$$

We are to find the probability of 142 or more successes in 200 trials. The mean and standard deviation of the binomial distribution are

$$\mu = np = 200\,(.67) = 134$$

$$\sigma = \sqrt{npq} = \sqrt{200\,(.67)\,(.33)} = 6.64981203$$

For the continuity correction, we subtract .5 from 142, which gives 141.5. Thus, the probability that 142 or more adults in a random sample of 200 would hold the given view is approximated by the area under the normal distribution curve to the right of $x = 141.5$, as shown in Figure 6.55.

$$\text{For } x = 141.5: \qquad z = \frac{141.5 - 134}{6.64981203} = 1.13$$

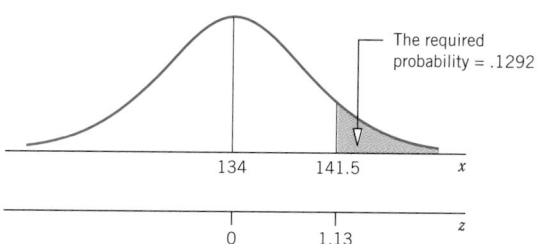

Figure 6.55 Area to the right of $x = 141.5$.

To find the required probability, we find the area between $z = 0$ and $z = 1.13$ and then subtract this area from .5, which is the total area to the right of $z = 0$. From Table VII of Appendix C, the area between $z = 0$ and $z = 1.13$ is .3708. Hence,

$$P(x \ge 141.5) = P(z \ge 1.13) = .5 - .3708 = \mathbf{.1292}$$

Thus, the probability that 142 or more adults in a random sample of 200 would hold the given view is approximately .1292. ■

EXERCISES

Concepts and Procedures

6.65 Under what conditions is the normal distribution usually used as an approximation to the binomial distribution?

6.66 For a binomial probability distribution, $n = 20$ and $p = .60$.
    a. Find the probability $P(x = 11)$ by using the table of binomial probabilities (Table IV of Appendix C).
    b. Find the probability $P(x = 11)$ by using the normal distribution as an approximation to the binomial distribution. What is the difference between this approximation and the exact probability calculated in part a?

6.67 For a binomial probability distribution, $n = 25$ and $p = .40$.
    a. Find the probability $P(8 \le x \le 12)$ by using the table of binomial probabilities (Table IV of Appendix C).
    b. Find the probability $P(8 \le x \le 12)$ by using the normal distribution as an approximation to the binomial distribution. What is the difference between this approximation and the exact probability calculated in part a?

6.68   For a binomial probability distribution, $n = 80$ and $p = .50$. Let x be the number of successes in 80 trials.

    a.   Find the mean and standard deviation of this binomial distribution.

    b.   Find $P(x \geq 37)$ using the normal approximation.

    c.   Find $P(41 \leq x \leq 44)$ using the normal approximation.

6.69   For a binomial probability distribution, $n = 120$ and $p = .60$. Let x be the number of successes in 120 trials.

    a.   Find the mean and standard deviation of this binomial distribution.

    b.   Find $P(x \leq 70)$ using the normal approximation.

    c.   Find $P(67 \leq x \leq 71)$ using the normal approximation.

6.70   Find the following binomial probabilities using the normal approximation.

    a. $n = 140$, $p = .45$, $P(x = 67)$

    b. $n = 100$, $p = .55$, $P(52 \leq x \leq 60)$

    c. $n = 90$, $p = .42$, $P(x \geq 40)$

    d. $n = 104$, $p = .75$, $P(x \leq 72)$

6.71   Find the following binomial probabilities using the normal approximation.

    a. $n = 70$, $p = .30$, $P(x = 18)$

    b. $n = 200$, $p = .70$, $P(133 \leq x \leq 145)$

    c. $n = 85$, $p = .40$, $P(x \geq 30)$

    d. $n = 150$, $p = .38$, $P(x \leq 62)$.

Applications

6.72   According to Labor Canada, 29.2% of Canada's civilian workers were members of labor unions in 1994 *(The 1996 Canadian Global Almanac)*. Suppose this result holds true for the current population of Canadian civilian workers. Find the probability that in a random sample of 750 Canadian civilian workers 215 to 222 are union members.

6.73   In a recent study of American high school students, about 20% of them said that they do not try as hard as they could in school because they fear the disapproval of their peers (Laurence Steinberg, et al., ''Beyond the Classroom: Why School Reform Has Failed and What Parents Need to Do.'' Cited in *Education Week* 15[37], June 5, 1996). Assume that 20% of all American high school students feel this way. Find the probability that in a random sample of 100 American high school students, 12 to 18 will feel this way.

6.74   According to a survey conducted by the Gallup Organization for Phi Delta Kappa, 60% of adults in the United States favor increasing the amount of time high school students spend in school by lengthening the school year or the school day *(Phi Delta Kappan, September 1996)*. Assume that this result is true for the current population of all American adults. Find the probability that the number of adults in a random sample of 250 who hold this view is

    a.   exactly 145   **b.**   at most 135   **c.**   155 to 160

6.75   A study published in *The Journal of the American Medical Association* reported that 42.3% of physicians in the United States are employed by others as opposed to being self-employed as solo or group practitioners *(U.S. News & World Report, September 2, 1996)*. Assume that this result is true for the current population of all U.S. physicians. Find the probability that in a random sample of 480 U.S. physicians, the number who are employed by others is

    a.   exactly 200   **b.**   205 or more   **c.**   195 to 210

6.76   According to the Americans' Use of Time Project for 1995, 54% of American adults surveyed reported feeling *a great deal of* or *moderate* stress in the past two weeks (John P. Robinson and Geoffrey Godbey, ''The Great American Slowdown,'' *American Demographics* 18[6], June 1996). Assume that this result holds true for the current population of all American adults.

    a.   Find the probability that in a random sample of 200 adults, exactly 100 have felt such stress in the past two weeks.

    b.   Find the probability that in a random sample of 200 adults, at most 110 have felt such stress in the past two weeks.

c. What is the probability that in a random sample of 200 adults, 105 to 115 have felt such stress in the past two weeks?

6.77 A fast food chain store conducted a taste survey before marketing a new hamburger. The results of the survey showed that 70% of the people who tried this hamburger liked it. Encouraged by this result, the company decided to market the new hamburger. Assume that 70% of all people like this hamburger. On a certain day, 100 customers bought this hamburger.

a. Find the probability that exactly 65 of the 100 customers will like this hamburger.
b. What is the probability that 60 or less of the 100 customers will like this hamburger?
c. What is the probability that 75 to 80 of the 100 customers will like this hamburger?

6.78 Johnson Electronics makes calculators. Consumer satisfaction is one of the top priorities of the company's management. The company guarantees the refund of money or a replacement for any calculator that malfunctions within two years from the date of purchase. It is known from past data that despite all efforts, 5% of the calculators manufactured by this company malfunction within a two-year period. The company recently mailed 500 such calculators to its customers.

a. Find the probability that exactly 28 of the 500 calculators will be returned for refund or replacement within a two-year period.
b. What is the probability that 26 or more of the 500 calculators will be returned for refund or replacement within a two-year period?
c. What is the probability that 15 to 20 of the 500 calculators will be returned for refund or replacement within a two-year period?

6.79 Hurbert Corporation makes font cartridges for laser printers that it sells to Alpha Electronics Inc. The cartridges are shipped to Alpha Electronics in large volumes. The quality control department at Alpha Electronics randomly selects 100 cartridges from each shipment and inspects them for being good or defective. If this sample contains seven or more defective cartridges, the entire shipment is rejected. Hurbert Corporation promises that of all the cartridges, only 5% are defective.

a. Find the probability that a given shipment of cartridges received by Alpha Electronics will be accepted.
b. Find the probability that a given shipment of cartridges received by Alpha Electronics will not be accepted.

GLOSSARY

Continuity correction factor   Addition of .5 and/or subtraction of .5 from the value(s) of x when the normal distribution is used as an approximation to the binomial distribution, where x is the number of successes in n trials.

Continuous random variable   A random variable that can assume any value in one or more intervals.

Normal probability distribution   The probability distribution of a continuous random variable that, when plotted, gives a specific bell-shaped curve. The parameters of the normal distribution are the mean μ and the standard deviation σ.

Standard normal distribution   The normal distribution with $\mu = 0$ and $\sigma = 1$. The units of the standard normal distribution are denoted by z.

z value or **z score**   The units of the standard normal distribution that are denoted by z.

KEY FORMULAS

1. **The z value for an x value to standardize a normal distribution**

$$z = \frac{x - \mu}{\sigma}$$

2. **The value of x for a normal distribution, when the values of μ, σ, and z are known**

$$x = \mu + z\sigma$$

3. **Normal approximation to the binomial distribution**

The normal distribution can be used as an approximation to the binomial distribution when

$$np > 5 \quad \text{and} \quad nq > 5$$

4. **Mean and standard deviation of the binomial distribution**

$$\mu = np \quad \text{and} \quad \sigma = \sqrt{npq}$$

SUPPLEMENTARY EXERCISES

6.80   The management at Ohio National Bank does not want its customers to wait in line for service for too long. The manager of a branch of this bank estimated that the customers currently have to wait an average of 8 minutes for service. Assume that the waiting times for all customers at this branch have a normal distribution with a mean of 8 minutes and a standard deviation of 2 minutes.

    a.   Find the probability that a randomly selected customer will have to wait for less than three minutes.

    b.   What percentage of the customers have to wait for 10 to 13 minutes?

    c.   What percentage of the customers have to wait for 6 to 12 minutes?

    d.   Is it possible that a customer may have to wait for more than 16 minutes for service? Explain.

6.81   The research department at Colorado Bank estimated that the mean daily balance of all checking accounts at this bank is $870. Suppose the balances of all checking accounts at this bank have a normal distribution with a mean of $870 and a standard deviation of $350.

    a.   Find the probability that on a certain day a randomly selected checking account will have a balance of more than $1500.

    b.   What percentage of the checking accounts at this bank will have a balance of $200 to $300 on a certain day?

    c.   What percentage of the checking accounts at this bank will have a balance of $1000 to $1400 on a certain day?

    d.   Is it possible that a checking account will have a balance of more than $2500 on a certain day? Explain.

6.82   At Jen and Perry Ice Cream Company, the machine that fills one-pound cartons of Top Flavor ice cream is set to dispense 16 ounces of ice cream in every carton. However, some cartons contain slightly less than and some contain slightly more than 16 ounces of ice cream. The amounts of ice cream in all such cartons have a normal distribution with a mean of 16 ounces and a standard deviation of .18 ounces.

    a.   Find the probability that a randomly selected carton contains 16.25 to 16.50 ounces of ice cream.

    b.   What percentage of such cartons contain less than 15.75 ounces of ice cream?

    c.   Is it possible for a carton to contain less than 15.25 ounces of ice cream? Explain.

6.83   A machine at Kasem Steel Corporation makes iron rods that are supposed to be 50 inches long. However, the machine does not make all rods of exactly the same length. It is known that the probability distribution of the lengths of rods made on this machine is normal with a mean of 50 inches and a standard deviation of .06 inch. The rods that are either shorter than 49.85 inches or longer than 50.15 inches are discarded. What percentage of the rods made on this machine are discarded?

6.84   The time taken by new employees at Shia Corporation to learn a packaging procedure is normally distributed with a mean of 24 hours and a standard deviation of 2.5 hours.

   a.   Find the time beyond which only 1% of the new employees take to learn this procedure.
   b.   What is the time below which only 5% of the new employees take to learn this procedure?

6.85   The print on the package of 100-watt General Electric soft-white light bulbs states that these bulbs have an average life of 750 hours. Assume that the lives of all such bulbs have a normal distribution with a mean of 750 hours and a standard deviation of 50 hours.

   a.   Let x be the life of such a light bulb. Find x so that only 2.5% of such light bulbs have lives longer than this value.
   b.   Let x be the life of such a light bulb. Find x so that about 80% of such light bulbs have lives shorter than this value.

*__6.86__   It is known that 15% of all homeowners pay a monthly mortgage of more than $2400 and that the standard deviation of the monthly mortgage payments of all homeowners is $350. Suppose that the monthly mortgage payments of all homeowners have a normal distribution. What is the mean monthly mortgage paid by all homeowners?

*__6.87__   At Jen and Perry Ice Cream Company, a machine fills one-pound cartons of Top Flavor ice cream. The machine can be set to dispense, on average, any amount of ice cream into these cartons. However, the machine does not put exactly the same amount of ice cream in each carton; it varies from carton to carton. It is known that the amount of ice cream put in each such carton has a normal distribution with a standard deviation of .18 ounces. The quality control inspector wants to set the machine such that at least 90% of the cartons have more than 16 ounces of ice cream. What should be the mean amount of ice cream put into these cartons by this machine?

6.88   Mong Corporation makes auto batteries. The company claims that 80% of its LL70 batteries are good for 70 months or more.

   a.   What is the probability that in a sample of 100 such batteries, exactly 85 will be good for 70 months or more?
   b.   Find the probability that in a sample of 100 such batteries, at most 74 will be good for 70 months or more.
   c.   What is the probability that in a sample of 100 such batteries, 75 to 87 will be good for 70 months or more?
   d.   Find the probability that in a sample of 100 such batteries, 72 to 77 will be good for 70 months or more.

 6.89   Stress on the job is a major concern of a large number of people who go into managerial positions. It is estimated that 80% of managers of all companies suffer from job-related stress.

   a.   What is the probability that in a sample of 200 managers of companies, exactly 150 suffer from job-related stress?
   b.   Find the probability that in a sample of 200 managers of companies, at least 170 suffer from job-related stress.
   c.   What is the probability that in a sample of 200 managers of companies, 165 or less suffer from job-related stress?
   d.   Find the probability that in a sample of 200 managers of companies, 164 to 172 suffer from job-related stress.

*__6.90__   Two companies, A and B, drill wells in a rural area. Company A charges a flat fee of $3500 to drill a well regardless of its depth. Company B charges $1000 plus $12 per foot to drill a well. The depths of wells drilled in this area have a normal distribution with a mean of 250 feet and a standard deviation of 40 feet.

   a.   What is the probability that Company B would charge more than Company A to drill a well?
   b.   Find the mean amount charged by Company B to drill a well.

*6.91   Otto is trying out for the javelin throw to compete in the Olympics. The lengths of his javelin throws are normally distributed with a mean of 290 feet and a standard deviation of 10 feet. What is the probability that the longest of 3 of his throws is 320 feet or more?

*6.92   Lori just bought a new set of 4 tires for her car. The life of each tire is normally distributed with a mean of 45,000 miles and a standard deviation of 2000 miles. Find the probability that all 4 tires will last at least 46,000 miles. Assume that the life of each of these tires is independent of the lives of other tires.

 *6.93   The Jen and Perry Ice Cream company makes a gourmet ice cream. Although the law allows ice cream to contain up to 50% air, this product is designed to contain only 20% air. Because of variability inherent in the manufacturing process, management is satisfied if each pint contains between 18% and 22% air. Currently two of Jen and Perry's plants are making gourmet ice cream. At Plant A, the mean amount of air per pint is 20% with a standard deviation of 2%. At Plant B, the mean amount of air per pint is 19% with a standard deviation of 1%. Assuming the amount of air is normally distributed at both plants, which plant is producing the greater proportion of pints that contain between 18% and 22% air?

*6.94   The highway police in a certain state are using aerial surveillance to control speeding on a highway with a posted speed limit of 55 miles per hour. Police officers watch cars from helicopters above a straight segment of this highway that has large marks painted on the pavement at 1-mile intervals. After the police officers observe how long a car takes to cover one mile, a computer estimates that car's speed. Assume that the errors of these estimates are normally distributed with a mean of 0 and a standard deviation of 2 miles per hour.

  a. The state police chief has directed his officers not to issue a speeding citation unless the aerial unit's estimate of speed is at least 65 miles per hour. What is the probability that a car traveling at 60 miles per hour or less will be cited for speeding?

  b. Suppose the chief does not want his officers to cite a car for speeding unless they are 99% sure that it is traveling at 60 miles per hour or faster. What is the minimum estimate of speed at which a car should be cited for speeding?

 *6.95   Ashley knows that the time it takes her to commute to work is approximately normally distributed with a mean of 45 minutes and a standard deviation of 3 minutes. What time must she leave home in the morning so that she is 95% sure of arriving at work by 9 A.M.?

*6.96   A soft-drink vending machine is supposed to pour 8 ounces of the drink into a paper cup. However, the actual amount poured into a cup varies. The amount poured into a cup follows a normal distribution with a mean that can be set to any desired amount by adjusting the machine. The standard deviation of the amount poured is always .07 ounces regardless of the mean amount. If the owner of the machine wants to be 99% sure that the amount in each cup is 8 ounces or more, to what level should she set the mean?

*6.97   A newspaper article reported that the mean mathematics score on SAT tests for local high school students was 480 and that 20% of the students scored below 400. Assume that the SAT scores for students from this school follow a normal distribution.

  a. Find the standard deviation of the mathematics scores for students from this school.

  b. Find the percentage of students at this school whose mathematics scores were above 500.

*6.98   Alpha Corporation is considering two suppliers to secure the large amounts of steel rods that it uses. Company A produces rods with a mean diameter of 8 mm and a standard deviation of .15 mm and sells 10,000 rods for $400. Company B produces rods with a mean diameter of 8 mm and a standard deviation of .12 mm and sells 10,000 rods for $460. A rod is usable only if its diameter is between 7.8 mm and 8.2 mm. Assume that the diameters of the rods produced by each company have a normal distribution. Which of the two companies should Alpha Corporation use as a supplier? Justify your answer with appropriate calculations.

*6.99   A gambler is planning to make a sequence of bets on a roulette wheel. Note that a roulette wheel has 38 numbers of which 18 are red, 18 are black, and 2 are green. Each time the wheel is spun, each of the 38 numbers is equally likely to occur. The gambler will choose one of the following two sequences.

Single-number bet: The gambler will bet $5 on a particular number before each spin. He will win a net amount of $175 if that number comes up and lose $5 otherwise.

Color bet: The gambler will bet $5 on the red color before each spin. He will win a net amount of $5 if a red number comes up and lose $5 otherwise.

   a.   If the gambler makes a sequence of 25 bets, which of the two betting schemes do you think offers him a better chance of coming out ahead (winning more money than losing) after the 25 bets?

   b.   Now compute the probability of coming out ahead after 25 single-number bets of $5 each and after 25 color bets of $5 each. Do these results confirm your guess in part a? (Before using an approximation to find either probability, be sure to check to see if it is appropriate.)

SELF-REVIEW TEST

1.   The normal probability distribution is applied to

   a.   a continuous random variable   **b.**   a discrete random variable   **c.**   any random variable

2.   For a continuous random variable, the probability of a single value of x is always

   a.   0   **b.**   1.0   **c.**   between 0 and 1

3.   Which of the following is not a characteristic of the normal distribution?

   a.   The total area under the curve is 1.0.
   b.   The curve is symmetric about the mean.
   c.   The two tails of the curve extend indefinitely.
   d.   The value of the mean is always greater than the value of the standard deviation.

4.   The parameters of a normal distribution are

   a. μ, z, and σ **b.** μ and σ **c.** μ, x, and σ

5.   For the standard normal distribution

   a. $\mu = 0$ and $\sigma = 1$ **b.** $\mu = 1$ and $\sigma = 0$ **c.** $\mu = 100$ and $\sigma = 10$

6.   The z value for μ for a normal distribution curve is always

   a.   positive   **b.**   negative   **c.**   0

7.   For a normal distribution curve, the z value for an x value that is less than μ is always

   a.   positive   **b.**   negative   **c.**   0

8.   Usually the normal distribution is used as an approximation to the binomial distribution when

   a. $n \geq 30$ **b.** $np > 5$ and $nq > 5$ **c.** $n > 20$ and $p = .50$

9.   Find the following probabilities for the standard normal distribution.

   a. $P(.87 \leq z \leq 2.33)$ **b.** $P(-2.97 \leq z \leq 1.46)$ **c.** $P(z \leq -1.19)$ **d.** $P(z > -.71)$

10.   Find the value of z for the standard normal curve such that the area

   a.   in the left tail is .1000
   b.   between 0 and z is .2291 and z is positive
   c.   in the right tail is .0500
   d.   between 0 and z is .3571 and z is negative

11.   A 1995 Gallup poll, sponsored by the National Sleep Foundation, found that people in the United States were averaging 7 hours of sleep per night (*The Chronicle of Higher Education*, August 9, 1996). Assume that the amount of sleep per night for all people in the United States is normally distributed with a mean of 7 hours and a standard deviation of 1.2 hours.

    a.   Find the probability that a randomly selected person gets between 5 and 8 hours of sleep per night.

    b.   What is the probability that a randomly selected person gets 8 hours or more of sleep per night?

    c.   What is the probability that a randomly selected person gets 6.5 hours or less sleep per night?

    d.   Find the probability that a randomly selected person gets between 7.5 and 9 hours of sleep per night.

12. Refer to Problem 11.

    a.   The 5% of Americans who sleep the least get fewer than how many hours of sleep per night?

    b.   The 10% of Americans who sleep the most get more than how many hours of sleep per night?

13. According to a report by the American College Testing Program, 26.9% of college freshmen who enrolled in Fall 1994 did not return for their sophomore year (*The Chronicle of Higher Education*, July 19, 1996). Assume that this percentage is true for the population of all college freshmen who enroll every Fall. A random sample of 400 college freshmen is selected.

    a.   Find the probability that 110 of them will not return for their sophomore year.

    b.   What is the probability that between 85 and 100 of them will not return for their sophomore year?

    c.   Find the probability that at least 120 of them will not return for their sophomore year.

    d.   What is the probability that at most 95 of them will not return for their sophomore year?

    e.   Find the probability that between 100 and 120 of them will not return for their sophomore year.

    f.   Find the probability that at most 275 of them will return for their sophomore year.

    g.   Find the probability that 280 to 330 of them will return for their sophomore year.

USING MINITAB

THE NORMAL PROBABILITY DISTRIBUTION

This section explains the procedure for finding an area under the normal distribution curve using MINITAB. Illustration M6–1 describes the steps involved in finding such areas.

ILLUSTRATION M6–1   According to the records of an electric company serving the Boston area, the mean electric consumption during winter for all households is 1650 kilowatt hours per month. Assume that the monthly electric consumption during winter by all households in this area has a normal distribution with a mean of 1650 kilowatt hours and a standard deviation of 320 kilowatt hours. Using MINITAB, find the probability that the monthly electric consumption during winter of a randomly selected household from this area is less than 1800 kilowatt hours.

Solution   The probability that the monthly electric consumption during winter of a randomly selected household from this area is less than 1800 kilowatt hours will be given by the area under the normal distribution curve to the left of $x = 1800$. The normal distribution of x has $\mu = 1650$ and $\sigma = 320$.

If you are using MINITAB FOR WINDOWS, perform the following steps to find the probability $P(x < 1800)$ when $\mu = 1650$ and $\sigma = 320$.

Step 1.   Click the **Calc** pull-down menu at the top of the screen.
Step 2.   Click **Probability Distributions** from the selections available in the **Calc** menu.
Step 3.   Click **Normal** from the selections available in the **Probability Distributions** menu.
Step 4.   You will see a dialog box entitled **Normal Distribution** appear on the screen. Click inside the circle next to **Cumulative probability**. Then, type **1650** (the value of μ) in the box next to **Mean** and **320** (the value of σ) in the box next to **Standard deviation**. Then, click inside the circle next to **Input constant** and type **1800** (the value of x) in the box next to it.
Step 5.   Click the **OK** button at the bottom of this dialog box. The required probability will appear on the screen.

If you are using the MINITAB COMMAND LANGUAGE, type the following MINITAB commands to obtain the probability $P(x < 1800)$.

```
MTB  > CDF x = 1800;
SUBC > NORMAL MEAN = 1650 SD = 320.
```

Whether you use MINITAB FOR WINDOWS or the MINITAB COMMAND LANGUAGE, you will obtain the output given in Figure 6.56. From this output, the required probability is .6804. The number 1.80E + 03 below x in Figure 6.56 refers to 1800. Here, +03 means that the decimal in 1.80 should be moved three digits to the right, which makes it 1800.

Note that when you use the **Cumulative probability** option in the above commands, MINITAB gives the whole area under the normal curve to the left of the given x-value. For example, in Step 4 of

321

the MINITAB FOR WINDOWS commands just presented, you clicked **Cumulative probability**. Also, in the MINITAB COMMAND LANGUAGE you used **CDF** in the first MINITAB command. Both of these commands helped us to find the cumulative probability to the left of $x = 1800$.

Figure 6.56   Normal probability $P(x < 1800)$ when $\mu = 1650$ and $\sigma = 320$.

```
Cumulative Distribution Function

Normal with mean = 1650.00 and standard deviation = 320.000

        x      P(X <= x)
  1.80E+03      0.6804
```

If you want to find $P(x > a)$ for a normal distribution, you first find $P(x \leq a)$ and then subtract this probability from 1.0, which is the total area under the normal curve. For example, to find the probability that the monthly electric consumption during winter of a randomly selected household from this area is more than 1800 kilowatt hours, you first find $P(x \leq 1800)$ and then subtract this probability from 1.0. Thus,

$$P(x > 1800) = 1 - P(x \leq 1800) = 1 - .6804 = .3196$$

To find the probability that x is between two numbers such as $P(a \leq x \leq b)$, we first find $P(x \leq a)$ and $P(x \leq b)$. Then you subtract the smaller probability from the larger probability. For example, suppose you need to find the probability that the monthly electric consumption during winter of a randomly selected household from this area is between 1500 and 1900 kilowatt hours. Using the procedures just described, you first find $P(x \leq 1500)$ and $P(x \leq 1900)$ and then take the difference between these two probabilities. Using the MINITAB procedures explained earlier, you will find

$$P(x \leq 1500) = .3196 \qquad \text{and} \qquad P(x \leq 1900) = .7827$$

Hence,

$$P(1500 \leq x \leq 1900) = P(x \leq 1900) - P(x \leq 1500) = .7827 - .3196 = .4631$$

The procedure for finding an area under the standard normal curve is exactly the same. In this case you enter the value of z instead of the value of x and enter the values of the mean and standard deviation as 0 and 1, respectively, in the MINITAB FOR WINDOWS steps and the MINITAB COMMAND LANGUAGE steps.

COMPUTER ASSIGNMENTS

M6.1   Using MINITAB, find the area under the standard normal curve

    a.   to the left of $z = -1.94$ **b.**   to the left of $z = .83$

    c.   to the right of $z = 1.45$ **d.**   to the right of $z = -1.65$

    e.   between $z = .75$ and $z = 1.90$ **f.**   between $z = -1.20$ and $z = 1.55$

M6.2   Using MINITAB, find the following areas under a normal curve with $\mu = 86$ and $\sigma = 14$.

    a.   Area to the left of $x = 71$ **b.**   Area to the left of $x = 96$

    c.   Area to the right of $x = 90$ **d.**   Area to the right of $x = 75$

    e.   Area between $x = 65$ and $x = 75$ **f.**   Area between $x = 72$ and $x = 95$

M6.3   According to the American Association of University Professors, the mean 1995–96 salary of full-time instructional faculty at two-year colleges was $41,640. Suppose that these salaries have a normal distribution with a mean of $41,640 and a standard deviation of $4000.

    a.   Using MINITAB, find the probability that the 1995–96 salary of a randomly selected faculty member from such colleges was less than $38,000.

    b.   Using MINITAB, find the probability that the 1995–96 salary of a randomly selected faculty member from such colleges was more than $35,000.

    c.   Using MINITAB, find the probability that the 1995–96 salary of a randomly selected faculty member from such colleges was between $43,000 and $51,000.

M6.4   According to the U.S. Bureau of the Census, Americans aged 26 to 44 spent an average of 22 minutes traveling to work in 1995 (*USA Today*, November 20, 1995). Assume that such travel times for workers aged 26 to 44 are normally distributed with a mean of 22 minutes and a standard deviation of 5 minutes.

    a.   Find the probability that a randomly selected U.S. worker aged 26 to 44 spent more than 18 minutes traveling to work in 1995.

    b.   What percentage of such workers spent between 10 and 18 minutes traveling to work in 1995?

    c.   Find the probability that a randomly selected U.S. worker aged 26 to 44 spent less than 28 minutes traveling to work in 1995.

M6.5   The transmission on a particular model of car has a warranty for 40,000 miles. It is known that the life of such a transmission has a normal distribution with a mean of 72,000 miles and a standard deviation of 12,000 miles. Answer the following questions using MINITAB.

    a.   What percentage of the transmissions will fail before the end of the warranty period?

    b.   What percentage of the transmissions will be good for more than 100,000 miles?

    c.   What percentage of the transmissions will be good for 80,000 to 100,000 miles?

MORE CHALLENGING EXERCISES (Optional)
CHAPTERS 4 TO 6

1. A hotel owner has determined that 83% of the hotel's guests eat either dinner or breakfast in the hotel restaurant. Further investigation reveals that 30% of the guests eat dinner and 60% of the guests eat breakfast in the hotel restaurant.

   a. What proportion of the hotel guests eat both dinner and breakfast in the hotel restaurant?
   b. What proportion of the hotel guests eat neither dinner nor breakfast in the hotel restaurant?
   c. What proportion of the hotel guests eat dinner but not breakfast in the hotel restaurant?

2. Eighty percent of all the apples picked on Andy's Apple Acres are satisfactory, but the other 20% are not suitable for market. Andy inspects each apple his pickers pick. Even though he has been in the apple business for 40 years, Andy is prone to make mistakes. He judges as suitable for market 5% of all unsatisfactory apples and judges as unsuitable for market 1% of all satisfactory apples.

   a. What proportion of all apples picked on Andy's Apple Acres are both satisfactory and judged by Andy as being satisfactory for market?
   b. What proportion of all apples picked are marketed?
   c. What proportion of apples that are marketed are actually satisfactory?

3. The principal of a large high school would like to determine the proportion of students in her school who used drugs during the past week. Since the results of directly asking each student "Have you used drugs during the past week?" would be unreliable, the principal might use the following randomized response scheme. Each student rolls a fair die once, the outcome of which only he or she knows. If a 1 or 2 is rolled, the student must answer the sensitive question truthfully. However, if a 3, 4, 5, or 6 is rolled, the student must answer the question with the opposite of the true answer. In this way, the principal would not know whether a *yes* response means the student used drugs and answered truthfully or the student did not use drugs and answered untruthfully. If 60% of the students in the school respond *yes*, what proportion of the students actually did use drugs during the past week? (Note: Splitting the outcomes of the roll of the die as described represents only one possibility. The outcomes could have been split differently but not in such a way that the probability of answering truthfully is .5. In this case, it would not be possible to estimate the proportion of all students who actually used drugs over the past week.)

4. Consider the following three games. Which of these would you most like to play? Which one would you least like to play? Explain your answers mathematically.

   Game I: You toss a fair coin once. If a head appears you receive $3 but if a tail appears you have to pay $1.
   Game II: You receive a single ticket for a raffle that has a total of 500 tickets. Two tickets are chosen without replacement from the 500. The holder of the first ticket selected receives $300 and the holder of the second ticket selected receives $150.
   Game III: You toss a fair coin once. If a head appears you receive $1,000,002 but if a tail appears you have to pay $1,000,000.

5. The random variable x represents the number of exams a randomly selected student has on any given day. The probability distribution function of x is given by the formula

$$P(x) = \frac{x^2 - x + 2}{8} \qquad \text{for } x = 0, 1, 2$$

   a. Find the probability a student has at least one exam on a given day.
   b. On average, how many exams do you expect a randomly selected student will have on a given day?
   c. Find the standard deviation of x.

6. Brad Henry is a stone products salesman. Let x be the number of contacts visited by him on a particular day. The following table gives the probability distribution of x.

x	1	2	3	4
$P(x)$	.12	.25	.56	.07

Let y be the total number of contacts Brad visits on two randomly selected days. Write the probability distribution of y.

7. A local radio station plans to give away tickets as a promotional campaign for a singles' riverboat cruise to random callers who are 21 years of age or older. From past experience, station managers know that 80% of the tickets given away are actually used. If the riverboat can hold at most 300 persons, how many tickets should the station give away so that there is a 99% chance that all ticket holders showing up for the cruise will be accommodated?

8. The number of calls coming into a small mail order company follow a Poisson distribution. Currently, these calls are serviced by a single operator. The manager knows from past experience that an additional operator will be necessary if the rate of calls exceeds 20 per hour. The manager observes that 9 calls came into the mail order company during a randomly selected 15-minute period.

   a. If the rate of calls is actually 20 per hour, what is the probability that 9 or more calls will come in during a given 15-minute period?

   b. If the rate of calls is really 30 per hour, what is the probability that 9 or more calls will come in during a given 15-minute period?

   c. Would you advise the manager to hire a second operator? Explain.

9. The amount of time taken by a bank teller to serve a randomly selected customer has a normal distribution with a mean of 2 minutes and a standard deviation of .5 minutes.

   a. What is the probability that both of two randomly selected customers will take less than 1 minute each to be served?

   b. What is the probability that at least one of four randomly selected customers will need more than 2.25 minutes to be served?

   c. Find the probability that exactly three of ten randomly selected customers will need between 1.15 and 1.65 minutes each to be served.

   d. Find the probability that between 75 and 80 of 100 randomly selected customers will need less than 2.45 minutes each to be served.

10. At a casino, gamblers may bet on the spin of a wheel that is divided into 54 equal pieces; 24 of the sections are marked $1, 15 are marked $2, 7 are marked $5, 4 are marked $10, 2 are marked $20, 1 is marked $40 green, and 1 is marked $40 red. If a gambler pays a dollar to play a denomination of his choice ($1, $2, $5, $10, $20, $40 green, or $40 red) and the denomination is spun on the wheel, his dollar plus the amount spun is returned to him. Otherwise, he loses his dollar. Based on expected values, which denomination is the best bet? ($40 green and $40 red are considered separate bets.) Interpret the expected values you compute.

11. A clothing distributor is considering marketing a new line of fashions that she thinks has a 70% chance of being successful. The distributor can test market the fashions to gain insight into whether or not the new fashion line will be successful. Unfortunately, test marketing is not 100% accurate in predicting the success of a product. If the new fashion line will actually be successful, test marketing will predict it to be successful 80% of the time. If the new fashion line will actually be unsuccessful, test marketing will predict it to be unsuccessful 75% of the time. The clothing distributor would not want to market the new fashion line unless she can be at least 85% certain that it will be successful.

   a. Should the clothing distributor test market the new fashion line?

   b. What decision should the clothing distributor make if the test marketing predicts a success?

   c. How sure can the clothing distributor be that the new fashion line will be unsuccessful if the test marketing predicts it to be unsuccessful?

7 | SAMPLING DISTRIBUTIONS

Chapters 5 and 6 discussed probability distributions of discrete and continuous random variables. This chapter extends the concept of probability distribution to that of a sample statistic. As we discussed in Chapter 3, a sample statistic is a numerical summary measure calculated for sample data. The mean, median, mode, and standard deviation calculated for sample data are called *sample statistics*. On the other hand, the same numerical summary measures calculated for population data are called *population parameters*. A population parameter is always a constant, whereas a sample statistic is always a random variable. Since every random variable must possess a probability distribution, each sample statistic possesses a probability distribution. The probability distribution of a sample statistic is more commonly called its *sampling distribution*. This chapter discusses the sampling distributions of the sample mean and the sample proportion. The concepts covered in this chapter are the foundation of the inferential statistics discussed in succeeding chapters.

7.1 POPULATION AND SAMPLING DISTRIBUTIONS

This section introduces the concept of population distribution and sampling distribution. Subsection 7.1.1 explains the population distribution, and Subsection 7.1.2 describes the sampling distribution of $\bar{x}$.

7.1.1 POPULATION DISTRIBUTION

The **population distribution** is the probability distribution derived from the information on all elements of a population.

POPULATION DISTRIBUTION

The *population distribution* is the probability distribution of the population data.

Suppose there are only five students in an advanced statistics class and the midterm scores of these five students are

$$70 \quad 78 \quad 80 \quad 80 \quad 95$$

Let x denote the score of a student. Using single-valued classes (as there are only five data values, there is no need to group them), we can write the frequency distribution of scores as in Table 7.1 along with the relative frequencies of classes, which are obtained by dividing the frequencies of classes by the population size. Table 7.2, which lists the probabilities of various x values, presents the probability distribution of the population. Note that these probabilities are the same as relative frequencies.

Table 7.1 Population Frequency and Relative Frequency Distributions

x	f	Relative Frequency
70	1	$1/5 = .20$
78	1	$1/5 = .20$
80	2	$2/5 = .40$
95	1	$1/5 = .20$
	$N = 5$	Sum $= 1.00$

Table 7.2 Population Probability Distribution

x	$P(x)$
70	.20
78	.20
80	.40
95	.20
	$\Sigma P(x) = 1.00$

The values of the mean and standard deviation calculated for the probability distribution of Table 7.2 give the values of the population parameters μ and σ. These values are $\mu = 80.60$ and $\sigma = 8.09$. The values of μ and σ for the probability distribution of Table 7.2 can be calculated using the formulas given in Sections 5.3 and 5.4 of Chapter 5 (see Exercise 7.6).

7.1.2 SAMPLING DISTRIBUTION

As mentioned at the beginning of this chapter, the value of a population parameter is always constant. For example, for any population data set, there is only one value of the population mean, μ. However, we cannot say the same about the sample mean, $\bar{x}$. We would expect different samples of the same size drawn from the same population to yield different values of the sample mean, $\bar{x}$. The value of the sample mean for any one sample will depend on the elements included in that sample. Consequently, *the sample mean, $\bar{x}$, is a random variable.* Therefore, like other random variables, the sample mean possesses a probability distribution, which is more commonly called the **sampling distribution of $\bar{x}$**. Other sample statistics such as the median, mode, and standard deviation also possess sampling distributions.

SAMPLING DISTRIBUTION OF $\bar{x}$

The probability distribution of $\bar{x}$ is called its sampling distribution. It lists the various values that $\bar{x}$ can assume and the probability of each value of $\bar{x}$.

In general, the probability distribution of a sample statistic is called its *sampling distribution.*

Reconsider the population of midterm scores of five students given in Table 7.1. Consider all possible samples of three scores each that can be selected, without replacement, from that population. The total number of possible samples, given by the combinations formula discussed in Chapter 5, is 10, that is,

$$\text{Total number of samples} = \binom{5}{3} = \frac{5!}{3! \, (5-3)!} = \frac{5 \cdot 4 \cdot 3 \cdot 2 \cdot 1}{3 \cdot 2 \cdot 1 \cdot 2 \cdot 1} = 10$$

Suppose we assign letters A, B, C, D, and E to the scores of five students so that

$$A = 70, \quad B = 78, \quad C = 80, \quad D = 80, \quad E = 95$$

Then the 10 possible samples of three scores each are

$$\text{ABC}, \quad \text{ABD}, \quad \text{ABE}, \quad \text{ACD}, \quad \text{ACE}, \quad \text{ADE}, \quad \text{BCD}, \quad \text{BCE}, \quad \text{BDE}, \quad \text{CDE}$$

These 10 samples and their respective means are listed in Table 7.3. Note that the first two samples have the same three scores. The reason for this is that two of the students (C and D) have the same score and, hence, the samples ABC and ABD contain the same values. The mean of each sample is obtained by dividing the sum of the three scores included in that sample by 3. For instance, the mean of the first sample is $(70 + 78 + 80)/3 = 76$. Note that the values of the means of samples in Table 7.3 are rounded to two decimal places.

Table 7.3 All Possible Samples and Their Means When the Sample Size Is 3

Sample	Scores in the Sample	$\bar{x}$
ABC	70, 78, 80	76.00
ABD	70, 78, 80	76.00
ABE	70, 78, 95	81.00
ACD	70, 80, 80	76.67
ACE	70, 80, 95	81.67
ADE	70, 80, 95	81.67
BCD	78, 80, 80	79.33
BCE	78, 80, 95	84.33
BDE	78, 80, 95	84.33
CDE	80, 80, 95	85.00

By using the values of $\bar{x}$ given in Table 7.3, we record the frequency distribution of $\bar{x}$ in Table 7.4. By dividing the frequencies of the various values of $\bar{x}$ by the sum of all frequencies, we obtain the relative frequencies of classes, which are listed in the third column of Table 7.4. These relative frequencies are used as probabilities and listed in Table 7.5. This table gives the sampling distribution of $\bar{x}$.

If we select just one sample of three scores from the population of five scores, we may draw any of the 10 possible samples. Hence, the sample mean, $\bar{x}$, can assume any of the values listed in Table 7.5 with the corresponding probability. For instance, the probability that the mean of a randomly selected sample of three scores is 81.67 is .20. This probability can be written as

$$P(\bar{x} = 81.67) = .20$$

Table 7.4 Frequency and Relative Frequency Distributions of $\bar{x}$ When the Sample Size Is 3

$\bar{x}$	f	Relative Frequency
76.00	2	2/10 = .20
76.67	1	1/10 = .10
79.33	1	1/10 = .10
81.00	1	1/10 = .10
81.67	2	2/10 = .20
84.33	2	2/10 = .20
85.00	1	1/10 = .10
	$\Sigma f = 10$	Sum = 1.00

Table 7.5 Sampling Distribution of $\bar{x}$ When the Sample Size Is 3

$\bar{x}$	$P(\bar{x})$
76.00	.20
76.67	.10
79.33	.10
81.00	.10
81.67	.20
84.33	.20
85.00	.10
	$\Sigma P(\bar{x}) = 1.00$

7.2 SAMPLING AND NONSAMPLING ERRORS

Usually, different samples selected from the same population will give different results because they contain different elements. This is obvious from Table 7.3, which shows that the mean of a sample of three scores depends on which three of the five scores are included in the sample. The result obtained from any one sample will generally be different from the result obtained from the corresponding population. The difference between the value of a sample statistic obtained from a sample and the value of the corresponding population parameter obtained from the population is called the **sampling error**. Note that this difference represents the sampling error only if the sample is random and no nonsampling error has been made. Otherwise only a part of this difference will be due to the sampling error.

SAMPLING ERROR

Sampling error is the difference between the value of a sample statistic and the value of the corresponding population parameter. In the case of the mean,

$$\text{Sampling error} = \bar{x} - \mu$$

assuming that the sample is random and no nonsampling error has been made.

It is important to remember that *a sampling error occurs because of chance*. The errors that occur for other reasons, such as errors made during collection, recording, and tabulation of data, are called **nonsampling errors**. Such errors occur because of human mistakes, and not chance. Note that there is only one kind of sampling error—the error that occurs due to chance. However, there is not just one nonsampling error but many nonsampling errors that may occur due to different reasons.

NONSAMPLING ERRORS

The errors that occur in the collection, recording, and tabulation of data are called *nonsampling errors*.

The following paragraph, reproduced from the *Current Population Reports* of the U.S. Bureau of the Census, explains how nonsampling errors can occur.

> Nonsampling errors can be attributed to many sources, e.g., inability to obtain information about all cases in the sample, definitional difficulties, differences in the interpretation of questions, inability or unwillingness on the part of the respondents to provide correct information, inability to recall information, errors made in collection such as in recording or coding the data, errors made in processing the data, errors made in estimating values for missing data, biases resulting from the differing recall periods caused by the interviewing pattern used, and failure of all units in the universe to have some probability of being selected for the sample (undercoverage).

The following are the main reasons for the occurrence of nonsampling errors.

1. If a sample is nonrandom (and, hence, nonrepresentative), the sample results may be too different from the census results. The following quote from *U.S. News & World Report* describes how even a randomly selected sample can become nonrandom if some of the members included in the sample cannot be contacted.

> A test poll conducted in the 1984 presidential election found that if the poll were halted after interviewing only those subjects who could be reached on the first try, Reagan showed a 3-percentage-point lead over Mondale. But when interviewers made a determined effort to reach everyone on their lists of randomly selected subjects—calling some as many as 30 times before finally reaching them—Reagan showed a 13 percent lead, much closer to the actual election result. As it turned out, people who were planning to vote Republican were simply less likely to be at home. ("The Numbers Racket: How Polls and Statistics Lie," *U.S. News & World Report*, July 11, 1988. Copyright © 1988 by U.S. News & World Report, Inc. Reprinted with permission.)

2. The questions may be phrased in such a way that they are not fully understood by the members of the sample or population. As a result, the answers obtained are not accurate.

3. The respondents may intentionally give false information in response to some sensitive questions. For example, people may not tell the truth about drinking habits, incomes, or opinions about minorities. Sometimes the respondents may give wrong answers because of ignorance. For example, a person may not remember the exact amount he spent on clothes during the last year. If asked in a survey, he may give an inaccurate answer.

4. The poll taker may make a mistake and enter a wrong number in the records or make an error while entering the data on a computer.

Note that nonsampling errors can occur both in a sample survey and in a census, whereas sampling error occurs only when a sample survey is conducted. Nonsampling errors can be minimized by preparing the survey questionnaire carefully and handling the data cautiously. However, it is impossible to avoid sampling error.

Example 7–1 illustrates the sampling and nonsampling errors using the mean.

Illustrating sampling and nonsampling errors.

EXAMPLE 7-1   Reconsider the population of five scores given in Table 7.1. The scores of the five students are 70, 78, 80, 80, and 95. The population mean is

$$\mu = (70 + 78 + 80 + 80 + 95)/5 = 80.60$$

Now suppose we take a random sample of three scores from this population. Assume that this sample includes the scores 70, 80, and 95. The mean for this sample is

$$\bar{x} = (70 + 80 + 95)/3 = 81.67$$

Consequently,

$$\text{Sampling error} = \bar{x} - \mu = 81.67 - 80.60 = \mathbf{1.07}$$

That is, the mean score estimated from the sample is 1.07 higher than the mean score of the population. Note that this difference occurred due to chance, that is, because we used a sample instead of the population.

Now suppose, when we select the above mentioned sample, we mistakenly record the second score as 82 instead of 80. As a result, we calculate the sample mean as

$$\bar{x} = (70 + 82 + 95)/3 = 82.33$$

Consequently, the difference between this sample mean and the population mean is

$$\bar{x} - \mu = 82.33 - 80.60 = 1.73$$

However, this difference between the sample mean and the population mean does not represent the sampling error. As we calculated earlier, only 1.07 of this difference is due to the sampling error. The remaining portion, which is equal to $1.73 - 1.07 = .66$, represents the nonsampling error because it occurred due to the error we made in recording the second score in the sample. Thus, in this case,

$$\text{Sampling error} = \mathbf{1.07}$$

$$\text{Nonsampling error} = \mathbf{.66}$$

Figure 7.1 shows the sampling and nonsampling errors for these calculations.

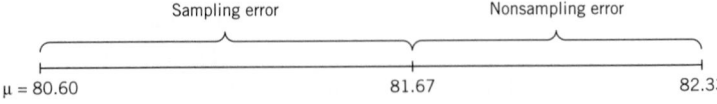

Figure 7.1   Sampling and nonsampling errors.

Thus, the sampling error is the difference between the correct value of $\bar{x}$ and μ, where the correct value of $\bar{x}$ is the value of $\bar{x}$ that does not contain any nonsampling errors. On the other hand, the nonsampling error(s) is (are) obtained by subtracting the correct value of $\bar{x}$ from the incorrect value of $\bar{x}$, where the incorrect value of $\bar{x}$ is the value that contains the nonsampling error(s). For our example 7–1,

$$\text{Sampling error} = \bar{x} - \mu = 81.67 - 80.60 = 1.07$$

$$\text{Nonsampling error} = \text{Incorrect } \bar{x} - \text{Correct } \bar{x} = 82.33 - 81.67 = .66 \quad \blacksquare$$

Note that in the real world we do not know the mean of a population. Hence, we select a sample to use the sample mean as an estimate of the population mean. Consequently, we never know the size of the sampling error.

Case Study 7–1, excerpted from an article that appeared in the *Business Week* of October 14, 1996, describes the sampling and nonsampling errors that may have occurred in the political polls conducted by many agencies before the Presidential election of November 1996. Although the election was over a while ago, this article makes a good reading to understand how different polls may give different results due to sampling and nonsampling errors.

CASE STUDY 7-1   LIES, DAMN LIES, AND POLITICAL POLLS?

Since Labor Day (1996), Presidential polls have been consistent about one thing: Bill Clinton is ahead of Bob Dole. But when it comes to the size of the lead, surveys have fluctuated wildly, ranging from a 6% Dole deficit to a Clinton blowout by 25%. A CNN/*USA Today*/Gallup poll boosted Clinton's lead from 9% to 22% in just two days. "It kind of shakes your faith in polling a little bit," sighs Thomas H. Silver, publisher of *The Polling Report*, a newsletter on political trends.

The erratic polls have some Republicans grousing that the public is getting a distorted picture of a race that many pundits are already declaring over. . . .

Polling pros concede there's some truth to GOP critics' gripes. A proliferation of error-prone quickie surveys, methodological differences between polls, and rising voter volatility all can skew results. "It's standard for every loser to complain about how a poll is conducted," says Claibourne Darden, an independent pollster in Atlanta. "But sometimes, the complaint is valid."

. . . A problem for all pollsters is questioning a valid cross-section of voters. It's an even greater challenge for the one-night telephone polls increasingly trumpeted by the media. Among the hazards: Some people don't have telephones or are not at home; others refuse to respond. Moreover, small polls have a high sampling error, so a candidate's purported 12% lead could be as large as 19% or as small as 5%.

Then there's the built-in polling bias that favors any incumbent. Larry Hugick, media polling director at Princeton Survey Research, says some voters don't like to tell pollsters they oppose the incumbent, preferring to say they are undecided. On Election Day, these voters tend to break in favor of the challenger. "Dole is going to do well among people who tell us they're undecided," Hugick predicts.

FAULTY MATH
The GOP's big complaint concerns formulas predicting how likely a respondent is to vote. These turnout models, and other weighting factors that compensate for the overrepresented or undersampled

groups, are fraught with potential for error. In the September 21 (1996) Louisiana open Senate primary, polls showed Republican Woody Jenkins running a poor third. But he ended up first because of heavier-than-predicted turnout by GOP activists.

Republicans say many turnout models favor Democrats. For 1996, most republicans are using as a predictive model the 1994 congressional elections, when GOP voters were far more likely to vote than Democrats. But press pollsters prefer the 1992 election model, when Democratic turnout was higher. "We won't know until the election who is right," says Democratic pollster Ann Bennett. . . .

Source: Excerpted from Richard S. Dunham, "Lies, Damn Lies, and Political Polls?" *Business Week*, October 14, 1996. Reprinted with permission. Copyright © McGraw-Hill, Inc., 1996.

EXERCISES

Concepts and Procedures

7.1 Briefly explain the meaning of a population distribution and a sampling distribution. Give an example of each.

7.2 Explain briefly the meaning of sampling error. Give an example. Does such an error occur only in a sample survey or can it occur both in a sample survey and a census?

7.3 Explain briefly the meaning of nonsampling errors. Give an example. Do such errors occur only in a sample survey or can they occur both in a sample survey and a census?

7.4 Consider the following population of six numbers.

$$15 \quad 13 \quad 8 \quad 17 \quad 9 \quad 10$$

a. Find the population mean.
b. Liza selected one sample of four numbers from this population. The sample included the numbers 13, 8, 9, and 10. Calculate the sample mean and sampling error for this sample.
c. Refer to part b. When Liza calculated the sample mean, she mistakenly used the numbers 13, 8, 6, and 10 to calculate the sample mean. Find the sampling and nonsampling errors in this case.
d. List all samples of four numbers (without replacement) that can be selected from this population. Calculate the sample mean and sampling error for each of these samples.

7.5 Consider the following population of 10 numbers.

$$20 \quad 25 \quad 13 \quad 19 \quad 9 \quad 15 \quad 11 \quad 7 \quad 17 \quad 23$$

a. Find the population mean.
b. Rich selected one sample of nine numbers from this population. The sample included the numbers 20, 25, 13, 9, 15, 11, 7, 17, and 23. Calculate the sample mean and sampling error for this sample.
c. Refer to part b. When Rich calculated the sample mean, he mistakenly used the numbers 20, 25, 13, 9, 15, 11, 17, 17, and 23 to calculate the sample mean. Find the sampling and nonsampling errors in this case.
d. List all samples of nine numbers (without replacement) that can be selected from this population. Calculate the sample mean and sampling error for each of these samples.

Applications

7.6 Using the formulas of Sections 5.3 and 5.4 of Chapter 5 for the mean and standard deviation of a discrete random variable, verify that the mean and standard deviation for the population probability distribution of Table 7.2 are 80.60 and 8.09, respectively.

7.7 The following data give the ages of all six members of a family.

$$55 \quad 53 \quad 28 \quad 25 \quad 21 \quad 15$$

  a. Let x denote the age of a member of this family. Write the population distribution of x.
  b. List all the possible samples of size five (without replacement) that can be selected from this population. Calculate the mean for each of these samples. Write the sampling distribution of $\bar{x}$.
  c. Calculate the mean for the population data. Select one random sample of size five and calculate the sample mean $\bar{x}$. Compute the sampling error.

7.8 The following data give the years of teaching experience for all five faculty members of a department at a university.

$$7 \quad 8 \quad 12 \quad 7 \quad 20$$

  a. Let x denote the years of teaching experience for a faculty member of this department. Write the population distribution of x.
  b. List all the possible samples of size four (without replacement) that can be selected from this population. Calculate the mean for each of these samples. Write the sampling distribution of $\bar{x}$.
  c. Calculate the mean for the population data. Select one random sample of size four and calculate the sample mean $\bar{x}$. Compute the sampling error.

7.3 MEAN AND STANDARD DEVIATION OF $\bar{x}$

The mean and standard deviation calculated for the sampling distribution of $\bar{x}$ are called the **mean** and **standard deviation of $\bar{x}$**. Actually, the mean and standard deviation of $\bar{x}$ are, respectively, the mean and standard deviation of the means of all samples of the same size selected from a population. The standard deviation of $\bar{x}$ is also called the *standard error of $\bar{x}$*.

> **MEAN AND STANDARD DEVIATION OF $\bar{x}$**
>
> The mean and standard deviation of the sampling distribution of $\bar{x}$ are called the *mean and standard deviation of $\bar{x}$* and are denoted by $\mu_{\bar{x}}$ and $\sigma_{\bar{x}}$, respectively.

If we calculate the mean and standard deviation of the 10 values of $\bar{x}$ listed in Table 7.3, we obtain the mean, $\mu_{\bar{x}}$, and the standard deviation, $\sigma_{\bar{x}}$, of $\bar{x}$. Alternatively, we can calculate the mean and standard deviation of the sampling distribution of $\bar{x}$ listed in Table 7.5. These will also be the values of $\mu_{\bar{x}}$ and $\sigma_{\bar{x}}$. From these calculations, we will obtain $\mu_{\bar{x}} = 80.60$ and $\sigma_{\bar{x}} = 3.30$ (see Exercise 7.25 at the end of this section).

The mean of the sampling distribution of $\bar{x}$ is always equal to the mean of the population.

> **MEAN OF THE SAMPLING DISTRIBUTION OF $\bar{x}$**
>
> The *mean of the sampling distribution of $\bar{x}$* is always equal to the mean of the population. Thus,
>
> $$\mu_{\bar{x}} = \mu$$

Hence, if we select all possible samples (of the same size) from a population and calculate their means, the mean ($\mu_{\bar{x}}$) of all these sample means will be the same as the mean (μ) of the population. If we calculate the mean for the population probability distribution of Table 7.2 and the mean for the sampling distribution of Table 7.5 by using the formula learned in Section 5.3 of Chapter 5, we get the same value of 80.60 for μ and $\mu_{\bar{x}}$ (see Exercise 7.25).

The sample mean, $\bar{x}$, is called an **estimator** of the population mean, μ. When the expected value (or mean) of a sample statistic is equal to the value of the corresponding population parameter, that sample statistic is said to be an **unbiased estimator**. For the sample mean $\bar{x}$, $\mu_{\bar{x}} = \mu$. Hence, $\bar{x}$ is an unbiased estimator of μ. This is a very important property that an estimator should possess.

However, the standard deviation, $\sigma_{\bar{x}}$, of $\bar{x}$ is not equal to the standard deviation, σ, of the population distribution (unless $n = 1$). The standard deviation of $\bar{x}$ is equal to the standard deviation of the population divided by the square root of the sample size. That is,

$$\sigma_{\bar{x}} = \frac{\sigma}{\sqrt{n}}$$

This formula for the standard deviation of $\bar{x}$ holds true only when the sampling is done either with replacement from a finite population or with or without replacement from an infinite population. These two conditions can be replaced by the condition that the above formula holds true if the sample size is small in comparison to the population size. The sample size is considered to be small compared to the population size if the sample size is equal to or less than 5% of the population size, that is, if

$$\frac{n}{N} \leq .05$$

If this condition is not satisfied, we use the following formula to calculate $\sigma_{\bar{x}}$.

$$\sigma_{\bar{x}} = \frac{\sigma}{\sqrt{n}} \sqrt{\frac{N - n}{N - 1}}$$

where the factor

$$\sqrt{\frac{N - n}{N - 1}}$$

is called the finite population correction factor.

In most practical applications, the sample size is usually small compared to the population size. Consequently, in most cases the formula used to calculate $\sigma_{\bar{x}}$ is $\sigma_{\bar{x}} = \sigma/\sqrt{n}$.

> **STANDARD DEVIATION OF THE SAMPLING DISTRIBUTION OF $\bar{x}$**
>
> The *standard deviation of the sampling distribution of $\bar{x}$* is
>
> $$\sigma_{\bar{x}} = \frac{\sigma}{\sqrt{n}}$$
>
> where σ is the standard deviation of the population and n is the sample size. This formula is used when $n/N \leq .05$, where N is the population size.

Following are two important observations regarding the sampling distribution of $\bar{x}$.

1. *The spread of the sampling distribution of $\bar{x}$ is smaller than the spread of the corresponding population distribution.* In other words, $\sigma_{\bar{x}} < \sigma$. This is obvious from the formula for $\sigma_{\bar{x}}$. When n is greater than 1, which is usually true, the denominator in $\sigma/\sqrt{n}$ is greater than 1. Hence, $\sigma_{\bar{x}}$ is smaller than σ.

2. *The standard deviation of the sampling distribution of $\bar{x}$ decreases as the sample size increases.* This feature of the sampling distribution of $\bar{x}$ is also obvious from the formula

$$\sigma_{\bar{x}} = \frac{\sigma}{\sqrt{n}}$$

If the standard deviation of a sample statistic decreases as the sample size is increased, that statistic is said to be a **consistent estimator**. This is another important property that an estimator should possess. It is obvious from the above formula for $\sigma_{\bar{x}}$ that as n increases, the value of $\sqrt{n}$ also increases and, consequently, the value of $\sigma/\sqrt{n}$ decreases. Thus, the sample mean $\bar{x}$ is a consistent estimator of the population mean μ. Example 7–2 illustrates this feature.

Mean and standard deviation of $\bar{x}$.

EXAMPLE 7–2 The mean wage per hour for all 5000 employees working at a large company is $13.50 and the standard deviation is $2.90. Let $\bar{x}$ be the mean wage per hour for a random sample of certain employees selected from this company. Find the mean and standard deviation of $\bar{x}$ for a sample size of

(a) 30   (b) 75   (c) 200

Solution   From the given information, for the population of all employees,

$$N = 5000, \quad \mu = \$13.50, \quad \text{and} \quad \sigma = \$2.90$$

(a)   The mean, $\mu_{\bar{x}}$, of the sampling distribution of $\bar{x}$ is

$$\mu_{\bar{x}} = \mu = \mathbf{\$13.50}$$

In this case, $n = 30$, $N = 5000$, and $n/N = 30/5000 = .006$. As n/N is less than .05, the standard deviation of $\bar{x}$ is obtained by using the formula $\sigma/\sqrt{n}$. Hence,

$$\sigma_{\bar{x}} = \frac{\sigma}{\sqrt{n}} = \frac{2.90}{\sqrt{30}} = \mathbf{\$.529}$$

Thus, we can state that if we take all possible samples of size 30 from the population of all employees of this company and prepare the sampling distribution of $\bar{x}$, the mean and standard deviation of this sampling distribution of $\bar{x}$ will be $13.50 and $.529, respectively.

(b)   In this case, $n = 75$ and $n/N = 75/5000 = .015$, which is less than .05. The mean and standard deviation of $\bar{x}$ are

$$\mu_{\bar{x}} = \mu = \textbf{\$13.50} \quad \text{and} \quad \sigma_{\bar{x}} = \frac{\sigma}{\sqrt{n}} = \frac{2.90}{\sqrt{75}} = \textbf{\$.335}$$

(c)   In this case, $n = 200$ and $n/N = 200/5000 = .04$, which is less than .05. Therefore, the mean and standard deviation of $\bar{x}$ are

$$\mu_{\bar{x}} = \mu = \textbf{\$13.50} \quad \text{and} \quad \sigma_{\bar{x}} = \frac{\sigma}{\sqrt{n}} = \frac{2.90}{\sqrt{200}} = \textbf{\$.205}$$

From the above calculations we observe that the mean of the sampling distribution of $\bar{x}$ is always equal to the mean of the population whatever the size of the sample. However, the value of the standard deviation of $\bar{x}$ decreases from $.529 to $.335 and then to $.205 as the sample size increases from 30 to 75 and then to 200.   ▬

EXERCISES

Concepts and Procedures

7.9   Let $\bar{x}$ be the mean of a sample selected from a population.
   a.   What is the mean of the sampling distribution of $\bar{x}$ equal to?
   b.   What is the standard deviation of the sampling distribution of $\bar{x}$ equal to? Assume $n/N \le .05$.

7.10   What is an estimator? When is an estimator unbiased? Is the sample mean, $\bar{x}$, an unbiased estimator of μ? Explain.

7.11   When is an estimator said to be consistent? Is the sample mean, $\bar{x}$, a consistent estimator of μ? Explain.

7.12   How does the value of $\sigma_{\bar{x}}$ change as the sample size increases? Explain.

7.13   Consider a large population with $\mu = 60$ and $\sigma = 12$. Assuming $n/N \le .05$, find the mean and standard deviation of the sample mean, $\bar{x}$, for a sample size of
   a.   18   b.   90

7.14   Consider a large population with $\mu = 90$ and $\sigma = 16$. Assuming $n/N \le .05$, find the mean and standard deviation of the sample mean, $\bar{x}$, for a sample size of
   a.   10   b.   35

7.15   A population of $N = 5000$ has $\sigma = 20$. In each of the following cases which formula will you use to calculate $\sigma_{\bar{x}}$ and why? Using the appropriate formula, calculate $\sigma_{\bar{x}}$ for each of these cases.
   a. $n = 300$ b. $n = 100$

7.16   A population of $N = 100,000$ has $\sigma = 35$. In each of the following cases which formula will you use to calculate $\sigma_{\bar{x}}$ and why? Using the appropriate formula, calculate $\sigma_{\bar{x}}$ for each of these cases.
   a. $n = 2500$ b. $n = 7000$

*7.17   For a population, $\mu = 125$ and $\sigma = 18$.

    a.   For a sample selected from this population, $\mu_{\bar{x}} = 125$ and $\sigma_{\bar{x}} = 3.6$. Find the sample size. Assume $n/N \leq .05$.

    b.   For a sample selected from this population, $\mu_{\bar{x}} = 125$ and $\sigma_{\bar{x}} = 2.25$. Find the sample size. Assume $n/N \leq .05$.

*7.18   For a population, $\mu = 46$ and $\sigma = 8$.

    a.   For a sample selected from this population, $\mu_{\bar{x}} = 46$ and $\sigma_{\bar{x}} = 2.0$. Find the sample size. Assume $n/N \leq .05$.

    b.   For a sample selected from this population, $\mu_{\bar{x}} = 46$ and $\sigma_{\bar{x}} = 1.6$. Find the sample size. Assume $n/N \leq .05$.

Applications

7.19   According to the U.S. Bureau of Labor Statistics, the mean number of hours worked per week by workers employed in the private (nonagricultural) industrial sector was 34.2 in 1996. Assume that for the hours worked per week by the current population of all workers employed in the private (nonagricultural) industrial sector, $\mu = 34.2$ and $\sigma = 4$. Let $\bar{x}$ be the mean hours worked per week by a random sample of 40 such workers. Find the mean and standard deviation of the sampling distribution of $\bar{x}$.

7.20   According to the U.S. Bureau of Labor Statistics, the mean weekly earnings of workers in the manufacturing sector were \$517 in February 1996. Assume that the current mean weekly earnings of all workers in the manufacturing sector are \$517 with a standard deviation of \$65. Let $\bar{x}$ be the mean weekly earnings of a sample of 400 workers selected from the manufacturing sector. Find the mean and standard deviation of the sampling distribution of $\bar{x}$.

7.21   The mean annual salary for all 1050 professors at a university is \$51,400 and the standard deviation of their annual salaries is \$7400. Let $\bar{x}$ be the mean salary of a random sample of 16 professors selected from this university. Find the mean and standard deviation of the sampling distribution of $\bar{x}$.

7.22   Data from the food industry indicate that Americans under the age of 18 consume an average of 14 pounds of cereal per year (*U.S. News & World Report*, April 29, 1996). Assume that for the annual cereal consumption of all Americans under the age of 18, $\mu = 14$ pounds and $\sigma = 3.8$ pounds. Let $\bar{x}$ be the mean annual cereal consumption for a random sample of 700 Americans under the age of 18. Find the mean and standard deviation of $\bar{x}$.

*7.23   The standard deviation of prices of all cars is known to be \$3600. Let $\bar{x}$ be the mean price of a sample of cars. What sample size will produce the standard deviation of $\bar{x}$ equal to \$180?

*7.24   The standard deviation of the 1996 gross sales of all corporations is known to be \$139.50 million. Let $\bar{x}$ be the mean of the 1996 gross sales of a sample of corporations. What sample size will produce the standard deviation of $\bar{x}$ equal to \$15.50 million?

*7.25   Consider the sampling distribution of $\bar{x}$ given in Table 7.5.

    a.   Calculate the value of $\mu_{\bar{x}}$ using the formula: $\mu_{\bar{x}} = \Sigma \bar{x} \, P(\bar{x})$. Is the value of μ calculated in Exercise 7.6 the same as the value of $\mu_{\bar{x}}$ calculated here?

    b.   Calculate the value of $\sigma_{\bar{x}}$ by using the formula

$$\sigma_{\bar{x}} = \sqrt{\Sigma \bar{x}^2 P(\bar{x}) - (\mu_{\bar{x}})^2}$$

    c.   From Exercise 7.6, $\sigma = 8.09$. Also, our sample size is 3 so that $n = 3$. Therefore, $\sigma/\sqrt{n} = 8.09/\sqrt{3} = 4.67$. From part b, you should get $\sigma_{\bar{x}} = 3.30$. Why does $\sigma/\sqrt{n}$ not equal $\sigma_{\bar{x}}$ in this case?

    d.   In our example (given in the beginning of Section 7.1.1) on scores, $N = 5$ and $n = 3$. Hence, $n/N = 3/5 = .60$. Because n/N is greater than .05, the appropriate formula to find $\sigma_{\bar{x}}$ is

$$\sigma_{\bar{x}} = \frac{\sigma}{\sqrt{n}} \sqrt{\frac{N-n}{N-1}}$$

Show that the value of $\sigma_{\bar{x}}$ calculated by using this formula gives the same value as the one calculated in part b.

7.4 SHAPE OF THE SAMPLING DISTRIBUTION OF $\bar{x}$

The shape of the sampling distribution of $\bar{x}$ relates to the following two cases.

1. The population from which samples are drawn has a normal distribution.
2. The population from which samples are drawn does not have a normal distribution.

7.4.1 SAMPLING FROM A NORMALLY DISTRIBUTED POPULATION

When the population from which samples are drawn is normally distributed with its mean equal to μ and standard deviation equal to σ, then

1. The mean of $\bar{x}$, $\mu_{\bar{x}}$, is equal to the mean of the population, μ.
2. The standard deviation of $\bar{x}$, $\sigma_{\bar{x}}$, is equal to $\sigma/\sqrt{n}$, assuming $n/N \leq .05$.
3. The shape of the sampling distribution of $\bar{x}$ is normal, whatever the value of n.

SAMPLING DISTRIBUTION OF $\bar{x}$ WHEN THE POPULATION HAS A NORMAL DISTRIBUTION

If the population from which the samples are drawn is normally distributed with mean μ and standard deviation σ, then the sampling distribution of the sample mean, $\bar{x}$, will also be normally distributed with the following mean and standard deviation, irrespective of the sample size.

$$\mu_{\bar{x}} = \mu \quad \text{and} \quad \sigma_{\bar{x}} = \frac{\sigma}{\sqrt{n}}$$

Remember For $\sigma_{\bar{x}} = \sigma/\sqrt{n}$ to be true, n/N must be less than or equal to .05.

Figure 7.2*a* shows the probability distribution curve for a population. The distribution curves in Figure 7.2*b* through Figure 7.2*e* show the sampling distributions of $\bar{x}$ for different sample sizes taken from the population of Figure 7.2*a*. As we can observe, the population has a normal distribution. Because of this, the sampling distribution of $\bar{x}$ is normal for each of the four cases illustrated in parts *b* through *e* of Figure 7.2. Also notice from Figure 7.2*b* through Figure 7.2*e* that the spread of the sampling distribution of $\bar{x}$ decreases as the sample size increases.

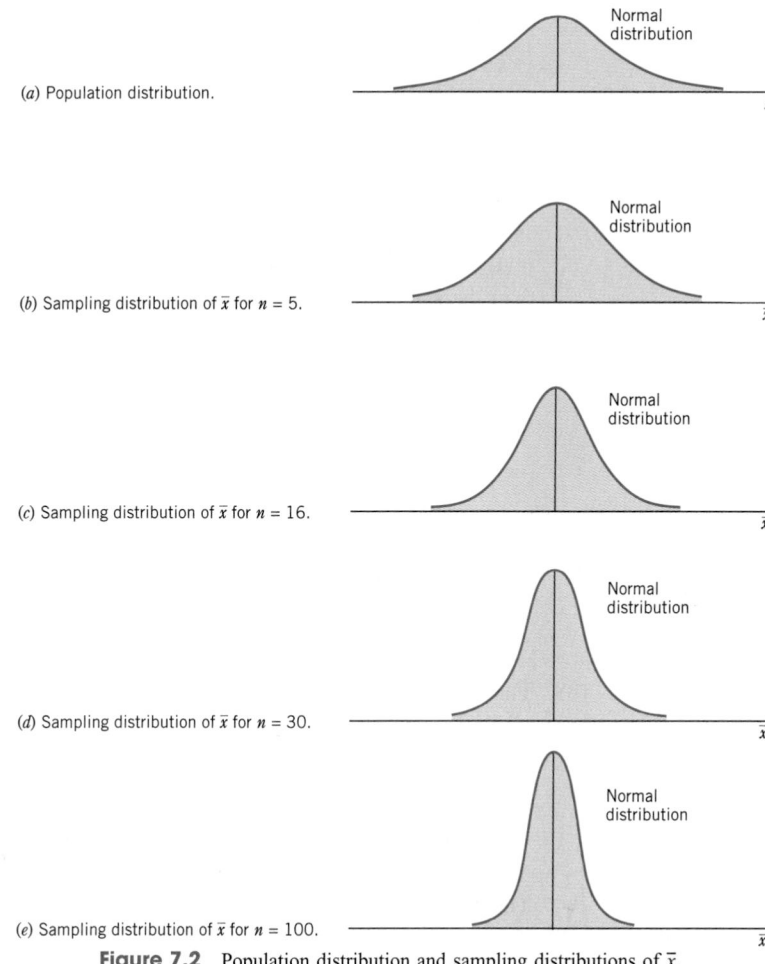

(a) Population distribution.

(b) Sampling distribution of $\bar{x}$ for $n = 5$.

(c) Sampling distribution of $\bar{x}$ for $n = 16$.

(d) Sampling distribution of $\bar{x}$ for $n = 30$.

(e) Sampling distribution of $\bar{x}$ for $n = 100$.

Figure 7.2 Population distribution and sampling distributions of $\bar{x}$.

Example 7–3 illustrates the calculation of the mean and standard deviation of $\bar{x}$ and the description of the shape of its sampling distribution.

Mean, standard deviation, and the sampling distribution of $\bar{x}$: normal population.

EXAMPLE 7–3 In a recent SAT test, the mean score for all examinees was 904. Assume that the distribution of SAT scores of all examinees is normal with a mean of 904 and a standard deviation of 153. Let $\bar{x}$ be the mean SAT score of a random sample of certain examinees. Calculate the mean and standard deviation of $\bar{x}$ and describe the shape of its sampling distribution when the sample size is

(a) 16   (b) 50   (c) 1000

Solution Let μ and σ be the mean and standard deviation of SAT scores of all examinees, and $\mu_{\bar{x}}$ and $\sigma_{\bar{x}}$ be the mean and standard deviation of the sampling distribution of $\bar{x}$. Then, from the given information,

$$\mu = 904 \quad \text{and} \quad \sigma = 153$$

(a)   The mean and standard deviation of $\bar{x}$ are

$$\mu_{\bar{x}} = \mu = \textbf{904} \quad \text{and} \quad \sigma_{\bar{x}} = \frac{\sigma}{\sqrt{n}} = \frac{153}{\sqrt{16}} = \textbf{38.250}$$

As the SAT scores of all examinees are assumed to be normally distributed, the sampling distribution of $\bar{x}$ for samples of 16 examinees is also normal. Figure 7.3 shows the population distribution and the sampling distribution of $\bar{x}$. Note that because σ is greater than $\sigma_{\bar{x}}$, the population distribution has a wider spread and a smaller height than the sampling distribution of $\bar{x}$ in Figure 7.3.

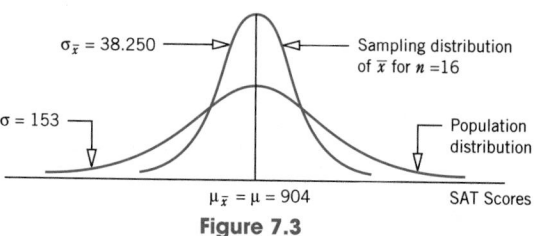

Figure 7.3

(b)   The mean and standard deviation of $\bar{x}$ are

$$\mu_{\bar{x}} = \mu = \textbf{904} \quad \text{and} \quad \sigma_{\bar{x}} = \frac{\sigma}{\sqrt{n}} = \frac{153}{\sqrt{50}} = \textbf{21.637}$$

Again, because the SAT scores of all examinees are assumed to be normally distributed, the sampling distribution of $\bar{x}$ for samples of 50 examinees is also normal. The population distribution and the sampling distribution of $\bar{x}$ are shown in Figure 7.4.

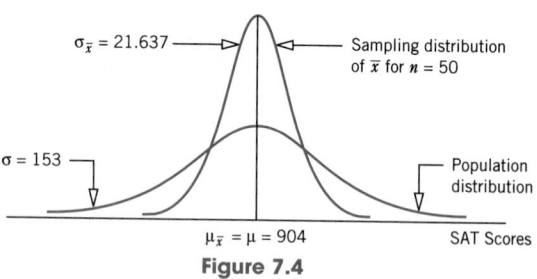

Figure 7.4

(c)   The mean and standard deviation of $\bar{x}$ are

$$\mu_{\bar{x}} = \mu = \textbf{904} \quad \text{and} \quad \sigma_{\bar{x}} = \frac{\sigma}{\sqrt{n}} = \frac{153}{\sqrt{1000}} = \textbf{4.838}$$

Again, because the SAT scores of all examinees are assumed to be normally distributed, the sampling distribution of $\bar{x}$ for samples of 1000 examinees is also normal. The two distributions are shown in Figure 7.5.

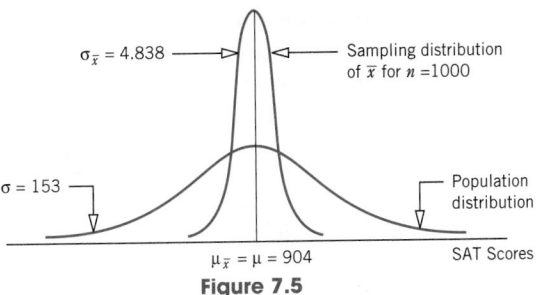

Figure 7.5

Thus, whatever the sample size, the sampling distribution of $\bar{x}$ is normal when the population from which the samples are drawn is normally distributed.

7.4.2 SAMPLING FROM A POPULATION THAT IS NOT NORMALLY DISTRIBUTED

Most of the time the population from which the samples are selected is not normally distributed. In such cases, the shape of the sampling distribution of $\bar{x}$ is inferred from a very important theorem called the **central limit theorem**.

CENTRAL LIMIT THEOREM

According to the *central limit theorem*, for a large sample size, the sampling distribution of $\bar{x}$ is approximately normal, irrespective of the shape of the population distribution. The mean and standard deviation of the sampling distribution of $\bar{x}$ are

$$\mu_{\bar{x}} = \mu \quad \text{and} \quad \sigma_{\bar{x}} = \frac{\sigma}{\sqrt{n}}$$

The sample size is usually considered to be large if $n \geq 30$.

Note that when the population does not have a normal distribution, the shape of the sampling distribution is not exactly normal but approximately normal for a large sample size. The approximation becomes more accurate as the sample size increases. Another point to remember is that the central limit theorem applies to *large* samples only. Usually, if the sample size is 30 or more, it is considered sufficiently large to apply the central limit theorem to the sampling distribution of $\bar{x}$. Thus, according to the central limit theorem,

1. When $n \geq 30$, the shape of the sampling distribution of $\bar{x}$ is approximately normal irrespective of the shape of the population.
2. The mean of $\bar{x}$, $\mu_{\bar{x}}$, is equal to the mean of the population, μ.
3. The standard deviation of $\bar{x}$, $\sigma_{\bar{x}}$, is equal to $\sigma/\sqrt{n}$.

Again, remember that for $\sigma_{\bar{x}} = \sigma/\sqrt{n}$ to be true, n/N must be less than or equal to .05.

Figure 7.6*a* shows the probability distribution curve for a population. The distribution curves in Figure 7.6*b* through Figure 7.6*e* show the sampling distributions of $\bar{x}$ for different sample sizes taken from the population of Figure 7.6*a*. As we can observe, the population is not normally distributed. The sampling distributions of $\bar{x}$ shown in parts *b* and *c*, when $n < 30$, are not normal. However, the sampling distributions of $\bar{x}$ shown in parts *d* and *e*, when $n \geq 30$, are (approximately) normal. Also notice that the spread of the sampling distribution of $\bar{x}$ decreases as the sample size increases.

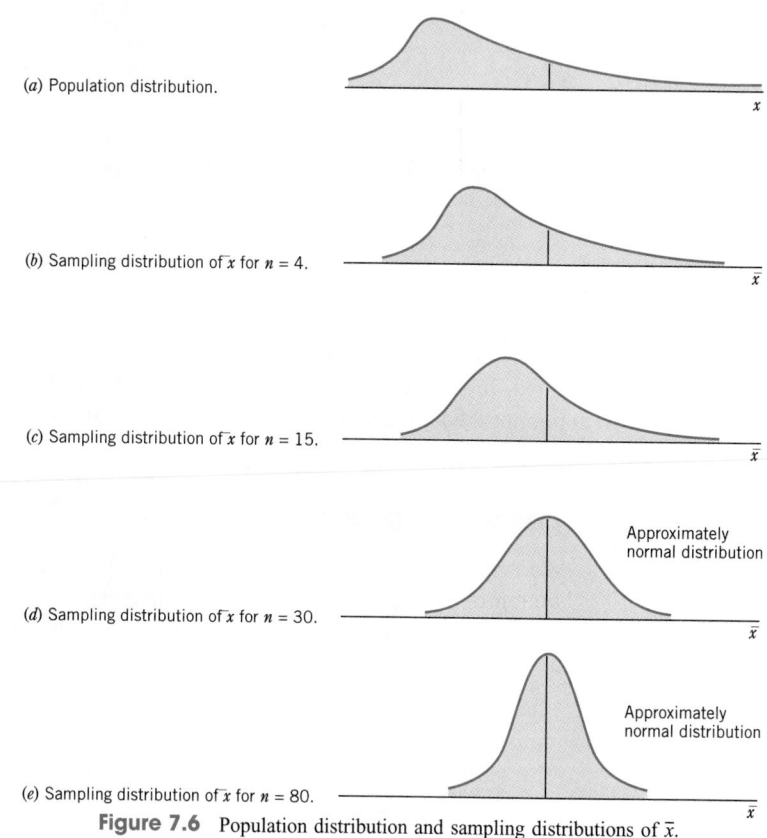

(*a*) Population distribution.

(*b*) Sampling distribution of $\bar{x}$ for $n = 4$.

(*c*) Sampling distribution of $\bar{x}$ for $n = 15$.

(*d*) Sampling distribution of $\bar{x}$ for $n = 30$.

Approximately normal distribution

(*e*) Sampling distribution of $\bar{x}$ for $n = 80$.

Approximately normal distribution

Figure 7.6   Population distribution and sampling distributions of $\bar{x}$.

Example 7–4 illustrates the calculation of the mean and standard deviation of $\bar{x}$ and describes the shape of the sampling distribution of $\bar{x}$ when the sample size is large.

Mean, standard deviation, and the sampling distribution of $\bar{x}$: nonnormal population.

EXAMPLE 7–4   The mean rent paid by all tenants in a large city is $950 with a standard deviation of $225. However, the population distribution of rents for all tenants in this city is skewed to the right. Calculate the mean and standard deviation of $\bar{x}$ and describe the shape of its sampling distribution when the sample size is

(a)   30   (b)   100

Solution   Although the population distribution of rents paid by all tenants is not normal, in each case the sample size is large ($n \geq 30$). Hence, the central limit theorem can be applied to infer the shape of the sampling distribution of $\bar{x}$.

(a) Let $\bar{x}$ be the mean rent paid by a sample of 30 tenants. Then, the sampling distribution of $\bar{x}$ is approximately normal with the values of the mean and standard deviation as

$$\mu_{\bar{x}} = \mu = \mathbf{\$950} \quad \text{and} \quad \sigma_{\bar{x}} = \frac{\sigma}{\sqrt{n}} = \frac{225}{\sqrt{30}} = \mathbf{\$41.079}$$

Figure 7.7 shows the population distribution and the sampling distribution of $\bar{x}$.

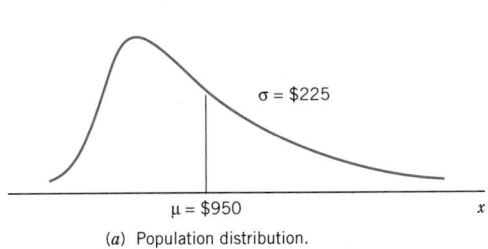
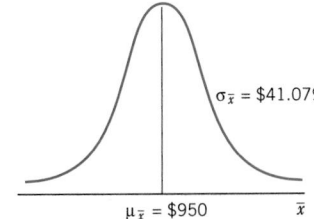

(a) Population distribution.   (b) Sampling distribution of $\bar{x}$ for $n = 30$.

Figure 7.7

(b) Let $\bar{x}$ be the mean rent paid by a sample of 100 tenants. Then, the sampling distribution of $\bar{x}$ is approximately normal with the values of the mean and standard deviation as

$$\mu_{\bar{x}} = \mu = \mathbf{\$950} \quad \text{and} \quad \sigma_{\bar{x}} = \frac{\sigma}{\sqrt{n}} = \frac{225}{\sqrt{100}} = \mathbf{\$22.500}$$

Figure 7.8 shows the population distribution and the sampling distribution of $\bar{x}$.

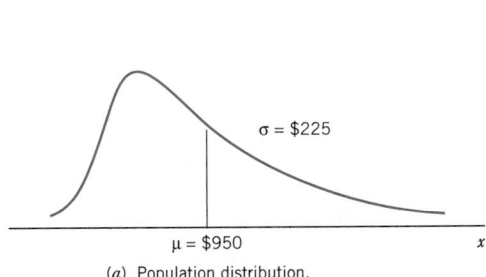
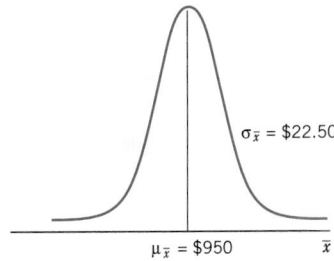

(a) Population distribution.   (b) Sampling distribution of $\bar{x}$ for $n = 100$.

Figure 7.8

EXERCISES

Concepts and Procedures

7.26 What condition or conditions must hold true for the sampling distribution of the sample mean to be normal when the sample size is less than 30?

7.27 Explain the central limit theorem.

7.28 A population has a distribution that is skewed to the left. Indicate in which of the following cases the central limit theorem will apply to describe the sampling distribution of the sample mean.

      a. $n = 400$ **b.** $n = 25$ **c.** $n = 36$

7.29 A population has a distribution that is skewed to the right. A sample of size n is selected from this population. Describe the shape of the sampling distribution of the sample mean for each of the following cases.

      a. $n = 25$ **b.** $n = 80$ **c.** $n = 29$

7.30 A population has a normal distribution. A sample of size n is selected from this population. Describe the shape of the sampling distribution of the sample mean for each of the following cases.

      a. $n = 94$ **b.** $n = 11$

7.31 A population has a normal distribution. A sample of size n is selected from this population. Describe the shape of the sampling distribution of the sample mean for each of the following cases.

      a. $n = 23$ **b.** $n = 450$

Applications

7.32 The time taken to complete a statistics test by all students is normally distributed with a mean of 120 minutes and a standard deviation of 10 minutes. Let $\bar{x}$ be the mean time taken to complete this test by a random sample of 16 students. Calculate the mean and standard deviation of $\bar{x}$ and describe the shape of its sampling distribution.

7.33 The speeds of all cars traveling on a stretch of Interstate Highway I-95 are normally distributed with a mean of 68 miles per hour and a standard deviation of 3 miles per hour. Let $\bar{x}$ be the mean speed of a random sample of 20 cars traveling on this highway. Calculate the mean and standard deviation of $\bar{x}$ and describe the shape of its sampling distribution.

7.34 The amounts of electric bills for all households in a city have an approximate normal distribution with a mean of $42 and a standard deviation of $7. Let $\bar{x}$ be the mean amount of electric bills for a random sample of 25 households selected from this city. Find the mean and standard deviation of $\bar{x}$ and comment on the shape of its sampling distribution.

7.35 The GPAs of all 5540 students enrolled at a university have an approximate normal distribution with a mean of 3.02 and a standard deviation of .29. Let $\bar{x}$ be the mean GPA of a random sample of 48 students selected from this university. Find the mean and standard deviation of $\bar{x}$ and comment on the shape of its sampling distribution.

7.36 The weights of all people living in a town have a distribution that is skewed to the right with a mean of 133 pounds and a standard deviation of 24 pounds. Let $\bar{x}$ be the mean weight of a random sample of 45 persons selected from this town. Find the mean and standard deviation of $\bar{x}$ and comment on the shape of its sampling distribution.

7.37 The amounts of telephone bills for all households in a large city have a distribution that is skewed to the right with a mean of $70 and a standard deviation of $25. Let $\bar{x}$ be the mean amount of telephone bills for a random sample of 90 households selected from this city. Calculate the mean and standard deviation of $\bar{x}$ and describe the shape of its sampling distribution.

7.38 The balances of checking accounts at a local bank have a distribution that is skewed to the right with its mean equal to $350 and standard deviation equal to $140. Let $\bar{x}$ be the mean balance of a random sample of 60 checking accounts selected from this bank. Calculate the mean and standard deviation of $\bar{x}$ and describe the shape of its sampling distribution.

7.39 According to the American Association of University Professors, the mean 1995–96 salary of full-time instructional faculty at institutions of higher education was $50,980. Suppose the distribution of 1995–96 salaries of these faculty is skewed to the right with its mean equal to $50,980 and standard deviation equal to $6300. Let $\bar{x}$ be the mean 1995–96 salary of a random sample of 80 faculty members selected from institutions of higher education. Calculate the mean and standard deviation of $\bar{x}$ and describe the shape of its sampling distribution.

7.5   APPLICATIONS OF THE SAMPLING DISTRIBUTION OF $\bar{x}$

From the central limit theorem, for large samples, the sampling distribution of $\bar{x}$ is approximately normal. Based on this result, we can make the following statements about $\bar{x}$ for large samples. The areas under the curve of $\bar{x}$ mentioned in these statements are found from the normal distribution table.

1. *If we take all possible samples of the same (large) size from a population and calculate the mean for each of these samples, then about 68.26% of the sample means will be within one standard deviation of the population mean.* Or we can state that if we take one sample (of n $\geq$ 30) from a population and calculate the mean for this sample, the probability that this sample mean will be within one standard deviation of the population mean is .6826. That is,

$$P(\mu - 1\sigma_{\bar{x}} \leq \bar{x} \leq \mu + 1\sigma_{\bar{x}}) = .6826$$

This probability is shown in Figure 7.9.

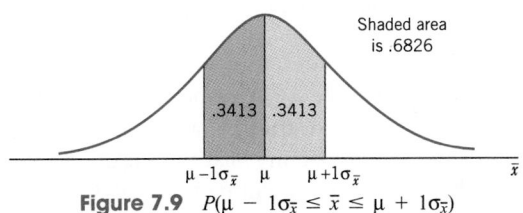

Figure 7.9 $P(\mu - 1\sigma_{\bar{x}} \leq \bar{x} \leq \mu + 1\sigma_{\bar{x}})$

2. *If we take all possible samples of the same (large) size from a population and calculate the mean for each of these samples, then about 95.44% of the sample means will be within two standard deviations of the population mean.* Or we can state that if we take one sample (of n $\geq$ 30) from a population and calculate the mean for this sample, the probability that this sample mean will be within two standard deviations of the population mean is .9544. That is,

$$P(\mu - 2\sigma_{\bar{x}} \leq \bar{x} \leq \mu + 2\sigma_{\bar{x}}) = .9544$$

This probability is shown in Figure 7.10.

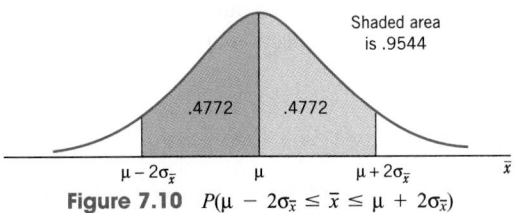

Figure 7.10 $P(\mu - 2\sigma_{\bar{x}} \leq \bar{x} \leq \mu + 2\sigma_{\bar{x}})$

3. *If we take all possible samples of the same (large) size from a population and calculate the mean for each of these samples, then about 99.74% of the sample means will be within three standard deviations of the population mean.* Or we can state that if we take one sample (of $n \geq 30$) from a population and calculate the mean for this sample, the probability that this sample mean will be within three standard deviations of the

population mean is .9974. That is,

$$P(\mu - 3\sigma_{\bar{x}} \le \bar{x} \le \mu + 3\sigma_{\bar{x}}) = .9974$$

This probability is shown in Figure 7.11.

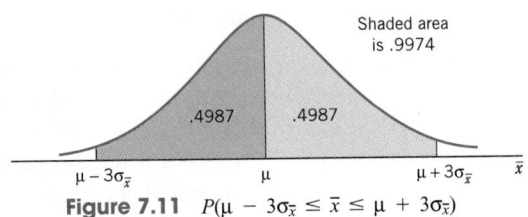

Figure 7.11 $P(\mu - 3\sigma_{\bar{x}} \le \bar{x} \le \mu + 3\sigma_{\bar{x}})$

When conducting a survey, we usually select one sample and compute the value of $\bar{x}$ based on that sample. We never select all possible samples of the same size and then prepare the sampling distribution of $\bar{x}$. Rather, we are more interested in finding the probability that the value of $\bar{x}$ computed from one sample falls within a given interval. Examples 7–5 and 7–6 illustrate this procedure.

Calculating probability of $\bar{x}$ in an interval: normal population.

EXAMPLE 7–5   Assume that the weights of all packages of a certain brand of cookies are normally distributed with a mean of 32 ounces and a standard deviation of .3 ounces. Find the probability that the mean weight, $\bar{x}$, of a random sample of 20 packages of this brand of cookies will be between 31.8 and 31.9 ounces.

Solution   Although the sample size is small ($n < 30$) the shape of the sampling distribution of $\bar{x}$ is normal because the population is normally distributed. The mean and standard deviation of $\bar{x}$ are

$$\mu_{\bar{x}} = \mu = 32 \text{ ounces} \qquad \text{and} \qquad \sigma_{\bar{x}} = \frac{\sigma}{\sqrt{n}} = \frac{.3}{\sqrt{20}} = .06708204 \text{ ounces}$$

We are to compute the probability that the value of $\bar{x}$ calculated for one randomly drawn sample of 20 packages is between 31.8 and 31.9 ounces, that is,

$$P(31.8 < \bar{x} < 31.9)$$

This probability is given by the area under the normal distribution curve for $\bar{x}$ between the points $\bar{x} = 31.8$ and $\bar{x} = 31.9$. The first step in finding this area is to convert the two $\bar{x}$ values to respective z values.

---

z VALUE FOR A VALUE OF $\bar{x}$

The *z value for a value of $\bar{x}$* is calculated as

$$z = \frac{\bar{x} - \mu}{\sigma_{\bar{x}}}$$

The z values for $\bar{x} = 31.8$ and $\bar{x} = 31.9$ are computed below and they are shown on the z scale below the normal distribution curve for $\bar{x}$ in Figure 7.12.

For $\bar{x} = 31.8$: $z = \dfrac{31.8 - 32}{.06708204} = -2.98$

For $\bar{x} = 31.9$: $z = \dfrac{31.9 - 32}{.06708204} = -1.49$

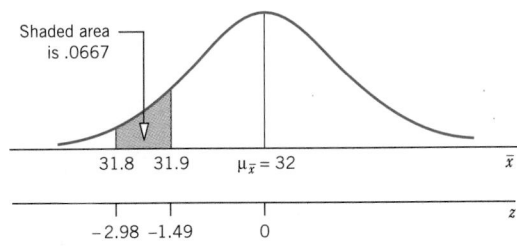

Figure 7.12 $P(31.8 < \bar{x} < 31.9)$

The probability that $\bar{x}$ is between 31.8 and 31.9 is given by the area under the standard normal curve between $z = -2.98$ and $z = -1.49$. Thus, the required probability is

$$P(31.8 < \bar{x} < 31.9) = P(-2.98 < z < -1.49)$$
$$= P(-2.98 < z < 0) - P(-1.49 < z < 0)$$
$$= .4986 - .4319 = \mathbf{.0667}$$

Therefore, the probability is .0667 that the mean weight of a sample of 20 packages will be between 31.8 and 31.9 ounces.   ■

Calculating probability of $\bar{x}$ in an interval: $n > 30$.

EXAMPLE 7–6   According to the National Association of Realtors, the mean price of houses in the Northeast region of the United States was $164,300 in 1995 (*Real Estate Outlook* 3[6], June 1996). Suppose the current prices of all houses in the Northeast region have a probability distribution that is skewed to the right with a mean of $164,300 and a standard deviation of $29,500. Let $\bar{x}$ be the mean price of a sample of 400 houses selected from this region.

(a)   What is the probability that the mean price obtained from this sample will be within $3000 of the population mean?

(b)   What is the probability that the mean price obtained from this sample will be lower than the population mean by $2500 or more?

Solution   From the given information, for the prices of all houses in the Northeast region,

$$\mu = \$164{,}300 \quad \text{and} \quad \sigma = \$29{,}500$$

Although the shape of the probability distribution of the population (prices of all houses) is skewed to the right, the sampling distribution of $\bar{x}$ is approximately normal because the sample size is large ($n > 30$). Remember that when the sample size is large, the central limit theorem applies. The mean and standard deviation of the sampling distribution of $\bar{x}$ are

$$\mu_{\bar{x}} = \mu = \$164{,}300 \quad \text{and} \quad \sigma_{\bar{x}} = \frac{\sigma}{\sqrt{n}} = \frac{29{,}500}{\sqrt{400}} = \$1475$$

(a) The probability that the mean price obtained from a sample of 400 houses will be within \$3000 of the population mean is written as

$$P(161{,}300 \leq \bar{x} \leq 167{,}300)$$

This probability is given by the area under the normal curve for $\bar{x}$ between $\bar{x} = \$161{,}300$ and $\bar{x} = \$167{,}300$, as shown in Figure 7.13. We find this area as follows.

For $\bar{x} = \$161{,}300$: $z = \dfrac{161{,}300 - 164{,}300}{1475} = -2.03$

For $\bar{x} = \$167{,}300$: $z = \dfrac{167{,}300 - 164{,}300}{1475} = 2.03$

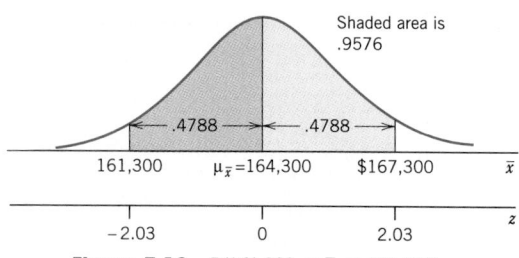

Figure 7.13 $P(161{,}300 \leq \bar{x} \leq 167{,}300)$

Hence, the required probability is

$$P(161{,}300 \leq \bar{x} \leq 167{,}300) = P(-2.03 \leq z \leq 2.03)$$
$$= P(-2.03 \leq z \leq 0) + P(0 \leq z \leq 2.03)$$
$$= .4788 + .4788 = \mathbf{.9576}$$

Therefore, the probability that the mean price of a sample of 400 houses selected from the Northeast region is within \$3000 of the population mean is .9576.

(b) The probability that the mean price obtained from a sample of 400 houses will be lower than the population mean by \$2500 or more is written as

$$P(\bar{x} \leq 161{,}800)$$

This probability is given by the area under the normal curve for $\bar{x}$ to the left of $\bar{x} = \$161{,}800$, as shown in Figure 7.14. We find this area as follows.

For $\bar{x} = 161{,}800$: $z = \dfrac{161{,}800 - 164{,}300}{1475} = -1.69$

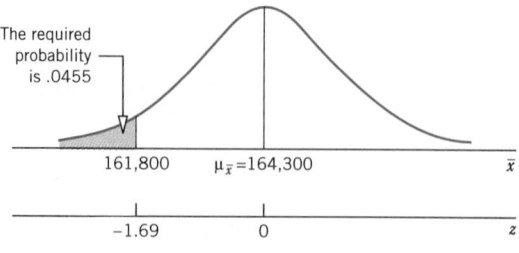

Figure 7.14 $P(\bar{x} \leq 161{,}800)$

Hence, the required probability is

$$P(\bar{x} \leq 161{,}800) = P(z \leq -1.69)$$
$$= .5 - P(-1.69 < z < 0)$$
$$= .5 - .4545 = \mathbf{.0455}$$

Therefore, the probability that the mean price of a sample of 400 houses selected from the Northeast region is lower than the population mean by \$2500 or more is .0455.

EXERCISES

Concepts and Procedures

7.40   If all possible samples of the same (large) size are selected from a population, what percent of all the sample means will be within 2.5 standard deviations of the population mean?

7.41   If all possible samples of the same (large) size are selected from a population, what percent of all the sample means will be within 1.5 standard deviations of the population mean?

 7.42   For a population, $N = 10{,}000$, $\mu = 124$, and $\sigma = 18$. Find the z value for each of the following for $n = 36$.
   a. $\bar{x} = 128.60$ b. $\bar{x} = 119.30$ c. $\bar{x} = 116.88$ d. $\bar{x} = 132.05$

7.43   For a population, $N = 205{,}000$, $\mu = 66$, and $\sigma = 7$. Find the z value for each of the following for $n = 49$.
   a. $\bar{x} = 68.44$ b. $\bar{x} = 58.75$ c. $\bar{x} = 62.35$ d. $\bar{x} = 71.82$

 7.44   Let x be a continuous random variable that has a normal distribution with $\mu = 75$ and $\sigma = 15$. Assuming $n/N \leq .05$, find the probability that the sample mean, $\bar{x}$, for a random sample of 20 taken from this population will be
   a.   between 68.5 and 77.3   b.   less than 72.4

7.45   Let x be a continuous random variable that has a normal distribution with $\mu = 48$ and $\sigma = 8$. Assuming $n/N \leq .05$, find the probability that the sample mean, $\bar{x}$, for a random sample of 16 taken from this population will be
   a.   between 49.6 and 52.2   b.   more than 45.7

7.46   Let x be a continuous random variable that has a distribution skewed to the right with $\mu = 60$ and $\sigma = 10$. Assuming $n/N \leq .05$, find the probability that the sample mean, $\bar{x}$, for a random sample of 40 taken from this population will be
   a.   less than 62.20   b.   between 61.4 and 64.2

7.47   Let x be a continuous random variable that follows a distribution skewed to the left with $\mu = 90$ and $\sigma = 16$. Assuming $n/N \leq .05$, find the probability that the sample mean, $\bar{x}$, for a random sample of 64 taken from this population will be
   a.   less than 82.3   b.   more than 86.7

Applications

 7.48   The speeds of all cars traveling on a stretch of Interstate Highway I-95 are normally distributed with a mean of 68 miles per hour and a standard deviation of 3 miles per hour. Find the probability that the mean speed of a random sample of 16 cars traveling on this stretch of this Interstate Highway is
   a.   less than 66 miles per hour
   b.   more than 69 miles per hour
   c.   67 to 69 miles per hour

7.49 The GPAs of all students enrolled at a large university have an approximate normal distribution with a mean of 3.02 and a standard deviation of .29. Find the probability that the mean GPA of a random sample of 20 students selected from this university is

      a. 3.10 or higher   **b.** 2.90 or lower   **c.** 2.95 to 3.11

7.50 The time taken to complete a statistics test by all students is normally distributed with a mean of 120 minutes and a standard deviation of 10 minutes. Find the probability that the mean time taken to complete this test by a random sample of 16 students would be

    a.   between 122 and 126 minutes

    b.   within 4 minutes of the population mean

    c.   lower than the population mean by 3 minutes or more

7.51 According to the U.S. Bureau of Labor Statistics, the mean weekly wage for manufacturing workers was $517 in February 1996. Assume that the weekly wages for all manufacturing workers for February 1996 are normally distributed with a mean of $517 and a standard deviation of $65. Find the probability that the mean weekly wage for a random sample of 25 manufacturing workers taken from this population would be

    a.   between $527 and $537

    b.   within $15 of the population mean weekly wage

    c.   lower than the population mean by $20 or more

7.52 The time that college students spend studying per week has a distribution that is skewed to the right with a mean of 8.4 hours and a standard deviation of 2.7 hours. Find the probability that the mean time spent studying per week for a random sample of 45 students would be

      a. between 8 and 9 hours   **b.** less than 8 hours

7.53 The ages of all college students follow a distribution that is skewed to the right with a mean of 26 years and a standard deviation of 4 years. Find the probability that the mean age for a random sample of 36 students would be

      a. between 25 and 27 years   **b.** less than 25 years

7.54 Total knee replacement surgery has become increasingly common in the past few years. In 1994, claims to the Metropolitan Life Insurance Company averaged $28,340 for the combined hospital and physicians' costs for this surgery (*Statistical Bulletin*, April–June 1996). Assume that the current costs of all such surgeries have a distribution that is skewed to the right with a mean of $28,340 and a standard deviation of $8000. Find the probability that the mean cost for a random sample of 100 such surgeries will lie

    a.   between $28,800 and $29,600

    b.   within $700 of the population mean

    c.   lower than the population mean by $800 or more

7.55 The amounts of electric bills for all households in a city have a skewed probability distribution with a mean of $65 and a standard deviation of $25. Find the probability that the mean amount of electric bills for a random sample of 75 households selected from this city will be

    a.   between $58 and $63

    b.   within $6 of the population mean

    c.   more than the population mean by at least $5

7.56 The balances of all savings accounts at a local bank have a distribution that is skewed to the right with its mean equal to $12,450 and standard deviation equal to $4160. Find the probability that the mean balance of a sample of 50 savings accounts selected from this bank will be

    a.   more than $11,500

    b.   between $12,000 and $13,800

    c.   within $1500 of the population mean

    d.   more than the population mean by at least $1000

7.57 The heights of all adults living in a large city have a distribution that is skewed to the right with a mean of 68 inches and a standard deviation of 4 inches. Find the probability that the mean height of a random sample of 100 adults selected from this city would be

    a.   less than 67.8 inches

    b.   between 67.5 inches and 68.7 inches

    c.   within .6 inches of the population mean

    d.   lower than the population mean by .5 inches or more

7.58   Johnson Electronics Corporation makes electric tubes. It is known that the standard deviation of the lives of these tubes is 150 hours. The company's research department takes a sample of 100 such tubes and finds that the mean life of these tubes is 2250 hours. What is the probability that this sample mean is within 25 hours of the mean life of all tubes produced by this company?

7.59   A machine at Katz Steel Corporation makes 3-inch-long nails. The probability distribution of the lengths of these nails is normal with a mean of 3 inches and a standard deviation of .1 inch. The quality control inspector takes a sample of 25 nails once a week and calculates the mean length of these nails. If the mean of this sample is either less than 2.95 inches or greater than 3.05 inches, the inspector concludes that the machine needs an adjustment. What is the probability that based on a sample of 25 nails the inspector will conclude that the machine needs an adjustment?

7.6 POPULATION AND SAMPLE PROPORTIONS

The concept of proportion is the same as the concept of relative frequency discussed in Chapter 2 and the concept of probability of success in a binomial experiment. The relative frequency of a category or class gives the proportion of the sample or population that belongs to that category or class. Similarly, the probability of success in a binomial experiment represents the proportion of the sample or population that possesses a given characteristic.

    The **population proportion**, denoted by **p**, is obtained by taking the ratio of the number of elements in a population with a specific characteristic to the total number of elements in the population. The **sample proportion**, denoted by $\hat{p}$ (pronounced *p hat*), gives a similar ratio for a sample.

POPULATION AND SAMPLE PROPORTIONS

The *population* and *sample proportions,* denoted by p and $\hat{p}$, respectively, are calculated as

$$p = \frac{X}{N} \quad \text{and} \quad \hat{p} = \frac{x}{n}$$

where

        N = Total number of elements in the population

        n = Total number of elements in the sample

        X = Number of elements in the population that possess a specific characteristic

        x = Number of elements in the sample that possess a specific characteristic

---

Example 7–7 illustrates the calculation of the population and sample proportions.

Calculating population and sample proportions.

EXAMPLE 7–7 Suppose a total of 789,654 families live in a city and 563,282 of them own homes. Then,

$$N = \text{Population size} = 789{,}654$$

$$X = \text{Families in the population who own homes} = 563{,}282$$

The proportion of all families in this city who own homes is

$$p = \frac{X}{N} = \frac{563{,}282}{789{,}654} = .71$$

Now, suppose a sample of 240 families is taken from this city and 158 of them are homeowners. Then,

$$n = \text{Sample size} = 240$$

$$x = \text{Families in the sample who own homes} = 158$$

The sample proportion is

$$\hat{p} = \frac{x}{n} = \frac{158}{240} = .66$$

As in the case of the mean, the difference between the sample proportion and the corresponding population proportion gives the sampling error, assuming that the sample is random and no nonsampling error has been made. That is, in case of the proportion,

$$\text{Sampling error} = \hat{p} - p$$

For instance, for Example 7–7,

$$\text{Sampling error} = \hat{p} - p = .66 - .71 = -.05$$

7.7 MEAN, STANDARD DEVIATION, AND SHAPE OF THE SAMPLING DISTRIBUTION OF $\hat{p}$

This section discusses the sampling distribution of the sample proportion, and the mean, standard deviation, and shape of this sampling distribution.

7.7.1 SAMPLING DISTRIBUTION OF $\hat{p}$

Just like the sample mean $\bar{x}$, the sample proportion, $\hat{p}$, is also a random variable. Hence, it possesses a probability distribution, which is called its **sampling distribution**.

SAMPLING DISTRIBUTION OF THE SAMPLE PROPORTION, $\hat{p}$

The probability distribution of the sample proportion, $\hat{p}$, is called its *sampling distribution*. It gives the various values that $\hat{p}$ can assume and their probabilities.

The value of $\hat{p}$ calculated for a particular sample depends on what elements of the population are included in that sample. Example 7–8 illustrates the concept of the sampling distribution of $\hat{p}$.

Illustrating the sampling distribution of $\hat{p}$.

EXAMPLE 7–8   Boe Consultant Associates has five employees. Table 7.6 gives the names of these five employees and information concerning their knowledge of statistics.

Table 7.6   Information on the Five Employees of Boe Consultant Associates

Name	Knows Statistics
Ally	yes
John	no
Susan	no
Lee	yes
Tom	yes

If we define the population proportion, p, as the proportion of employees who know statistics, then,

$$p = 3/5 = .60$$

Now, suppose we draw all possible samples of three employees each and compute the proportion of employees, for each sample, who know statistics. The total number of samples of size three that can be drawn from the population of five employees is

$$\text{Total number of samples} = \binom{5}{3} = \frac{5!}{3!\,(5-3)!} = \frac{5 \cdot 4 \cdot 3 \cdot 2 \cdot 1}{3 \cdot 2 \cdot 1 \cdot 2 \cdot 1} = 10$$

Table 7.7 lists these 10 possible samples and the proportion of employees who know statistics for each of those samples. Note that we have rounded the values of $\hat{p}$ to two decimal places.

Table 7.7   All Possible Samples of Size 3 and the Value of $\hat{p}$ for Each Sample

Sample	Proportion Who Know Statistics $\hat{p}$
Ally, John, Susan	1/3 = .33
Ally, John, Lee	2/3 = .67
Ally, John, Tom	2/3 = .67
Ally, Susan, Lee	2/3 = .67
Ally, Susan, Tom	2/3 = .67
Ally, Lee, Tom	3/3 = 1.00
John, Susan, Lee	1/3 = .33
John, Susan, Tom	1/3 = .33
John, Lee, Tom	2/3 = .67
Susan, Lee, Tom	2/3 = .67

Using Table 7.7, we prepare the frequency distribution of $\hat{p}$ as recorded in Table 7.8, along with the relative frequencies of classes, which are obtained by dividing the frequencies of classes by the population size. The relative frequencies are used as probabilities and listed in Table 7.9. This table gives the sampling distribution of $\hat{p}$.

Table 7.8 Frequency and Relative Frequency Distributions of $\hat{p}$ When the Sample Size is 3

$\hat{p}$	f	Relative Frequency
.33	3	$3/10 = .30$
.67	6	$6/10 = .60$
1.00	1	$1/10 = .10$
	$\Sigma f = 10$	Sum $= 1.00$

Table 7.9 Sampling Distribution of $\hat{p}$ When the Sample Size is 3

$\hat{p}$	$P(\hat{p})$
.33	.30
.67	.60
1.00	.10
	$\Sigma P(\hat{p}) = 1.00$

7.7.2 MEAN AND STANDARD DEVIATION OF $\hat{p}$

The **mean of $\hat{p}$**, which is the same as the mean of the sampling distribution of $\hat{p}$, is always equal to the population proportion, p, just as the mean of the sampling distribution of $\bar{x}$ is always equal to the population mean, μ.

MEAN OF THE SAMPLE PROPORTION

The *mean of the sample proportion, $\hat{p}$,* is denoted by $\mu_{\hat{p}}$ and is equal to the population proportion, p. Thus,

$$\mu_{\hat{p}} = p$$

The sample proportion, $\hat{p}$, is called an **estimator** of the population proportion, p. As mentioned earlier in this chapter, when the expected value (or mean) of a sample statistic is equal to the value of the corresponding population parameter, that sample statistic is said to be an **unbiased estimator**. Since for the sample proportion, $\mu_{\hat{p}} = p$, $\hat{p}$ is an unbiased estimator of p.

The **standard deviation of $\hat{p}$**, denoted by $\sigma_{\hat{p}}$, is given by the following formula. This formula is true only when the sample size is small compared to the population size. As we know from Section 7.3, the sample size is said to be small compared to the population size if $n/N \leq .05$.

STANDARD DEVIATION OF THE SAMPLE PROPORTION

The *standard deviation of the sample proportion,* $\hat{p}$, is denoted by $\sigma_{\hat{p}}$ and is given by the formula

$$\sigma_{\hat{p}} = \sqrt{\frac{pq}{n}}$$

where p is the population proportion, $q = 1 - p$, and n is the sample size.
This formula is used when $n/N \leq .05$ where N is the population size.

However, if n/N is greater than .05, then $\sigma_{\hat{p}}$ is calculated as follows.

$$\sigma_{\hat{p}} = \sqrt{\frac{pq}{n}} \sqrt{\frac{N - n}{N - 1}}$$

where the factor

$$\sqrt{\frac{N - n}{N - 1}}$$

is called the finite population correction factor.

In almost all cases, the sample size is small compared to the population size and, consequently, the formula used to calculate $\sigma_{\hat{p}}$ is $\sqrt{pq/n}$.

As mentioned earlier in this chapter, if the standard deviation of a sample statistic decreases as the sample size is increased, that statistic is said to be a **consistent estimator**. It is obvious from the above formula for $\sigma_{\hat{p}}$, that as n increases, the value of $\sqrt{pq/n}$ decreases. Thus, the sample proportion, $\hat{p}$, is a consistent estimator of the population proportion, p.

7.7.3 SHAPE OF THE SAMPLING DISTRIBUTION OF $\hat{p}$

The shape of the sampling distribution of $\hat{p}$ is inferred from the central limit theorem.

CENTRAL LIMIT THEOREM FOR SAMPLE PROPORTION

According to the central limit theorem, the *sampling distribution of* $\hat{p}$ is approximately normal for a sufficiently large sample size. In the case of proportion, the sample size is considered to be sufficiently large if np and nq are both greater than 5, that is, if

$$np > 5 \quad \text{and} \quad nq > 5$$

Note that the sampling distribution of $\hat{p}$ will be approximately normal if $np > 5$ and $nq > 5$. This is the same condition that was required for the application of the normal approximation to the binomial probability distribution in Chapter 6.

Example 7–9 shows the calculation of the mean and standard deviation of $\hat{p}$ and describes the shape of its sampling distribution.

Mean, standard deviation, and the sampling distribution of $\hat{p}$.

EXAMPLE 7–9   Forty-four percent of U.S. adults polled by the Gallup Organization for *U.S. News*/CNN said that teenage boys are the worst drivers (*U.S. News & World Report*, July 31, 1995). Assume that this result is true for the current population of all U.S. adults. Let $\hat{p}$ be the proportion in a random sample of 100 U.S. adults who hold this view. Find the mean and standard deviation of $\hat{p}$ and describe the shape of its sampling distribution.

Solution   Let p be the proportion of all U.S. adults who think that teenage boys are the worst drivers. Then,

$$p = .44 \quad \text{and} \quad q = 1 - p = 1 - .44 = .56$$

The mean of the sampling distribution of $\hat{p}$ is

$$\mu_{\hat{p}} = p = \mathbf{.44}$$

The standard deviation of $\hat{p}$ is

$$\sigma_{\hat{p}} = \sqrt{\frac{pq}{n}} = \sqrt{\frac{(.44)(.56)}{100}} = \mathbf{.050}$$

The values of np and nq are

$$np = 100\,(.44) = 44 \quad \text{and} \quad nq = 100\,(.56) = 56$$

As np and nq are both greater than 5, we can apply the central limit theorem to make an inference about the shape of the sampling distribution of $\hat{p}$. Therefore, the sampling distribution of $\hat{p}$ is approximately normal with a mean of .44 and a standard deviation of .050, as shown in Figure 7.15.

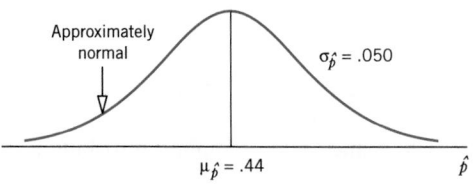

Figure 7.15

EXERCISES

Concepts and Procedures

7.60   In a population of 1000 subjects, 640 possess a certain characteristic. A sample of 40 subjects selected from this population has 24 subjects who possess the same characteristic. What are the values of the population and sample proportions?

7.61   In a population of 5000 subjects, 600 possess a certain characteristic. A sample of 120 subjects selected from this population contains 18 subjects who possess the same characteristic. What are the values of the population and sample proportions?

7.62 In a population of 18,700 subjects, 30% possess a certain characteristic. In a sample of 250 subjects selected from this population, 25% possess the same characteristic. How many subjects in the population and sample, respectively, possess this characteristic?

7.63 In a population of 9500 subjects, 75% possess a certain characteristic. In a sample of 400 subjects selected from this population, 78% possess the same characteristic. How many subjects in the population and sample, respectively, possess this characteristic?

7.64 Let $\hat{p}$ be the proportion of elements in a sample that possess a characteristic.

    a. What is the mean of $\hat{p}$?

    b. What is the standard deviation of $\hat{p}$? Assume $n/N \leq .05$.

    c. What condition(s) must hold true for the sampling distribution of $\hat{p}$ to be approximately normal?

7.65 For a population, $N = 12,000$ and $p = .72$. A random sample of 900 elements selected from this population gave $\hat{p} = .66$. Find the sampling error.

7.66 For a population, $N = 2800$ and $p = .28$. A random sample of 80 elements selected from this population gave $\hat{p} = .33$. Find the sampling error.

7.67 What is the estimator of the population proportion? Is this estimator an unbiased estimator of p? Explain why or why not.

7.68 Is the sample proportion a consistent estimator of the population proportion? Explain why or why not.

7.69 How does the value of $\sigma_{\hat{p}}$ change as the sample size increases? Explain. Assume $n/N \leq .05$.

7.70 Consider a large population with $p = .65$. Assuming $n/N \leq .05$, find the mean and standard deviation of the sample proportion $\hat{p}$ for a sample size of

    a. 100   **b.** 900

7.71 Consider a large population with $p = .18$. Assuming $n/N \leq .05$, find the mean and standard deviation of the sample proportion $\hat{p}$ for a sample size of

    a. 400   **b.** 750

7.72 A population of $N = 4000$ has a population proportion equal to .10. In each of the following cases which formula will you use to calculate $\sigma_{\hat{p}}$ and why? Using the appropriate formula, calculate $\sigma_{\hat{p}}$ for each of these cases.

    a. $n = 800$ **b.** $n = 30$

7.73 A population of $N = 1400$ has a population proportion equal to .45. In each of the following cases which formula will you use to calculate $\sigma_{\hat{p}}$ and why? Using the appropriate formula, calculate $\sigma_{\hat{p}}$ for each of these cases.

    a. $n = 90$ **b.** $n = 50$

7.74 According to the central limit theorem, the sampling distribution of $\hat{p}$ is approximately normal when the sample is large. What is considered a large sample in the case of the proportion? Briefly explain.

7.75 Indicate in which of the following cases the central limit theorem will apply to describe the sampling distribution of the sample proportion.

    a. $n = 400$ and $p = .28$ **b.** $n = 80$ and $p = .05$

    c. $n = 60$ and $p = .12$ **d.** $n = 100$ and $p = .035$

7.76 Indicate in which of the following cases the central limit theorem will apply to describe the sampling distribution of the sample proportion.

    a. $n = 20$ and $p = .45$ **b.** $n = 75$ and $p = .22$

    c. $n = 350$ and $p = .01$ **d.** $n = 200$ and $p = .022$

Applications

7.77 A company manufactured six television sets on a given day and these television sets were inspected for being good or defective. The results of the inspection are as follows.

Good   Good   Defective   Defective   Good   Good

    a.   What proportion of these television sets are good?

    b.   How many total samples (without replacement) of size 5 can be selected from this population?

    c.   List all the possible samples of size 5 that can be selected from this population and calculate the sample proportion, $\hat{p}$, of television sets that are good for each sample. Prepare the sampling distribution of $\hat{p}$.

    d.   For each sample listed in part c, calculate the sampling error.

7.78   The following data give the information on all 5 employees of a company.

Male   Female   Female   Male   Female

    a.   What proportion of employees of this company are female?

    b.   How many total samples (without replacement) of size 3 can be drawn from this population?

    c.   List all the possible samples of size 3 that can be selected from this population and calculate the sample proportion, $\hat{p}$, of the employees who are female for each sample. Prepare the sampling distribution of $\hat{p}$.

    d.   For each sample listed in part c, calculate the sampling error.

7.79   The U.S. public has become increasingly concerned about the ill-mannered behavior of its fellow citizens. A poll conducted in February 1996 by *U.S. News*/Bozell Worldwide found that 73% of adults in the United States feel that mean-spirited political campaigns contribute to this problem (*U.S. News & World Report*, April 22, 1996). Assume that 73% of all adults in the United States feel that mean-spirited political campaigns contribute to the problem of uncivil behavior. Let $\hat{p}$ be the proportion of U.S. adults in a sample of 500 who hold this view. Find the mean and standard deviation of $\hat{p}$ and describe the shape of its sampling distribution.

7.80   The Gallup Organization conducted a poll for Phi Delta Kappa on the attitudes of U.S. adults toward public schools (*Phi Delta Kappan*, September 1996). Sixty-six percent of these adults favored a community service requirement for high school graduation. Assume that this percentage is true for the current population of all U.S. adults. Let $\hat{p}$ be the proportion of adults in a random sample of 200 U.S. adults who hold this view. Calculate the mean and standard deviation of $\hat{p}$ and describe the shape of its sampling distribution.

7.81   Between December 1995 and February 1996, Louis Harris and Associates conducted a national survey of students in grades 7 through 12 for the Metropolitan Life Insurance Company (*Statistical Bulletin*, July–September 1996). Nearly 20% of these students were at least somewhat afraid of being attacked in or near their schools. Suppose that 20% of all students in grades 7 through 12 are afraid of such attacks. Let $\hat{p}$ be the proportion of students in a random sample of 50 seventh- through-twelfth-grade students who fear such attacks. Calculate the mean and standard deviation of $\hat{p}$ and comment on the shape of its sampling distribution.

7.82   According to a U.S. Bureau of the Census survey in 1995, 75.1% of the *baby boomers* (aged 26 to 44 years) drive alone to work (*USA Today*, November 20, 1995). Assume that this percentage is true for the current population of all baby boomers. Let $\hat{p}$ be the proportion of baby boomers in a random sample of 100 who drive alone to work. Calculate the mean and standard deviation of $\hat{p}$ and describe the shape of its sampling distribution.

7.8 APPLICATIONS OF THE SAMPLING DISTRIBUTION OF $\hat{p}$

As mentioned in Section 7.5, when we conduct a study we usually take only one sample and make all decisions or inferences on the basis of the results of that one sample. We use the concepts of the mean, standard deviation, and shape of the sampling distribution of $\hat{p}$

to determine the probability that the value of $\hat{p}$ computed from one sample falls within a given interval. Examples 7–10 and 7–11 illustrate this application.

Calculating probability that $\hat{p}$ is in an interval.

EXAMPLE 7–10   According to a poll commissioned by Merck Family Fund, 85% of women said that as a nation Americans are "addicted to shopping" (*Working Woman*, August 1995). Assume that this result is true for the current population of all U.S. women. Let $\hat{p}$ be the proportion in a random sample of 200 U.S. women who hold this view. Find the probability that the value of $\hat{p}$ is between .87 and .90.

Solution   From the given information,

$$n = 200, \qquad p = .85, \qquad \text{and} \qquad q = 1 - p = 1 - .85 = .15$$

where p is the proportion of all women who would say that as a nation Americans are "addicted to shopping."

The mean of the sample proportion $\hat{p}$ is

$$\mu_{\hat{p}} = p = .85$$

The standard deviation of $\hat{p}$ is

$$\sigma_{\hat{p}} = \sqrt{\frac{pq}{n}} = \sqrt{\frac{(.85)(.15)}{200}} = .02524876$$

The values of np and nq are

$$np = 200 \,(.85) = 170 \qquad \text{and} \qquad nq = 200 \,(.15) = 30$$

As np and nq are both greater than 5, we can infer from the central limit theorem that the sampling distribution of $\hat{p}$ is approximately normal. The probability that $\hat{p}$ is between .87 and .90 is given by the area under the normal curve for $\hat{p}$ between $\hat{p} = .87$ and $\hat{p} = .90$, as shown in Figure 7.16.

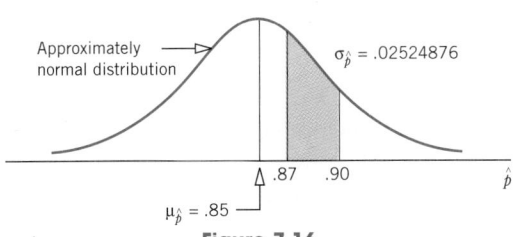

Figure 7.16

The first step in finding the area under the normal curve between $\hat{p} = .87$ and $\hat{p} = .90$ is to convert these two values to respective z values. The z value for $\hat{p}$ is computed using the following formula.

> **z VALUE FOR A VALUE OF $\hat{p}$**
>
> The z *value for a value of* $\hat{p}$ is calculated as
>
> $$z = \frac{\hat{p} - p}{\sigma_{\hat{p}}}$$

Next, the two values of $\hat{p}$ are converted to their respective z values and then the area under the normal curve between these two points is found using the normal distribution table.

$$\text{For } \hat{p} = .87: \quad z = \frac{.87 - .85}{.02524876} = .79$$

$$\text{For } \hat{p} = .90: \quad z = \frac{.90 - .85}{.02524876} = 1.98$$

Thus, the probability that $\hat{p}$ is between .87 and .90 is given by the area under the standard normal curve between $z = .79$ and $z = 1.98$. This area is shown in Figure 7.17. The required probability is

$$P(.87 < p < .90) = P(.79 < z < 1.98)$$
$$= P(0 < z < 1.98) - P(0 < z < .79)$$
$$= .4761 - .2852 = \textbf{.1909}$$

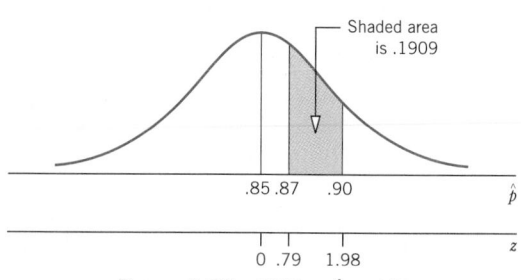

Figure 7.17 $P(.87 < \hat{p} < .90)$

Thus, the probability is .1909 that the proportion of women in a random sample of 200 who hold the given view is between .87 and .90. ■

Probability of $\hat{p}$ being less than a certain value.

EXAMPLE 7–11 Maureen Webster, who is contesting for the mayor's position in a large city, claims that she is favored by 53% of all eligible voters of that city. Assume that this claim is true. What is the probability that in a random sample of 400 registered voters taken from this city, less than 49% will favor Maureen Webster?

Solution Let p be the proportion of all eligible voters who favor Maureen Webster. Then,

$$p = .53 \quad \text{and} \quad q = 1 - p = 1 - .53 = .47$$

Then, the mean of the sampling distribution of the sample proportion, $\hat{p}$, is

$$\mu_{\hat{p}} = p = .53$$

The population of all voters is large (because of the city being large) and the sample size is small compared to the population. Consequently, we can assume that $n/N \le .05$. Hence, the standard deviation of $\hat{p}$ is calculated as

$$\sigma_{\hat{p}} = \sqrt{\frac{pq}{n}} = \sqrt{\frac{(.53)\,(.47)}{400}} = .02495496$$

From the central limit theorem, the shape of the sampling distribution of $\hat{p}$ is approximately normal. The probability that $\hat{p}$ is less than .49 is given by the area under the normal distribution curve for $\hat{p}$ to the left of $\hat{p} = .49$, as shown in Figure 7.18. The z value for $\hat{p} = .49$ is

$$z = \frac{\hat{p} - p}{\sigma_{\hat{p}}} = \frac{.49 - .53}{.02495496} = -1.60$$

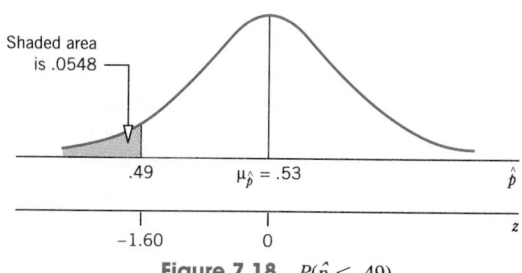

Shaded area is .0548

.49 $\mu_{\hat{p}} = .53$ $\hat{p}$

−1.60 0 z

Figure 7.18 $P(\hat{p} < .49)$

Thus, the required probability is

$$P(\hat{p} < .49) = P(z < -1.60) = .5 - P(-1.60 < z < 0)$$
$$= .5 - .4452 = \mathbf{.0548}$$

Hence, the probability that less than 49% of the voters in a random sample of 400 will favor Maureen Webster is .0548. ▬

EXERCISES

Concepts and Procedures

7.83 If all possible samples of the same (large) size are selected from a population, what percentage of all sample proportions will be within 2.0 standard deviations of the population proportion?

7.84 If all possible samples of the same (large) size are selected from a population, what percentage of all sample proportions will be within 3.0 standard deviations of the population proportion?

7.85 For a population, $N = 30{,}000$ and $p = .60$. Find the z value for each of the following for $n = 100$.

    a. $\hat{p} = .56$ **b.** $\hat{p} = .68$ **c.** $\hat{p} = .53$ **d.** $\hat{p} = .65$

7.86 For a population, $N = 18{,}000$ and $p = .24$. Find the z value for each of the following for $n = 70$.

    a. $\hat{p} = .26$ **b.** $\hat{p} = .32$ **c.** $\hat{p} = .17$ **d.** $\hat{p} = .20$

Applications

7.87 In a recent *U.S. News*/CNN poll conducted by the Gallup Organization, 32% of adults said that it was *permissible* to take home pens and pencils from their employers (*U.S. News & World Report*, July 17, 1995). Assume that this percentage is true for the current population of all U.S. adults and let $\hat{p}$ be the proportion in a random sample of 400 adults who hold this view. Find the probability that the value of $\hat{p}$ will be

    a. between .27 and .36   **b.** less than .30

7.88   A survey of all medium- and large-sized corporations showed that 65% of them offer retirement plans to their employees. Let $\hat{p}$ be the proportion in a random sample of 50 such corporations that offer retirement plans to their employees. Find the probability that the value of $\hat{p}$ will be

    a.   between .54 and .61   **b.**   more than .71

7.89   According to a Centers for Disease Control and Prevention estimate, 70% of the adult cigarette smokers in the United States want to quit smoking (*Statistical Bulletin*, April–June 1996). Assume that this result holds true for the current population of all adult smokers and that $\hat{p}$ is the proportion in a random sample of 100 smokers who want to quit smoking. Find the probability that the value of $\hat{p}$ will be

    a.   between .68 and .73   **b.**   more than .67

7.90   Dartmouth Distribution Warehouse makes deliveries of a large number of products to its customers. It is known that 85% of all the orders it receives from its customers are delivered on time. Let $\hat{p}$ be the proportion of orders in a random sample of 100 that are delivered on time. Find the probability that the value of $\hat{p}$ will be

    a.   between .81 and .88   **b.**   less than .87

7.91   Brooklyn Corporation manufactures computer diskettes. The machine that is used to make these diskettes is known to produce 6% defective diskettes. The quality control inspector selects a sample of 100 diskettes every week and inspects them for being good or defective. If 8% or more of the diskettes in the sample are defective, the process is stopped and the machine is readjusted. What is the probability that based on a sample of 100 diskettes the process will be stopped to readjust the machine?

7.92   Mong Corporation makes auto batteries. The company claims that 80% of its LL70 batteries are good for 70 months or more. Assume that this claim is true. Let $\hat{p}$ be the proportion in a sample of 100 such batteries that are good for 70 months or more.

    a.   What is the probability that this sample proportion is within .05 of the population proportion?

    b.   What is the probability that this sample proportion is lower than the population proportion by .06 or more?

    c.   What is the probability that this sample proportion is greater than the population proportion by .07 or more?

GLOSSARY

Central limit theorem   The theorem from which it is inferred that for a large sample size ($n \geq 30$), the shape of the sampling distribution of $\bar{x}$ is approximately normal. Also, by the same theorem, the shape of the sampling distribution of $\hat{p}$ is approximately normal for a sample for which $np > 5$ and $nq > 5$.

Consistent estimator   A sample statistic with a standard deviation that decreases as the sample size increases.

Estimator   The sample statistic that is used to estimate a population parameter.

Mean of $\hat{p}$ The mean of the sampling distribution of $\hat{p}$, denoted by $\mu_{\hat{p}}$, is equal to the population proportion p.

Mean of $\bar{x}$ The mean of the sampling distribution of $\bar{x}$, denoted by $\mu_{\bar{x}}$, is equal to the population mean μ.

Nonsampling errors   The errors that occur during the collection, recording, and tabulation of data.

Population distribution   The probability distribution of the population data.

Population proportion p   The ratio of the number of elements in a population with a specific characteristic to the total number of elements in the population.

Sample proportion $\hat{p}$ The ratio of the number of elements in a sample with a specific characteristic to the total number of elements in that sample.

Sampling distribution of $\hat{p}$ The probability distribution of all the values of $\hat{p}$ calculated from all possible samples of the same size selected from a population.

Sampling distribution of $\bar{x}$ The probability distribution of all the values of $\bar{x}$ calculated from all possible samples of the same size selected from a population.

Sampling error   The difference between the value of a sample statistic calculated from a random sample and the value of the corresponding population parameter. This type of error occurs due to chance.

Standard deviation of $\hat{p}$ The standard deviation of the sampling distribution of $\hat{p}$, denoted by $\sigma_{\hat{p}}$, is equal to $\sqrt{pq/n}$ when $n/N \leq .05$.

Standard deviation of $\bar{x}$ The standard deviation of the sampling distribution of $\bar{x}$, denoted by $\sigma_{\bar{x}}$, is equal to $\sigma/\sqrt{n}$ when $n/N \leq .05$.

Unbiased estimator   An estimator with an expected value (or mean) that is equal to the value of the corresponding population parameter.

KEY FORMULAS

1. **Mean of the sampling distribution of $\bar{x}$**

$$\mu_{\bar{x}} = \mu$$

2. **Standard deviation of the sampling distribution of $\bar{x}$ when $n/N \leq .05$**

$$\sigma_{\bar{x}} = \frac{\sigma}{\sqrt{n}}$$

3. **The z value for a value of $\bar{x}$**

$$z = \frac{\bar{x} - \mu}{\sigma_{\bar{x}}} \quad \text{where } \sigma_{\bar{x}} = \frac{\sigma}{\sqrt{n}}$$

4. **Population proportion**

$$p = \frac{X}{N}$$

where

      N = Total number of elements in the population

      X = Number of elements in the population that possess a specific characteristic

5. **Sample proportion**

$$\hat{p} = \frac{x}{n}$$

where

      n = Total number of elements in the sample

      x = Number of elements in the sample that possess a specific characteristic

6. **Mean of the sampling distribution of $\hat{p}$**

$$\mu_{\hat{p}} = p$$

7. **Standard deviation of the sampling distribution of $\hat{p}$ when $n/N \leq .05$**

$$\sigma_{\hat{p}} = \sqrt{\frac{pq}{n}}$$

8. **The z value for a value of $\hat{p}$**

$$z = \frac{\hat{p} - p}{\sigma_{\hat{p}}} \quad \text{where } \sigma_{\hat{p}} = \sqrt{\frac{pq}{n}}$$

SUPPLEMENTARY EXERCISES

7.93 The print on the package of 100-watt General Electric soft-white light bulbs claims that these bulbs have an average life of 750 hours. Assume that the lives of all such bulbs have a normal distribution with a mean of 750 hours and a standard deviation of 50 hours. Let $\bar{x}$ be the mean life of a random sample of 25 such bulbs. Find the mean and standard deviation of $\bar{x}$ and describe the shape of its sampling distribution.

7.94 The weekly earnings of all 2480 employees of a company have a distribution that is skewed to the right with its mean equal to $438 and standard deviation equal to $40. Let $\bar{x}$ be the mean weekly earnings of a random sample of 100 employees selected from this company. Calculate the mean and standard deviation of $\bar{x}$ and comment on the shape of its sampling distribution.

7.95 Refer to Exercise 7.93. The print on the package of 100-watt General Electric soft-white light bulbs says that these bulbs have an average life of 750 hours. Assume that the lives of all such bulbs have a normal distribution with a mean of 750 hours and a standard deviation of 50 hours. Find the probability that the mean life of a random sample of 25 such bulbs will be

    a. greater than 735 hours   **b.** between 725 and 740 hours
    c. within 15 hours of the population mean
    d. lower than the population mean by 20 hours or more

7.96 Refer to Exercise 7.94. The weekly earnings of all 2480 employees of a company have a distribution that is skewed to the right with its mean equal to $438 and standard deviation equal to $40. Find the probability that the mean weekly earnings of a random sample of 100 employees selected from this company will be

    a. less than $433   **b.** between $435 and $440
    c. within $7 of the population mean
    d. greater than the population mean by $2.50 or more

7.97 According to a Priority Management survey, adults spend an average of 10 hours a day at work and commuting. Let the daily work and commute times for all adults have a mean of 10 hours and a standard deviation of 1.8 hours. Find the probability that the mean of the daily work and commute times for a random sample of 80 adults will be

    a. greater than 10.45 hours   **b.** between 9.75 and 10.50 hours
    c. within .25 hours of the population mean
    d. lower than the population mean by .50 hours or more

7.98 A machine at Keats Corporation fills 64-ounce detergent jugs. The probability distribution of the amount of detergent in these jugs is normal with a mean of 64 ounces and a standard deviation of .4 ounces. The quality control inspector takes a sample of 16 jugs once a week and measures the amount of detergent in these jugs. If the mean of this sample is either less than 63.75 ounces or greater than 64.25 ounces, the inspector concludes that the machine needs an adjustment. What is the probability that based on a sample of 16 jugs the inspector will conclude that the machine needs an adjustment when actually it does not?

7.99 Ten percent of all items produced on a machine are defective. Let $\hat{p}$ be the proportion of defective items in a random sample of 80 items selected from the production line. Calculate the mean and standard deviation of $\hat{p}$ and describe the shape of its sampling distribution.

7.100 Seventy percent of adults favor some kind of government control on the prices of medicines. Assume that this percentage is true for the current population of all adults. Let $\hat{p}$ be the proportion of adults in a random sample of 400 who favor government control on the prices of medicines. Calculate the mean and standard deviation of $\hat{p}$ and describe the shape of its sampling distribution.

7.101 Refer to Exercise 7.100. Seventy percent of adults favor some kind of government control on the prices of medicines. Assume that this percentage is true for the current population of all adults.
    a. Find the probability that the proportion of adults in a random sample of 400 who favor some kind of government control on the prices of medicines is
        i.   less than .65   ii.   between .73 and .76
    b. What is the probability that the proportion of adults in a random sample of 400 who favor some kind of government control is within .06 of the population proportion?
    c. What is the probability that the sample proportion is greater than the population proportion by .05 or more?

7.102 According to a recent *U.S. News*/CNN poll conducted by the Gallup Organization, 62% of American adults are frightened always, most of the time, or sometimes when they fly on small commuter airplanes (*U.S. News and World Report*, April 17, 1995). Assume that this result holds true for the current population of all American adults. Let $\hat{p}$ be the proportion in a random sample of 100 adults who are frightened of flying on small commuter airplanes.
    a. What is the probability that this sample proportion is within .08 of the population proportion?
    b. What is the probability that this sample proportion is not within .08 of the population proportion?
    c. What is the probability that this sample proportion is lower than the population proportion by .10 or more?
    d. What is the probability that this sample proportion is greater than the population proportion by .09 or more?

*7.103 Let μ be the mean annual salary of major league baseball players for 1996. Assume that the standard deviation of the salaries of these players is $50,000. What is the probability that the 1996 mean salary of a random sample of 100 baseball players was within $10,000 of the population mean, μ? Assume that $n/N \leq .05$.

*7.104 The test scores for 300 students were entered into a computer, analyzed, and stored in a file. Unfortunately, someone accidentally erased a major portion of this file from the computer. The only information that is available is that 30% of the scores were below 65 and 15% of the scores were above 90. Assuming the scores are normally distributed, find their mean and standard deviation.

*7.105 A chemist has a 10-gallon sample of river water taken just downstream from the outflow of a chemical plant. He is concerned about the concentration, c (in parts per million), of a certain toxic substance in the water. He wants to take several measurements, find the mean concentration of the toxic substance for this sample, and have a 95% chance of being within .5 parts per million of the true mean value of c. If the concentration of the toxic substance in all measurements is normally distributed with $\sigma = .8$ parts per million, how many measurements are necessary to achieve this goal?

*7.106 A television reporter is covering the election for mayor of a large city and will conduct an exit poll (interviews with voters immediately after they vote) to make an early prediction of the outcome. Assume that the eventual winner of the election will get 60% of the votes.
    a. What is the probability that a prediction based on an exit poll of a random sample of 25 voters will be correct? In other words, what is the probability that 13 or more of the 25 voters in the sample will have voted for the eventual winner?
    b. How large a sample would the reporter have to take so that the probability of correctly predicting the outcome would be .95 or more?

*7.107   A city is planning to build a hydroelectric power plant. A local newspaper found that 53% of the voters in this city favor the construction of this plant. Assume that this result holds true for the population of all voters in this city.

    a.   What is the probability that more than 50% of the voters in a random sample of 200 voters selected from this city will favor the construction of this plant?

    b.   A politician would like to take a random sample of voters in which over 50% would favor the plant construction. How large a sample should be selected so that the politician is 95% sure of this outcome?

SELF-REVIEW TEST

1.   A sampling distribution is the probability distribution of

    a.   a population parameter   **b.**   a sample statistic   **c.**   any random variable

2.   Nonsampling errors are

    a.   the errors that occur because the sample size is too large in relation to the population size
    b.   the errors made while collecting, recording, and tabulating data
    c.   the errors that occur because an untrained person conducts the survey

3.   A sampling error is

    a.   the difference between the value of a sample statistic based on a random sample and the value of the corresponding population parameter
    b.   the error made while collecting, recording, and tabulating data
    c.   the error that occurs because the sample is too small

4.   The mean of the sampling distribution of $\bar{x}$ is always equal to

    a. μ **b.** $\mu - 5$ **c.** $\sigma/\sqrt{n}$

5.   The condition for the standard deviation of the sample mean to be $\sigma/\sqrt{n}$ is that

    a. $np > 5$ **b.** $n/N \leq .05$ **c.** $n > 30$

6.   The standard deviation of the sampling distribution of the sample mean decreases when

    a. x increases   **b.** n increases   **c.** n decreases

7.   When samples are selected from a normally distributed population, the sampling distribution of the sample mean has a normal distribution

    a.   if $n \geq 30$ **b.**   if $n/N \leq .05$ **c.**   all the time

8.   When samples are selected from a nonnormally distributed population, the sampling distribution of the sample mean has an approximate normal distribution

    a.   if $n \geq 30$ **b.**   if $n/N \leq .05$ **c.**   always

9.   In a sample of 200 customers of a mail-order company, 148 are found to be satisfied with the service they receive from the company. The proportion of customers in this sample who are satisfied with the company's service is

    a.   .26   **b.**   .74   **c.**   .148

10.   The mean of the sampling distribution of $\hat{p}$ is always equal to

    a. p **b.** μ **c.** $\hat{p}$

11. The condition for the standard deviation of the sampling distribution of the sample proportion to be $\sqrt{pq/n}$ is

    a. $np > 5$ and $nq > 5$ **b.** $n > 30$ **c.** $n/N \leq .05$

12. The sampling distribution of $\hat{p}$ is (approximately) normal if

    a. $np > 5$ and $nq > 5$ **b.** $n > 30$ **c.** $n/N \leq .05$

13. Briefly state and explain the central limit theorem.

14. The weights of all students at a large university have an approximate normal distribution with a mean of 145 pounds and a standard deviation of 18 pounds. Let $\bar{x}$ be the mean weight of a random sample of certain students selected from this university. Calculate the mean and standard deviation of $\bar{x}$ and describe the shape of its sampling distribution for a sample size of

    a. 25   **b.** 100

15. The time taken to run a certain road race for all participants has an unknown distribution with a mean of 47 minutes and a standard deviation of 8.4 minutes. Let $\bar{x}$ be the mean time taken to run this road race for a random sample of certain participants. Find the mean and standard deviation of $\bar{x}$ and describe the shape of its sampling distribution for a sample size of

    a. 20   **b.** 70

16. According to an IRS study, it takes an average of 336 minutes for taxpayers to prepare, copy, and mail a 1040 tax form. Assume that the time taken by all taxpayers to prepare, copy, and mail the 1040 tax form has an unknown distribution with a mean of 336 minutes and a standard deviation of 70 minutes. Find the probability that the mean time taken to prepare, copy, and mail this tax form for a random sample of 60 taxpayers would be

    a.   Between 320 and 330 minutes
    b.   more than 355 minutes
    c.   less than 345 minutes
    d.   between 325 and 355 minutes

17. At Jen and Perry Ice Cream Company, the machine that fills 1-pound cartons of Top Flavor ice cream is set to dispense 16 ounces of ice cream in every carton. However, some cartons contain slightly less than and some contain slightly more than 16 ounces of ice cream. The amounts of ice cream in all such cartons have a normal distribution with a mean of 16 ounces and a standard deviation of .18 ounces.

    a.   Find the probability that the mean amount of ice cream in a random sample of 16 such cartons will be
        i.   between 15.90 and 15.95 ounces
        ii.   less than 15.95 ounces
        iii.   more than 15.97 ounces
    b.   What is the probability that the mean amount of ice cream in a random sample of 16 such cartons will be within .10 ounces of the population mean?
    c.   What is the probability that the mean amount of ice cream in a random sample of 16 such cartons will be lower than the population mean by .135 ounces or more?

18. According to a poll conducted by Celinda Lake of Lake Research and Ed Goeas of the Tarrance Group for *U.S. News & World Report* in March 1996, 92% of adults think that television contributes to violence in the United States (*U.S. News & World Report*, April 15, 1996). Suppose this result holds true for the current population of all U.S. adults. Let $\hat{p}$ be the proportion in a random sample of U.S. adults who hold this opinion. Calculate the mean and standard deviation of $\hat{p}$ and describe the shape of its sampling distribution when the sample size is

    a. 50   **b.** 200   **c.** 900

19. The Gallup Organization conducted a poll for Phi Delta Kappa on the attitudes of U.S. adults toward public schools (*Phi Delta Kappan*, September 1996). Sixty-six percent of these adults favored

a community service requirement for high school graduation. Assume that this percentage is true for the current population of all U.S. adults.

a. Find the probability that the proportion of adults in a random sample of 400 who will favor the community service requirement for high school graduation is

    i.   more than .68   ii.   between .63 and .70

    iii. less than .68   iv. between .61 and .64

b. What is the probability that the proportion of adults in a random sample of 400 who will hold this view is within .05 of the population proportion?

c. What is the probability that the sample proportion for a random sample of 400 adults is lower than the population proportion by .04 or more?

d. What is the probability that the sample proportion for a random sample of 400 adults is greater than the population proportion by .03 or more?

USING MINITAB

This MINITAB section describes how to construct the sampling distribution of the sample mean by simulating the sampling procedure. Illustration M7–1 describes the procedure to do so. Note that MINITAB FOR WINDOWS does not contain this procedure. It can only be done by using the MINITAB COMMAND LANGUAGE as shown in the following illustration.

ILLUSTRATION M7–1   Consider the experiment of rolling a die. Obtain 1000 samples, each containing the results of 40 rolls of the die. Then find the sample mean for each of these 1000 samples and prepare a sampling distribution of $\bar{x}$ by using these 1000 sample means.

Solution   Each roll of the die has six possible outcomes: 1 spot, 2 spots, 3 spots, 4 spots, 5 spots, and 6 spots. Rolling a die can be simulated by randomly selecting a number from 1, 2, 3, 4, 5, and 6. The MINITAB commands given below will produce 1000 samples, each sample containing the results of 40 rolls. Note that a roll of the die is simulated by randomly selecting a number from 1 through 6. The set of values in the first row of the 40 columns gives the first sample, the set of values in the second row of the 40 columns produces the second sample, and so on. Thus, you obtain 1000 samples, each containing the results of 40 rolls of the die.

```
MTB  > RANDOM 1000 C1–C40;
SUBC > INTEGER 1 6.
```

To calculate the sample means for the 1000 samples you obtained, use the **RMEANS** command, as shown below. The **RMEANS** command calculates the means of rows. The following command will calculate the means of 1000 samples given in 1000 rows of the 40 columns and will put these means in column C41.

```
MTB > RMEANS C1–C40 PUT IN C41
```

Now, by using the following MINITAB commands, you can calculate the mean and standard deviation and construct the histogram of the sample means given in column C41.

```
MTB > MEAN C41
MTB > STDEV C41
MTB > HISTOGRAM C41
```

By combining all the above commands, the mean, the standard deviation, and the histogram of the sample means of 1000 samples (each sample containing the results of 40 rolls of a die) are obtained. One such MINITAB output is shown in Figure 7.19 as are all the MINITAB commands mentioned in this section.

The frequency histogram of Figure 7.19 can be used to construct the sampling distribution of the sample mean $\bar{x}$. The shape of the sampling distribution of $\bar{x}$ for these 1000 samples will look exactly

371

like the histogram of Figure 7.19. The mean and standard deviation of this sampling distribution of $\bar{x}$ are 3.4857 and .26677, respectively. That is,

$$\mu_{\bar{x}} = 3.4857 \qquad \text{and} \qquad \sigma_{\bar{x}} = .26677$$

Figure 7.19 Mean, standard deviation, and histogram of $\bar{x}$.

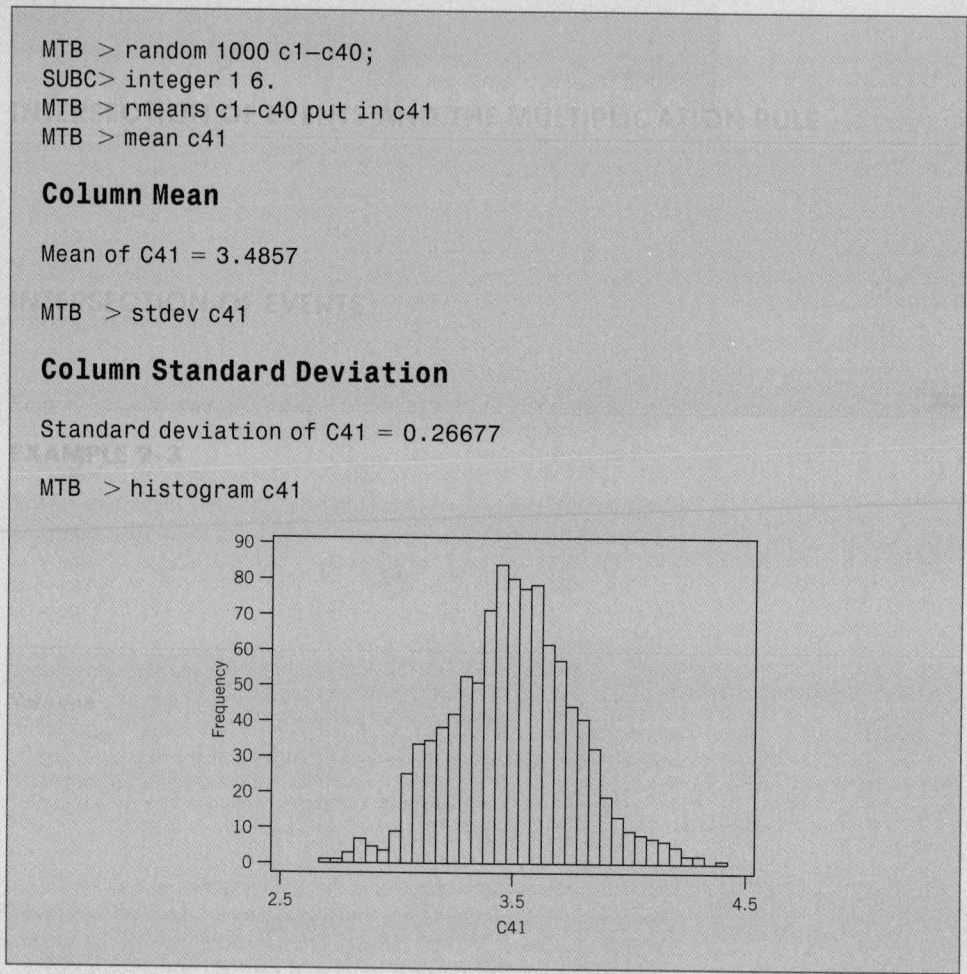

```
MTB > random 1000 c1–c40;
SUBC> integer 1 6.
MTB > rmeans c1–c40 put in c41
MTB > mean c41
```

Column Mean

```
Mean of C41 = 3.4857

MTB  > stdev c41
```

Column Standard Deviation

```
Standard deviation of C41 = 0.26677

MTB  > histogram c41
```

Note that each time you repeat this experiment, you will obtain different values of $\mu_{\bar{x}}$ and $\sigma_{\bar{x}}$ and a different histogram. ▇

COMPUTER ASSIGNMENTS

M7.1 Using MINITAB, create 200 samples, each containing the results of 30 rolls of a die. Calculate the means of these 200 samples. Construct the histogram and calculate the mean and standard deviation of these 200 sample means.

M7.2 Using the following MINITAB commands, create 150 samples each containing the results of selecting 35 numbers from 1 through 100.

```
MTB  > RANDOM 150 C1–C35;
SUBC > INTEGER 1 100.
```

Calculate the means of these 150 samples and put those in column C36. Construct the histogram and calculate the mean and standard deviation of these 150 sample means.

8

ESTIMATION OF THE MEAN AND PROPORTION

N ow we are entering that part of statistics called *inferential statistics*. In Chapter 1 inferential statistics was defined as the part of statistics that helps us to make decisions about some characteristics of a population based on sample information. In other words, inferential statistics uses the sample results to make decisions and draw conclusions about the population from which the sample is drawn. Estimation is the first topic to be considered in our discussion of inferential statistics. Estimation and hypothesis testing (discussed in Chapter 9) taken together are usually referred to as inference making. This chapter explains how to estimate the population mean and population proportion for a single population.

8.1 ESTIMATION: AN INTRODUCTION

Estimation is a procedure by which numerical value or values are assigned to a population parameter based on the information collected from a sample.

> **ESTIMATION**
>
> The assignment of value(s) to a population parameter based on a value of the corresponding sample statistic is called *estimation*.

In inferential statistics, μ is called the *true population mean* and p is called the *true population proportion*. There are many other population parameters such as the median, mode, variance, and standard deviation.

The following are a few examples of estimation: an auto company may want to estimate the mean fuel consumption for a particular model of a car; a manager may want to estimate the average time taken by new employees to learn a job; the U.S. Census Bureau may want to find the mean housing expenditure per month incurred by households; and the AWAH (Association of Wives of Alcoholic Husbands) may want to find the proportion (or percentage) of all husbands who are alcoholic.

The examples about estimating the mean fuel consumption, estimating the average time taken to learn a job by new employees, and estimating the mean housing expenditure per month incurred by households are illustrations of estimating the *true population mean*, μ. The example about estimating the proportion (or percentage) of all husbands who are alcoholic is an illustration of estimating the *true population proportion, p*.

If we can conduct a *census* (a survey that includes the entire population) each time we want to find the value of a population parameter, then the estimation procedures explained in this and subsequent chapters are not needed. For example, if the U.S. Census Bureau can contact every household living in the United States to find the mean housing expenditure incurred by households, the result of the survey (which will actually be a census) will give the value of μ and the procedures learned in this chapter will not be needed. However, it is too expensive, very time consuming, or virtually impossible to contact every member of a population to collect information to find the true value of a population parameter. Therefore, we usually take a sample from the population and calculate the value of the appropriate sample statistic. Then we assign a value or values to the corresponding population parameter based on the value of the sample statistic. This chapter (and subsequent chapters) explains how to assign values to population parameters based on the values of sample statistics.

For example, to estimate the mean time taken to learn a certain job by new employees, the manager will take a sample of new employees and record the time taken by each of these employees to learn the job. Using this information, he or she will calculate the sample mean, $\bar{x}$. Then, based on the value of $\bar{x}$, he or she will assign certain values to μ. As another example, to estimate the mean housing expenditure per month incurred by all households in the United States, the U.S. Census Bureau will take a sample of certain households, collect the information on the housing expenditure that each of these households incurs per month, and compute the value of the sample mean, $\bar{x}$. Based on this value of $\bar{x}$, the bureau will then assign values to the population mean, μ. Similarly, the AWAH will take a sample of husbands and determine the value of the sample proportion, $\hat{p}$, which represents the proportion of husbands in the sample who are alcoholic. Then, using this value of the sample proportion, $\hat{p}$, AWAH will assign values to the population proportion, p.

The value(s) assigned to a population parameter based on the value of a sample statistic is called an **estimate** of the population parameter. For example, suppose the manager takes a sample of 40 new employees and finds that the mean time, $\bar{x}$, taken to learn this job for these employees is 5.5 hours. If he or she assigns this value to the population mean, then 5.5 hours will be called an estimate of μ. The sample statistic used to estimate a population parameter is called an **estimator**. Thus, the sample mean, $\bar{x}$, is an estimator of the population mean, μ, and the sample proportion, $\hat{p}$, is an estimator of the population proportion, p.

ESTIMATE AND ESTIMATOR

The value(s) assigned to a population parameter based on the value of a sample statistic is called an *estimate*. The sample statistic used to estimate a population parameter is called an *estimator*.

The estimation procedure involves the following steps.

1. Select a sample.
2. Collect the required information from the members of the sample.
3. Calculate the value of the sample statistic.
4. Assign value(s) to the corresponding population parameter.

8.2 POINT AND INTERVAL ESTIMATES

An estimate may be a point estimate or an interval estimate. These two types of estimates are described in this section.

8.2.1 A POINT ESTIMATE

If we select a sample and compute the value of the sample statistic for this sample, this value gives the **point estimate** of the corresponding population parameter.

> **POINT ESTIMATE**
>
> The value of a sample statistic that is used to estimate a population parameter is called a *point estimate*.

Thus, the value computed for the sample mean, $\bar{x}$, from a sample is a point estimate of the corresponding population mean, μ. For the example mentioned earlier, suppose the U.S. Census Bureau takes a sample of 10,000 households and determines that the mean housing expenditure per month, $\bar{x}$, for this sample is $874. Then, using $\bar{x}$ as a point estimate of μ, the bureau can state that the mean housing expenditure per month, μ, for all households is about $874. This procedure is called **point estimation**.

Usually, whenever we use point estimation, we calculate the **margin of error** associated with that point estimation. For the estimation of the population mean, the margin of error is calculated as follows.

$$\text{Margin of error} = \pm 1.96 \, \sigma_{\bar{x}} \qquad \text{or} \qquad \pm 1.96 \, s_{\bar{x}}$$

That is, we find the standard deviation of the sample mean and multiply it by 1.96. Here $s_{\bar{x}}$ is a point estimator of $\sigma_{\bar{x}}$, and it will be discussed later in this chapter.

Each sample selected from a population is expected to yield a different value of the sample statistic. Thus, the value assigned to a population mean, μ, based on a point estimate depends on which of the samples is drawn. Consequently, the point estimate assigns a value to μ that almost always differs from the true value of the population mean.

8.2.2 AN INTERVAL ESTIMATE

In the case of **interval estimation**, instead of assigning a single value to a population parameter, an interval is constructed around the point estimate and then a probabilistic statement that this interval contains the corresponding population parameter is made.

> **INTERVAL ESTIMATION**
>
> In *interval estimation,* an interval is constructed around the point estimate, and it is stated that this interval is likely to contain the corresponding population parameter.

For the example about the mean housing expenditure (Section 8.2.1), instead of saying that the mean housing expenditure per month for all households is $874, we obtain an interval by subtracting a number from $874 and adding the same number to $874. Then we state that this interval contains the population mean, μ. For purposes of illustration, suppose we subtract $110 from $874 and add $110 to $874. Consequently, we obtain the interval ($874 − $110) to ($874 + $110) or $764 to $984. Then we state that the interval $764 to $984 is likely to contain the population mean, μ, and that the mean housing expenditure per month for all households in the United States is between $764 and $984. This procedure is called *interval estimation*. The value $764 is called the *lower limit* of the interval and $984 is called the *upper limit* of the interval. Figure 8.1 illustrates the concept of interval estimation.

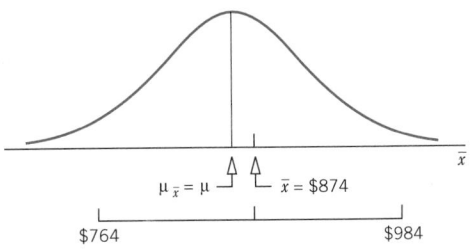

Figure 8.1 Interval estimation.

The question arises, what number should we subtract from and add to a point estimate to obtain an interval estimate? The answer to this question depends on two considerations:

1. The standard deviation $\sigma_{\bar{x}}$ of the sample mean, $\bar{x}$
2. The level of confidence to be attached to the interval

First, the larger the standard deviation of $\bar{x}$, the greater the number subtracted from and added to the point estimate. Thus, it is obvious that if the range over which $\bar{x}$ can assume values is larger, the interval constructed around $\bar{x}$ must be wider to include μ.

Second, the quantity subtracted and added must be larger if we want to have a higher confidence in our interval. We always attach a probabilistic statement to the interval estimation. This probabilistic statement is given by the **confidence level**. An interval that is constructed based on this confidence level is called a **confidence interval**.

CONFIDENCE LEVEL AND CONFIDENCE INTERVAL

Each interval is constructed with regard to a given *confidence level* and is called a *confidence interval*. The confidence level associated with a confidence interval states how much confidence we have that this interval contains the true population parameter. The confidence level is denoted by $(1 - \alpha)100\%$.

The confidence level is denoted by $(1 - \alpha)100\%$ (α is the Greek letter *alpha*). When expressed as probability, it is called the *confidence coefficient* and is denoted by $1 - \alpha$. In passing, note that α is called the *significance level*, which will be explained in detail in Chapter 9.

Although any value of the confidence level can be chosen to construct a confidence interval, the more common values are 90%, 95%, and 99%. The corresponding confidence coefficients are .90, .95, and .99. The next section describes how to actually construct a confidence interval for the population mean for a large sample.

8.3 INTERVAL ESTIMATION OF A POPULATION MEAN: LARGE SAMPLES

This section explains how to construct a confidence interval for the population mean μ when the sample size is large.[1] Recall from the discussion in Chapter 7 that in the case of $\bar{x}$, the

---

[1]Some statisticians prefer to discuss estimation and tests of hypotheses based on whether or not the population standard deviation σ is known. However, we prefer to use the large sample and small sample criteria. The reason is that σ is almost always unknown. Hence, discussing estimation and tests of hypotheses based on large and small samples makes more sense than whether or not σ is known.

sample size is considered to be large when n is 30 or larger. According to the central limit theorem, for a large sample the sampling distribution of the sample mean, $\bar{x}$, is (approximately) normal irrespective of the shape of the population from which the sample is drawn. Therefore, *when the sample size is 30 or larger, we will use the normal distribution to construct a confidence interval for* μ. We also know from Chapter 7 that the standard deviation of $\bar{x}$ is $\sigma_{\bar{x}} = \sigma/\sqrt{n}$. However, if the population standard deviation, σ, is not known, then we use the sample standard deviation, s, for σ. Consequently, we use

$$s_{\bar{x}} = \frac{s}{\sqrt{n}}$$

for $\sigma_{\bar{x}} = \sigma/\sqrt{n}$. Note that the value of $s_{\bar{x}}$ is a point estimate of $\sigma_{\bar{x}}$.

CONFIDENCE INTERVAL FOR μ FOR LARGE SAMPLES

The $(1 - \alpha)100\%$ *confidence interval* for μ is

$$\bar{x} \pm z\sigma_{\bar{x}} \qquad \text{if } \sigma \text{ is known}$$

$$\bar{x} \pm zs_{\bar{x}} \qquad \text{if } \sigma \text{ is not known}$$

where $\sigma_{\bar{x}} = \sigma/\sqrt{n}$ and $s_{\bar{x}} = s/\sqrt{n}$.

The value of z used here is read from the standard normal distribution table for the given confidence level.

The quantity $z\sigma_{\bar{x}}$ (or $zs_{\bar{x}}$ when σ is not known) in the confidence interval formula is called the **maximum error of estimate** and is denoted by E.

MAXIMUM ERROR OF ESTIMATE FOR μ

The *maximum error of estimate for* μ, denoted by E, is the quantity that is subtracted from and added to the value of $\bar{x}$ to obtain a confidence interval for μ. Thus,

$$E = z\sigma_{\bar{x}} \qquad \text{or} \qquad zs_{\bar{x}}$$

The value of z in the confidence interval formula is obtained from the standard normal distribution table (Table VII of Appendix C) for the given confidence level. To illustrate, suppose we want to construct a 95% confidence interval for μ. A 95% confidence level means that the total area under the normal curve for $\bar{x}$ between two points (at the same distance) on different sides of μ is 95% or .95, as shown in Figure 8.2. To find the value of z, we first divide the given confidence coefficient by 2. Then we look for this number in the body of the normal table. The corresponding value of z is the value we use in the confidence interval. Thus, to find the z value for a 95% confidence level, we perform the following two steps.

1. First we divide .95 by 2, which gives .4750.
2. Then we locate .4750 in the body of the normal distribution table and record the corresponding value of z. This value of z is 1.96.

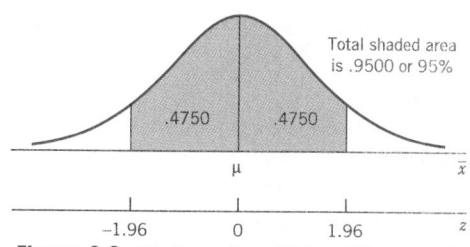

Figure 8.2 Finding z for a 95% confidence level.

For a $(1 - \alpha)100\%$ confidence level, the area between $-z$ and z is $1 - \alpha$. Because the total area under the normal curve is 1.0, the total area under the curve in the two tails is α. This, as mentioned earlier, is called the significance level. In the example of Figure 8.2, $\alpha = 1 - .95 = .05$. Therefore, as shown in Figure 8.3, the area under the curve in each of the two tails is $\alpha/2$. Thus, the value of z associated with a $(1 - \alpha)100\%$ confidence level is sometimes denoted by $z_{\alpha/2}$. However, this text will denote this value simply by z.

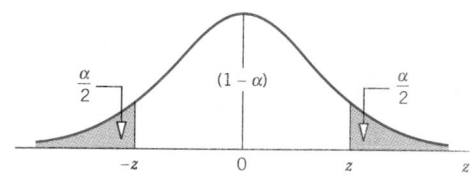

Figure 8.3 Area in the tails.

Example 8–1 describes the procedure to construct a confidence interval for μ for a large sample.

Point estimate and confidence interval for μ: σ known and $n > 30$.

EXAMPLE 8-1 A publishing company has just published a new college textbook. Before the company decides the price at which to sell this textbook, it wants to know the average price of all such textbooks in the market. The research department at the company took a sample of 36 such textbooks and collected information on their prices. This information produced a mean price of $54.40 for this sample. It is known that the standard deviation of the prices of all such textbooks is $4.50.

(a) What is the point estimate of the mean price of all such college textbooks? What is the margin of error for this estimate?

(b) Construct a 90% confidence interval for the mean price of all such college textbooks.

Solution From the given information,

$$n = 36, \quad \bar{x} = \$54.40, \quad \text{and} \quad \sigma = \$4.50$$

The standard deviation of $\bar{x}$ is

$$\sigma_{\bar{x}} = \frac{\sigma}{\sqrt{n}} = \frac{4.50}{\sqrt{36}} = \$.75$$

(a) The point estimate of the mean price of all such college textbooks is $54.40, that is,

$$\text{Point estimate of } \mu = \bar{x} = \$54.40$$

The margin of error associated with this point estimate of μ is

$$\text{Margin of error} = \pm 1.96 \ \sigma_{\bar{x}} = \pm 1.96 \ (.75) = \pm \mathbf{\$1.47}$$

The margin of error states that the mean price of all such college textbooks is $54.40, give or take $1.47. Note that the margin of error is simply the maximum error of estimate for a 95% confidence interval.

(b) The confidence level is 90% or .90. First we find the z value for a 90% confidence level. To do so, we divide .90 by 2 to obtain .4500. Then we locate .4500 in the body of the normal distribution table (Table VII of Appendix C). Because .4500 is not in the normal table, we can use the number closest to .4500, which is either .4495 or .4505. If we use .4505 as an approximation for .4500, the value of z for this number is 1.65.[2]

Next, we substitute all the values in the confidence interval formula for μ. The 90% confidence interval for μ is

$$\bar{x} \pm z\sigma_{\bar{x}} = 54.40 \pm 1.65 \ (.75) = 54.40 \pm 1.24$$

$$= (54.40 - 1.24) \text{ to } (54.40 + 1.24) = \mathbf{\$53.16 \text{ to } \$55.64}$$

Thus, we are 90% confident that the mean price of all such college textbooks is between $53.16 and $55.64. Note that we cannot say for sure whether the interval $53.16 to $55.64 contains the true population mean or not. Since μ is a constant, we cannot say that the probability is .90 that this interval contains μ because either it contains μ or it does not. Consequently, the probability is either 1.0 or 0 that this interval contains μ. All we can say is that we are 90% confident that the mean price of all such college textbooks is between $53.16 and $55.64.   ▪

How do we interpret a 90% confidence level? In terms of Example 8–1, if we take all possible samples of 36 such college textbooks each and construct a 90% confidence interval for μ around each sample mean, we can expect that 90% of these intervals will include μ and 10% will not. In Figure 8.4 we show means $\bar{x}_1$, $\bar{x}_2$, and $\bar{x}_3$ of three different samples of the same size drawn from the same population. Also shown in this figure are the 90%

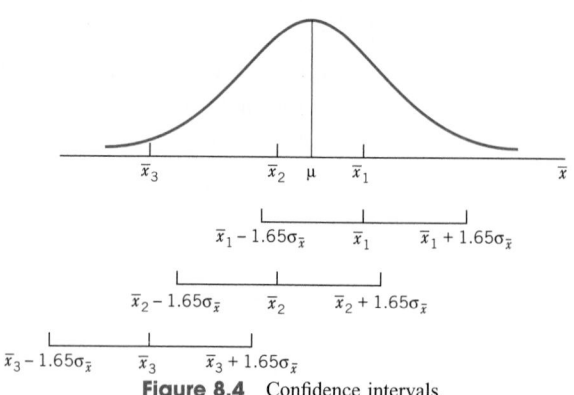

Figure 8.4 Confidence intervals.

---

[2]Note that there is no apparent reason for choosing .4505 and not choosing .4495. If we choose .4495, the z value will be 1.64. An alternative is to use the average of 1.64 and 1.65, 1.645, which we will not do in this text.

confidence intervals constructed around these three sample means. As we observe, the 90% confidence intervals constructed around $\bar{x}_1$ and $\bar{x}_2$ include μ, but the one constructed around $\bar{x}_3$ does not. We can state for a 90% confidence level that if we take many samples of the same size from a population and construct 90% confidence intervals around the means of these samples, then 90% of these confidence intervals will be like the ones around $\bar{x}_1$ and $\bar{x}_2$ in Figure 8.4, which include μ, and 10% will be like the one around $\bar{x}_3$, which does not include μ.

In Example 8–1, the value of the population standard deviation, σ, was known. However, more often we do not know the value of σ. In such cases, we estimate the population standard deviation, σ, by the sample standard deviation, s, and estimate the standard deviation, $\sigma_{\bar{x}}$, of $\bar{x}$ by $s_{\bar{x}}$. Then we use $s_{\bar{x}}$ for $\sigma_{\bar{x}}$ in the formula for the confidence interval for μ. Owing to the central limit theorem, as long as the sample size is large ($n \geq 30$), even if we do not know σ, we can use the normal distribution.

Example 8–2 illustrates the construction of a confidence interval for μ when σ is not known.

Constructing confidence interval for μ*:* σ *not known and* $n > 30$*.*

EXAMPLE 8–2   The U.S. Bureau of Labor often conducts surveys to collect information on the labor market. According to one such recent survey, the workers employed in manufacturing industries in the United States earned an average of $546 per week in September 1996 (*Bureau of Labor Statistics News*, December 6, 1996). Assume that this mean is based on a random sample of 1000 workers selected from the manufacturing industries and that the standard deviation of weekly earnings for this sample is $75. Find a 99% confidence interval for the mean weekly earnings of all workers employed in manufacturing industries in September 1996.

Solution   From the given information,

$$n = 1000, \qquad \bar{x} = \$546, \qquad s = \$75,$$

and $$\text{Confidence level} = 99\% \text{ or } .99$$

First we find the standard deviation of $\bar{x}$. Because σ is not known, we will use $s_{\bar{x}}$ as an estimator of $\sigma_{\bar{x}}$. The value of $s_{\bar{x}}$ is

$$s_{\bar{x}} = \frac{s}{\sqrt{n}} = \frac{75}{\sqrt{1000}} = \$2.37170825$$

Note that we have rounded $s_{\bar{x}}$ to eight decimal places. We will continue with this rounding-off policy for all intermediary calculations.

Because the sample size is large ($n > 30$), we will use the normal distribution to determine the confidence interval for μ. To find z for a 99% confidence level, we divide .99 by 2 to obtain .4950. From the normal distribution table, the z value for .4950 is approximately 2.58. Substituting all the values in the formula, the 99% confidence interval for μ is

$$\bar{x} \pm z s_{\bar{x}} = 546 \pm 2.58 \,(2.37170825) = 546 \pm 6.12 = \textbf{\$539.88 to \$552.12}$$

Thus, we can state with 99% confidence that the average weekly earnings of workers employed in manufacturing industries in September 1996 were between $539.88 and $552.12. ▬

The **width of a confidence interval** depends on the size of the maximum error, $z\sigma_{\bar{x}}$, which depends on the values of z, σ, and n because $\sigma_{\bar{x}} = \sigma/\sqrt{n}$. However, the value of σ is not within the control of the investigator. Hence, the width of a confidence interval depends on

1. The value of z, which depends on the confidence level
2. The sample size n

The confidence level determines the value of z, which in turn determines the size of the maximum error. The value of z increases as the confidence level increases, and it decreases as the confidence level decreases. For example, the value of z is approximately 1.65 for a 90% confidence level, 1.96 for a 95%, and approximately 2.58 for a 99% confidence level. Hence, the higher the confidence level, the larger the width of the confidence interval, other things remaining the same.

For the same value of σ, an increase in the sample size decreases the value of $\sigma_{\bar{x}}$, which in turn decreases the size of the maximum error when the confidence level remains unchanged. Therefore, an increase in the sample size decreases the width of the confidence interval.

Thus, if we want to decrease the width of a confidence interval, we have two choices:

1. Lower the confidence level
2. Increase the sample size

However, lowering the confidence level is not a good choice because a lower confidence level may give less reliable results. Therefore, we should always prefer to increase the sample size if we want to decrease the width of a confidence interval. Next we illustrate, using Example 8–2, how either a decrease in the confidence level or an increase in the sample size decreases the width of the confidence interval.

1. CONFIDENCE LEVEL AND THE WIDTH OF CONFIDENCE INTERVAL

Reconsider Example 8–2. Suppose all the information given in that example remains the same. First, let us decrease the confidence level to 95%. From the normal distribution table, $z = 1.96$ for a 95% confidence level. Then, using $z = 1.96$ in the confidence interval for Example 8–2, we obtain:

$$\bar{x} \pm zs_{\bar{x}} = 546 \pm 1.96\,(2.37170825) = 546 \pm 4.65 = \textbf{\$541.35 to \$550.65}$$

Comparing this confidence interval to the one obtained in Example 8–2, we observe that the width of the confidence interval for a 95% confidence level is smaller than the one for a 99% confidence level.

2. SAMPLE SIZE AND THE WIDTH OF CONFIDENCE INTERVAL

Consider Example 8–2 again. Now suppose the information given in that example is based on a sample size of 1600. Further assume that all other information given in that example, including the confidence level, remains the same. First we calculate the standard deviation of the sample mean using $n = 1600$.

$$s_{\bar{x}} = \frac{s}{\sqrt{n}} = \frac{75}{\sqrt{1600}} = \$1.875$$

Then, the 99% confidence interval for μ is

$$\bar{x} \pm zs_{\bar{x}} = 546 \pm 2.58 \,(1.875) = 546 \pm 4.84 = \textbf{\$541.16 to \$550.84}$$

Comparing this confidence interval to the one obtained in Example 8–2, we observe that the width of the 99% confidence interval for $n = 1600$ is smaller than the 99% confidence interval for $n = 1000$.

CASE STUDY 8-1 CRYING BEHAVIOR IN THE HUMAN ADULT

Dr. William H. Frey, II, of the St. Paul-Ramsey Medical Center at St. Paul, Minnesota, and his associates studied crying behavior in 286 women and 45 men aged 18 to 75 years. The subjects included in the survey were asked to keep all records for each episode of emotional and irritant crying "including date, time, duration, reason for crying (situation, thought, emotions), and components of crying episode (lump in throat, watery eyes, flowing tears, sobbing)" for a period of 30 days. The authors divided all the subjects into two groups, "those subjects who met all of the psychiatric status criteria and those who failed to meet one or more of those criteria. The criteria were as follows: no diagnosed psychiatric illness, medication for psychiatric illness, or mental health counseling in the last six months; no episode of depression lasting at least 1 week in the last six months; no evidence of depression . . . ; and no evidence of labile or histrionic personality disorders as indicated by answers to 11 questions. . . ." The subjects included in the first group (who met all the criteria) were called "normal subjects" in the study. Let us refer to the second group (who failed to meet one or more of the criteria) as a "special group."

Based on the records of subjects in the samples, the authors calculated the mean number of times women and men cried per month, the mean duration of crying per episode, and the corresponding standard deviations. Using the means and the standard deviations of the means calculated by the authors, we have computed the 95% confidence intervals for some of the population parameters in the following table.

Subjects	Group	Population Parameter	95% Confidence Interval
Women	Normal	Mean number of emotional crying episodes per month	$5.3 \pm 1.96 \,(.3)$
Men	Normal	Mean number of emotional crying episodes per month	$1.4 \pm 1.96 \,(.4)$
Women	Special	Mean number of emotional crying episodes per month	$7.1 \pm 1.96 \,(.6)$
Women	Normal	Mean duration of emotional crying episodes (in minutes)	$6.0 \pm 1.96 \,(1.0)$
Men	Normal	Mean duration of emotional crying episodes (in minutes)	$6.0 \pm 1.96 \,(2.0)$
Women	Special	Mean duration of emotional crying episodes (in minutes)	$11.0 \pm 1.96 \,(2.0)$

(*Note*: The number of subjects in different groups were: 175 in normal group for women, 111 in special group for women, and 30 in normal group for men. The special group for men is not included in the above table because it included only 15 subjects, which does not make a large sample.)

The values within parentheses in the fourth column of the table are the values of $s_{\bar{x}}$.

For example, according to the authors the mean number of emotional crying episodes per month for normal women (the first group in the table) was 5.3, and the standard deviation of the mean was .3. Hence, a 95% confidence interval for the mean number of emotional crying episodes per month for women belonging to the normal group is 5.3 $\pm$ 1.96 (.3) or 4.71 to 5.89. In other words, at the 95% confidence level, we can state that all the women belonging to the normal group are expected to cry on average 4.71 to 5.89 times per month. We can interpret the other intervals for the remaining two groups the same way.

Source: William H. Frey II, Carrie Hoffman-Ahern et al., "Crying Behavior in the Human Adult," *Integrative Psychiatry*, September-October 1983, 94–100. Copyright © 1983 by Elsevier Science Publishing Co., Inc. Data and excerpts within quotes reprinted with permission of the publisher.

EXERCISES

Concepts and Procedures

8.1 Briefly explain the meaning of an estimator and an estimate.

8.2 Explain the meaning of a point estimate and an interval estimate.

8.3 What is the point estimator of the population mean, μ? How is the margin of error for a point estimate of μ calculated?

8.4 Explain the various alternatives for decreasing the width of a confidence interval. Which of them is the best alternative?

8.5 Briefly explain how the width of a confidence interval decreases with an increase in the sample size. Give an example.

8.6 Briefly explain how the width of a confidence interval decreases with a decrease in the confidence level. Give an example.

8.7 Briefly explain the difference between a confidence level and a confidence interval.

8.8 What is the maximum error of estimate for μ for a large sample? How is it calculated?

8.9 How will you interpret a 99% confidence interval for μ for a large sample? Explain.

8.10 Find z for each of the following confidence levels.
    a. 90%   **b.** 95%   **c.** 96%   **d.** 97%   **e.** 98%   **f.** 99%

8.11 For a data set obtained from a sample, $n = 64$, $\bar{x} = 22.5$, and $s = 3.4$.
    a. What is the point estimate of μ?
    b. What is the margin of error associated with the point estimate of μ?
    c. Make a 99% confidence interval for μ.
    d. What is the maximum error of estimate for part c?

8.12 For a data set obtained from a sample, $n = 81$, $\bar{x} = 44.25$, and $s = 4.5$.
    a. What is the point estimate of μ?
    b. What is the margin of error associated with the point estimate of μ?
    c. Make a 95% confidence interval for μ.
    d. What is the maximum error of estimate for part c?

8.13 The standard deviation for a population is $\sigma = 12.6$. A sample of 36 observations selected from this population gave a mean equal to 74.8.

a. Make a 90% confidence interval for μ.
b. Construct a 95% confidence interval for μ.
c. Determine a 99% confidence interval for μ.
d. Does the width of the confidence intervals constructed in parts a through c increase as the confidence level increases? Explain your answer.

8.14 The standard deviation for a population is $\sigma = 16.4$. A sample of 100 observations selected from this population gave a mean equal to 143.72.

a. Make a 99% confidence interval for μ.
b. Construct a 95% confidence interval for μ.
c. Determine a 90% confidence interval for μ.
d. Does the width of the confidence intervals constructed in parts a through c decrease as the confidence level decreases? Explain your answer.

8.15 The standard deviation for a population is $\sigma = 6.30$. A random sample selected from this population gave a mean equal to 78.90.

a. Make a 99% confidence interval for μ assuming $n = 36$.
b. Construct a 99% confidence interval for μ assuming $n = 81$.
c. Determine a 99% confidence interval for μ assuming $n = 100$.
d. Does the width of the confidence intervals constructed in parts a through c decrease as the sample size increases? Explain.

8.16 The standard deviation for a population is $\sigma = 7.14$. A random sample selected from this population gave a mean equal to 55.63.

a. Make a 95% confidence interval for μ assuming $n = 196$.
b. Construct a 95% confidence interval for μ assuming $n = 100$.
c. Determine a 95% confidence interval for μ assuming $n = 49$.
d. Does the width of the confidence intervals constructed in parts a through c increase as the sample size decreases? Explain.

8.17 a. A sample of 100 observations taken from a population produced a sample mean equal to 55.32 and a standard deviation equal to 8.4. Make a 90% confidence interval for μ.
b. Another sample of 100 observations taken from the same population produced a sample mean equal to 57.40 and a standard deviation equal to 7.5. Make a 90% confidence interval for μ.
c. A third sample of 100 observations taken from the same population produced a sample mean equal to 56.25 and a standard deviation equal to 7.9. Make a 90% confidence interval for μ.
d. The true population mean for this population is 55.80. Which of the confidence intervals constructed in parts a through c cover this population mean and which do not?

8.18 a. A sample of 400 observations taken from a population produced a sample mean equal to 92.45 and a standard deviation equal to 12.20. Make a 97% confidence interval for μ.
b. Another sample of 400 observations taken from the same population produced a sample mean equal to 91.75 and a standard deviation equal to 14.50. Make a 97% confidence interval for μ.
c. A third sample of 400 observations taken from the same population produced a sample mean equal to 89.63 and a standard deviation equal to 13.40. Make a 97% confidence interval for μ.
d. The true population mean for this population is 90.65. Which of the confidence intervals constructed in parts a through c cover this population mean and which do not?

8.19 For a population, the value of the standard deviation is 2.45. A sample of 35 observations taken from this population produced the following data.

42	51	42	31	28	36	49
29	46	37	32	27	33	41
44	41	28	46	34	39	48
26	35	37	38	46	48	37
29	31	44	41	37	38	46

　　a. What is the point estimate of μ?
　　b. What is the margin of error associated with the point estimate of μ?
　　c. Make a 98% confidence interval for μ.
　　d. What is the maximum error of estimate for part c?

8.20 For a population, the value of the standard deviation is 4.56. A sample of 32 observations taken from this population produced the following data.

74	85	72	73	86	81	77	80
83	78	79	88	76	73	84	78
81	72	82	81	79	83	88	86
78	83	87	82	80	84	76	74

　　a. What is the point estimate of μ?
　　b. What is the margin of error associated with the point estimate of μ?
　　c. Make a 99% confidence interval for μ.
　　d. What is the maximum error of estimate for part c?

Applications

8.21 A survey of Americans planning *long* summer vacations in 1995 revealed a mean planned expenditure of $1076 (*U.S. News & World Report*, June 12, 1995). Assume that this mean is based on a random sample of 300 Americans who were planning long summer vacations in 1995 and that the sample standard deviation was $345. Construct a 99% confidence interval for the mean planned expenditure by all Americans taking long summer vacations in 1995.

8.22 According to an estimate by the Fraser Institute, Canadian families paid an average of $28,127 (in Canadian dollars) in taxes (sum of income, property, sales, social security taxes, etc.) during 1995 (*The 1996 Canadian Global Almanac*). Assume that this estimate is based on a random sample of 500 Canadian families and that the standard deviation of taxes paid by these 500 families was $8100. Make a 90% confidence interval for the corresponding population mean.

8.23 According to a study done by Dr. Martha S. Linet and others, the mean duration of the most recent headache was 8.2 hours for a sample of 5055 females aged 12 through 29. (Martha S. Linet et al., "An Epidemiologic Study of Headache Among Adolescents and Young Adults," *The Journal of the American Medical Association* 261[15], April 21, 1989.) Assume that this sample represents the current population of all headaches for all females aged 12 through 29 and that the standard deviation for this sample is 2.4 hours. Make a 95% confidence interval for the mean duration of all headaches for all 12- to 29-year-old females.

8.24 According to a study done by Professor Adam Drewnowski and others on weight loss among college students, a sample of 507 female college students indicated that they desired to lose on average 3.6 kilograms. (Adam Drewnowski et al., "The Prevalence of Bulimia Nervosa in the U.S. College Student Population," *American Journal of Public Health* 78[10], October 1988.) Assume that this sample represents the current population of all female college students and that the standard deviation of desired weight loss for this sample is .97 kilograms. Construct a 95% confidence interval for the mean desired weight loss for all female college students.

8.25 Do people who stop smoking tend to gain weight? A study of American adults with different smoking histories was undertaken to answer this question (Katherine M. Flegal et al., "The Influence of Smoking Cessation on the Prevalence of Overweight in the United States," *The New England Journal of Medicine* 333[18], November 2, 1995). The authors of this study collected data on a random sample of 315 men over the age of 35 who had quit smoking during the past 10 years and found that these men had gained an average weight of 5.28 kilograms since quitting smoking with a standard deviation of .59 kilogram.

　　a. What is the point estimate of the corresponding population mean? What is the margin of error for this estimate?
　　b. Make a 98% confidence interval for the corresponding population mean.

8.26 According to a study by the Organization for Economic Cooperation and Development released in 1996, the mean starting salary of public elementary school teachers in the United States was $22,753

(*The Willimantic Chronicle*, December 13, 1996). Assume that this result is based on a random sample of 900 public elementary school teachers selected from the United States and that the sample standard deviation of the starting salaries of these teachers was $2700.

    a. What is the point estimate of the mean starting salary of all public elementary school teachers in the United States based on this sample result? What is the margin of error for this estimate?

    b. Make a 90% confidence interval for the mean starting salary of all public elementary school teachers in the United States.

8.27 The U.S. Travel Industry estimated that Americans planned to spend an average of 4.8 nights away on vacations in 1995 (*U.S. News & World Report*, June 12, 1995). Suppose that this mean was based on a random sample of 500 Americans who planned vacations and that the sample standard deviation was 1.5 nights.

    a. Make a 97% confidence interval for the mean length of vacations Americans planned in 1995.

    b. Explain why we need to make the confidence interval. Why can we not say that the mean length of vacations planned by all American people was 4.8 nights in 1995?

8.28 Computer Action Company sells computers and computer parts by mail. The company assures its customers that products are mailed as soon as possible after an order is placed with the company. A sample of 50 recent orders showed that the mean time taken to mail products for these orders was 70 hours and the standard deviation was 14 hours.

    a. Construct a 95% confidence interval for the mean time taken to mail products for all orders received at the office of this company.

    b. Explain why we need to make the confidence interval. Why can we not say that the mean time taken to mail products for all orders received at the office of this company is 70 hours?

8.29 Lazurus Steel Corporation produces iron rods that are supposed to be 36 inches long. The machine that makes these rods does not produce each rod exactly 36 inches long. The lengths of these rods vary slightly. It is known that when the machine is working properly, the mean length of the rods made on this machine is 36 inches. The standard deviation of the lengths of all rods produced on this machine is always equal to .10 inch. The quality control department takes a sample of 40 such rods every week, calculates the mean length of these rods, and makes a 99% confidence interval for the population mean. If either the upper limit of this confidence interval is greater than 36.05 inches or the lower limit of this confidence interval is less than 35.95 inches, the machine is stopped and adjusted. A recent such sample of 40 rods produced a mean length of 36.02 inches. Based on this sample, will you conclude that the machine needs an adjustment?

8.30 At Farmer's Dairy, a machine is set to fill 32-ounce milk cartons. However, this machine does not put exactly 32 ounces of milk in each carton; the amount varies slightly from carton to carton. It is known that when the machine is working properly, the mean net weight of these cartons is 32 ounces. The standard deviation of the amount of milk in all such cartons is always equal to .15 ounce. The quality control department takes a sample of 35 such cartons every week, calculates the mean net weight of these cartons, and makes a 99% confidence interval for the population mean. If either the upper limit of this confidence interval is greater than 32.15 ounces or the lower limit of this confidence interval is less than 31.85 ounces, the machine is stopped and adjusted. A recent sample of 35 such cartons produced a mean net weight of 31.94 ounces. Based on this sample, will you conclude that the machine needs an adjustment?

8.31 A consumer agency that proposes that lawyers' rates are too high wanted to estimate the mean hourly rate for all lawyers in New York City. A sample of 70 lawyers taken from New York City showed that the mean hourly rate charged by them is $223 and the standard deviation of hourly charges is $55.

    a. Construct a 99% confidence interval for the mean hourly charges for all lawyers in New York City.

    b. Suppose the confidence interval obtained in part a is too wide. How can the width of this interval be reduced? Discuss all possible alternatives. Which of these alternatives is the best?

8.32   A bank manager wants to know the mean amount of mortgage paid per month by homeowners in an area. A random sample of 40 homeowners selected from this area showed that they pay an average of $1350 per month for their mortgage with a standard deviation of $215.

   a.   Find a 97% confidence interval for the mean amount of mortgage paid per month by all homeowners in this area.

   b.   Suppose the confidence interval obtained in part a is too wide. How can the width of this interval be reduced? Discuss all possible alternatives. Which of these alternatives is the best?

8.33   You are interested in estimating the mean commuting time from home to school for all commuter students at your school. Briefly explain the procedure you will follow to conduct this study. Collect the required data from a sample of 30 or more such students and then estimate the population mean at a 99% confidence level.

8.34   You are interested in estimating the mean age of cars owned by all people in the United States. Briefly explain the procedure you will follow to conduct this study. Collect the required data on a sample of 30 or more cars and then estimate the population mean at a 95% confidence level.

8.4   INTERVAL ESTIMATION OF A POPULATION MEAN: SMALL SAMPLES

Recall from Section 8.3 that for large samples ($n \geq 30$), whether or not σ is known, the normal distribution is used to estimate the population mean, μ. We use the normal distribution in such cases because, according to the central limit theorem, the sampling distribution of $\bar{x}$ is approximately normal for large samples irrespective of the shape of the population distribution.

However, many times we can select only small samples. This may be due either to the nature of the experiment or to the cost involved in taking a sample. For example, to test a new drug on patients, research may have to be based on a small sample either because there are not many patients available or willing to participate or because it is too expensive to include enough patients in the research to have a large sample.

If the sample size is small, the normal distribution can still be used to construct a confidence interval for μ if (1) the population from which the sample is drawn is normally distributed, and (2) the value of σ is known. But more often we do not know σ and, consequently, we have to use the sample standard deviation, s, as an estimator of σ. In such cases, the normal distribution cannot be used to make confidence intervals about μ. When (1) the population from which the sample is selected is (approximately) normally distributed, (2) the sample size is small (that is, $n < 30$), and (3) the population standard deviation, σ, is not known, the normal distribution is replaced by the *t distribution* to construct confidence intervals about μ. The *t* distribution is described in the next subsection.

CONDITIONS UNDER WHICH THE *t* DISTRIBUTION IS USED TO MAKE A CONFIDENCE INTERVAL ABOUT μ

The *t distribution* is used to make a confidence interval about μ if

1.   The population from which the sample is drawn is (approximately) normally distributed

2.   The sample size is small (that is, $n < 30$)

3.   The population standard deviation, σ, is not known

8.4.1 THE *t* DISTRIBUTION

The **t distribution** was developed by W. S. Gossett in 1908 and published under the pseudonym *Student*. As a result, the *t* distribution is also called *Student's t distribution*. The *t* distribution is similar to the normal distribution in some respects. Like the normal distribution curve, the *t* distribution curve is symmetric (bell-shaped) about the mean, and it never meets the horizontal axis. The total area under a *t* distribution curve is 1.0 or 100%. However, the *t* distribution curve is flatter than the standard normal distribution curve. In other words, the *t* distribution curve has a lower height and a wider spread (or, we can say, larger standard deviation) than the standard normal distribution. However, as the sample size increases, the *t* distribution approaches the standard normal distribution. The units of a *t* distribution are denoted by *t*.

The shape of a particular *t* distribution curve depends on the number of **degrees of freedom (*df*)**. For the purpose of Chapters 8 and 9, the number of degrees of freedom for a *t* distribution is equal to the sample size minus one, that is,

$$df = n - 1$$

The number of degrees of freedom is the only parameter of the *t* distribution. There is a different *t* distribution for each number of degrees of freedom. Like the standard normal distribution, the mean of the *t* distribution is 0. But unlike the standard normal distribution, whose standard deviation is 1, the standard deviation of a *t* distribution is $\sqrt{df/(df-2)}$, which is always greater than 1. Thus, the standard deviation of a *t* distribution is larger than the standard deviation of the standard normal distribution.

THE *t* DISTRIBUTION

The *t distribution* is a specific type of bell-shaped distribution with a lower height and a wider spread than the standard normal distribution. As the sample size becomes larger, the *t* distribution approaches the standard normal distribution. The *t* distribution has only one parameter, called the degrees of freedom (*df*). The mean of the *t* distribution is equal to 0 and its standard deviation is $\sqrt{df/(df-2)}$.

Figure 8.5 shows the standard normal distribution and the *t* distribution for 9 degrees of freedom. The standard deviation of the standard normal distribution is 1.0, and the standard deviation of the *t* distribution is $\sqrt{df/(df-2)} = \sqrt{9/(9-2)} = 1.134$.

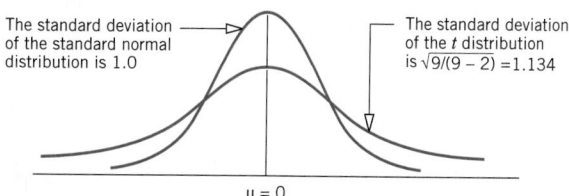

The standard deviation of the standard normal distribution is 1.0

The standard deviation of the *t* distribution is $\sqrt{9/(9-2)} = 1.134$

$\mu = 0$

Figure 8.5 The *t* distribution for *df* = 9, and the standard normal distribution.

As stated earlier, the number of degrees of freedom for a *t* distribution for the purpose of this chapter is *n* − 1. **The number of degrees of freedom is defined as the number of observations that can be chosen freely.** As an example, suppose we know that the mean of 4 values is 20. Consequently, the sum of these 4 values is 20(4) = 80. Now, how many values out of 4 can we choose freely so that the sum of these 4 values is 80? The answer is that we can freely choose $4 - 1 = 3$ values. Suppose we choose 27, 8, and 19 as the 3 values. Given these 3 values and the information that the mean of the 4 values is 20, the fourth value is $80 - 27 - 8 - 19 = 26$. Thus, once we have chosen 3 values, the fourth value is automatically determined. Consequently, the number of degrees of freedom for this example is

$$df = n - 1 = 4 - 1 = 3$$

We subtract 1 from n because we lose 1 degree of freedom to calculate the mean.

Table VIII of Appendix C lists the values of t for the given number of degrees of freedom and areas in the right tail of a t distribution. Because the t distribution is symmetric, these are also the values of $-t$ for the same number of degrees of freedom and the same areas in the left tail of the t distribution. Example 8–3 describes how to read Table VIII of Appendix C.

Reading the t distribution table.

EXAMPLE 8–3   Find the value of t for 16 degrees of freedom and .05 area in the right tail of a t distribution curve.

Solution   In Table VIII of Appendix C, we locate 16 in the column of degrees of freedom (labeled df) and .05 in the row of *area in the right tail under the t distribution curve* at the top of the table. The entry at the intersection of the row of 16 and the column of .05, which is 1.746, gives the required value of t. The relevant portion of Table VIII of Appendix C is shown here as Table 8.1. The value of t read from the t distribution table is shown in Figure 8.6.

Table 8.1   Determining t for 16 df and .05 Area in the Right Tail

Area in the right tail

| df | Area in the Right Tail Under the t Distribution Curve | | | | |
	.10	.05	.025	...	.001
1	3.078	6.314	12.706	...	318.309
2	1.886	2.920	4.303	...	22.327
3	1.638	2.353	3.182	...	10.215
.	...	...	...	...	...
.	...	...	...	...	...
.	...	...	...	...	...
16	1.337	**1.746**	2.120	...	3.686
.	...	...	...	...	...
.	...	...	...	...	...
.	...	...	...	...	...
75	1.293	1.665	1.992	...	3.202
∞	1.282	1.645	1.960	...	3.090

The required value of t for 16 df and .05 area in the right tail

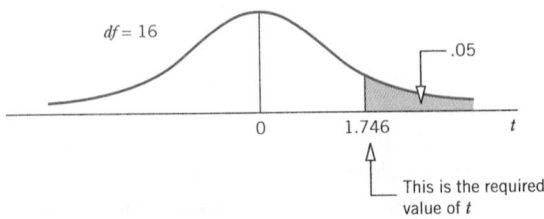

Figure 8.6   The value of t for 16 df and .05 area in the right tail.

Because of the symmetric shape of the t distribution curve, the value of t for 16 degrees of freedom and .05 area in the left tail is -1.746. Figure 8.7 illustrates this case.

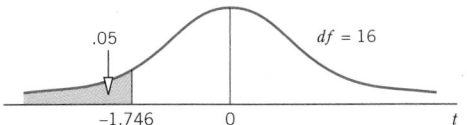

Figure 8.7 The value of t for 16 df and .05 area in the left tail.

8.4.2 CONFIDENCE INTERVAL FOR μ USING THE t DISTRIBUTION

To reiterate, when the following three conditions hold true, we use the t distribution to construct a confidence interval for the population mean, μ.

1. The population from which the sample is drawn is (approximately) normally distributed
2. The sample size is small (that is, $n < 30$)
3. The population standard deviation, σ, is not known

CONFIDENCE INTERVAL FOR μ FOR SMALL SAMPLES

The $(1 - \alpha)100\%$ *confidence interval* for μ is

$$\bar{x} \pm ts_{\bar{x}} \qquad \text{where} \qquad s_{\bar{x}} = \frac{s}{\sqrt{n}}$$

The value of t is obtained from the t distribution table for $n - 1$ degrees of freedom and the given confidence level.

Examples 8–4 and 8–5 describe the procedure of constructing a confidence interval for μ using the t distribution.

Constructing a 95% confidence interval for μ using the t distribution.

EXAMPLE 8–4 Dr. Moore wanted to estimate the mean cholesterol level for all adult males living in Hartford. He took a sample of 25 adult males from Hartford and found that the mean cholesterol level for this sample is 186 with a standard deviation of 12. Assume that the cholesterol levels for all adult males in Hartford are (approximately) normally distributed. Construct a 95% confidence interval for the population mean μ.

Solution From the given information,

$$n = 25, \qquad \bar{x} = 186, \qquad s = 12,$$

and

$$\text{Confidence level} = 95\% \text{ or } .95$$

The value of $s_{\bar{x}}$ is

$$s_{\bar{x}} = \frac{s}{\sqrt{n}} = \frac{12}{\sqrt{25}} = 2.40$$

To find the value of t, we need to know the degrees of freedom and the area under the t distribution curve in each tail.

$$\text{Degrees of freedom} = n - 1 = 25 - 1 = 24$$

To find the area in each tail, we divide the confidence level by 2 and subtract the number obtained from .5. Thus,

$$\text{Area in each tail} = .5 - (.95/2) = .5 - .4750 = .025$$

From the t distribution table, Table VIII of Appendix C, the value of t for $df = 24$ and .025 area in the right tail is 2.064. The value of t is shown in Figure 8.8.

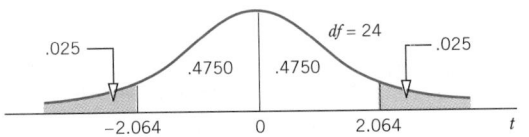

Figure 8.8 The value of t.

Substituting all values in the formula for the confidence interval for μ, the 95% confidence interval is

$$\bar{x} \pm ts_{\bar{x}} = 186 \pm 2.064 \, (2.40) = 186 \pm 4.95 = \textbf{181.05 to 190.95}$$

Thus, we can state with 95% confidence that the mean cholesterol level for all adult males living in Hartford lies between 181.05 and 190.95.

Note that $\bar{x} = 186$ is a point estimate of μ in this example. ▄

Constructing a 99% confidence interval for μ using the t distribution.

EXAMPLE 8–5 According to a survey by *Working Woman* magazine, female pharmacists at chain stores earn an average of $53,600 per year (*Working Woman*, January 1997). Assume that this survey is based on a random sample of 25 female pharmacists working at chain stores. Further assume that the current annual salaries of all such female pharmacists have an approximate normal distribution and the sample standard deviation is $6300. Determine a 99% confidence interval for the corresponding population mean.

Solution From the given information,

$$n = 25, \qquad \bar{x} = \$53,600, \qquad s = \$6300,$$

and   Confidence level = 99% or .99

First we calculate the standard deviation of $\bar{x}$, the number of degrees of freedom, and the area in each tail of the t distribution.

$$s_{\bar{x}} = \frac{s}{\sqrt{n}} = \frac{6300}{\sqrt{25}} = \$1260$$

$$df = n - 1 = 25 - 1 = 24$$

$$\text{Area in each tail} = .5 - (.99/2) = .5 - .4950 = .005$$

From the t distribution table, $t = 2.797$ for 24 degrees of freedom and .005 area in the right tail. The 99% confidence interval for μ is

$$\bar{x} \pm ts_{\bar{x}} = \$53{,}600 \pm 2.797 \ (1260)$$

$$= \$53{,}600 \pm \$3524.22 = \textbf{\$50{,}075.78 to \$57{,}124.22}$$

Thus, we can state with 99% confidence that based on this sample the mean annual salary of all female pharmacists working at chain stores is between $50,075.78 and $57,124.22. ◼

Again, we can decrease the width of a confidence interval for μ either by lowering the confidence level or by increasing the sample size, as was done in Section 8.3. However, increasing the sample size is the better alternative.

CASE STUDY 8-2 CARDIAC DEMANDS OF HEAVY SNOW SHOVELING

During the winter months, newspapers often report cases of sudden deaths occurring during snow shoveling. Dr. Barry A. Franklin et al. conducted a controlled experiment to measure the effects of snow shoveling on men. After screening for health problems, the doctors included 10 men, with a mean age of 32.4 years, in the experiment. Each of these men completed exercise tests on a treadmill and on a device known as an arm-crank ergometer. Then, each of these men participated in the snow removal tests. Each man cleared two 15-meter (49.2-feet) long tracts of heavy, wet snow ranging from 3 to 5 inches in depth, using a snow shovel in one phase and an electric snow thrower in the other. The order of these two phases was determined randomly for each man. Each man shoveled snow for 10 minutes at a time, with 10- to 15-minute rest periods. During each 10-minute period of work, heart rates were recorded at 2-minute intervals. Four variables were measured for each man during the experiment: heart rate, systolic blood pressure, oxygen consumption, and the subject's own rating of his level of exertion.

The following table compares the men's reaction to snow removal with shovels and with electric snow throwers. The means and standard deviations are based on the maximum value for each variable for each man during the experiment, and the confidence intervals are calculated using these values of the mean and the standard deviation. It is assumed that the population of each variable has a normal distribution.

Here, heart rate is measured in beats per minute, systolic blood pressure is in millimeters of mercury, oxygen consumption is in metabolic equivalents, and rating of perceived exertion is on a numerical scale ranging from 6 to 20.

Variable	Snow Shoveling			Automated Snow Removal		
	$\bar{x}$	s	90% Confidence Interval	$\bar{x}$	s	90% Confidence Interval
Heart Rate	175	15	175 ± 1.833 (4.7434)	124	18	124 ± 1.833 (5.6921)
Systolic Blood Pressure	198	17	198 ± 1.833 (5.3759)	161	14	161 ± 1.833 (4.4272)
Oxygen Consumption	5.7	.8	5.7 ± 1.833 (.2530)	2.4	.7	2.4 ± 1.833 (.2214)
Own Rating of Exertion	16.7	1.7	16.7 ± 1.833 (.5376)	9.9	1.0	9.9 ± 1.833 (.3162)

The confidence intervals, given in the table, are constructed by using the method studied in Section 8.4.2 of this chapter. For example, for the variable on heart rate while shoveling snow manually,

$$\bar{x} = 175, \quad s = 15, \quad \text{and} \quad n = 10$$

Because $n < 30$, we will use the t distribution to make the confidence interval, assuming that the corresponding population has a normal distribution. For the 90% confidence level and $df = 9$, the t value from the t distribution table is 1.833. The standard deviation of $\bar{x}$ is:

$$s_{\bar{x}} = \frac{s}{\sqrt{n}} = \frac{15}{\sqrt{10}} = 4.7434$$

Thus, the 90% confidence interval for the corresponding population mean is:

$$\bar{x} \pm ts_{\bar{x}} = 175 \pm 1.833\,(4.7434) = 175 \pm 8.69 = 166.31 \text{ to } 183.69$$

Based on this result, we can state with 90% confidence that the mean heart rate for healthy men in this age group while shoveling snow manually is between 166.31 and 183.69 beats per minute.

Source: Barry A. Franklin et al., "Cardiac Demands of Heavy Snow Shoveling," *The Journal of the American Medical Association* 273[11], March 15, 1995. Copyright © The American Medical Association, 1995. Excerpts and data reproduced with permission.

EXERCISES

Concepts and Procedures

8.35 Briefly explain the similarities and the differences between the standard normal distribution and the t distribution.

8.36 What are the parameters of a normal distribution and a t distribution? Explain.

8.37 Briefly explain the meaning of the degrees of freedom for a t distribution. Give one example.

8.38 What assumptions must hold true to use the t distribution to make a confidence interval for μ?

8.39 Find the value of t for the t distribution for each of the following.
  a. Area in the right tail = .05 and $df = 13$
  b. Area in the left tail = .025 and $n = 22$
  c. Area in the left tail = .001 and $df = 17$
  d. Area in the right tail = .005 and $n = 26$

8.40 a. Find the value of t for a t distribution with a sample size of 20 and area in the left tail equal to .10.
  b. Find the value of t for a t distribution with a sample size of 12 and area in the right tail equal to .025.
  c. Find the value of t for a t distribution with 14 degrees of freedom and .001 area in the right tail.
  d. Find the value of t for a t distribution with 24 degrees of freedom and .005 area in the left tail.

8.41 For each of the following, find the area in the appropriate tail of the t distribution.

    a. $t = 2.060$ and $df = 25$ **b.** $t = -3.686$ and $df = 16$
    c. $t = -2.650$ and $n = 14$ **d.** $t = 2.845$ and $n = 21$

8.42 For each of the following, find the area in the appropriate tail of the t distribution.

    a. $t = -1.323$ and $df = 21$ **b.** $t = 2.467$ and $n = 29$
    c. $t = 3.250$ and $df = 9$ **d.** $t = -2.947$ and $df = 15$

8.43 Find the value of t from the t distribution table for each of the following.

    a. Confidence level $= 99\%$ and $df = 16$
    b. Confidence level $= 95\%$ and $n = 24$
    c. Confidence level $= 90\%$ and $df = 19$

8.44 **a.** Find the value of t from the t distribution table for a sample size of 20 and a confidence level of 95%.

    b. Find the value of t from the t distribution table for 11 degrees of freedom and a 90% confidence level.

    c. Find the value of t from the t distribution table for a sample size of 26 and a confidence level of 99%.

8.45 A sample of 12 observations taken from a normally distributed population produced the following data.

13	15	9	11	8	16
14	9	10	14	16	12

    a. What is the point estimate of μ?
    b. Make a 99% confidence interval for μ.
    c. What is the maximum error of estimate for part b?

8.46 A sample of 10 observations taken from a normally distributed population produced the following data.

44	52	31	48	46	39	43	36	41	49

    a. What is the point estimate of μ?
    b. Make a 95% confidence interval for μ.
    c. What is the maximum error of estimate for part b?

8.47 Suppose, for a sample selected from a normally distributed population, $\bar{x} = 65.50$ and $s = 8.6$.

    a. Construct a 95% confidence interval for μ assuming $n = 16$.
    b. Construct a 90% confidence interval for μ assuming $n = 16$. Is the width of the 90% confidence interval smaller than the width of the 95% confidence interval calculated in part a? If yes, explain why.
    c. Find a 95% confidence interval for μ assuming $n = 25$. Is the width of the 95% confidence interval for μ with $n = 25$ smaller than the width of the 95% confidence interval for μ with $n = 16$ calculated in part a? If so, why? Explain.

8.48 Suppose, for a sample selected from a normally distributed population, $\bar{x} = 21.5$ and $s = 3.9$.

    a. Construct a 95% confidence interval for μ assuming $n = 27$.
    b. Construct a 99% confidence interval for μ assuming $n = 27$. Is the width of the 99% confidence interval larger than the width of the 95% confidence interval calculated in part a? If yes, explain why.
    c. Find a 95% confidence interval for μ assuming $n = 20$. Is the width of the 95% confidence interval for μ with $n = 20$ larger than the width of the 95% confidence interval for μ with $n = 27$ calculated in part a? If so, why? Explain.

Applications

8.49 A random sample of 15 statistics students showed that the mean time taken to solve a computer assignment is 19 minutes with a standard deviation of 3 minutes. Construct a 99% confidence interval for the mean time taken by all statistics students to solve this computer assignment. Assume that the time taken to solve this computer assignment by all students follows a normal distribution.

8.50 A random sample of 20 acres gave a mean yield of wheat equal to 39.4 bushels per acre with a standard deviation of 3 bushels. Assuming that the yield of wheat per acre is normally distributed, construct a 90% confidence interval for the population mean μ.

8.51   Reports of increased death rates due to heart attacks while shoveling snow are common in medical literature. In order to assess the effect of snow shoveling on the heart, a group of medical researchers conducted a study of the physiologic responses of 10 healthy men aged 22 to 35 to manual snow removal (B. A. Franklin et al., ''Cardiac Demands of Heavy Snow Shoveling,'' *The Journal of the American Medical Association* 273[11], March 15, 1995). The heart rates of these 10 men averaged 175 beats per minute during snow shoveling with a sample standard deviation of 15 beats. Assume that the heart rates during snow shoveling are normally distributed for healthy men in this age group. Construct a 95% confidence interval for the mean heartbeat rate during snow shoveling for all healthy men in this age group.

8.52   A company wants to estimate the mean net weight of its Top Taste cereal boxes. A sample of 16 such boxes produced the mean net weight of 31.98 ounces with a standard deviation of .26 ounces. Make a 95% confidence interval for the mean net weight of all Top Taste cereal boxes. Assume that the net weights of all such cereal boxes have a normal distribution.

8.53   The high cost of health care is a matter of major concern for a large number of families. A random sample of 25 families selected from an area showed that they spend an average of $120 per month on health care with a standard deviation of $28. Make a 98% confidence interval for the mean health care expenditure per month incurred by all families in this area. Assume that the monthly health care expenditures of all families in this area have a normal distribution.

8.54   A random sample of 27 auto claims filed with an insurance company produced a mean amount for these claims equal to $2100 with a standard deviation of $460. Construct a 99% confidence interval for the corresponding population mean. Assume that the amounts of all such claims filed with this company have a normal distribution.

8.55   A random sample of 16 mid-sized cars, which were tested for fuel consumption, gave a mean of 26.4 miles per gallon with a standard deviation of 2.3 miles per gallon.

    a. Assuming that the miles per gallon given by all mid-sized cars have a normal distribution, find a 99% confidence interval for the population mean, μ.

    b. Suppose the confidence interval obtained in part a is too wide. How can the width of this interval be reduced? Describe all possible alternatives. Which alternative is the best and why?

8.56   The mean time taken to design a home plan by 20 architects was found to be 21 hours with a standard deviation of 3.75 hours.

    a. Assume that the time taken by all architects to design this home plan is normally distributed. Construct a 98% confidence interval for the population mean μ.

    b. Suppose the confidence interval obtained in part a is too wide. How can the width of this interval be reduced? Describe all possible alternatives. Which alternative is the best and why?

8.57   The following data give the speeds (in miles per hour), as measured by radar, of 10 cars traveling on Interstate Highway I-15.

66	72	62	68	76
74	65	78	67	69

Assuming that the speeds of all cars traveling on this highway have a normal distribution, construct a 90% confidence interval for the mean speed of all cars traveling on this highway.

8.58 A sample of eight adults was taken, and these adults were asked about the time they spend per week on leisure activities. Their responses (in hours) are as follows.

$$45 \quad 12 \quad 31 \quad 16 \quad 28 \quad 14 \quad 18 \quad 26$$

Assuming that the times spent on leisure activities by all adults are normally distributed, make a 95% confidence interval for the mean time spent per week on leisure activities by all adults.

*8.59 You are working for a supermarket. The manager has asked you to estimate the mean time taken by a cashier to serve customers at this supermarket. Unfortunately, he has asked you to take a small sample. Briefly explain how you will conduct this study. Collect data on time taken by any supermarket cashier to serve 10 customers. Then estimate the population mean. Choose your own confidence level.

*8.60 You are working for a bank. The bank manager wants to know the mean waiting time for all customers who visit this bank. She has asked you to estimate this mean by taking a small sample. Briefly explain how you will conduct this study. Collect data on waiting times for 15 customers who visit a bank. Then estimate the population mean. Choose your own confidence level.

8.5 INTERVAL ESTIMATION OF A POPULATION PROPORTION: LARGE SAMPLES

Often we want to estimate the population proportion or percentage. (Recall that a percentage is obtained by multiplying the proportion by 100.) For example, the production manager of a company may want to estimate the proportion of defective items produced on a machine. A bank manager may want to find the percentage of customers who are satisfied with the service provided by the bank.

Again, if we can conduct a census each time we want to find the value of a population proportion, there is no need to learn the procedures discussed in this section. However, we usually derive our results from sample surveys. Hence, to take into account the variability in the results obtained from different sample surveys, we need to know the procedures, discussed in this section, for estimating a population proportion.

Recall from Chapter 7 that the population proportion is denoted by p and the sample proportion is denoted by $\hat{p}$. This section explains how to estimate the population proportion, p, using the sample proportion, $\hat{p}$. The sample proportion, $\hat{p}$, is a sample statistic, and it possesses a sampling distribution. From Chapter 7, we know that for large samples:

1. The sampling distribution of the sample proportion, $\hat{p}$, is (approximately) normal
2. The mean, $\mu_{\hat{p}}$, of the sampling distribution of $\hat{p}$ is equal to the population proportion, p
3. The standard deviation, $\sigma_{\hat{p}}$, of the sampling distribution of the sample proportion, $\hat{p}$, is $\sqrt{pq/n}$ where $q = 1 - p$

☞ *Remember* In the case of proportion, a sample is considered to be large if np and nq are both greater than 5. If p and q are not known, then $n\hat{p}$ and $n\hat{q}$ should each be greater than 5 for the sample to be large.

When estimating the value of a population proportion, we do not know the values of p and q. Consequently, we cannot compute $\sigma_{\hat{p}}$. Therefore, in the estimation of a population proportion, we use the value of $s_{\hat{p}}$ as an estimate of $\sigma_{\hat{p}}$. The value of $s_{\hat{p}}$ is calculated using the following formula.

ESTIMATOR OF THE STANDARD DEVIATION OF $\hat{p}$

The value of $s_{\hat{p}}$, which gives a point estimate of $\sigma_{\hat{p}}$, is calculated as

$$s_{\hat{p}} = \sqrt{\frac{\hat{p}\,\hat{q}}{n}}$$

The sample proportion, $\hat{p}$, is the point estimator of the corresponding population proportion, p. As in the case of the mean, whenever we use point estimation for the population proportion, we calculate the **margin of error** associated with that point estimation. For the estimation of the population proportion, the margin of error is calculated as follows.

$$\text{Margin of error} = \pm 1.96 \, s_{\hat{p}}$$

That is, we find the standard deviation of the sample proportion and multiply it by 1.96. The $(1 - \alpha)100\%$ confidence interval for p is constructed using the following formula.

CONFIDENCE INTERVAL FOR THE POPULATION PROPORTION, p

The $(1 - \alpha)100\%$ *confidence interval for the population proportion, p*, is

$$\hat{p} \pm zs_{\hat{p}}$$

The value of z used here is obtained from the standard normal distribution table for the given confidence level, and $s_{\hat{p}} = \sqrt{\hat{p}\,\hat{q}/n}$.

Examples 8–6 and 8–7 illustrate the procedure for constructing a confidence interval for p.

Point estimate and a 99% confidence interval for p: large sample.

EXAMPLE 8-6 In a poll of 1000 U.S. adults conducted by Celinda Lake of Lake Research and Ed Goeas of Tarrance Group for *U.S. News & World Report*, 81% of adults said that television contributes to the decline in family values (*U.S. News & World Report*, April 15, 1996).

(a) What is the point estimate of the population proportion? What is the margin of error of this estimate?

(b) Find, with a 99% confidence level, what percentage of all adults would hold this opinion.

Solution Let p be the proportion of all adults who hold the opinion that television contributes to the decline in family values and let $\hat{p}$ be the corresponding sample proportion. From the given information,

$$n = 1000, \qquad \hat{p} = .81, \qquad \text{and} \qquad \hat{q} = 1 - \hat{p} = 1 - .81 = .19$$

First, we calculate the value of the standard deviation of the sample proportion as follows.

$$s_{\hat{p}} = \sqrt{\frac{\hat{p}\,\hat{q}}{n}} = \sqrt{\frac{(.81)\,(.19)}{1000}} = .01240564$$

Note that $n\hat{p}$ and $n\hat{q}$ are both greater than 5. (The reader should check that this condition is satisfied.) Consequently the sampling distribution of $\hat{p}$ is approximately normal, and we will use the normal distribution to calculate the margin of error for the point estimate of p and to make a confidence interval about p.

(a) The point estimate of the proportion of all adults who hold the opinion that television contributes to the decline in family values is equal to .81, that is,

$$\text{Point estimate of } p = \hat{p} = \mathbf{.81}$$

The margin of error associated with this point estimate of p is

$$\text{Margin of error} = \pm 1.96\ s_{\hat{p}} = \pm 1.96\ (.01240564) = \pm\mathbf{.024}\text{ or }\mathbf{2.4\%}$$

The margin of error states that the proportion of all adults who hold the opinion that television contributes to the decline in family values is .81 (or 81%), give or take .024 (or 2.4%). As in the case of the mean, the margin of error is simply the maximum error of estimate for a 95% confidence interval for the population proportion.

(b) The confidence level is 99% or .99. To find z, we divide the confidence level by 2 and look for this number in the body of the normal distribution table and record the corresponding value of z.

$$.99/2 = .4950$$

The z value for .4950 is approximately 2.58 from the normal distribution table. Substituting all the values in the confidence interval formula for p, we obtain:

$$\hat{p} \pm zs_{\hat{p}} = .81 \pm 2.58\ (.01240564) = .81 \pm .032$$

$$= \mathbf{.778\text{ to }.842}\text{ or }\mathbf{77.8\%\text{ to }84.2\%}$$

Thus, we can state with 99% confidence that .778 to .842 or 77.8% to 84.2% of all adults hold the opinion that television contributes to the decline in family values. ▬

Constructing a 95% confidence interval for p: large sample.

EXAMPLE 8-7 In a poll of 2003 U.S. adults conducted by Peter Hart and Robert Teeter for *The Wall Street Journal*/NBC News, 57% of adults said that public education needs to be improved (*The Wall Street Journal*, December 13, 1996). Construct a 95% confidence interval for the proportion of all U.S. adults who hold this opinion.

Solution Let p be the proportion of all U.S. adults who hold the opinion that public education needs to be improved, and let $\hat{p}$ be the corresponding sample proportion. From the given information,

$$n = 2003, \quad \hat{p} = .57, \quad \hat{q} = 1 - \hat{p} = 1 - .57 = .43$$

and $$\text{Confidence level} = 95\%\text{ or }.95$$

The standard deviation of the sample proportion is

$$s_{\hat{p}} = \sqrt{\frac{\hat{p}\,\hat{q}}{n}} = \sqrt{\frac{(.57)\,(.43)}{2003}} = .01106194$$

From the normal distribution table, the value of z for $.95/2 = .4750$ is 1.96. The 95% confidence interval for p is

$$\hat{p} + zs_{\hat{p}} = .57 \pm 1.96 \,(.01106194) = .57 \pm .022$$

$$= \textbf{.548 to .592 or 54.8\% to 59.2\%}$$

Thus, we can state with 95% confidence that the proportion of all U.S. adults who hold the opinion that public education needs to be improved is between .548 and .592. The confidence interval can be converted to a percentage interval as 54.8% to 59.2%.　　■

Again, we can decrease the width of a confidence interval for p either by lowering the confidence level or by increasing the sample size. However, lowering the confidence level is not a good choice because it simply decreases the likelihood that the confidence interval contains p. Hence, to decrease the width of a confidence interval for p, we should always increase the sample size.

Case Study 8–3 discusses the philosophy and meaning of confidence intervals using the example of projecting election results.

CASE STUDY 8-3　ASK MR. STATISTICS

Dear Meisterstatser:

Being preternaturally partial to candidates of conservative coloration, I gloatfully stayed up late on election night and found myself brooding anew on the mystery of anchorpersons' projecting this or that candidate a winner even though half or more of the ballots are yet to be counted. How do they do this? is my first interrogatory, and the follow-up is, How can it be that on the morning after the election, the *New York Times* still didn't seem to know that the Republicans had gained control of the House of Representatives even though the television networks had all reported this fact relatively early in the evening? Don't the paper's editors have reporters or at least mothers who call in to tell them what's going on out there on the tube? Or was this invincibly liberal broadsheet simply unable to face the emerging rightist reality?

SUSPECT THE LATTER

Dear Gloatist:

One reluctantly judges your suspicion groundless. Based on knowledge and belief, and one actual interview with a *Times* man who was working election night, one would say that the paper missed the biggest political story in years for two other reasons: (1) Unlike AP and the networks, it had not made the huge investments required to make early calls of close races, and (2) it couldn't bear to give AP or the despised networks as the *source* of the GOP House victory—and at the time it was going to press, there wasn't any other source.

As to how the networks project winners on incomplete returns, the answer is that they rely on the magic of statistical sampling, which enables one to take highly incomplete data and convert them into useful probabilistic statements about final outcomes. Imagine a state with six million voters, 55% of whom favor a ballot initiative outlawing blue cheese. If you polled only 300 randomly selected individuals—1/20,000th of the electorate—on the blue cheese question, you would be right 19 times out of 20 about which side had the majority.

The networks' election night projection problems are mainly managed by an amazing enterprise called Voter News Service. Headquartered in a dowdy office building on New York City's West 34th Street, VNS is a consortium owned by NBC, CBS, ABC, CNN, and AP. Its main purpose is to biennially make possible the calling of elections before the votes are counted. On election day, its normal head count of 22 swells to over 30,000.

The data are fed into a sophisticated computer model enabling Murray Edelman, VNS "editorial director" (i.e., statistics boss), to forecast winners with various levels of confidence. In most electoral contests, Edelman calls a winner if he can be 99.5% certain that the candidate showing up ahead on his computer monitor will finally prevail. But Edelman called the House Republican majority with only a 95% confidence level, in part because VNS also had abundant political analysis confirming the strength of the Republican tide. The five principals in the consortium have instantaneous access to all the data pouring into VNS, and occasionally one of the networks will opt to accept a lower level of confidence and make a projection before Edelman.

The model itself has been built up over the years by Warren Mitofsky and Edelman, who had earlier worked together at the Census Bureau. (Mitofsky now has his own polling service, which tries to stay busy between U.S. elections by taking on elections abroad.) Before the polls close, the model is being bombarded by exit-poll data from preselected sample precincts; the data might show responses from 100 or so voters per precinct, with each voter typically responding to 25 questions about age, income, education, and other personal data, and assorted political views. For a single fairly large state in which, say, 50 precincts are being tracked, the computer program must rapidly assimilate 125,000 bits of exit-poll information. After the polls close, the computers begin receiving actual vote results from the sample precincts. Last to come in are actual data from all other precincts.

The sample precinct data are matched against past voting behavior, and the model kicks out an estimate of the likely outcome along with a "standard error" of the estimate. When the night's data are similar to those for past elections, the standard error will be small and an early call will be more likely. A figure of supreme importance to Edelman is the race's "critical value," which is the percentage difference between the two candidates' votes divided by the standard error. Edelman is usually ready to call a race when the critical value reaches 3. The Lautenberg-Haytaian senatorial race in New Jersey looked close to viewers getting the vote totals, but it had an unusually small standard error and a critical value of 3.8, which made it an easy call, and VNS declared Lautenberg the winner before 9 P.M. The final tally gave him a four-point edge over Haytaian.

Question: Should VNS and the networks share more information with viewers? Many moderately sophisticated viewers would like to know when a candidate is 90% or even 75% certain to win. VNS shrinks from imparting such data because they would result in some likely winners ending up as losers, which unsophisticated viewers would then construe as network screwups. Being sensitive to such opinions, VNS and the networks are unlikely to change course.

Source: Daniel Seligman, "Ask Mr. Statistics," *Fortune,* December 26, 1994. Copyright © 1994, Time Inc. Reproduced with permission. All rights reserved.

EXERCISES

Concepts and Procedures

8.61 What assumption(s) must hold true to use the normal distribution to make a confidence interval for the population proportion, p?

8.62 What is the point estimator of the population proportion, p? How is the margin of error for a point estimate of p calculated?

8.63 Check if the sample size is large enough to use the normal distribution to make a confidence interval for p for each of the following cases.

    a. $n = 50$ and $\hat{p} = .25$ **b.** $n = 160$ and $\hat{p} = .03$

    c. $n = 400$ and $\hat{p} = .65$ **d.** $n = 75$ and $\hat{p} = .06$

8.64 Check if the sample size is large enough to use the normal distribution to make a confidence interval for p for each of the following cases.

    a. $n = 120$ and $\hat{p} = .04$ **b.** $n = 60$ and $\hat{p} = .08$

    c. $n = 40$ and $\hat{p} = .50$ **d.** $n = 900$ and $\hat{p} = .15$

8.65   **a.** A sample of 400 observations taken from a population produced a sample proportion of .63. Make a 95% confidence interval for p.

    b. Another sample of 400 observations taken from the same population produced a sample proportion of .59. Make a 95% confidence interval for p.

    c. A third sample of 400 observations taken from the same population produced a sample proportion of .67. Make a 95% confidence interval for p.

    d. The true population proportion for this population is .65. Which of the confidence intervals constructed in parts a through c cover this population proportion and which do not?

8.66   **a.** A sample of 900 observations taken from a population produced a sample proportion of .32. Make a 90% confidence interval for p.

    b. Another sample of 900 observations taken from the same population produced a sample proportion of .36. Make a 90% confidence interval for p.

    c. A third sample of 900 observations taken from the same population produced a sample proportion of .30. Make a 90% confidence interval for p.

    d. The true population proportion for this population is .34. Which of the confidence intervals constructed in parts a through c cover this population proportion and which do not?

8.67 A sample of 500 observations selected from a population produced a sample proportion equal to .72.

    a. Make a 90% confidence interval for p.

    b. Construct a 95% confidence interval for p.

    c. Make a 99% confidence interval for p.

    d. Does the width of the confidence intervals constructed in parts a through c increase as the confidence level increases? If yes, explain why.

8.68 A sample of 200 observations selected from a population gave a sample proportion equal to .25.

    a. Make a 99% confidence interval for p.

    b. Construct a 97% confidence interval for p.

    c. Make a 90% confidence interval for p.

    d. Does the width of the confidence intervals constructed in parts a through c decrease as the confidence level decreases? If yes, explain why.

8.69 A sample selected from a population gave a sample proportion equal to .67.

    a. Make a 99% confidence interval for p assuming $n = 100$.

    b. Construct a 99% confidence interval for p assuming $n = 600$.

    c. Make a 99% confidence interval for p assuming $n = 1500$.

    d. Does the width of the confidence intervals constructed in parts a through c decrease as the sample size increases? If yes, explain why.

8.70 A sample selected from a population gave a sample proportion equal to .34.

    a. Make a 95% confidence interval for p assuming $n = 1200$.

    b. Construct a 95% confidence interval for p assuming $n = 500$.

    c. Make a 95% confidence interval for p assuming $n = 80$.

    d. Does the width of the confidence intervals constructed in parts a through c increase as the sample size decreases? If yes, explain why.

Applications

8.71 In 1995, Fairbanks, Maslin, Maullin, and Associates interviewed 750 American children aged 10 to 16 years for the Children Now advocacy group to assess television's influence on the children in this age group. Of these children, 49% felt that television makes them think that people are mostly dishonest. Assume that these 750 children make a random sample of all American children in the 10-to-16-year age group.

    a. What is the point estimate of the corresponding population proportion? What is the margin of error associated with this point estimate?

    b. Construct a 99% confidence interval for the proportion of all U.S. children in this age group who feel that television makes them think that people are mostly dishonest.

8.72 In *virtual reality* a person views a computer-generated scene that changes as if the viewer's body were in motion. Some individuals experience unpleasant side effects from virtual reality, such as nausea, dizziness, or disorientation. In a recent study by Clare Tegan of Britain's Defense Research Agency, each of the 150 people included in the study spent 20 minutes wearing a head-mounted virtual reality system through which he or she explored a virtual environment consisting of a series of rooms (Jane Seymour, "Virtually Real, Really Sick," *New Scientist* 129[2014], January 27, 1996). Either during their time in the virtual environment or in the 10 minutes immediately afterward, 61% of these 150 persons reported some side effects. Suppose that these 150 persons make a random sample of all users of virtual reality.

    a. What is the point estimate of the corresponding population proportion? What is the margin of error associated with this point estimate?

    b. Find the 95% confidence interval for the proportion of all virtual reality users who would suffer side effects.

8.73 In a 1995 *U.S. News*/CNN poll conducted by the Gallup organization, 92% of U.S. male drivers rated their driving as excellent or good (*U.S. News & World Report*, March 27, 1995). Suppose that this percentage was based on a random sample of 400 U.S. male drivers.

    a. What is the point estimate of the corresponding population proportion? What is the margin of error associated with this estimate?

    b. Find a 95% confidence interval for the corresponding population proportion.

8.74 It is said that happy and healthy workers are more efficient and productive. A company that manufactures exercising machines wanted to know the percentage of large companies that provide on-site health club facilities. A sample of 240 such companies showed that 80 of them provide such facilities on site.

    a. What is the point estimate of the percentage of all such companies that provide such facilities on site? What is the margin of error associated with this point estimate?

    b. Construct a 97% confidence interval for the percentage of all such companies that provide such facilities on site.

8.75 A mail-order company promises its customers that the products ordered will be mailed within 72 hours after an order is placed. The quality control department at the company checks from time to time to see if this promise is fulfilled. Recently the quality control department took a sample of 50 orders and found that 42 of them were mailed within 72 hours of the placement of the orders.

    a. Construct a 98% confidence interval for the percentage of all orders that are mailed within 72 hours of their placement.

    b. Suppose the confidence interval obtained in part a is too wide. How can the width of this interval be reduced? Discuss all possible alternatives. Which of these alternatives is the best?

8.76 One of the major problems faced by department stores is a high percentage of returns. The manager of a department store wanted to estimate the percentage of all sales that result in returns. A sample of 40 sales showed that 8 of them had products returned within the time allowed for returns.

    a. Make a 99% confidence interval for the percentage of all sales that result in returns.

b. Suppose the confidence interval obtained in part a is too wide. How can the width of this interval be reduced? Discuss all possible alternatives. Which of these alternatives is the best?

8.77 Market Facts' Telenation conducted a survey in August 1996 for *U.S. News & World Report* to examine parents' concerns about the effect of television on their children (*U.S. News & World Report*, September 9, 1996). Parents were especially uncomfortable about cartoons in which the characters solve their problems by violence. A total of 373 parents of children under 18 years of age were polled. Of them, 73% were concerned about such use of violence on television.

a. Determine a 99% confidence interval for the percentage of all parents with children under 18 years of age who are concerned about such use of violence on television.
b. Explain why we need to make the confidence interval. Why cannot we simply say that 73% of all such parents are concerned about such use of violence on television?

8.78 According to a 1995 survey by Roper Starch Worldwide, 45% of women in the United States prefer to work outside their homes. Suppose this percentage is based on a random sample of 1200 women.

a. Make a 98% confidence interval for the proportion of all U.S. women who prefer to work outside their homes.
b. Explain why we need to make the confidence interval. Why cannot we simply say that 45% of all U.S. women prefer to work outside their homes?

8.79 A researcher wanted to know the percentage of judges who are in favor of the death penalty. He took a random sample of 15 judges and asked them whether or not they favor the death penalty. The responses of these judges are given below.

| Yes | No | Yes | Yes | No | No | No | |
| Yes | No | Yes | Yes | Yes | No | Yes | Yes |

a. What is the point estimate of the population proportion? What is the margin of error associated with this point estimate?
b. Make a 95% confidence interval for the percentage of all judges who are in favor of the death penalty.

8.80 The manager of a supermarket wanted to know the percentage of shoppers who prefer to buy name brand products. A random sample of 20 shoppers who shopped at this store were asked this question. The following are the responses of these shoppers.

No	No	No	Yes	Yes	No	No
Yes	No	Yes	No	No	Yes	No
No	No	No	Yes	No	Yes	

a. What is the point estimate of the population proportion? What is the margin of error associated with this point estimate?
b. Construct a 99% confidence interval for the percentage of all shoppers who prefer to buy name brand products.

*8.81 You want to estimate the proportion of students at your college who hold off-campus (part-time or full-time) jobs. Briefly explain how you will make such an estimate. Collect data from 40 students at your college on whether or not they hold off-campus jobs. Then, calculate the proportion of students in this sample who hold off-campus jobs. Using this information, estimate the population proportion. Select your own confidence level.

*8.82 You want to estimate the percentage of people who are satisfied with the services provided by their banks. Briefly explain how you will make such an estimate. Collect data from 30 people on whether or not they are satisfied with the services provided by their banks. Then, calculate the percentage of people in this sample who are satisfied. Using this information, estimate the population percentage. Select your own confidence level.

8.6 SAMPLE SIZE DETERMINATION FOR THE ESTIMATION OF MEAN

One reason why we usually conduct a sample survey and not a census is that almost always we have limited resources at our disposal. In light of this, if a smaller sample can serve our purpose, then we will be wasting our resources by taking a larger sample. For instance, suppose we want to estimate the mean life of a certain auto battery. If a sample of 40 batteries can give us the type of confidence interval we are looking for, then we will be wasting money and time if we take a sample of a much larger size, say, 500 batteries. In such cases, if we know the confidence level and the width of the confidence interval that we want, then we can find the (approximate) size of the sample that will produce the required result.

In Section 8.3 we learned that $E = z\sigma_{\bar{x}}$ is called the maximum error of estimate for μ. As we know, the standard deviation of the sample mean is equal to $\sigma/\sqrt{n}$. Therefore, we can write the maximum error of estimate for μ as

$$E = z \cdot \frac{\sigma}{\sqrt{n}}$$

Suppose we predetermine the size of the maximum error, E, and want to find the size of the sample that will yield this maximum error. From the above expression, the following formula is obtained that determines the required sample size n.

DETERMINING THE SAMPLE SIZE FOR THE ESTIMATION OF μ

Given the confidence level and the standard deviation of the population, the sample size that will produce a predetermined maximum error E of the confidence interval *estimate of* μ is

$$n = \frac{z^2 \sigma^2}{E^2}$$

If we do not know σ, we can take a preliminary sample (of any arbitrarily determined size) and find the sample standard deviation, s. Then we can use s for σ in the formula. However, note that using s for σ may give a sample size that eventually may produce an error much larger (or smaller) than the predetermined maximum error. This will depend on how close s and σ are.

Example 8–8 illustrates how we determine the sample size that will produce the maximum error of estimate for μ within a certain limit.

Determining sample size for the estimation of μ.

EXAMPLE 8–8 Suppose the U.S. Bureau of the Census wants to estimate the mean family size for all U.S. families at a 99% confidence level. It is known that the standard deviation σ for the sizes of all families in the United States is .6. How large a sample should the bureau select if it wants its estimate to be within .01 of the population mean?

Solution The Bureau of the Census wants the 99% confidence interval for the mean family size to be

$$\bar{x} \pm .01$$

Hence, the maximum size of the error of estimate is to be .01, that is

$$E = .01$$

The value of z for a 99% confidence level is 2.58. The value of σ is given to be .6. Therefore, substituting all values in the formula and simplifying

$$n = \frac{z^2 \, \sigma^2}{E^2} = \frac{(2.58)^2 \, (.6)^2}{(.01)^2} = \frac{(6.6564) \, (.36)}{(.0001)} = 23,963.04 \approx \mathbf{23,964}$$

Thus, the required sample size is 23,964. If the Bureau of the Census takes a sample of 23,964 families, computes the mean family size for this sample, and then constructs a 99% confidence interval around this sample mean, the maximum error of the estimate will be approximately .01. Note that we have rounded the final answer for the sample size to the next higher integer. This is always the case when determining the sample size. ▄▄

EXERCISES

Concepts and Procedures

8.83   For data on a variable for a population, $\sigma = 10.5$.

    a.   How large a sample should be selected so that the maximum error of estimate for a 99% confidence interval for μ is 2.50?

    b.   How large a sample should be selected so that the maximum error of estimate for a 96% confidence interval for μ is 3.20?

8.84   For data on a variable for a population, $\sigma = 16.42$.

    a.   What should the sample size be for a 98% confidence interval for μ to have a maximum error of estimate equal to 5.50?

    b.   What should the sample size be for a 95% confidence interval for μ to have a maximum error of estimate equal to 4.25?

8.85   Determine the sample size for the estimate of μ for the following.

    a. $E = 2.3$, $\sigma = 15.40$,   confidence level = 99%

    b. $E = 4.1$, $\sigma = 23.45$,   confidence level = 95%

    c. $E = 25.9$, $\sigma = 122.25$,   confidence level = 90%

8.86   Determine the sample size for the estimate of μ for the following.

    a. $E = .17$, $\sigma = .90$,   confidence level = 99%

    b. $E = 1.45$, $\sigma = 5.82$,   confidence level = 95%

    c. $E = 5.65$, $\sigma = 18.20$,   confidence level = 90%

Applications

8.87   A researcher wants to determine a 95% confidence interval for the mean number of hours that high school students spend doing homework per week. She knows that the standard deviation for hours spent per week by all high school students doing homework is 7. How large a sample should the researcher select so that the estimate will be within 1.5 hours of the population mean?

8.88   A company that produces detergents wants to estimate the mean amount of detergent in 64-ounce jugs at a 99% confidence level. The company knows that the standard deviation of amounts of detergent in such jugs is .20 ounces. How large a sample should the company take so that the estimate is within .04 ounces of the population mean?

8.89   A department store manager wants to estimate at a 90% confidence level the mean amount spent by all customers at this store. From an earlier study, the manager knows that the standard deviation of amounts spent by customers at this store is $27. What sample size should he choose so that the estimate is within $3 of the population mean?

8.90 A U.S. government agency wants to estimate at a 95% confidence level the mean speed for all cars traveling on Interstate Highway I-95. From a previous study, the agency knows that the standard deviation of speeds of cars traveling on this highway is 3.5 miles per hour. What sample size should the agency choose for the estimate to be within 1.5 miles per hour of the population mean?

8.7 SAMPLE SIZE DETERMINATION FOR THE ESTIMATION OF PROPORTION

Just as we did with the mean, we can also determine the sample size for estimating the population proportion, p. This sample size will yield an error of estimate that may not be larger than a predetermined maximum error. By knowing the sample size that can give us the required results, we can save our scarce resources by not taking an unnecessarily large sample. From Section 8.5, the maximum error, E, of the interval estimation of the population proportion is

$$E = z\, \sigma_{\hat{p}} = z \times \sqrt{\frac{pq}{n}}$$

By manipulating this expression algebraically, we obtain the following formula to find the required sample size given E, p, q, and z.

DETERMINING THE SAMPLE SIZE FOR THE ESTIMATION OF p

Given the confidence level and the values of p and q, the sample size that will produce a predetermined maximum error E of the confidence interval *estimate of p* is

$$n = \frac{z^2\, p\, q}{E^2}$$

We can observe from this formula that to find n, we need to know the values of p and q. However, the values of p and q are not known to us. In such a situation, we can choose one of the following alternatives.

1. We make the *most conservative estimate* of the sample size n by using $p = .50$ and $q = .50$. For a given E, these values of p and q will give us the largest sample size by comparison to any other pair of values of p and q because the product of $p = .50$ and $q = .50$ is greater than the product of any other pair of values for p and q.

2. We take a *preliminary sample* (of arbitrarily determined size) and calculate $\hat{p}$ and $\hat{q}$ for this sample. Then, we use these values of $\hat{p}$ and $\hat{q}$ to find n.

Examples 8–9 and 8–10 illustrate how to determine the sample size that will produce the error of estimation for the population proportion within a predetermined maximum value. Example 8–9 gives the most conservative estimate of n, and Example 8–10 uses the results from a preliminary sample to determine the required sample size.

Most conservative estimate of n for the estimation of p.

EXAMPLE 8–9 Lombard Electronics Company has just installed a new machine that makes a part that is used in clocks. The company wants to estimate the proportion of these parts produced by this machine that are defective. The company manager wants this estimate to be within .02 of the population proportion for a 95% confidence level. What is the most conservative estimate of the sample size that will limit the maximum error to within .02 of the population proportion?

Solution   The company manager wants the 95% confidence interval to be

$$\hat{p} \pm .02$$

Therefore,

$$E = .02$$

The value of z for a 95% confidence level is 1.96. For the most conservative estimate of the sample size, we will use $p = .50$ and $q = .50$. Hence, the required sample size is

$$n = \frac{z^2 \, p \, q}{E^2} = \frac{(1.96)^2 \, (.50) \, (.50)}{(.02)^2} = \mathbf{2401}$$

Thus, if the company takes a sample of 2401 parts, the estimate of p will be within .02 of the population proportion. ▬

Determining n for the estimation of p using the preliminary sample results.

EXAMPLE 8–10 Consider Example 8–9 again. Suppose a preliminary sample of 200 parts produced by this machine showed that 7% of them are defective. How large a sample should the company select so that the 95% confidence interval for p is within .02 of the population proportion?

Solution   Again, the company wants the 95% confidence interval for p to be

$$\hat{p} \pm .02$$

Hence,

$$E = .02$$

The value of z for a 95% confidence level is 1.96. From the preliminary sample,

$$\hat{p} = .07 \quad \text{and} \quad \hat{q} = 1 - .07 = .93$$

Using these values of $\hat{p}$ and $\hat{q}$ as estimates of p and q, we obtain:

$$n = \frac{z^2 \, \hat{p} \, \hat{q}}{E^2} = \frac{(1.96)^2 \, (.07) \, (.93)}{(.02)^2} = \frac{(3.8416) \, (.07) \, (.93)}{.0004} = 625.22 \approx \mathbf{626}$$

Thus, if the company takes a sample of 626 items, the estimate of p will be within .02 of the population proportion. However, we should note that this sample size will produce the maximum error within .02 points only if $\hat{p}$ is .07 or less for the new sample. But if $\hat{p}$ for the new sample happens to be much higher than .07, the maximum error will not be within .02. Therefore, to avoid such a situation, we may be more conservative and take a much larger sample than 626 items. ▬

EXERCISES

Concepts and Procedures

8.91 **a.** How large a sample should be selected so that the maximum error of estimate for a 99% confidence interval for p is .035 when the value of the sample proportion obtained from a preliminary sample is .26?

b. Find the most conservative sample size that will produce the maximum error for a 99% confidence interval for p equal to .035.

8.92 **a.** How large a sample should be selected so that the maximum error of estimate for a 98% confidence interval for p is .045 when the value of the sample proportion obtained from a preliminary sample is .57?

b. Find the most conservative sample size that will produce the maximum error for a 98% confidence interval for p equal to .045.

8.93 Determine the most conservative sample size for the estimation of the population proportion for the following.

    a. $E = .03$,   confidence level = 99%
    b. $E = .04$,   confidence level = 95%
    c. $E = .01$,   confidence level = 90%

8.94 Determine the sample size for the estimation of the population proportion for the following where $\hat{p}$ is the sample proportion based on a preliminary sample.

    a. $E = .03$, $\hat{p} = .32$,   confidence level = 99%
    b. $E = .04$, $\hat{p} = .78$,   confidence level = 95%
    c. $E = .02$, $\hat{p} = .64$,   confidence level = 90%

Applications

8.95 Tony's Pizza guarantees all pizza deliveries within 30 minutes of the placement of orders. An agency wants to estimate the proportion of all pizzas delivered within 30 minutes by Tony's. What is the most conservative estimate of the sample size that would limit the maximum error to within .02 of the population proportion for a 99% confidence interval?

8.96 Refer to Exercise 8.95. Assume that a preliminary study has shown that 93% of all Tony's pizzas are delivered within 30 minutes. How large should the sample size be so that the 99% confidence interval for the population proportion has a maximum error of .02?

8.97 A consumer agency wants to estimate the proportion of all drivers who wear seat belts while driving. Assume that a preliminary study has shown that 76% of drivers wear seat belts while driving. How large should the sample size be so that the 99% confidence interval for the population proportion has a maximum error of .03?

8.98 Refer to Exercise 8.97. What is the most conservative estimate of the sample size that would limit the maximum error to within .03 of the population proportion for a 99% confidence interval?

GLOSSARY

Confidence interval An interval constructed around the value of a sample statistic to estimate the corresponding population parameter.

Confidence level Confidence level, denoted by $(1 - \alpha)100\%$, states how much confidence we have that a confidence interval contains the true population parameter.

Degrees of freedom (*df*) The number of observations that can be chosen freely. For the estimation of μ using the t distribution, the degrees of freedom are $n - 1$.

Estimate The value of a sample statistic that is used to find the corresponding population parameter.

Estimation   A procedure by which numerical value or values are assigned to a population parameter based on the information collected from a sample.

Estimator   The sample statistic that is used to estimate a population parameter.

Interval estimate   An interval constructed around the point estimate that is likely to contain the corresponding population parameter. Each interval estimate has a confidence level.

Maximum error of estimate   The quantity that is subtracted from and added to the value of a sample statistic to obtain a confidence interval for the corresponding population parameter.

Point estimate   The value of a sample statistic assigned to the corresponding population parameter.

***t* distribution**   A continuous distribution with a specific type of bell-shaped curve with its mean equal to 0 and standard deviation equal to $\sqrt{df/(df-2)}$.

KEY FORMULAS

1. **Margin of error associated with the point estimation of μ**

    $$\pm 1.96 \ \sigma_{\bar{x}} \quad \text{or} \quad \pm 1.96 \ s_{\bar{x}}$$

    where $\qquad \sigma_{\bar{x}} = \sigma/\sqrt{n} \quad \text{and} \quad s_{\bar{x}} = s/\sqrt{n}$

2. **The $(1 - \alpha)100\%$ confidence interval for μ for a large sample ($n \geq 30$)**

    $$\bar{x} \pm z\sigma_{\bar{x}} \quad \text{if } \sigma \text{ is known}$$
    $$\bar{x} \pm zs_{\bar{x}} \quad \text{if } \sigma \text{ is not known}$$

    where $\qquad \sigma_{\bar{x}} = \sigma/\sqrt{n} \quad \text{and} \quad s_{\bar{x}} = s/\sqrt{n}$

3. **The $(1 - \alpha)100\%$ confidence interval for μ for a small sample ($n < 30$) when population is (approximately) normally distributed and σ is not known**

    $$\bar{x} \pm ts_{\bar{x}} \quad \text{where} \quad s_{\bar{x}} = s/\sqrt{n}$$

4. **Margin of error associated with the point estimation of p**

    $$\pm 1.96 \ s_{\hat{p}} \quad \text{where} \quad s_{\hat{p}} = \sqrt{\frac{\hat{p}\,\hat{q}}{n}}$$

5. **The $(1 - \alpha)100\%$ confidence interval for p for a large sample**

    $$\hat{p} \pm zs_{\hat{p}} \quad \text{where} \quad s_{\hat{p}} = \sqrt{\frac{\hat{p}\,\hat{q}}{n}}$$

6. **Maximum error, E, of the estimate for μ**

    $$E = z\sigma_{\bar{x}} \quad \text{or} \quad zs_{\bar{x}}$$

7. **Required sample size for a predetermined maximum error for estimating μ**

    $$n = \frac{z^2 \sigma^2}{E^2}$$

8. **Maximum error, E, of the estimate for the population proportion**

    $$E = zs_{\hat{p}} \quad \text{where} \quad s_{\hat{p}} = \sqrt{\frac{\hat{p}\,\hat{q}}{n}}$$

9. **Required sample size for a predetermined maximum error for estimating *p***

$$n = \frac{z^2\, p\, q}{E^2}$$

Use $p = .50$ and $q = .50$ for the most conservative estimate and the values of $\hat{p}$ and $\hat{q}$ if the estimate is to be based on a preliminary sample.

SUPPLEMENTARY EXERCISES

8.99 A company opened a new movie theater. Before setting the price of a movie ticket, the company wants to find the average price charged by other movie theaters. A random sample of 100 movie theaters taken by the company showed that the mean price of a movie ticket is $6.75 with a standard deviation of $.80.

    a.   What is the point estimate of the mean price of movie tickets for all theaters? What is the margin of error associated with this estimate?

    b.   Make a 95% confidence interval for the population mean, μ.

8.100 A bank manager wants to know the mean amount owed on credit card accounts that become delinquent. A random sample of 100 delinquent credit card accounts taken by the manager produced a mean amount owed on these accounts equal to $2130 with a standard deviation of $578.

    a.   What is the point estimate of the mean amount owed on all delinquent credit card accounts at this bank? What is the margin of error associated with this estimate?

    b.   Construct a 97% confidence interval for the mean amount owed on all delinquent credit card accounts for this bank.

8.101 York Steel Corporation produces iron rings that are supplied to other companies. These rings are supposed to have a diameter of 24 inches. The machine that makes these rings does not produce each ring with a diameter of exactly 24 inches. The diameter of each of these rings varies slightly. It is known that when the machine is working properly, the rings made on this machine have a mean diameter of 24 inches. The standard deviation of the diameters of all rings produced on this machine is always equal to .06 inches. The quality control department takes a sample of 36 such rings every week, calculates the mean of diameters for these rings, and makes a 99% confidence interval for the population mean. If either the lower limit of this confidence interval is less than 23.975 inches or the upper limit of this confidence interval is greater than 24.025 inches, the machine is stopped and adjusted. A recent such sample of 36 rings produced a mean diameter of 24.015 inches. Based on this sample, can you conclude that the machine needs an adjustment? Explain.

8.102 Yunan Corporation produces bolts that are supplied to other companies. These bolts are supposed to be 4 inches long. The machine that makes these bolts does not produce each bolt exactly 4 inches long. It is known that when the machine is working properly, the mean length of the bolts made on this machine is 4 inches. The standard deviation of the lengths of all bolts produced on this machine is always equal to .04 inches. The quality control department takes a sample of 50 such bolts every week, calculates the mean length of these bolts, and makes a 98% confidence interval for the population mean. If either the upper limit of this confidence interval is greater than 4.02 inches or the lower limit of this confidence interval is less than 3.98 inches, the machine is stopped and adjusted. A recent such sample of 50 bolts produced a mean length of 3.99 inches. Based on this sample, will you conclude that the machine needs an adjustment?

8.103 A magazine wanted to estimate the mean daily cost for a room for all major hotels in the United States. The research department at the magazine took a sample of 33 hotels and obtained the following data on the daily charges per room.

$124	157	105	185	86	210	130	120	195	230	160
145	177	180	205	153	240	210	75	115	165	189
205	143	125	112	254	142	179	190	224	90	125

    a. What is the point estimate of the mean daily cost per room for all hotels in the United States? What is the margin of error associated with this estimate?

    b. Construct a 98% confidence interval for the mean daily cost per room for all hotels in the United States.

8.104 A travel magazine wanted to estimate the mean leisure time per week enjoyed by adults. The research department at the magazine took a sample of 36 adults and obtained the following data on the weekly leisure time (in hours).

15	12	18	23	11	21	16	13	9	19	26	11
7	18	11	15	23	26	10	8	17	21	12	7
19	17	11	13	21	16	14	9	15	12	10	14

    a. What is the point estimate of the mean leisure time per week enjoyed by all adults? What is the margin of error associated with this estimate?

    b. Construct a 99% confidence interval for the mean leisure time per week enjoyed by all adults.

8.105 A random sample of 25 life insurance policyholders showed that the average premium they pay on their life insurance policies is $420 per year with a standard deviation of $62. Assuming that the life insurance policy premiums for all life insurance policyholders have a normal distribution, make a 99% confidence interval for the population mean, μ.

8.106 A drug that provides relief from headaches was tried on 18 randomly selected patients. The experiment showed that the mean time to get relief from headache for these patients after taking this drug was 26 minutes with a standard deviation of 4.5 minutes. Assuming that the time taken to get relief from a headache after taking this drug is (approximately) normally distributed, determine a 95% confidence interval for the mean relief time for this drug for all patients.

8.107 A survey of 20 randomly selected adult males showed that the mean time they spend per week watching sports on television is 10.25 hours with a standard deviation of 2.2 hours. Assuming that the time spent per week watching sports on television by all adult males is (approximately) normally distributed, construct a 90% confidence interval for the population mean, μ.

8.108 A random sample of 20 female members of health clubs in Los Angeles showed that they spend, on average, 4.5 hours per week doing physical exercise with a standard deviation of .75 hours. Assume that the time spent doing physical exercise by all female members of health clubs in Los Angeles is (approximately) normally distributed. Find a 98% confidence interval for the population mean.

8.109 A computer company that recently developed a new software product wanted to estimate the mean time taken to learn how to use this software by people who are somewhat familiar with computers. A random sample of 12 such persons was selected. The following data give the time taken (in hours) by these persons to learn how to use this software.

1.75	2.25	2.40	1.90	1.50	2.75
2.15	2.25	1.80	2.20	3.25	2.60

Construct a 95% confidence interval for the population mean. Assume that the time taken by all persons who are somewhat familiar with computers to learn how to use this software is approximately normally distributed.

8.110 A company that produces 8-ounce low-fat yogurt cups wanted to estimate the mean number of calories for such cups. A random sample of 10 such cups produced the following data on calories.

147	159	153	146	144	161	163	153	143	158

Construct a 99% confidence interval for the population mean. Assume that the number of calories for such cups of yogurt produced by this company has an approximately normal distribution.

8.111 An insurance company selected a sample of 50 auto claims filed with it and investigated those claims carefully. The company found that 12% of those claims were fraudulent.

    a. What is the point estimate of the percentage of all auto claims filed with this company that are fraudulent? What is the margin of error associated with this estimate?

    b. Make a 99% confidence interval for the percentage of all auto claims filed with this company that are fraudulent.

8.112 An auto company wanted to know the percentage of people who prefer to own safer cars (that is, cars that possess more safety features) even if they have to pay a few thousand dollars more. A random sample of 500 persons showed that 47% of them will not mind paying a few thousand dollars more to have safer cars.

    a. What is the point estimate of the percentage of all people who will not mind paying a few thousand dollars more to have safer cars? What is the margin of error associated with this estimate?

    b. Construct a 90% confidence interval for the percentage of all people who will not mind paying a few thousand dollars more to have safer cars.

8.113 A sample of 20 managers was taken, and they were asked whether or not they usually take work home. The responses of these managers are given below where *yes* indicates they usually take work home and *no* means they do not.

Yes	Yes	No	No	No	Yes	No	No
No	No	Yes	Yes	No	Yes	Yes	No
No	No	No	Yes				

Make a 99% confidence interval for the percentage of all managers who take work home.

8.114 A sample of 16 salespersons was taken and they were asked whether or not they have ever been fired from their jobs. The following are the responses of these salespersons.

No	No	Yes	No	Yes	No	Yes	Yes
No	No	No	No	Yes	No	Yes	No

Construct a 97% confidence interval for the percentage of all salespersons who have ever been fired from their jobs.

8.115 A researcher wants to determine a 99% confidence interval for the mean number of hours that adults spend per week doing community service. How large a sample should the researcher select so that the estimate is within 1.2 hours of the population mean? Assume that the standard deviation for time spent per week doing community service by all adults is 3 hours.

8.116 An economist wants to find a 90% confidence interval for the mean sale price of houses in a state. How large a sample should he or she select so that the estimate is within $3500 of the population mean? Assume that the standard deviation for the sale prices of all houses in this state is $31,500.

8.117 A telephone company wants to estimate the proportion of all households who own telephone answering machines. What is the most conservative estimate of the sample size that would limit the maximum error to be within .03 of the population proportion for a 95% confidence interval?

8.118 Refer to Exercise 8.117. Assume that a preliminary sample has shown that 67% of the households in the sample own telephone answering machines. How large should the sample size be so that the 95% confidence interval for the population proportion has a maximum error of .03?

*8.119 Let μ be the mean hourly wage of carpenters in the state of Idaho. A random sample of carpenters from Idaho yielded a 95% confidence interval for μ of $11.45 to $14.35. Assume that this sample included more than 30 carpenters.

    a. Find the value of $\bar{x}$ for this sample.

    b. Find the 99% confidence interval for μ based on this same sample.

*8.120 The director of Channel 66, an independent television station, claims that this channel devotes an average of 12 minutes or less per hour to commercials. Students in a communications course at a local college are asked to estimate the mean time that Channel 66 actually devotes per hour to commercials. Since the amount of commercial time seems to vary from hour to hour, each student is asked to watch Channel 66 for a preassigned 20-minute time interval and use a stopwatch to time the commercials. Thirty students participated in this study, which showed that a mean of 4.68 minutes of commercials were aired on this channel per 20-minute interval with a standard deviation of 1.30 minutes.

    a. Using this information, find the 90% confidence interval for μ, the mean time devoted to commercials per hour on Channel 66.

    b. A student named Waldo was absent when the project was assigned. The instructor has asked him to make his own estimate of μ. Waldo does not have much time to watch television, and, hence, he decides to turn on Channel 66 at randomly chosen times and record whether or not a commercial is on at that time. He does this 30 times and observes a commercial on 7 of those occasions. Use this information to find a point estimate of μ.

    c. Compare Waldo's point estimate of μ to the one based on the information collected by the other 30 students.

    d. Use Waldo's data to find a 90% confidence interval for μ. (Hint: First find a 90% confidence interval for p, the proportion of Channel 66's broadcast time that is devoted to commercials. Then convert this interval to a confidence interval for μ.)

    e. Compare the maximum error of Waldo's estimate of μ to that of the class.

    f. To achieve the same accuracy as that of the class's estimate, how many times would Waldo have to turn on Channel 66? Do you think this is feasible?

*8.121 In an international Gallup poll released in June 1995, residents of 18 countries were asked several questions concerning their satisfaction with life in their own countries (*USA Today*, June 12, 1995). Among the results from the United States, 83% of adults are satisfied with their personal lives, 64% are satisfied with the way democracy works in the United States, 52% feel that the world is worse now than when their parents were growing up, and 11% would like to relocate permanently to another country. Suppose these results are based on a sample of 400 adults. Using each of these four percentages, find the 95% confidence interval for the corresponding true percentage of all U.S. adults who feel that way. Write a one-page report to present to a group of college students who have not taken statistics. Your report should answer questions such as: What is a confidence interval? Why is a range of values more informative than a single percentage? What does ''95%'' mean in this context? What assumptions, if any, are we making when we construct the confidence intervals?

*8.122 A group of veterinarians wants to test a new canine vaccine for Lyme disease. (Lyme disease is transmitted by the bite of an infected tick.) In an area having a high incidence of Lyme disease, 100 dogs are randomly selected (with their owners' permission) to receive the vaccine. Over a 12-month period, these dogs are periodically examined by veterinarians for symptoms of Lyme disease. At the end of 12 months, 10 of these 100 dogs are diagnosed with the disease. During the same 12-month period, 18% of the unvaccinated dogs in the area have been found to have Lyme disease. Let p be the proportion of all potential vaccinated dogs who would contract Lyme disease in this area.

    a. Find a 95% confidence interval for p.

    b. Does 18% lie within your confidence interval of part a? Does this suggest the vaccine might or might not be effective to some degree?

    c. Write a brief critique of this experiment pointing out anything that may have distorted the results or conclusions.

SELF-REVIEW TEST

1. Complete the following sentences using the terms *population parameter* and *sample statistic*.
    a. Estimation means assigning values to a _____ based on the value of a _____.
    b. An estimator is the _____ used to estimate a _____.
    c. The value of a _____ is called the point estimate of the corresponding _____.

2. A 95% confidence interval for μ can be interpreted to mean that if we take 100 samples of the same size and construct 100 such confidence intervals for μ then
    a. 95 of them will not include μ   **b.** 95 will include μ   **c.** 95 will include $\bar{x}$

3. The confidence level is denoted by
    a. $(1 - \alpha)100\%$ **b.** $100\alpha\%$ **c.** α

4. The maximum error of the estimate for μ is
    a. $z\sigma_{\bar{x}}$ (or $zs_{\bar{x}}$)   **b.** $\sigma/\sqrt{n}$ (or $s/\sqrt{n}$)   **c.** $\sigma_{\bar{x}}$ (or $s_{\bar{x}}$)

5. Which of the following assumptions is not required to use the *t* distribution to make a confidence interval for μ?
    a. The population from which the sample is taken is (approximately) normally distributed
    b. $n < 30$
    c. The population standard deviation, σ, is not known
    d. The sample size is at least 10

6. The parameter(s) of the *t* distribution is (are)
    a. n **b.** degrees of freedom   **c.** μ and degrees of freedom

7. A sample of 50 packages mailed from a specific post office showed a mean mailing charge of $2.85 with a standard deviation of $.72.
    a. What is the point estimate of the population mean? What is the margin of error associated with this estimate?
    b. Construct a 99% confidence interval for the mean mailing charge for all packages mailed from this post office.

8. A sample of 25 malpractice lawsuits filed against doctors showed that the mean compensation awarded to the plaintiffs was $297,364 with a standard deviation of $74,820. Find a 95% confidence interval for the mean compensation awarded to plaintiffs of all such lawsuits. Assume that the compensations awarded to plaintiffs of all such lawsuits are normally distributed.

9. A telephone poll of U.S. adults conducted by Yankelovich Partners for *Time*/CNN found that 58% of these adults felt that they were happier than their parents (*Time*, January 16, 1995). Suppose that this percentage is based on a random sample of 550 U.S. adults.
    a. What is the point estimate of the corresponding population proportion? What is the margin of error associated with this estimate?
    b. Construct a 95% confidence interval for the proportion of all U.S. adults who feel that they are happier than their parents.

10. A statistician is interested in estimating at a 95% confidence level the mean number of houses sold per month by all real estate agents in a large city. From an earlier study, it is known that the standard deviation of the number of houses sold per month by all real estate agents in this city is 1.9. How large a sample should be taken so that the estimate is within .65 of the population mean?

11. A company wants to estimate the proportion of all workers who hold more than one job. What is the most conservative estimate of the sample size that would limit the maximum error to be within .025 of the population proportion for a 99% confidence interval?

12. Refer to Problem 11. Assume that a preliminary study has shown that 12% of adults hold more than one job. How large a sample should be taken in this case so that the maximum error is within .025 of the population proportion for a 99% confidence interval?

13. Dr. Garcia estimated the mean stress score before a statistics test for a random sample of 25 students. She found the mean and standard deviation for this sample to be 6.8 (on a scale of 1 to 10) and 1.2, respectively. She used a 97% confidence level. However, she thinks that the confidence interval is too wide. How can she reduce the width of the confidence interval? Describe all possible alternatives. Which alternative do you think is best and why?

∗14. You want to estimate the mean number of hours that students at your college work per week. Briefly explain how you will conduct this study using a small sample. Take a sample of 12 students from your college who hold a job. Collect data on the number of hours that these students spent working last week. Then estimate the population mean. Choose your own confidence level. What assumptions will you make to estimate this population mean?

∗15. You want to estimate the proportion of people who are happy with their current jobs. Briefly explain how you will conduct this study. Take a sample of 35 persons and collect data on whether or not they are happy with their current jobs. Then estimate the population proportion. Choose your own confidence level.

USING MINITAB

This section describes how to use MINITAB to make confidence intervals for the population mean for large and small samples. Remember that when the sample size is large ($n \geq 30$), whether σ is known or not, the normal distribution can be used to make a confidence interval for the population mean. However, when the sample size is small ($n < 30$), the population standard deviation is not known, and the population has a normal distribution, then the t distribution is used to make a confidence interval for the population mean.

INTERVAL ESTIMATION OF A POPULATION MEAN: LARGE SAMPLES

Illustration M8–1 shows how to make a confidence interval for the population mean for a large sample.

ILLUSTRATION M8–1   The following data give the heights (in inches) of 36 randomly selected adults.

65	68	71	66	69	60	62	69	74	68	70	63
62	72	75	69	65	62	71	68	63	60	66	68
70	68	63	67	71	61	66	61	69	72	71	67

Assuming that the population standard deviation is not known, use MINITAB to construct a 95% confidence interval for the mean height of all adults.

Solution   If you are using MINITAB FOR WINDOWS, perform the following steps to find the 95% confidence interval for the mean height of all adults.

Step 1.   Enter the data on heights of 36 adults in column C1 of the Data window.

Step 2.   Calculate the standard deviation of the sample data using the MINITAB procedure explained in Chapter 3. You will obtain $s = 4.0356$.

Step 3.   Click the **Stat** pull-down menu at the top of the screen.

Step 4.   Click **Basic Statistics** from the selections available in the **Stat** menu.

Step 5.   Click **1-Sample z** from the selections available in the **Basic Statistics** menu.

Step 6.   You will see a dialog box entitled **1-Sample z** appear on the screen. Type **C1** in the box below **Variables**. Click inside the circle next to **Confidence interval** and type **95%** in the box next to **Level**. Enter **4.0356** (the value of s obtained in Step 2) in the box next to **Sigma**. Note that you are using the value of s for σ here.

Step 7.   Click the **OK** button at the bottom of this dialog box. The MINITAB output will appear on the screen.

If you are using the MINITAB COMMAND LANGUAGE, first enter the given data on the heights of 36 adults into column C1 using the **SET** command, find the standard deviation of the sample data (which will be 4.0356), and then type the following MINITAB command to obtain the required confidence interval for μ.

```
MTB > ZINTERVAL 95% SIGMA = 4.0356 C1
```

Here, **ZINTERVAL** instructs MINITAB to use the normal distribution, 95% indicates the confidence level in percent, SIGMA = 4.0356 represents the value of the standard deviation, and C1 instructs MINITAB to construct a confidence interval for μ using the value of $\bar{x}$ calculated for the data entered in column C1.

Whether you use MINITAB FOR WINDOWS or the MINITAB COMMAND LANGUAGE, you will obtain the output given in Figure 8.9.

Figure 8.9   Confidence interval for μ for a large sample.

```
Confidence Intervals

The assumed sigma = 4.04

Variable    N     Mean   StDev   SE Mean      95.0 % CI
C1          36   67.000  4.036    0.673    ( 65.682, 68.318)
```

In the MINITAB solution of Figure 8.9, *The assumed sigma* = 4.04 is the value of *s* used for σ, 36 gives the number of values in the data entered in column C1, 67.000 is the sample mean ($\bar{x}$) of that data set, 4.036 is the standard deviation (*s*) of the sample data, 0.673 is the standard error (or standard deviation) of the sample mean calculated as $s/\sqrt{n} = 4.036/\sqrt{36} = .673$, 65.682 is the lower limit of the 95% confidence interval for μ, and 68.318 is the upper limit of that confidence interval.

Thus, the 95% confidence interval for the mean height (in inches) of all adults based on the heights of adults included in this sample is

$$65.682 \text{ to } 68.318$$ ■

If you know the population standard deviation when estimating the population mean for a large sample, skip Step 2 and use the given value of σ in Step 6 of the MINITAB FOR WINDOWS procedure explained above. In the MINITAB COMMAND LANGUAGE as well, use the given value of σ instead of using the value of *s*.

INTERVAL ESTIMATION OF A POPULATION MEAN: SMALL SAMPLES

We know from the discussion in this chapter that when the sample size is small, the population is normally distributed, and the population standard deviation is not known, the *t* distribution is used to construct a confidence interval for μ. Illustration M8–2 shows how to find the confidence interval for the population mean in such a case.

ILLUSTRATION M8-2   The following data give the yearly earnings (in thousands of dollars) of 12 randomly selected households from a small town.

36.75	52.43	18.82	28.45	39.50	22.65
14.30	46.75	24.48	31.70	17.25	40.27

Using MINITAB, construct a 99% confidence interval for the mean yearly earnings of all households in this town. Assume that the distribution of yearly earnings of all households in this town is approximately normal.

Solution   Because all three conditions of the *t* distribution are satisfied, the *t* distribution will be used to make the confidence interval for μ. If you are using MINITAB FOR WINDOWS, perform the following steps to find the 99% confidence interval for the mean yearly earnings of all households in this town using the *t* distribution.

Step 1.　Enter the given data on yearly earnings of 12 households in column C1 of the Data window.

Step 2.　Click the **Stat** pull-down menu at the top of the screen.

Step 3.　Click **Basic Statistics** from the selections available in the **Stat** menu.

Step 4.　Click **1-Sample t** from the selections available in the **Basic Statistics** menu.

Step 5.　You will see a dialog box entitled **1-Sample t** appear on the screen. Type **C1** in the box below **Variables.** Click inside the circle next to **Confidence interval** and type **99%** in the box next to **Level.**

Step 6.　Click the **OK** button at the bottom of this dialog box. The MINITAB output will appear on the screen.

If you are using the MINITAB COMMAND LANGUAGE, first enter the given data on the yearly earnings of 12 households into column C1 using the **SET** command and then type the following MINITAB command to obtain the required confidence interval for μ.

```
MTB > TINTERVAL 99% C1
```

Here, **TINTERVAL** instructs MINITAB to use the *t* distribution, 99% indicates the confidence level as a percentage, and C1 instructs MINITAB to construct a confidence interval for μ using the value of $\bar{x}$ calculated for the data entered in column C1.

Whether you use MINITAB FOR WINDOWS or the MINITAB COMMAND LANGUAGE, you will obtain the output given in Figure 8.10.

Figure 8.10　Confidence interval for μ for a small sample.

```
Confidence Intervals

Variable    N    Mean   StDev   SE Mean      99.0 % CI
C1         12   31.11   12.19     3.52    ( 20.18, 42.05)
```

Thus, the 99% confidence interval for the yearly earnings of all households in this town based on the earnings of 12 households included in this sample is

$$20.18 \text{ to } 42.05$$

Because the given data are in thousands of dollars, the confidence interval can be written as

$$\$20,180 \text{ to } \$42,050$$

INTERVAL ESTIMATION OF A POPULATION PROPORTION: LARGE SAMPLES

The current version of MINITAB FOR WINDOWS does not include procedures to make a confidence interval for the population proportion. However, the MINITAB COMMAND LANGUAGE can be used to make such a confidence interval. This is done in combination with the test of hypothesis about *p*, which we will study in Chapter 9. The use of the MINITAB COMMAND LANGUAGE to make a confidence interval for *p* will be explained in the section on hypothesis tests about a population proportion in the MINITAB section of Chapter 9.

COMPUTER ASSIGNMENTS

M8.1 The following data give the annual income (in thousands of dollars) before taxes for a sample of 36 randomly selected families from a city.

21.6	13.0	25.6	27.9	50.0	18.1
10.1	21.5	10.0	72.8	58.2	15.4
7.2	27.0	32.2	45.0	95.0	27.8
92.8	8.4	45.3	76.0	28.6	9.3
30.6	19.0	25.5	27.5	9.7	15.1
6.3	44.5	24.0	13.0	61.7	16.0

Using MINITAB, construct a 99% confidence interval for μ assuming that the population standard deviation is $13.75 thousand.

M8.2 The following data give the checking account balances on a certain day for a randomly selected sample of 30 households.

500	100	650	1917	2200	500	180	3000	1500	1300
319	1500	1102	405	124	1000	134	2000	150	800
200	750	300	2300	40	1200	500	900	20	160

Using MINITAB, construct a 97% confidence interval for μ assuming that the population standard deviation is unknown.

M8.3 Refer to Data Set I of Appendix B on prices of various products in different cities across the country. Using the data on monthly telephone charges given in column C7, make a 98% confidence interval for the population mean μ.

M8.4 Refer to the Manchester Road Race data for all participants on the floppy diskette provided by the publisher. Using MINITAB, take a sample of 100 observations from this data set.
   a.   Using the sample data, make a 95% confidence interval for the mean time taken to complete this race by all participants.
   b.   The mean time taken to run this race by all participants is 48.12 minutes. Does the confidence interval made in part a include this population mean?

M8.5 Repeat Computer Assignment M8.4 for a sample of 25 observations. Assume that the distribution of time taken to run this race by all participants is approximately normal.

M8.6 The following data give the prices (in thousands of dollars) of 16 recently sold houses in an area.

141	163	127	104	197	203	113	179
256	228	183	119	133	199	271	191

Using MINITAB, construct a 99% confidence interval for the mean price of all houses in this area. Assume that the distribution of prices of all houses in the given area is normal.

M8.7 A researcher wanted to estimate the mean contributions made to charitable causes by major companies. A random sample of 18 companies produced the following data on contributions (in millions of dollars) made by them.

1.8	.6	1.2	.3	2.6	1.9	3.4	2.6	.2
2.4	1.4	2.5	3.1	.9	1.2	2.0	.8	1.1

Using MINITAB, make a 98% confidence interval for the mean contributions made to charitable causes by all major companies. Assume that the contributions made to charitable causes by all major companies have a normal distribution.

9

HYPOTHESIS TESTS ABOUT THE MEAN AND PROPORTION

This chapter introduces the second topic in inferential statistics: tests of hypotheses. In a test of hypothesis, we test a certain given theory or belief about a population parameter. We may want to find out, using some sample information, whether or not a given claim (or statement) about a population parameter is true. This chapter discusses how to make such tests of hypotheses about the population mean, μ, and the population proportion, p.

As an example, a soft-drink company may claim that, on average, its cans contain 12 ounces of soda. A government agency may want to test whether or not such cans contain, on average, 12 ounces of soda. As another example, according to the U.S. Bureau of the Census, 15.4% of the population in the United States lacked health insurance in 1995. An economist may want to check if this percentage is still true for this year. In the first of these two examples we are to test a hypothesis about the population mean, μ, and in the second example we are to test a hypothesis about the population proportion, p.

9.1 HYPOTHESIS TESTS: AN INTRODUCTION

Why do we need to perform a test of hypothesis? Reconsider the example about soft-drink cans. Suppose we take a sample of 100 cans of the soft drink under investigation. We then find out that the mean amount of soda in these 100 cans is 11.89 ounces. Based on this result, can we state that, on average, all such cans contain less than 12 ounces of soda and that the company is lying to the public? Not until we perform a test of hypothesis can we make such an accusation. The reason is that the mean, $\bar{x} = 11.89$ ounces, is obtained from a sample. The difference between 12 ounces (the required average amount for the population) and 11.89 ounces (the observed average amount for the sample) may have occurred only because of the sampling error. Another sample of 100 cans may give us a mean of 12.04 ounces. Therefore, we make a test of hypothesis to find out how large the difference between 12 ounces and 11.89 ounces is and to investigate whether or not this difference has occurred as a result of chance alone. Now, if 11.89 ounces is the mean for all cans and not for just 100 cans, then we do not need to make a test of hypothesis. Instead, we can immediately state that the mean amount of soda in all such cans is less than 12 ounces. We perform a test of hypothesis only when we are making a decision about a population parameter based on the value of a sample statistic.

9.1.1 TWO HYPOTHESES

Consider a nonstatistical example of a person who has been indicted for committing a crime and is being tried in a court. Based on the available evidence, the judge or jury will make one of two possible decisions:

1. The person is not guilty.
2. The person is guilty.

At the outset of the trial, the person is presumed not guilty. The prosecutor's efforts are to prove that the person has committed the crime and, hence, is guilty. In statistics, *the person is not guilty* is called the **null hypothesis** and *the person is guilty* is called the **alternative hypothesis**. The null hypothesis is denoted by H_0 and the alternative hypothesis is denoted by H_1. In the beginning of the trial it is assumed that the person is not guilty. The null hypothesis is usually the hypothesis that is assumed to be true to begin with. The two hypotheses for the court case are written as follows (notice the colon after H_0 and H_1).

Null hypothesis: H_0: The person is not guilty

Alternative hypothesis: H_1: The person is guilty

In a statistics example, the null hypothesis states that a given claim (or statement) about a population parameter is true. Reconsider the example of the soft-drink company's claim that, on average, its cans contain 12 ounces of soda. In reality, this claim may or may not be true. However, we will initially assume that the company's claim is true (that is, the company is not guilty of cheating and lying). To test the claim of the soft-drink company, the null hypothesis will be that the company's claim is true. Let μ be the mean amount of soda in all cans. The company's claim will be true if $\mu = 12$ ounces. Thus, the null hypothesis will be written as

$$H_0: \mu = 12 \text{ ounces} \qquad \text{(The company's claim is true)}$$

In this example, the null hypothesis can also be written as $\mu \geq 12$ ounces because the claim of the company will still be true if the cans contain, on average, more than 12 ounces of soda. The company will be accused of cheating the public only if the cans contain, on average, less than 12 ounces of soda. However, it will not affect the test whether we use an $=$ or a $\geq$ sign in the null hypothesis as long as the alternative hypothesis has a $<$ sign. Remember that in the null hypothesis (and in the alternative hypothesis also) we use the population parameter (such as μ or p), and not the sample statistic (such as $\bar{x}$ or $\hat{p}$).

NULL HYPOTHESIS

A *null hypothesis* is a claim (or statement) about a population parameter that is assumed to be true until it is declared false.

The alternative hypothesis in our statistics example will be that the company's claim is false and its soft-drink cans contain, on average, less than 12 ounces of soda, that is, $\mu < 12$ ounces. The alternative hypothesis will be written as

$$H_1: \mu < 12 \text{ ounces} \qquad \text{(The company's claim is false)}$$

ALTERNATIVE HYPOTHESIS

An *alternative hypothesis* is a claim about a population parameter that will be true if the null hypothesis is false.

Let us return to the example of the court trial. The trial begins with the assumption that the null hypothesis is true, that is, the person is not guilty. The prosecutor assembles all the possible evidence and presents it in the court to prove that the null hypothesis is false and the alternative hypothesis is true (that is, the person is guilty). In the case of our statistics example, the information obtained from a sample will be used as evidence to decide whether or not the claim of the company is true. In the court case, the decision made by the judge (or jury) depends on the amount of evidence presented by the prosecutor. At the end of the trial, the judge (or jury) will consider whether or not the evidence presented by the prosecutor is sufficient to declare the person guilty. The amount of evidence that will be considered to be sufficient to declare the person guilty depends on the discretion of the judge (or jury).

9.1.2   REJECTION AND NONREJECTION REGIONS

In Figure 9.1, which represents the court case, the point marked ''0'' indicates that there is no evidence against the person being tried. The farther the point is to the right on the horizontal axis, the more convincing the evidence is that the person has committed the crime. We have arbitrarily marked a point C on the horizontal axis. Let us assume that a judge (or jury) considers any amount of evidence to the right of point C to be sufficient and any amount of evidence to the left of C to be insufficient to declare the person guilty. Point C is called the **critical value** or **critical point** in statistics. If the amount of evidence presented by the prosecutor falls in the area to the left of point C, the verdict will reflect that there is not enough evidence to declare the person guilty. Consequently, the accused person will be declared *not guilty*. In statistics, this decision is stated as *do not reject H_0*. It is equivalent to saying that there is not enough evidence to declare the null hypothesis false. The area to the left of point C is called the *nonrejection region*, that is, this is the region where the null hypothesis is not rejected. However, if the amount of evidence falls in the area to the right of point C, the verdict will be that there is sufficient evidence to declare the person guilty. In statistics, this decision is stated as *reject H_0* or *the null hypothesis is false*. Rejecting H_0 is equivalent to saying that *the alternative hypothesis is true*. The area to the right of point C is called the *rejection region*, that is, this is the region where the null hypothesis is rejected.

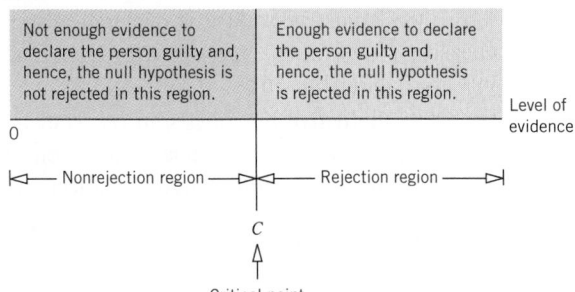

Figure 9.1   Nonrejection and rejection regions for the court case.

9.1.3   TWO TYPES OF ERRORS

We all know that a court's verdict is not always correct. If a person is declared guilty at the end of a trial, there are two possibilities.

1. The person has *not* committed the crime but is declared guilty (because of what may be false evidence).

2. The person *has* committed the crime and is rightfully declared guilty.

In the first case, the court has made an error by punishing an innocent person. In statistics, this kind of error is called a **Type I** or an **α** *(alpha)* **error**. In the second case, because the guilty person has been punished, the court has made the correct decision. The second row in the shaded portion of Table 9.1 shows these two cases. The two columns of Table 9.1, corresponding to *the person is not guilty* and *the person is guilty*, give the two actual situations. Which one of these is true is known only to the person being tried. The two rows in this table, corresponding to *the person is not guilty* and *the person is guilty*, show the two possible court decisions.

Table 9.1

		Actual Situation	
		The Person Is Not Guilty	**The Person Is Guilty**
Court's decision	The person is not guilty	Correct decision	Type II or β error
	The person is guilty	Type I or α error	Correct decision

In our statistics example, a Type I error will occur when H_0 is actually true (that is, the cans do contain, on average, 12 ounces of soda), but it just happens that we draw a sample with a mean that is well below 12 ounces and we wrongfully reject the null hypothesis, H_0. The value of **α**, called the **significance level** of the test, represents the probability of making a Type I error. In other words, α is the probability of rejecting the null hypothesis, H_0, when in fact it is true.

TYPE I ERROR

A *Type I error* occurs when a true null hypothesis is rejected. The value of α represents the probability of committing this type of error, that is,

$$\alpha = P(H_0 \text{ is rejected} \mid H_0 \text{ is true})$$

The value of α represents the significance level of the test.

The size of the rejection region in a statistics problem of a test of hypothesis depends on the value assigned to α. In a test of hypothesis, we usually assign a value to α before making the test. Although any value can be assigned to α, the commonly used values of α are .01, .025, .05, and .10. Usually the value assigned to α does not exceed .10 (or 10%).

Now, suppose that in the court trial case the person is declared not guilty at the end of the trial. Such a verdict does not indicate that the person has indeed *not* committed the crime. It is possible that the person is guilty but there is not enough evidence to prove the guilt. Consequently, in this situation there are again two possibilities.

1. The person has *not* committed the crime and is declared not guilty.
2. The person *has* committed the crime but, *because of the lack of enough evidence*, is declared not guilty.

In the first case, the court's decision is correct. But in the second case the court has committed an error by setting a guilty person free. In statistics, this type of error is called a

Type II or a β (the Greek letter *beta*) **error**. These two cases are shown in the first row of the shaded portion of Table 9.1.

In our statistics example, a Type II error will occur when the null hypothesis H_0 is actually false (that is, the soda contained in all cans, on average, is less than 12 ounces), but it happens by chance that we draw a sample with a mean that is close to or larger than 12 ounces and we wrongfully conclude *do not reject H_0*. The value of β represents the probability of making a Type II error. It represents the probability that H_0 is not rejected when actually H_0 is false. The value of $1 - \beta$ is called the **power of the test**. It represents the probability of not making a Type II error.

TYPE II ERROR

A *Type II error* occurs when a false null hypothesis is not rejected. The value of β represents the probability of committing a Type II error, that is,

$$\beta = P(H_0 \text{ is not rejected} \mid H_0 \text{ is false})$$

The value of $1 - \beta$ is called the power of the test. It represents the probability of not making a Type II error.

The two types of errors that occur in tests of hypotheses depend on each other. We cannot lower the values of α and β simultaneously for a test of hypothesis for a fixed sample size. Lowering the value of α will raise the value of β, and lowering the value of β will raise the value of α. However, we can decrease both α and β simultaneously by increasing the sample size. The explanation of how α and β are related and the computation of β are not within the scope of this text.

Table 9.2, which is similar to Table 9.1, is written for the statistics problem of a test of hypothesis. In Table 9.2 *the person is not guilty* is replaced by H_0 *is true*, *the person is guilty* by H_0 *is false*, and the *court's decision* by *decision*.

Table 9.2

		Actual Situation	
		H_0 **is true**	H_0 **is false**
Decision	Do not reject H_0	Correct decision	Type II or β error
	Reject H_0	Type I or α error	Correct decision

9.1.4   TAILS OF A TEST

The statistical hypothesis-testing procedure is similar to the trial of a person in the court but with two major differences. The first major difference is that in a statistical test of hypothesis, the partition of the total region into rejection and nonrejection regions is not arbitrary. Instead, it depends on the value assigned to α (Type I error). As mentioned earlier, α is also called the significance level of the test.

The second major difference relates to the rejection region. In the court case, the rejection region is on the right side of the critical point, as shown in Figure 9.1. However, in statistics, the rejection region for a hypothesis-testing problem can be on both sides with the nonrejection region in the middle, or it can be on the left side or on the right side of the nonrejection region. These possibilities are explained in the next three parts of this section. A test with two rejection regions is called a **two-tailed test**, and a test with one rejection region is called a **one-tailed test**. The one-tailed test is called a **left-tailed test** if the rejection region is in the left tail of the distribution curve, and it is called a **right-tailed test** if the rejection region is in the right tail of the distribution curve.

TAILS OF THE TEST

A *two-tailed test* has rejection regions in both tails, a *left-tailed test* has the rejection region in the left tail, and a *right-tailed test* has the rejection region in the right tail of the distribution curve.

A Two-tailed Test

According to the U.S. Bureau of the Census, the mean family size in the United States was 3.19 in 1995. A researcher wants to check whether or not this mean has changed since 1995. The key word here is *changed*. The mean family size has changed if it has either increased or decreased during the period since 1995. This is an example of a two-tailed test. Let μ be the current mean family size for all families. The two possible decisions are

1. The mean family size has not changed, that is, $\mu = 3.19$.
2. The mean family size has changed, that is, $\mu \neq 3.19$.

We write the null and alternative hypotheses for this test as

$$H_0: \mu = 3.19 \qquad \text{(The mean family size has not changed)}$$

$$H_1: \mu \neq 3.19 \qquad \text{(The mean family size has changed)}$$

Whether a test is two-tailed or one-tailed is determined by the sign in the alternative hypothesis. If the alternative hypothesis has a *not equal to* ($\neq$) sign, as in this example, it is a two-tailed test. As shown in Figure 9.2, a two-tailed test has two rejection regions, one in each tail of the distribution curve. Figure 9.2 shows the sampling distribution of $\bar{x}$ for a large sample. Assuming H_0 is true, $\bar{x}$ has a normal distribution with its mean equal to 3.19 (the value of μ in H_0). In Figure 9.2, the area of each of the two rejection regions is $\alpha/2$ and the total area of both rejection regions is α (the significance level). As shown in this figure, a two-tailed test of hypothesis has two critical values that separate the two rejection regions from the nonrejection region. We will reject H_0 if the value of $\bar{x}$ obtained from the sample falls in either of the two rejection regions. We will not reject H_0 if the value of $\bar{x}$ lies in the nonrejection region. By rejecting H_0, we are saying that the difference between the value of μ stated in H_0 and the value of $\bar{x}$ obtained from the sample is too large to have occurred because of the sampling error alone. Consequently, this difference is real. By not rejecting H_0, we are saying that the difference between the value of μ stated in H_0 and the value of $\bar{x}$ obtained from the sample is small and it may have occurred because of the sampling error alone.

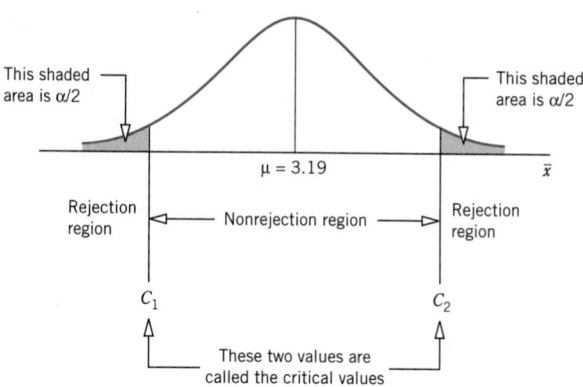

Figure 9.2 A two-tailed test.

A Left-tailed Test

Reconsider the example of the mean amount of soda in all soft-drink cans produced by a company. The company claims that these cans, on average, contain 12 ounces of soda. However, if these cans contain less than the claimed amount of soda, then the company can be accused of cheating. Suppose a consumer agency wants to test whether the mean amount of soda per can is less than 12 ounces. Note that the key phrase this time is *less than*, which indicates a left-tailed test. Let μ be the mean amount of soda in all cans. The two possible decisions are

1. The mean amount of soda in all cans is not less than 12 ounces, that is, $\mu = 12$ ounces.
2. The mean amount of soda in all cans is less than 12 ounces, that is, $\mu < 12$ ounces.

The null and alternative hypotheses for this test are written as

$$H_0: \mu = 12 \text{ ounces} \quad \text{(The mean is not less than 12 ounces)}$$

$$H_1: \mu < 12 \text{ ounces} \quad \text{(The mean is less than 12 ounces)}$$

In this case, we can also write the null hypothesis as $H_0: \mu \geq 12$. This will not affect the result of the test as long as the sign in H_1 is *less than* ($<$).

When the alternative hypothesis has a *less than* ($<$) sign, as in this case, the test is always left-tailed. In a left-tailed test, the rejection region is always in the left tail of the distribution curve, as shown in Figure 9.3, and the area of this rejection region is equal to α (the significance level). We can observe from this figure that there is only one critical value in a left-tailed test.

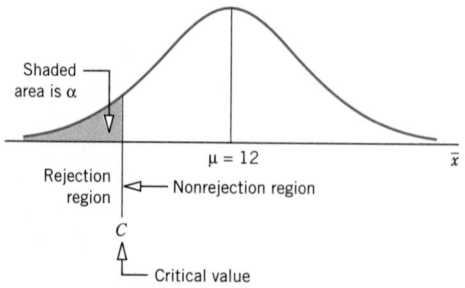

Figure 9.3 A left-tailed test.

Assuming H_0 is true, $\bar{x}$ has a normal distribution for a large sample with its mean equal to 12 ounces (the value of μ in H_0). We will reject H_0 if the value of $\bar{x}$ obtained from the sample falls in the rejection region; we will not reject H_0 otherwise.

A Right-tailed Test

To illustrate the third case, according to a study by the Organization for Economic Cooperation and Development released in 1996, the mean starting salary of public elementary school teachers in the United States was $22,753. Suppose we want to test if the current mean starting salary of all public elementary school teachers in the United States is higher than $22,753. The key phrase in this case is *higher than*, which indicates a right-tailed test. Let μ be the current mean starting salary of public elementary school teachers in the United States. The two possible decisions this time are

1. The current mean starting salary of all public elementary school teachers in the United States is not higher than $22,753, that is, $\mu = \$22,753$.
2. The current mean starting salary of all public elementary school teachers in the United States is higher than $22,753, that is, $\mu > \$22,753$.

We write the null and alternative hypotheses for this test as

H_0: $\mu = \$22,753$ (The current mean starting salary is not higher than $22,753)

H_1: $\mu > \$22,753$ (The current mean starting salary is higher than $22,753)

In this case, we can also write the null hypothesis as H_0: $\mu \leq \$22,753$, which states that the current mean starting salary of all public elementary school teachers in the United States is either equal to or less than $22,753. Again, the result of the test will not be affected whether we use an *equal to* ($=$) or a *less than or equal to* ($\leq$) sign in H_0 as long as the alternative hypothesis has a *greater than* ($>$) sign.

When the alternative hypothesis has a *greater than* ($>$) sign, the test is always right-tailed. As shown in Figure 9.4, in a right-tailed test, the rejection region is in the right tail of the distribution curve. The area of this rejection region is equal to α, the significance level. Like a left-tailed test, a right-tailed test has only one critical value.

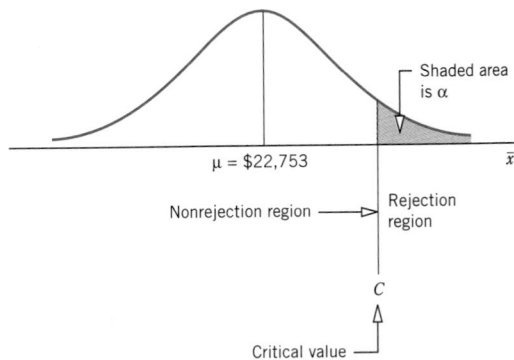

Figure 9.4 A right-tailed test.

Again, assuming H_0 is true, $\bar{x}$ has a normal distribution for a large sample with its mean equal to $22,753 (the value of μ in H_0). We will reject H_0 if the value of $\bar{x}$ obtained from the sample falls in the rejection region. Otherwise, we will not reject H_0.

Table 9.3 summarizes the foregoing discussion about the relationship between the signs in H_0 and H_1 and the tails of a test.

Table 9.3

	Two-tailed Test	Left-tailed Test	Right-tailed Test
Sign in the null hypothesis H_0	=	= or ≥	= or ≤
Sign in the alternative hypothesis H_1	≠	<	>
Rejection region	In both tails	In the left tail	In the right tail

Note that the null hypothesis always has an *equal to* (=) or a *greater than or equal to* (≥) or a *less than or equal to* (≤) sign, and the alternative hypothesis always has a *not equal to* (≠) or a *less than* (<) or a *greater than* (>) sign.

A test of hypothesis involves five steps, which are listed below.

STEPS TO PERFORM A TEST OF HYPOTHESIS

A statistical test of hypothesis procedure contains the following five steps.

1. State the null and alternative hypotheses
2. Select the distribution to use
3. Determine the rejection and nonrejection regions
4. Calculate the value of the test statistic
5. Make a decision

With the help of examples, these steps will be described in the next section.

EXERCISES

Concepts and Procedures

9.1 Briefly explain the meaning of each of the following terms.
   - a. Null hypothesis
   - b. Alternative hypothesis
   - c. Critical point(s)
   - d. Significance level
   - e. Nonrejection region
   - f. Rejection region
   - g. Tails of a test
   - h. Two types of errors

9.2 What are the four possible outcomes for a test of hypothesis? Show these outcomes by writing a table. Briefly describe the Type I and Type II errors.

9.3 Explain how the tails of a test depend on the sign in the alternative hypothesis. Describe the signs in the null and alternative hypotheses for a two-tailed, a left-tailed, and a right-tailed test, respectively.

9.4 Explain which of the following is a two-tailed test, a left-tailed test, or a right-tailed test.
   - a. H_0: $\mu = 45$, H_1: $\mu > 45$
   - b. H_0: $\mu = 23$, H_1: $\mu \neq 23$
   - c. H_0: $\mu \geq 75$, H_1: $\mu < 75$

Show the rejection and nonrejection regions for each of these cases by drawing a sampling distribution curve for the sample mean, assuming that the sample size is large in each case.

9.5 Explain which of the following is a two-tailed test, a left-tailed test, or a right-tailed test.

    a. H_0: $\mu = 12$, H_1: $\mu < 12$

    b. H_0: $\mu \leq 85$, H_1: $\mu > 85$

    c. H_0: $\mu = 33$, H_1: $\mu \neq 33$

Show the rejection and nonrejection regions for each of these cases by drawing a sampling distribution curve for the sample mean, assuming that the sample size is large in each case.

9.6 Which of the two hypotheses (null and alternative) is initially assumed to be true in a test of hypothesis?

9.7 Consider H_0: $\mu = 20$ versus H_1: $\mu < 20$.

    a. What type of error would you make if the null hypothesis is actually false and you fail to reject it?

    b. What type of error would you make if the null hypothesis is actually true and you reject it?

9.8 Consider H_0: $\mu = 55$ versus H_1: $\mu \neq 55$.

    a. What type of error would you make if the null hypothesis is actually false and you fail to reject it?

    b. What type of error would you make if the null hypothesis is actually true and you reject it?

Applications

9.9 Write the null and alternative hypotheses for each of the following examples. Determine if each is a case of a two-tailed, a left-tailed, or a right-tailed test.

    a. To test whether or not the mean price of houses in Connecticut is greater than $143,000

    b. To test if the mean number of hours spent working per week by college students who hold jobs is different from 15 hours

    c. To test whether the mean life of a particular brand of auto batteries is less than 45 months

    d. To test if the mean amount of time spent doing homework by all fourth-graders is different from 5 hours a week

    e. To test if the mean age of all college students is different from 24 years

9.10 Write the null and alternative hypotheses for each of the following examples. Determine if each is a case of a two-tailed, a left-tailed, or a right-tailed test.

    a. To test if the mean amount of time spent per week watching sports on television by all adult males is different from 9.5 hours

    b. To test if the mean amount of money spent by all customers at a supermarket is less than $85

    c. To test whether the mean starting salary of college graduates is higher than $29,000 per year

    d. To test if the mean GPA of all students at a university is lower than 2.9

    e. To test if the mean cholesterol level of all adult males in the United States is higher than 175

9.2 HYPOTHESIS TESTS ABOUT A POPULATION MEAN: LARGE SAMPLES

From the central limit theorem discussed in Chapter 7, the sampling distribution of $\bar{x}$ is approximately normal for large samples ($n \geq 30$). Consequently, whether or not σ is known, the normal distribution is used to test hypotheses about the population mean when a sample size is large.

TEST STATISTIC

In tests of hypotheses about μ for large samples, the random variable

$$z = \frac{\bar{x} - \mu}{\sigma_{\bar{x}}} \quad \text{or} \quad \frac{\bar{x} - \mu}{s_{\bar{x}}}$$

where $\qquad \sigma_{\bar{x}} = \sigma/\sqrt{n} \quad$ and $\quad s_{\bar{x}} = s/\sqrt{n}$

is called the *test statistic*. The test statistic can be defined as a rule or criterion that is used to make the decision whether or not to reject the null hypothesis.

At the end of Section 9.1, it was mentioned that a test of hypothesis procedure involves the following five steps.

1. State the null and alternative hypotheses
2. Select the distribution to use
3. Determine the rejection and nonrejection regions
4. Calculate the value of the test statistic
5. Make a decision

Examples 9–1 through 9–3 illustrate the use of these five steps to perform tests of hypotheses about the population mean μ. Example 9–1 is concerned with a two-tailed test and Examples 9–2 and 9–3 describe one-tailed tests.

Conducting a two-tailed test of hypothesis about μ for a large sample.

EXAMPLE 9–1   The TIV Telephone Company provides long-distance telephone service in an area. According to the company's records, the average length of all long-distance calls placed through this company in 1996 was 12.44 minutes. The company's management wanted to check if the mean length of the current long-distance calls is different from 12.44 minutes. A sample of 150 such calls placed through this company produced a mean length of 13.71 minutes with a standard deviation of 2.65 minutes. Using the 5% significance level, can you conclude that the mean length of all current long-distance calls is different from 12.44 minutes?

Solution   Let μ be the mean length of all current long-distance calls placed through this company and $\bar{x}$ be the corresponding mean for the sample. From the given information,

$$n = 150, \quad \bar{x} = 13.71 \text{ minutes}, \quad \text{and} \quad s = 2.65 \text{ minutes}$$

We are to test whether or not the mean length of all current long-distance calls is different from 12.44 minutes. The significance level α is .05. That is, the probability of rejecting the null hypothesis when it actually is true should not exceed .05. This is the probability of making a Type I error. We perform the test of hypothesis using the five steps as follows.

Step 1.   *State the null and alternative hypotheses*

Notice that we are testing to find whether or not the mean length of all current long-distance calls is different from 12.44 minutes. We write the null and alternative hypotheses as follows.

H_0: $\mu = 12.44$ (The mean length of all current long-distance calls is 12.44 minutes)

H_1: $\mu \neq 12.44$ (The mean length of all current long-distance calls is different from 12.44 minutes)

Step 2. *Select the distribution to use*

Because the sample size is large ($n > 30$), the sampling distribution of $\bar{x}$ is (approximately) normal. Consequently, we use the normal distribution to make the test.

Step 3. *Determine the rejection and nonrejection regions*

The significance level is .05. The $\neq$ sign in the alternative hypothesis indicates that the test is two-tailed with two rejection regions, one in each tail of the normal distribution curve of $\bar{x}$. Because the total area of both rejection regions is .05 (the significance level), the area of the rejection region in each tail is .025, that is,

Area in each tail $= \alpha/2 = .05/2 = .025$

These areas are shown in Figure 9.5. Two critical points in this figure separate the two rejection regions from the nonrejection region. Next we find the z values for the two critical points using the area of the rejection region. To find the z values for these critical points, we first find the area between the mean and one of the critical points. We obtain this area by subtracting .025 (the area in each tail) from .5, which gives .4750. Next we look for .4750 in the standard normal distribution table, Table VII of Appendix C. The value of z for .4750 is 1.96. Hence, the z values of the two critical points, as shown in Figure 9.5, are -1.96 and 1.96.

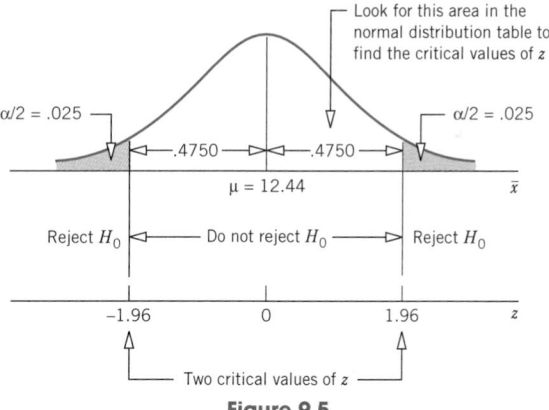

Figure 9.5

Step 4. *Calculate the value of the test statistic*

The decision to reject or not to reject the null hypothesis will depend on whether the evidence from the sample falls in the rejection or nonrejection region. If the value of $\bar{x}$ falls in either of the two rejection regions, we reject H_0. Otherwise, we do not reject H_0. The value of $\bar{x}$ obtained from the sample is called the *observed value of $\bar{x}$*. To locate the position of $\bar{x} = 13.71$ on the sampling distribution curve of $\bar{x}$ in Figure 9.5, we first calculate the z value for $\bar{x} = 13.71$. This is called the *value of the test statistic*. Then, we compare the value of the test statistic with the two critical values of z, -1.96 and 1.96, shown in Figure 9.5. If the value of the test statistic is between -1.96 and 1.96, we do not reject H_0. If the value of the test statistic is either greater than 1.96 or less than -1.96, we reject H_0.

CALCULATING THE VALUE OF THE TEST STATISTIC

For a large sample, *the value of the test statistic z for $\bar{x}$ for a test of hypothesis about μ* is computed as follows.

$$z = \frac{\bar{x} - \mu}{\sigma_{\bar{x}}} \quad \text{if } \sigma \text{ is known}$$

$$z = \frac{\bar{x} - \mu}{s_{\bar{x}}} \quad \text{if } \sigma \text{ is not known}$$

where $\sigma_{\bar{x}} = \sigma/\sqrt{n}$ and $s_{\bar{x}} = s/\sqrt{n}$

The value of z calculated for $\bar{x}$ using the above formula is also called the **observed value of z**.

The value of $\bar{x}$ from the sample is 13.71. As σ is not known, we calculate the z value using $s_{\bar{x}}$ as follows.

$$s_{\bar{x}} = \frac{s}{\sqrt{n}} = \frac{2.65}{\sqrt{150}} = .21637159$$

$$z = \frac{\bar{x} - \mu}{s_{\bar{x}}} = \frac{13.71 - 12.44}{.21637159} = 5.87$$

From H_0

The value of μ in the calculation of the z value is substituted from the null hypothesis. The value of $z = 5.87$ calculated for $\bar{x}$ is called the *computed value of the test statistic z*. This is the value of z that corresponds to the value of $\bar{x}$ observed from the sample. It is also called the *observed value of z*.

Step 5. Make a decision

In the final step we make a decision based on the location of the value of the test statistic z computed for $\bar{x}$ in Step 4. This value of $z = 5.87$ is greater than the critical value of $z = 1.96$, and it falls in the rejection region in the right tail in Figure 9.5. Hence, we reject H_0 and conclude that based on the sample information, it appears that the mean length of all such calls is not equal to 12.44 minutes.

By rejecting the null hypothesis we are stating that the difference between the sample mean, $\bar{x} = 13.71$ minutes, and the hypothesized value of the population mean, $\mu = 12.44$ minutes, is too large and may not have occurred because of chance or sampling error alone. This difference seems to be real and, hence, the mean length of all such calls is different from 12.44 minutes. Note that the rejection of the null hypothesis does not necessarily indicate that the mean length of all such calls is definitely different from 12.44 minutes. It simply indicates that there is strong evidence (from the sample) that the mean length of such calls is not equal to 12.44 minutes. There is a possibility that the mean length of all such calls is equal to 12.44 minutes but, by the luck of the draw, we selected a sample with a mean that is too far from the hypothesized mean of 12.44 minutes. If so, we have wrongfully rejected the null hypothesis H_0. This is a Type I error and its probability is .05 in this example.

Making a right-tailed test of hypothesis about μ for a large sample.

EXAMPLE 9-2   According to the National Association of Realtors, the mean sales price of existing single-family homes in the United States was $139,000 in 1995 (*Real Estate Outlook*, 3[6], June 1996). A random sample of 500 such homes that were recently sold gave a mean sales price of $146,690 with a standard deviation of $23,700. Test at the 1% significance level if the current mean sales price of such homes is greater than $139,000.

Solution   Let μ be the current mean sales price of all existing single-family homes in the United States and $\bar{x}$ the corresponding mean for the sample. From the given information,

$$n = 500, \qquad \bar{x} = \$146,690, \qquad \text{and} \qquad s = \$23,700$$

The significance level is $\alpha = .01$.

Step 1.   *State the null and alternative hypotheses*

We are to test if the current mean sales price of existing single-family homes is greater than $139,000. The null and alternative hypotheses are

H_0: $\mu = \$139,000$ (The current mean is not greater than $139,000)

H_1: $\mu > \$139,000$ (The current mean is greater than $139,000)

Step 2.   *Select the distribution to use*

Because the sample size is large ($n > 30$), the sampling distribution of $\bar{x}$ is (approximately) normal. Consequently, we use the normal distribution to make the test.

Step 3.   *Determine the rejection and nonrejection regions*

The significance level is .01. The $>$ sign in the alternative hypothesis indicates that the test is right-tailed with its rejection region in the right tail of the sampling distribution curve of $\bar{x}$. Because there is only one rejection region, its area is $\alpha = .01$. As shown in Figure 9.6, the critical value of z, obtained from Table VII of Appendix C for .4900, is approximately 2.33.

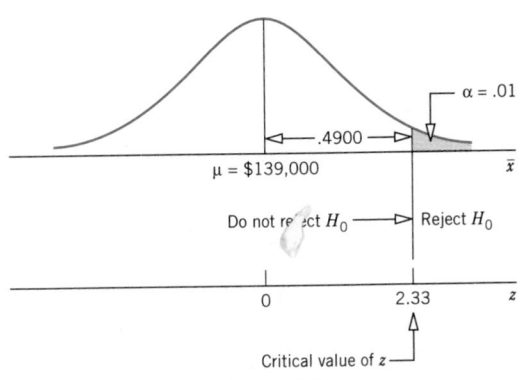

Figure 9.6

Step 4. *Calculate the value of the test statistic*

The value of the test statistic z for $\bar{x} = \$146,690$ is computed as follows.

$$s_{\bar{x}} = \frac{s}{\sqrt{n}} = \frac{23,700}{\sqrt{500}} = \$1059.89622133$$

$$z = \frac{\bar{x} - \mu}{s_{\bar{x}}} = \frac{146,690 - 139,000}{1059.89622133} = 7.26$$

(From H_0)

Step 5. *Make a decision*

Because the value of the test statistic $z = 7.26$ is larger than the critical value of $z = 2.33$ and it falls in the rejection region, we reject H_0. Consequently, we can state that the sample mean $\bar{x} = \$146,690$ is too far from the hypothesized population mean $\mu = \$139,000$. The difference between the two may not be attributed to chance or sampling error alone. Therefore, the current mean sales price of existing single-family homes in the United States is greater than $139,000.

Making a left-tailed test of hypothesis about μ *for a large sample.*

EXAMPLE 9–3 Because couples are deciding to have fewer children, the family size in the United States has declined continuously during the past few decades. According to the U.S. Bureau of the Census, the mean family size was 3.19 in 1995. A researcher wanted to check if the current mean family size is less than 3.19. A sample of 900 families taken this year by this researcher produced a mean family size of 3.16 with a standard deviation of .70. Using the .025 significance level, can we conclude that the mean family size has declined since 1995?

Solution Let μ be the current mean size of all families and $\bar{x}$ the mean family size for the sample. From the given information,

$$n = 900, \quad \bar{x} = 3.16, \quad \text{and} \quad s = .70$$

The mean family size for 1995 is given to be 3.19. The significance level α is .025.

Step 1. *State the null and alternative hypotheses*

Notice that we are testing for a *decline* in the mean family size. The null and alternative hypotheses are written as follows.

$$H_0: \mu = 3.19 \quad \text{(The mean family size has not declined)}$$
$$H_1: \mu < 3.19 \quad \text{(The mean family size has declined)}$$

Step 2. *Select the distribution to use*

Because the sample size is large ($n > 30$), the sampling distribution of $\bar{x}$ is (approximately) normal. Consequently, we use the normal distribution to make the test.

Step 3. *Determine the rejection and nonrejection regions*

The significance level is .025. The $<$ sign in the alternative hypothesis indicates that the test is left-tailed with the rejection region in the left tail of the sampling distribution curve of $\bar{x}$. The critical value of z, obtained from the normal table for .4750, is -1.96, as shown in Figure 9.7.

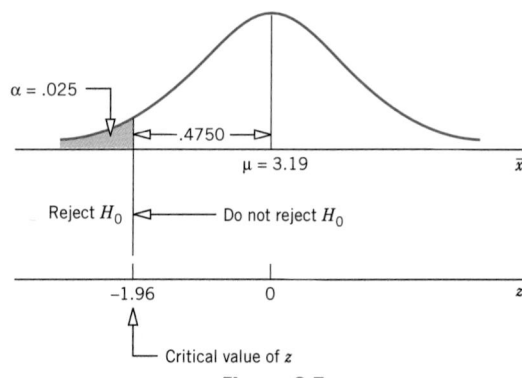

Figure 9.7

Step 4. *Calculate the value of the test statistic*

The value of the test statistic z for $\bar{x} = 3.16$ is calculated as follows.

$$s_{\bar{x}} = \frac{s}{\sqrt{n}} = \frac{.70}{\sqrt{900}} = .02333333$$

From H_0

$$z = \frac{\bar{x} - \mu}{s_{\bar{x}}} = \frac{3.16 - 3.19}{.02333333} = -1.29$$

Step 5. *Make a decision*

The value of the test statistic $z = -1.29$ is greater than the critical value of $z = -1.96$, and it falls in the nonrejection region. As a result, we fail to reject H_0. Consequently, we can state that based on the sample information, it appears that the mean family size has not declined since 1995. Note that we are not concluding that the mean family size has definitely not declined. By not rejecting the null hypothesis, we are saying that the information obtained from the sample is not strong enough to reject the null hypothesis and to conclude that the family size has declined since 1995. ▬

In studies published in various journals, authors usually use the terms *significantly different* and *not significantly different* when deriving conclusions based on hypothesis tests. These terms are short versions of the terms *statistically significantly different* and *statistically not significantly different*. The statement *significantly different* means that the difference between the observed value of the sample mean $\bar{x}$ and the hypothesized value of the population mean μ is so large that it probably did not occur because of the sampling error alone. Consequently, the null hypothesis is rejected. In other words, the difference between $\bar{x}$ and μ is statistically significant. Thus, the statement *significantly different* is equivalent to saying that the *null hypothesis is rejected*. In Example 9–2, we can state as a conclusion that the observed value of $\bar{x} = \$146,690$ is significantly different from the hypothesized value of $\mu = \$139,000$. That is, the current mean sales price of existing single-family homes is significantly different from $139,000.

On the other hand, the statement *not significantly different* means that the difference between the observed value of the sample mean $\bar{x}$ and the hypothesized value of the population mean μ is so small that it may have occurred just because of chance. Consequently, the null hypothesis is not rejected. Thus, the statement *not significantly different* is equivalent to saying that we *fail to reject the null hypothesis*. In Example 9–3, we can state as a conclusion that the observed value of $\bar{x} = 3.16$ is not significantly different from the hypothesized value of $\mu = 3.19$. In other words, the current mean family size does not seem to be significantly different from 3.19.

EXERCISES

Concepts and Procedures

9.11   What are the five steps of a test of hypothesis? Explain briefly.

9.12   What does the level of significance represent in a test of hypothesis? Explain.

9.13   By rejecting the null hypothesis in a test of hypothesis example, are you stating that the alternative hypothesis is true?

9.14   What is the difference between the critical value of z and the observed value of z?

9.15   For each of the following examples of tests of hypotheses about μ, show the rejection and nonrejection regions on the sampling distribution of the sample mean.
    a.   A two-tailed test with $\alpha = .05$ and $n = 40$
    b.   A left-tailed test with $\alpha = .01$ and $n = 67$
    c.   A right-tailed test with $\alpha = .02$ and $n = 55$

9.16   For each of the following examples of tests of hypotheses about μ, show the rejection and nonrejection regions on the sampling distribution of the sample mean.
    a.   A two-tailed test with $\alpha = .01$ and $n = 100$
    b.   A left-tailed test with $\alpha = .005$ and $n = 60$
    c.   A right-tailed test with $\alpha = .025$ and $n = 36$

9.17   Consider the following null and alternative hypotheses.

$$H_0: \mu = 25 \quad \text{versus} \quad H_1: \mu \neq 25$$

Suppose you perform this test at $\alpha = .05$ and reject the null hypothesis. Would you state that the difference between the hypothesized value of the population mean and the observed value of the sample mean is "statistically significant" or would you state that this difference is "statistically not significant"? Explain.

9.18   Consider the following null and alternative hypotheses.

$$H_0: \mu = 60 \quad \text{versus} \quad H_1: \mu > 60$$

Suppose you perform this test at $\alpha = .01$ and fail to reject the null hypothesis. Would you state that the difference between the hypothesized value of the population mean and the observed value of the sample mean is "statistically significant" or would you state that this difference is "statistically not significant"? Explain.

9.19   For each of the following significance levels, what is the probability of making a Type I error?
    a. $\alpha = .025$ b. $\alpha = .05$ c. $\alpha = .01$

9.20   For each of the following significance levels, what is the probability of making a Type I error?
    a. $\alpha = .10$ b. $\alpha = .02$ c. $\alpha = .005$

9.21   A random sample of 100 observations produced a sample mean of 32 and a standard deviation of 6. Find the critical and observed values of z for each of the following tests of hypotheses using $\alpha = .05$.
    a. $H_0: \mu = 28$ versus $H_1: \mu > 28$
    b. $H_0: \mu = 28$ versus $H_1: \mu \neq 28$

9.22   A random sample of 80 observations produced a sample mean of 15 and a standard deviation of 4. Find the critical and observed values of z for each of the following tests of hypotheses using $\alpha = .01$.

a. H_0: $\mu = 20$ versus H_1: $\mu < 20$
b. H_0: $\mu = 20$ versus H_1: $\mu \neq 20$

9.23 Consider the null hypothesis H_0: $\mu = 50$. Suppose a random sample of 100 observations is taken to perform this test. Using $\alpha = .05$, show the rejection and nonrejection regions on the sampling distribution curve of the sample mean and find the critical value(s) of z when the alternative hypothesis is

    a. H_1: $\mu < 50$ b. H_1: $\mu \neq 50$ c. H_1: $\mu > 50$

9.24 Consider the null hypothesis H_0: $\mu = 35$. Suppose a random sample of 60 observations is taken to perform this test. Using $\alpha = .01$, show the rejection and nonrejection regions on the sampling distribution curve of the sample mean and find the critical value(s) of z for a

    a. left-tailed test   b. two-tailed test   c. right-tailed test

9.25 Consider H_0: $\mu = 100$ versus H_1: $\mu \neq 100$.

    a. A random sample of 64 observations produced a sample mean of 98 and a standard deviation of 12. Using $\alpha = .01$, would you reject the null hypothesis?
    b. Another random sample of 64 observations taken from the same population produced a sample mean of 104 and a standard deviation of 10. Using $\alpha = .01$, would you reject the null hypothesis?

Comment on the results of parts a and b.

9.26 Consider H_0: $\mu = 45$ versus H_1: $\mu < 45$.

    a. A random sample of 100 observations produced a sample mean of 43 and a standard deviation of 5. Using $\alpha = .025$, would you reject the null hypothesis?
    b. Another random sample of 100 observations taken from the same population produced a sample mean of 43.8 and a standard deviation of 7. Using $\alpha = .025$, would you reject the null hypothesis?

Comment on the results of parts a and b.

9.27 Make the following tests of hypotheses.

    a. H_0: $\mu = 25$, H_1: $\mu \neq 25$, $n = 81$, $\bar{x} = 28$, $s = 3$, $\alpha = .01$
    b. H_0: $\mu = 12$, H_1: $\mu < 12$, $n = 45$, $\bar{x} = 11$, $\sigma = 4.5$, $\alpha = .05$
    c. H_0: $\mu = 40$, H_1: $\mu > 40$, $n = 100$, $\bar{x} = 46$, $s = 7$, $\alpha = .10$

9.28 Make the following tests of hypotheses.

    a. H_0: $\mu = 80$, H_1: $\mu \neq 80$, $n = 33$, $\bar{x} = 76$, $\sigma = 15$, $\alpha = .10$
    b. H_0: $\mu = 32$, H_1: $\mu < 32$, $n = 75$, $\bar{x} = 27$, $s = 7.4$, $\alpha = .01$
    c. H_0: $\mu = 55$, H_1: $\mu > 55$, $n = 40$, $\bar{x} = 60$, $s = 4$, $\alpha = .05$

Applications

9.29 According to a Ciba Geneva Pharmacy Benefit Report, Americans spent an average of $220 per person on prescription drugs in 1994 (*Kiplinger's Personal Finance Magazine*, May 1996). A recent survey of 300 randomly chosen Americans showed that they spent an average of $235 per person on prescription drugs with a standard deviation of $90. Test at the 2.5% significance level whether the mean amount currently spent on prescription drugs by all Americans exceeds $220 per person.

9.30 According to an estimate, the mean income of attorneys was $66,271 in 1991–92 (*USA Today*, December 20, 1993). A researcher wanted to check if the current mean income of attorneys is greater than $66,271. A random sample of 64 attorneys taken by this researcher produced a mean income of $69,484 with a standard deviation of $11,500. Test at the 1% significance level whether the current mean income of all attorneys is greater than $66,271.

9.31 According to *Statistics Canada*, the average family income in Canada was $53,459 in 1993. A recently taken random sample of 1200 Canadian families yielded a mean income of $54,900 with a sample standard deviation of $16,850. Using the 2% significance level, can you conclude that the mean family income in Canada has increased since 1993? Explain your conclusion in words.

9.32 The U.S. Bureau of Labor Statistics often conducts surveys to collect information on the labor market. According to the bureau, workers in the private sector earned an average of $11.66 an hour in February 1996. A labor economist took a random sample of 1000 private sector workers recently that produced a mean hourly wage of $11.88 with a standard deviation of $1.90. Test at the 1% significance level if the current mean hourly wage for private sector workers is greater than $11.66. Explain your conclusion in words.

9.33 According to the U.S. Bureau of Labor Statistics, the mean hourly earnings for production workers in mining in January 1996 were $15.66. The owner of a coal mine claims that his workers earn, on average, a wage equal to the U.S. mean hourly wage. However, a random sample of 36 of his workers yielded a mean hourly rate of $15.35 with a standard deviation of $.45.

    a. At the 1% level of significance, can you conclude that the claim made by this coal mine owner is false?

    b. What is the Type I error in this case? Explain in words. What is the probability of making this error?

9.34 The U.S. Travel Industry estimated that the average amount Americans planned to spend on *long* vacations was $1076 in 1995 (*U.S. News & World Report,* June 12, 1995). A recently taken random sample of 300 Americans planning long vacations yielded a mean planned expenditure of $1150 with a sample standard deviation of $350.

    a. Testing at the 5% significance level, do you think that the mean amount Americans plan to spend this year on long vacations is different from $1076?

    b. What is the Type I error in this case? Explain in words. What is the probability of making this error?

9.35 A study conducted a few years ago claims that adult males spend an average of 11 hours a week watching sports on television. A recent sample of 100 adult males showed that the mean time they spend per week watching sports on television is 9.50 hours with a standard deviation of 2.2 hours.

    a. Test at the 1% significance level if currently all adult males spend less than 11 hours watching sports on television.

    b. What will your decision be in part a if the probability of making a Type I error is zero? Explain.

9.36 A restaurant franchise company has a policy of opening new restaurants only in those areas that have a mean household income of at least $35,000 per year. The company is currently considering an area in which to open a new restaurant. The company's research department took a sample of 150 households from this area and found that the mean income of these households is $33,564 per year with a standard deviation of $5400.

    a. Using the 1% significance level, would you conclude that the company should not open a restaurant in this area?

    b. What will your decision be in part a if the probability of making a Type I error is zero? Explain.

9.37 The manufacturer of a certain brand of auto batteries claims that the mean life of these batteries is 45 months. A consumer protection agency that wants to check this claim took a random sample of 36 such batteries and found that the mean life for this sample is 43.75 months with a standard deviation of 4 months.

    a. Using the 2.5% significance level, would you conclude that the mean life of these batteries is less than 45 months?

    b. Make the test of part a using a 5% significance level. Is your decision different from the one in part a? Comment on the results of parts a and b.

9.38 A study claims that all adults spend an average of 8 hours or more on chores during a weekend. A researcher wanted to check if this claim is true. A random sample of 200 adults taken by this researcher showed that these adults spend an average of 7.70 hours on chores during a weekend with a standard deviation of 2.1 hours.

    a. Using the 1% significance level, can you conclude that the claim that all adults spend an average of 8 hours or more on chores during a weekend is false?

    b. Make the test of part a using a 2.5% significance level. Is your decision different from the one in part a? Comment on the results of parts a and b.

9.39 Lazurus Steel Corporation produces iron rods that are supposed to be 36 inches long. The machine that makes these rods does not produce each rod exactly 36 inches long. The lengths of these rods vary slightly. It is known that when the machine is working properly, the mean length of the rods is 36 inches. The standard deviation of the lengths of all rods produced on this machine is always equal to .05 inches. The quality control department at the company takes a sample of 40 such rods each week, calculates the mean length of these rods, and tests the null hypothesis $\mu = 36$ inches against the alternative hypothesis $\mu \neq 36$ inches using a 1% significance level. If the null hypothesis is rejected, the machine is stopped and adjusted. A recent sample of 40 such rods produced a mean length of 36.015 inches. Based on this sample, would you conclude that the machine needs an adjustment?

9.40 At Farmer's Dairy, a machine is set to fill 32-ounce milk cartons. However, this machine does not put exactly 32 ounces of milk in each carton; the amount varies slightly from carton to carton. It is known that when the machine is working properly, the mean net weight of these cartons is 32 ounces. The standard deviation of the milk in all such cartons is always equal to .15 ounces. The quality control inspector at this dairy takes a sample of 35 such cartons each week, calculates the mean net weight of these cartons, and tests the null hypothesis $\mu = 32$ ounces against the alternative hypothesis $\mu \neq 32$ ounces using a 2% significance level. If the null hypothesis is rejected, the machine is stopped and adjusted. A recent sample of 35 such cartons produced a mean net weight of 31.90 ounces. Based on this sample, would you conclude that the machine needs to be adjusted?

9.41 A company claims that the mean net weight of the contents of its All Taste cereal boxes is at least 18 ounces. Suppose you want to test whether or not the claim of the company is true. Explain briefly how you would conduct this test using a large sample.

9.42 A researcher claims that college students spend an average of 45 minutes per week on community service. You want to test if the mean time spent per week on community service by college students is different from 45 minutes. Explain briefly how you would conduct this test using a large sample.

9.3 HYPOTHESIS TESTS USING THE *p*-VALUE APPROACH

In the discussion of tests of hypotheses in Section 9.2, the value of the significance level α was selected before the test was performed. Sometimes we may prefer not to predetermine α. Instead, we may want to find a value such that a given null hypothesis will be rejected for any α greater than this value and it will not be rejected for any α smaller than this value. The **probability-value approach**, more commonly called the *p-value approach*, gives such a value. In this approach, we calculate the *p*-value for the test, which is defined as the smallest level of significance at which the given null hypothesis is rejected.

> **p-VALUE**
>
> The *p-value* is the smallest significance level at which the null hypothesis is rejected.

Using the *p*-value approach, we reject the null hypothesis if

$$p\text{-value} < \alpha$$

and we do not reject the null hypothesis if

$$p\text{-value} \geq \alpha$$

For a one-tailed test, the p-value is given by the area in the tail of the sampling distribution curve beyond the observed value of the sample statistic. Figure 9.8 shows the p-value for a right-tailed test about μ.

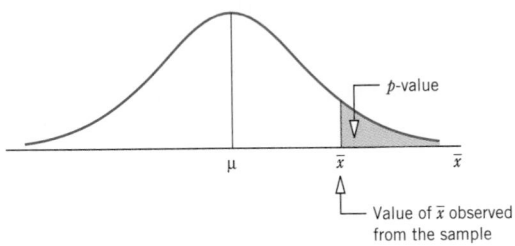

Figure 9.8 The p-value for a right-tailed test.

For a two-tailed test, the p-value is twice the area in the tail of the sampling distribution curve beyond the observed value of the sample statistic. Figure 9.9 shows the p-value for a two-tailed test. Each of the areas in the two tails gives one-half the p-value.

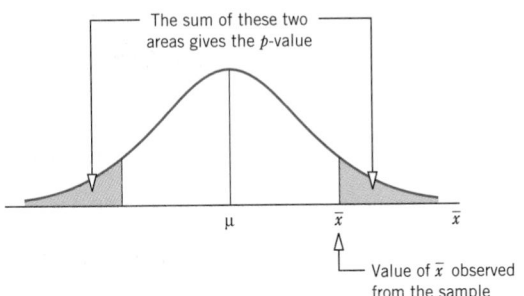

Figure 9.9 The p-value for a two-tailed test.

Examples 9–4 and 9–5 illustrate the calculation and use of the p-value.

Calculating p-value for a one-tailed test of hypothesis.

EXAMPLE 9–4 The management of Priority Health Club claims that its members lose an average of 10 pounds or more within the first month after joining the club. A consumer agency that wanted to check this claim took a random sample of 36 members of this health club and found that they lost an average of 9.2 pounds within the first month of membership with a standard deviation of 2.4 pounds. Find the p-value for this test.

Solution Let μ be the mean weight lost during the first month of membership by all members of this health club and $\bar{x}$ be the corresponding mean for the sample. From the given information,

$$n = 36, \quad \bar{x} = 9.2 \text{ pounds}, \quad \text{and} \quad s = 2.4 \text{ pounds}$$

The claim of the club is that its members lose, on average, 10 pounds or more within the first month of membership. To calculate the p-value, we apply the following three steps.

Step 1. *State the null and alternative hypotheses*

H_0: $\mu \geq 10$ (The mean weight lost is 10 pounds or more)

H_1: $\mu < 10$ (The mean weight lost is less than 10 pounds)

Step 2. *Select the distribution to use*

Because the sample size is large, we use the normal distribution to make the test and to calculate the *p*-value.

Step 3. *Calculate the p-value*

The $<$ sign in the alternative hypothesis indicates that the test is left-tailed. The *p*-value is given by the area to the left of $\bar{x} = 9.2$ under the sampling distribution curve of $\bar{x}$, as shown in Figure 9.10. To find this area, we first find the *z* value for $\bar{x} = 9.2$ as follows.

$$s_{\bar{x}} = \frac{s}{\sqrt{n}} = \frac{2.4}{\sqrt{36}} = .40$$

$$z = \frac{\bar{x} - \mu}{s_{\bar{x}}} = \frac{9.2 - 10}{.40} = -2.00$$

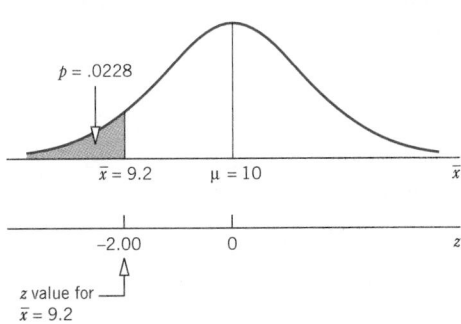

Figure 9.10   The *p*-value for a left-tailed test.

The area to the left of $\bar{x} = 9.2$ under the sampling distribution of $\bar{x}$ is equal to the area under the standard normal curve to the left of $z = -2.00$. From the normal distribution table, the area between the mean and $z = -2.00$ is .4772. Hence, the area to the left of $z = -2.00$ is $.5 - .4772 = .0228$. Consequently,

$$p\text{-value} = \mathbf{.0228}$$

Thus, based on the *p*-value of .0228 we can state that for any α (significance level) greater than .0228 we will reject the null hypothesis stated in Step 1 and for any α less than .0228 we will not reject the null hypothesis. Suppose we make the test for this example at $\alpha = .01$. Because $\alpha = .01$ is less than the *p*-value of .0228, we will not reject the null hypothesis. Now, suppose we make the test at $\alpha = .05$. This time, because $\alpha = .05$ is greater than the *p*-value of .0228, we will reject the null hypothesis. ■

The reader should make the test of hypothesis for Example 9–4 at $\alpha = .01$ and at $\alpha = .05$ by using the five steps learned in Section 9.2. The null hypothesis will not be rejected at $\alpha = .01$ (as .01 is less than $p = .0228$), and the null hypothesis will be rejected at $\alpha = .05$ (as .05 is greater than $p = .0228$).

*Calculating p-value
for a two-tailed test
of hypothesis.*

EXAMPLE 9–5   At Canon Food Corporation, it used to take an average of 50 minutes for new workers to learn a food processing job. Recently the company installed a new food processing machine. The supervisor at the company wants to find if the mean time taken by new workers to learn the food processing procedure on this new machine is different from 50 minutes. A sample of 40 workers showed that it took, on average, 47 minutes for them to learn the food processing procedure on the new machine with a standard deviation of 7 minutes. Find the *p*-value for the test that the mean learning time for the food processing procedure on the new machine is different from 50 minutes.

Solution   Let μ be the mean time (in minutes) taken to learn the food processing procedure on the new machine by all workers and $\bar{x}$ the corresponding sample mean. From the given information,

$$n = 40, \quad \bar{x} = 47 \text{ minutes}, \quad \text{and} \quad s = 7 \text{ minutes}$$

To calculate the *p*-value, we apply the following three steps.

Step 1.   *State the null and alternative hypotheses*

$$H_0: \mu = 50 \text{ minutes}$$

$$H_1: \mu \neq 50 \text{ minutes}$$

Note that the null hypothesis states that the mean time for learning the food processing procedure on the new machine is 50 minutes, and the alternative hypothesis states that this time is different from 50 minutes.

Step 2.   *Select the distribution to use*

Because the sample size is large, we use the normal distribution to make the test and to calculate the *p*-value.

Step 3.   *Calculate the p-value*

The $\neq$ sign in the alternative hypothesis indicates that the test is two-tailed. The *p*-value is equal to twice the area in the tail of the sampling distribution curve of $\bar{x}$ to the left of $\bar{x} = 47$, as shown in Figure 9.11. To find this area, we first find the *z* value for $\bar{x} = 47$ as follows.

$$s_{\bar{x}} = \frac{s}{\sqrt{n}} = \frac{7}{\sqrt{40}} = 1.10679718 \text{ minutes}$$

$$z = \frac{\bar{x} - \mu}{s_{\bar{x}}} = \frac{47 - 50}{1.10679718} = -2.71$$

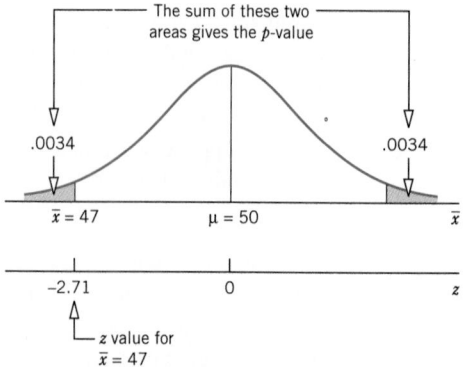

Figure 9.11   The *p*-value for a two-tailed test.

The area to the left of $\bar{x} = 47$ is equal to the area under the standard normal curve to the left of $z = -2.71$. From the normal distribution table, the area between the mean and $z = -2.71$ is .4966. Hence, the area to the left of $z = -2.71$ is

$$.5 - .4966 = .0034$$

Consequently, the p-value is

$$p\text{-value} = 2\ (.0034) = \mathbf{.0068}$$

Thus, based on the p-value of .0068, we conclude that for any α (significance level) greater than .0068 we will reject the null hypothesis and for any α less than .0068 we will not reject the null hypothesis. ■

EXERCISES

Concepts and Procedures

9.43 Briefly explain the procedure used to calculate the p-value for a two-tailed and for a one-tailed test, respectively.

9.44 Find the p-value for each of the following hypothesis tests.

    a. H_0: $\mu = 23$, H_1: $\mu \neq 23$, $n = 50$, $\bar{x} = 21$, $s = 5$
    b. H_0: $\mu = 15$, H_1: $\mu < 15$, $n = 80$, $\bar{x} = 13.2$, $s = 5.5$
    c. H_0: $\mu = 38$, H_1: $\mu > 38$, $n = 35$, $\bar{x} = 40.6$, $s = 7.2$

9.45 Find the p-value for each of the following hypothesis tests.

    a. H_0: $\mu = 46$, H_1: $\mu \neq 46$, $n = 40$, $\bar{x} = 49.43$, $s = 9.7$
    b. H_0: $\mu = 26$, H_1: $\mu < 26$, $n = 33$, $\bar{x} = 24.2$, $s = 4.3$
    c. H_0: $\mu = 18$, H_1: $\mu > 18$, $n = 55$, $\bar{x} = 20.4$, $s = 7.8$

9.46 Consider H_0: $\mu = 29$ versus H_1: $\mu \neq 29$. A random sample of 60 observations taken from this population produced a sample mean of 31.4 and a standard deviation of 8.

    a. Calculate the p-value.
    b. Considering the p-value of part a, would you reject the null hypothesis if the test were made at the significance level of .05?
    c. Considering the p-value of part a, would you reject the null hypothesis if the test were made at the significance level of .01?

9.47 Consider H_0: $\mu = 72$ versus H_1: $\mu > 72$. A random sample of 36 observations taken from this population produced a sample mean of 74.07 and a standard deviation of 6.

    a. Calculate the p-value.
    b. Considering the p-value of part a, would you reject the null hypothesis if the test were made at the significance level of .01?
    c. Considering the p-value of part a, would you reject the null hypothesis if the test were made at the significance level of .025?

Applications

9.48 According to the U.S. Department of Agriculture, the average coffee consumption in the United States in 1993 was 26.0 gallons per person. A recently taken random sample of 400 Americans found a mean coffee consumption of 26.7 gallons with a sample standard deviation of 6.8 gallons. Find the p-value for the test of hypothesis with the alternative hypothesis that the mean coffee consumption now exceeds 26.0 gallons per person.

9.49 Is giving birth becoming a *drive-through* experience at American hospitals? According to the U.S. National Center for Health Statistics, the mean hospital stay for delivery of a baby was 2.4 days in 1993. A recently taken random sample of 150 births found a mean hospital stay of 2.2 days with a standard deviation of .9 days. Find the *p*-value for the hypothesis test with the alternative hypothesis that the current mean time in the hospital for childbirth is less than 2.4 days.

9.50 The manufacturer of a certain brand of auto batteries claims that the mean life of these batteries is 45 months. A consumer protection agency that wants to check this claim took a random sample of 36 such batteries and found that the mean life for this sample is 43.75 months with a standard deviation of 4 months. Find the *p*-value for the test of hypothesis with the alternative hypothesis that the mean life of these batteries is less than 45 months.

9.51 A study claims that all adults spend an average of 14 hours or more on chores during a weekend. A researcher wanted to check if this claim is true. A random sample of 200 adults taken by this researcher showed that these adults spend an average of 13.55 hours on chores during a weekend with a standard deviation of 3.1 hours. Find the *p*-value for the hypothesis test with the alternative hypothesis that all adults spend less than 14 hours on chores during a weekend.

9.52 According to the U.S. Bureau of the Census, the mean monthly salary of people with a professional degree was $5534 in 1993. Assume that this result holds true for the 1993 population of all people with a professional degree. A random sample of 400 people with a professional degree taken recently showed that their mean monthly salary is $5640 with a standard deviation of $990.

    a.   Find the *p*-value for the test of hypothesis with the alternative hypothesis that the current mean monthly salary of all people with a professional degree is greater than $5534.

    b.   If $\alpha = .01$, based on the *p*-value calculated in part a, would you reject the null hypothesis? Explain.

    c.   If $\alpha = .025$, based on the *p*-value calculated in part a, would you reject the null hypothesis? Explain.

9.53 A telephone company claims that the mean duration of all long-distance phone calls made by its residential customers is 10 minutes. A random sample of 100 long-distance calls made by its residential customers taken from the records of this company showed that the mean duration of calls for this sample is 9.0 minutes with a standard deviation of 5.2 minutes.

    a.   Find the *p*-value for the test that the mean duration of all long-distance calls made by residential customers is less than 10 minutes.

    b.   If $\alpha = .02$, based on the *p*-value calculated in part a, would you reject the null hypothesis? Explain.

    c.   If $\alpha = .05$, based on the *p*-value calculated in part a, would you reject the null hypothesis? Explain.

9.54 Lazurus Steel Corporation produces iron rods that are supposed to be 36 inches long. The machine that makes these rods does not produce each rod exactly 36 inches long. The lengths of these rods vary slightly. It is known that when the machine is working properly, the mean length of the rods is 36 inches. The standard deviation of the lengths of all rods produced on this machine is always equal to .05 inches. The quality control department at the company takes a sample of 40 such rods every week, calculates the mean length of these rods, and tests the null hypothesis $\mu = 36$ inches against the alternative hypothesis $\mu \neq 36$ inches. If the null hypothesis is rejected, the machine is stopped and adjusted. A recent such sample of 40 rods produced a mean length of 36.015 inches.

    a.   Calculate the *p*-value for this test of hypothesis.

    b.   Based on the *p*-value calculated in part a, will the quality control inspector decide to stop the machine and adjust it if he chooses the maximum probability of a Type I error to be .02? What if the maximum probability of a Type I error is .10?

9.55 At Farmer's Dairy, a machine is set to fill 32-ounce milk cartons. However, this machine does not put exactly 32 ounces of milk in each carton; the amount varies slightly from carton to carton. It is known that when the machine is working properly, the mean net weight of these cartons is 32 ounces. The standard deviation of the milk in all such cartons is always equal to .15 ounces. The quality control inspector at this company takes a sample of 35 such cartons every week, calculates the mean net weight of these cartons, and tests the null hypothesis $\mu = 32$ ounces against the alternative hypothesis $\mu \neq 32$ ounces. If the null hypothesis is rejected, the machine is stopped and adjusted. A recent sample of 35 such cartons produced a mean net weight of 31.90 ounces.

a. Calculate the *p*-value for this test of hypothesis.
b. Based on the *p*-value calculated in part a, will the quality control inspector decide to stop the machine and readjust it if she chooses the maximum probability of a Type I error to be .01? What if the maximum probability of a Type I error is .05?

9.4 HYPOTHESIS TESTS ABOUT A POPULATION MEAN: SMALL SAMPLES

Many times the size of a sample that is used to make a test of hypothesis about μ is small, that is, $n < 30$. This may be the case because we have limited resources and cannot afford to take a large sample or it may be because of the nature of the experiment itself. For example, to test a new model of a car for fuel efficiency (miles per gallon), the company may prefer to use a small sample. All cars included in such a test must be sold as used cars. In the case of a small sample, if the population from which the sample is drawn is (approximately) normally distributed and the population standard deviation σ is known, we can still use the normal distribution to make a test of hypothesis about μ. However, if the population is (approximately) normally distributed, the population standard deviation σ is not known, and the sample size is small ($n < 30$), then the normal distribution is replaced by the t distribution to make a test of hypothesis about μ. In such a case the random variable

$$t = \frac{\bar{x} - \mu}{s_{\bar{x}}} \qquad \text{where} \qquad s_{\bar{x}} = \frac{s}{\sqrt{n}}$$

has a t distribution. The t is called the **test statistic** to make a hypothesis test about a population mean for small samples.

CONDITIONS UNDER WHICH THE *t* DISTRIBUTION IS USED TO MAKE TESTS OF HYPOTHESIS ABOUT μ

The t distribution is used to conduct a *test of hypothesis about μ* if

1. The sample size is small ($n < 30$)
2. The population from which the sample is drawn is (approximately) normally distributed
3. The population standard deviation σ is unknown

The procedure that is used to make hypothesis tests about μ in the case of small samples is similar to the one for large samples. We perform the same five steps with the only difference being the use of the t distribution in place of the normal distribution.

TEST STATISTIC

The value of the *test statistic t* for the sample mean $\bar{x}$ is computed as

$$t = \frac{\bar{x} - \mu}{s_{\bar{x}}} \qquad \text{where} \qquad s_{\bar{x}} = \frac{s}{\sqrt{n}}$$

The value of t calculated for $\bar{x}$ by using the above formula is also called the **observed value of t**.

Examples 9–6, 9–7, and 9–8 describe the procedure of testing hypotheses about the population mean using the *t* distribution.

Conducting a two-tailed test of hypothesis about μ: n < 30.

EXAMPLE 9-6 A psychologist claims that the mean age at which children start walking is 12.5 months. Carol wanted to check if this claim is true. She took a random sample of 18 children and found that the mean age at which these children started walking was 12.9 months with a standard deviation of .80 months. Using the 1% significance level, can you conclude that the mean age at which all children start walking is different from 12.5 months? Assume that the ages at which all children start walking have an approximate normal distribution.

Solution Let μ be the mean age at which all children start walking and $\bar{x}$ the corresponding mean for the sample. Then, from the given information,

$$n = 18, \quad \bar{x} = 12.9 \text{ months}, \quad s = .80 \text{ months}, \quad \text{and} \quad \alpha = .01$$

Step 1. *State the null and alternative hypotheses*

We are to test if the mean age at which all children start walking is different from 12.5 months. The null and alternative hypotheses are

H_0: μ = 12.5   (The mean walking age is 12.5 months)

H_1: μ ≠ 12.5   (The mean walking age is different from 12.5 months)

Step 2. *Select the distribution to use*

The sample size is small, and the population is approximately normally distributed. However, we do not know the population standard deviation σ. Hence, we use the *t* distribution to make the test.

Step 3. *Determine the rejection and nonrejection regions*

The significance level is .01. The ≠ sign in the alternative hypothesis indicates that the test is two-tailed and the rejection region lies in both tails. The area of the rejection region in each tail of the *t* distribution curve is

$$\text{Area in each tail} = \alpha/2 = .01/2 = .005$$

$$df = n - 1 = 18 - 1 = 17$$

From the *t* distribution table, the critical values of *t* for 17 degrees of freedom and .005 area in each tail of the *t* distribution curve are −2.898 and 2.898. These values are shown in Figure 9.12.

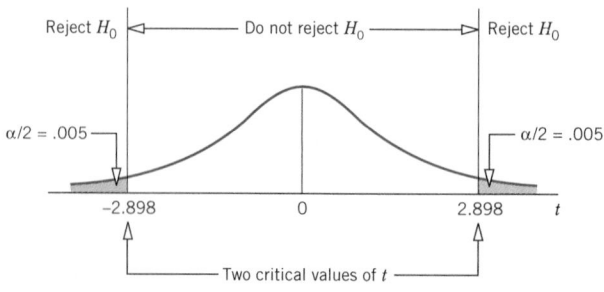

Figure 9.12

Step 4. *Calculate the value of the test statistic*

We calculate the value of the test statistic t for $\bar{x} = 12.9$ as follows.

$$s_{\bar{x}} = \frac{s}{\sqrt{n}} = \frac{.80}{\sqrt{18}} = .18856181$$

$$t = \frac{\bar{x} - \mu}{s_{\bar{x}}} = \frac{12.9 - 12.5}{.18856181} = 2.121$$

Step 5. *Make a decision*

The value of the test statistic $t = 2.121$ falls between the two critical points, -2.898 and 2.898, which is the nonrejection region. Consequently, we fail to reject H_0. As a result, we can state that the difference between the hypothesized population mean and the sample mean is so small that it may have occurred because of sampling error. The mean age at which children start walking is not different from 12.5 months. ■

Using the *p*-Value Approach in Example 9–6

Note that using the procedure discussed in Section 9.3, we can use the *p*-value approach to make a decision in all problems relating to hypothesis testing in this and succeeding chapters. For instance, we can find the *p*-value for $\bar{x} = 12.9$ in Example 9–6 and compare it with the given significance level to make a decision. As shown above in Step 4 of Example 9–6, the t value for $\bar{x} = 12.9$ is 2.121. From the t distribution table, for df = 17 and $t = 2.121$, the *p*-value for a two-tailed test is approximately .05. (Note that this *p*-value for df = 17 and $t = 2.110$ is $2 \times .025 = .05$. However, for the same degrees of freedom but $t = 2.121$, this *p*-value will be slightly lower than .05.) Since $\alpha = .01$ is less than the *p*-value of .05, we fail to reject the null hypothesis.

The same procedure can be used to obtain the *p*-value for the test of hypothesis problems in Section 9.5 and in succeeding chapters. To do so, we find the value of the test statistic for the given value of the sample statistic obtained from the sample and then obtain the *p*-value from the corresponding probability distribution table for that value of the test statistic. Finally, we compare that *p*-value with the significance level and make a decision.

All the statistical software packages, including MINITAB, give the *p*-value in the solution to a test of hypothesis problem. Thus, if you are using a statistical software to solve a test of hypothesis problem, you can compare the *p*-value given in the computer solution to the significance level and make a decision.

Conducting a left-tailed test of hypothesis about μ: n < 30.

EXAMPLE 9–7 Grand Auto Corporation produces auto batteries. The company claims that its top-of-the-line Never Die batteries are good, on average, for at least 65 months. A consumer protection agency tested 15 such batteries to check this claim. It found the mean life of these 15 batteries to be 63 months with a standard deviation of 2 months. At the 5% significance level, can you conclude that the claim of the company is true? Assume that the life of such a battery has an approximate normal distribution.

Solution Let μ be the mean life of all Never Die batteries and $\bar{x}$ the corresponding mean for the sample. Then, from the given information,

$$n = 15, \quad \bar{x} = 63 \text{ months}, \quad \text{and} \quad s = 2 \text{ months}$$

The significance level is $\alpha = .05$. The company's claim is that the mean life of these batteries is at least 65 months.

Step 1. *State the null and alternative hypotheses*

We are to test whether or not the mean life of Never Die batteries is at least 65 months. The null and alternative hypotheses are as follows.

$$H_0: \mu \geq 65 \quad \text{(The mean life is at least 65 months)}$$

$$H_1: \mu < 65 \quad \text{(The mean life is less than 65 months)}$$

Step 2. *Select the distribution to use*

The sample size is small ($n < 30$), and the life of a battery is approximately normally distributed. However, the population standard deviation is not known. Hence, we use the t distribution to make the test.

Step 3. *Determine the rejection and nonrejection regions*

The significance level is .05. The $<$ sign in the alternative hypothesis indicates that the test is left-tailed with the rejection region in the left tail of the t distribution curve. To find the critical value of t, we need to know the area in the left tail and the degrees of freedom.

$$\text{Area in the left tail} = \alpha = .05$$

$$df = n - 1 = 15 - 1 = 14$$

From the t distribution table, the critical value of t for 14 degrees of freedom and an area of .05 in the left tail is -1.761. This value is shown in Figure 9.13.

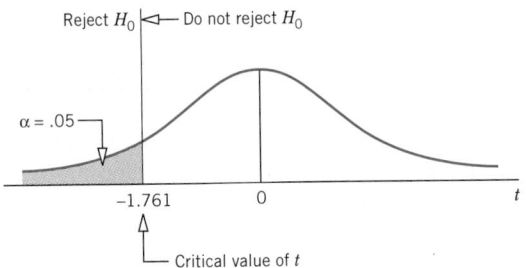

Figure 9.13

Step 4. *Calculate the value of the test statistic*

The value of the test statistic t for $\bar{x} = 63$ is calculated as follows.

$$s_{\bar{x}} = \frac{s}{\sqrt{n}} = \frac{2}{\sqrt{15}} = .51639778$$

$$t = \frac{\bar{x} - \mu}{s_{\bar{x}}} = \frac{63 - 65}{.51639778} = -3.873 \qquad \text{From } H_0$$

Step 5. *Make a decision*

The value of the test statistic $t = -3.873$ is less than the critical value of $t = -1.761$, and it falls in the rejection region. Therefore, we reject H_0 and conclude that the sample mean is too small compared to 65 (company's claimed value of μ) and the difference between the two may not be attributed to chance alone. We can conclude that the mean life of the company's Never Die batteries is less than 65 months. ■

Making a right-tailed test of hypothesis about μ: $n < 30$.

EXAMPLE 9-8 The management at Massachusetts Savings Bank is always concerned about the quality of service provided to its customers. With the old computer system, a teller at this bank could serve, on average, 22 customers per hour. The management noticed that with this service rate, the waiting time for customers was too long. Recently the management of the bank installed a new computer system in the bank expecting that it would increase the service rate and consequently make the customers happier by reducing the waiting time. To check if the new computer system is more efficient than the old system, the management of the bank took a random sample of 18 hours and found that during these hours the mean number of customers served by tellers was 28 per hour with a standard deviation of 2.5. Testing at the 1% significance level, would you conclude that the new computer system is more efficient than the old computer system? Assume that the number of customers served per hour by a teller on this computer system has an approximate normal distribution.

Solution Let μ be the mean number of customers served per hour by a teller using the new system and $\bar{x}$ be the corresponding mean for the sample. Then, from the given information,

$$n = 18 \text{ hours,} \qquad \bar{x} = 28 \text{ customers,} \qquad s = 2.5 \text{ customers,} \qquad \text{and} \qquad \alpha = .01$$

Step 1. *State the null and alternative hypotheses*

We are to test whether or not the new computer system is more efficient than the old system. The new computer system will be more efficient than the old system if the mean number of customers served per hour by using the new computer system is significantly more than 22; otherwise, it will not be more efficient. The null and alternative hypotheses are

$$H_0: \mu = 22 \qquad \text{(The new computer system is not more efficient)}$$
$$H_1: \mu > 22 \qquad \text{(The new computer system is more efficient)}$$

Step 2. *Select the distribution to use*

The sample size is small, and the population is approximately normally distributed. However, we do not know the population standard deviation σ. Hence, we use the t distribution to make the test.

Step 3. *Determine the rejection and nonrejection regions*

The significance level is .01. The $>$ sign in the alternative hypothesis indicates that the test is right-tailed and the rejection region lies in the right tail of the t distribution curve.

$$\text{Area in the right tail} = \alpha = .01$$
$$df = n - 1 = 18 - 1 = 17$$

From the t distribution table, the critical value of t for 17 degrees of freedom and .01 area in the right tail is 2.567. This value is shown in Figure 9.14.

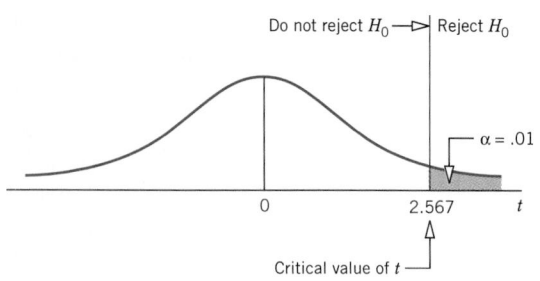

Figure 9.14

Step 4. *Calculate the value of the test statistic*

The value of the test statistic t for $\bar{x} = 28$ is calculated as follows.

$$s_{\bar{x}} = \frac{s}{\sqrt{n}} = \frac{2.5}{\sqrt{18}} = .58925565$$

$$t = \frac{\bar{x} - \mu}{s_{\bar{x}}} = \frac{28 - 22}{.58925565} = 10.182$$

(arrow labeled —From H_0)

Step 5. *Make a decision*

The value of the test statistic $t = 10.182$ is larger than the critical value of $t = 2.567$, and it falls in the rejection region. Consequently, we reject H_0. As a result, we conclude that the value of the sample mean is too large compared to the hypothesized value of the population mean and the difference between the two may not be attributed to chance alone. The mean number of customers served per hour using the new computer system is more than 22. The new computer system is more efficient than the old computer system. ■

EXERCISES

Concepts and Procedures

9.56   Briefly explain the conditions that must hold true to use the t distribution to make a test of hypothesis about the population mean.

9.57   For each of the following examples of tests of hypotheses about μ, show the rejection and nonrejection regions on the t distribution curve.

    a.   A two-tailed test with $\alpha = .02$ and $n = 20$
    b.   A left-tailed test with $\alpha = .01$ and $n = 16$
    c.   A right-tailed test with $\alpha = .05$ and $n = 18$

9.58   For each of the following examples of tests of hypotheses about μ, show the rejection and nonrejection regions on the t distribution curve.

    a.   A two-tailed test with $\alpha = .01$ and $n = 15$
    b.   A left-tailed test with $\alpha = .005$ and $n = 25$
    c.   A right-tailed test with $\alpha = .025$ and $n = 22$

9.59   A random sample of 25 observations taken from a population that is normally distributed produced a sample mean of 58.5 and a standard deviation of 7.5. Find the critical and observed values of t for each of the following tests of hypotheses using $\alpha = .01$.

    a. H_0: $\mu = 55$ versus H_1: $\mu > 55$

    b. H_0: $\mu = 55$ versus H_1: $\mu \neq 55$

9.60 A random sample of 16 observations taken from a population that is normally distributed produced a sample mean of 42.4 and a standard deviation of 8. Find the critical and observed values of t for each of the following tests of hypotheses using $\alpha = .05$.

    a. H_0: $\mu = 46$ versus H_1: $\mu < 46$

    b. H_0: $\mu = 46$ versus H_1: $\mu \neq 46$

9.61 Consider the null hypothesis H_0: $\mu = 70$ about the mean of a population that is normally distributed. Suppose a random sample of 20 observations is taken from this population to make this test. Using $\alpha = .01$, show the rejection and nonrejection regions and find the critical value(s) of t for a

    a. left-tailed test   b. two-tailed test   c. right-tailed test

9.62 Consider the null hypothesis H_0: $\mu = 35$ about the mean of a population that is normally distributed. Suppose a random sample of 22 observations is taken from this population to make this test. Using $\alpha = .05$, show the rejection and nonrejection regions and find the critical value(s) of t for a

    a. left-tailed test   b. two-tailed test   c. right-tailed test

9.63 Consider H_0: $\mu = 80$ versus H_1: $\mu \neq 80$ for a population that is normally distributed.

    a. A random sample of 25 observations taken from this population produced a sample mean of 77 and a standard deviation of 8. Using $\alpha = .01$, would you reject the null hypothesis?

    b. Another random sample of 25 observations taken from the same population produced a sample mean of 86 and a standard deviation of 6. Using $\alpha = .01$, would you reject the null hypothesis?

Comment on the results of parts a and b.

9.64 Consider H_0: $\mu = 40$ versus H_1: $\mu > 40$ for a population that is normally distributed.

    a. A random sample of 16 observations taken from this population produced a sample mean of 45 and a standard deviation of 5. Using $\alpha = .025$, would you reject the null hypothesis?

    b. Another random sample of 16 observations taken from the same population produced a sample mean of 41.9 and a standard deviation of 7. Using $\alpha = .025$, would you reject the null hypothesis?

Comment on the results of parts a and b.

9.65 Assuming that the respective populations are normally distributed, perform the following hypothesis tests.

    a. H_0: $\mu = 24$, H_1: $\mu \neq 24$, $n = 25$, $\bar{x} = 28$, $s = 4.9$, $\alpha = .01$

    b. H_0: $\mu = 30$, H_1: $\mu < 30$, $n = 16$, $\bar{x} = 27$, $s = 6.6$, $\alpha = .025$

    c. H_0: $\mu = 18$, H_1: $\mu > 18$, $n = 20$, $\bar{x} = 22$, $s = 8$, $\alpha = .10$

9.66 Assuming that the respective populations are normally distributed, perform the following hypothesis tests.

    a. H_0: $\mu = 60$, H_1: $\mu \neq 60$, $n = 14$, $\bar{x} = 56$, $s = 9$, $\alpha = .05$

    b. H_0: $\mu = 35$, H_1: $\mu < 35$, $n = 24$, $\bar{x} = 29$, $s = 5.4$, $\alpha = .005$

    c. H_0: $\mu = 47$, H_1: $\mu > 47$, $n = 18$, $\bar{x} = 51$, $s = 6$, $\alpha = .001$

Applications

9.67 According to a basketball coach, the mean height of all female college basketball players is 69.5 inches. A random sample of 25 such players produced a mean height of 70.2 inches with a standard deviation of 2.1 inches. Assuming that the heights of all female college basketball players are normally distributed, test at the 1% significance level if their mean height is different from 69.5 inches.

9.68 According to a U.S. Bureau of the Census survey in 1995, American workers aged 26 to 44 spent an average of 22 minutes traveling to work (*USA Today*, November 20, 1995). A recently taken random sample of 20 American workers in this age group yielded a mean commuting time to work equal to 25 minutes with a standard deviation of 8 minutes. Assume that the times to commute to work for all such workers are normally distributed. Using the 5% significance level, can you conclude that the current mean travel time to work for all U.S. workers aged 26 to 44 is different than 22 minutes?

9.69 The president of a university claims that the mean time spent partying by all students at this university is not more than 7 hours per week. A random sample of 20 students taken from this university showed that they spent an average of 10.7 hours partying the previous week with a standard deviation of 2.3 hours. Assuming that the time spent partying by all students at this university is approximately normally distributed, test at the 2.5% significance level if the president's claim is true. Explain your conclusion in words.

9.70 The mean balance of all checking accounts at a bank on December 31, 1996 was $850. A random sample of 25 checking accounts taken recently from this bank gave a mean balance of $775 with a standard deviation of $230. Assume that the balances of all checking accounts at this bank are normally distributed. Using the 1% significance level, can you conclude that the mean balance of such accounts has decreased during this period? Explain your conclusion in words.

9.71 A soft-drink manufacturer claims that its 12-ounce cans do not contain, on average, more than 30 calories. A random sample of 16 cans of this soft drink, which were checked for calories, contained a mean of 31.8 calories with a standard deviation of 3 calories. Assume that the number of calories in 12-ounce soda cans is normally distributed. Does the sample information support the alternative hypothesis that the manufacturer's claim is false? Use a significance level of 5%. Explain your conclusion in words.

9.72 According to the U.S. Bureau of the Census data for 1993, students required an average of 6.29 years beyond high school to obtain a bachelor's degree. A recently taken random sample of 27 newly awarded bachelor's degree holders showed that it took them a mean of 6.90 years to obtain the degree after finishing high school, with a sample standard deviation of 1.1 years. Using the 1% significance level, can you conclude that the mean time required for all students to obtain a bachelor's degree is currently greater than 6.29 years? Assume that the time taken by all students to obtain a bachelor's degree after finishing high school is normally distributed.

9.73 A paint manufacturing company claims that the mean drying time for its paints is not more than 45 minutes. A random sample of 20 gallons of paints selected from the production line of this company showed that the mean drying time for this sample is 50 minutes with a standard deviation of 3 minutes. Assume that the drying times for these paints have a normal distribution.

    a. Using the 1% significance level, would you conclude that the company's claim is true?
    b. What is the Type I error in this exercise? Explain in words. What is the probability of making such an error?

9.74 A 1995 Michigan State University study examined employment prospects for new college graduates (*The Hartford Courant*, December 3, 1995). A survey of businesses, industries, and government agencies revealed that chemical engineering promised to be the most lucrative field with an average annual starting salary of $41,183. A recent sample of 22 newly hired college graduates in chemical engineering yielded a mean annual starting salary of $40,400 with a sample standard deviation of $3000. Assume that the annual starting salaries of all new college graduates in chemical engineering are normally distributed.

    a. Using the 2.5% significance level, can you conclude that the current mean starting annual salary for all new college graduates in chemical engineering is less than $41,183?
    b. What is the Type I error in this exercise? Explain in words. What is the probability of making such an error?

9.75 A business school claims that students who complete a three-month typing course can type, on average, at least 1200 words an hour. A random sample of 25 students who completed this course typed, on average, 1130 words an hour with a standard deviation of 85 words. Assume that the typing speeds for all students who complete this course have an approximate normal distribution.

a. Suppose the probability of making a Type I error is selected to be zero. Can you conclude that the claim of the business school is true? Answer without performing the five steps of a test of hypothesis.

b. Using the 5% significance level, can you conclude that the claim of the business school is true?

9.76 *Fortune* magazine compiled salary information for various occupational groups in 1995 (*Fortune*, June 26, 1995). According to this information, television news anchorpersons earn an average annual salary of $65,824. Assume that this mean was true for all television news anchorpersons in 1995. A recent sample of 20 television news anchorpersons showed that they earn a mean annual salary of $73,500 with a sample standard deviation of $9500. Assume that the annual salaries of all television news anchorpersons are normally distributed.

a. Suppose the probability of making a Type I error is selected to be zero. Would you accept or reject the null hypothesis that the current mean annual salary of all television news anchorpersons is $65,824? Answer without performing the five steps of a test of hypothesis.

b. Using the 5% significance level, can you conclude that the current mean annual salary of television news anchorpersons is different from $65,824?

9.77 A past study claims that adults in America spend an average of 18 hours a week on leisure activities. A researcher wanted to test this claim. She took a sample of 10 adults and asked them about the time they spend per week on leisure activities. Their responses (in hours) are as follows.

14	25	22	38	16	26	19	23	41	33

Assume that the time spent on leisure activities by all adults is normally distributed. Using the 5% significance level, can you conclude that the claim of the earlier study is true? (*Hint:* First, calculate the sample mean and the sample standard deviation for these data using the formulas learned in Sections 3.1.1 and 3.2.2 of Chapter 3. Then make the test of hypothesis about μ.)

9.78 The past records of a supermarket show that its customers spend an average of $65 per visit at this store. Recently the management of the store initiated a promotional campaign according to which each customer receives points based on the total money spent at the store and these points can be used to buy products at the store. The management expects that as a result of this campaign, the customers should be encouraged to spend more money at the store. To check whether this is true, the manager of the store took a sample of 12 customers who visited the store. The following data give the money (in dollars) spent by these customers at this supermarket during their visits.

$88	69	141	28	106	45
32	51	78	54	110	83

Assume that the money spent by all customers at this supermarket has a normal distribution. Using the 1% significance level, can you conclude that the mean amount of money spent by all customers at this supermarket after the campaign was started is higher than $65? (*Hint:* First, calculate the sample mean and the sample standard deviation for these data using the formulas learned in Sections 3.1.1 and 3.2.2 of Chapter 3. Then make the test of hypothesis about μ.)

9.79 The manager of a service station claims that the mean amount spent on gas by its customers is $10.90. You want to test if the mean amount spent on gas at this station is different from $10.90. Briefly explain how you would conduct this test by taking a small sample.

9.80 A tool manufacturing company claims that its top-of-the-line machine that is used to manufacture bolts produces an average of 88 or more bolts per hour. A company that is interested in buying this machine wants to check this claim. Suppose you are asked to conduct this test. Briefly explain how you would do so by taking a small sample.

9.5   HYPOTHESIS TESTS ABOUT A POPULATION PROPORTION: LARGE SAMPLES

Often we want to conduct a test of hypothesis about a population proportion. For example, 33% of the students listed in *Who's Who Among American High School Students* said that drugs and alcohol are the most serious problems facing their high schools. A sociologist may want to check if this percentage still holds. As another example, a mail-order company claims that 90% of all orders it receives are shipped within 72 hours. The company's management may want to determine from time to time whether or not this claim is true.

This section presents the procedure to perform tests of hypotheses about the population proportion, p, for large samples. The procedure to make such tests is similar in many respects to the one for the population mean, μ. The procedure includes the same five steps. Again, the test can be two-tailed or one-tailed. We know from Chapter 7 that when the sample size is large, the sample proportion, $\hat{p}$, is approximately normally distributed with its mean equal to p and standard deviation equal to $\sqrt{pq/n}$. Hence, we use the normal distribution to perform a test of hypothesis about the population proportion, p, for a large sample. As was mentioned in Chapters 7 and 8, in the case of a proportion, the sample size is considered to be large when np and nq are both greater than 5.

TEST STATISTIC

The value of the *test statistic z* for the sample proportion, $\hat{p}$, is computed as

$$z = \frac{\hat{p} - p}{\sigma_{\hat{p}}} \qquad \text{where} \qquad \sigma_{\hat{p}} = \sqrt{\frac{pq}{n}}$$

The value of p used in this formula is the one used in the null hypothesis. The value of q is equal to $1 - p$.

The value of z calculated for $\hat{p}$ using the above formula is also called the **observed value of z**.

Examples 9–9, 9–10, and 9–11 describe the procedure to make tests of hypotheses about the population proportion, p.

Conducting a two-tailed test of hypothesis about p: large sample.

EXAMPLE 9–9   According to a *Wall Street Journal*/NBC News nationwide poll, 69% of adults interviewed said that they are in favor of tighter handgun control (*The Wall Street Journal*, December 13, 1996). For convenience, assume that 69% of all adults in the United States held this view in 1996. A researcher took a nationwide poll of 900 adults recently and found that 71% of them are in favor of tighter handgun control. Using the .01 significance level, can you conclude that the current percentage of adults in the United States who are in favor of tighter handgun control is different from that for 1996?

Solution   Let p be the proportion of the current population of all adults in the United States who are in favor of tighter handgun control and $\hat{p}$ the corresponding sample proportion. Then, from the given information,

$$n = 900, \qquad \hat{p} = .71, \qquad \text{and} \qquad \alpha = .01$$

Based on the 1996 poll, 69% of adults were in favor of tighter handgun control. Assuming this is true for the current population of adults,

$$p = .69 \quad \text{and} \quad q = 1 - p = 1 - .69 = .31$$

Note that we have changed all percentages to proportions.

Step 1. *State the null and alternative hypotheses*

The current percentage of adults in the United States who are in favor of tighter handgun control is not different from that for 1996 if $p = .69$ and the current percentage is different from 1996 if $p \neq .69$. The null and alternative hypotheses are as follows.

$$H_0: p = .69 \quad \text{(The current percentage is not different from 1996)}$$
$$H_1: p \neq .69 \quad \text{(The current percentage is different from 1996)}$$

Step 2. *Select the distribution to use*

The values of np and nq are

$$np = 900 \, (.69) = 621 \quad \text{and} \quad nq = 900 \, (.31) = 279$$

Because both np and nq are greater than 5, the sample size is large. Consequently, we use the normal distribution to make the hypothesis test about p.

Step 3. *Determine the rejection and nonrejection regions*

The $\neq$ sign in the alternative hypothesis indicates that the test is two-tailed. The significance level is .01. Therefore, the total area of the two rejection regions is .01 and the rejection region in each tail of the sampling distribution of $\hat{p}$ is $\alpha/2 = .01/2 = .005$. The critical values of z, obtained from the standard normal distribution table, are -2.58 and 2.58, as shown in Figure 9.15.

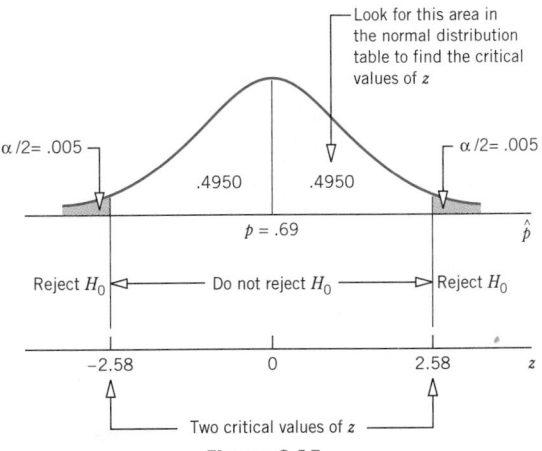

Figure 9.15

Step 4. *Calculate the value of the test statistic*

The value of the test statistic z for $\hat{p} = .71$ is calculated as follows.

$$\sigma_{\hat{p}} = \sqrt{\frac{pq}{n}} = \sqrt{\frac{(.69)\,(.31)}{900}} = .01541644$$

From H_0

$$z = \frac{\hat{p} - p}{\sigma_{\hat{p}}} = \frac{.71 - .69}{.01541644} = 1.30$$

Step 5. *Make a decision*

The value of the test statistic $z = 1.30$ for $\hat{p}$ lies in the nonrejection region. Consequently, we fail to reject H_0. Therefore, we can state that the sample proportion is not too far from the hypothesized value of the population proportion, and the difference between the two can be attributed to chance. We conclude that the current percentage of adults in the United States who are in favor of tighter handgun control is not different from that for 1996. ■

Making a right-tailed test of hypothesis about p: large sample.

EXAMPLE 9-10 When working properly, a machine that is used to make chips for calculators does not produce more than 4% defective chips. Whenever the machine produces more than 4% defective chips, it needs an adjustment. To check if the machine is working properly, the quality control department at the company often takes samples of chips and inspects them to determine if they are good or defective. One such random sample of 200 chips taken recently from the production line contained 14 defective chips. Test at the 5% significance level whether or not the machine needs an adjustment.

Solution Let p be the proportion of defective chips in all chips produced by this machine and $\hat{p}$ be the corresponding sample proportion. Then, from the given information,

$$n = 200, \quad \hat{p} = 14/200 = .07, \quad \text{and} \quad \alpha = .05$$

When the machine is working properly it does not produce more than 4% defective chips. Consequently, assuming that the machine is working properly,

$$p = .04 \quad \text{and} \quad q = 1 - p = 1 - .04 = .96$$

Step 1. *State the null and alternative hypotheses*

The machine will not need an adjustment if the percentage of defective chips is 4% or less, and it will need an adjustment if this percentage is greater than 4%. Hence, the null and alternative hypotheses are

$H_0: p \leq .04$ (The machine does not need an adjustment)

$H_1: p > .04$ (The machine needs an adjustment)

Step 2. *Select the distribution to use*

The values of np and nq are

$$np = 200\,(.04) = 8 > 5 \quad \text{and} \quad nq = 200\,(.96) = 192 > 5$$

Because the sample size is large, we use the normal distribution to make the hypothesis test about p.

Step 3. *Determine the rejection and nonrejection regions*

The significance level is .05. The $>$ sign in the alternative hypothesis indicates that the test is right-tailed and the rejection region lies in the right tail of the sampling distribution of $\hat{p}$ with its area equal to .05. As shown in Figure 9.16, the critical value of z, obtained from the normal distribution table for .4500, is approximately 1.65.

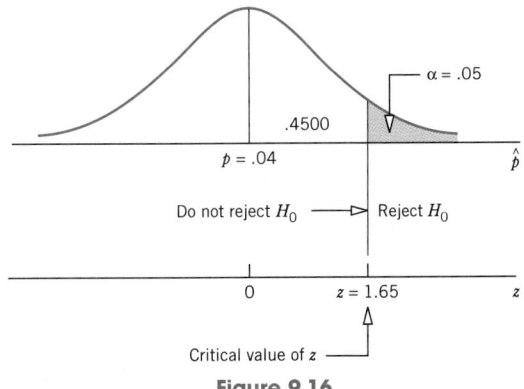

Figure 9.16

Step 4. *Calculate the value of the test statistic*

The value of the test statistic z for $\hat{p} = .07$ is calculated as follows.

$$\sigma_{\hat{p}} = \sqrt{\frac{pq}{n}} = \sqrt{\frac{(.04)\,(.96)}{200}} = .01385641$$

$$z = \frac{\hat{p} - p}{\sigma_{\hat{p}}} = \frac{.07 - .04}{.01385641} = 2.17$$

From H_0

Step 5. *Make a decision*

Because the value of the test statistic $z = 2.17$ is greater than the critical value of $z = 1.65$ and it falls in the rejection region, we reject H_0. We conclude that the sample proportion is too far from the hypothesized value of the population proportion and the difference between the two cannot be attributed to chance alone. Therefore, based on the sample information, we conclude that the machine needs an adjustment. ◾

Conducting a left-tailed test of hypothesis about p: large sample.

EXAMPLE 9–11   Direct Mailing Company sells computers and computer parts by mail. The company claims that at least 90% of all orders are mailed within 72 hours after they are received. The quality control department at the company often takes samples to check if this claim is valid. A recently taken sample of 150 orders showed that 129 of them were mailed within 72 hours. Do you think the company's claim is true? Use a 2.5% significance level.

Solution   Let p be the proportion of all orders that are mailed by the company within 72 hours and $\hat{p}$ the corresponding sample proportion. Then, from the given information,

$$n = 150, \qquad \hat{p} = 129/150 = .86, \qquad \text{and} \qquad \alpha = .025$$

The company claims that at least 90% of all orders are mailed within 72 hours. Assuming that this claim is true, the values of p and q are

$$p = .90 \quad \text{and} \quad q = 1 - p = 1 - .90 = .10$$

Step 1.   *State the null and alternative hypotheses*

The null and alternative hypotheses are

$$H_0\colon p \geq .90 \quad \text{(The company's claim is true)}$$

$$H_1\colon p < .90 \quad \text{(The company's claim is false)}$$

Step 2.   *Select the distribution to use*

We first check whether both np and nq are greater than 5.

$$np = 150\,(.90) = 135 > 5 \quad \text{and} \quad nq = 150\,(.10) = 15 > 5$$

Consequently, the sample size is large. Therefore, we use the normal distribution to make the hypothesis test about p.

Step 3.   *Determine the rejection and nonrejection regions*

The significance level is .025. The $<$ sign in the alternative hypothesis indicates that the test is left-tailed and the rejection region lies in the left tail of the sampling distribution of $\hat{p}$ with its area equal to .025. As shown in Figure 9.17, the critical value of z, obtained from the normal distribution table for .4750, is (approximately) -1.96.

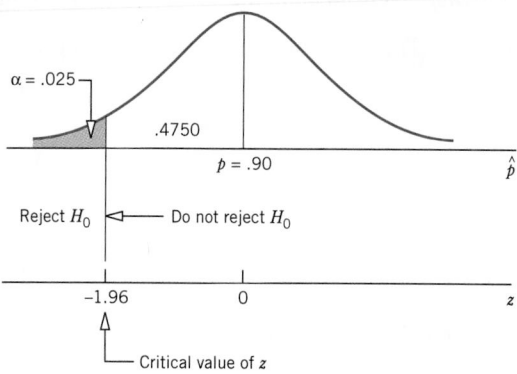

Figure 9.17

Step 4.   *Calculate the value of the test statistic*

The value of the test statistic z for $\hat{p} = .86$ is calculated as follows.

$$\sigma_{\hat{p}} = \sqrt{\frac{pq}{n}} = \sqrt{\frac{(.90)\,(.10)}{150}} = .02449490$$

$$z = \frac{\hat{p} - p}{\sigma_{\hat{p}}} = \frac{.86 - .90}{.02449490} = -1.63$$

(From H_0)

Step 5.   *Make a decision*

The value of the test statistic $z = -1.63$ is greater than the critical value of $z = -1.96$, and it falls in the nonrejection region. Therefore, we fail to reject H_0. We can state that the difference between the sample proportion and the hypothesized value of the population proportion is small and this difference may have occurred owing to chance alone. Therefore, the proportion of all orders that are mailed within 72 hours is at least 90%, and the company's claim is true. ▬

We can also use the *p*-value approach to make tests of hypotheses about the population proportion *p*. The procedure to calculate the *p*-value for the sample proportion is similar to the one applied to the sample mean in Section 9.3.

Case Study 9–1 discusses the risk of lung cancer for passive smokers. This case study presents a confidence interval for the increase in risk of lung cancer for those who are exposed to environmental smoke and also states that the results obtained for the test of hypothesis performed to check this increased risk were significant.

CASE STUDY 9-1   STATISTICIANS OCCUPY FRONT LINES IN BATTLE OVER PASSIVE SMOKING

In the controversy over passive smoking, the difference between 90% and 95% has become a matter of life and death.

The U.S. Environmental Protection Agency says there is a 90% probability that the risk of lung cancer for passive smokers is somewhere between 4% and 35% higher than for those who aren't exposed to environmental smoke. To statisticians, this calculation is called the ''90% confidence interval.''

And that, say tobacco-company statisticians, is the rub. ''Ninety-nine percent of all epidemiological studies use a 95% confidence interval,'' says Gio B. Gori, director of the Health Policy Center in Bethesda, MD, who has frequently served as a consultant and an expert witness for the tobacco industry.

. . . [T]he validity of the evidence already is being weighed by individuals, managers, city and state legislatures and others who have the power to ban smoking in their immediate environs. . . . [T]he EPA, citing its earlier conclusions, urged schools, day-care centers, parents, party hosts, and others to voluntarily ban smoking in their respective areas or to at least increase ventilation.

. . . Dr. Burns cites the EPA report as offering a completed chain of evidence convicting environmental smoke of causing cancer. One link in the chain is the fact that the same proven cancer-causing chemicals found in directly inhaled cigarette smoke—mainstream smoke—are found in the exhaled or environmental smoke. The final link is the evidence that nonsmokers exposed to environmental smoke have a higher-than-normal incidence of lung cancer.

This last link of evidence is based in large part on studies of nonsmoking women who lived with longtime smokers. Adding up the results from 11 such studies, the EPA concluded that the nonsmoking women who live with smokers have, on the average, a 19% higher risk of developing lung cancer than comparable women who live in a smoke-free home. The risk is higher for wives of heavy smokers and lower for wives of light smokers.

This 19% higher risk translates into 1,500 to 1,760 women dying each year of lung cancer caused by breathing other people's cigarette smoke, the EPA statisticians calculated. An equal number of non-smoking men also die of lung cancer from environmental smoke, for a total of more than 3,000 deaths a year, the EPA report declares.

... When the 19%-higher-risk figure was first calculated, a statistical test determined its "statistical significance." The term refers to the odds that the answer was the result of chance, explains Kenneth G. Brown, an independent statistician consultant in Chapel Hill, N.C., who did the risk calculations.

This latter calculation showed that there were only two chances out of 100—a probability of 0.02—that the 19% figure was a matter of happenstance. This more than meets the standard of 0.05 (five chances out of 100) at which most scientific studies are considered statistically significant.

Mr. Brown says that it was during the reviews of the final drafts that a second reliability calculation was added to give reviewers a better feeling for the reliability of the calculations.

This second calculation produced the controversial 90% confidence interval, or 90% probability that the lung-cancer-risk range is between 4% and 35% higher for passive smokers than those who aren't so exposed.

The Health Policy Center's Dr. Gori explains that the standard in such studies is to calculate the range within which it is 95% certain that the true answer lies, rather than the range for a 90% certainty. The reason the EPA didn't use the standard 95% confidence interval, Dr. Gori says, is that it would be so wide it might even hint that passive smoking actually reduced the risk of lung cancer. Although such a calculation wasn't made, it might show, for instance, that passive smokers' risk of cancer ranges from, say, 15% lower to 160% higher than the risk run by those in a smoke-free environment.

... Dr. Wood, the EPA consultant, says that Dr. Gori is correct in saying that using a 95% confidence interval would hint that passive smoking might reduce the risk of cancer. But, he says, this is exactly why it wasn't used. The EPA believes it is inconceivable that breathing in smoke containing known cancer-causing substances could be healthy and any hint in the report that it might be would be meaningless and confusing, he explains.

"I could have presented any level of confidence interval you wanted and it still wouldn't change the conclusion" that passive smoking boosts the risk of lung cancer an average of 19%, he says.

"The confidence interval isn't a substantive issue," Mr. Wood says. The 90% confidence interval used in the report was added for the convenience of scientifically oriented readers. The tobacco industry's harping on it, "is just to confuse the public."

Source: The Wall Street Journal, July 28, 1993. Copyright © Dow Jones Company 1993. Excerpts reprinted with permission.

EXERCISES

Concepts and Procedures

9.81   Explain when a sample is large enough to use the normal distribution to make a test of hypothesis about the population proportion.

9.82   In each of the following cases, do you think the sample size is large enough to use the normal distribution to make a test of hypothesis about the population proportion? Explain why or why not.

a. $n = 40$ and $p = .11$ b. $n = 100$ and $p = .73$
c. $n = 80$ and $p = .05$ d. $n = 50$ and $p = .14$

9.83 In each of the following cases, do you think the sample size is large enough to use the normal distribution to make a test of hypothesis about the population proportion? Explain why or why not.

a. $n = 30$ and $p = .65$ b. $n = 70$ and $p = .05$
c. $n = 60$ and $p = .06$ d. $n = 900$ and $p = .17$

9.84 For each of the following examples of tests of hypotheses about the population proportion, show the rejection and nonrejection regions on the graph of the sampling distribution of the sample proportion.

a. A two-tailed test with $\alpha = .10$
b. A left-tailed test with $\alpha = .01$
c. A right-tailed test with $\alpha = .05$

9.85 For each of the following examples of tests of hypotheses about the population proportion, show the rejection and nonrejection regions on the graph of the sampling distribution of the sample proportion.

a. A two-tailed test with $\alpha = .05$
b. A left-tailed test with $\alpha = .02$
c. A right-tailed test with $\alpha = .025$

9.86 A random sample of 500 observations produced a sample proportion equal to .37. Find the critical and observed values of z for each of the following tests of hypotheses using $\alpha = .05$.

a. $H_0: p = .30$ versus $H_1: p > .30$
b. $H_0: p = .30$ versus $H_1: p \neq .30$

9.87 A random sample of 200 observations produced a sample proportion equal to .59. Find the critical and observed values of z for each of the following tests of hypotheses using $\alpha = .01$.

a. $H_0: p = .63$ versus $H_1: p < .63$
b. $H_0: p = .63$ versus $H_1: p \neq .63$

9.88 Consider the null hypothesis $H_0: p = .65$. Suppose a random sample of 1000 observations is taken to make this test about the population proportion. Using $\alpha = .05$, show the rejection and nonrejection regions and find the critical value(s) of z for a

a. left-tailed test   b. two-tailed test   c. right-tailed test

9.89 Consider the null hypothesis $H_0: p = .25$. Suppose a random sample of 400 observations is taken to make this test about the population proportion. Using $\alpha = .01$, show the rejection and nonrejection regions and find the critical value(s) of z for a

a. left-tailed test   b. two-tailed test   c. right-tailed test

9.90 Consider $H_0: p = .70$ versus $H_1: p \neq .70$.

a. A random sample of 600 observations produced a sample proportion equal to .67. Using $\alpha = .01$, would you reject the null hypothesis?
b. Another random sample of 600 observations taken from the same population produced a sample proportion equal to .76. Using $\alpha = .01$, would you reject the null hypothesis?

Comment on the results of parts a and b.

9.91 Consider $H_0: p = .45$ versus $H_1: p < .45$.

a. A random sample of 400 observations produced a sample proportion equal to .41. Using $\alpha = .025$, would you reject the null hypothesis?
b. Another random sample of 400 observations taken from the same population produced a sample proportion of .395. Using $\alpha = .025$, would you reject the null hypothesis?

Comment on the results of parts a and b.

9.92 Make the following hypothesis tests about p.

a. $H_0: p = .45$, $H_1: p \neq .45$, $n = 100$, $\hat{p} = .48$, $\alpha = .10$

   b. H_0: $p = .72$, H_1: $p < .72$, $n = 700$, $\hat{p} = .65$, $\alpha = .05$
   c. H_0: $p = .30$, H_1: $p > .30$, $n = 200$, $\hat{p} = .34$, $\alpha = .01$

9.93   Make the following hypothesis tests about p.

   a. H_0: $p = .57$, H_1: $p \neq .57$, $n = 800$, $\hat{p} = .51$, $\alpha = .05$
   b. H_0: $p = .26$, H_1: $p < .26$, $n = 400$, $\hat{p} = .22$, $\alpha = .01$
   c. H_0: $p = .84$, H_1: $p > .84$, $n = 250$, $\hat{p} = .86$, $\alpha = .025$

Applications

9.94   In an earlier *USA Today*/CNN/Gallup poll of American adults, 45% of adults said that cigarette advertising should be banned (*USA Today*, March 16, 1994). Assume that this result is true for the population of all American adults at the time of that poll. A researcher wanted to check whether this result holds true for the current population of American adults. A sample of 700 adults taken recently by this researcher showed that 42% of them hold this view. Using the 2.5% significance level, can you conclude that the current percentage of American adults who think that cigarette advertising should be banned is less than 45%?

9.95   According to a University of Michigan study released in July 1995, 49% of eighth-graders in the United States believe that smoking one pack of cigarettes a day does not pose a serious risk to their health (*USA Today*, July 20, 1995). A recent sample survey of 400 eighth-graders from the United States showed that 176 of them believe that smoking one pack of cigarettes a day does not pose a serious risk to their health. Test at the 1% significance level whether the current proportion of all U.S. eighth-graders who hold this opinion is less than 49%.

9.96   A study of skiing injuries was conducted at a large ski resort in Jackson Hole, Wyoming, from 1981 to 1993 (*The American Journal of Sports Medicine* 23[5], September/October, 1995). This study analyzed a total of 9749 skiing injuries and classified 2935 of them as knee sprains. The article also states that past research indicated that about 20% of skiing injuries are knee sprains. Assume that the data from Jackson Hole represent a random sample of current skiing injuries. Test the alternative hypothesis that the percentage of knee sprains among current skiing injuries based on this study exceeds 20%. Use a 1% significance level.

9.97   In 1994, Harrah's Survey of Casino Entertainment estimated that 59% of adult Americans approved of gambling (*USA Today*, March 23, 1995). Assume that this percentage is true for the population of all adult Americans at the time of the survey. In a recent random sample of 1000 adult Americans, 630 approved of gambling. At the 5% level of significance, can you conclude that the proportion of all adult Americans who approve of gambling is currently different from 59%?

9.98   In 1996, 32% of U.S. households owned a personal computer (*Fortune*, March 18, 1996). In a recent sample of 850 U.S. households, 305 own personal computers.

   a.   Test at the 2% significance level whether the current percentage of all U.S. households who own personal computers is different from 32%.
   b.   What is the Type I error in this case? What is the probability of making this error?

9.99   A 1995 poll, conducted by Louis Harris & Associates for *Business Week* magazine, examined the perception of American adults of their government (*Business Week*, March 13, 1995). When asked whether the U.S. government represents the concerns of the American people or those of special interest groups, 63% of those polled felt that the U.S. government primarily represents the concerns of special interest groups. Assume that this percentage was true for the population of all American adults at the time the poll was conducted. A recent random sample of 1200 American adults found that 61% of them feel that the U.S. government represents the concerns of special interest groups.

   a.   Test at the 5% significance level whether the percentage of Americans who hold this view has changed since 1995.
   b.   What is the Type I error in this case? What is the probability of making this error?

9.100   A food company is planning to market a new type of frozen yogurt. However, before marketing this yogurt, the company wants to find what percentage of the people like it. The company's management has decided that it will market this yogurt only if at least 35% of the people like it. The company's research department selected a random sample of 400 persons and asked them to taste this yogurt. Of these 400 persons, 112 said they liked it.

    a. Testing at the 2.5% significance level, can you conclude that the company should market this yogurt?

    b. What will your decision be in part a if the probability of making a Type I error is zero? Explain.

9.101 A mail-order company claims that at least 60% of all orders are mailed within 48 hours. From time to time the quality control department at the company checks if this promise is fulfilled. Recently the quality control department at this company took a sample of 400 orders and found that 212 of them were mailed within 48 hours of the placement of the orders.

    a. Testing at the 1% significance level, can you conclude that the company's claim is true?

    b. What will your decision be in part a if the probability of making a Type I error is zero? Explain.

9.102 Brooklyn Corporation manufactures computer diskettes. The machine that is used to make these diskettes is known to produce not more than 5% defective diskettes. The quality control inspector selects a sample of 200 diskettes each week and inspects them for being good or defective. Using the sample proportion, the quality control inspector tests the null hypothesis $p \leq .05$ against the alternative hypothesis $p > .05$, where p is the proportion of diskettes that are defective. She always uses a 2.5% significance level. If the null hypothesis is rejected, the production process is stopped to make any necessary adjustments. A recent such sample of 200 diskettes contained 17 defective diskettes.

    a. Using the 2.5% significance level, would you conclude that the production process should be stopped to make necessary adjustments?

    b. Perform the test of part a using a 1% significance level. Is your decision different from the one in part a?

Comment on the results of parts a and b.

9.103 Shulman Steel Corporation makes bearings that are supplied to other companies. One of the machines makes bearings that are supposed to have a diameter of four inches. The bearings that have a diameter of either more or less than four inches are considered defective and are discarded. When working properly, the machine does not produce more than 7% of bearings that are defective. The quality control inspector selects a sample of 200 bearings each week and inspects them for the size of their diameters. Using the sample proportion, the quality control inspector tests the null hypothesis $p \leq .07$ against the alternative hypothesis $p > .07$, where p is the proportion of bearings that are defective. He always uses a 2% significance level. If the null hypothesis is rejected, the machine is stopped to make any necessary adjustments. One such sample of 200 bearings taken recently contained 22 defective bearings.

    a. Using the 2% significance level, will you conclude that the machine should be stopped to make necessary adjustments?

    b. Perform the test of part a using a 1% significance level. Is your decision different from the one in part a?

Comment on the results of parts a and b.

9.104 Two years ago, 75% of the customers of a bank said that they were satisfied with the services provided by the bank. The manager of the bank wants to know if this percentage of satisfied customers has changed since then. She assigns this responsibility to you. Briefly explain how you would conduct such a test.

9.105 A study claims that 65% of students at all colleges and universities hold off-campus (part-time or full-time) jobs. You want to check if the percentage of students at your school who hold off-campus jobs is different from 65%. Briefly explain how you would conduct such a test. Collect data from 40 students at your school on whether or not they hold off-campus jobs. Then, calculate the proportion of students in this sample who hold off-campus jobs. Using this information, test the hypothesis. Select your own significance level.

GLOSSARY

α The significance level of a test of hypothesis that denotes the probability of rejecting a null hypothesis when it actually is true. (The probability of committing a Type I error.)

Alternative hypothesis   A claim about a population parameter that will be true if the null hypothesis is false.

β The probability of not rejecting a null hypothesis when it actually is false. (The probability of committing a Type II error.)

Critical value or **critical point**   One or two values that divide the whole region under the sampling distribution of a sample statistic into rejection and nonrejection regions.

Left-tailed test   A test in which the rejection region lies in the left tail of the distribution curve.

Null hypothesis   A claim about a population parameter that is assumed to be true until proven otherwise.

Observed value of z or t   The value of z or t calculated for a sample statistic such as the sample mean or the sample proportion.

One-tailed test   A test in which there is only one rejection region, either in the left tail or in the right tail of the distribution curve.

***p*-value**   The smallest significance level at which a null hypothesis can be rejected.

Right-tailed test   A test in which the rejection region lies in the right tail of the distribution curve.

Significance level   The value of α that gives the probability of committing a Type I error.

Test statistic   The value of z or t calculated for a sample statistic such as the sample mean or the sample proportion.

Two-tailed test   A test in which there are two rejection regions, one in each tail of the distribution curve.

Type I error   An error that occurs when a true null hypothesis is rejected.

Type II error   An error that occurs when a false null hypothesis is not rejected.

KEY FORMULAS

1. **Value of the test statistic z for $\bar{x}$ in a test of hypothesis about μ for a large sample**

$$z = \frac{\bar{x} - \mu}{\sigma_{\bar{x}}} \quad \text{if } \sigma \text{ is known, where } \sigma_{\bar{x}} = \frac{\sigma}{\sqrt{n}}$$

or

$$z = \frac{\bar{x} - \mu}{s_{\bar{x}}} \quad \text{if } \sigma \text{ is not known, where } s_{\bar{x}} = \frac{s}{\sqrt{n}}$$

2. **Value of the test statistic t for $\bar{x}$ in a test of hypothesis about μ for a small sample**

$$t = \frac{\bar{x} - \mu}{s_{\bar{x}}} \quad \text{where} \quad s_{\bar{x}} = \frac{s}{\sqrt{n}}$$

3. **Value of the test statistic z for $\hat{p}$ in a test of hypothesis about p for a large sample**

$$z = \frac{\hat{p} - p}{\sigma_{\hat{p}}} \quad \text{where} \quad \sigma_{\hat{p}} = \sqrt{\frac{pq}{n}}$$

SUPPLEMENTARY EXERCISES

9.106 Consider the following null and alternative hypotheses.

$$H_0: \mu = 120 \text{ versus } H_1: \mu > 120$$

A random sample of 81 observations taken from this population produced a sample mean of 123.5 and a sample standard deviation of 15.

    a. If this test is made at the 2.5% significance level, would you reject the null hypothesis?
    b. What is the probability of making a Type I error in part a?
    c. Calculate the p-value for the test. Based on this p-value, would you reject the null hypothesis if $\alpha = .01$? What if $\alpha = .05$?

9.107 Consider the following null and alternative hypotheses.

$$H_0: \mu = 40 \text{ versus } H_1: \mu \neq 40$$

A random sample of 64 observations taken from this population produced a sample mean of 38.4 and a sample standard deviation of 6.

    a. If this test is made at the 2% significance level, would you reject the null hypothesis?
    b. What is the probability of making a Type I error in part a?
    c. Calculate the p-value for the test. Based on this p-value, would you reject the null hypothesis if $\alpha = .01$? What if $\alpha = .05$?

9.108 Consider the following null and alternative hypotheses.

$$H_0: p = .82 \text{ versus } H_1: p \neq .82$$

A random sample of 600 observations taken from this population produced a sample proportion of .86.

    a. If this test is made at the 2% significance level, would you reject the null hypothesis?
    b. What is the probability of making a Type I error in part a?
    c. Calculate the p-value for the test. Based on this p-value, would you reject the null hypothesis if $\alpha = .025$? What if $\alpha = .005$?

9.109 Consider the following null and alternative hypotheses.

$$H_0: p = .44 \text{ versus } H_1: p < .44$$

A random sample of 450 observations taken from this population produced a sample proportion of .39.

    a. If this test is made at the 2% significance level, would you reject the null hypothesis?
    b. What is the probability of making a Type I error in part a?
    c. Calculate the p-value for the test. Based on this p-value, would you reject the null hypothesis if $\alpha = .01$? What if $\alpha = .025$?

9.110 A manufacturer of fluorescent light bulbs claims that the mean life of these bulbs is at least 2500 hours. A consumer agency wanted to check whether or not this claim is true. The agency took a random sample of 36 such bulbs and tested them. The mean life for the sample was found to be 2447 hours with a standard deviation of 180 hours.

    a. Do you think the sample information supports the company's claim? Use $\alpha = .025$.
    b. What is the Type I error in this case? Explain. What is the probability of making this error?
    c. Will your conclusion of part a change if the probability of making a Type I error is zero?

9.111 The rapid increases in college tuition and fees over the past few years have been a source of great concern for college students and their parents. According to a recent study by the U.S. General Accounting Office, the average annual cost of in-state tuition and fees for four-year public colleges in the United States was $2865 in 1996, which represents approximately 9% of the median household income (*Willimantic Chronicle*, September 20, 1996). A recent sample of 40 four-year U.S. public colleges yielded a mean in-state tuition and fees of $3000 with a sample standard deviation of $300.

    a. Does the sample information support the alternative hypothesis that the current mean cost of in-state tuition and fees for all four-year public colleges is greater than $2865? Use $\alpha = .01$.

    b. What is the Type I error in this case? Explain. What is the probability of making this error?

    c. Will your conclusion for part a change if the probability of making a Type I error is zero?

9.112 According to the U.S. Bureau of Labor Statistics, U.S. households spent an average of $1567 on entertainment in 1994. A recent sample of 400 U.S. households showed that they spent an average of $1618 on entertainment in 1996 with a standard deviation of $435.

    a. Using $\alpha = .05$, can you conclude that the mean amount spent on entertainment by all U.S. households in 1996 is different from $1567?

    b. Using $\alpha = .01$, can you conclude that the mean amount spent on entertainment in 1996 by all U.S. households is different from $1567?

Comment on the results of parts a and b.

9.113 During the past few years people have become more health conscious, especially in regard to the consumption of red meat. In 1993, the average consumption of red meat per person was 111.9 pounds in the United States. A sample of 100 persons showed that they consumed, on average, 106.5 pounds of red meat in 1996 with a standard deviation of 26.5 pounds.

    a. Using $\alpha = .05$, does the sample information support the alternative hypothesis that the 1996 mean consumption of red meat is different from 111.9 pounds?

    b. Using $\alpha = .01$, does the sample information support the alternative hypothesis that the 1996 mean consumption of red meat is different from 111.9 pounds?

Comment on the results of parts a and b.

9.114 Customers often complain about long waiting times at restaurants before the food is served. A restaurant claims that it serves food to its customers, on average, within 15 minutes after the order is placed. A local newspaper journalist wanted to check if the restaurant's claim is true. A sample of 36 customers showed that the mean time taken to serve food to them was 15.9 minutes with a standard deviation of 2.4 minutes. Using the sample mean, the journalist says that the restaurant's claim is false. Do you think the journalist's conclusion is fair to the restaurant? Use the 1% significance level to answer this question.

9.115 The customers at a bank complained about long lines and the time they had to spend waiting for service. It is known that the customers at this bank had to wait 8 minutes, on average, before being served. The management made some changes to reduce the waiting time for its customers. A sample of 32 customers taken after these changes were made produced a mean waiting time of 7.4 minutes with a standard deviation of 2.1 minutes. Using this sample mean, the bank manager displayed a huge banner inside the bank mentioning that the mean waiting time for customers has been reduced by new changes. Do you think the bank manager's claim is justifiable? Use the 2.5% significance level to answer this question.

9.116 A 1995 study by the International Mass Retailers Association found that young shoppers aged 8 to 17 averaged 12.7 shopping trips per month (*USA Today*, July 25, 1995). A recent sample of 300 shoppers aged 8 to 17 averaged 12.3 shopping trips per month with a standard deviation of 3.2. Find the *p*-value for the test with the alternative hypothesis that the current mean number of shopping trips by young shoppers aged 8 to 17 is different from 12.7 per month.

9.117 The mean consumption of water per household in a city was 1245 cubic feet per month. Due to a water shortage because of a drought, the city council campaigned for water usage conservation by households. A few months after the campaign was started, the mean consumption of water for a sample of 100 households was found to be 1175 cubic feet per month with a standard deviation of 250 cubic feet. Find the *p*-value for the hypothesis test that the mean consumption of water per household has decreased due to the campaign by the city council.

9.118 A recent journal article examined the amount of housework done by adult children living with their parents (Glenna Spitzee and Russell Ward, ''Household Labor in Intergenerational Households,'' *Journal of Marriage and the Family*, May 1995, pp. 355–361). Among the data gathered were times spent per week on housework as reported by the adult children and as reported by their parents. The average time claimed by children was 14.4 hours per week, while their parents reported an average of only 6.9 hours per week. A researcher wanted to do a follow-up study so she installed video cameras in a random sample of 16 such homes to observe the actual length of time each of the 16 adult children spent per week on household tasks. This sample yielded a mean of 9.0 hours per week spent by these children on housework with a standard deviation of 3.3 hours. Assume that the times spent per week on housework by all adult children in such homes are normally distributed.

a. At the 1% level of significance, can you conclude that the mean time spent on housework per week by all adult children is less than 14.4 hours?

b. At the 1% level of significance, can you conclude that the mean time spent on housework per week by all adult children is greater than 6.9 hours?

c. Based on the results of parts a and b, whose estimate of the mean time spent on housework by adult children is the more credible, the adult children's or the parents'?

9.119 The administrative office of a hospital claims that the mean waiting time for patients to get treatment in its emergency ward is 25 minutes. A random sample of 16 patients who received treatment in the emergency ward of this hospital produced a mean waiting time of 27.5 minutes with a standard deviation of 4.8 minutes. Using the 1% significance level, test whether the mean waiting time at the emergency ward is different from 25 minutes. Assume that the waiting times for all patients at this emergency ward have a normal distribution.

9.120 According to the American Medical Association, physicians in the United States work an average of 59 hours a week (*The Wall Street Journal*, March 19, 1993). A researcher wanted to check whether or not this result holds true for the current population of all physicians. A random sample of 25 physicians showed that they work an average of 54 hours a week with a standard deviation of 7 hours. Assume that the number of hours worked per week by all physicians has a normal distribution.

a. Using $\alpha = .025$, can you conclude that the mean number of hours worked per week by all physicians is less than 59 hours?

b. Suppose the probability of making a Type I error is zero. Can you make a decision for the test of part a without going through the five steps of hypothesis testing? If yes, what is your decision? Explain.

9.121 An earlier study claims that U.S. adults spend an average of 114 minutes with their families per day. A recently taken sample of 25 adults showed that they spend an average of 109 minutes per day with their families. The sample standard deviation is 11 minutes. Assume that the time spent by adults with their families has an approximate normal distribution.

a. Using the 1% significance level, test whether the mean time spent currently by all adults with their families is less than 114 minutes a day.

b. Suppose the probability of making a Type I error is zero. Can you make a decision for the test of part a without going through the five steps of hypothesis testing? If yes, what is your decision? Explain.

9.122 A computer company that recently introduced a new software product claims that the mean time taken to learn how to use this software is not more than two hours for people who are somewhat familiar with computers. A random sample of 12 such persons was selected. The following data give the time taken (in hours) by these persons to learn how to use this software.

1.75	2.25	2.40	1.90	1.50	2.75
2.15	2.25	1.80	2.20	3.25	2.60

Test at the 1% significance level if the company's claim is true. Assume that the time taken by all persons who are somewhat familiar with computers to learn how to use this software is approximately normally distributed.

 9.123 A company claims that its 8-ounce low-fat yogurt cups contain, on average, at most 150 calories per cup. A consumer agency wanted to check whether or not this claim is true. A random sample of 10 such cups produced the following data on calories.

147	159	153	146	144	161	163	153	143	158

Test at the 2.5% significance level if the company's claim is true. Assume that the number of calories for such cups of yogurt produced by this company has an approximate normal distribution.

9.124 In 1996, *Fortune* magazine conducted a poll of professional money managers to assess their opinion of economists (*Fortune*, September 9, 1996). When asked how well economists were able to forecast the level of economic activity, 67% of the money managers rated economists' performance as *fair* or *poor*. Assume that 67% is the true percentage of all professional money managers who felt this way when the poll was taken in 1996. In a recent sample of 200 professional money managers, 126 rated economists fair or poor at forecasting the level of economic activity.

   a.   Test at the 2.5% significance level whether the current proportion of all professional money managers who would rate economists fair or poor in forecasting the level of economic activity is lower than 67%.
   b.   How will you explain the Type I error in this case? What is the probability of making this error in part a?

9.125 In January 1995, Yankelovich Partners, Inc. took a telephone poll of 800 adult Americans for *Time*/CNN to examine their attitude toward environmental issues (*Time*, February 27, 1995). Eleven percent of the respondents felt that protection of the environment was not very important. Assume that this result was true for the population of all adult Americans at the time of the poll. In a recent random sample of 600 adult Americans, 13.5% stated that protection of the environment was not very important.

   a.   Test at the 1% level of significance if the current percentage of American adults who believe that protection of the environment is not very important is higher than 11%.
   b.   How will you explain the Type I error in this case? What is the probability of making this error in part a?

9.126 More and more people are abandoning national brand products and buying store brand products to save money. The president of a company that produces national brand coffee claims that 40% of the people prefer to buy national brand coffee. A random sample of 700 people who buy coffee showed that 252 of them buy national brand coffee. Using $\alpha = .01$, can you conclude that the percentage of people who buy national brand coffee is different from 40%?

 9.127 In recent years, the U.S. media have called attention to the increasingly ill-mannered behavior of people in the United States. A *U.S. News*/Bozell poll of American adults, conducted by KRC Research & Consulting in February 1996, found that 78% of those polled felt that incivility has worsened in the United States in the past 10 years (*U.S. News & World Report*, April 22, 1996). A recent poll of 500 American adults found that 81% of them hold this view. At the 5% level of significance, can you conclude that the current percentage of adults who feel that incivility has worsened in the past 10 years is different from 78%?

9.128 Mong Corporation makes auto batteries. The company claims that 80% of its LL70 batteries are good for 70 months or more. A consumer agency wanted to check if this claim is true. The agency took a random sample of 40 such batteries and found that 75% of them were good for 70 months or more.

   a.   Using the 1% significance level, can you conclude that the company's claim is false?
   b.   What will your decision be in part a if the probability of making a Type I error is zero? Explain.

9.129 Dartmouth Distribution Warehouse makes deliveries of a large number of products to its customers. To keep its customers happy and satisfied, the company's policy is to deliver on time at least 90% of all the orders it receives from its customers. The quality control inspector at the company quite often takes samples of orders delivered and checks if this policy is maintained. A recent such sample of 90 orders taken by this inspector showed that 75 of them were delivered on time.

    a. Using the 2% significance level, can you conclude that the company's policy is maintained?

    b. What will your decision be in part a if the probability of making a Type I error is zero? Explain.

*__9.130__ Refer to Exercise 9.125. Find the p-value for the test of hypothesis mentioned in that exercise. Using this p-value, would you reject the null hypothesis at $\alpha = .05$? What if $\alpha = .01$?

*__9.131__ Refer to Exercise 9.129. Find the p-value for the test of hypothesis mentioned in that exercise. Using this p-value, would you reject the null hypothesis at $\alpha = .05$? What if $\alpha = .01$?

*__9.132__ Professor Hansen believes that some people have the ability to predict in advance the outcome of a spin of a roulette wheel. He takes 100 student volunteers to a casino. The roulette wheel has 38 numbers, each of which is equally likely to occur. Of these 38 numbers, 18 are red, 18 are black, and 2 are green. Each student is to place a series of five bets, choosing either a red or a black number before each spin of the wheel. Thus, a student who bets on red has an 18/38 chance of winning that bet. The same is true of betting on black.

    a. Assuming random guessing, what is the probability that a particular student will win all five of his or her bets?

    b. Suppose for each student we formulate the hypothesis test

        H_0: The student is guessing

        H_1: The student has some predictive ability

    Suppose we reject H_0 only if the student wins all five bets. What is the significance level?

    c. Suppose that two of the 100 students win all five of their bets. Professor Hansen says "For these two students we can reject H_0 and conclude that we have found two students with some ability to predict." What might you make of Professor Hansen's conclusion?

*__9.133__ Acme Bicycle Company makes derailleurs for mountain bikes. Usually, no more than 4% of these parts are defective, but occasionally the machines that make them get out of adjustment and the rate of defectives exceeds 4%. To guard against this, the chief quality control inspector takes a random sample of 100 derailleurs each week and checks each one for defects. If too many of these parts are defective, the machines are shut down and adjusted. To decide how many parts must be defective in order to shut down the machines, the company's statistician has set up the hypothesis test

$$H_0: p \leq .04 \qquad \text{versus} \qquad H_1: p > .04$$

where p is the proportion of defectives among all derailleurs being made currently. Rejection of H_0 would call for shutting down the machines. For the inspector's convenience, the statistician would like the rejection region to have the form, "Reject H_0 if the number of defective parts is C or more." Find the value of C that will make the significance level (approximately) .05.

*__9.134__ Alpha Airlines claims that only 15% of its flights arrive more than 10 minutes late. Let p be the proportion of all of Alpha's flights that arrive more than 10 minutes late. Consider the hypothesis test

$$H_0: p \leq .15 \qquad \text{versus} \qquad H_1: p > .15$$

Suppose we take a random sample of 50 flights by Alpha Airlines and agree to reject H_0 if 9 or more of them arrive late. Find the significance level for this test.

SELF-REVIEW TEST

1. A test of hypothesis is always about

   a. a population parameter   **b.** a sample statistic   **c.** a test statistic

2. A Type I error is committed when

   a. a null hypothesis is not rejected when it is actually false
   b. a null hypothesis is rejected when it is actually true
   c. an alternative hypothesis is rejected when it is actually true

3. A Type II error is committed when

   a. a null hypothesis is not rejected when it is actually false
   b. a null hypothesis is rejected when it is actually true
   c. an alternative hypothesis is rejected when it is actually true

4. A critical value is the value

   a. calculated from sample data
   b. determined from a table (e.g., the normal distribution table or other such tables)
   c. neither a nor b

5. The computed value of a test statistic is the value

   a. calculated for a sample statistic
   b. determined from a table (e.g., the normal distribution table or other such tables)
   c. neither a nor b

6. The observed value of a test statistic is the value

   a. calculated for a sample statistic
   b. determined from a table (e.g., the normal distribution table or other such tables)
   c. neither a nor b

7. The significance level, denoted by α, is

   a. the probability of committing a Type I error
   b. the probability of committing a Type II error
   c. neither a nor b

8. The value of β gives the

   a. probability of committing a Type I error
   b. probability of committing a Type II error
   c. power of the test

9. The value of $1 - \beta$ gives the

   a. probability of committing a Type I error
   b. probability of committing a Type II error
   c. power of the test

10. A two-tailed test is a test with

   a. two rejection regions   **b.** two nonrejection regions   **c.** two test statistics

11. A one-tailed test

   a. has one rejection region   **b.** has one nonrejection region   **c.** both a and b

12. The smallest level of significance at which a null hypothesis is rejected is called

   a. α **b.** p-value   **c.** β

13. Which of the following is not required to apply the t distribution to make a test of hypothesis about μ?

    a. $n < 30$ **b.** population is normally distributed   **c.** σ is unknown   **d.** β is known

14. The sign in the alternative hypothesis in a two-tailed test is always

    a. $<$ **b.** $>$ **c.** $\neq$

15. The sign in the alternative hypothesis in a left-tailed test is always

    a. $<$ **b.** $>$ **c.** $\neq$

16. The sign in the alternative hypothesis in a right-tailed test is always

    a. $<$ **b.** $>$ **c.** $\neq$

17. A bank loan officer claims that the mean monthly mortgage payment made by all homeowners in a certain city is $1365. A housing magazine wanted to test this claim. A random sample of 100 homeowners taken by this magazine produced the mean monthly mortgage of $1489 with a standard deviation of $278.

    a. Testing at the 1% significance level, would you conclude that the mean monthly mortgage payment made by all homeowners in this city is different from $1365?

    b. What is the Type I error in part a? What is the probability of making this error?

    c. What will your decision be in part a if the probability of making a Type I error is zero? Explain.

18. An editor of a New York publishing company claims that the mean time taken to write a textbook is at least 30 months. A sample of 16 textbook authors showed that the mean time taken by them to write a textbook was 25 months with a standard deviation of 7.2 months.

    a. Using the 2.5% significance level, would you conclude that the editor's claim is true? Assume that the time taken to write a textbook is normally distributed for all textbook authors.

    b. What is the Type I error in part a? What is the probability of making this error?

    c. What will your decision be in part a if the probability of making a Type I error is .001?

19. Twenty-three percent of the physicians in the United States were women in 1994 (*U.S. News & World Report*, May 1, 1995). A recent sample of 300 U.S. physicians found that 51 of them were women.

    a. Using a 5% significance level, can you conclude that the current percentage of women in the population of all physicians in the United States is less than 23%?

    b. What is the Type I error in part a? What is the probability of making this error?

    c. What will your decision be in part a if the probability of making a Type I error is zero? Explain.

20. According to an IRS estimate in the 1995 instruction booklet for income tax Form 1040, it takes an average of 336 minutes to prepare, copy, assemble, and mail this form. (Note that this estimate does not include the time spent on record keeping or learning about the laws or the form.) A random sample of 100 taxpayers who filed Form 1040 in 1996 showed that they took, on average, 349 minutes to prepare, copy, assemble, and mail this form. The standard deviation for the sample was 62 minutes.

    a. Find the p-value for the test with the alternative hypothesis that the mean time taken to prepare, copy, assemble, and mail Form 1040 is different from 336 minutes.

    b. Using the p-value calculated in part a, would you reject the null hypothesis if $\alpha = .01$? What if $\alpha = .05$?

*21. Refer to Problem 19.

    a. Find the p-value for the test of hypothesis mentioned in part a of that problem.

    b. Using this p-value, will you reject the null hypothesis if $\alpha = .01$? What if $\alpha = .005$?

USING MINITAB

This section describes how to use MINITAB to test hypotheses about the population mean for large and small samples and about the population proportion for large samples. Remember that when the sample size is large ($n \geq 30$), whether σ is known or not, the normal distribution is used to make a hypothesis test about a population mean. However, when the sample size is small ($n < 30$), the population standard deviation is not known, and the population has a normal distribution, then the t distribution is used to make a hypothesis test about a population mean.

HYPOTHESIS TESTS ABOUT A POPULATION MEAN: LARGE SAMPLES

Illustration M9–1 shows how to test a hypothesis about the population mean for a large sample.

ILLUSTRATION M9–1 An earlier study showed that the mean time spent by all college students on community service is 50 minutes a week. The following data give the time spent on community service by each of 34 college students based on a recently taken random sample.

34	56	74	23	12	89	87	56	48	13	9	85
76	56	17	28	66	38	46	81	29	33	41	78
11	57	17	22	91	54	19	35	65	47		

Using MINITAB, test at the 5% significance level whether the mean time spent doing community service by all college students is less than 50 minutes.

Solution You are to test whether or not the mean time spent doing community service by all college students is less than 50 minutes. Hence, the null and alternative hypotheses are

$$H_0: \mu = 50$$

$$H_1: \mu < 50$$

The significance level is .05 and the test is left-tailed.

   If you are using MINITAB FOR WINDOWS, perform the following steps to make the test of hypothesis about the population mean for this example.

Step 1. Enter the given data on community service in column C1 of the Data window.
Step 2. Calculate the standard deviation of the sample data using the MINITAB procedure explained in Chapter 3. You will obtain $s = 25.659$.
Step 3. Click the **Stat** pull-down menu at the top of the screen.
Step 4. Click **Basic Statistics** from the selections available in the **Stat** menu.
Step 5. Click **1-Sample z** from the selections available in the **Basic Statistics** menu.
Step 6. You will see a dialog box entitled **1-Sample z** appear on the screen. Type **C1** in the box below **Variables**. Click inside the circle next to **Test mean** and enter **50** in the box next to it. Click the arrow in the corner of the box next to **Alternative**. You will see three alternatives appear on the screen. Since the test is left-tailed, click the **less than** alternative. Enter **25.659** (the value of s from Step 2 above) in the box next to **Sigma**. Note that you are using the value of s for σ here.

Step 7. Click the **OK** button at the bottom of this dialog box. The MINITAB output will appear on the screen.

If you are using the MINITAB COMMAND LANGUAGE, first enter the given data on community service into column C1 using the **SET** command, find the standard deviation of the sample data (which will be 25.659), and then type the following MINITAB commands to make the required test of hypothesis about μ.

```
MTB > ZTEST MEAN = 50 SIGMA = 25.659 C1;
SUBC > ALTERNATIVE = -1.
```

Here, in the first MINITAB command, *ZTEST* instructs MINITAB to use the normal distribution, *MEAN* = 50 indicates the value of μ in the null hypothesis, *SIGMA* = 25.659 represents the value of the sample standard deviation, which is used as a value of σ, and C1 instructs MINITAB to make the test about μ using the value of $\bar{x}$ calculated for the data entered in column C1. In the MINITAB subcommand, *ALTERNATIVE* = −1 tells MINITAB that it is a left-tailed test.

Note that the value of **Alternative** used in the subcommand would be −1, 0, or 1 depending on whether the test is *left-tailed*, *two-tailed*, or *right-tailed*, respectively. In other words,

ALTERNATIVE = −1   if the test is left-tailed

ALTERNATIVE = 0   if the test is two-tailed

ALTERNATIVE = 1   if the test is right-tailed

Whether you use MINITAB FOR WINDOWS or the MINITAB COMMAND LANGUAGE, you will obtain the output given in Figure 9.18.

Figure 9.18   MINITAB output for a test of hypothesis about μ for a large sample.

```
Z-Test

Test of mu = 50.00 vs mu < 50.00
The assumed sigma = 25.7

Variable    N    Mean   StDev   SE Mean      Z     P
C1         34   46.85   25.66     4.40   -0.72  0.24
```

In the MINITAB solution given in Figure 9.18, *Test of mu* = 50.00 *vs mu* < 50.00 indicates that the null hypothesis is μ = 50 and the alternative hypothesis is μ < 50. In the row of C1 in MINITAB output, 34 is the sample size, 46.85 is the mean of the sample data, 25.66 is the standard deviation of the sample data, 4.40 is the standard deviation (or standard error) of the sample mean, −0.72 is the observed value of the test statistic *z*, and 0.24 is the *p*-value discussed in Section 9.3 of this chapter.

To make a decision, find the critical value of *z* from the normal distribution table for a left-tailed test with α = .05. This value (for .4500 area between the mean and *z*) is approximately −1.65. Because the observed value of the test statistic, *z* = −.72, is greater than the critical value of *z* = −1.65 and it falls in the nonrejection region, the null hypothesis is not rejected. (The reader is advised to draw a graph that shows the rejection and nonrejection regions.) Based on this sample information, it can be concluded that the mean time spent doing community service by college students is not less than 50 minutes a week.

You can also use the *p*-value, printed in the MINITAB solution, to make the decision. As you know from Section 9.3, you will reject H_0 if the value of α is larger than the *p*-value, and you will fail to reject H_0 if the value of α is smaller than the *p*-value. In the MINITAB solution in Figure 9.18, the *p*-value is .24. As α = .05 is smaller than the *p*-value of .24, you fail to reject H_0. ■

In Illustration M9–1, the population standard deviation was unknown. However, if the population standard deviation is known, you will use that value of σ in the test of hypothesis, and you do not need to calculate the sample standard deviation before making the test.

HYPOTHESIS TESTS ABOUT A POPULATION MEAN: SMALL SAMPLES

We know from the discussion in this chapter that we apply the t distribution to make a hypothesis test about the population mean when

1. the sample size is small, that is, n $<$ 30
2. the population is (approximately) normally distributed
3. the population standard deviation, σ, is not known

Illustration M9–2 shows how to test a hypothesis about the population mean for a small sample.

ILLUSTRATION M9–2 A psychologist claims that the mean age at which children start walking is 12.5 months. The following data give the age (in months) at which 18 randomly selected children started walking.

15	11	13	14	15	12	15	10	16
17	14	16	13	15	15	14	11	13

Using MINITAB, test at the 1% significance level if the mean age at which children start walking is different from 12.5 months. Assume that the age at which all children start walking is normally distributed.

Solution You are to test if the mean age at which children start walking is different from 12.5 months. The null and alternative hypotheses are

$$H_0: \mu = 12.5$$

$$H_1: \mu \neq 12.5$$

The significance level is .01 and the test is two-tailed.

If you are using MINITAB FOR WINDOWS, perform the following steps to make the test of hypothesis about the population mean for this example.

Step 1. Enter the given data on 18 children in column C1 of the Data window.
Step 2. Click the **Stat** pull-down menu at the top of the screen.
Step 3. Click **Basic Statistics** from the selections available in the **Stat** menu.
Step 4. Click **1-Sample t** from the selections available in the **Basic Statistics** menu.
Step 5. You will see a dialog box entitled **1-Sample t** appear on the screen. Type **C1** in the box below **Variables**. Click inside the circle next to **Test mean** and enter **12.5** in the box next to it. Click the arrow in the corner of the box next to **Alternative**. You will see three alternatives appear on the screen. Since the test is two-tailed, click the **not equal** alternative.
Step 6. Click the **OK** button at the bottom of this dialog box. The MINITAB output will appear on the screen.

If you are using the MINITAB COMMAND LANGUAGE, first enter the given data on 18 children into column C1 using the **SET** command and, then, type the following MINITAB commands to make the required test of hypothesis about μ.

```
MTB > TTEST MEAN = 12.5 C1;
SUBC > ALTERNATIVE = 0.
```

Here, in the first MINITAB command, *TTEST* instructs MINITAB to use the t distribution, *MEAN* = 12.5 indicates the value of μ in the null hypothesis, and *C1* instructs MINITAB to make the test about μ using the value of $\bar{x}$ calculated for the data entered in column C1. In the MINITAB subcommand, *ALTERNATIVE* = 0 tells MINITAB that it is a two-tailed test.

Whether you use MINITAB FOR WINDOWS or the MINITAB COMMAND LANGUAGE, you will obtain the output given in Figure 9.19.

Figure 9.19 MINITAB output for a test of hypothesis about μ for a small sample.

T-Test of the Mean

Test of mu = 12.500 vs mu not = 12.500

```
Variable    N    Mean   StDev  SE Mean    T       P
C1         18  13.833   1.917   0.452   2.95   0.0090
```

In the MINITAB solution given in Figure 9.19, *Test of mu* = 12.500 *vs mu not* = 12.500 indicates that the null hypothesis is $\mu = 12.5$ and the alternative hypothesis is $\mu \neq 12.5$. In the row of C1 in MINITAB output, 18 is the sample size, 13.833 is the mean of the sample data, 1.917 is the standard deviation of the sample data, 0.452 is the standard deviation (or standard error) of the sample mean, 2.95 is the observed value of the test statistic *t*, and 0.0090 is the *p*-value discussed in Section 9.3 of this chapter.

To make a decision, find the critical values of *t* from the *t* distribution table for a two-tailed test with $\alpha/2 = .005$ and $df = 18 - 1 = 17$. These values of *t* are -2.898 and 2.898. Because the value of the test statistic, $t = 2.95$, is greater than 2.898, it falls in the rejection region. (The reader is advised to draw a graph showing the rejection and nonrejection regions.) Consequently, the null hypothesis is rejected. Thus, based on this sample information, it is concluded that the mean age at which all children start walking is different from 12.5 months.

You can reach the same conclusion by using the *p*-value. As $\alpha = .01$ is larger than the *p*-value of .0090, you reject the null hypothesis.

HYPOTHESIS TESTS ABOUT A POPULATION PROPORTION: LARGE SAMPLES

The current version of MINITAB FOR WINDOWS does not include procedures to make a test of hypothesis about the population proportion. However, the MINITAB COMMAND LANGUAGE can be used to make such a test. The use of the MINITAB COMMAND LANGUAGE to make a test of hypothesis about *p* will be explained with the help of Illustration M9–3. This illustration also explains how a confidence interval for *p* can be obtained using MINITAB.

ILLUSTRATION M9–3 Refer to Example 9–10. When working properly, a machine that is used to make chips for calculators does not produce more than 4% defective chips. Whenever the machine produces more than 4% defective chips, it needs an adjustment. To check if the machine is working properly, the quality control department at the company often takes samples of chips and inspects them to determine if they are good or defective. One such random sample of 200 chips taken recently from the production line contained 14 defective chips. Using MINITAB, test at the 5% significance level whether or not the machine needs an adjustment. Based on this sample information, make a 99% confidence interval for the population proportion.

Solution Let *n* be the number of chips in the sample and *x* be the number of defective chips in the sample. Then, from the given information,

$$n = 200 \quad \text{and} \quad x = 14$$

For the test of hypothesis, $\alpha = .05$. The null and alternative hypotheses are

$$H_0: p \leq .04 \quad \text{(The machine does not need an adjustment)}$$

$$H_1: p > .04 \quad \text{(The machine needs an adjustment)}$$

Figure 9.20 shows the MINITAB commands and the MINITAB output for both the above-mentioned test of hypothesis and a 99% confidence interval for *p*.

Figure 9.20   Test of hypothesis and a confidence interval for p.

```
MTB > exec 'pinf'
Executing from file: C:\MTBWIN\MACROS\pinf.MTB

  At the ''DATA>'' prompt below, enter the following values:
  1. p-naught, the hypothesized value of the population proportion;
  2. x, the observed number of successes; and
  3. n, the sample size.

DATA> .04
DATA> 14
DATA> 200
```

Data Display

```
Input:
    0.04   14.00   200.00

  At the ''DATA>'' prompt below, enter a
      1 to perform an upper-tailed test, or
      0 to perform a two-tailed test, or
     -1 to perform a lower-tailed test.

DATA> 1
```

Data Display

```
      Ho: p = 0.0400   vs.   Ha: p  >   0.0400

      z = 2.16506   p-value = 0.01519


  A confidence interval for the true population proportion will be
  produced below. At the ''DATA>'' prompt below, enter the
  confidence level (95 for a 95% confidence interval) for the
  confidence interval.

DATA> 99
```

Data Display

```
CI Level
    99
```

Data Display

```
  The 99% confidence interval is (0.0235 to 0.1165)
```

As shown in Figure 9.20, to make a test of hypothesis about p, first enter the following MINITAB command.

```
MTB > EXEC 'PINF'
```

Then, MINITAB will ask you to enter the values of p, x, and n. Here, the value of p will be entered from the null hypothesis. You enter these values, one at a time, next to the **DATA** prompt and hit the Enter/Return key. Note that you must know the value of x. If you know $\hat{p}$ instead of x, then multiply n by $\hat{p}$ to obtain x. After you have entered the values of p, x, and n, MINITAB asks you to enter the value of **Alternative**. Type -1, 0, or 1, whichever is applicable, at the **DATA** prompt and hit the Enter/Return key. MINITAB prints the values of H_0, H_1, z, and p. For our test, $z = 2.165$ and $p = .01519$. Since $\alpha = .05$, the critical value of z is 1.65. Hence, the null hypothesis is rejected. Also, $\alpha = .05$ is greater than $p = .01519$, which again helps to reject the null hypothesis.

After MINITAB gives the output for the test of hypothesis, it asks you for the confidence level to find the confidence interval for p. Enter **99** next to the **DATA** prompt and hit the Enter/Return key. MINITAB prints the confidence interval. For our example, this 99% confidence interval for p is .0235 to .1165 or 2.35% to 11.65%. In other words, based on this sample, this machine is currently producing between 2.35% and 11.65% defective chips. ▆

COMPUTER ASSIGNMENTS

M9.1 According to an earlier study, the mean amount spent on clothes by American women is $575 per year. A researcher wanted to check if this result still holds true. A random sample of 39 women taken recently by this researcher produced the following data on the amount they spent on clothes last year.

671	584	328	498	827	921	425	204	382	539
1070	854	669	328	537	849	930	1234	695	738
341	189	867	923	721	125	298	473	876	932
573	931	460	1430	391	887	958	674	782	

Using MINITAB, test at the 1% significance level if the mean expenditure on clothes for American women for last year is different from $575. Assume that the population standard deviation is $132.

M9.2 The mean weight of all babies born at a hospital last year was 7.6 pounds. A random sample of 35 babies born at this hospital this year produced the following data.

8.2	9.1	6.9	5.8	6.4	10.3	12.1	9.1	5.9	7.3
11.2	8.3	6.5	7.1	8.0	9.2	5.7	9.5	8.3	6.3
4.9	7.6	10.1	9.2	8.4	7.5	7.2	8.3	7.2	9.7
6.0	8.1	6.1	8.3	6.7					

Using MINITAB, test at the 2.5% significance level if the mean weight of babies born at this hospital this year is higher than 7.6 pounds.

M9.3 The president of a large university claims that the mean time spent partying by all students at the university is not more than seven hours per week. The following data give the time spent partying during the previous week by a random sample of 16 students taken from this university.

12	9	5	15	11	13	10	6
4	11	6	9	13	6	16	8

Using MINITAB, test at the 1% significance level if the president's claim is true. Assume that the time spent partying by all students at this university has an approximate normal distribution.

M9.4   According to a basketball coach, the mean height of all male college basketball players is 74 inches. A random sample of 25 such players produced the following data on their heights.

68	76	74	83	77	76	69	67	71	74	79	85	69
78	75	78	68	72	83	79	82	76	69	70	81	

Using MINITAB, test at the 2% significance level if the mean height of all male college basketball players is different from 74 inches. Assume that the heights of all male college basketball players are (approximately) normally distributed.

M9.5   A past study claims that adults in America spend an average of 18 hours a week on leisure activities. A researcher wanted to test this claim. She took a sample of 10 adults and asked them about the time they spend per week on leisure activities. Their responses (in hours) are as follows.

14	25	22	38	16	26	19	23	41	33

Assume that the times spent on leisure activities by all adults are normally distributed. Using the 5% significance level, can you conclude that the claim of the earlier study is true? Use MINITAB to answer this question.

M9.6   In an earlier *USA Today*/CNN/Gallup poll of American adults, 45% of adults said that cigarette advertising should be banned (*USA Today*, March 16, 1994). Assume that this result was true for the population of all American adults at the time of that poll. A researcher wanted to check whether this result holds true for the current population of American adults. A sample of 700 adults taken recently by this researcher showed that 294 of them hold this view. Using MINITAB, testing at the 2.5% significance level, can you conclude that the current percentage of American adults who think that cigarette advertising should be banned is different from 45%? Using MINITAB, make a 95% confidence interval for the current percentage of American adults who think that cigarette advertising should be banned.

M9.7   A mail-order company claims that at least 60% of all orders it receives are mailed within 48 hours. From time to time the quality control department at the company checks if this promise is fulfilled. Recently, the quality control department at this company took a sample of 400 orders and found that 224 of them were mailed within 48 hours of the placement of the orders. Using MINITAB, test at the 1% significance level whether or not the company's claim is true. Using MINITAB, make a 97% confidence interval for the population proportion.

MORE CHALLENGING EXERCISES (Optional)
CHAPTERS 7 TO 9

1. The weights of the packages handled by a parcel delivery service have a mean of 20 pounds and a standard deviation of 6 pounds. The van that is used to pick up these parcels has a load limit of 1500 pounds. On a particular day, the van must pick up 68 packages from different locations.
    a. Find the probability the load limit of the van will not be exceeded.
    b. What is the highest number of packages the van can pick up so that the probability the load limit of the van will be exceeded is no more than .05?

2. You are to conduct the experiment of sampling 10 times (with replacement) from the digits 0, 1, 2, 3, 4, 5, 6, 7, 8, and 9. You can do this in a variety of ways. One way to do so is to write each digit on a separate slip of paper, place all the slips in a hat, and select 10 times from the hat, returning each selected slip before the next pick. As alternatives, you can use a random numbers table such as Table I of Appendix C (see Section A.2.4 of Appendix A for instructions on how to use this table), you can use a 10-sided die, MINITAB (see Minitab section of Chapter 7), or a calculator that generates random numbers. Perform this experiment using any of these methods and compute the sample mean $\bar{x}$ for the 10 numbers obtained. Now repeat the above procedure 49 more times. When you are done you will have 50 sample means.
    a. Make a table of the population distribution and display it using a graph.
    b. Make a stem-and-leaf display of your 50 sample means. What shape does it have?
    c. What does the central limit theorem say about the shape of the sampling distribution of $\bar{x}$? What mean and standard deviation does the sampling distribution of $\bar{x}$ have in this problem?

3. Determine which of the following sampling schemes properly sample from the corresponding populations and which do not. In each case, write the population distribution and explain how the sampling procedure does or does not result in a proper observation from the population. In cases of improper sampling, how can the procedure be modified to get a proper sample value?
    a. You would like to observe the toss of a true coin but you have only a fair die. You roll this die once and observe the outcome. If an even number is rolled you say a head appears, and if an odd number is rolled you say a tail appears.
    b. You would like to observe the outcome of randomly picking a number from 0 to 9 but you have only a deck of cards. You randomly select a card from the deck. If an ace is selected you say the outcome is 1, if a two is selected you say the outcome is 2, . . . , if a nine is selected you say the outcome is 9, and if any other card is selected you say the outcome is 0.
    c. You would like to observe the outcome of the roll of a fair die but you only have access to the random numbers table, Table I of Appendix C. You obtain a three-digit number as described in Section A.2.4 of Appendix A. If the number is in the 100s (i.e., 100 to 199) you say a 1 is rolled, if the number is in the 200s you say a 2 is rolled, . . . , and if the number is in the 600s a 6 is rolled. However, if the number is in the 700s, 800s, or 900s you start over by obtaining a new three-digit number.
    d. You would like to observe the number of heads in two tosses of a true coin but you have only a deck of cards. You randomly select a card from the deck. If a heart is selected you say no heads were tossed, if a club is selected you say exactly one head was tossed, if a spade is selected you say two heads were tossed, but if a diamond is selected you start all over again by picking a new card.
    e. You would like to observe the number of heads in two tosses of a true coin but have only a deck of cards. You randomly select two cards from the deck. If the first card is red you say the first toss is a head, otherwise the first toss is a tail. If the second card is

red the second toss is a head, otherwise the second toss is a tail. You say the number of heads tossed equals the number of heads on the simulated tosses.

4. A gas station attendant would like to estimate p, the proportion of all households who own more than two vehicles. To obtain an estimate, the attendant decides to ask the next 200 gasoline customers how many vehicles their households own. To obtain an estimate of p, the attendant counts the number of customers who say there are more than two vehicles in their households and then divides this number by 200. How would you critique this estimation procedure? Is there anything wrong with this procedure that would result in sampling or nonsampling error? If so, can you suggest a procedure that would reduce this error?

5. A couple considering the purchase of a new home would like to estimate the average number of cars that go past the location per day. The couple guesses that the number of cars passing this location per day has a standard deviation of 170.

    a. On how many randomly selected days should the number of cars passing the location be observed so that the couple can be 99% certain the estimate will be within 100 cars of the true average?

    b. Suppose the couple finds out that the standard deviation of the number of cars passing the location per day is not 170 but actually 272. If they have already taken a sample of the size computed in part a, what confidence does the couple have that their point estimate is within 100 cars of the true average?

    c. If the couple has already taken a sample of the size computed in part a and later finds out that the standard deviation of the number of cars passing the location per day is actually 130, they can be 99% confident their point estimate is within how much of the true average?

6. A thumbtack that is tossed on a desk can land in one of the following two ways.

Heads   Tails

Brad and Dan cannot agree on the likelihood of obtaining a head or a tail. Brad argues that obtaining a tail is more likely than obtaining a head because of the shape of the tack. If the tack had no point at all it would resemble a coin that has the same probability of coming up heads or tails when tossed. And, the longer the point, the less likely it is that the tack will stand up on its head when tossed. Dan believes that as the tack lands tails, the point causes the tack to jump around and come to rest in the heads position. Brad and Dan need you to settle their dispute. Do you think the tack is equally likely to land heads or tails? To investigate this question, find an ordinary thumbtack and toss it a large number of times (say 100 times).

    a. What is the meaning, in words, of the unknown parameter in this problem?

    b. Set up the null and alternative hypotheses and compute the p-value based on your results from tossing the tack.

    c. How would you answer the original question now? If you decide the tack is not fair, do you side with Brad or Dan?

    d. What would you estimate the value of the parameter in part a to be? Find a 90% confidence interval for this parameter.

    e. After doing this experiment, do you feel 100 tosses are enough to infer about the nature of your tack? Using your result as a preliminary estimate, determine how many tosses would be necessary to be 95% certain of having 4% accuracy; that is, the maximum error of estimate is $\pm 4\%$. Have you observed enough tosses?

7. Since 1984, all automobiles have been manufactured with a middle taillight. You have been hired to answer the question, "Is the middle taillight effective in reducing rear-end collisions?" You have

available to you any information you could possibly want about all rear-end collisions involving cars built before 1984. How would you conduct an experiment to answer the above question? In your answer, include things like: (i) the precise meaning of the unknown parameter you are testing; (ii) H_0 and H_1; (iii) a detailed explanation of what sample data you would collect to draw a conclusion; (iv) any assumptions you would make, particularly about the characteristics of cars built before 1984 versus those built since 1984.

8. The Parks and Recreation Department has determined that the thickness of the ice on a town pond must average 5 inches to be safe for ice skating. Consider the decision of whether or not to allow ice skating as a hypothesis-testing problem,

    a. What is the meaning of the unknown parameter that is being tested?

    b. Set up the appropriate null and alternative hypotheses for this situation.

    c. What would it mean to commit a Type I error for your test in part b? What about a Type II error?

    d. Would you want the significance level for your test in part b to be larger or smaller? Suggest a significance level and explain what it means in this problem.

9. Thirty percent of all people who are inoculated with the current vaccine that is used to prevent a disease contract the disease within a year. The developer of a new vaccine that is intended to prevent this disease wishes to test for significant evidence that the new vaccine is more effective.

    a. Determine the appropriate null and alternative hypotheses.

    b. The developer decides to study 100 randomly selected people by inoculating them with the new vaccine. If 84 or more of them do not contract the disease within a year, the developer will conclude that the new vaccine is superior to the old one. What significance level is the developer using for the test?

    c. If 20 people inoculated with the new vaccine are studied and the new vaccine is concluded to be better than the old one if fewer than 3 people contract the disease within a year, what is the significance level of the test?

10. The print on the package of 100-watt General Electric soft-white light bulbs states that these light bulbs have an average life of 750 hours. Assume that the standard deviation of the lengths of lives of these light bulbs is 50 hours. A skeptical consumer does not think these light bulbs last as long as the manufacturer says and she decides to test 64 randomly selected light bulbs. She has set up the decision rule that if the average life of these 64 light bulbs is less than 735 hours then she will conclude that G.E. has printed too high an average length of life on the package and will write them a letter to that effect. Approximately what significance level is the consumer using? Approximately what significance level is she using if she decides G.E. has printed too high an average length of life on the package if the average life of the 64 light bulbs is less than 700 hours? Interpret the values you get.

11. The standard therapy used to treat a disorder cures 60% of all patients in an average of 140 visits. A health care provider considers supporting a new therapy regime for the disorder if it is effective in reducing the number of visits while still retaining the cure rate of the standard therapy. A study of 200 patients with the disorder who were treated by the new therapy regime reveals that 108 were cured in an average of 132 visits with a standard deviation of 38 visits. What decision should be made using a .01 level of significance?

10 ESTIMATION AND HYPOTHESIS TESTING: TWO POPULATIONS

Chapters 8 and 9 discussed the estimation and hypothesis-testing procedures for μ and p involving a single population. This chapter extends the discussion of estimation and hypothesis-testing procedures to the difference between two population means and the difference between two population proportions. For example, we may want to make a confidence interval for the difference between mean prices of houses in California and in New York. Or we may want to test the hypothesis that the mean price of houses in California is different from that in New York. As another example, we may want to make a confidence interval for the difference between the proportions of all male and female adults who abstain from drinking. Or we may want to test the hypothesis that the proportion of all adult males who abstain from drinking is different from the proportion of all adult females who abstain from drinking. Constructing confidence intervals and testing hypotheses about population parameters are referred to as *making inferences*.

10.1 INFERENCES ABOUT THE DIFFERENCE BETWEEN TWO POPULATION MEANS FOR LARGE AND INDEPENDENT SAMPLES

Let μ_1 be the mean of the first population and μ_2 be the mean of the second population. Suppose we want to make a confidence interval and test a hypothesis about the difference between these two population means, that is, $\mu_1 - \mu_2$. Let $\bar{x}_1$ be the mean of a sample taken from the first population and $\bar{x}_2$ be the mean of a sample taken from the second population. Then, $\bar{x}_1 - \bar{x}_2$ is the sample statistic that is used to make an interval estimate and to test a hypothesis about $\mu_1 - \mu_2$. This section discusses how to make confidence intervals and test hypotheses about $\mu_1 - \mu_2$ when the two samples are large and independent. As discussed in earlier chapters, in the case of μ, a sample is considered to be large if it contains 30 or more observations. The concept of independent and dependent samples is explained next.

10.1.1 INDEPENDENT VERSUS DEPENDENT SAMPLES

Two samples are **independent** if they are drawn from two different populations and the elements of one sample have no relationship to the elements of the second sample. If the elements of the two samples are somehow related, then the samples are said to be **dependent**. Thus, in two independent samples, the selection of one sample has no effect on the selection of the second sample.

> **INDEPENDENT VERSUS DEPENDENT SAMPLES**
>
> Two samples drawn from two populations are *independent* if the selection of one sample from one population does not affect the selection of the second sample from the second population. Otherwise, the samples are *dependent*.

Examples 10–1 and 10–2 illustrate independent and dependent samples, respectively.

Illustrating two independent samples.

EXAMPLE 10-1 Suppose we want to estimate the difference between the mean salaries of all male and all female executives. To do so, we draw two samples, one from the population of male executives and another from the population of female executives. These two samples are independent because they are drawn from two different populations, and the samples have no effect on each other. ▪

Illustrating two dependent samples.

EXAMPLE 10-2 Suppose we want to estimate the difference between the mean weights of all participants before and after a weight loss program. To accomplish this, suppose we take a sample of 40 participants and measure their weights before and after the completion of this program. Note that these two samples include the same 40 participants. This is an example of two dependent samples. Such samples are also called *paired* or *matched samples*. ▪

This section and Sections 10.2, 10.3, and 10.5 discuss how to make confidence intervals and test hypotheses about the difference between two population parameters when samples are independent. Section 10.4 discusses how to make confidence intervals and test hypotheses about the difference between two population means when samples are dependent.

10.1.2 MEAN, STANDARD DEVIATION, AND SAMPLING DISTRIBUTION OF $\bar{x}_1 - \bar{x}_2$

Suppose we draw two (independent) large samples from two different populations that are referred to as population 1 and population 2. Let

μ_1 = the mean of population 1

μ_2 = the mean of population 2

σ_1 = the standard deviation of population 1

σ_2 = the standard deviation of population 2

n_1 = the size of the sample drawn from population 1 ($n_1 \geq 30$)

n_2 = the size of the sample drawn from population 2 ($n_2 \geq 30$)

$\bar{x}_1$ = the mean of the sample drawn from population 1

$\bar{x}_2$ = the mean of the sample drawn from population 2

Then, from the central limit theorem, $\bar{x}_1$ is approximately normally distributed with mean μ_1 and standard deviation $\sigma_1/\sqrt{n_1}$, and $\bar{x}_2$ is approximately normally distributed with mean μ_2 and standard deviation $\sigma_2/\sqrt{n_2}$.

Using these results, we can now make the following statements about the mean, the standard deviation, and the shape of the sampling distribution of $\bar{x}_1 - \bar{x}_2$. Figure 10.1 shows the sampling distribution of $\bar{x}_1 - \bar{x}_2$.

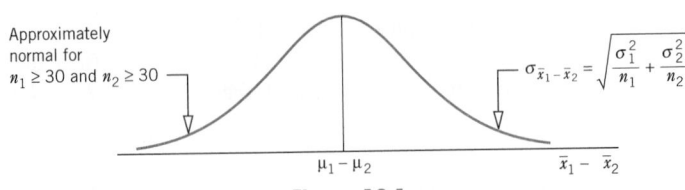

Approximately normal for $n_1 \geq 30$ and $n_2 \geq 30$

$$\sigma_{\bar{x}_1 - \bar{x}_2} = \sqrt{\frac{\sigma_1^2}{n_1} + \frac{\sigma_2^2}{n_2}}$$

$\mu_1 - \mu_2$

$\bar{x}_1 - \bar{x}_2$

Figure 10.1

1. The mean of $\bar{x}_1 - \bar{x}_2$, denoted by $\mu_{\bar{x}_1-\bar{x}_2}$, is

$$\mu_{\bar{x}_1-\bar{x}_2} = \mu_1 - \mu_2$$

2. The standard deviation of $\bar{x}_1 - \bar{x}_2$, denoted by $\sigma_{\bar{x}_1-\bar{x}_2}$, is[1]

$$\sigma_{\bar{x}_1-\bar{x}_2} = \sqrt{\frac{\sigma_1^2}{n_1} + \frac{\sigma_2^2}{n_2}}$$

3. Regardless of the shapes of the two populations, the shape of the sampling distribution of $\bar{x}_1 - \bar{x}_2$ is approximately normal. This is so because the difference between two normally distributed random variables is also normally distributed. Note again that for this to hold true, both samples must be large.

THE SAMPLING DISTRIBUTION, MEAN, AND STANDARD DEVIATION OF $\bar{x}_1 - \bar{x}_2$

For two large and independent samples, selected from two different populations, the *sampling distribution* of $\bar{x}_1 - \bar{x}_2$ is (approximately) normal with its *mean* and *standard deviation* as follows.

$$\mu_{\bar{x}_1-\bar{x}_2} = \mu_1 - \mu_2 \quad \text{and} \quad \sigma_{\bar{x}_1-\bar{x}_2} = \sqrt{\frac{\sigma_1^2}{n_1} + \frac{\sigma_2^2}{n_2}}$$

However, we usually do not know the standard deviations σ_1 and σ_2 of the two populations. In such cases, we replace $\sigma_{\bar{x}_1-\bar{x}_2}$ by its point estimator $s_{\bar{x}_1-\bar{x}_2}$, which is calculated as follows.

AN ESTIMATE OF THE STANDARD DEVIATION OF $\bar{x}_1 - \bar{x}_2$

The value of $s_{\bar{x}_1-\bar{x}_2}$, which gives an *estimate* of $\sigma_{\bar{x}_1-\bar{x}_2}$, is calculated as

$$s_{\bar{x}_1-\bar{x}_2} = \sqrt{\frac{s_1^2}{n_1} + \frac{s_2^2}{n_2}}$$

where s_1 and s_2 are the standard deviations of the two samples selected from the two populations.

Thus, when both samples are large, the sampling distribution of $\bar{x}_1 - \bar{x}_2$ is approximately normal. Consequently, in such cases, we use the normal distribution to make a confidence interval and to test a hypothesis about $\mu_1 - \mu_2$.

---

[1]The formula for the standard deviation of $\bar{x}_1 - \bar{x}_2$ can also be written as

$$\sigma_{\bar{x}_1-\bar{x}_2} = \sqrt{\sigma_{\bar{x}_1}^2 + \sigma_{\bar{x}_2}^2}$$

where $\sigma_{\bar{x}_1} = \sigma_1/\sqrt{n_1}$ and $\sigma_{\bar{x}_2} = \sigma_2/\sqrt{n_2}$.

10.1.3 INTERVAL ESTIMATION OF $\mu_1 - \mu_2$

By constructing a confidence interval for $\mu_1 - \mu_2$ we find the difference between the means of two populations. For example, we may want to find the difference between the mean heights of male and female adults. The difference between the two sample means, $\bar{x}_1 - \bar{x}_2$, is the point estimator of the difference between the two population means, $\mu_1 - \mu_2$. Again, in this section we assume that the two samples are large and independent. When these assumptions hold true, we use the normal distribution to make a confidence interval for the difference between the two population means. The following formula gives the interval estimation for $\mu_1 - \mu_2$.

CONFIDENCE INTERVAL FOR $\mu_1 - \mu_2$

The $(1 - \alpha)100\%$ *confidence interval for* $\mu_1 - \mu_2$ is

$$(\bar{x}_1 - \bar{x}_2) \pm z\sigma_{\bar{x}_1-\bar{x}_2} \qquad \text{if } \sigma_1 \text{ and } \sigma_2 \text{ are known}$$

$$(\bar{x}_1 - \bar{x}_2) \pm zs_{\bar{x}_1-\bar{x}_2} \qquad \text{if } \sigma_1 \text{ and } \sigma_2 \text{ are not known}$$

The value of z is obtained from the normal distribution table for the given confidence level. The values of $\sigma_{\bar{x}_1-\bar{x}_2}$ and $s_{\bar{x}_1-\bar{x}_2}$ are calculated as explained earlier.

Examples 10–3 and 10–4 illustrate the procedure to construct a confidence interval for $\mu_1 - \mu_2$ for large samples. In Example 10–3 the population standard deviations are known, and in Example 10–4 they are not known.

Constructing a confidence interval for $\mu_1 - \mu_2$: σ_1 and σ_2 known.

EXAMPLE 10-3 A survey of credit card holders revealed that Americans carried an average credit card balance of $3900 in 1995 and $3300 in 1994 (*U.S. News & World Report*, January 1, 1996). Suppose that these averages are based on random samples of 400 credit card holders in 1995 and 450 credit card holders in 1994 and that the population standard deviations of the balances were $880 in 1995 and $810 in 1994.

(a) What is the point estimate of $\mu_1 - \mu_2$?

(b) Construct a 95% confidence interval for the difference between the mean credit card balances for all credit card holders in 1995 and 1994.

Solution Refer to all credit card holders in 1995 as population 1 and all credit card holders in 1994 as population 2. The respective samples, then, are samples 1 and 2. Let μ_1 and μ_2 be the mean credit card balances for populations 1 and 2, and let $\bar{x}_1$ and $\bar{x}_2$ be the means of the respective samples. From the given information,

$$\text{For 1995:} \quad n_1 = 400, \quad \bar{x}_1 = \$3900, \quad \sigma_1 = \$880$$

$$\text{For 1994:} \quad n_2 = 450, \quad \bar{x}_2 = \$3300, \quad \sigma_2 = \$810$$

(a) The point estimate of $\mu_1 - \mu_2$ is given by the value of $\bar{x}_1 - \bar{x}_2$. Thus,

$$\text{Point estimate of } \mu_1 - \mu_2 = \$3900 - \$3300 = \$600$$

(b) The confidence level is $1 - \alpha = .95$.

First, we calculate the standard deviation of $\bar{x}_1 - \bar{x}_2$ as follows.

$$\sigma_{\bar{x}_1 - \bar{x}_2} = \sqrt{\frac{\sigma_1^2}{n_1} + \frac{\sigma_2^2}{n_2}} = \sqrt{\frac{(880)^2}{400} + \frac{(810)^2}{450}} = \$58.25804665$$

Next, we find the z value for the 95% confidence level. From the normal distribution table, this value of z is 1.96.

Finally, substituting all the values in the confidence interval formula, we obtain the 95% confidence interval for $\mu_1 - \mu_2$ as

$$(\bar{x}_1 - \bar{x}_2) \pm z\sigma_{\bar{x}_1 - \bar{x}_2} = (3900 - 3300) \pm 1.96 \, (58.25804665)$$

$$= 600 \pm 114.19 = \textbf{\$485.81 to \$714.19}$$

Thus, with 95% confidence we can state that the difference in the mean credit card balances for all credit card holders in 1995 and 1994 was between $485.81 and $714.19.

Constructing a confidence interval for $\mu_1 - \mu_2$: σ_1 and σ_2 not known.

EXAMPLE 10–4 In 1996, *Money* magazine conducted an experiment in which one-dollar coins and quarters were left on busy sidewalks in six major U.S. cities (*Money*, October 1996). An observer then recorded the length of time each coin remained on the sidewalk before being picked up by a pedestrian. According to the results of the experiment, the mean length of time one-dollar coins remained on the sidewalks before being picked up by pedestrians was 6.50 minutes and the mean length of time quarters stayed on the sidewalks was 5.75 minutes. Assume that these means are based on samples of 50 one-dollar coins and 45 quarters and that the two sample standard deviations are 1.75 minutes and 1.20 minutes, respectively. Find a 99% confidence interval for the difference between the corresponding population means.

Solution Let μ_1 be the mean length of time all one-dollar coins will remain on the sidewalk before being picked up by pedestrians and μ_2 the mean length of time all quarters will remain on the sidewalk before being picked up. Let $\bar{x}_1$ and $\bar{x}_2$ be the means of the respective samples. From the given information,

For $1 coins: $n_1 = 50$, $\bar{x}_1 = 6.50$, $s_1 = 1.75$

For quarters: $n_2 = 45$, $\bar{x}_2 = 5.75$, $s_2 = 1.20$

The confidence level is $1 - \alpha = .99$.

Because σ_1 and σ_2 are not known, we use $s_{\bar{x}_1 - \bar{x}_2}$ in the confidence interval formula. The value of $s_{\bar{x}_1 - \bar{x}_2}$ is

$$s_{\bar{x}_1 - \bar{x}_2} = \sqrt{\frac{s_1^2}{n_1} + \frac{s_2^2}{n_2}} = \sqrt{\frac{(1.75)^2}{50} + \frac{(1.20)^2}{45}} = .30536863$$

From the normal distribution table, the z value for a 99% confidence level is (approximately) 2.58. The 99% confidence interval for $\mu_1 - \mu_2$ is

$$(\bar{x}_1 - \bar{x}_2) \pm z s_{\bar{x}_1 - \bar{x}_2} = (6.50 - 5.75) \pm 2.58 \, (.30536863)$$

$$= .75 \pm .79 = \textbf{-.04 to 1.54 minutes}$$

Thus, with 99% confidence we can state that the difference between the two population means is $-.04$ to 1.54 minutes.

10.1.4 HYPOTHESIS TESTING ABOUT $\mu_1 - \mu_2$

It is often necessary to compare the means of two populations. For example, we may want to know if the mean price of houses in Chicago is the same as that in Los Angeles. Similarly, we may be interested in knowing if, on average, American children spend fewer hours in school than Japanese children. In both these cases we will perform a test of hypothesis about $\mu_1 - \mu_2$. The alternative hypothesis in a test of hypothesis may be that the means of the two populations are different, or that the mean of the first population is greater than the mean of the second population, or that the mean of the first population is smaller than the mean of the second population. These three situations are described below.

1. Testing an alternative hypothesis that the means of two populations are different is equivalent to $\mu_1 \neq \mu_2$, which is the same as $\mu_1 - \mu_2 \neq 0$.
2. Testing an alternative hypothesis that the mean of the first population is greater than the mean of the second population is equivalent to $\mu_1 > \mu_2$, which is the same as $\mu_1 - \mu_2 > 0$.
3. Testing an alternative hypothesis that the mean of the first population is smaller than the mean of the second population is equivalent to $\mu_1 < \mu_2$, which is the same as $\mu_1 - \mu_2 < 0$.

The procedure followed to perform a test of hypothesis about the difference between two population means is similar to the one used to test hypotheses about single population parameters in Chapter 9. The procedure involves the same five steps that were used in Chapter 9 to test hypotheses about μ and p. Because we are dealing with large (and independent) samples in this section, we will use the normal distribution to conduct a test of hypothesis about $\mu_1 - \mu_2$.

TEST STATISTIC z FOR $\bar{x}_1 - \bar{x}_2$

The value of the *test statistic z for $\bar{x}_1 - \bar{x}_2$* is computed as

$$z = \frac{(\bar{x}_1 - \bar{x}_2) - (\mu_1 - \mu_2)}{\sigma_{\bar{x}_1 - \bar{x}_2}}$$

The value of $\mu_1 - \mu_2$ is substituted from H_0. If the values of σ_1 and σ_2 are not known, we replace $\sigma_{\bar{x}_1 - \bar{x}_2}$ by $s_{\bar{x}_1 - \bar{x}_2}$ in the formula.

Making a two-tailed test of hypothesis about $\mu_1 - \mu_2$: large samples.

EXAMPLE 10–5 Refer to Example 10–3 about the mean credit card balances maintained by credit card holders in America in 1995 and 1994. Test at the 1% significance level if the mean credit card balances for all credit card holders in America in 1995 and 1994 were different.

Solution From the information given in Example 10–3,

$$\text{For 1995:} \quad n_1 = 400, \quad \bar{x}_1 = \$3900, \quad \sigma_1 = \$880$$

$$\text{For 1994:} \quad n_2 = 450, \quad \bar{x}_2 = \$3300, \quad \sigma_2 = \$810$$

Let μ_1 and μ_2 be the mean credit card balances for all credit card holders in America in 1995 and 1994, respectively.

Step 1. *State the null and alternative hypotheses*

We are to test if the two population means are different. The two possibilities are
(i) The mean credit card balances for all credit card holders in America in 1995 and 1994 are not different. In other words, $\mu_1 = \mu_2$, which can be written as $\mu_1 - \mu_2 = 0$.
(ii) The mean credit card balances for all credit card holders in America in 1995 and 1994 are different. That is, $\mu_1 \neq \mu_2$, which can be written as $\mu_1 - \mu_2 \neq 0$.

Considering these two possibilities, the null and alternative hypotheses are

$$H_0\text{: } \mu_1 - \mu_2 = 0 \qquad \text{(the two population means are not different)}$$

$$H_1\text{: } \mu_1 - \mu_2 \neq 0 \qquad \text{(the two population means are different)}$$

Step 2. *Select the distribution to use*

Because $n_1 > 30$ and $n_2 > 30$, both sample sizes are large. Therefore, the sampling distribution of $\bar{x}_1 - \bar{x}_2$ is approximately normal, and we use the normal distribution to make the hypothesis test.

Step 3. *Determine the rejection and nonrejection regions*

The significance level is given to be .01. The $\neq$ sign in the alternative hypothesis indicates that the test is two-tailed. The area in each tail of the normal distribution curve is $\alpha/2 = .01/2 = .005$. The critical values of z for .005 area in each tail of the normal distribution curve are (approximately) 2.58 and -2.58. These values are shown in Figure 10.2.

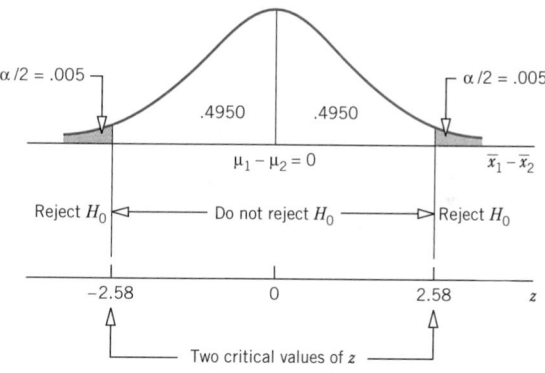

Figure 10.2

Step 4. *Calculate the value of the test statistic*

The value of the test statistic z for $\bar{x}_1 - \bar{x}_2$ is computed as follows.

$$\sigma_{\bar{x}_1-\bar{x}_2} = \sqrt{\frac{\sigma_1^2}{n_1} + \frac{\sigma_2^2}{n_2}} = \sqrt{\frac{(880)^2}{400} + \frac{(810)^2}{450}} = \$58.25804665$$

$$z = \frac{(\bar{x}_1 - \bar{x}_2) - (\mu_1 - \mu_2)}{\sigma_{\bar{x}_1-\bar{x}_2}} = \frac{(3900 - 3300) - 0}{58.25804665} = 10.30$$

From H_0

Step 5. *Make a decision*

Because the value of the test statistic $z = 10.30$ falls in the rejection region, we reject the null hypothesis H_0. Therefore, we conclude that the mean credit card balances for all credit card holders in America in 1995 and 1994 were different. Note that we cannot say for sure that the two population means are different. All we can say is that the evidence from the two samples is very strong that the corresponding population means are different. ■

Making a right-tailed
test of hypothesis about
$\mu_1 - \mu_2$: *large samples.*

EXAMPLE 10-6   Refer to Example 10–4 about the mean length of times one-dollar coins and quarters remained on the sidewalks before being picked up by pedestrians. Test at the 2.5% significance level if the mean length of time all one-dollar coins will remain on the sidewalks before being picked up is higher than the mean length of time all quarters will stay on the sidewalks.

Solution   From the information given in Example 10–4,

$$\text{For \$1 coins:} \qquad n_1 = 50, \qquad \bar{x}_1 = 6.50, \qquad s_1 = 1.75 \text{ minutes}$$

$$\text{For quarters:} \qquad n_2 = 45, \qquad \bar{x}_2 = 5.75, \qquad s_2 = 1.20 \text{ minutes}$$

Let μ_1 and μ_2 be the mean lengths of time all one-dollar coins and all quarters will remain on the sidewalks before being picked up, respectively.

Step 1.   *State the null and alternative hypotheses*

The two possibilities are
(i)   The mean length of time all one-dollar coins will remain on the sidewalks before being picked up is not higher than the mean length of time all quarters will stay on the sidewalks, which can be written as $\mu_1 = \mu_2$ or $\mu_1 - \mu_2 = 0$.
(ii)   The mean length of time all one-dollar coins will remain on the sidewalks before being picked up is higher than the mean length of time all quarters will stay on the sidewalks, which can be written as $\mu_1 > \mu_2$ or $\mu_1 - \mu_2 > 0$.

The null and alternative hypotheses are

$$H_0: \mu_1 - \mu_2 = 0 \qquad (\mu_1 \text{ is equal to } \mu_2)$$

$$H_1: \mu_1 - \mu_2 > 0 \qquad (\mu_1 \text{ is higher than } \mu_2)$$

Step 2.   *Select the distribution to use*

Because $n_1 > 30$ and $n_2 > 30$, both sample sizes are large. Therefore, the sampling distribution of $\bar{x}_1 - \bar{x}_2$ is approximately normal, and we use the normal distribution to make the hypothesis test.

Step 3.   *Determine the rejection and nonrejection regions*

The significance level is given to be .025. The $>$ sign in the alternative hypothesis indicates that the test is right-tailed. Consequently, the critical value of z is 1.96, as shown in Figure 10.3.

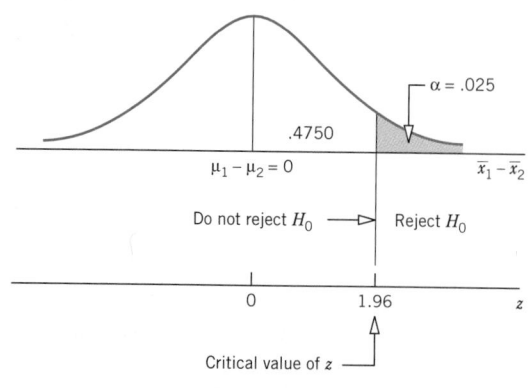

Figure 10.3

Step 4. *Calculate the value of the test statistic*

The value of the test statistic z for $\bar{x}_1 - \bar{x}_2$ is computed as follows.

$$s_{\bar{x}_1 - \bar{x}_2} = \sqrt{\frac{s_1^2}{n_1} + \frac{s_2^2}{n_2}} = \sqrt{\frac{(1.75)^2}{50} + \frac{(1.20)^2}{45}} = .30536863$$

From H_0

$$z = \frac{(\bar{x}_1 - \bar{x}_2) - (\mu_1 - \mu_2)}{s_{\bar{x}_1 - \bar{x}_2}} = \frac{(6.50 - 5.75) - 0}{.30536863} = 2.46$$

Step 5. *Make a decision*

Because the value of the test statistic $z = 2.46$ falls in the rejection region, we reject the null hypothesis H_0. Therefore, we conclude that the mean length of time all one-dollar coins will remain on the sidewalks before being picked up is higher than the mean length of time all quarters will stay on the sidewalks. ▄

Case Study 10–1 further illustrates the concept of confidence intervals and tests of hypotheses. Note that a part of this case study discusses the test of hypothesis and the confidence interval for the two population medians. Although we have not discussed the procedures for confidence intervals and tests of hypotheses for the medians, this case study shows that we can make such inferences for population parameters other than the mean.

CASE STUDY 10-1   INFERENCES MADE BY THE BUREAU OF THE CENSUS

The U.S. Bureau of the Census publishes a large number of reports every year that contain results of various surveys conducted by the bureau. Using these survey results, the bureau estimates many population parameters such as the mean, the median, and the proportion. It also performs the tests of hypotheses about the changes in the values of these parameters over time. Following are two statements quoted from two of the bureau's recent publications. The figures within parentheses in the second statement give the 90% confidence intervals of the estimates.

1. ''Per capita income did not change in real terms between 1994 and 1995 for the total population. . . .''
2. ''The annual real median earnings of women working year round, full time declined . . . between 1994 and 1995—going from \$22,834 ($\pm$\$193) to \$22,497 ($\pm$\$225). . . .''

According to the first statement, the real per capita income of the U.S. population did not change between 1994 and 1995. In other words, the real per capita income for the year 1995 was not (significantly) different from the real per capita income for the year 1994. (Note that the real per capita income is actually the mean income per person adjusted for inflation.) Let μ_1 and μ_2, respectively, be the real per capita incomes of the total population for the years 1994 and 1995. The bureau's researchers performed the following test of hypothesis.

$$H_0: \mu_1 = \mu_2$$

$$H_1: \mu_1 \neq \mu_2$$

Based on the information from sample surveys for the years 1994 and 1995, the bureau's researchers failed to reject the null hypothesis and concluded that μ_1 is not different from μ_2.

The second statement has two parts. According to the first part of this statement, the real median annual earnings of full time women workers decreased between 1994 and 1995. Let M_1 and M_2 be, respectively, the real median annual earnings of full time women workers for the years 1994 and 1995. The bureau's researchers performed the following test of hypothesis.

$$H_0: M_2 = M_1$$

$$H_1: M_2 < M_1$$

Using the information from sample surveys for the years 1994 and 1995, the bureau's researchers rejected the null hypothesis and concluded that M_2 is less than M_1. Note that we have not discussed the test of hypothesis and confidence interval procedures for the median in this text.

The second part of the second statement gives the 90% confidence intervals for the real median annual earnings of full time women workers for the years 1994 and 1995. According to this information, the 90% confidence interval for the real median annual earnings of full time women workers for the year 1994 is $22,834 ± $193, and the corresponding 90% confidence interval for 1995 is $22,497 ± $225.

Source: Money Income in the United States: 1995. Series P-60, 193. U.S. Department of Commerce, Economics and Statistics Administration, Bureau of the Census, September 1996.

EXERCISES

Concepts and Procedures

10.1   Briefly explain the meaning of independent and dependent samples. Give one example of each.

10.2   Describe the sampling distribution of $\bar{x}_1 - \bar{x}_2$ for large and independent samples. What are the mean and standard deviation of this sampling distribution?

10.3   The following information is obtained from two independent samples selected from two populations.

$$n_1 = 150 \quad \bar{x}_1 = 5.56 \quad s_1 = 1.65$$
$$n_2 = 170 \quad \bar{x}_2 = 4.80 \quad s_2 = 1.58$$

    a.   What is the point estimate of $\mu_1 - \mu_2$?

    b.   Construct a 99% confidence interval for $\mu_1 - \mu_2$.

 10.4   The following information is obtained from two independent samples selected from two populations.

$$n_1 = 300 \quad \bar{x}_1 = 25.0 \quad s_1 = 4.9$$
$$n_2 = 250 \quad \bar{x}_2 = 28.5 \quad s_2 = 4.5$$

a. What is the point estimate of $\mu_1 - \mu_2$?
b. Construct a 95% confidence interval for $\mu_1 - \mu_2$.

10.5 Refer to the information given in Exercise 10.3. Test at the 5% significance level if the two population means are different.

10.6 Refer to the information given in Exercise 10.4. Test at the 1% significance level if the two population means are different.

10.7 Refer to the information given in Exercise 10.3. Test at the 1% significance level if μ_1 is greater than μ_2.

10.8 Refer to the information given in Exercise 10.4. Test at the 5% significance level if μ_1 is less than μ_2.

Applications

10.9 According to the Canadian Real Estate Association data reproduced in the *1996 Canadian Global Almanac*, the average sale price of homes in Toronto was $208,922 in 1994 and $206,490 in 1993. Suppose that these means were based on random samples of 420 Toronto homes in 1994 and 400 homes in 1993 and that the sample standard deviations of the sale prices are $34,000 for 1994 and $32,000 for 1993.

a. Let μ_1 and μ_2 be the mean sale prices of all homes in Toronto in 1994 and 1993, respectively. Find the point estimate of $\mu_1 - \mu_2$.
b. Find a 99% confidence interval for the difference in the two population means.
c. Using the 5% level of significance, can you conclude that the mean sale prices of all homes in Toronto were different in 1994 and 1993?

10.10 During a one-year period ending in June 1992, state courts in the 75 largest counties in the United States settled an estimated 366,000 civil contract disputes (*Bureau of Justice Statistics: Special Report*, February 1996). These cases involved disputes between buyers and sellers, lenders and borrowers, and so on. The processing time for such cases varied widely from county to county. During this one-year period, the mean processing time for contract cases was 50 months in Fresno County (California) and 37 months in San Bernardino County (California). Suppose that these means are based on random samples of 30 and 35 cases for the two counties, respectively. Assume that the population standard deviations for the processing times for all such cases are 12 months for Fresno County and 10 months for San Bernardino County.

a. Let μ_1 and μ_2 be the mean processing times for all such cases in Fresno and San Bernardino counties, respectively. What is the point estimate of $\mu_1 - \mu_2$?
b. Construct a 97% confidence interval for the difference between the mean processing times for all such cases in Fresno and San Bernardino counties.
c. Test at the 2% significance level whether the mean processing times for Fresno County and San Bernardino County are different.

10.11 According to a survey, the mean price of gasoline in the United States was $1.20 per gallon in 1995 and $1.10 per gallon in 1994 (*U.S. News & World Report*, June 12, 1995). Suppose these means were based on random samples of 100 gas stations for 1995 and 120 gas stations for 1994. Also, assume that the sample standard deviations were $.11 for 1995 and $.10 for 1994.

a. Let μ_1 and μ_2 be the mean prices per gallon of gasoline in the United States in 1995 and 1994, respectively. What is the point estimate of $\mu_1 - \mu_2$?
b. Find a 90% confidence interval for the difference between the mean gasoline prices for 1995 and 1994.
c. Test at the 1% significance level if the mean gasoline price was higher in 1995 than in 1994.

10.12 According to the authors of a study about employee reliability, "Although theft is a major problem, it is only one factor of the larger construct of organizational delinquency. Excessive absences, tardiness, malingering, equipment damage, drug and alcohol abuse, grievances, suspensions from work, insubordination, and ordinary rule infractions are all components of the delinquency syndrome" (Joyce Hogan and Robert Hogan, "How to Measure Employee Reliability," *Journal of Applied Psy-*

chology 74[2], April 1989). The authors measured the employee reliability scores for male and female employees. A sample of 1637 male employees produced a mean score of 45.4 with a standard deviation of 8.0, and a sample of 590 female employees gave a mean score of 46.5 with a standard deviation of 8.3.

a. Let μ_1 and μ_2 be the mean reliability scores of male and female employees, respectively. What is the point estimate of $\mu_1 - \mu_2$?

b. Construct a 97% confidence interval for the difference between the mean reliability scores of all male and all female employees.

c. Test at the 2% significance level if the mean reliability scores for all male and all female employees are different.

10.13 A business consultant wanted to investigate if providing day care facilities on premises by companies reduces the absentee rate of working mothers with 6-year-old or younger children. She took a sample of 45 such mothers from companies that provide day care facilities on premises. These mothers missed an average of 6.4 days from work last year with a standard deviation of 1.20 days. Another sample of 50 such mothers taken from companies that do not provide day care facilities on premises showed that these mothers missed an average of 9.3 days last year with a standard deviation of 1.85 days.

a. Construct a 98% confidence interval for the difference between the two population means.

b. Using the 2.5% significance level, can you conclude that the mean number of days missed per year by mothers working for companies that provide day care facilities on premises is less than the mean number of days missed per year by mothers working for companies that do not provide day care facilities on premises?

c. What is the Type I error and its probability for the test of hypothesis in part b? Explain.

10.14 According to the U.S. Bureau of Justice Statistics, the mean time actually served in prison for crimes of violence in 1994 was 46 months in the Southern states and 54 months in the Northeastern states (*Bureau of Justice Statistics: Selected Findings*, July 1995). Assume that the population standard deviation of time served for such crimes is 15 months for the Southern states and 18 months for the Northeastern states. Also assume that these means are based on a random sample of 200 cases for the Southern States and 240 cases for the Northeastern states.

a. Construct a 99% confidence interval for the difference in mean times served in prison for such crimes in the two regions.

b. Using the 1% significance level, can you conclude that the mean time served for violent crimes is lower in the Southern states than in the Northeastern states?

c. What is the Type I error and its probability for the test of hypothesis in part b? Explain.

10.15 According to the Bureau of Labor Statistics, the mean hourly wage in April 1996 was $14.49 for transportation and public utility workers and $12.70 for manufacturing workers. Assume that these two estimates are based on random samples of 1000 and 1200 workers taken, respectively, from the two populations. Further assume that the standard deviations of the two populations are $1.85 and $1.40, respectively.

a. Construct a 95% confidence interval for the difference between the mean hourly wages of the two populations.

b. Test at the 2.5% significance level if the mean hourly wage for transportation and public utility workers in April 1996 was higher than that of manufacturing workers.

c. What will your decision be in part b if the probability of making a Type I error is zero? Explain.

10.16 Professors Paul W. Kingston and Steven L. Nock studied the time spent together by single- and dual-earner couples. According to the records kept by wives in this study, the mean time spent together by a husband and wife watching television was 61.6 minutes per day for single-earner couples and 44.4 minutes per day for dual-earner couples. The respective sample sizes were 144 and 177, and the sample standard deviations were 79.0 and 56.8 minutes, respectively (Paul William Kingston and Steven L. Nock, "Time Together Among Dual-earner Couples," *American Sociological Review* 52[3], pp. 391–400).

a. Construct a 99% confidence interval for the difference between the two population means.

b. Using the 1% significance level, can you conclude that the mean time spent together watching television by single-earner couples is higher than that of dual-earner couples?

c. What will your decision be in part b if the probability of making a Type I error is zero? Explain.

 10.17 The management at the New Century Bank claims that the mean waiting time for all customers at its branches is less than that at the Public Bank, which is its main competitor. A business consulting firm took a sample of 200 customers from the New Century Bank and found that they waited an average of 4.75 minutes with a standard deviation of 1.2 minutes before being served. Another sample of 300 customers taken from the Public Bank showed that these customers waited an average of 5.00 minutes with a standard deviation of 1.5 minutes before being served.

a. Make a 97% confidence interval for the difference between the two population means.

b. Test at the 2.5% significance level if the claim of the management of the New Century Bank is true.

*c. Calculate the *p*-value for the test of part b. Based on this *p*-value, would you reject the null hypothesis if $\alpha = .01$? What if $\alpha = .05$?

10.18 Maine Mountain Dairy claims that its 8-ounce low-fat yogurt cups contain, on average, fewer calories than the 8-ounce low-fat yogurt cups produced by a competitor. A consumer agency wanted to check this claim. A sample of 50 such yogurt cups produced by this company showed that they contained an average of 141 calories per cup with a standard deviation of 5.4 calories. A sample of 40 such yogurt cups produced by its competitor showed that they contained an average of 144 calories per cup with a standard deviation of 6.3 calories.

a. Make a 98% confidence interval for the difference between the mean number of calories in the 8-ounce low-fat yogurt cups produced by the two companies.

b. Test at the 1% significance level if Maine Mountain Dairy's claim is true.

*c. Calculate the *p*-value for the test of part b. Based on this *p*-value, would you reject the null hypothesis if $\alpha = .005$? What if $\alpha = .025$?

10.2 INFERENCES ABOUT THE DIFFERENCE BETWEEN TWO POPULATION MEANS FOR SMALL AND INDEPENDENT SAMPLES: EQUAL STANDARD DEVIATIONS

Many times, due to either budget constraint or the nature of the populations, it may not be possible to take large samples to make inferences about the difference between two population means. This section discusses how to make a confidence interval and test a hypothesis about the difference between two population means when the samples are small ($n_1 < 30$ and $n_2 < 30$) and independent. Our main assumption in this case is that the two populations from which the two samples are drawn are (approximately) normally distributed. If this assumption is true, and we know the population standard deviations, we can still use the normal distribution to make inferences about $\mu_1 - \mu_2$ when samples are small and independent. However, we usually do not know the population standard deviations σ_1 and σ_2. In such cases, we replace the normal distribution by the t distribution to make inferences about $\mu_1 - \mu_2$ for small and independent samples. We will make one more assumption in this section that the standard deviations of the two populations are equal. In other words, we assume that although σ_1 and σ_2 are unknown, they are equal. The case when σ_1 and σ_2 are not equal will be discussed in Section 10.3.

WHEN TO USE THE t DISTRIBUTION TO MAKE INFERENCES ABOUT $\mu_1 - \mu_2$

The t distribution is used to make inferences about $\mu_1 - \mu_2$ when the following assumptions hold true.

1.   The two populations from which the two samples are drawn are (approximately) normally distributed.
2.   The samples are small ($n_1 < 30$ and $n_2 < 30$) and independent.
3.   The standard deviations σ_1 and σ_2 of the two populations are unknown but they are equal, that is, $\sigma_1 = \sigma_2$.

When the standard deviations of the two populations are equal, we can use σ for both σ_1 and σ_2. Since σ is unknown, we replace it by its point estimator s_p, which is called the **pooled sample standard deviation** (hence, the subscript p). The value of s_p is computed by using the information from the two samples as follows.

THE POOLED STANDARD DEVIATION FOR TWO SAMPLES

The *pooled standard deviation for two samples* is computed as

$$s_p = \sqrt{\frac{(n_1 - 1)s_1^2 + (n_2 - 1)s_2^2}{n_1 + n_2 - 2}}$$

where n_1 and n_2 are the sizes of the two samples and s_1^2 and s_2^2 are the variances of the two samples.

In this formula, $n_1 - 1$ are the degrees of freedom for sample 1, $n_2 - 1$ are the degrees of freedom for sample 2, and $n_1 + n_2 - 2$ *are the degrees of freedom for the two samples taken together*.

When s_p is used as an estimator of σ, the standard deviation $\sigma_{\bar{x}_1 - \bar{x}_2}$ of $\bar{x}_1 - \bar{x}_2$ is estimated by $s_{\bar{x}_1 - \bar{x}_2}$. The value of $s_{\bar{x}_1 - \bar{x}_2}$ is calculated by using the following formula.

ESTIMATOR OF THE STANDARD DEVIATION OF $\bar{x}_1 - \bar{x}_2$

The *estimator of the standard deviation of $\bar{x}_1 - \bar{x}_2$* is

$$s_{\bar{x}_1 - \bar{x}_2} = s_p \sqrt{\frac{1}{n_1} + \frac{1}{n_2}}$$

Now we are ready to discuss the procedures that are used to make confidence intervals and test hypotheses about $\mu_1 - \mu_2$ for small and independent samples selected from two populations with unknown but equal standard deviations.

10.2.1   INTERVAL ESTIMATION OF $\mu_1 - \mu_2$

As was mentioned earlier, the difference between the two sample means, $\bar{x}_1 - \bar{x}_2$, is the point estimator of the difference between the two population means, $\mu_1 - \mu_2$. The following formula gives the confidence interval for $\mu_1 - \mu_2$ when the t distribution is used.

CONFIDENCE INTERVAL FOR $\mu_1 - \mu_2$

The $(1 - \alpha)100\%$ *confidence interval for* $\mu_1 - \mu_2$ is

$$(\bar{x}_1 - \bar{x}_2) \pm ts_{\bar{x}_1 - \bar{x}_2}$$

where the value of t is obtained from the t distribution table for the given confidence level and $n_1 + n_2 - 2$ degrees of freedom, and $s_{\bar{x}_1 - \bar{x}_2}$ is calculated as explained earlier in Section 10.2.

Example 10–7 describes the procedure to make a confidence interval for $\mu_1 - \mu_2$ using the t distribution.

Constructing a confidence interval for $\mu_1 - \mu_2$: small and independent samples, and $\sigma_1 = \sigma_2$.

EXAMPLE 10-7   A consumer agency wanted to estimate the difference in the mean amounts of caffeine in two brands of coffee. The agency took a sample of 15 one-pound jars of Brand I coffee that showed the mean amount of caffeine in these jars to be 80 milligrams per jar with a standard deviation of 5 milligrams. Another sample of 12 one-pound jars of Brand II coffee gave a mean amount of caffeine equal to 77 milligrams per jar with a standard deviation of 6 milligrams. Construct a 95% confidence interval for the difference between the mean amounts of caffeine in one-pound jars of these two brands of coffee. Assume that the two populations are normally distributed and that the standard deviations of the two populations are equal.

Solution   Let μ_1 and μ_2 be the mean amounts of caffeine per jar in all one-pound jars of Brands I and II, respectively, and let $\bar{x}_1$ and $\bar{x}_2$ be the means of the two respective samples. From the given information,

$$n_1 = 15, \quad \bar{x}_1 = 80 \text{ milligrams}, \quad s_1 = 5 \text{ milligrams}$$

$$n_2 = 12, \quad \bar{x}_2 = 77 \text{ milligrams}, \quad s_2 = 6 \text{ milligrams}$$

The confidence level is $1 - \alpha = .95$.
First we calculate the standard deviation of $\bar{x}_1 - \bar{x}_2$ as follows.

$$s_p = \sqrt{\frac{(n_1 - 1)s_1^2 + (n_2 - 1)s_2^2}{n_1 + n_2 - 2}} = \sqrt{\frac{(15 - 1)(5)^2 + (12 - 1)(6)^2}{15 + 12 - 2}} = 5.46260011$$

$$s_{\bar{x}_1 - \bar{x}_2} = s_p \sqrt{\frac{1}{n_1} + \frac{1}{n_2}} = (5.46260011) \sqrt{\frac{1}{15} + \frac{1}{12}} = 2.11565593$$

Next, to find the t value from the t distribution table, we need to know the area in each tail of the t distribution curve and the degrees of freedom.

$$\text{Area in each tail} = \alpha/2 = .5 - (.95/2) = .025$$

$$\text{Degrees of freedom} = n_1 + n_2 - 2 = 15 + 12 - 2 = 25$$

The t value for $df = 25$ and .025 area in the right tail of the t distribution curve is 2.060. The 95% confidence interval for $\mu_1 - \mu_2$ is

$$(\bar{x}_1 - \bar{x}_2) \pm t s_{\bar{x}_1 - \bar{x}_2} = (80 - 77) \pm 2.060\,(2.11565593)$$

$$= 3 \pm 4.36 = \mathbf{-1.36 \text{ to } 7.36 \text{ milligrams}}$$

Thus, with 95% confidence we can state that based on these two sample results, the difference in the mean amounts of caffeine in one-pound jars of these two brands of coffee lies between -1.36 and 7.36 milligrams. Because the lower limit of the interval is negative, it is possible that the mean amount of caffeine is greater in the second brand than in the first brand of coffee.

Note that the value of $\bar{x}_1 - \bar{x}_2$, which is $80 - 77 = 3$, gives the point estimate of $\mu_1 - \mu_2$.

10.2.2 HYPOTHESIS TESTING ABOUT $\mu_1 - \mu_2$

When the three assumptions mentioned in Section 10.2 are satisfied, then the t distribution is applied to make a hypothesis test about the difference between two population means. The test statistic in this case is t, which is calculated as follows.

TEST STATISTIC t FOR $\bar{x}_1 - \bar{x}_2$

The value of the *test statistic t for $\bar{x}_1 - \bar{x}_2$* is computed as

$$t = \frac{(\bar{x}_1 - \bar{x}_2) - (\mu_1 - \mu_2)}{s_{\bar{x}_1 - \bar{x}_2}}$$

The value of $\mu_1 - \mu_2$ in this formula is substituted from the null hypothesis and $s_{\bar{x}_1 - \bar{x}_2}$ is calculated as explained in Section 10.2.

Examples 10–8 and 10–9 illustrate how a test of hypothesis about the difference between two population means for small and independent samples that are selected from two populations with equal standard deviations is conducted using the t distribution.

Making a two-tailed test of hypothesis about $\mu_1 - \mu_2$: small and independent samples, and $\sigma_1 = \sigma_2$.

EXAMPLE 10–8   A sample of 14 cans of Brand I diet soda gave the mean number of calories of 23 per can with a standard deviation of 3 calories. Another sample of 16 cans of Brand II diet soda gave the mean number of calories of 25 per can with a standard deviation of 4 calories. At the 1% significance level, can you conclude that the mean number of calories per can are different for these two brands of diet soda? Assume that the calories per can of diet soda are normally distributed for each of the two brands and that the standard deviations for the two populations are equal.

Solution   Let μ_1 and μ_2 be the mean number of calories per can for diet soda of Brand I and Brand II, respectively, and let $\bar{x}_1$ and $\bar{x}_2$ be the means of the respective samples. From the given information,

$$n_1 = 14, \quad \bar{x}_1 = 23, \quad s_1 = 3$$

$$n_2 = 16, \quad \bar{x}_2 = 25, \quad s_2 = 4$$

The significance level is $\alpha = .01$.

Step 1. *State the null and alternative hypotheses*

We are to test for the difference in the mean number of calories per can for the two brands. The null and alternative hypotheses are

$$H_0: \mu_1 - \mu_2 = 0 \qquad \text{(the mean number of calories are not different)}$$
$$H_1: \mu_1 - \mu_2 \neq 0 \qquad \text{(the mean number of calories are different)}$$

Step 2. *Select the distribution to use*

The two populations are normally distributed, the samples are small and independent, and the standard deviations of the two populations are unknown but equal. Consequently, we will use the t distribution.

Step 3. *Determine the rejection and nonrejection regions*

The $\neq$ sign in the alternative hypothesis indicates that the test is two-tailed. The significance level is .01. Hence,

$$\text{Area in each tail} = \alpha/2 = .01/2 = .005$$
$$\text{Degrees of freedom} = n_1 + n_2 - 2 = 14 + 16 - 2 = 28$$

The critical values of t for $df = 28$ and .005 area in each tail of the t distribution curve are -2.763 and 2.763, as shown in Figure 10.4.

Reject H_0 ◄─── Do not reject H_0 ───► Reject H_0

$\alpha/2 = .005$ $\alpha/2 = .005$

$-2.763 \qquad 0 \qquad 2.763 \qquad t$

Two critical values of t

Figure 10.4

Step 4. *Calculate the value of the test statistic*

The value of the test statistic t for $\bar{x}_1 - \bar{x}_2$ is computed as follows.

$$s_p = \sqrt{\frac{(n_1 - 1)s_1^2 + (n_2 - 1)s_2^2}{n_1 + n_2 - 2}} = \sqrt{\frac{(14 - 1)(3)^2 + (16 - 1)(4)^2}{14 + 16 - 2}} = 3.57071421$$

$$s_{\bar{x}_1 - \bar{x}_2} = s_p \sqrt{\frac{1}{n_1} + \frac{1}{n_2}} = (3.57071421) \sqrt{\frac{1}{14} + \frac{1}{16}} = 1.30674760$$

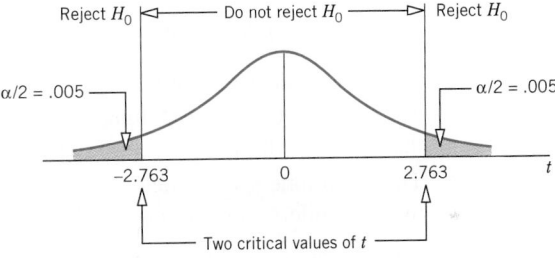

$$t = \frac{(\bar{x}_1 - \bar{x}_2) - (\mu_1 - \mu_2)}{s_{\bar{x}_1 - \bar{x}_2}} = \frac{(23 - 25) - 0}{1.30674760} = -1.531$$

Step 5. *Make a decision*

Because the value of the test statistic $t = -1.531$ for $\bar{x}_1 - \bar{x}_2$ falls in the nonrejection region, we fail to reject the null hypothesis. Consequently we conclude that there is no difference in the mean number of calories per can for the two brands of diet soda. The difference in $\bar{x}_1$ and $\bar{x}_2$ observed for the two samples may have occurred due to sampling error only. ■

Making a right-tailed test of hypothesis about $\mu_1 - \mu_2$: small and independent samples, and $\sigma_1 = \sigma_2$.

EXAMPLE 10-9   A sample of 15 children from New York State showed that the mean time they spend watching television is 28.5 hours per week with a standard deviation of 4 hours. Another sample of 16 children from California showed that the mean time spent by them watching television is 23.25 hours per week with a standard deviation of 5 hours. Using a 2.5% significance level, can you conclude that the mean time spent watching television by children in New York State is higher than that for children in California? Assume that the time spent watching television by children has a normal distribution for both populations and that the standard deviations for the two populations are equal.

Solution   Let the children from New York State be referred to as population 1 and those from California as population 2. Let μ_1 and μ_2 be the mean time spent watching television by children in populations 1 and 2, respectively, and let $\bar{x}_1$ and $\bar{x}_2$ be the mean time spent watching televison by children in the respective samples. From the given information,

For New York State:	$n_1 = 15$,	$\bar{x}_1 = 28.5$ hours,	$s_1 = 4$ hours
For California:	$n_2 = 16$,	$\bar{x}_2 = 23.25$ hours,	$s_2 = 5$ hours

The significance level is $\alpha = .025$.

Step 1.   *State the null and alternative hypotheses*

The two possible decisions are
(i)   The mean time spent watching television by children in New York State is not higher than that for children in California. This can be written as $\mu_1 = \mu_2$ or $\mu_1 - \mu_2 = 0$.
(ii) The mean time spent watching television by children in New York State is higher than that for children in California. This can be written as $\mu_1 > \mu_2$ or $\mu_1 - \mu_2 > 0$.

Hence, the null and alternative hypotheses are

$$H_0: \mu_1 - \mu_2 = 0$$
$$H_1: \mu_1 - \mu_2 > 0$$

Note that the null hypothesis can also be written as $\mu_1 - \mu_2 \leq 0$.

Step 2.   *Select the distribution to use*

The two populations are normally distributed, the samples are small and independent, and the standard deviations of the two populations are unknown but equal. Consequently, we use the t distribution to make the test.

Step 3.   *Determine the rejection and nonrejection regions*

The $>$ sign in the alternative hypothesis indicates that the test is right-tailed. The significance level is .025.

Area in the right tail of the t distribution $= \alpha = .025$

Degrees of freedom $= n_1 + n_2 - 2 = 15 + 16 - 2 = 29$

From the t distribution table, the critical value of t for $df = 29$ and .025 area in the right tail of the t distribution is 2.045. This value is shown in Figure 10.5.

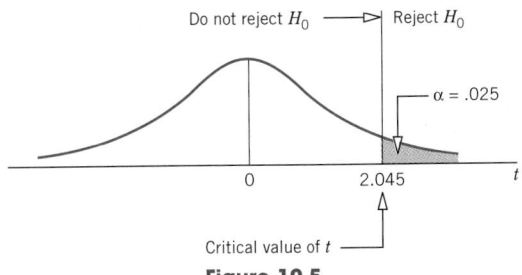

Figure 10.5

Step 4. *Calculate the value of the test statistic*

The value of the test statistic t for $\bar{x}_1 - \bar{x}_2$ is computed as follows.

$$s_p = \sqrt{\frac{(n_1 - 1)s_1^2 + (n_2 - 1)s_2^2}{n_1 + n_2 - 2}} = \sqrt{\frac{(15 - 1)(4)^2 + (16 - 1)(5)^2}{15 + 16 - 2}} = 4.54479619$$

$$s_{\bar{x}_1 - \bar{x}_2} = s_p \sqrt{\frac{1}{n_1} + \frac{1}{n_2}} = (4.54479619) \sqrt{\frac{1}{15} + \frac{1}{16}} = 1.63338904$$

$$t = \frac{(\bar{x}_1 - \bar{x}_2) - (\mu_1 - \mu_2)}{s_{\bar{x}_1 - \bar{x}_2}} = \frac{(28.5 - 23.25) - 0}{1.63338904} = 3.214$$

From H_0

Step 5. *Make a decision*

Because the value of the test statistic $t = 3.214$ for $\bar{x}_1 - \bar{x}_2$ falls in the rejection region, we reject the null hypothesis H_0. Hence, we conclude that children in New York State spend, on average, more time watching TV than children in California. ▬

EXERCISES

Concepts and Procedures

10.19 Explain what conditions must hold true to use the t distribution to make a confidence interval and to test a hypothesis about $\mu_1 - \mu_2$ for two independent samples selected from two populations with unknown but equal standard deviations.

10.20 The following information was obtained from two independent samples selected from two normally distributed populations with unknown but equal standard deviations.

$$n_1 = 22 \qquad \bar{x}_1 = 12.50 \qquad s_1 = 3.75$$
$$n_2 = 18 \qquad \bar{x}_2 = 14.60 \qquad s_2 = 3.10$$

a. What is the point estimate of $\mu_1 - \mu_2$?
b. Construct a 95% confidence interval for $\mu_1 - \mu_2$.

10.21 The following information was obtained from two independent samples selected from two normally distributed populations with unknown but equal standard deviations.

$$n_1 = 20 \qquad \bar{x}_1 = 33.75 \qquad s_1 = 5.25$$
$$n_2 = 23 \qquad \bar{x}_2 = 28.50 \qquad s_2 = 4.55$$

a. What is the point estimate of $\mu_1 - \mu_2$?
b. Construct a 99% confidence interval for $\mu_1 - \mu_2$.

10.22   Refer to the information given in Exercise 10.20. Test at the 5% significance level if the two population means are different.

10.23   Refer to the information given in Exercise 10.21. Test at the 1% significance level if the two population means are different.

10.24   Refer to the information given in Exercise 10.20. Test at the 1% significance level if μ_1 is less than μ_2.

10.25   Refer to the information given in Exercise 10.21. Test at the 5% significance level if μ_1 is greater than μ_2.

10.26   The following information was obtained from two independent samples selected from two normally distributed populations with unknown but equal standard deviations.

Sample 1:	27	31	25	33	21	35	30	26	25	31	33	30	28
Sample 2:	24	28	22	25	24	22	29	26	25	28	19	29	

    a.   Let μ_1 be the mean of population 1 and μ_2 be the mean of population 2. What is the point estimate of $\mu_1 - \mu_2$?
    b.   Construct a 98% confidence interval for $\mu_1 - \mu_2$.
    c.   Test at the 1% significance level if μ_1 is greater than μ_2.

10.27   The following information was obtained from two independent samples selected from two normally distributed populations with unknown but equal standard deviations.

Sample 1:	13	11	9	12	8	10	5	10	9	12	13	
Sample 2:	16	14	11	19	14	17	13	16	17	18	19	12

    a.   Let μ_1 be the mean of population 1 and μ_2 be the mean of population 2. What is the point estimate of $\mu_1 - \mu_2$?
    b.   Construct a 99% confidence interval for $\mu_1 - \mu_2$.
    c.   Test at the 2.5% significance level if μ_1 is lower than μ_2.

Applications

10.28   The management of a supermarket wanted to investigate if the male customers spend less money, on average, than the female customers. A sample of 25 male customers who shopped at this supermarket showed that they spent an average of $75 with a standard deviation of $17.50. Another sample of 20 female customers who shopped at the same supermarket showed that they spent an average of $89 with a standard deviation of $14.40. Assume that the amounts spent at this supermarket by all male and all female customers are normally distributed with equal but unknown population standard deviations.

    a.   Construct a 99% confidence interval for the difference between the mean amounts spent by all male and all female customers at this supermarket.
    b.   Using the 5% significance level, can you conclude that the mean amount spent by all male customers at this supermarket is less than that of all female customers?

10.29   A manufacturing company is interested in buying one of two different kinds of machines. The company tested the two machines for production purposes. The first machine was run for 8 hours. It produced an average of 126 items per hour with a standard deviation of 9 items. The second machine was run for 10 hours. It produced an average of 117 items per hour with a standard deviation of 6 items. Assume that the production per hour for each machine is (approximately) normally distributed. Further assume that the standard deviations of the hourly productions of the two populations are equal.

    a.   Make a 95% confidence interval for the difference between the two population means.
    b.   Using the 2.5% significance level, can you conclude that the mean number of items produced per hour by the first machine is higher than that of the second machine?

10.30   An insurance company wants to know if the average speed at which men drive cars is higher than that of women drivers. The company took a random sample of 27 cars driven by men on a highway and found the mean speed to be 68 miles per hour with a standard deviation of 2.2 miles.

Another sample of 18 cars driven by women on the same highway gave a mean speed of 65 miles per hour with a standard deviation of 2.5 miles. Assume that the speeds at which all men and all women drive cars on this highway are both normally distributed with the same population standard deviation.

    a. Construct a 98% confidence interval for the difference between the mean speeds of cars driven by all men and all women drivers on this highway.

    b. Test at the 1% significance level if the mean speed of cars driven by all men drivers on this highway is higher than that of cars driven by all women drivers.

10.31 According to Metropolitan Life Insurance Company claims data, the mean hospital and physician's charges for prostate surgery are $14,460 for Maryland and $17,570 for Virginia (*Statistical Bulletin*, July–September 1996). These two means are based on 19 claims for Maryland and 16 claims for Virginia. Assume that the standard deviations for these sets of claims for the two states are $1920 and $2250, respectively. Further assume that all such charges have a normal distribution with the same population standard deviation for each of these two states.

    a. Construct a 90% confidence interval for the difference between the mean hospital and physician's charges for all prostate surgeries for these two states.

    b. Test at the 5% significance level if the mean hospital and physician's charges for all prostate surgeries for these two states are different.

10.32 Quadro Corporation has two supermarket stores in a city. The company's quality control department wanted to check if the customers are equally satisfied with the service provided at these two stores. A sample of 25 customers selected from Supermarket I produced a mean satisfaction index of 7.8 (on a scale of 1 to 10, 1 being the lowest and 10 being the highest) with a standard deviation of .75. Another sample of 28 customers selected from Supermarket II produced a mean satisfaction index of 8.3 with a standard deviation of .59. Assume that the customer satisfaction index for each supermarket has a normal distribution with the same population standard deviation.

    a. Construct a 98% confidence interval for the difference between the mean satisfaction indexes for all customers for the two supermarkets.

    b. Test at the 1% significance level if the mean satisfaction indexes for all customers for the two supermarkets are different.

10.33 A company claims that its medicine, Brand A, provides faster relief from pain than another company's medicine, Brand B. A researcher tested both brands of medicine on two groups of randomly selected patients. The results of the test are given in the following table. The mean and standard deviation of relief times are in minutes.

Brand	Sample Size	Mean of Relief Times	Standard Deviation of Relief Times
A	25	44	13
B	22	49	11

    a. Construct a 99% confidence interval for the difference between the mean relief times for the two brands of medicine.

    b. Test at the 1% significance level if the mean relief time for Brand A is less than that for Brand B.

Assume that the two populations are normally distributed with equal standard deviations.

10.34 In 1995, Runzheimer International studied camera film prices in several large cities in the world (*USA Today*, September 1, 1995). The average price for a roll of 35mm film with 24 exposures was $6.55 in Tokyo and $4.33 in New York City. Suppose that these means are based on random samples of 20 shops in Tokyo and 25 shops in New York City and that the sample standard deviations of the prices are $1.05 for Tokyo and $1.00 for New York. Further assume that the two populations from which these two samples are selected are normally distributed with equal but unknown standard deviations.

    a. Construct a 90% confidence interval for the difference between the population mean prices of such films in the two cities.

    b. Test at the 5% significance level if the population mean price of such films is higher in Tokyo than in New York City.

10.35 According to the American Association of University Professors, the mean salary of male professors at American colleges and universities was $66,740 and that of female professors was $58,990 in 1995–96. For convenience, assume that these two means are based on random samples of 28 male professors and 26 female professors. Further assume that the standard deviations for the two samples are $3700 and $3200, respectively.

    a. Construct a 90% confidence interval for the difference between the two population means.

    b. Using the 1% significance level, can you conclude that the mean salary of all male professors for 1995–96 was higher than that of all female professors?

Assume that the salaries of all male and all female professors are both normally distributed with equal standard deviations.

10.3 INFERENCES ABOUT THE DIFFERENCE BETWEEN TWO POPULATION MEANS FOR SMALL AND INDEPENDENT SAMPLES: UNEQUAL STANDARD DEVIATIONS

Section 10.2 explained how to make inferences about the difference between two population means using the t distribution when the standard deviations of the two populations are unknown but equal and certain other assumptions hold true. Now, what if all other assumptions of Section 10.2 hold true but the population standard deviations are not only unknown but also unequal? In this case, the procedures used to make confidence intervals and to test hypotheses about $\mu_1 - \mu_2$ remain similar to the ones we learned in Sections 10.2.1 and 10.2.2 except for two differences. When the population standard deviations are unknown and not equal, then the degrees of freedom are no longer given by $n_1 + n_2 - 2$ and the standard deviation of $\bar{x}_1 - \bar{x}_2$ is not calculated using the pooled standard deviation s_p.

DEGREES OF FREEDOM

When

1. The two populations from which the samples are drawn are (approximately) normally distributed
2. The two samples are small and independent
3. The two population standard deviations are unknown and unequal

then the t distribution is used to make inferences about $\mu_1 - \mu_2$ and the *degrees of freedom* for the t distribution are given by

$$df = \frac{\left(\dfrac{s_1^2}{n_1} + \dfrac{s_2^2}{n_2} \right)^2}{\dfrac{\left(\dfrac{s_1^2}{n_1} \right)^2}{n_1 - 1} + \dfrac{\left(\dfrac{s_2^2}{n_2} \right)^2}{n_2 - 1}}$$

The number given by this formula is always rounded down for df.

Because the standard deviations of the two populations are not known, we use $s_{\bar{x}_1 - \bar{x}_2}$ as a point estimator of $\sigma_{\bar{x}_1 - \bar{x}_2}$. The following formula is used to calculate the standard deviation $s_{\bar{x}_1 - \bar{x}_2}$ of $\bar{x}_1 - \bar{x}_2$.

ESTIMATE OF THE STANDARD DEVIATION OF $\bar{x}_1 - \bar{x}_2$

The value of $s_{\bar{x}_1 - \bar{x}_2}$ is calculated as

$$s_{\bar{x}_1 - \bar{x}_2} = \sqrt{\frac{s_1^2}{n_1} + \frac{s_2^2}{n_2}}$$

10.3.1 INTERVAL ESTIMATION OF $\mu_1 - \mu_2$

Again, the difference between the two sample means, $\bar{x}_1 - \bar{x}_2$, is the point estimator of the difference between the two population means, $\mu_1 - \mu_2$. The following formula gives the confidence interval for $\mu_1 - \mu_2$ when the t distribution is used and the population standard deviations are unknown and presumed to be unequal.

CONFIDENCE INTERVAL FOR $\mu_1 - \mu_2$

The $(1 - \alpha)100\%$ *confidence interval for $\mu_1 - \mu_2$* is

$$(\bar{x}_1 - \bar{x}_2) \pm t \, s_{\bar{x}_1 - \bar{x}_2}$$

where the value of t is obtained from the t distribution table for a given confidence level and the degrees of freedom given by the formula mentioned in Section 10.3, and $s_{\bar{x}_1 - \bar{x}_2}$ is calculated as explained earlier.

Example 10–10 describes how to construct a confidence interval for $\mu_1 - \mu_2$ when the standard deviations of the two populations are unknown and unequal.

Constructing a confidence interval for $\mu_1 - \mu_2$: small and independent samples, and $\sigma_1 \neq \sigma_2$.

EXAMPLE 10–10 According to Example 10–7 of Section 10.2.1, a sample of 15 one-pound jars of coffee of Brand I showed that the mean amount of caffeine in these jars is 80 milligrams per jar with a standard deviation of 5 milligrams. Another sample of 12 one-pound coffee jars of Brand II gave a mean amount of caffeine equal to 77 milligrams per jar with a standard deviation of 6 milligrams. Construct a 95% confidence interval for the difference between the mean amounts of caffeine in one-pound coffee jars of these two brands. Assume that the two populations are normally distributed and that the standard deviations of the two populations are not equal.

Solution Let μ_1 and μ_2 be the mean amount of caffeine per jar in all one-pound jars of Brand I and II, respectively, and let $\bar{x}_1$ and $\bar{x}_2$ be the means of the two respective samples.

From the given information,

$$n_1 = 15, \quad \bar{x}_1 = 80 \text{ milligrams}, \quad s_1 = 5 \text{ milligrams}$$
$$n_2 = 12, \quad \bar{x}_2 = 77 \text{ milligrams}, \quad s_2 = 6 \text{ milligrams}$$

The confidence level is $1 - \alpha = .95$.

First, we calculate the standard deviation of $\bar{x}_1 - \bar{x}_2$ as follows.

$$s_{\bar{x}_1 - \bar{x}_2} = \sqrt{\frac{s_1^2}{n_1} + \frac{s_2^2}{n_2}} = \sqrt{\frac{(5)^2}{15} + \frac{(6)^2}{12}} = 2.16024690$$

Next, to find the t value from the t distribution table, we need to know the area in each tail of the t distribution curve and the degrees of freedom.

$$\text{Area in each tail} = \alpha/2 = .5 - (.95/2) = .025$$

$$df = \frac{\left(\frac{s_1^2}{n_1} + \frac{s_2^2}{n_2}\right)^2}{\frac{\left(\frac{s_1^2}{n_1}\right)^2}{n_1 - 1} + \frac{\left(\frac{s_2^2}{n_2}\right)^2}{n_2 - 1}} = \frac{\left(\frac{(5)^2}{15} + \frac{(6)^2}{12}\right)^2}{\frac{\left(\frac{(5)^2}{15}\right)^2}{(15 - 1)} + \frac{\left(\frac{(6)^2}{12}\right)^2}{(12 - 1)}} = 21.42 \approx 21$$

Note that the degrees of freedom are always rounded down as in this calculation.

From the t distribution table, the t value for $df = 21$ and .025 area in the right tail of the t distribution curve is 2.080. The 95% confidence interval for $\mu_1 - \mu_2$ is

$$(\bar{x}_1 - \bar{x}_2) \pm t\, s_{\bar{x}_1 - \bar{x}_2} = (80 - 77) \pm 2.080\,(2.16024690)$$
$$= 3 \pm 4.49 = \mathbf{-1.49 \text{ to } 7.49}$$

Thus, with 95% confidence we can state that based on these two sample results, the difference in the mean amounts of caffeine in one-pound jars of these two brands of coffee is between -1.49 and 7.49 milligrams. ▬

Comparing this confidence interval with the one obtained in Example 10–7, we observe that the two confidence intervals are very close. From this we can conclude that even if the standard deviations of the two populations are not equal and we use the procedure of Section 10.2.1 to make a confidence interval for $\mu_1 - \mu_2$, the margin of error will be small as long as the difference between the two standard deviations is not too large.

10.3.2 HYPOTHESIS TESTING ABOUT $\mu_1 - \mu_2$

When the standard deviations of the two populations are unknown and unequal, with the other conditons of Section 10.2.2 holding true, we use the t distribution to make a test of hypothesis about $\mu_1 - \mu_2$. This procedure differs from the one in Section 10.2.2 only in the calculation of degrees of freedom for the t distribution and the standard deviation of $\bar{x}_1 - \bar{x}_2$. The df and the standard deviation of $\bar{x}_1 - \bar{x}_2$ in this case are given by the formulas given earlier in Section 10.3.

TEST STATISTIC t FOR $\bar{x}_1 - \bar{x}_2$

The value of the *test statistic t* for $\bar{x}_1 - \bar{x}_2$ is computed as

$$t = \frac{(\bar{x}_1 - \bar{x}_2) - (\mu_1 - \mu_2)}{s_{\bar{x}_1 - \bar{x}_2}}$$

The value of $\mu_1 - \mu_2$ in the above formula is substituted from the null hypothesis and $s_{\bar{x}_1 - \bar{x}_2}$ is calculated as explained earlier.

Example 10–11 illustrates the procedure to conduct a test of hypothesis about $\mu_1 - \mu_2$ when the standard deviations of the two populations are unknown and unequal.

Making a two-tailed test of hypothesis about $\mu_1 - \mu_2$: small and independent samples, and $\sigma_1 \neq \sigma_2$.

EXAMPLE 10-11   According to Example 10–8 of Section 10.2.2, a sample of 14 cans of Brand I diet soda gave the mean number of calories per can as 23 with a standard deviation of 3 calories. Another sample of 16 cans of Brand II diet soda gave the mean number of calories of 25 per can with a standard deviation of 4 calories. Test at the 1% significance level if the mean number of calories per can of diet soda are different for these two brands. Assume that the calories per can of diet soda are normally distributed for each of these two brands and that the standard deviations for the two populations are not equal.

Solution   Let μ_1 and μ_2 be the mean number of calories for all cans of diet soda of Brand I and Brand II, respectively, and let $\bar{x}_1$ and $\bar{x}_2$ be the means of the respective samples. From the given information,

$$n_1 = 14, \quad \bar{x}_1 = 23, \quad s_1 = 3$$
$$n_2 = 16, \quad \bar{x}_2 = 25, \quad s_2 = 4$$

The significance level is $\alpha = .01$.

Step 1.   *State the null and alternative hypotheses*

We are to test for the difference in the mean number of calories per can for the two brands. The null and alternative hypotheses are

H_0: $\mu_1 - \mu_2 = 0$ (the mean number of calories are not different)

H_1: $\mu_1 - \mu_2 \neq 0$ (the mean number of calories are different)

Step 2.   *Select the distribution to use*

The two populations are normally distributed, the samples are small and independent, and the standard deviations of the two populations are unknown and unequal. Consequently, we use the t distribution to make the test.

Step 3.   *Determine the rejection and nonrejection regions*

The $\neq$ sign in the alternative hypothesis indicates that the test is two-tailed. The significance level is .01. Hence,

$$\text{Area in each tail} = \alpha/2 = .01/2 = .005$$

The degrees of freedom are calculated as follows.

$$df = \frac{\left(\dfrac{s_1^2}{n_1} + \dfrac{s_2^2}{n_2}\right)^2}{\dfrac{\left(\dfrac{s_1^2}{n_1}\right)^2}{n_1 - 1} + \dfrac{\left(\dfrac{s_2^2}{n_2}\right)^2}{n_2 - 1}} = \frac{\left(\dfrac{(3)^2}{14} + \dfrac{(4)^2}{16}\right)^2}{\dfrac{\left(\dfrac{(3)^2}{14}\right)^2}{(14 - 1)} + \dfrac{\left(\dfrac{(4)^2}{16}\right)^2}{(16 - 1)}} = 27.41 \approx 27$$

From the t distribution table, the critical values of t for $df = 27$ and .005 area in each tail of the t distribution curve are -2.771 and 2.771. These values are shown in Figure 10.6.

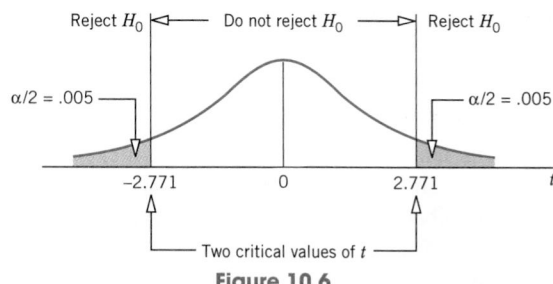

Figure 10.6

Step 4. *Calculate the value of the test statistic*

The value of the test statistic t for $\bar{x}_1 - \bar{x}_2$ is computed as follows.

$$s_{\bar{x}_1 - \bar{x}_2} = \sqrt{\frac{s_1^2}{n_1} + \frac{s_2^2}{n_2}} = \sqrt{\frac{(3)^2}{14} + \frac{(4)^2}{16}} = 1.28173989$$

$$t = \frac{(\bar{x}_1 - \bar{x}_2) - (\mu_1 - \mu_2)}{s_{\bar{x}_1 - \bar{x}_2}} = \frac{(23 - 25) - 0}{1.28173989} = -1.560$$

— From H_0

Step 5. *Make a decision*

Because the value of the test statistic $t = -1.560$ for $\bar{x}_1 - \bar{x}_2$ falls in the nonrejection region, we fail to reject the null hypothesis. Hence, there is no difference in the mean number of calories per can for the two brands of diet soda. The difference in $\bar{x}_1$ and $\bar{x}_2$ observed for the two samples may have occurred due to sampling error only. ■

☞ *Remember* The degrees of freedom for the procedures to make a confidence interval and to test a hypothesis about $\mu_1 - \mu_2$ learned in Section 10.3.1 and 10.3.2 are always rounded down.

EXERCISES

Concepts and Procedures

10.36 Assuming that the two populations are normally distributed with unequal and unknown population standard deviations, construct a 95% confidence interval for $\mu_1 - \mu_2$ for the following.

$$n_1 = 24 \qquad \bar{x}_1 = 17.20 \qquad s_1 = 3.90$$
$$n_2 = 16 \qquad \bar{x}_2 = 19.40 \qquad s_2 = 5.15$$

10.37 Assuming that the two populations are normally distributed with unequal and unknown population standard deviations, construct a 99% confidence interval for $\mu_1 - \mu_2$ for the following.

$$n_1 = 15 \qquad \bar{x}_1 = 52.61 \qquad s_1 = 3.27$$
$$n_2 = 19 \qquad \bar{x}_2 = 43.75 \qquad s_2 = 5.85$$

10.38 Refer to Exercise 10.36. Test at the 5% significance level if the two population means are different.

10.39 Refer to Exercise 10.37. Test at the 1% significance level if the two population means are different.

10.40 Refer to Exercise 10.36. Test at the 1% significance level if μ_1 is less than μ_2.

10.41 Refer to Exercise 10.37. Test at the 5% significance level if μ_1 is greater than μ_2.

Applications

10.42 According to the information given in Exercise 10.28, a sample of 25 male customers who shopped at a supermarket showed that they spent an average of $75 with a standard deviation of $17.50. Another sample of 20 female customers who shopped at the same supermarket showed that they spent an average of $89 with a standard deviation of $14.40. Assume that the amounts spent at this supermarket by all male and all female customers are normally distributed with unequal and unknown population standard deviations.

    a. Construct a 99% confidence interval for the difference between the mean amounts spent by all male and all female customers at this supermarket.

    b. Using the 5% significance level, can you conclude that the mean amount spent by all male customers at this supermarket is less than that of all female customers?

10.43 According to Exercise 10.29, a manufacturing company is interested in buying one of two different kinds of machines. The company tested the two machines for production purposes. The first machine was run for 8 hours. It produced an average of 126 items per hour with a standard deviation of 9 items. The second machine was run for 10 hours. It produced an average of 117 items per hour with a standard deviation of 6 items. Assume that the production per hour for each machine is (approximately) normally distributed. Further assume that the standard deviations of the hourly productions of the two populations are unequal.

    a. Make a 95% confidence interval for the difference between the two population means.

    b. Using the 2.5% significance level, can you conclude that the mean number of items produced per hour by the first machine is higher than that of the second machine?

10.44 According to Exercise 10.30, an insurance company wants to know if the average speed at which men drive cars is higher than that of women drivers. The company took a random sample of 27 cars driven by men on a highway and found the mean speed to be 68 miles per hour with a standard deviation of 2.2 miles. Another sample of 18 cars driven by women on the same highway gave a mean speed of 65 miles per hour with a standard deviation of 2.5 miles. Assume that the speeds at which all men and all women drive cars on this highway are both normally distributed with unequal population standard deviations.

    a. Construct a 98% confidence interval for the difference between the mean speeds of cars driven by all men and all women drivers on this highway.

    b. Test at the 1% significance level if the mean speed of cars driven by all men drivers on this highway is higher than that of cars driven by all women drivers.

10.45 Refer to Exercise 10.31. According to Metropolitan Life Insurance Company claims data, the mean hospital and physician's charges for prostate surgery are $14,460 for Maryland and $17,570 for Virginia (*Statistical Bulletin*, July–September 1996). These two means are based on 19 claims for Maryland and 16 claims for Virginia. Assume that the standard deviations for these sets of claims for the two states are $1920 and $2250, respectively. Further assume that such charges have normal distributions with different population standard deviations for these two states.

a. Construct a 90% confidence interval for the difference between the mean hospital and physician's charges for all prostate surgeries for these two states.

b. Test at the 5% significance level if the mean hospital and physician's charges for all prostate surgeries for these two states are different.

10.46 As mentioned in Exercise 10.32, Quadro Corporation has two supermarket stores in a city. The company's quality control department wanted to check if the customers are equally satisfied with the service provided at these two stores. A sample of 25 customers selected from Supermarket I produced a mean satisfaction index of 7.8 (on a scale of 1 to 10, 1 being the lowest and 10 being the highest) with a standard deviation of .75. Another sample of 28 customers selected from Supermarket II produced a mean satisfaction index of 8.3 with a standard deviation of .59. Assume that the customer satisfaction index for each supermarket has a normal distribution with a different population standard deviation.

a. Construct a 98% confidence interval for the difference between the mean satisfaction indexes for all customers for the two supermarkets.

b. Test at the 1% significance level if the mean satisfaction indexes for all customers for the two supermarkets are different.

10.47 As mentioned in Exercise 10.33, a company claims that its medicine, Brand A, provides faster relief from pain than another company's medicine, Brand B. A researcher tested both brands of medicine on two groups of randomly selected patients. The results of the test are given in the following table. The mean and standard deviation of relief times are in minutes.

Brand	Sample Size	Mean of Relief Times	Standard Deviation of Relief Times
A	25	44	13
B	22	49	11

a. Construct a 99% confidence interval for the difference between the mean relief times for the two brands of medicine.

b. Test at the 1% significance level if the mean relief time for Brand A is less than that for Brand B.

Assume that the two populations are normally distributed with unequal standard deviations.

10.48 In 1995, Runzheimer International studied camera film prices in several large cities in the world (*USA Today*, September 1, 1995). The average price for a roll of 35mm film with 24 exposures was $6.55 in Tokyo and $4.33 in New York City. Suppose that these means are based on random samples of 20 shops in Tokyo and 25 shops in New York City, and that the sample standard deviations of the prices are $1.05 for Tokyo and $1.00 for New York. Further assume that the two populations from which these two samples are selected are normally distributed with unequal and unknown standard deviations.

a. Construct a 90% confidence interval for the difference between the population mean prices of such films in the two cities.

b. Test at the 5% significance level if the population mean price of such films is higher in Tokyo than in New York City.

10.49 Refer to Exercise 10.35. According to the American Association of University Professors, the mean salary of male professors at American colleges and universities was $66,740 and that of female professors was $58,990 in 1995–96. For convenience, assume that these two means are based on random samples of 28 male professors and 26 female professors. Further assume that the standard deviations for the two samples are $3700 and $3200, respectively.

a. Construct a 90% confidence interval for the difference between the two population means.

b. Using the 1% significance level, can you conclude that the mean salary of all male professors for 1995–96 was higher than that of all female professors?

Assume that the salaries of all male and all female professors are both normally distributed with unequal standard deviations.

10.4 INFERENCES ABOUT THE DIFFERENCE BETWEEN TWO POPULATION MEANS FOR PAIRED SAMPLES

Sections 10.1, 10.2, and 10.3 were concerned with estimation and hypothesis testing about the difference between two population means when the two samples were drawn independently from two different populations. This section describes estimation and hypothesis-testing procedures for the difference between two population means when the samples are dependent.

In a case of two dependent samples, two data values—one for each sample—are collected from the same source (or element) and, hence, these are also called **paired** or **matched samples**. For example, we may want to make inferences about the mean weight loss for members of a health club after they have gone through an exercise program for a certain period. To do so, suppose we select a sample of 15 members of this health club and record their weights before and after the program. In this example, both sets of data are collected from the same 15 persons, once before and once after the program. Thus, although there are two samples, they contain the same 15 persons. This is an example of paired (or dependent or matched) samples. The procedures to make confidence intervals and test hypotheses in the case of paired samples are different from the ones for independent samples discussed in earlier sections of this chapter.

PAIRED OR MATCHED SAMPLES

Two samples are said to be *paired* or *matched samples* when for each data value collected from one sample there is a corresponding data value collected from the second sample, and both these data values are collected from the same source.

As another example of paired samples, suppose an agronomist wants to measure the effect of a new brand of fertilizer on the yield of potatoes. To do so, he selects 10 pieces of land and divides each piece of land into two portions. Then he randomly assigns one of the two portions from each piece of land to grow potatoes without using fertilizer (or using some other brand of fertilizer). The second portion from each piece of land is used to grow potatoes using the new brand of fertilizer. Thus, he will have 10 pairs of data values. Then, using the procedure to be discussed in this section, he will make inferences about the difference in the mean yields of potatoes with the new fertilizer and without it.

The question arises, why does the agronomist not choose 10 pieces of land on which to grow potatoes without using the new brand of fertilizer and another 10 pieces of land to grow potatoes by using the new brand of fertilizer? If he does so, the effect of the fertilizer might be confused with the effects due to soil differences at different locations. Thus, he will not be able to isolate the effect of the new brand of fertilizer on the yield of potatoes. Consequently, the results will not be reliable. By choosing 10 pieces of land and then dividing each of them into two portions, the researcher decreases the possibility that the difference in the productivities of different pieces of land affects the results.

In paired samples, the difference between the two data values for each element of the two samples is denoted by **d**. This value of d is called the **paired difference**. We then treat all the values of d as one sample and make inferences applying procedures similar to the ones used for one-sample cases in Chapters 8 and 9. Note that as each source (or element) gives a pair of values (one for each of the two data sets), each sample contains the same number of values. That is, both samples are of the same size. Therefore, we denote the (common) **sample size** by n, which gives the number of paired difference values denoted by d. The **degrees of freedom** for the paired samples are $n - 1$. Let

μ_d = the mean of the paired differences for the population

σ_d = the standard deviation of the paired differences for the population

$\bar{d}$ = the mean of the paired differences for the sample

s_d = the standard deviation of the paired differences for the sample

n = the number of paired difference values

MEAN AND STANDARD DEVIATION OF THE PAIRED DIFFERENCES FOR SAMPLES

The values of $\bar{d}$ and s_d are calculated as[2]

$$\bar{d} = \frac{\Sigma d}{n}$$

$$s_d = \sqrt{\frac{\Sigma d^2 - \dfrac{(\Sigma d)^2}{n}}{n - 1}}$$

In paired samples, instead of using $\bar{x}_1 - \bar{x}_2$ as the sample statistic to make inferences about $\mu_1 - \mu_2$, we use the sample statistic $\bar{d}$ to make inferences about μ_d. Actually the value of $\bar{d}$ is always equal to $\bar{x}_1 - \bar{x}_2$, and the value of μ_d is always equal to $\mu_1 - \mu_2$.

SAMPLING DISTRIBUTION, MEAN, AND STANDARD DEVIATION OF $\bar{d}$

If the number of paired values is large ($n \geq 30$), because of the central limit theorem, the *sampling distribution* of $\bar{d}$ is approximately normal with its *mean* and *standard deviation* as

$$\mu_{\bar{d}} = \mu_d \quad \text{and} \quad \sigma_{\bar{d}} = \frac{\sigma_d}{\sqrt{n}}$$

---

[2]The basic formula to calculate s_d is

$$s_d = \sqrt{\frac{\Sigma(d - \bar{d})^2}{n - 1}}$$

However, we will not use this formula to make calculations in this chapter.

In cases when $n \geq 30$, the normal distribution can be used to make inferences about μ_d.

However, in cases of paired samples, the sample sizes are usually small and σ_d is unknown. In such cases, assuming that the paired differences for the population are (approximately) normally distributed, the normal distribution is replaced by the t distribution to make inferences about μ_d. When σ_d is not known, the standard deviation of $\overline{d}$ is estimated by $s_{\overline{d}} = s_d/\sqrt{n}$.

ESTIMATE OF THE STANDARD DEVIATION OF PAIRED DIFFERENCES

If

1. n is less than 30
2. σ_d is not known
3. the population of paired differences is (approximately) normally distributed

then the t distribution is used to make inferences about μ_d. The standard deviation $\sigma_{\overline{d}}$ of $\overline{d}$ is estimated by $s_{\overline{d}}$, which is calculated as

$$s_{\overline{d}} = \frac{s_d}{\sqrt{n}}$$

Sections 10.4.1 and 10.4.2 describe the procedures to make a confidence interval and test a hypothesis about μ_d when σ_d is unknown and n is small. The inferences are made using the t distribution. However, if n is large, even if σ_d is unknown, the normal distribution can be used to make inferences about μ_d.

10.4.1 INTERVAL ESTIMATION OF μ_d

The mean $\overline{d}$ of paired differences for paired samples is the point estimator of μ_d. The following formula is used to construct a confidence interval for μ_d in the case of (approximately) normally distributed populations.

CONFIDENCE INTERVAL FOR μ_d

The $(1 - \alpha)100\%$ *confidence interval for* μ_d is

$$\overline{d} \pm t s_{\overline{d}}$$

where the value of t is obtained from the t distribution table for the given confidence level and $n - 1$ degrees of freedom, and $s_{\overline{d}}$ is calculated as explained earlier.

Example 10–12 illustrates the procedure to construct a confidence interval for μ_d.

Constructing a
confidence interval
for μ_d: paired samples.

EXAMPLE 10-12 A researcher wanted to find the effect of a special diet on systolic blood pressure. She selected a sample of seven adults and put them on this dietary plan for three months. The following table gives the systolic blood pressures of these seven adults before and after the completion of this plan.

Before	210	180	195	220	231	199	224
After	193	186	186	223	220	183	233

Let μ_d be the mean reduction in the systolic blood pressures due to this special dietary plan for the population of all adults. Construct a 95% confidence interval for μ_d. Assume that the population of paired differences is (approximately) normally distributed.

Solution Because the information obtained is from paired samples, we will make the confidence interval for the paired difference mean μ_d of the population using the paired difference mean $\bar{d}$ of the sample. Let d be the difference in the systolic blood pressure of an adult before and after this special dietary plan. Then, d is obtained by subtracting the systolic blood pressure after the plan from the systolic blood pressure before the plan. The third column of Table 10.1 lists the values of d for the seven adults. The fourth column of the table records the values of d^2, which are obtained by squaring each of the d values.

Table 10.1

		Difference	
Before	**After**	**d**	**d^2**
210	193	17	289
180	186	−6	36
195	186	9	81
220	223	−3	9
231	220	11	121
199	183	16	256
224	233	−9	81
		$\Sigma d = 35$	$\Sigma d^2 = 873$

The values of $\bar{d}$ and s_d are calculated as follows.

$$\bar{d} = \frac{\Sigma d}{n} = \frac{35}{7} = 5.00$$

$$s_d = \sqrt{\frac{\Sigma d^2 - \frac{(\Sigma d)^2}{n}}{n - 1}} = \sqrt{\frac{873 - \frac{(35)^2}{7}}{7 - 1}} = 10.78579312$$

Hence, the standard deviation of $\bar{d}$ is

$$s_{\bar{d}} = \frac{s_d}{\sqrt{n}} = \frac{10.78579312}{\sqrt{7}} = 4.07664661$$

For the 95% confidence interval, the area in each tail of the t distribution curve is

$$\text{Area in each tail} = \alpha/2 = .5 - (.95/2) = .025$$

The degrees of freedom are

$$df = n - 1 = 7 - 1 = 6$$

From the t distribution table, the t value for $df = 6$ and .025 area in the right tail of the t distribution curve is 2.447. Therefore, the 95% confidence interval for μ_d is

$$\bar{d} \pm ts_{\bar{d}} = 5.00 \pm 2.447 \, (4.07664661) = 5.00 \pm 9.98 = -\textbf{4.98 to } \textbf{14.98}$$

Thus, we can state with 95% confidence that the mean difference between systolic blood pressures before and after the given dietary plan for all adult participants is between -4.98 and 14.98. ■

10.4.2 HYPOTHESIS TESTING ABOUT μ_d

A hypothesis about μ_d is tested by using the sample statistic $\bar{d}$. If n is 30 or larger, we can use the normal distibution to test a hypothesis about μ_d. However, if n is less than 30, we replace the normal distribution by the t distribution. To use the t distribution, we assume that the population of all paired differences is (approximately) normally distributed and that the population standard deviation, σ_d, of paired differences in not known. This section illustrates the case of the t distribution only. The following formula is used to calculate the value of the test statistic t when testing a hypothesis about μ_d.

TEST STATISTIC t FOR $\bar{d}$

The value of the *test statistic t* for $\bar{d}$ is computed as follows.

$$t = \frac{\bar{d} - \mu_d}{s_{\bar{d}}}$$

The critical value of t is found from the t distribution table for the given significance level and $n - 1$ degrees of freedom.

Examples 10–13 and 10–14 illustrate the hypothesis-testing procedure for μ_d.

Conducting a left-tailed test of hypothesis about μ_d: paired samples.

EXAMPLE 10–13 A company wanted to know if attending a course on ''how to be a successful salesperson'' can increase the average sales of its employees. The company sent six of its salespersons to attend this course. The following table gives the one-week sales of these salespersons before and after they attended this course.

Before	12	18	25	9	14	16
After	18	24	24	14	19	20

Using the 1% significance level, can you conclude that the mean weekly sales for all salespersons increase as a result of attending this course? Assume that the population of paired differences has a normal distribution.

Solution Because the data are for paired samples, we test a hypothesis about the paired differences mean μ_d of the population using the paired differences mean $\bar{d}$ of the sample.

Let

$$d = \text{(weekly sales before the course)} - \text{(weekly sales after the course)}$$

In Table 10.2, we calculate d for each of the six salespersons by subtracting the sales after the course from the sales before the course. The fourth column of the table lists the values of d^2.

Table 10.2

Before	After	Difference d	d^2
12	18	−6	36
18	24	−6	36
25	24	1	1
9	14	−5	25
14	19	−5	25
16	20	−4	16
		$\Sigma d = -25$	$\Sigma d^2 = 139$

The values of $\bar{d}$ and s_d are calculated as follows.

$$\bar{d} = \frac{\Sigma d}{n} = \frac{-25}{6} = -4.17$$

$$s_d = \sqrt{\frac{\Sigma d^2 - \frac{(\Sigma d)^2}{n}}{n-1}} = \sqrt{\frac{139 - \frac{(-25)^2}{6}}{6-1}} = 2.63944439$$

The standard deviation of $\bar{d}$ is

$$s_{\bar{d}} = \frac{s_d}{\sqrt{n}} = \frac{2.63944439}{\sqrt{6}} = 1.07754866$$

Step 1. *State the null and alternative hypotheses*

We are to test if the mean weekly sales for all salespersons increase as a result of taking the course. Let μ_1 be the mean weekly sales for all salespersons before the course and μ_2 the mean weekly sales for all salespersons after the course. Then $\mu_d = \mu_1 - \mu_2$. The mean weekly sales for all salespersons will increase due to attending the course if μ_1 is less than μ_2, which can be written as $\mu_1 - \mu_2 < 0$ or $\mu_d < 0$. Consequently, the null and alternative hypotheses are

H_0: $\mu_d = 0$ ($\mu_1 - \mu_2 = 0$ or the mean weekly sales do not increase)

H_1: $\mu_d < 0$ ($\mu_1 - \mu_2 < 0$ or the mean weekly sales do increase)

Note that we can also write the null hypothesis as $\mu_d \geq 0$.

Step 2. *Select the distribution to use*

The sample size is small ($n < 30$), the population of paired differences is normal, and σ_d is unknown. Therefore, we use the t distribution to conduct the test.

Step 3. *Determine the rejection and nonrejection regions*

The $<$ sign in the alternative hypothesis indicates that the test is left-tailed. The significance level is .01. Hence,

$$\text{Area in left tail} = \alpha = .01$$

$$\text{Degrees of freedom} = n - 1 = 6 - 1 = 5$$

The critical value of t for $df = 5$ and .01 area in left tail of the t distribution curve is -3.365. This value is shown in Figure 10.7.

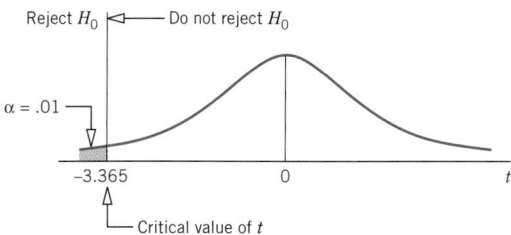

Figure 10.7

Step 4. *Calculate the value of the test statistic*

The value of the test statistic t for $\bar{d}$ is computed as follows.

$$t = \frac{\bar{d} - \mu_d}{s_{\bar{d}}} = \frac{-4.17 - 0}{1.07754866} = -3.870$$

From H_0

Step 5. *Make a decision*

Because the value of the test statistic $t = -3.870$ for $\bar{d}$ falls in the rejection region, we reject the null hypothesis. Consequently, we conclude that the mean weekly sales for all salespersons increase as a result of this course. ▬

Making a two-tailed test of hypothesis about μ_d: paired samples.

EXAMPLE 10-14 Refer to Example 10–12. The table that gives the blood pressures of seven adults before and after the completion of a special dietary plan is reproduced below.

Before	210	180	195	220	231	199	224
After	193	186	186	223	220	183	233

Let μ_d be the mean of the differences between the systolic blood pressures before and after completing this special dietary plan for the population of all adults. Using the 5% significance level, can we conclude that the mean of the paired differences μ_d is different from zero? Assume that the population of paired differences is (approximately) normally distributed.

Solution Table 10.3 gives d and d^2 for each of the seven adults.

Table 10.3

Before	After	Difference d	d^2
210	193	17	289
180	186	−6	36
195	186	9	81
220	223	−3	9
231	220	11	121
199	183	16	256
224	233	−9	81
		$\Sigma d = 35$	$\Sigma d^2 = 873$

The values of $\bar{d}$ and s_d are calculated as follows.

$$\bar{d} = \frac{\Sigma d}{n} = \frac{35}{7} = 5.00$$

$$s_d = \sqrt{\frac{\Sigma d^2 - \frac{(\Sigma d)^2}{n}}{n-1}} = \sqrt{\frac{873 - \frac{(35)^2}{7}}{7-1}} = 10.78579312$$

Hence, the standard deviation of $\bar{d}$ is

$$s_{\bar{d}} = \frac{s_d}{\sqrt{n}} = \frac{10.78579312}{\sqrt{7}} = 4.07664661$$

Step 1. *State the null and alternative hypotheses*

H_0: $\mu_d = 0$ (the mean of the paired differences is not different from zero)

H_1: $\mu_d \neq 0$ (the mean of the paired differences is different from zero)

Step 2. *Select the distribution to use*

Because the sample size is small, the population of paired differences is (approximately) normal, and σ_d is not known. Therefore, we use the t distribution to make the test.

Step 3. *Determine the rejection and nonrejection regions*

The $\neq$ sign in the alternative hypothesis indicates that the test is two-tailed. The significance level is .05.

$$\text{Area in each tail of the curve} = \alpha/2 = .05/2 = .025$$

$$\text{Degrees of freedom} = n - 1 = 7 - 1 = 6$$

The two critical values of t for $df = 6$ and .025 area in each tail of the t distribution curve are −2.447 and 2.447. These values are shown in Figure 10.8.

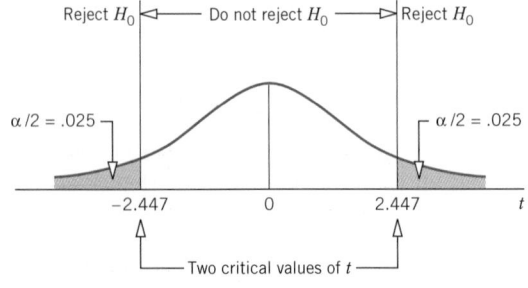

Figure 10.8

Step 4. *Calculate the value of the test statistic*

The value of the test statistic t for $\bar{d}$ is computed as follows.

$$t = \frac{\bar{d} - \mu_d}{s_{\bar{d}}} = \frac{5.00 - \overset{\text{From } H_0}{0}}{4.07664661} = 1.226$$

Step 5. *Make a decision*

Because the value of the test statistic $t = 1.226$ for $\bar{d}$ falls in the nonrejection region, we fail to reject the null hypothesis. Hence, we conclude that the mean of the population paired differences is not different from zero. In other words, we can state that the mean of the differences between the systolic blood pressures before and after completing this special dietary plan for the population of all adults is not different from zero. ▬

EXERCISES

Concepts and Procedures

10.50 Explain when you would use the paired samples procedure to make confidence intervals and test hypotheses.

10.51 Find the following confidence intervals for μ_d assuming that the populations of paired differences are normally distributed.

  a. $n = 9$, $\bar{d} = 25.4$, $s_d = 13.5$, confidence level = 99%
  b. $n = 26$, $\bar{d} = 13.2$, $s_d = 4.8$, confidence level = 95%
  c. $n = 12$, $\bar{d} = 34.6$, $s_d = 11.7$, confidence level = 90%

10.52 Find the following confidence intervals for μ_d assuming that the populations of paired differences are normally distributed.

  a. $n = 10$, $\bar{d} = 17.5$, $s_d = 6.3$, confidence level = 99%
  b. $n = 24$, $\bar{d} = 55.9$, $s_d = 14.7$, confidence level = 95%
  c. $n = 14$, $\bar{d} = 29.3$, $s_d = 8.3$, confidence level = 90%

10.53 Perform the following tests of hypotheses assuming that the populations of paired differences are normally distributed.

  a. $H_0: \mu_d = 0$, $H_1: \mu_d \neq 0$, $n = 9$, $\bar{d} = 6.7$, $s_d = 2.5$, $\alpha = .10$
  b. $H_0: \mu_d = 0$, $H_1: \mu_d > 0$, $n = 22$, $\bar{d} = 14.8$, $s_d = 6.4$, $\alpha = .05$
  c. $H_0: \mu_d = 0$, $H_1: \mu_d < 0$, $n = 17$, $\bar{d} = -9.3$, $s_d = 4.8$, $\alpha = .01$

10.54 Conduct the following tests of hypotheses assuming that the populations of paired differences are normally distributed.

  a. $H_0: \mu_d = 0$, $H_1: \mu_d \neq 0$, $n = 26$, $\bar{d} = 9.6$, $s_d = 3.9$, $\alpha = .05$
  b. $H_0: \mu_d = 0$, $H_1: \mu_d > 0$, $n = 15$, $\bar{d} = 8.8$, $s_d = 4.7$, $\alpha = .01$
  c. $H_0: \mu_d = 0$, $H_1: \mu_d < 0$, $n = 20$, $\bar{d} = -7.4$, $s_d = 2.3$, $\alpha = .10$

Applications

10.55 A company sent seven of its employees to attend a course in building self-confidence. These employees were evaluated for their self-confidence before and after attending this course. The following table gives the scores (on a scale of 1 to 15, 1 being the lowest and 15 being the highest score) of these employees before and after they attended the course.

Before	8	5	4	9	6	8	5
After	10	7	5	11	6	7	9

    a. Construct a 95% confidence interval for the mean μ_d of the population paired differences where a paired difference is equal to the score of an employee before attending the course minus the score of the same employee after attending the course.

    b. Test at the 1% significance level if attending this course increases the mean score of employees.

Assume that the population of paired differences has a normal distribution.

 10.56 Many students suffer from math anxiety. A professor who teaches statistics offered her students a two-hour lecture on math anxiety and ways to overcome it. The following table gives the test scores in statistics of seven students before and after they attended this lecture.

Before	56	69	48	74	65	71	58
After	62	73	44	85	71	70	69

    a. Construct a 99% confidence interval for the mean μ_d of the population paired differences where a paired difference is equal to the score before attending this lecture minus the score after attending this lecture.

    b. Test at the 2.5% significance level if attending this lecture increases the average score in statistics.

Assume that the population of paired differences is (approximately) normally distributed.

10.57 A private agency claims that the crash course it offers significantly increases the writing speed of secretaries. The following table gives the scores of eight secretaries before and after they attended this course.

Before	81	75	89	91	65	70	90	69
After	97	72	93	110	78	69	115	75

    a. Make a 90% confidence interval for the mean μ_d of the population paired differences where a paired difference is equal to the score before attending the course minus the score after attending the course.

    b. Using the 5% significance level, can you conclude that attending this course increases the writing speed of secretaries?

Assume that the population of paired differences is (approximately) normally distributed.

10.58 A company claims that its 12-week special exercise program significantly reduces weight. A random sample of six persons was selected, and these persons were put on this exercise program for 12 weeks. The following table gives the weights (in pounds) of those six persons before and after the program.

Before	180	195	177	221	208	199
After	183	187	161	204	197	189

    a. Make a 95% confidence interval for the mean μ_d of the population paired differences where a paired difference is equal to the weight before joining this exercise program minus the weight at the end of the 12-week program.

    b. Using the 1% significance level, can you conclude that the mean weight loss for all persons due to this special exercise program is greater than zero?

Assume that the population of all paired differences is (approximately) normally distributed.

10.59 The manufacturer of a gasoline additive claims that the use of this additive increases gasoline mileage. A random sample of six cars was selected and these cars were driven for one week without the gasoline additive and then for one week with the gasoline additive. The following table gives the miles per gallon for these cars without and with the gasoline additive.

Without	24.6	28.3	18.9	23.7	15.4	29.5
With	26.3	31.7	18.2	25.3	18.3	30.9

a. Construct a 99% confidence interval for the mean μ_d of the population paired differences where a paired difference is equal to the miles per gallon without the gasoline additive minus the miles per gallon with the gasoline additive.

b. Using the 2.5% significance level, can you conclude that the use of the gasoline additive increases the gasoline mileage?

Assume that the population of paired differences is (approximately) normally distributed.

 10.60 A company is considering installing new machines to assemble its products. The company is considering two types of machines, but it will buy only one type. The company selected eight assembly workers and asked them to use these two types of machines to assemble products. The following table gives the time taken (in minutes) to assemble one unit of the product on each type of machine for each of these eight workers.

Machine I	23	26	19	24	26	22	20	18
Machine II	21	24	23	25	24	25	24	23

a. Construct a 98% confidence interval for the mean μ_d of the population paired differences where a paired difference is equal to the time taken to assemble a unit of the product on Machine I minus the time taken to assemble a unit of the product on Machine II.

b. Test at the 5% significance level if the mean time taken to assemble a unit of the product is different for the two types of machines.

Assume that the population of paired differences is (approximately) normally distributed.

10.5 INFERENCES ABOUT THE DIFFERENCE BETWEEN TWO POPULATION PROPORTIONS FOR LARGE AND INDEPENDENT SAMPLES

Quite often we need to construct a confidence interval and test a hypothesis about the difference between two population proportions. For instance, we may want to estimate the difference between the proportion of defective items produced on two different machines. If p_1 and p_2 are the proportions of defective items produced on the first and second machine, respectively, then we are to make a confidence interval for $p_1 - p_2$. Or we may want to test the hypothesis that the proportion of defective items produced on Machine I is different from the proportion of defective items produced on Machine II. In this case, we are to test the null hypothesis $p_1 - p_2 = 0$ against the alternative hypothesis $p_1 - p_2 \neq 0$.

This section discusses how to make a confidence interval and test a hypothesis about $p_1 - p_2$ for two large and independent samples. The sample statistic that is used to make inferences about $p_1 - p_2$ is $\hat{p}_1 - \hat{p}_2$ where $\hat{p}_1$ and $\hat{p}_2$ are the proportions for two large and independent samples. As discussed in Section 7.6 of Chapter 7, we determine a sample proportion by dividing the number of elements in the sample that possess a given attribute by the sample size. Thus,

$$\hat{p}_1 = x_1/n_1 \qquad \text{and} \qquad \hat{p}_2 = x_2/n_2$$

where x_1 and x_2 are the number of elements that possess a given characteristic in the two samples and n_1 and n_2 are the sizes of the two samples, respectively.

10.5.1 MEAN, STANDARD DEVIATION, AND SAMPLING DISTRIBUTION OF $\hat{p}_1 - \hat{p}_2$

As discussed in Chapter 7, for a large sample the sample proportion $\hat{p}$ is (approximately) normally distributed with mean p and standard deviation $\sqrt{pq/n}$. Hence, for two large and independent samples of sizes n_1 and n_2, respectively, their sample proportions $\hat{p}_1$ and $\hat{p}_2$ are (approximately) normally distributed with means p_1 and p_2 and standard deviations $\sqrt{p_1 q_1/n_1}$ and $\sqrt{p_2 q_2/n_2}$, respectively. Using these results, we can make the following statements about the shape of the sampling distribution of $\hat{p}_1 - \hat{p}_2$ and its mean and standard deviation.

MEAN, STANDARD DEVIATION, AND SAMPLING DISTRIBUTION OF $\hat{p}_1 - \hat{p}_2$

For two large and independent samples, the *sampling distribution of $\hat{p}_1 - \hat{p}_2$ is* (approximately) normal with its *mean* and *standard deviation* as

$$\mu_{\hat{p}_1 - \hat{p}_2} = p_1 - p_2$$

and

$$\sigma_{\hat{p}_1 - \hat{p}_2} = \sqrt{\frac{p_1 q_1}{n_1} + \frac{p_2 q_2}{n_2}}$$

respectively, where $q_1 = 1 - p_1$ and $q_2 = 1 - p_2$.

Thus, to construct a confidence interval and test a hypothesis about $p_1 - p_2$ for large and independent samples, we use the normal distribution. As was indicated in Chapter 7, in the case of proportion, the sample is large if np and nq are both greater than 5. In the case of two samples, both sample sizes will be large if $n_1 p_1$, $n_1 q_1$, $n_2 p_2$, and $n_2 q_2$ are all greater than 5.

10.5.2 INTERVAL ESTIMATION OF $p_1 - p_2$

The difference between two sample proportions $\hat{p}_1 - \hat{p}_2$ is the point estimator for the difference between two population proportions $p_1 - p_2$. Because we do not know p_1 and p_2 when we are making a confidence interval for $p_1 - p_2$, we cannot calculate the value of $\sigma_{\hat{p}_1 - \hat{p}_2}$. Therefore, we use $s_{\hat{p}_1 - \hat{p}_2}$ as the point estimator of $\sigma_{\hat{p}_1 - \hat{p}_2}$ in the interval estimation. We construct the confidence interval for $p_1 - p_2$ using the following formula.

CONFIDENCE INTERVAL FOR $p_1 - p_2$

The $(1 - \alpha)100\%$ *confidence interval for $p_1 - p_2$* is

$$(\hat{p}_1 - \hat{p}_2) \pm z s_{\hat{p}_1 - \hat{p}_2}$$

where the value of z is read from the normal distribution table for the given confidence level, and $s_{\hat{p}_1 - \hat{p}_2}$ is calculated as

$$s_{\hat{p}_1 - \hat{p}_2} = \sqrt{\frac{\hat{p}_1 \hat{q}_1}{n_1} + \frac{\hat{p}_2 \hat{q}_2}{n_2}}$$

Example 10–15 describes the procedure to make a confidence interval for the difference between two population proportions for large samples.

Constructing a confidence interval for $p_1 - p_2$: large and independent samples.

EXAMPLE 10-15   A researcher wanted to estimate the difference between the percentages of users of two toothpastes who will never switch to another toothpaste. In a sample of 500 users of Toothpaste A taken by this researcher, 100 said that they will never switch to another toothpaste. In another sample of 400 users of Toothpaste B taken by the same researcher, 68 said that they will never switch to another toothpaste.

(a)   Let p_1 and p_2 be the proportions of all users of Toothpastes A and B, respectively, who will never switch to another toothpaste. What is the point estimate of $p_1 - p_2$?

(b)   Construct a 97% confidence interval for the difference between the proportions of all users of the two toothpastes who will never switch.

Solution   Let p_1 and p_2 be the proportions of all users of Toothpastes A and B, respectively, who will never switch to another toothpaste and let $\hat{p}_1$ and $\hat{p}_2$ be the respective sample proportions. Let x_1 and x_2 be the number of users of Toothpastes A and B, respectively, in the two samples who said that they will never switch to another toothpaste. From the given information,

$$\text{Toothpaste A:} \quad n_1 = 500 \quad \text{and} \quad x_1 = 100$$

$$\text{Toothpaste B:} \quad n_2 = 400 \quad \text{and} \quad x_2 = 68$$

The two sample proportions are calculated as follows.

$$\hat{p}_1 = x_1/n_1 = 100/500 = .20$$

$$\hat{p}_2 = x_2/n_2 = 68/400 = .17$$

Then,

$$\hat{q}_1 = 1 - .20 = .80 \quad \text{and} \quad \hat{q}_2 = 1 - .17 = .83$$

(a)   The point estimate of $p_1 - p_2$ is as follows.

$$\text{Point estimate of } p_1 - p_2 = \hat{p}_1 - \hat{p}_2 = .20 - .17 = \mathbf{.03}$$

(b)   The values of $n_1\hat{p}_1$, $n_1\hat{q}_1$, $n_2\hat{p}_2$, and $n_2\hat{q}_2$ are

$$n_1\hat{p}_1 = 500\,(.20) = 100, \qquad n_1\hat{q}_1 = 500\,(.80) = 400,$$

$$n_2\hat{p}_2 = 400\,(.17) = 68, \quad \text{and} \quad n_2\hat{q}_2 = 400\,(.83) = 332$$

Since each of these values is greater than 5, both sample sizes are large. Consequently we use the normal distribution to make a confidence interval for $p_1 - p_2$.
The standard deviation of $\hat{p}_1 - \hat{p}_2$ is

$$s_{\hat{p}_1-\hat{p}_2} = \sqrt{\frac{\hat{p}_1\hat{q}_1}{n_1} + \frac{\hat{p}_2\hat{q}_2}{n_2}} = \sqrt{\frac{(.20)\,(.80)}{500} + \frac{(.17)\,(.83)}{400}} = .02593742$$

The z value for a 97% confidence level, obtained from the normal distribution table for $.97/2 = .4850$, is 2.17. The 97% confidence interval for $p_1 - p_2$ is

$$(\hat{p}_1 - \hat{p}_2) \pm z s_{\hat{p}_1-\hat{p}_2} = (.20 - .17) \pm 2.17\,(.02593742)$$

$$= .03 \pm .056 = \mathbf{-.026 \text{ to } .086}$$

Thus, with 97% confidence we can state that the difference between the two population proportions is between $-.026$ and $.086$.

10.5.3 HYPOTHESIS TESTING ABOUT $p_1 - p_2$

In this section we learn how to test a hypothesis about $p_1 - p_2$ for two large and independent samples. The procedure involves the same five steps that we have used previously. Once again, we calculate the standard deviation of $\hat{p}_1 - \hat{p}_2$ as

$$\sigma_{\hat{p}_1 - \hat{p}_2} = \sqrt{\frac{p_1 q_1}{n_1} + \frac{p_2 q_2}{n_2}}$$

When a test of hypothesis about $p_1 - p_2$ is performed, usually the null hypothesis is $p_1 = p_2$ and the values of p_1 and p_2 are not known. Assuming that the null hypothesis is true and $p_1 = p_2$, a common value of p_1 and p_2, denoted by $\bar{p}$, is calculated by using one of the following formulas.

$$\bar{p} = \frac{x_1 + x_2}{n_1 + n_2} \quad \text{or} \quad \frac{n_1 \hat{p}_1 + n_2 \hat{p}_2}{n_1 + n_2}$$

Which of these formulas is used depends on whether the values of x_1 and x_2 or the values of $\hat{p}_1$ and $\hat{p}_2$ are known. Note that x_1 and x_2 are the number of elements in each of the two samples that possess a certain characteristic. This value of $\bar{p}$ is called the **pooled sample proportion**. Using the value of the pooled sample proportion, we compute an estimate of the standard deviation of $\hat{p}_1 - \hat{p}_2$ as follows.

$$s_{\hat{p}_1 - \hat{p}_2} = \sqrt{\bar{p}\,\bar{q}\left(\frac{1}{n_1} + \frac{1}{n_2}\right)}$$

where $\bar{q} = 1 - \bar{p}$.

TEST STATISTIC z FOR $\hat{p}_1 - \hat{p}_2$

The value of the *test statistic z for $\hat{p}_1 - \hat{p}_2$* is calculated as

$$z = \frac{(\hat{p}_1 - \hat{p}_2) - (p_1 - p_2)}{s_{\hat{p}_1 - \hat{p}_2}}$$

The value of $p_1 - p_2$ is substituted from H_0, which usually is zero.

Examples 10–16 and 10–17 illustrate the procedure to test hypotheses about the difference between two population proportions for large samples.

Making a right-tailed test of hypothesis about $p_1 - p_2$: large and independent samples.

EXAMPLE 10–16 Reconsider Example 10–15 about the percentages of users of two toothpastes who will never switch to another toothpaste. At the 1% significance level, can we conclude that the proportion of users of Toothpaste A who will never switch to another toothpaste is higher than the proportion of users of Toothpaste B who will never switch to another toothpaste?

Solution Let p_1 and p_2 be the proportions of all users of Toothpastes A and B, respectively, who will never switch to another toothpaste and let $\hat{p}_1$ and $\hat{p}_2$ be the corresponding sample proportions. Let x_1 and x_2 be the number of users of Toothpastes A and B, respectively, in the two samples who said that they will never switch to another toothpaste. From the given information,

$$\text{Toothpaste A:} \quad n_1 = 500 \quad \text{and} \quad x_1 = 100$$

$$\text{Toothpaste B:} \quad n_2 = 400 \quad \text{and} \quad x_2 = 68$$

The significance level is $\alpha = .01$.

The two sample proportions are calculated as follows.

$$\hat{p}_1 = x_1/n_1 = 100/500 = .20$$

$$\hat{p}_2 = x_2/n_2 = 68/400 = .17$$

Step 1. *State the null and alternative hypotheses*

We are to test if the proportion of users of Toothpaste A who will never switch to another toothpaste is higher than the proportion of users of Toothpaste B who will never switch to another toothpaste. In other words, we are to test whether p_1 is greater than p_2. This can be written as $p_1 - p_2 > 0$. Thus, the two hypotheses are

$$H_0: p_1 - p_2 = 0 \quad (p_1 \text{ is not greater than } p_2)$$

$$H_1: p_1 - p_2 > 0 \quad (p_1 \text{ is greater than } p_2)$$

Step 2. *Select the distribution to use*

As shown in Example 10–15, $n_1\hat{p}_1$, $n_1\hat{q}_1$, $n_2\hat{p}_2$, and $n_2\hat{q}_2$ are all greater than 5. Consequently both samples are large, and we apply the normal distribution to make the test.

Step 3. *Determine the rejection and nonrejection regions*

The $>$ sign in the alternative hypothesis indicates that the test is right-tailed. From the normal distribution table, for a .01 significance level, the critical value of z is 2.33. This is shown in Figure 10.9.

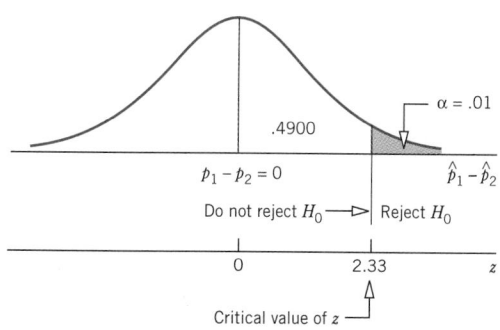

Figure 10.9

Step 4. *Calculate the value of the test statistic*

The pooled sample proportion is

$$\bar{p} = \frac{x_1 + x_2}{n_1 + n_2} = \frac{100 + 68}{500 + 400} = .187$$

and

$$\bar{q} = 1 - \bar{p} = 1 - .187 = .813$$

The estimate of the standard deviation of $\hat{p}_1 - \hat{p}_2$ is

$$s_{\hat{p}_1-\hat{p}_2} = \sqrt{\bar{p}\,\bar{q}\left(\frac{1}{n_1} + \frac{1}{n_2}\right)} = \sqrt{(.187)\,(.813)\left(\frac{1}{500} + \frac{1}{400}\right)} = .02615606$$

The value of the test statistic z for $\hat{p}_1 - \hat{p}_2$ is

From H_0

$$z = \frac{(\hat{p}_1 - \hat{p}_2) - (p_1 - p_2)}{s_{\hat{p}_1-\hat{p}_2}} = \frac{(.20 - .17) - 0}{.02615606} = 1.15$$

Step 5. *Make a decision*

Since the value of the test statistic $z = 1.15$ for $\hat{p}_1 - \hat{p}_2$ falls in the nonrejection region, we fail to reject the null hypothesis. Therefore, we conclude that the proportion of users of Toothpaste A who will never switch to another toothpaste is not greater than the proportion of users of Toothpaste B who will never switch to another toothpaste. ▄

Conducting a two-tailed test of hypothesis about $p_1 - p_2$: large and independent samples.

EXAMPLE 10-17 A clinical trial was conducted to test the effectiveness of "nicotine patch" therapy for cessation of smoking (Douglas E. Jorenby, Ph.D. et al., "Varying Nicotine Patch Dose and Type of Smoking Cessation Counseling," *The Journal of the American Medical Association* 274[17], November 1, 1995). Two dosage levels of nicotine patch, 22-mg and 44-mg, were tested. Of the 504 smokers included in the study, 252 received the 22-mg dose, and 10% of them suffered from nausea. The remaining 252 smokers were given the 44-mg dose, and 28% of them reported feeling nauseous. Test whether all smokers receiving the 22-mg dose nicotine patches would experience a different rate of nausea than all smokers receiving the 44-mg dose nicotine patches. Use the 1% significance level.

Solution Let p_1 and p_2 be the proportions of smokers who experience nausea among all smokers who receive 22-mg dose and 44-mg dose nicotine patches, respectively. Let $\hat{p}_1$ and $\hat{p}_2$ be the corresponding sample proportions. From the given information,

For 22-mg dose nicotine patch receivers: $\quad n_1 = 252 \quad$ and $\quad \hat{p}_1 = .10$

For 44-mg dose nicotine patch receivers: $\quad n_2 = 252 \quad$ and $\quad \hat{p}_2 = .28$

The significance level is $\alpha = .01$.

Step 1. *State the null and alternative hypotheses*

The null and alternative hypotheses are

$\quad H_0: p_1 - p_2 = 0 \quad$ (the two population proportions are not different)

$\quad H_1: p_1 - p_2 \neq 0 \quad$ (the two population proportions are different)

Step 2. *Select the distribution to use*

Because the samples are large and independent, we apply the normal distribution to make the test. (The reader should check that $n_1\hat{p}_1$, $n_1\hat{q}_1$, $n_2\hat{p}_2$, and $n_2\hat{q}_2$ are all greater than 5.)

Step 3. *Determine the rejection and nonrejection regions*

The $\neq$ sign in the alternative hypothesis indicates that the test is two-tailed. For a 1% significance level, the critical values of z are -2.58 and 2.58. These values are shown in Figure 10.10.

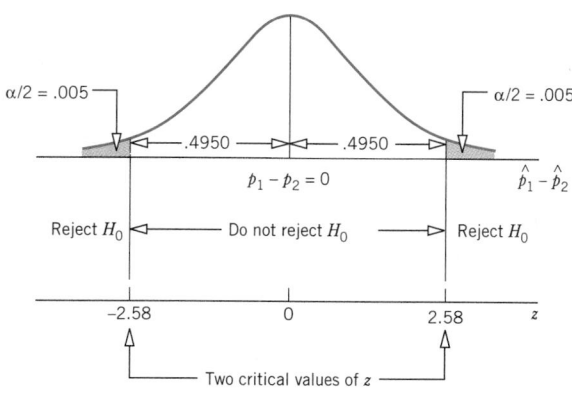

Figure 10.10

Step 4. *Calculate the value of the test statistic*

The pooled sample proportion is

$$\bar{p} = \frac{n_1 \, \hat{p}_1 + n_2 \, \hat{p}_2}{n_1 + n_2} = \frac{252 \, (.10) + 252 \, (.28)}{252 + 252} = .19$$

and
$$\bar{q} = 1 - \bar{p} = 1 - .19 = .81$$

The estimate of the standard deviation of $\hat{p}_1 - \hat{p}_2$ is

$$s_{\hat{p}_1-\hat{p}_2} = \sqrt{\bar{p}\,\bar{q}\left(\frac{1}{n_1} + \frac{1}{n_2}\right)} = \sqrt{(.19)\,(.81)\left(\frac{1}{252} + \frac{1}{252}\right)} = .03494894$$

The value of the test statistic z for $\hat{p}_1 - \hat{p}_2$ is

$$z = \frac{(\hat{p}_1 - \hat{p}_2) - (p_1 - p_2)}{s_{\hat{p}_1-\hat{p}_2}} = \frac{(.10 - .28) - 0}{.03494894} = -5.15$$

Step 5. *Make a decision*

The value of the test statistic $z = -5.15$ for $\hat{p}_1 - \hat{p}_2$ falls in the rejection region. Consequently, we reject the null hypothesis. As a result, we conclude that the proportions of smokers who experience nausea among all smokers who receive 22-mg dose and 44-mg dose nicotine patches are different. ■

CASE STUDY 10-2   MORE ON THE INFERENCES MADE BY THE BUREAU OF THE CENSUS

As we mentioned in Case Study 10–1, the U.S. Bureau of the Census publishes a large number of reports every year that summarize results of various surveys done by the bureau. The following statement is quoted from one of the bureau's recent publications. The figures in parentheses in this statement give the 90% confidence interval of the estimate.

The poverty rate was 13.8 (± 0.3) percent in 1995, significantly lower than the 14.5 (± 0.3) percent poverty rate in 1994.

A part of this statement gives 90% confidence intervals for the poverty rates for the years 1994 and 1995. According to the sample surveys conducted by the Bureau of the Census, the 90% confidence interval for the percentage of persons in the United States who were living below the poverty level in 1995 is 13.8 $\pm$ 0.3. The corresponding 90% confidence interval for 1994 is 14.5 $\pm$ 0.3.

A part of this statement also gives the decision of a test of hypothesis. It states that the percentage of all persons in the United States who were living below the poverty level in 1995 is significantly lower than the corresponding percentage for the year 1994. Let p_1 and p_2 be, respectively, the proportions of all persons in the United States who were living below the poverty level in 1994 and 1995. The Bureau's researchers tested the following hypothesis.

$$H_0: p_2 = p_1$$

$$H_1: p_2 < p_1$$

Using the information from sample surveys for the years 1994 and 1995, the bureau's researchers rejected the null hypothesis. Consequently, they concluded that the percentage of all persons in the United States who were living below the poverty level in 1995 is significantly lower than the corresponding percentage for the year 1994.

Source: Poverty in the United States: 1995. Series P-60, 194. U.S. Department of Commerce, Economics and Statistics Administration, Bureau of the Census, September 1996.

EXERCISES

Concepts and Procedures

10.61   What is the shape of the sampling distribution of $\hat{p}_1 - \hat{p}_2$ for two large samples? What are the mean and standard deviation of this sampling distribution?

10.62   When are the samples considered large enough for the sampling distribution of the difference between two sample proportions to be (approximately) normal?

10.63   Construct a 99% confidence interval for $p_1 - p_2$ for the following.

$$n_1 = 300 \qquad \hat{p}_1 = .53 \qquad n_2 = 200 \qquad \hat{p}_2 = .59$$

10.64   Construct a 95% confidence interval for $p_1 - p_2$ for the following.

$$n_1 = 100 \qquad \hat{p}_1 = .81 \qquad n_2 = 150 \qquad \hat{p}_2 = .76$$

10.65   Refer to the information given in Exercise 10.63. Test at the 1% significance level if the two population proportions are different.

10.66   Refer to the information given in Exercise 10.64. Test at the 5% significance level if $p_1 - p_2$ is different from zero.

10.67   Refer to the information given in Exercise 10.63. Test at the 1% significance level if p_1 is less than p_2.

10.68 Refer to the information given in Exercise 10.64. Test at the 2.5% significance level if p_1 is greater than p_2.

10.69 A sample of 500 observations taken from the first population gave $x_1 = 310$. Another sample of 600 observations taken from the second population gave $x_2 = 348$.

    a. Find the point estimate of $p_1 - p_2$.

    b. Make a 97% confidence interval for $p_1 - p_2$.

    c. Show the rejection and nonrejection regions on the sampling distribution of $\hat{p}_1 - \hat{p}_2$ for H_0: $p_1 = p_2$ versus H_1: $p_1 > p_2$. Use a significance level of 2.5%.

    d. Find the value of the test statistic z for the test of part c.

    e. Will you reject the null hypothesis mentioned in part c at a significance level of 2.5%?

10.70 A sample of 1000 observations taken from the first population gave $x_1 = 280$. Another sample of 1200 observations taken from the second population gave $x_2 = 396$.

    a. Find the point estimate of $p_1 - p_2$.

    b. Make a 98% confidence interval for $p_1 - p_2$.

    c. Show the rejection and nonrejection regions on the sampling distribution of $\hat{p}_1 - \hat{p}_2$ for H_0: $p_1 = p_2$ versus H_1: $p_1 < p_2$. Use a significance level of 1%.

    d. Find the value of the test statistic z for the test of part c.

    e. Will you reject the null hypothesis mentioned in part c at a significance level of 1%?

Applications

10.71 According to the National Household Survey on Drug Abuse, 8.2% of the 12- to 17-year-olds interviewed admitted to illicit drug use during the month before the survey in 1994, and the corresponding percentage was 10.9% for 1995 (*USA Today*, August 21, 1996). Assume that these estimates are based on random samples of 1800 and 2000 such young persons for 1994 and 1995, respectively.

    a. Determine a 99% confidence interval for the difference between the population proportions of 12- to 17-year-olds who used illicit drugs in 1994 and 1995.

    b. At the 1% significance level, can you conclude that the proportion of all 12- to 17-year-olds using illicit drugs in 1994 was less than that in 1995?

10.72 A company has two restaurants in two different areas of New York City. The company wants to estimate the percentages of patrons who think that the food and service at each of these restaurants are excellent. A sample of 200 patrons taken from the restaurant in area A showed that 118 of them think that the food and service are excellent at this restaurant. Another sample of 250 patrons taken from the restaurant in area B showed that 160 of them think that the food and service are excellent at this restaurant.

    a. Construct a 97% confidence interval for the difference between the two population proportions.

    b. Testing at the 2.5% significance level, can you conclude that the proportion of patrons at the restaurant in area A who think that the food and service are excellent is lower than the corresponding proportion at the restaurant in area B?

10.73 In a *Prevention* magazine survey released in 1994, Princeton Survey Research Associates examined the weights of children aged 3 to 17 (*USA Today*, October 27, 1994). According to this study, 24% of children in this age group were overweight in 1984, and 31% were considered overweight in 1993. Suppose that these percentages are based on random samples of 400 and 500 children in the given age group in 1984 and 1993, respectively.

    a. Construct a 98% confidence interval for the difference between the proportions of the overweight 3- to 17-year-olds in 1984 and 1993.

    b. Test at the 1% level of significance if the proportion of overweight children in this age group was less in 1984 than in 1993.

 10.74 The Institute for Social Inquiry at the University of Connecticut conducted an opinion poll on college/university athletics in June 1996 in which 500 randomly selected Connecticut residents were contacted by telephone (*The Hartford Courant*, July 6, 1996). Sixty percent of the males and

54% of the females in the poll felt that athletics at larger colleges/universities are overemphasized. Assume that this poll included 250 men and 250 women.

- a. Construct a 95% confidence interval for the difference between the proportions of all men and all women in Connecticut who feel that college/university athletics are over-emphasized.
- b. Testing at the 5% significance level, can you conclude that the proportions of all men and all women in Connecticut who feel that college/university athletics are overemphasized are different?

10.75 The management of a supermarket wanted to investigate if the percentages of men and women who prefer to buy national brand products over the store brand products are different. A sample of 600 men shoppers at the company's supermarkets showed that 246 of them prefer to buy national brand products over the store brand products. Another sample of 700 women shoppers at the company's supermarkets showed that 266 of them prefer to buy national brand products over the store brand products.

- a. What is the point estimate of the difference between the two population proportions?
- b. Construct a 95% confidence interval for the difference between the proportions of all men and all women shoppers at these supermarkets who prefer to buy national brand products over the store brand products.
- c. Testing at the 5% significance level, can you conclude that the proportions of all men and all women shoppers at these supermarkets who prefer to buy national brand products over the store brand products are different?

10.76 The lottery commissioner's office in a state wanted to find if the percentages of men and women who play the lottery often are different. A sample of 500 men taken by the commissioner's office showed that 165 of them play the lottery often. Another sample of 300 women showed that 69 of them play the lottery often.

- a. What is the point estimate of the difference between the two population proportions?
- b. Construct a 99% confidence interval for the difference between the proportions of all men and all women who play the lottery often.
- c. Testing at the 1% significance level, can you conclude that the proportions of all men and all women who play the lottery often are different?

10.77 A mail-order company has two warehouses, one on the West Coast and the second on the East Coast. The company's policy is to mail all orders placed with it within 72 hours. The company's quality control department checks quite often whether or not this policy is maintained at the two warehouses. A recently taken sample of 400 orders placed with the warehouse on the West Coast showed that 368 of them were mailed within 72 hours. Another sample of 300 orders placed with the warehouse on the East Coast showed that 285 of them were mailed within 72 hours.

- a. Construct a 97% confidence interval for the difference between the proportions of all orders placed at the two warehouses that are mailed within 72 hours.
- b. Using the 2.5% significance level, can you conclude that the proportion of all orders placed at the warehouse on the West Coast that are mailed within 72 hours is lower than the corresponding proportion for the warehouse on the East Coast?
- *c. Find the p-value for the test mentioned in part b.

10.78 A company that has many department stores in the Southern states wanted to find the percentage of sales at two such stores for which at least one of the items was returned. A sample of 800 sales randomly selected from Store A showed that for 280 of them at least one item was returned. Another sample of 900 sales randomly selected from Store B showed that for 324 of them at least one item was returned.

- a. Construct a 98% confidence interval for the difference between the proportions of all sales at the two stores for which at least one item is returned.
- b. Using the 1% significance level, can you conclude that the proportions of all sales at the two stores for which at least one item is returned are different?
- *c. Find the p-value for the test mentioned in part b.

GLOSSARY

d The difference between two matched values in two samples collected from the same source. It is called the paired difference.

$\bar{d}$ The mean of the paired differences for a sample.

Independent samples   Two samples drawn from two populations such that the selection of one does not affect the selection of the other.

Paired or **matched samples**   Two samples drawn in such a way that they include the same elements and two data values are obtained from each element, one for each sample. Also called **dependent samples**.

μ_d The mean of the paired differences for the population.

s_d The standard deviation of the paired differences for a sample.

σ_d The standard deviation of the paired differences for the population.

KEY FORMULAS

1.   **Mean of the sampling distribution of $\bar{x}_1 - \bar{x}_2$**

$$\mu_{\bar{x}_1-\bar{x}_2} = \mu_1 - \mu_2$$

2.   **Standard deviation of $\bar{x}_1 - \bar{x}_2$ for two large and independent samples**

$$\sigma_{\bar{x}_1-\bar{x}_2} = \sqrt{\frac{\sigma_1^2}{n_1} + \frac{\sigma_2^2}{n_2}}$$

3.   **The $(1 - \alpha)100\%$ confidence interval for $\mu_1 - \mu_2$ for two large and independent samples**

$$(\bar{x}_1 - \bar{x}_2) \pm z\sigma_{\bar{x}_1-\bar{x}_2}$$

If σ_1 and σ_2 are not known, then $\sigma_{\bar{x}_1-\bar{x}_2}$ is replaced by its point estimator $s_{\bar{x}_1-\bar{x}_2}$, which is calculated as

$$s_{\bar{x}_1-\bar{x}_2} = \sqrt{\frac{s_1^2}{n_1} + \frac{s_2^2}{n_2}}$$

4.   **Value of the test statistic z for $\bar{x}_1 - \bar{x}_2$ for two large and independent samples**

$$z = \frac{(\bar{x}_1 - \bar{x}_2) - (\mu_1 - \mu_2)}{\sigma_{\bar{x}_1-\bar{x}_2}}$$

If σ_1 and σ_2 are not known, then $\sigma_{\bar{x}_1-\bar{x}_2}$ is replaced by $s_{\bar{x}_1-\bar{x}_2}$.

5.   **Pooled standard deviation for two small and independent samples taken from two populations with equal standard deviations**

$$s_p = \sqrt{\frac{(n_1 - 1)s_1^2 + (n_2 - 1)s_2^2}{n_1 + n_2 - 2}}$$

6.   **Estimate of the standard deviation of $\bar{x}_1 - \bar{x}_2$ for two small and independent samples taken from two populations with equal standard deviations**

$$s_{\bar{x}_1-\bar{x}_2} = s_p \sqrt{\frac{1}{n_1} + \frac{1}{n_2}}$$

7. The $(1 - \alpha)100\%$ confidence interval for $\mu_1 - \mu_2$ for two small and independent samples taken from two normally distributed populations with equal standard deviations

$$(\bar{x}_1 - \bar{x}_2) \pm t s_{\bar{x}_1 - \bar{x}_2}$$

8. Value of the test statistic t for $\bar{x}_1 - \bar{x}_2$ for two small and independent samples taken from two normally distributed populations with equal standard deviations

$$t = \frac{(\bar{x}_1 - \bar{x}_2) - (\mu_1 - \mu_2)}{s_{\bar{x}_1 - \bar{x}_2}}$$

9. Degrees of freedom to make inferences about $\mu_1 - \mu_2$ for two small and independent samples taken from two normally distributed populations with unequal standard deviations

$$df = \frac{\left(\dfrac{s_1^2}{n_1} + \dfrac{s_2^2}{n_2}\right)^2}{\dfrac{\left(\dfrac{s_1^2}{n_1}\right)^2}{n_1 - 1} + \dfrac{\left(\dfrac{s_2^2}{n_2}\right)^2}{n_2 - 1}}$$

10. Estimate of the standard deviation of $\bar{x}_1 - \bar{x}_2$ for two small and independent samples taken from two normally distributed populations with unequal standard deviations

$$s_{\bar{x}_1 - \bar{x}_2} = \sqrt{\frac{s_1^2}{n_1} + \frac{s_2^2}{n_2}}$$

11. The $(1 - \alpha)100\%$ confidence interval for $\mu_1 - \mu_2$ for two small and independent samples taken from two normally distributed populations with unequal standard deviations

$$(\bar{x}_1 - \bar{x}_2) \pm t s_{\bar{x}_1 - \bar{x}_2}$$

12. Value of the test statistic t for $\bar{x}_1 - \bar{x}_2$ for two small and independent samples taken from two normally distributed populations with unequal standard deviations

$$t = \frac{(\bar{x}_1 - \bar{x}_2) - (\mu_1 - \mu_2)}{s_{\bar{x}_1 - \bar{x}_2}}$$

13. Sample mean for paired differences

$$\bar{d} = \frac{\Sigma d}{n}$$

14. Sample standard deviation for paired differences

$$s_d = \sqrt{\frac{\Sigma d^2 - \dfrac{(\Sigma d)^2}{n}}{n - 1}}$$

15. Mean and standard deviation of the sampling distribution of $\bar{d}$

$$\mu_{\bar{d}} = \mu_d \qquad \text{and} \qquad s_{\bar{d}} = \frac{s_d}{\sqrt{n}}$$

16. **The $(1 - \alpha)100\%$ confidence interval for μ_d**

$$\bar{d} \pm ts_{\bar{d}}$$

17. **Value of the test statistic t for $\bar{d}$**

$$t = \frac{\bar{d} - \mu_d}{s_{\bar{d}}}$$

18. **Mean of the sampling distribution of $\hat{p}_1 - \hat{p}_2$**

$$\mu_{\hat{p}_1-\hat{p}_2} = p_1 - p_2$$

19. **Estimate of the standard deviation of $\hat{p}_1 - \hat{p}_2$ for two large and independent samples**

$$s_{\hat{p}_1-\hat{p}_2} = \sqrt{\frac{\hat{p}_1\hat{q}_1}{n_1} + \frac{\hat{p}_2\hat{q}_2}{n_2}}$$

20. **The $(1 - \alpha)100\%$ confidence interval for $p_1 - p_2$ for two large and independent samples**

$$(\hat{p}_1 - \hat{p}_2) \pm zs_{\hat{p}_1-\hat{p}_2}$$

21. **Pooled sample proportion for two samples**

$$\bar{p} = \frac{x_1 + x_2}{n_1 + n_2} \quad \text{or} \quad \frac{n_1\hat{p}_1 + n_2\hat{p}_2}{n_1 + n_2}$$

22. **Estimator of the standard deviation of $\hat{p}_1 - \hat{p}_2$ using the pooled sample proportion**

$$s_{\hat{p}_1-\hat{p}_2} = \sqrt{\bar{p}\,\bar{q}\left(\frac{1}{n_1} + \frac{1}{n_2}\right)}$$

23. **Value of the test statistic z for $\hat{p}_1 - \hat{p}_2$ for large and independent samples**

$$z = \frac{(\hat{p}_1 - \hat{p}_2) - (p_1 - p_2)}{s_{\hat{p}_1-\hat{p}_2}}$$

SUPPLEMENTARY EXERCISES

10.79 According to a 1995 survey by Nielsen Media Research, American women spend an average of 5 hours per day and American men spend an average of 4.28 hours per day watching television (*USA Today*, August 21, 1995). Assume that these results are based on random samples of 500 women and 500 men and that the sample standard deviation was 1.1 hours for women and 1.2 hours for men.

    a. Construct a 99% confidence interval for the difference between the mean time spent watching television per day by all women and by all men in the United States.

    b. Using the 1% significance level, can you conclude that the mean time spent watching television per day is higher for women than for men in the United States?

    c. What will your decision be in part b if the probability of making a Type I error is zero? Explain.

10.80 During the second quarter (April through June) of 1995, about 1657 mass layoff events occurred in the United States (*U.S. Bureau of Labor Statistics News*, March 22, 1996). A *mass layoff* event involves a layoff of at least 50 workers by one employer. Layoffs that occurred due to a change in business ownership averaged 356 workers laid off per event, while those caused by financial difficulties averaged 282 workers laid off per event. Suppose that these means are based on random samples of 30 mass layoffs caused by ownership changes and 45 caused by financial difficulties.

Assume that the sample standard deviations in cases of ownership changes and financial difficulties were 90 and 75, respectively.

a. Construct a 99% confidence interval for the difference between the two population means.
b. Test at the 5% significance level if the mean number of workers laid off in all mass layoffs due to ownership changes is different from the mean number of workers laid off in all mass layoff cases involving financial difficulty.
c. What will your decision be in part b if the probability of making a Type I error is zero? Explain.

10.81 A consulting agency was asked by a large insurance company to investigate if business majors were better salespersons. A sample of 40 salespersons with a business degree showed that they sold an average of 10 insurance policies per week with a standard deviation of 1.80 policies. Another sample of 45 salespersons with a degree other than business showed that they sold an average of 8.5 insurance policies per week with a standard deviation of 1.35 policies.

a. Construct a 99% confidence interval for the difference between the two population means.
b. Using the 1% significance level, can you conclude that persons with a business degree are better salespersons than those who have a degree in another area?

10.82 According to an estimate, the average earnings of female workers who are not union members are $348 per week and those of female workers who are union members are $467 per week. Suppose that these average earnings are calculated based on random samples of 1500 female workers who are not union members and 2000 female workers who are union members. Further assume that the standard deviations for these two samples are $30 and $35, respectively.

a. Construct a 95% confidence interval for the difference between the two population means.
b. Test at the 2.5% significance level if the mean weekly earnings of female workers who are not union members are less than those of female workers who are union members.

10.83 A researcher wants to test if the mean GPAs (grade point averages) of all male and all female college students who actively participate in sports are different. She took a random sample of 28 male students and 24 female students who are actively involved in sports. She found the mean GPAs of the two groups to be 2.62 and 2.74, respectively, with the corresponding standard deviations equal to .43 and .38.

a. Test at the 5% significance level if the mean GPAs of the two populations are different.
b. Construct a 90% confidence interval for the difference between the two population means.

Assume that the GPAs of all male and all female student athletes both have a normal distribution with the same standard deviation.

10.84 In 1995, Opinion Research Corporation conducted a survey for Teledyne Water Pik to study the amount of time Americans spend in showers (*USA Today*, July 11, 1995). The survey showed that persons in the 18- to 24-year age group spend an average of 16.4 minutes in the shower, while persons aged 45 to 54 spend an average of 10.4 minutes. Suppose that these results are based on random samples of 25 individuals in each age group and that the sample standard deviations for the time spent in showers are 5.0 minutes for the 18- to 24-year-olds and 3.7 minutes for the 45- to 54-year-olds. Assume that the two populations are normally distributed with equal standard deviations.

a. Construct a 90% confidence interval for the difference in the mean time spent in showers by all persons in these two age groups.
b. Test at the 5% significance level if the mean time spent in showers by all persons in the two age groups are different.

10.85 An agency wanted to estimate the difference between the auto insurance premiums paid by drivers insured with two different insurance companies. A random sample of 25 drivers insured with insurance company A showed that they paid an average monthly insurance premium of $83 with a standard deviation of $14. Another random sample of 20 drivers insured with insurance company B showed that these drivers paid an average monthly insurance premium of $76 with a standard deviation of $12. Assume that the insurance premiums paid by all drivers insured with companies A and B are both normally distributed with equal standard deviations.

a. Construct a 99% confidence interval for the difference between the two population means.
b. Test at the 1% significance level if the mean monthly insurance premium paid by drivers insured with company A is higher than that of drivers insured with company B.

10.86 A random sample of 28 children selected from families with only one child gave a mean tolerance level of 2.4 (on a scale of 1 to 8) with a standard deviation of .62. Another random sample of 25 children selected from families with more than one child gave a mean tolerance level of 3.5 with a standard deviation of .47.

a. Construct a 99% confidence interval for the difference between the two population means.
b. Test at the 5% significance level if the mean tolerance level for children from families with only one child is lower than that for children from families with more than one child.

Assume that the tolerance levels for all children in both groups have normal distributions with the same standard deviation.

10.87 Repeat Exercise 10.83, but now assume that the GPAs of all male and all female student athletes are both normally distributed with unequal standard deviations.

10.88 Repeat Exercise 10.84, but now assume that the two populations have normal distributions with unequal standard deviations.

10.89 Repeat Exercise 10.85, but now assume that the insurance premiums paid by all drivers insured with companies A and B both are normally distributed with unequal standard deviations.

10.90 Repeat Exercise 10.86, but now assume that the tolerance levels for all children in both groups are normally distributed with unequal standard deviations.

10.91 A random sample of eight students was selected to test for the effectiveness of hypnosis on their academic performances. The following table gives the GPAs for the semester before and the semester after the students tried hypnosis.

Before	2.3	2.8	3.1	2.7	3.4	2.6	2.8	2.5
After	2.6	3.2	3.0	3.5	3.7	2.4	2.9	2.9

a. Construct a 99% confidence interval for the mean μ_d of the population paired differences where a paired difference is defined as the difference between the GPA of a student before and after trying hypnosis.
b. Using a 2.5% significance level, can you conclude that the academic performance of all students improves due to hypnotism?

Assume that the population of paired differences is (approximately) normally distributed.

10.92 A random sample of nine students was selected to test for the effectiveness of a special course designed to improve memory. The following table gives the results of a memory test given to these students before and after this course.

Before	43	57	48	65	71	49	38	69	58
After	49	56	55	77	79	57	36	64	69

a. Construct a 95% confidence interval for the mean μ_d of the population paired differences where a paired difference is defined as the difference between the memory test scores of a student before and after attending this course.
b. Test at the 1% significance level if this course makes any statistically significant improvement in the memory of all students.

Assume that the population of the paired differences has a normal distribution.

10.93 According to a survey of 1000 men and 900 women, 21% of men and 28% of women read for fun almost every day.

    a. Construct a 95% confidence interval for the difference between the proportions of all men and all women who read for fun every day.

    b. Test at the 2% significance level if the proportions of all men and all women who read for fun every day are different.

10.94 The U.S. Bureau of Justice Statistics conducts periodic surveys to estimate the percentages of various types of crimes that are reported to the police (*Bureau of Justice Statistics Bulletin*, April 1996). In 1994, the bureau estimated that 66.7% of robberies in which the victims were injured were reported to the police and 63.1% of robberies in which the victims were not injured were reported to the police. Suppose that these estimates are based on 700 robberies in which the victims were injured and 1100 in which the victims were not injured.

    a. Construct a 99% confidence interval for the difference between the two population proportions.

    b. Test at the 2.5% significance level if the proportion of robberies reported to police in which the victims were injured is higher than the proportion of robberies reported to police in which the victims were not injured.

10.95 A 1995 survey by *Leisure Trends* magazine found that 44% of men and 31% of women had more than four hours of leisure time per day (*USA Today*, April 10, 1995). Suppose that these percentages are based on random samples of 395 men and 410 women.

    a. Find a 95% confidence interval for the difference between the two population proportions.

    b. At the 1% significance level, can you conclude that the percentages of all men and of all women who have more than four hours of leisure time per day are different?

10.96 According to the U.S. Bureau of Labor Statistics, 14.3% of employed female wage and salary workers aged 16 years and over were represented by labor unions in 1995 (*Bureau of Labor Statistics News*, February 9, 1996). The corresponding percentage for men was 18.8% for the same year. Suppose that these percentages are based on random samples of 1200 female and 1100 male wage and salary workers.

    a. Construct a 97% confidence interval for the difference between the true proportions of such female and male workers represented by unions in 1995.

    b. Using the 1% significance level, can you conclude that the proportion of all such female workers represented by unions in 1995 was less than the proportion of all such male workers represented by unions?

∗10.97 In a recent study on the effectiveness of flu vaccination in healthy, working adults, the authors randomly assigned 409 such adults to receive the influenza vaccine and 416 to receive placebo injections, hereafter called the nonvaccinated adults (K. L. Nichol et al., ''The Effectiveness of Vaccination against Influenza in Healthy, Working Adults,'' *The New England Journal of Medicine* 333[14], October 5, 1995). These adults were monitored for the flu season from December 1, 1994, to March 31, 1995. The total number of episodes of upper respiratory illness for each group was determined and converted to the number of episodes per 100 adults. Let μ_1 be the mean number of episodes per 100 for all potential nonvaccinated adults and μ_2 the mean number of episodes per 100 for all potential vaccine recipients. The experiment yielded a 95% confidence interval for $\mu_1 - \mu_2$ of 17 to 53.

    a. Find $\bar{x}_1 - \bar{x}_2$ for this example.

    b. Find a 99% confidence interval for $\mu_1 - \mu_2$ for this example.

∗10.98 Manufacturers of two competing automobile models, Gofer and Diplomat, each claim to have the lowest mean fuel consumption. Let μ_1 be the mean fuel consumption in miles per gallon (mpg) for the Gofer and μ_2 the mean fuel consumption in mpg for the Diplomat. The two manufacturers have agreed to a test in which several cars of each model will be driven on a 100-mile test run. Then the fuel consumption, in mpg, will be calculated for each test run. The average of the mpg for all 100-mile test runs for each model gives the corresponding mean. Assume that for each model the gas mileages for the test runs are normally distributed with $\sigma = 2$ mpg. Note that each car is driven for one and only one 100-mile test run.

    a. How many cars (i.e., sample size) for each model are required to estimate $\mu_1 - \mu_2$ with a 90% confidence level and with a maximum error of estimate of 1.5 mpg? Use the same number of cars (i.e., sample size) for each model.

b. If μ_1 is actually 33 mpg and μ_2 is actually 30 mpg, what is the probability that five cars for each model would yield $\bar{x}_1 \geq \bar{x}_2$?

*10.99   Maria and Ellen both specialize in throwing the javelin. Maria throws the javelin a mean distance of 200 feet with a standard deviation of 10 feet, whereas Ellen throws the javelin a mean distance of 210 feet with a standard deviation of 12 feet. Assume that the distances each of these athletes throws the javelin are normally distributed with these means and standard deviations. If Maria and Ellen each throw the javelin once, what is the probability that Maria's throw is longer than Ellen's?

*10.100   A new type of sleeping pill is tested against an older standard pill. Two thousand insomniacs are randomly divided into two equal groups. The first group is given the old standard pill, while the second group receives the new pill. The time required to fall asleep after the pill is administered is recorded for each person. The results of the experiment are given in the following table, where $\bar{x}$ and s represent the mean and standard deviation for the times required to fall asleep for people in each group after the pill is taken.

	Group 1 (Old Pill)	Group 2 (New Pill)
n	1000	1000
$\bar{x}$	15.4 minutes	15.0 minutes
s	3.5 minutes	3.0 minutes

Consider the test of hypothesis H_0: $\mu_1 - \mu_2 = 0$ versus H_1: $\mu_1 - \mu_2 > 0$, where μ_1 and μ_2 are the mean times required for all potential users to fall asleep using the old pill and the new pill, respectively.

a.   Find the p-value for this test.

b.   Does your answer to part a indicate that the result is statistically significant?

c.   Find the 95% confidence interval for $\mu_1 - \mu_2$.

d.   Does your answer to part c imply that this result is of great *practical* significance?

*10.101   In a study released in 1996, Karen Allen of the University of Buffalo Medical School worked with 240 couples who owned dogs (*USA Today*, March 3, 1996). These couples were assigned stressful tasks and then monitored for physical signs of stress. When the subjects worked math problems in the presence of their spouses, the average increase in pulse rate was 37 beats per minute. However, when the same subjects worked the same math problems with only the dog present, the average increase in pulse rate was just 7 beats per minute. Note that the samples are paired since each person is measured twice, once with the spouse and once with the dog present (not necessarily in that order). Assume that $s_d = 18$ beats per minute and note that $n = 480$.

a.   Find a 99% confidence interval for the difference between the mean increase in the pulse rates for all such couples with the spouse and with the dog present.

b.   At the 1% level of significance, can you conclude that the mean increase in the pulse rate is higher with the spouse present than with the dog present for all such couples?

*10.102   Gamma Corporation is considering the installation of governors on cars driven by its sales staff. These devices would limit the car speeds to a preset level, which is expected to improve fuel economy. The company is planning to test several cars for fuel consumption without governors for one week. Then governors would be installed in the same cars, and fuel consumption will be monitored for another week. Gamma Corporation wants to estimate the mean difference in fuel consumption with a maximum error of estimate of 2 mpg with a 90% confidence level. Assume that the differences in fuel consumption are normally distributed and that previous studies suggest that an estimate of $s_d = 3$ mpg is reasonable. How many cars should be tested? (Note that the critical value of t will depend on n, so it will be necessary to use trial and error.)

*10.103   Refer to Exercise 10.102. Suppose the Gamma Corporation decides to test governors on seven cars. However, the management is afraid that the speed limit imposed by the governors will reduce the number of contacts the sales persons can make each day. Thus, both the fuel consumption and the number of contacts made are recorded for each car/salesperson for each week of the testing period, both before and after the installation of governors.

Salesperson	Number of Contacts		Fuel Consumption (mpg)	
	Before	After	Before	After
A	50	49	25	26
B	63	60	21	24
C	42	47	27	26
D	55	51	23	25
E	44	50	19	24
F	65	60	18	22
G	66	58	20	23

Suppose that as a statistical analyst with the company you are directed to prepare a brief report that includes statistical analysis and interpretation of the data. Management will use your report to help it decide whether or not to install governors on all salespersons' cars. Use the 90% confidence intervals and .05 significance levels for any hypothesis tests to make suggestions. Assume that the differences in fuel consumption and the differences in the number of contacts are both normally distributed.

*10.104　Two competing airlines, Alpha and Beta, fly a route between Des Moines, Iowa, and Wichita, Kansas. Each airline claims to have a lower percentage of flights that arrive late. Let p_1 be the proportion of Alpha's flights that arrive late and p_2 the proportion of Beta's flights that arrive late.

　　a.　You are asked to observe a random sample of arrivals for each airline to estimate $p_1 - p_2$ with a 90% confidence level and a maximum error of estimate of .05. How many arrivals for each airline would you have to observe? (Assume that you will observe the same number of arrivals, n, for each airline. To be sure of taking a large enough sample, use $p_1 = p_2 = .50$ in your calculations for n.)

　　b.　Suppose that p_1 is actually .30 and p_2 is actually .23. What is the probability that a sample of 100 flights for each airline (200 in all) would yield $\hat{p}_1 \geq \hat{p}_2$?

SELF-REVIEW TEST

1.　To test the hypothesis that the mean blood pressure of university professors is lower than that of company executives, which of the following would you use?

　　a.　A left-tailed test　　b.　A two-tailed test　　c.　A right-tailed test

2.　Briefly explain the meaning of independent and dependent samples. Give one example of each of these cases.

3.　A company psychologist wanted to test if company executives have job-related stress scores higher than those of university professors. He took a sample of 40 executives and 50 professors and tested them for job-related stress. The sample of 40 executives gave a mean stress score of 7.6 with a standard deviation of .8. The sample of 50 professors produced a mean stress score of 5.4 with a standard deviation of 1.3.

　　a.　Construct a 99% confidence interval for the difference between the mean stress scores of all executives and all professors.

　　b.　Test at the 2.5% significance level if the mean stress score of all executives is higher than that of all professors.

4. A sample of 20 alcoholic fathers showed that they spend an average of 2.3 hours per week playing with their children with a standard deviation of .54 hours. A sample of 25 nonalcoholic fathers gave a mean of 4.6 hours per week with a standard deviation of .8 hours.

   a. Construct a 95% confidence interval for the difference between the mean time spent per week playing with their children by all alcoholic and all nonalcoholic fathers.

   b. Test at the 1% significance level if the mean time spent per week playing with their children by all alcoholic fathers is less than that of nonalcoholic fathers.

Assume that the times spent per week playing with their children by all alcoholic and all nonalcoholic fathers both are normally distributed with equal but unknown standard deviations.

5. Repeat Problem 4 assuming that the times spent per week playing with their chidren by all alcoholic and all nonalcoholic fathers both are normally distributed with unequal and unknown standard deviations.

6. The following table gives the number of items made in one hour by seven randomly selected workers on two different machines.

Worker	1	2	3	4	5	6	7
Machine I	15	18	14	20	16	18	21
Machine II	16	20	13	23	19	18	20

   a. Construct a 99% confidence interval for the mean μ_d of the population paired differences where a paired difference is equal to the number of items made by an employee in one hour on Machine I minus the number of items made by the same employee in one hour on Machine II.

   b. Test at the 5% significance level if the mean μ_d of the population paired differences is different from zero.

Assume that the population of paired differences is (approximately) normally distributed.

7. A sample of 500 male registered voters showed that 57% of them voted in the last presidential election. Another sample of 400 female registered voters showed that 55% of them voted in the same election.

   a. Construct a 97% confidence interval for the difference between the proportion of all male and all female registered voters who voted in the last presidential election.

   b. Test at the 1% significance level if the proportion of all male voters who voted in the last presidential election is different from that of all female voters.

USING MINITAB

INFERENCES ABOUT THE DIFFERENCE BETWEEN TWO POPULATION MEANS FOR LARGE AND INDEPENDENT SAMPLES

MINITAB does not contain a procedure that can be used to make a confidence interval and a test of hypothesis about the difference between two population means for large and independent samples using the normal distribution. The simplest way to make such inferences with MINITAB is to use the t distribution irrespective of the sample sizes. This procedure is explained next. MINITAB likewise does not have any procedure to make a confidence interval and test a hypothesis about the difference between two population proportions for large and independent samples using the normal distribution. Hence, this procedure is not discussed in this MINITAB section.

INFERENCES ABOUT THE DIFFERENCE BETWEEN TWO POPULATION MEANS FOR SMALL AND INDEPENDENT SAMPLES: EQUAL POPULATION STANDARD DEVIATIONS

Illustration M10–1 shows how to use MINITAB to make a confidence interval and test a hypothesis for the difference between two population means for small and independent samples when the two population standard deviations are equal and the two populations are normally distributed. Note that we can also use this procedure to make a confidence interval and test a hypothesis for the difference between two population means for large and independent samples. In such a case, we do not have to assume that the two populations are normally distributed and that the two population standard deviations are equal.

ILLUSTRATION M10–1   An insurance company wanted to find the difference between the mean speeds of cars driven by all men drivers and by all women drivers on a highway. A random sample of 16 men who were driving on this highway produced the following data on the speeds of their cars at the time of the survey.

$$
\begin{array}{cccccccc}
70 & 67 & 65 & 72 & 71 & 54 & 74 & 69 \\
63 & 57 & 64 & 76 & 60 & 55 & 63 & 69
\end{array}
$$

Another random sample of 14 women who were driving on the same highway produced the following data on the speeds of their cars at the time of the survey.

$$
\begin{array}{ccccccc}
61 & 55 & 58 & 66 & 70 & 54 & 57 \\
60 & 63 & 72 & 65 & 63 & 59 & 67
\end{array}
$$

Construct a 99% confidence interval for the difference between the mean speeds of cars driven by all men drivers and by all women drivers on this highway. Also, test at the 1% significance level if the mean speed of cars driven by all men drivers on this highway is greater than that of cars driven by all women drivers. Assume that the speeds at which all men and all women drive cars on this highway are normally distributed with equal but unknown population standard deviations.

Solution Let μ_1 and μ_2 be the mean speeds of cars driven by all men and all women, respectively, on this highway, and let $\bar{x}_1$ and $\bar{x}_2$ be the means of the respective samples.

You are to make a 99% confidence interval for $\mu_1 - \mu_2$ and test at the 1% significance level whether or not μ_1 is greater than μ_2. Thus, the null and alternative hypotheses are

$$H_0: \mu_1 - \mu_2 = 0$$

$$H_1: \mu_1 - \mu_2 > 0$$

The significance level is .01 and the test is right-tailed.

If you are using MINITAB FOR WINDOWS, perform the following steps to make the required confidence interval and test of hypothesis.

Step 1. Enter the given data in columns C1 and C2 of the Data window. Suppose C1 contains data on men drivers and C2 on women drivers.
Step 2. Click the **Stat** pull-down menu at the top of the screen.
Step 3. Click **Basic Statistics** from the selections available in the **Stat** menu.
Step 4. Click **2-Sample t** from the selections available in the **Basic Statistics** menu.
Step 5. You will see a dialog box entitled **2-Sample t** appear on the screen. Click inside the circle next to **Samples in different columns**. Enter **C1** in the box next to **First** and **C2** in the box next to **Second**. Click the arrow in the corner of the box next to **Alternative**. You will see three alternatives appear on the screen. Since the test is right-tailed, click the **greater than** alternative. Then enter **99** (which is the confidence level) in the box next to **Confidence level**. Click inside the box next to **Assume equal variances**.
Step 6. Click the **OK** button at the bottom of this dialog box. The MINITAB output will appear on the screen.

If you are using the MINITAB COMMAND LANGUAGE, first enter the given data on speeds into columns C1 and C2 using the **SET** command and then type the following MINITAB commands to make the required test of hypothesis about $\mu_1 - \mu_2$.

```
MTB  > TWOSAMPLE T C1 C2;
SUBC > POOLED;
SUBC > ALTERNATIVE = 1.
```

Here, in the first command, *TWOSAMPLE T C1 C2* instructs MINITAB that you are to make a test of hypothesis about the difference between the two population means using the *t* distribution for the data from two samples entered in columns C1 and C2. The subcommand *POOLED* indicates that the test is to be made using the pooled standard deviation (see Section 10.2) assuming that the standard deviations of the two populations are equal. The subcommand *ALTERNATIVE = 1* tells MINITAB that it is a right-tailed test. Remember that *ALTERNATIVE* will be equal to -1, 0, or 1 depending on whether the alternative hypothesis is $\mu_1 - \mu_2 < 0$, $\mu_1 - \mu_2 \neq 0$, or $\mu_1 - \mu_2 > 0$, respectively. Note that when you make a test of hypothesis about $\mu_1 - \mu_2$ using the MINITAB COMMAND LANGUAGE, you will also obtain a 95% confidence interval for $\mu_1 - \mu_2$.

The following MINITAB commands will give you a 99% confidence interval for $\mu_1 - \mu_2$.

```
MTB  > TWOSAMPLE 99% CONFIDENCE INTERVAL C1 C2;
SUBC > POOLED.
```

Here, the first command instructs MINITAB that you are to make a 99% confidence interval for the difference between the two population means using the data on two samples entered in columns C1 and C2. The subcommand *POOLED* instructs MINITAB that the standard deviations of the two populations are equal. In response to these commands, MINITAB will not only give a 99% confidence interval for $\mu_1 - \mu_2$ but it will also give the results for a test of hypothesis (using the t distribution) for H_0: $\mu_1 - \mu_2 = 0$ against H_1: $\mu_1 - \mu_2 \neq 0$. This test of hypothesis is based on the assumption that the standard deviations of the two populations are equal.

When you use MINITAB FOR WINDOWS to make a test of hypothesis and a confidence interval for $\mu_1 - \mu_2$, you will obtain the output given in Figure 10.11. Note that the MINITAB COMMAND LANGUAGE procedures will produce a slightly different looking output.

Figure 10.11   Test of hypothesis and a confidence interval for $\mu_1 - \mu_2$.

Two Sample T-Test and Confidence Interval

```
Two sample T for C1 vs C2
      N    Mean   StDev   SE Mean
C1   16   65.56    6.64     1.7
C2   14   62.14    5.43     1.5


99% CI for mu C1 − mu C2: ( −2.8, 9.6)
T-Test mu C1 = mu C2 (vs >): T= 1.53   P= 0.069   DF= 28
Both use Pooled StDev = 6.11
```

In Figure 10.11, *T-Test mu C1 = mu C2 (vs >)* indicates that it is a test about H_0: $\mu_1 - \mu_2 = 0$ against H_1: $\mu_1 - \mu_2 > 0$ using the t distribution. Also, from this figure,

The row of C1 gives: $n_1 = 16$, $\bar{x}_1 = 65.56$, $s_1 = 6.64$, and $s_{\bar{x}_1} = 1.7$

The row of C2 gives: $n_2 = 14$, $\bar{x}_2 = 62.14$, $s_2 = 5.43$, and $s_{\bar{x}_2} = 1.5$

The pooled standard deviation is: $s_p = 6.11$

The value of the test statistic, t, for $\bar{x}_1 - \bar{x}_2$ is: $t = 1.53$

The p-value is: $p = .069$

Degrees of freedom = DF = 28

The test is right-tailed. The significance level is given to be 1%.

From the t distribution table, the critical value of t for $df = 28$ and .01 area in the right tail is 2.467. Because the value of the test statistic, $t = 1.53$, is less than the critical value of $t = 2.467$, it falls in the nonrejection region. (The reader should draw a graph showing the rejection and nonrejection regions.) Consequently, the null hypothesis is not rejected.

You can reach the same conclusion using the p-value printed in the MINITAB solution. From the MINITAB solution, the p-value is .069. Since the value of $\alpha = .01$ is less than the p-value of .069, you fail to reject the null hypothesis.

The following portion of the MINITAB solution in Figure 10.11 gives the 99% confidence interval for $\mu_1 - \mu_2$.

99% CI for mu C1 − mu C2: (−2.8, 9.6)

From this, the 99% confidence interval for $\mu_1 - \mu_2$ is -2.8 to 9.6.

INFERENCES ABOUT THE DIFFERENCE BETWEEN TWO POPULATION MEANS FOR SMALL AND INDEPENDENT SAMPLES: UNEQUAL POPULATION STANDARD DEVIATIONS

The procedure to make a confidence interval and to test a hypothesis about $\mu_1 - \mu_2$ for small and independent samples when the standard deviations of the two populations are unknown and unequal is similar to the one used earlier in this section for the case of equal population standard deviations with one exception: we **do not click inside the box next to Assume equal variances** in MINITAB FOR WINDOWS and **do not include the subcommand POOLED** in the MINITAB COMMAND LANGUAGE of the previous section. Illustration M10–2 describes this procedure.

ILLUSTRATION M10-2 Reconsider the data on the speeds of cars of 16 men and 14 women drivers driving on a highway given in Illustration M10-1. Construct a 99% confidence interval for the difference between the mean speeds of cars driven by all men drivers and by all women drivers on this highway. Also, test at the 1% significance level if the mean speed of cars driven by all men drivers on this highway is greater than that of cars driven by all women drivers. Assume that the speeds at which all men and all women drive cars on this highway are normally distributed with unequal and unknown population standard deviations.

Solution Let μ_1 and μ_2 be the mean speeds of cars driven by all men and all women on this highway, and let $\bar{x}_1$ and $\bar{x}_2$ be the means of the respective samples. You are to make a 99% confidence interval for $\mu_1 - \mu_2$ and test at the 1% significance level whether or not μ_1 is greater than μ_2. Thus, the null and alternative hypotheses are

$$H_0: \mu_1 - \mu_2 = 0$$

$$H_1: \mu_1 - \mu_2 > 0$$

The significance level is .01 and the test is right-tailed.

If you are using MINITAB FOR WINDOWS, perform the following steps to make the required confidence interval and test of hypothesis.

Step 1. Enter the given data in columns C1 and C2 of the Data window.
Step 2. Click the **Stat** pull-down menu at the top of the screen.
Step 3. Click **Basic Statistics** from the selections available in the **Stat** menu.
Step 4. Click **2-Sample t** from the selections available in the **Basic Statistics** menu.
Step 5. You will see a dialog box entitled **2-Sample t** appear on the screen. Click inside the circle next to **Samples in different columns**. Enter **C1** in the box next to **First** and **C2** in the box next to **Second**. Click the arrow in the corner of the box next to **Alternative**. You will see three alternatives appear on the screen. As the test is right-tailed, click the **greater than** alternative. Then enter **99** (which is the confidence level) in the box next to **Confidence level**. Make sure the box next to **Assume equal variances** is **not checked**. If this box contains an X, click inside it to remove this X.
Step 6. Click the **OK** button at the bottom of this dialog box. The MINITAB output will appear on the screen.

If you are using the MINITAB COMMAND LANGUAGE, first enter the given data on speeds into columns C1 and C2 using the **SET** command and then type the following MINITAB commands to make the required test of hypothesis about $\mu_1 - \mu_2$.

```
MTB > TWOSAMPLE T C1 C2;
SUBC > ALTERNATIVE = 1.
```

The following MINITAB command will give you a 99% confidence interval for $\mu_1 - \mu_2$.

```
MTB > TWOSAMPLE 99% CONFIDENCE INTERVAL C1 C2
```

In response to the MINITAB FOR WINDOWS steps mentioned earlier, you will obtain the MINITAB output given in Figure 10.12. Note that the MINITAB COMMAND LANGUAGE procedures will produce a slightly different looking output.

Figure 10.12 Test of hypothesis and a confidence interval for $\mu_1 - \mu_2$.

```
Two Sample T-Test and Confidence Interval

Two sample T for C1 vs C2
     N    Mean   StDev   SE Mean
C1  16   65.56    6.64      1.7
C2  14   62.14    5.43      1.5


99% CI for mu C1 - mu C2: ( -2.7,  9.5)
T-Test mu C1 = mu C2  (vs >):  T= 1.55  P= 0.066  DF= 27
```

From the MINITAB output given in Figure 10.12, the value of the test statistic for $\bar{x}_1 - \bar{x}_2$ is $t = 1.55$. The p-value is .066.

As shown in the MINITAB solution of Figure 10.12, the degrees of freedom for the test are 27. The degrees of freedom are calculated using the formula given in Section 10.3 of this chapter. The test is right-tailed with $\alpha = .01$. From the t distribution table, the critical value of t for $df = 27$ and .01 area in the right tail is 2.473. Since the value of the test statistic, $t = 1.55$, is smaller than the critical value of $t = 2.473$, it falls in the nonrejection region. Consequently, the null hypothesis is not rejected.

You can reach the same conclusion using the p-value. From the MINITAB solution of Figure 10.12, the p-value is .066. Because the value of $\alpha = .01$ is smaller than the p-value of .066, the null hypothesis is not rejected.

The following portion of the MINITAB solution in Figure 10.12 gives the 99% confidence interval for $\mu_1 - \mu_2$.

99% CI for mu C1 − mu C2: (−2.7, 9.5)

From this, the 99% confidence interval for $\mu_1 - \mu_2$ is −2.7 to 9.5. ■

INFERENCES ABOUT THE DIFFERENCE BETWEEN TWO POPULATION MEANS FOR PAIRED SAMPLES

To make a confidence interval and to test a hypothesis about the difference between two population means for paired samples, we first enter the given data in columns C1 and C2. Then we take the difference between the corresponding data values of columns C1 and C2 and put them in column C3. The values recorded in column C3 are the paired differences denoted by d. Then, to make a confidence interval and to test a hypothesis about μ_d, we use the MINITAB procedures explained in the MINITAB sections of Chapters 8 and 9. Illustrations M10–3 and M10–4 explain how to make a confidence interval and test a hypothesis about μ_d.

ILLUSTRATION M10–3   Recall Example 10–12. A researcher wanted to find the effect of a special diet on systolic blood pressure. She selected a sample of seven adults and put them on this dietary program for three months. The following table gives the systolic blood pressures of these seven adults before and after the completion of the program.

Before	210	180	195	220	231	199	224
After	193	186	186	223	220	183	233

Construct a 95% confidence interval for μ_d where μ_d is the mean reduction in the systolic blood pressure due to this special dietary program for the population of all adults. Assume that the population of paired differences has a normal distribution.

Solution   Let d be the reduction in the systolic blood pressure of a person due to this special dietary program. Then,

$$d = \text{Systolic blood pressure before} - \text{Systolic blood pressure after}$$

Thus, the differences in the two blood pressures will be obtained by subtracting C2 from C1.

If you are using MINITAB FOR WINDOWS, perform the following steps to find the 95% confidence interval for μ_d.

Step 1. Enter the given data in columns C1 and C2 of the Data window. Note that C1 will contain the blood pressure before, and C2 will contain the blood pressure after the completion of the program.

Step 2. Calculate the difference C1–C2 and store it in C3. To do so, click the **Calc** pull-down menu at the top of the screen, then click **Matrices** from the selections available in the **Calc** menu, then click **Arithmetic** from the selections available in the **Matrices** menu. You will see a dialog box entitled **Matrix Arithmetic** appear on the screen. Click inside the circle next to **Subtract**. Type **C2** in the box next to it and **C1** in the box next to **from**. Then type C3 in the box next to **Store result in.** Click the **OK** button at the bottom of this dialog box. You will see the result appear in column C3 of the Data window. Alternatively, you can calculate the difference C1–C2 yourself and then enter those differences in C3.

Step 3. Click the **Stat** pull-down menu at the top of the screen.

Step 4. Click **Basic Statistics** from the selections available in the **Stat** menu.

Step 5. Click **1-Sample t** from the selections available in the **Basic Statistics** menu.

Step 6. You will see a dialog box entitled **1-Sample t** appear on the screen. Type **C3** in the box below **Variables**. Click inside the circle next to **Confidence interval** and type **95** in the box next to **Level**.

Step 7. Click the **OK** button at the bottom of this dialog box. The MINITAB output will appear on the screen.

If you are using the MINITAB COMMAND LANGUAGE, first enter the given data into columns C1 and C2 using the **SET** command. Then type the following MINITAB commands.

```
MTB > LET C3 = C1 − C2
MTB > TINTERVAL 95% C3
```

Here, the first MINITAB command calculates the difference between columns C1 and C2 and stores it in column C3. The second command instructs MINITAB to make a 95% confidence interval for μ_d using the t distribution and the value of $\overline{d}$ calculated for the data of column C3.

Whether you use MINITAB FOR WINDOWS or the MINITAB COMMAND LANGUAGE, you will obtain the output given in Figure 10.13.

Figure 10.13 Confidence interval for μ_d for small sample.

```
Confidence Intervals

Variable   N   Mean   StDev   SE Mean      95.0% CI
C3         7   5.00   10.79     4.08    ( −4.98,  14.98)
```

From Figure 10.13, $n = 7$, $\overline{d} = 5.00$, $s_d = 10.79$, $s_{\overline{d}} = 4.08$, and the 95% confidence interval for μ_d is -4.98 to 14.98.

ILLUSTRATION M10-4 Refer to Illustration M10–3. Test at the 5% significance level if the mean of the population paired differences, μ_d, is different from zero.

Solution Here, the null and alternative hypotheses are

$$H_0: \mu_d = 0$$

$$H_1: \mu_d \neq 0$$

The significance level is .05, and the test is two-tailed.

If you are using MINITAB FOR WINDOWS, perform the following steps to make a test of hypothesis about μ_d.

Step 1. Enter the given data in columns C1 and C2 of the Data window. Note that C1 will contain the blood pressure before, and C2 will contain the blood pressure after.

Step 2. Calculate the difference C1–C2 and put it in C3. To do so, click the **Calc** pull-down menu at the top of the screen, then click **Matrices** from the selections available in the **Calc** menu, then click **Arithmetic** from the selections available in the **Matrices** menu. You will see a dialog box entitled **Matrix Arithmetic** appear on the screen. Click inside the circle next to **Subtract**. Type **C2** in the box next to it and **C1** in the box next to **from**. Then type **C3** in the box next to **Store result in.** Click the **OK** button at the bottom of this dialog box. You will see the result appear in column C3 of the Data window. Alternatively, you can calculate the difference C1−C2 yourself and then enter those differences in C3.

Step 3.   Click the **Stat** pull-down menu at the top of the screen.

Step 4.   Click **Basic Statistics** from the selections available in the **Stat** menu.

Step 5.   Click **1-Sample t** from the selections available in the **Basic Statistics** menu.

Step 6.   You will see a dialog box entitled **1-Sample t** appear on the screen. Type **C3** in the box below **Variables**. Click inside the circle next to **Test mean** and enter 0 (which is the value of μ_d in H_0) in the box next to it. Click the arrow in the corner of the box next to **Alternative**. You will see three alternatives appear on the screen. Since the test is two-tailed, click the **not equal** alternative.

Step 7.   Click the **OK** button at the bottom of this dialog box. The MINITAB output will appear on the screen.

If you are using the MINITAB COMMAND LANGUAGE, first enter the given data into columns C1 and C2 using the **SET** command. Then type the following MINITAB commands.

```
MTB  > LET C3 = C1 − C2
MTB  > TTEST MEAN = 0 C3;
SUBC > ALTERNATIVE = 0.
```

Here, the first MINITAB command calculates the difference between columns C1 and C2 and stores it in column C3. The second command instructs MINITAB to make a test of hypothesis about μ_d, with the null hypothesis as H_0: $\mu_d = 0$, using the t distribution and the value of $\bar{d}$ calculated for the data of column C3. In the MINITAB subcommand, *ALTERNATIVE* = 0 tells MINITAB that it is a two-tailed test.

Whether you use MINITAB FOR WINDOWS or the MINITAB COMMAND LANGUAGE, you will obtain the output given in Figure 10.14.

Figure 10.14   MINITAB output for a test of hypothesis about μ_d for a small sample.

T-Test of the Mean

Test of mu = 0.00 vs mu not = 0.00

Variable	N	Mean	StDev	SE Mean	T	P
C3	7	5.00	10.79	4.08	1.23	0.27

In the MINITAB solution given in Figure 10.14, *Test of mu* = 0.00 *vs mu not* = 0.00 indicates that the null hypothesis is $\mu_d = 0$ and the alternative hypothesis is $\mu_d \neq 0$. From Figure 10.14, $n = 7$, $\bar{d} = 5.00$, $s_d = 10.79$, $s_{\bar{d}} = 4.08$, the observed value of the test statistic is $t = 1.23$, and the p-value is .27.

To make a decision, find the critical values of t from the t distribution table for a two-tailed test with $\alpha/2 = .025$ and $df = 7 - 1 = 6$. These values of t are -2.447 and 2.447. Because the value of the test statistic, $t = 1.23$, is between the two critical values, it falls in the nonrejection region. (The reader is advised to draw a graph showing the rejection and nonrejection regions.) Consequently, the null hypothesis is not rejected. Thus, based on this sample information, it is concluded that the mean of paired differences, μ_d, is equal to zero. In other words, it is concluded that the mean systolic blood pressure for all adults does not decrease due to this special dietary program.

You can reach the same conclusion by using the p-value. As $\alpha = .05$ is smaller than the p-value of .27, the null hypothesis is not rejected.   ■

COMPUTER ASSIGNMENTS

M10.1   A random sample of 13 male college students who hold jobs gave the following data on their GPAs.

3.12	2.84	2.43	2.15	3.92	2.45	2.73
3.06	2.36	1.93	2.81	3.27	1.83	

Another random sample of 16 female college students who also hold jobs gave the following data on their GPAs.

2.76	3.84	2.24	2.81	1.79	3.89	2.96	3.77
2.36	2.81	3.29	2.08	3.11	1.69	2.84	3.02

    a.   Using MINITAB, construct a 99% confidence interval for the difference between the mean GPAs of all male and all female college students who hold jobs.

    b.   Using MINITAB, test at the 5% significance level if the mean GPAs of all male and all female college students who hold jobs are different.

Assume that the GPAs of all such male and female college students are normally distributed with equal but unknown population standard deviations.

M10.2   A company recently opened two supermarkets in two different areas. The management wants to know if the mean sales per day for these two supermarkets are different. A sample of 10 days for Supermarket A produced the following data on daily sales (in thousand dollars).

47.56	57.66	51.23	58.29	43.71
49.33	52.35	50.13	47.45	53.86

A sample of 12 days for Supermarket B produced the following data on daily sales (in thousand dollars).

56.34	63.55	61.64	63.75	54.78	58.19
55.40	59.44	62.33	67.82	56.65	67.90

Assume that the daily sales of the two supermarkets are both normally distributed with equal but unknown standard deviations.

    a.   Using MINITAB, construct a 99% confidence interval for the difference between the mean daily sales for these two supermarkets.

    b.   Using MINITAB, test at the 1% significance level if the mean daily sales for these two supermarkets are different.

M10.3   Refer to Computer Assignment M10.1. Now do that assignment assuming the GPAs of all such male and female college students are normally distributed with unequal and unknown population standard deviations.

M10.4   Refer to Computer Assignment M10.2. Now do that assignment assuming the daily sales of the two supermarkets are both normally distributed with unequal and unknown standard deviations.

M10.5   Refer to Exercise 10.59. The manufacturer of a gasoline additive claims that the use of this additive increases gasoline mileage. A random sample of six cars was selected. These cars were driven for one week without the gasoline additive and then for one week with the gasoline additive. The following table gives the miles per gallon for these cars without and with the gasoline additive.

Without	24.6	28.3	18.9	23.7	15.4	29.5
With	26.3	31.7	18.2	25.3	18.3	30.9

a. Using MINITAB, construct a 99% confidence interval for the mean μ_d of the population paired differences.

b. Using MINITAB, test at the 1% significance level if the use of the gasoline additive increases the gasoline mileage.

Assume that the population of paired differences is (approximately) normally distributed.

M10.6   Refer to Exercise 10.60. A company is considering installing new machines to assemble its products. The company is considering two types of machines, but it will buy only one type. The company selected eight assembly workers and asked them to use these two types of machines to assemble products. The following table gives the time taken (in minutes) to assemble one unit of the product on each type of machine for each of these eight workers.

Machine I	23	26	19	24	26	22	20	18
Machine II	21	24	23	25	24	25	24	23

a. Using MINITAB, construct a 98% confidence interval for the mean μ_d of the population paired differences where a paired difference is equal to the time taken to assemble a unit of the product on Machine I minus the time taken to assemble a unit of the product on Machine II.

b. Using MINITAB, test at the 5% significance level if the mean time taken to assemble a unit of the product is different for the two types of machines.

Assume that the population of paired differences is (approximately) normally distributed.

11 CHI-SQUARE TESTS

he tests of hypotheses about the mean, the difference between two means, the proportion, and the difference between two proportions were discussed in Chapters 9 and 10. The tests about proportions dealt with countable or categorical data. In the case of a proportion and the difference between two proportions, the tests concerned experiments with only two categories. Recall from Chapter 5 that such experiments are called binomial experiments.

This chapter describes three types of tests:

1. Tests of hypotheses for experiments with more than two categories, called goodness-of-fit tests

2. Tests of hypotheses about contingency tables, called independence and homogeneity tests

3. Tests of hypotheses about the variance and standard deviation of a single population

All of these tests are performed by using the **chi-square distribution**. The chi-square distribution is sometimes written as χ^2 *distribution* and read as "chi-square distribution." The symbol χ is the Greek letter **chi**, pronounced "kī." The values of a chi-square distribution are denoted by the symbol χ^2 (read as *chi-square*), just as the values of the standard normal distribution and the t distribution are denoted by z and t, respectively. Section 11.1 describes the chi-square distribution.

11.1 THE CHI-SQUARE DISTRIBUTION

Like the t distribution, the chi-square distribution has only one parameter called the degrees of freedom (df). The shape of a specific chi-square distribution depends on the number of degrees of freedom.[1] (The degrees of freedom for a chi-square distribution are calculated by using different formulas for different tests. This will be explained when we discuss those tests.) The random variable χ^2 assumes nonnegative values only. Hence, a chi-square distribution curve starts at the origin (zero point) and lies entirely to the right of the vertical axis. Figure 11.1 shows three chi-square distribution curves. They are for 2, 7, and 12 degrees of freedom, respectively.

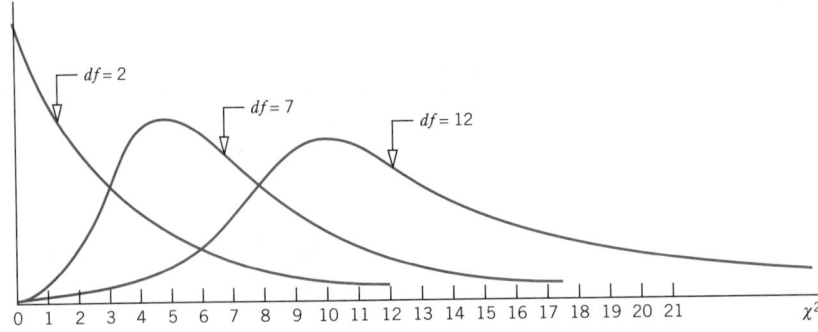

Figure 11.1 Three chi-square distribution curves.

[1]The mean of a chi-square distribution is equal to its df, and the standard deviation is equal to $\sqrt{2\ df}$.

As we can see from Figure 11.1, the shape of a chi-square distribution curve is skewed for very small degrees of freedom, and it changes drastically as the degrees of freedom increase. Eventually, for large degrees of freedom, the chi-square distribution curve looks like a normal distribution curve. The peak (or mode) of a chi-square distribution curve with 1 or 2 degrees of freedom occurs at zero and for a curve with 3 or more degrees of freedom at $df - 2$. For instance, the peak of the chi-square distribution curve with $df = 2$ in Figure 11.1 occurs at zero. The peak for the curve with $df = 7$ occurs at $7 - 2 = 5$. Finally, the peak for the curve with $df = 12$ occurs at $12 - 2 = 10$. Like all other continuous distribution curves, the total area under a chi-square distribution curve is 1.0.

THE CHI-SQUARE DISTRIBUTION

The *chi-square distribution* has only one parameter called the degrees of freedom. The shape of a chi-square distribution curve is skewed to the right for small df and becomes symmetric for large df. The entire chi-square distribution curve lies to the right of the vertical axis. The chi-square distribution assumes nonnegative values only, and these are denoted by the symbol χ^2 (read as *chi-square*).

If we know the degrees of freedom and the area in the right tail of a chi-square distribution curve, we can find the value of χ^2 from Table IX of Appendix C. Examples 11–1 and 11–2 show how to read that table.

Reading the chi-square distribution table: area in the right tail known.

EXAMPLE 11–1 Find the value of χ^2 for 7 degrees of freedom and an area of .10 in the right tail of the chi-square distribution curve.

Solution To find the required value of χ^2, we locate 7 in the column for df and .100 in the top row in Table IX of Appendix C. The required χ^2 value is given by the entry at the intersection of the row for 7 and the column for .100. This value is 12.017. The relevant portion of Table IX is presented as Table 11.1 below.

Table 11.1 χ^2 for $df = 7$ and .10 Area in the Right Tail

df	Area in the Right Tail Under the Chi-square Distribution Curve				
	.995	. . .	**.100**	. . .	**.005**
1	0.000	. . .	2.706	. . .	7.879
2	0.010	. . .	4.605	. . .	10.597
.	. . .	. . .	. . .	. . .	. . .
.	. . .	. . .	. . .	. . .	. . .
.	. . .	. . .	. . .	. . .	. . .
7	0.989	. . .	12.017 ←	. . .	20.278
.	. . .	. . .	. . .	. . .	. . .
.	. . .	. . .	. . .	. . .	. . .
100	67.328	. . .	118.498	. . .	140.169

Required value of χ^2

As shown in Figure 11.2, the χ^2 value for $df = 7$ and an area of .10 in the right tail of the chi-square distribution curve is 12.017.

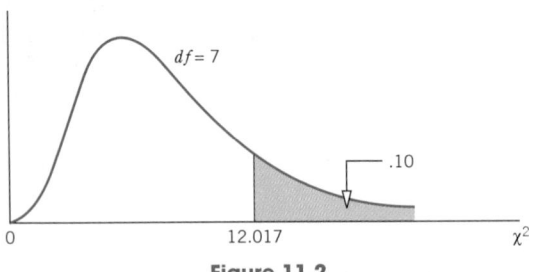

Figure 11.2

Reading the chi-square distribution table: area in the left tail known.

EXAMPLE 11–2   Find the value of χ^2 for 12 degrees of freedom and an area of .05 in the left tail of the chi-square distribution curve.

Solution   We can read Table IX of Appendix C only when an area in the right tail of the chi-square distribution curve is known. When the given area is in the left tail, as in this example, the first step is to find the area in the right tail of the chi-square distribution curve as follows.

$$\text{Area in the right tail} = 1 - \text{area in the left tail}$$

Therefore, for our example,

$$\text{Area in the right tail} = 1 - .05 = .95$$

Next, we locate 12 in the column for df and .950 in the top row in Table IX of Appendix C. The required value of χ^2, given by the entry at the intersection of the row for 12 and the column for .950, is 5.226. The relevant portion of Table IX is presented as Table 11.2.

Table 11.2 χ^2 for $df = 12$ and .95 Area in the Right Tail

	Area in the Right Tail Under the Chi-square Distribution Curve				
df	.995	...	.950	. . .	.005
1	0.000	. . .	0.004	. . .	7.879
2	0.010	. . .	0.103	. . .	10.597
.	. . .	. . .	. . .	. . .	. . .
.	. . .	. . .	. . .	. . .	. . .
.	. . .	. . .	. . .	. . .	. . .
12	3.074	. . .	5.226 ←	. . .	28.300
.	. . .	. . .	. . .	. . .	. . .
.	. . .	. . .	. . .	. . .	. . .
100	67.328	. . .	77.929	. . .	140.169

Required value of χ^2

As shown in Figure 11.3, the χ^2 value for $df = 12$ and .05 area in the left tail is 5.226.

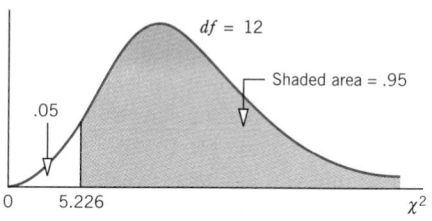

Figure 11.3

EXERCISES

Concepts and Procedures

11.1   Describe the chi-square distribution. What is the parameter (parameters) of such a distribution?

11.2   Find the value of χ^2 for 10 degrees of freedom and an area of .025 in the right tail of the chi-square distribution curve.

11.3   Find the value of χ^2 for 30 degrees of freedom and an area of .05 in the right tail of the chi-square distribution curve.

11.4   Determine the value of χ^2 for 15 degrees of freedom and an area of .10 in the left tail of the chi-square distribution curve.

11.5   Determine the value of χ^2 for 19 degrees of freedom and an area of .990 in the left tail of the chi-square distribution curve.

11.6   Find the value of χ^2 for 4 degrees of freedom and
  a.   .005 area in the right tail of the chi-square distribution curve
  b.   .05 area in the left tail of the chi-square distribution curve

11.7   Determine the value of χ^2 for 13 degrees of freedom and
  a.   .025 area in the left tail of the chi-square distribution curve
  b.   .995 area in the right tail of the chi-square distribution curve

11.2   A GOODNESS-OF-FIT TEST

This section explains how to make tests of hypotheses about experiments with more than two possible outcomes (or categories). Such experiments, called **multinomial experiments**, possess four characteristics. Note that a binomial experiment is a special case of a multinomial experiment.

A MULTINOMIAL EXPERIMENT

An experiment with the following characteristics is called a *multinomial experiment*.

1.   It consists of n identical trials (repetitions).
2.   Each trial results in one of k possible outcomes (or categories) where $k > 2$.
3.   The trials are independent.
4.   The probabilities of the various outcomes remain constant for each trial.

An experiment of many rolls of a die is an example of a multinomial experiment. It consists of many identical rolls (trials); each roll (trial) results in one of the six possible outcomes; each roll is independent of the other rolls; and the probabilities of the six outcomes remain constant for each roll.

As a second example of a multinomial experiment, suppose we select a random sample of people and ask them whether or not the quality of American cars is better than that of Japanese cars. The response of a person can be *yes*, *no*, or *do not know*. Each person included in the sample can be considered as one trial (repetition) of the experiment. There will be as many trials for this experiment as the number of persons selected. Each person can belong to any of the three categories—*yes*, *no*, or *do not know*. The response of each selected person is independent of the responses of other persons. Given that the population is large, the probabilities of a person belonging to the three categories remain the same for each trial. Consequently, this is an example of a multinomial experiment.

The frequencies obtained from the actual performance of an experiment are called the **observed frequencies**. In a **goodness-of-fit test**, we test the null hypothesis that the observed frequencies for an experiment follow a certain pattern or theoretical distribution. The test is called a goodness-of-fit test because the hypothesis tested is how *good* the observed frequencies *fit* a given pattern.

For our first example involving the experiment of many rolls of a die, we may test the null hypothesis that the given die is fair. The die will be fair if the observed frequency for each outcome is close to one-sixth of the total number of rolls.

For our second example involving opinions of people on the quality of American cars, suppose such a survey was conducted in 1995 and in that survey 41% of the people said *yes*, 48% said *no*, and 11% said *do not know*. We want to test if these percentages still hold true. Suppose we take a random sample of 1000 adults and observe that 536 of them think that the quality of American cars is better than that of Japanese cars, 362 say it is worse, and 102 have no opinion. The frequencies 536, 362, and 102 are the observed frequencies. These frequencies are obtained by actually performing the survey. Now, assuming that the 1995 percentages are still true (which will be our null hypothesis), in a sample of 1000 adults we will expect 410 to say *yes*, 480 to say *no*, and 110 to say *do not know*. These frequencies are obtained by multiplying the sample size (1000) by the 1995 proportions. These frequencies are called the **expected frequencies**. Then, we will make a decision to reject or not to reject the null hypothesis based on how large the difference between the observed frequencies and the expected frequencies is. To perform this test, we will use the chi-square distribution. Note that in this case, we are testing the null hypothesis that all three percentages (or proportions) are unchanged. However, if we want to make a test for only one of the three proportions, we use the procedure learned in Section 9.5 of Chapter 9. For example, if we are testing the hypothesis that the percentage of people who think the quality of American cars is better than that of the Japanese cars is different from 41%, then we will test the null hypothesis H_0: $p = .41$ against the alternative hypothesis H_1: $p \neq .41$. This test will be conducted using the procedure discussed in Section 9.5 of Chapter 9.

As mentioned earlier, the frequencies obtained from the performance of an experiment are called the observed frequencies. They are denoted by "*O*." To make a goodness-of-fit test, we calculate the expected frequencies for all categories of the experiment. The expected frequency for a category, denoted by "*E*," is given by the product of n and p where n is the total number of trials and p is the probability for that category.

OBSERVED AND EXPECTED FREQUENCIES

The frequencies obtained from the performance of an experiment are called the *observed frequencies* and are denoted by O. The *expected frequencies*, denoted by E, are the frequencies that we will expect to obtain if the null hypothesis is true. The expected frequency for a category is obtained as

$$E = np$$

where n is the sample size and p is the probability that an element belongs to that category if the null hypothesis is true.

DEGREES OF FREEDOM FOR A GOODNESS-OF-FIT TEST

In a goodness-of-fit test, the *degrees of freedom* are

$$df = k - 1$$

where k denotes the number of possible outcomes (or categories) for the experiment.

The procedure to make a goodness-of-fit test involves the same five steps that were used in the preceding chapters. *The chi-square goodness-of-fit test is always a right-tailed test.*

TEST STATISTIC FOR A GOODNESS-OF-FIT TEST

The *test statistic for a goodness-of-fit test* is χ^2 and its value is calculated as

$$\chi^2 = \sum \frac{(O - E)^2}{E}$$

where
$$O = \text{observed frequency for a category}$$
$$E = \text{expected frequency for a category} = np$$

Remember that a chi-square goodness-of-fit test is always a right-tailed test.

Whether or not the null hypothesis is rejected depends on how much the observed and expected frequencies differ from each other. To find how large the difference between the observed frequencies and the expected frequencies is, we do not just look at $\Sigma(O - E)$ because some of the $O - E$ values will be positive and others will be negative. The net result of the sum of these differences will always be zero. Therefore, we square each of the $O - E$ values to obtain $(O - E)^2$ and then weight them according to the reciprocals of their expected frequencies. The sum of the resulting numbers gives the computed value of the test statistic χ^2.

To make a goodness-of-fit test, the sample size should be large enough so that the expected frequency for each category is at least 5. If there is a category with an expected frequency of less than 5, either increase the sample size or combine two or more categories to make each expected frequency at least 5.

Examples 11–3 and 11–4 describe the procedure for performing goodness-of-fit tests using the chi-square distribution.

Goodness-of-fit test: equal proportions for all categories.

EXAMPLE 11–3   The following table lists the age distribution for a sample of 100 persons arrested for drunk driving.

Age	16–25	26–35	36–45	46–55	56 & older
Arrests	32	25	19	16	8

Using the 1% significance level, can we reject the null hypothesis that the proportion of people arrested for drunk driving is the same for all age groups?

Solution   To make this test, we proceed as follows.

Step 1.   *State the null and alternative hypotheses*

Because there are five categories listed in the table, the proportion of drunk drivers will be the same for all age groups if each group contains one-fifth of the total drunk drivers. The null and alternative hypotheses are as follows.

H_0: The proportion of people arrested for drunk driving is the same for all age groups

H_1: The proportion of people arrested for drunk driving is not the same for all age groups

If the proportion of people arrested for drunk driving is the same for all age groups, then the probability for any randomly selected drunk driver to belong to any of the five age groups listed in the table will be 1/5 or .20. Let p_1, p_2, p_3, p_4, and p_5 be the probabilities that any randomly selected drunk driver will belong to each of the five age groups, respectively. Then, the null hypothesis can also be written as

$$H_0: p_1 = p_2 = p_3 = p_4 = p_5 = .20$$

and the alternative hypothesis can be stated as

$$H_1: \text{At least two of the five probabilities are not equal to } .20$$

Step 2.   *Select the distribution to use*

Because there are five categories (i.e., five age groups listed in the table), this is a multinomial experiment. Consequently, we use the chi-square distribution to make this test.

Step 3.   *Determine the rejection and nonrejection regions*

The significance level is given to be .01, and the goodness-of-fit test is always right-tailed. Therefore, the area in the right tail of the chi-square distribution curve is

$$\text{Area in the right tail} = \alpha = .01$$

The number of degrees of freedom are calculated as follows.

$$k = \text{number of categories} = 5$$
$$df = k - 1 = 5 - 1 = 4$$

From the chi-square distribution table (Table IX of Appendix C), the critical value of χ^2 for $df = 4$ and .01 area in the right tail of the chi-square distribution curve is 13.277, as shown in Figure 11.4.

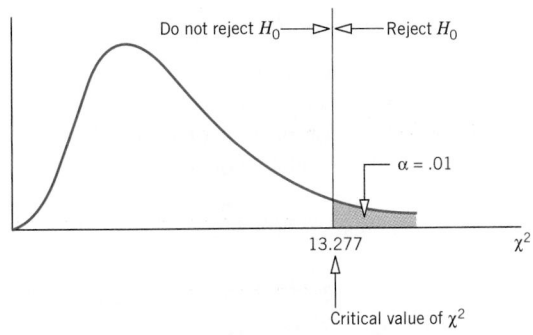

Figure 11.4

Step 4. *Calculate the value of the test statistic*

All the required calculations to find the value of the test statistic χ^2 are shown in Table 11.3.

Table 11.3

Category (age)	Observed Frequency O	p	Expected Frequency $E = np$	$(O - E)$	$(O - E)^2$	$\dfrac{(O - E)^2}{E}$
16–25	32	.20	$100(.20) = 20$	12	144	7.200
26–35	25	.20	$100(.20) = 20$	5	25	1.250
36–45	19	.20	$100(.20) = 20$	−1	1	.050
46–55	16	.20	$100(.20) = 20$	−4	16	.800
56 and over	8	.20	$100(.20) = 20$	−12	144	7.200
	$n = 100$					Sum = 16.500

The calculations made in Table 11.3 are explained below.

1. The first two columns in Table 11.3 list the five categories (age groups) and the observed frequencies for the sample of 100 drunk drivers, respectively. The third column contains the probabilities for the five categories assuming that the null hypothesis is true.

2. The fourth column contains the expected frequencies. These frequencies are obtained by multiplying the sample size ($n = 100$) by the probabilities listed in the third column. If the null hypothesis is true (i.e., the drunk drivers are equally distributed over all categories), then we will expect 20 out of 100 drunk drivers to belong to each category. Consequently, each category in the fourth column has the same expected frequency.

3. The fifth column lists the differences between the observed and expected frequencies, that is, $O - E$. These values are squared and recorded in the sixth column.

4. Finally, we divide the squared differences (of the sixth column) by the corresponding expected frequencies (listed in the fourth column) and write the resulting numbers in the seventh column.

5. The sum of the seventh column gives the value of the test statistic χ^2. Thus,

$$\chi^2 = \sum \frac{(O - E)^2}{E} = 16.500$$

Step 5.   *Make a decision*

The value of the test statistic $\chi^2 = 16.500$ is greater than the critical value of $\chi^2 = 13.277$ and it falls in the rejection region. Hence, we reject the null hypothesis and state that the proportion of drunk drivers is not the same for all age groups listed in the given table. In other words, we conclude that a higher percentage of drunk drivers belong to one or more of the age groups. However, based on this test we cannot say which age groups contain a higher percentage of drunk drivers. ■

Goodness-of-fit test: testing if results of a survey fit a given distribution.

EXAMPLE 11–4   In a 1995 American Express Financial Advisor sample survey, married women aged 45 and over were asked who makes the financial decisions in their households. The results of the survey are shown in the table below. Assume that these results hold true for the 1995 population of all such households.

Decisions made by	Wife	Husband	Both
Percentage of households	15	36	49

In a recent sample of 400 married women aged 45 and over, 68 indicated that in their households they make the financial decisions themselves, 124 said that their husbands make such decisions, and 208 stated that such decisions are made by both together. Test at the 2.5% significance level whether the current pattern of financial decision making is different from that for 1995.

Solution   We perform the following five steps to make this test of hypothesis.

Step 1.   *State the null and alternative hypotheses*

The null and alternative hypotheses are

H_0: The current percentage distribution of financial decision making in such households is the same as that for 1995

H_1: The current percentage distribution of financial decision making in such households is different from that for 1995

Step 2.   *Select the distribution to use*

Because this experiment has three categories (*Wife*, *Husband*, and *Both*), it is a multinomial experiment. Consequently, we use the chi-square distribution to make the test.

Step 3.   *Determine the rejection and nonrejection regions*

The significance level is .025. Because the goodness-of-fit test is right-tailed, the area in the right tail of the chi-square distribution curve is

$$\text{Area in the right tail} = \alpha = .025$$

The number of degrees of freedom is calculated as follows.

$$k = \text{number of categories} = 3$$
$$df = k - 1 = 3 - 1 = 2$$

From the chi-square table (Table IX of Appendix C), the critical value of χ^2 for $df = 2$ and .025 area in the right tail of the chi-square distribution curve is 7.378. This value is shown in Figure 11.5.

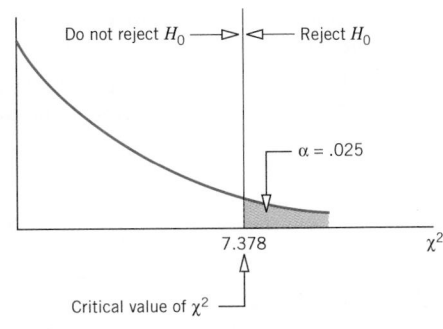

Figure 11.5

Step 4. *Calculate the value of the test statistic*

All the required calculations to find the value of the test statistic χ^2 are shown in Table 11.4.

Table 11.4

Category	Observed Frequency O	p	Expected Frequency $E = np$	$(O - E)$	$(O - E)^2$	$\dfrac{(O - E)^2}{E}$
Wife	68	.15	400(.15) = 60	8	64	1.067
Husband	124	.36	400(.36) = 144	−20	400	2.778
Both	208	.49	400(.49) = 196	12	144	.735
	$n = 400$					Sum = 4.580

Note that the given percentages have been converted to probabilities and recorded in the third column of Table 11.4. The value of the test statistic χ^2 is given by the sum of the last column. Thus,

$$\chi^2 = \sum \frac{(O - E)^2}{E} = 4.580$$

Step 5. *Make a decision*

The value of the test statistic $\chi^2 = 4.580$ is smaller than the critical value of $\chi^2 = 7.378$, and it falls in the nonrejection region. Hence, we fail to reject the null hypothesis and state that the current percentage distribution of financial decision making in households with married women aged 45 and over is not different from that for 1995. The difference between the observed frequencies and the expected frequencies seems to have occurred only because of the sampling error. ■

EXERCISES

Concepts and Procedures

11.8 Describe the four characteristics of a multinomial experiment.

11.9 What is a goodness-of-fit test and when is it applied? Explain.

11.10 Explain the difference between the observed and expected frequencies for a goodness-of-fit test.

11.11   How is the expected frequency of a category calculated for a goodness-of-fit test? What are the degrees of freedom for such a test?

11.12   To make a goodness-of-fit test, what should be the minimum expected frequency for each category? What are the alternatives if this condition is not satisfied?

11.13   The following table lists the frequency distribution for 60 rolls of a die.

Outcome	1-spot	2-spot	3-spot	4-spot	5-spot	6-spot
Frequency	7	12	8	15	11	7

Test at the 5% significance level if the null hypothesis that the given die is fair is true.

Applications

11.14   The following table gives the percentage distribution, according to marital status, of the 1994 Canadian female population aged 15 and over (*Source: 1996 Canadian Global Almanac*).

Marital status	Single	Married	Widowed	Divorced
Percentage	24.7	59.4	10.1	5.8

A recent random sample of 400 Canadian females, aged 15 and over, showed that 95 of them were single, 252 were married, 33 were widowed, and 20 were divorced. Test at the 5% significance level if the current percentage distribution of marital status of the Canadian female population aged 15 and over is different from that of 1994.

11.15   A defendant charged with a felony can hire an attorney if he or she can afford to. Otherwise he or she is represented by an attorney from a public defender's office or by a private attorney assigned by the court. The table below gives the percentages for various types of legal representation for felony defendants in the 75 largest counties in the United States in 1992 (*Source: Bureau of Justice Statistics: Selected Findings*, February 1996).

Counsel	Percentage of Felony Defendants
Public defender	59
Assigned counsel	22
Hired counsel	18
Other	1

A recent sample of 2000 felony defendants from the same 75 counties showed that 1198 of them had attorneys from a public defender's office, 404 were assigned counsel by the court, 377 hired their own counsel, and 21 had other forms of representation. Test at the 1% significance level if the percentage distribution of these representations for felony defendants has changed since 1992 in these 75 counties.

11.16   The following table gives the percentage distribution by educational attainment of U.S. full-time workers 25 years of age and over for the first quarter of 1996 (*Source: U.S. Bureau of Labor Statistics News*, April 19, 1996.)

Educational Attainment	Percentage of Workers
Less than a high school diploma	9.9
High school graduate, no college	32.5
Some college or associate's degree	27.8
Bachelor's degree	19.7
Advanced degree	10.1

A recent random sample of 1000 U.S. full-time workers in this age group showed that 112 have less than a high school diploma, 344 are high school graduates with no college, 252 have some college or an associate's degree, 193 hold a bachelor's degree, and 99 possess an advanced degree. Test at the 1% significance level if the percentage distribution by educational attainment of U.S. full-time workers in this age group has changed since the first quarter of 1996.

11.17 Home Mail Corporation sells products by mail. The company's management wants to find out if the number of orders received at the company's office on each of the five days of the week is the same. The company took a sample of 400 orders received during a four-week period. The following table lists the frequency distribution for these orders by the day of the week.

Day of the week	Mon	Tue	Wed	Thu	Fri
Number of orders received	92	68	65	86	89

Test at the 5% significance level if the null hypothesis that the orders are evenly distributed over all days of the week is true.

11.18 The following table lists the frequency distribution for a sample of 50 absences of college students from classes according to the day of occurrence.

Day of the week	Mon	Tue	Wed	Thu	Fri
Number of absences	14	6	4	10	16

Test at the 5% significance level if the null hypothesis that the absences are equally distributed over all days of the week is true.

11.19 The following table lists the frequency distribution of cars sold at an auto dealership during the past 12 months.

Month	Jan	Feb	Mar	Apr	May	Jun	Jul	Aug	Sep	Oct	Nov	Dec
Cars sold	21	17	15	12	14	12	13	15	23	26	27	29

Using the 10% significance level, will you reject the null hypothesis that the number of cars sold at this dealership is the same for each month?

11.20 Of all students enrolled at a large undergraduate university, 19% are seniors, 23% are juniors, 27% are sophomores, and 31% are freshmen. A sample of 200 students taken from this university by the student senate to conduct a survey includes 50 seniors, 46 juniors, 55 sophomores, and 49 freshmen. Using the 10% significance level, test the null hypothesis that this sample is a random sample. (*Hint*: This sample will be a random sample if it includes approximately 19% seniors, 23% juniors, 27% sophomores, and 31% freshmen.)

11.21 Chance Corporation produces beauty products. Two years ago the quality control department at the company conducted a survey of users of one of the company's products. The survey revealed that 53% of the users said the product was excellent, 31% said it was satisfactory, 7% said it was unsatisfactory, and 9% had no opinion. Assume that these percentages are true for the population of all users of this product at that time. After this survey was conducted, the company redesigned this product. A recent survey of 800 users of the redesigned product conducted by the quality control department at the company showed that 495 of the users think the product is excellent, 259 think it is satisfactory, 31 think it is unsatisfactory, and 15 have no opinion. Do you think the percentage distribution of the opinions of users of the redesigned product is different from the percentage distribution of users of this product before it was redesigned? Use $\alpha = .025$.

11.22 Henderson Corporation makes metal sheets among other products. When the process that is used to make metal sheets works properly, 92% of the metal sheets contain no defects, 5% have one defect each, and 3% have two or more defects each. The quality control inspectors at the company take samples of metal sheets quite often and check them for defects. If the distribution of defects for a sample is significantly different from the above mentioned percentage distribution, the process is stopped and adjusted. A recent sample of 300 sheets produced the frequency distribution of defects listed in the following table.

Number of defects	None	One	Two or More
Number of metal sheets	262	22	16

Does the evidence from this sample suggest that the process needs an adjustment? Use $\alpha = .01$.

11.3   CONTINGENCY TABLES

We often may have information on more than one variable for each element. Such information can be summarized and presented using a two-way classification table, which is also called a *contingency table* or *cross-tabulation*. Table 11.5 is an example of a contingency table, which contains data collected by the U.S. Department of Education. It gives information on the gender of all students enrolled in 1995 at institutions of higher education and on whether these students were full-time or part-time. Note that the numbers in the table give enrollments in thousands. Table 11.5 has two rows (one for males and the second for females) and two columns (one for full-time and the second for part-time students). Hence, it is also called a 2 × 2 (read as "two by two") contingency table.

Table 11.5   Total 1995 Enrollment (in thousands) at Institutions of Higher Education

	Full-time	**Part-time**
Male	3903	2478 ←
Female	4161	3668

Students who are male and enrolled part-time

Source: U.S. Department of Education, National Center for Education Statistics.

A contingency table can be of any size. For example, it can be 2 × 3, 3 × 2, 3 × 3, or 4 × 2. Note that in these notations, the first digit refers to the number of rows in the table, and the second digit refers to the number of columns. For example, a 3 × 2 table will contain three rows and two columns. In general, an $R \times C$ table contains R rows and C columns.

Each of the four boxes that contain numbers in Table 11.5 is called a *cell*. The number of cells for a contingency table is obtained by multiplying the number of rows by the number of columns. Thus, Table 11.5 contains 2 × 2 = 4 cells. The subjects that belong to a cell of a contingency table possess two characteristics. For example, 2478 thousand students listed in the second cell of the first row in Table 11.5 are *male* and *enrolled part-time*. The numbers written inside the cells are usually called the *joint frequencies*. For example, 2478 thousand students belong to the joint category of *male* and *enrolled part-time*. Hence, it is referred to as the joint frequency of this category.

11.4 A TEST OF INDEPENDENCE OR HOMOGENEITY

This section is concerned with tests of independence and homogeneity, which are performed using the contingency tables. Except for a few modifications, the procedure used to make such tests is almost the same as the one applied in Section 11.2 for a goodness-of-fit test.

11.4.1 A TEST OF INDEPENDENCE

In a **test of independence** for a contingency table, we test the null hypothesis that the two attributes (characteristics) of the elements of a given population are not related (that is, they are independent) against the alternative hypothesis that the two characteristics are related (that is, they are dependent). For example, we may want to test if the affiliation of people with the Democratic and Republican parties is independent of their income levels. We perform such a test by using the chi-square distribution. As another example, we may want to test if there is an association between being a male or female and having a preference for watching sports or soap operas on television.

DEGREES OF FREEDOM FOR A TEST OF INDEPENDENCE

A test of independence involves a test of the null hypothesis that two attributes of a population are not related. The *degrees of freedom for a test of independence* are

$$df = (R - 1)(C - 1)$$

where R and C are the number of rows and the number of columns, respectively, in the given contingency table.

The value of the test statistic χ^2 in the case of a test of independence is obtained using the same formula as in the goodness-of-fit test described in Section 11.2.

TEST STATISTIC FOR A TEST OF INDEPENDENCE

The value of the *test statistic χ^2 for a test of independence* is calculated as

$$\chi^2 = \sum \frac{(O - E)^2}{E}$$

where O and E are the observed and expected frequencies, respectively, for a cell.

The null hypothesis in a test of independence is always that the two attributes are not related. The alternative hypothesis is that the two attributes are related.

The frequencies obtained from the performance of an experiment for a contingency table are called the **observed frequencies**. The procedure to calculate the **expected frequencies** for a contingency table for a test of independence is different from the one for a goodness-of-fit test. Example 11–5 describes this procedure.

EXAMPLE 11-5 Violence and lack of discipline have become major problems in schools in the United States. A random sample of 300 adults was selected, and they were asked if they favor giving more freedom to schoolteachers to punish students for violence and lack of discipline. The two-way classification of the responses of these adults is presented in the following table.

	In Favor (F)	Against (A)	No Opinion (N)
Men (M)	93	70	12
Women (W)	87	32	6

Calculate the expected frequencies for this table assuming that the two attributes, gender and opinions on the issue, are independent.

Solution The table in Example 11–5 is reproduced below as Table 11.6. Note that Table 11.6 includes the row and column totals.

Table 11.6

	In Favor (F)	Against (A)	No Opinion (N)	Row Totals
Men (M)	93	70	12	175
Women (W)	87	32	6	125
Column totals	180	102	18	300

The numbers 93, 70, 12, 87, 32, and 6 listed inside the six cells of Table 11.6 are called the *observed frequencies* of the respective cells.

As mentioned earlier, the null hypothesis in a test of independence is that the two attributes (or classifications) are independent. In an independence test of hypothesis, first we assume that the null hypothesis is true and that the two attributes are independent. Assuming that the null hypothesis is true and that gender and opinions are not related in this example, the expected frequency for the cell corresponding to *Men* and *In Favor* is calculated as shown next. From Table 11.6,

$$P(\text{a person is a } Man) = P(M) = 175/300$$

$$P(\text{a person is } In\ Favor) = P(F) = 180/300$$

Because we are assuming that M and F are independent (by assuming that the null hypothesis is true), using the formula learned in Chapter 4, the joint probability of these two events is

$$P(M \text{ and } F) = P(M) \times P(F) = (175/300) \times (180/300)$$

Then, assuming that M and F are independent, the number of persons expected to be *Men* and *In Favor* in a sample of 300 is

$$E \text{ for } Men \text{ and } In\ Favor = 300 \times P(M \text{ and } F)$$

$$= 300 \times \frac{175}{300} \times \frac{180}{300} = \frac{175 \times 180}{300}$$

$$= \frac{(\text{Row total})(\text{Column total})}{\text{Sample size}}$$

Thus, the rule for obtaining the expected frequency for a cell is to divide the product of the corresponding row and column totals by the sample size.

EXPECTED FREQUENCIES FOR A TEST OF INDEPENDENCE

The expected frequency E for a cell is calculated as

$$E = \frac{(\text{Row total})(\text{Column total})}{\text{Sample size}}$$

Using this rule, we calculate the expected frequencies of the six cells of Table 11.6 as follows.

E for *Men* and *In Favor* cell $= (175)(180)/300 = 105.00$

E for *Men* and *Against* cell $= (175)(102)/300 = 59.50$

E for *Men* and *No Opinion* cell $= (175)(18)/300 = 10.50$

E for *Women* and *In Favor* cell $= (125)(180)/300 = 75.00$

E for *Women* and *Against* cell $= (125)(102)/300 = 42.50$

E for *Women* and *No Opinion* cell $= (125)(18)/300 = 7.50$

The expected frequencies are usually written in parentheses below the observed frequencies within the corresponding cells, as shown in Table 11.7.

Table 11.7

	In Favor (F)	Against (A)	No Opinion (N)	Row Totals
Men (M)	93 (105.00)	70 (59.50)	12 (10.50)	175
Women (W)	87 (75.00)	32 (42.50)	6 (7.50)	125
Column totals	180	102	18	300

Like a goodness-of-fit test, *a test of independence is always right-tailed*. To apply a chi-square test of independence, *the sample size should be large enough so that the expected frequency for each cell is at least 5*. If the expected frequency for a cell is not at least 5, we either increase the sample size or combine some categories. Examples 11–6 and 11–7 describe the procedure to make tests of independence using the chi-square distribution.

A test of independence: 2 × 3 table.

EXAMPLE 11–6 Reconsider the two-way classification table given in Example 11–5. In that example, a random sample of 300 adults was selected, and they were asked if they favor giving more freedom to schoolteachers to punish students for violence and lack of discipline. Based on the results of the survey, a two-way classification table was prepared and presented in Example 11–5. Does the sample provide sufficient information to conclude that the two attributes, gender and opinions of adults, are dependent? Use a 1% significance level.

Solution   The test involves the following five steps.

Step 1. *State the null and alternative hypotheses*

As mentioned earlier, the null hypothesis must be that the two attributes are independent. Consequently, the alternative hypothesis is that these attributes are dependent.

$$H_0: \text{Gender and opinions of adults are independent}$$

$$H_1: \text{Gender and opinions of adults are dependent}$$

Step 2. *Select the distribution to use*

We use the chi-square distribution to make a test of independence for a contingency table.

Step 3. *Determine the rejection and nonrejection regions*

The significance level is 1%. Because a test of independence is always right-tailed, the area of the rejection region is .01, and it falls in the right tail of the chi-square distribution curve. The contingency table contains two rows (*Men* and *Women*) and three columns (*In Favor*, *Against*, and *No Opinion*). Note that we do not count the row and column of totals. The degrees of freedom are

$$df = (R - 1)(C - 1) = (2 - 1)(3 - 1) = 2$$

From Table IX of Appendix C, the critical value of χ^2 for $df = 2$ and $\alpha = .01$ is 9.210. This value is shown in Figure 11.6.

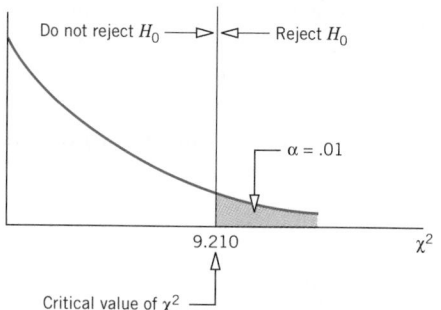

Figure 11.6

Step 4. *Calculate the value of the test statistic*

Table 11.7, with the observed and expected frequencies constructed in Example 11–5, is reproduced as Table 11.8 below.

Table 11.8

	In Favor (F)	Against (A)	No Opinion (N)	Row Totals
Men (M)	93 (105.00)	70 (59.50)	12 (10.50)	175
Women (W)	87 (75.00)	32 (42.50)	6 (7.50)	125
Column totals	180	102	18	300

To compute the value of the test statistic χ^2, we take the difference between each pair of observed and expected frequencies listed in Table 11.8, square those differences, and then divide each of the squared differences by the respective expected frequencies. The sum of the resulting numbers gives the value of the test statistic χ^2. All these calculations are made as follows.

$$\chi^2 = \sum \frac{(O - E)^2}{E}$$

$$= \frac{(93 - 105.00)^2}{105.00} + \frac{(70 - 59.50)^2}{59.50} + \frac{(12 - 10.50)^2}{10.50}$$

$$+ \frac{(87 - 75.00)^2}{75.00} + \frac{(32 - 42.50)^2}{42.50} + \frac{(6 - 7.50)^2}{7.50}$$

$$= 1.371 + 1.853 + .214 + 1.920 + 2.594 + .300 = 8.252$$

Step 5. *Make a decision*

The value of the test statistic $\chi^2 = 8.252$ is less than the critical value of $\chi^2 = 9.210$, and it falls in the nonrejection region. Hence, we fail to reject the null hypothesis and state that there is not enough evidence from the sample to conclude that the two characteristics, *gender* and *opinions of adults*, are dependent for this issue. ▄

A test of independence: 3 × 2 table.

EXAMPLE 11-7 In a recent study, students selected from all 50 states were interviewed in order to analyze drinking behavior of U.S. high school students (*Source:* Luis G. Escobedo, M.D., M.P.H., et al., "Patterns of Alcohol Use and the Risk of Drinking and Driving Among U.S. High School Students," *American Journal of Public Health* 85[7], July 1995). Each student included in the study was classified according to academic performance at school and was asked whether or not he/she had engaged in *binge drinking* (defined as five or more drinks in a row) at least once during the past month. The results obtained from the survey are given below.

		Binge Drinking	
		Yes	No
School	Excellent	1260	3588
Performance	Average	2157	4186
	Below average	441	497

Using the 5% level of significance, can you conclude that school performance and binge drinking are related for all high school students in the United States?

Solution We perform the following five steps to make this test of hypothesis.

Step 1. *State the null and alternative hypotheses*

The null and alternative hypotheses are

H_0: School performance and binge drinking are not related

H_1: School performance and binge drinking are related

Step 2. *Select the distribution to use*

Because we are performing a test of independence, we use the chi-square distribution to make the test.

Step 3.   *Determine the rejection and nonrejection regions*

With a significance level of 5%, the area of the rejection region is .05, and it falls in the right tail of the chi-square distribution curve. The contingency table contains three rows (*Excellent*, *Average*, and *Below average*) and two columns (*Yes* and *No*). The degrees of freedom are

$$df = (R - 1)(C - 1) = (3 - 1)(2 - 1) = 2$$

From Table IX of Appendix C, the critical value of χ^2 for $df = 2$ and $\alpha = .05$ is 5.991. This value is shown in Figure 11.7.

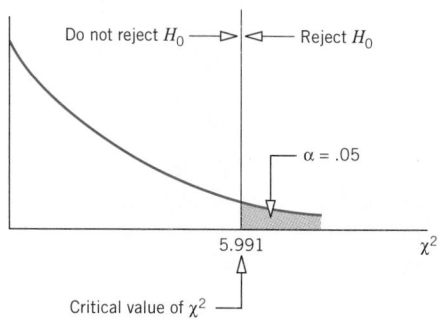

Figure 11.7

Step 4.   *Calculate the value of the test statistic*

The expected frequencies for the various cells are calculated as follows, and they are listed within parentheses in Table 11.9.

E for *Excellent* and *Yes* cell = (4848)(3858)/12129 = 1542.05

E for *Excellent* and *No* cell = (4848)(8271)/12129 = 3305.95

E for *Average* and *Yes* cell = (6343)(3858)/12129 = 2017.59

E for *Average* and *No* cell = (6343)(8271)/12129 = 4325.41

E for *Below average* and *Yes* cell = (938)(3858)/12129 = 298.36

E for *Below average* and *No* cell = (938)(8271)/12129 = 639.64

Table 11.9

	Binge Drinking		**Row**
	Yes (Y)	**No** (N)	**Totals**
Excellent (C)	1260 (1542.05)	3588 (3305.95)	4848
Average (A)	2157 (2017.59)	4186 (4325.41)	6343
Below average (B)	441 (298.36)	497 (639.64)	938
Column totals	3858	8271	12129

The value of the test statistic χ^2 is calculated as follows.

$$\chi^2 = \sum \frac{(O - E)^2}{E}$$

$$= \frac{(1260 - 1542.05)^2}{1542.05} + \frac{(3588 - 3305.95)^2}{3305.95} + \frac{(2157 - 2017.59)^2}{2017.59}$$

$$+ \frac{(4186 - 4325.41)^2}{4325.41} + \frac{(441 - 298.36)^2}{298.36} + \frac{(497 - 639.64)^2}{639.64}$$

$$= 51.589 + 24.063 + 9.633 + 4.493 + 68.193 + 31.809 = 189.780$$

Step 5. *Make a decision*

The value of the test statistic $\chi^2 = 189.780$ is greater than the critical value of $\chi^2 = 5.991$, and it falls in the rejection region. Hence, we reject the null hypothesis and state that there is strong evidence from the sample to conclude that the two characteristics, *school performance* and *binge drinking*, are related for all high school students in the United States.

11.4.2 A TEST OF HOMOGENEITY

In a **test of homogeneity**, we test if two (or more) populations are homogeneous (similar) with regard to the distribution of a certain characteristic. For example, we might be interested in testing the null hypothesis that the proportions of households that belong to different income groups are the same in California and Wisconsin. Or we may want to test whether or not the preferences of people in Florida, Arizona, and Vermont are similar with regard to Coke, Pepsi, and 7-Up.

A TEST OF HOMOGENEITY

A *test of homogeneity* involves testing the null hypothesis that the proportions of elements with certain characteristics in two or more different populations are the same against the alternative hypothesis that these proportions are not the same.

Let us consider the example of testing the null hypothesis that the proportions of households in California and Wisconsin who belong to various income groups are the same. (Note that in a test of homogeneity, the null hypothesis will always be that the proportions of elements with certain characteristics are the same in two or more populations. The alternative hypothesis will be that these proportions are not the same.) Suppose we define three income strata: high income group (with an income of more than $60,000), medium income group (with an income of $30,000 to $60,000), and low income group (with an income of less than $30,000). Furthermore, assume that we take one sample of 250 households from California and another sample of 150 households from Wisconsin, collect the information on incomes of these households, and prepare the contingency Table 11.10.

Table 11.10

	California	Wisconsin	Row Totals
High income	70	34	104
Medium income	80	40	120
Low income	100	76	176
Column totals	250	150	400

Note that in this example the column totals are fixed. That is, we decided in advance to take samples of 250 households from California and 150 from Wisconsin. However, the row totals (of 104, 120, and 176) are determined randomly by the outcomes of the two samples. If we compare this example to the one about violence and lack of discipline in schools taken in the previous section, we will notice that neither the column nor the row totals were fixed in that example. Instead, the researcher took just one sample of 300 adults, collected the information on gender and opinions, and prepared the contingency table. Thus, in that example, the row and column totals were all determined randomly. Thus, when both the row and column totals are determined randomly, we make a test of independence. However, when either the column totals or the row totals are fixed, we make a test of homogeneity. In the case of income groups in California and Wisconsin, we will make a test of homogeneity to test for the similarity of income groups in two states.

The procedure to make a test of homogeneity is similar to the procedure used to make a test of independence discussed earlier. Like a test of independence, a test of homogeneity is also right-tailed. Example 11–8 illustrates the procedure to make a homogeneity test.

A test of homogeneity.

EXAMPLE 11–8 Consider the data on income distribution for households in California and Wisconsin given in Table 11.10. Using the 2.5% significance level, test the null hypothesis that the distribution of households with regard to income levels is similar (homogeneous) for the two states.

Solution   We perform the following five steps to make this test of hypothesis.

Step 1. *State the null and alternative hypotheses*

The two hypotheses are[2]

H_0: The proportion of households that belong to each income group is the same in both states

H_1: The proportion of households that belong to each income group is not the same in both states

Step 2. *Select the distribution to use*

We use the chi-square distribution to make a homogeneity test.

---

[2]Let p_{HC}, p_{MC}, and p_{LC} be the proportions of households in California who belong to high, middle, and low income groups, respectively. Let p_{HW}, p_{MW}, and p_{LW} be the corresponding proportions for Wisconsin. Then, we can also write the null hypothesis as

$$H_0: p_{HC} = p_{HW}, p_{MC} = p_{MW}, \text{ and } p_{LC} = p_{LW}$$

and the alternative hypothesis as

$$H_1: \text{At least two of the equalities mentioned in } H_0 \text{ are not true}$$

Step 3. *Determine the rejection and nonrejection regions*

The significance level is 2.5%. Because the homogeneity test is right-tailed, the area of the rejection region is .025, and it lies in the right tail of the chi-square distribution curve. The contingency table for income groups in California and Wisconsin contains three rows and two columns. Hence, the degrees of freedom are

$$df = (R - 1)(C - 1) = (3 - 1)(2 - 1) = 2$$

From Table IX of Appendix C, the value of χ^2 for $df = 2$ and a .025 area in the right tail of the chi-square distribution curve is 7.378. This value is shown in Figure 11.8.

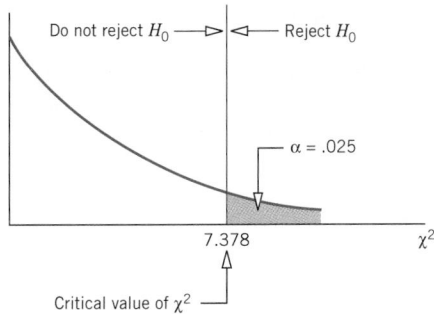

Figure 11.8

Step 4. *Calculate the value of the test statistic*

To compute the value of the test statistic χ^2, we need to calculate the expected frequencies first. Table 11.11 lists the observed as well as expected frequencies. The numbers in parentheses in this table are expected frequencies, which are calculated using the formula

$$E = \frac{(\text{Row total})(\text{Column total})}{\text{Total of both samples}}$$

Thus, for instance,

$$E \text{ for } \textit{High income} \text{ and } \textit{California} \text{ cell} = \frac{(104)(250)}{(400)} = 65$$

The remaining expected frequencies are calculated in the same way. Note that the expected frequencies in a test of homogeneity are calculated in the same way as in a test of independence.

Table 11.11

	California	Wisconsin	Row Totals
High income	70 (65)	34 (39)	104
Medium income	80 (75)	40 (45)	120
Low income	100 (110)	76 (66)	176
Column totals	250	150	400

The value of the test statistic χ^2 is computed as follows.

$$\chi^2 = \sum \frac{(O - E)^2}{E}$$

$$= \frac{(70 - 65)^2}{65} + \frac{(34 - 39)^2}{39} + \frac{(80 - 75)^2}{75} + \frac{(40 - 45)^2}{45}$$

$$+ \frac{(100 - 110)^2}{110} + \frac{(76 - 66)^2}{66}$$

$$= .385 + .641 + .333 + .556 + .909 + 1.515 = \mathbf{4.339}$$

Step 5. *Make a decision*

The value of the test statistic $\chi^2 = 4.339$ is less than the critical value of $\chi^2 = 7.378$, and it falls in the nonrejection region. Hence, we fail to reject the null hypothesis and state that the distribution of households with regard to income appears to be similar (homogeneous) in California and Wisconsin. ■

EXERCISES

Concepts and Procedures

11.23 Describe in your own words a test of independence and a test of homogeneity. Give one example of each.

11.24 Explain how the expected frequencies for cells of a contingency table are calculated in a test of independence or homogeneity. How do you find the degrees of freedom for such tests?

11.25 To make a test of independence or homogeneity, what should be the minimum expected frequency for each cell? What are the alternatives if this condition is not satisfied?

11.26 Consider the following contingency table.

	Column 1	Column 2	Column 3
Row 1	137	67	102
Row 2	98	71	65
Row 3	113	83	115

a. Write the null and alternative hypotheses for a test of independence for this table.
b. Calculate the expected frequencies for all cells assuming that the null hypothesis is true.
c. For $\alpha = .01$, find the critical value of χ^2. Show the rejection and nonrejection regions on the chi-square distribution curve.
d. Find the value of the test statistic χ^2.
e. Using $\alpha = .01$, would you reject the null hypothesis?

11.27 Consider the following contingency table that records the results obtained for four samples of fixed sizes selected from four populations.

	Sample Selected from			
	Population 1	Population 2	Population 3	Population 4
Row 1	24	81	58	123
Row 2	46	64	91	72
Row 3	18	39	105	93

a. Write the null and alternative hypotheses for a test of homogeneity for this table.
b. Calculate the expected frequencies for all cells assuming that the null hypothesis is true.
c. For $\alpha = .025$, find the critical value of χ^2. Show the rejection and nonrejection regions on the chi-square distribution curve.
d. Find the value of the test statistic χ^2.
e. Using $\alpha = .025$, would you reject the null hypothesis?

Applications

11.28 During the recession in the early 1990s, many families faced hard times financially. Some studies observed that more and more people stopped buying national brand products and started buying less expensive store brand products instead. Data produced by a recent sample of 700 adults on whether they usually buy store brand or name brand products are recorded in the following table.

	More Often Buy	
	National Brand	**Store Brand**
Men	172	143
Women	182	203

Using the 1% significance level, can you reject the null hypothesis that the two attributes, gender and buying national or store brand products, are independent?

11.29 One hundred auto drivers, who were stopped by police for some violation, were also checked to see if they were wearing their seat belts. The following table records the results of this survey.

	Wearing Seat Belt	**Not Wearing Seat Belt**
Male	34	21
Female	30	15

Test at the 2.5% significance level if being a male or a female and wearing or not wearing a seat belt are related.

11.30 A study conducted during 1993 and 1994 at a Los Angeles County medical center attempted to assess the effects of language barriers on the emergency care of patients (*Source:* David W. Baker et al., "Use and Effectiveness of Interpreters in an Emergency Department," *The Journal of the American Medical Association* 275[10], March 13, 1996). One aspect of the study involved interviews with 463 patients. Some of them spoke English and others were fluent only in Spanish. A few days after their emergency room visit, patients were questioned about interpreter use and about their own understanding of their diagnosis. The table below is a two-way classification of responses calculated using the percentages given in the article.

		Interpreter Use		
		No Need for Interpreter	**Interpreter Used**	**Interpreter Not Used**
Patients' Understanding of Diagnosis	Good-excellent	160	69	39
	Fair-poor	80	52	63

At the 10% level of significance, do patients' understanding of diagnosis and interpreter use appear to be related?

11.31 An article published recently in *The Journal of the American Medical Association* studied the relationship between smoking and household income (yuan per year) for males aged 15 years and older in the Minhang district of China (*Source:* Y. L. Gong et al., "Cigarette Smoking in China: Prevalence, Characteristics, and Attitudes in Minhang District," 274[15], October 18, 1995). The following two-way table gives the observed frequencies from this survey based on household incomes and smoking status.

		Current Smokers	Former Smokers	Never Smokers
Household Income (in yuan)	Less than 5000	359	18	186
	5000 to 6999	505	15	202
	7000 to 9999	655	16	278
	10,000 to 14,999	466	6	266
	15,000 or higher	294	6	151

Using the 5% significance level, can you conclude that household incomes and smoking status are dependent for people living in the Minhang district of China?

11.32 The following table gives the two-way classification of 400 randomly selected persons based on their status as a smoker or a nonsmoker and on the number of visits they made to their physicians last year.

	Visits to the Physician		
	0–1	2–4	≥ 5
Smoker	20	60	80
Nonsmoker	110	90	40

Test at the 5% significance level if smoking and visits to the physician are related for all persons.

 11.33 A study conducted in Wisconsin from 1981 to 1989 considered divorced fathers' compliance with child support orders (*Source:* Daniel R. Meyer and Judi Bartfeld, "Compliance with Child Support Orders in Divorce Cases," *Journal of Marriage and the Family* 58[1], February 1996). One of the variables affecting fathers' compliance considered in this study was the custody arrangement. The following table, adapted from the article, gives a two-way classification of divorced fathers according to legal custody and compliance with child support orders.

		Number of Fathers Who		
		Pay Nothing	**Pay in Part**	**Pay in Full**
Legal Custody	Mother	398	1069	1020
	Joint/Split	75	225	382

At the 1% significance level, can you conclude that the legal custody and compliance with child support orders are related?

11.34 National Electronics Company buys parts from two subsidiaries. The quality control department at this company wanted to check if the distribution of good and defective parts is the same for the supplies of parts received from both subsidiaries. The quality control inspector selected a sample of 300 parts received from Subsidiary A and a sample of 400 parts received from Subsidiary B. These parts were checked for being good or defective. The table on the next page records the results of this investigation.

	Subsidiary A	Subsidiary B
Good	284	381
Defective	16	19

Using the 5% significance level, test the null hypothesis that the distribution of good and defective parts is the same for both subsidiaries.

11.35 Two drugs were administered to two groups of randomly assigned patients to cure the same disease. One group of 60 patients and another group of 40 patients were selected. The following table gives information about the number of patients who were cured and not cured by each of the two drugs.

	Cured	Not Cured
Drug I	46	14
Drug II	18	22

Test at the 1% significance level whether or not the two drugs are similar in curing and not curing the patients.

11.36 A company introduced a new product in the market a few months ago. The management wants to determine the reaction of customers in different regions to this product. The research department at the company selected four different samples of 400 users of this product from four regions—the East, South, Midwest, and West. The users of the product were asked whether or not they like the product. The responses of these people are recorded in the following table.

	East	South	Midwest	West
Like	274	203	291	257
Do not like	126	197	109	143

Based on the evidence from these samples, can you conclude that the distribution of opinions of users of this product are not homogeneous for all four regions with regard to liking and not liking this product? Use $\alpha = .01$.

11.37 A poll of 500 adults conducted in 1996 by the Institute for Social Inquiry at the University of Connecticut sought their opinions on weather forecasts and the greenhouse effect (*Source: The Hartford Courant*, February 16, 1996). These adults were asked, "Do you think there have been big changes in world climate because of the greenhouse effect, or not?" Assuming that the pollsters selected 250 women and 250 men for this survey, the percentages given in the article would yield the following two-way classification table.

	Men	Women
Have been big changes	84	120
Have not been	112	65
Only small changes	39	30
Do not know	15	35

Is there sufficient evidence from these data to reject the null hypothesis that men and women have similar opinions about the greenhouse effect? Use $\alpha = .10$.

11.38 The following table gives the distribution of grades for three professors for a few randomly selected classes that each of them taught during the past two years.

		Professor		
		Miller	**Smith**	**Moore**
	A	18	36	20
	B	25	44	15
Grade	C	85	73	82
	D&F	17	12	8

Using the 2.5% significance level, test the null hypothesis that the grade distributions are homogeneous for these three professors.

11.39   Two random samples, one of 95 blue-collar workers and a second of 50 white-collar workers, were taken from a large company. These workers were asked about their views on a certain company issue. The following table gives the results of the survey.

	Opinion		
	Favor	**Oppose**	**Uncertain**
Blue-collar worker	47	39	9
White-collar worker	21	26	3

Using the 2.5% significance level, test the null hypothesis that the distributions of opinions are homogeneous for the two groups of workers.

11.5   INFERENCES ABOUT THE POPULATION VARIANCE

Earlier chapters explained how to make inferences (confidence intervals and hypothesis tests) about the population mean and population proportion. However, we may often need to control the variance (or standard deviation). Consequently, there may be a need to estimate and to test a hypothesis about the population variance σ^2. Section 11.5.1 describes how to make a confidence interval for the population variance (or standard deviation). Section 11.5.2 explains how to test a hypothesis about the population variance.

As an example, suppose a machine is set up to fill packages of cookies so that the net weight of cookies per package is 32 ounces. Note that the machine will not put exactly 32 ounces of cookies in each package. Some of the packages will contain less and some will contain more than 32 ounces. However, if the variance (and, hence, the standard deviation) is too large, some of the packages will contain quite a bit less than 32 ounces of cookies and some others will contain quite a bit more than 32 ounces. The manufacturer will not want a large variation in the amounts of cookies put in different packages. To keep this variation within some specified acceptable limit, the machine will be adjusted from time to time. Before the manager decides to adjust the machine at any time, he must estimate the variance or test a hypothesis or do both to find out if the variance exceeds the maximum acceptable value.

Like every sample statistic, the sample variance is a random variable, and it possesses a sampling distribution. If all the possible samples of a given size are taken from a population and their variances are calculated, the probability distribution of these variances is called the *sampling distribution of the sample variance.*

SAMPLING DISTRIBUTION OF $(n - 1)s^2/\sigma^2$

If the population from which the sample is selected is (approximately) normally distributed, then

$$\frac{(n - 1)\, s^2}{\sigma^2}$$

has a chi-square distribution with $n - 1$ degrees of freedom.

Thus, the chi-square distribution is used to construct a confidence interval and test a hypothesis about the population variance σ^2.

11.5.1 ESTIMATION OF THE POPULATION VARIANCE

The value of the sample variance s^2 is a point estimate of the population variance σ^2. The $(1 - \alpha)100\%$ confidence interval for σ^2 is given by the following formula.

CONFIDENCE INTERVAL FOR THE POPULATION VARIANCE σ^2

Assuming that the population from which the sample is selected is (approximately) normally distributed, the $(1 - \alpha)100\%$ *confidence interval for the population variance* σ^2 is

$$\frac{(n - 1)\, s^2}{\chi^2_{\alpha/2}} \quad \text{to} \quad \frac{(n - 1)\, s^2}{\chi^2_{1-\alpha/2}}$$

where $\chi^2_{\alpha/2}$ and $\chi^2_{1-\alpha/2}$ are obtained from the chi-square distribution table for $\alpha/2$ and $1 - \alpha/2$ areas in the right tail of the chi-square distribution curve, respectively, and for $n - 1$ degrees of freedom.

The confidence interval for the population standard deviation can be obtained by simply taking the positive square root of the two limits of the confidence interval for the population variance.

The procedure for making a confidence interval for σ^2 involves the following three steps.

1. Take a sample of size n and compute s^2 using the formula learned in Chapter 3. However, if n and s^2 are given, then perform only steps 2 and 3.
2. Calculate $\alpha/2$ and $1 - \alpha/2$. Find two values of χ^2 from the chi-square distribution table (Table IX of Appendix C): one for $\alpha/2$ area in the right tail of the chi-square distribution curve and $df = n - 1$, and the second for $1 - \alpha/2$ area in the right tail and $df = n - 1$.
3. Substitute all the values in the formula for the confidence interval for σ^2 and simplify.

Example 11–9 illustrates the estimation of the population variance and population standard deviation.

580

EXAMPLE 11-9 One type of cookie manufactured by Haddad Food Company is Cocoa Cookies. The machine that fills packages of these cookies is set up in such a way that the average net weight of these packages is 32 ounces with a variance of .015 square ounces. From time to time the quality control inspector at the company selects a sample of a few such packages, calculates the variance of the net weights of these packages, and constructs a 95% confidence interval for the population variance. If either both or one of the two limits of this confidence interval is not in the interval .008 to .030, the machine is stopped and adjusted. A recently taken random sample of 25 packages from the production line gave a sample variance of .029 square ounces. Based on this sample information, do you think the machine needs an adjustment? Assume that the net weights of cookies in all packages are normally distributed.

Solution The following three steps are performed to estimate the population variance and to make a decision.

Step 1. From the given information,

$$n = 25 \quad \text{and} \quad s^2 = .029$$

Step 2. The confidence level is $1 - \alpha = .95$. Hence,

$$\alpha = 1 - .95 = .05$$
$$\alpha/2 = .05/2 = .025$$
$$1 - \alpha/2 = 1 - .025 = .975$$
$$df = n - 1 = 25 - 1 = 24$$

From Table IX of Appendix C,

$$\chi^2 \text{ for 24 } df \text{ and .025 area in the right tail} = 39.364$$
$$\chi^2 \text{ for 24 } df \text{ and .975 area in the right tail} = 12.401$$

These values are shown in Figure 11.9.

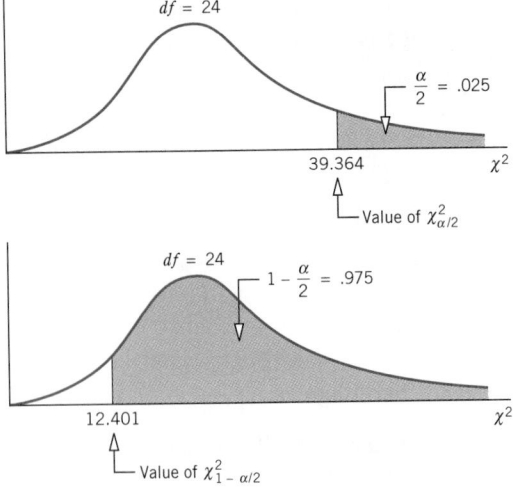

Figure 11.9

Step 3. The 95% confidence interval for σ^2 is

$$\frac{(n-1)\,s^2}{\chi^2_{\alpha/2}} \quad \text{to} \quad \frac{(n-1)\,s^2}{\chi^2_{1-\alpha/2}}$$

or

$$\frac{(25-1)\,(.029)}{39.364} \quad \text{to} \quad \frac{(25-1)\,(.029)}{12.401}$$

or

$$\textbf{.0177} \quad \textbf{to} \quad \textbf{.0561}$$

Thus, with 95% confidence, we can state that the variance for all packages of Cocoa Cookies lies between .0177 and .0561 square ounces. Note that the lower limit (.0177) of this confidence interval is between .008 and .030, but the upper limit (.0561) is larger than .030 and falls outside the interval .008 to .030. Since the upper limit is larger than .030, we can state that the machine needs to be stopped and adjusted.

We can obtain the confidence interval for the population standard deviation σ by taking the positive square root of the two limits of the above confidence interval for the population variance. Thus, a 95% confidence interval for the population standard deviation is

$$\sqrt{.0177} \quad \text{to} \quad \sqrt{.0561} \quad \text{or} \quad \textbf{.133 to .237}$$

Hence, the standard deviation of all packages of Cocoa Cookies is between .133 and .237 ounces at a 95% confidence level. ▪

11.5.2 HYPOTHESIS TESTS ABOUT THE POPULATION VARIANCE

A test of hypothesis about the population variance can be one-tailed or two-tailed. To make a test of hypothesis about σ^2, we perform the same five steps we have used earlier in hypothesis-testing examples. The procedure to test a hypothesis about σ^2 discussed in this section is applied only when the population from which a sample is selected is (approximately) normally distributed.

TEST STATISTIC FOR A TEST OF HYPOTHESIS ABOUT σ^2

The value of the *test statistic* χ^2 is calculated as

$$\chi^2 = \frac{(n-1)\,s^2}{\sigma^2}$$

where s^2 is the sample variance, σ^2 is the hypothesized value of the population variance, and $n-1$ represents the degrees of freedom. The population from which the sample is selected is assumed to be (approximately) normally distributed.

Examples 11–10 and 11–11 illustrate the procedure to make tests of hypothesis about σ^2.

Making a right-tailed test of hypothesis about σ^2.

EXAMPLE 11–10 One type of cookie manufactured by Haddad Food Company is Cocoa Cookies. The machine that fills packages of these cookies is set up in such a way that the average net weight of these packages is 32 ounces with a variance of .015 square ounces.

From time to time the quality control inspector at the company selects a sample of a few such packages, calculates the variance of the net weights of these packages, and makes a test of hypothesis about the population variance. She always uses $\alpha = .01$. The acceptable value of the population variance is .015 square ounces or less. If the conclusion from the test of hypothesis is that the population variance is not within the acceptable limit, the machine is stopped and adjusted. A recently taken random sample of 25 packages from the production line gave a sample variance of .029 square ounces. Based on this sample information, do you think the machine needs an adjustment? Assume that the net weights of cookies in all packages are normally distributed.

Solution   From the given information,

$$n = 25, \qquad \alpha = .01, \qquad \text{and} \qquad s^2 = .029$$

The population variance should not exceed .015 square ounces.

Step 1.   *State the null and alternative hypotheses*

We are to test whether or not the population variance is within the acceptable limit. The population variance is within the acceptable limit if it is less than or equal to .015; otherwise, it is not. Thus, the two hypotheses are

$H_0: \sigma^2 \leq .015$ (the population variance is within the acceptable limit)

$H_1: \sigma^2 > .015$ (the population variance exceeds the acceptable limit)

Step 2.   *Select the distribution to use*

We use the chi-square distribution to test a hypothesis about σ^2.

Step 3.   *Determine the rejection and nonrejection regions*

The significance level is 1% and, because of the $>$ sign in H_1, the test is right-tailed. The rejection region lies in the right tail of the chi-square distribution curve with its area equal to .01. The degrees of freedom for a chi-square test about σ^2 are $n - 1$, that is,

$$df = n - 1 = 25 - 1 = 24$$

From Table IX of Appendix C, the critical value of χ^2 for 24 degrees of freedom and .01 area in the right tail is 42.980. This value is shown in Figure 11.10.

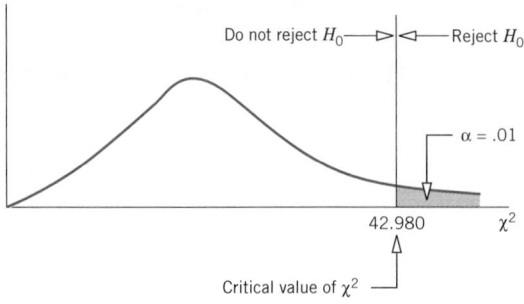

Figure 11.10

Step 4. *Calculate the value of the test statistic*

The value of the test statistic χ^2 for the sample variance is calculated as follows.

$$\chi^2 = \frac{(n-1)\,s^2}{\sigma^2} = \frac{(25-1)\,(.029)}{(.015)} = 46.400$$

From H_0

Step 5. *Make a decision*

The value of the test statistic $\chi^2 = 46.400$ is greater than the critical value of $\chi^2 = 42.980$, and it falls in the rejection region. Consequently, we reject H_0 and conclude that the population variance is not within the acceptable limit. The machine should be stopped and adjusted. ▪

Conducting a two-tailed test of hypothesis about σ^2.

EXAMPLE 11–11   The variance of scores on a standardized mathematics test for all high school seniors was 150 in 1996. A sample of scores for 20 high school seniors who took this test this year gave a variance of 170. Test at the 5% significance level if the variance of current scores of all high school seniors on this test is different from 150. Assume that the scores of all high school seniors on this test are (approximately) normally distributed.

Solution   From the given information,

$$n = 20, \qquad \alpha = .05, \qquad \text{and} \qquad s^2 = 170$$

The population variance was 150 in 1996.

Step 1. *State the null and alternative hypotheses*

The null and alternative hypotheses are

H_0: $\sigma^2 = 150$ (the population variance is not different from 150)

H_1: $\sigma^2 \neq 150$ (the population variance is different from 150)

Step 2. *Select the distribution to use*

We use the chi-square distribution to test a hypothesis about σ^2.

Step 3. *Determine the rejection and nonrejection regions*

The significance level is 5%. The $\neq$ sign in H_1 indicates that the test is two-tailed. The rejection region lies in both tails of the chi-square distribution curve with its total area equal to .05. Consequently, the area in each tail of the distribution curve is .025. The values of $\alpha/2$ and $1 - \alpha/2$ are

$$\frac{\alpha}{2} = \frac{.05}{2} = .025 \qquad \text{and} \qquad 1 - \frac{\alpha}{2} = 1 - .025 = .975$$

The degrees of freedom are

$$df = n - 1 = 20 - 1 = 19$$

From Table IX of Appendix C, the critical values of χ^2 for 19 degrees of freedom and for $\alpha/2$ and $1 - \alpha/2$ areas in the right tail are

χ^2 for 19 df and .025 area in the right tail = 32.852

χ^2 for 19 df and .025 area in the left tail

$\quad = \chi^2$ for 19 df and .975 area in the right tail = 8.907

These two values are shown in Figure 11.11.

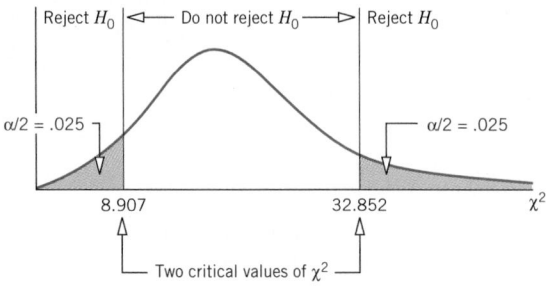

Figure 11.11

Step 4. *Calculate the value of the test statistic*

The value of the test statistic χ^2 for the sample variance is calculated as follows.

$$\chi^2 = \frac{(n-1)\,s^2}{\sigma^2} = \frac{(20-1)\,(170)}{(150)} = 21.533$$

From H_0

Step 5. *Make a decision*

The value of the test statistic $\chi^2 = 21.533$ is between the two critical values of χ^2, 8.907 and 32.852, and it falls in the nonrejection region. Consequently, we fail to reject H_0 and conclude that the population variance of the current scores of high school seniors on this standardized mathematics test does not appear to be different from 150. ▬

Note that we can make a test of hypothesis about the population standard deviation σ using the same procedure as that for the population variance σ^2. To make a test of hypothesis about σ, the only change will be mentioning the values of σ in H_0 and H_1. The rest of the procedure remains the same as in case of σ^2.

EXERCISES

Concepts and Procedures

11.40 A sample of certain observations selected from a normally distributed population produced a sample variance of 42. Construct a 95% confidence interval for σ^2 for each of the following cases and comment on what happens to the confidence interval of σ^2 when the sample size increases.

     a. $n = 12$ **b.** $n = 16$ **c.** $n = 25$

11.41 A sample of 25 observations selected from a normally distributed population produced a sample variance of 39. Construct a confidence interval for σ^2 for each of the following confidence levels and comment on what happens to the confidence interval of σ^2 when the confidence level decreases.

     a. $1 - \alpha = .99$ **b.** $1 - \alpha = .95$ **c.** $1 - \alpha = .90$

11.42 A sample of 20 observations selected from a normally distributed population produced a sample variance of 18.

     a. Write the null and alternative hypotheses to test whether the population variance is different from 14.

b. Using $\alpha = .05$, find the critical values of χ^2. Show the rejection and nonrejection regions on a chi-square distribution curve.

c. Find the value of the test statistic χ^2.

d. Using the 5% significance level, will you reject the null hypothesis stated in part a?

11.43 A sample of 16 observations selected from a normally distributed population produced a sample variance of 1.13.

a. Write the null and alternative hypotheses to test whether the population variance is greater than .80.

b. Using $\alpha = .01$, find the critical value of χ^2. Show the rejection and nonrejection regions on a chi-square distribution curve.

c. Find the value of the test statistic χ^2.

d. Using the 1% significance level, will you reject the null hypothesis stated in part a?

11.44 A sample of 25 observations selected from a normally distributed population produced a sample variance of .75.

a. Write the null and alternative hypotheses to test whether the population variance is less than 1.25.

b. Using $\alpha = .025$, find the critical value of χ^2. Show the rejection and nonrejection regions on a chi-square distribution curve.

c. Find the value of the test statistic χ^2.

d. Using the 2.5% significance level, will you reject the null hypothesis stated in part a?

11.45 A sample of 14 observations selected from a normally distributed population produced a sample variance of 4.6.

a. Write the null and alternative hypotheses to test whether the population variance is different from 2.2.

b. Using $\alpha = .05$, find the critical values of χ^2. Show the rejection and nonrejection regions on a chi-square distribution curve.

c. Find the value of the test statistic χ^2.

d. Using the 5% significance level, will you reject the null hypothesis stated in part a?

Applications

11.46 The management of a soft-drink company does not want the variance of the amount of soda in 12-ounce cans to be more than .01 square ounces. (Recall from Chapter 3 that the variance is always in square units.) The company manager takes a sample of certain cans and estimates the population variance quite often. A random sample of twenty 12-ounce cans taken from the production line of this company showed that the variance for this sample was .014 square ounces.

a. Construct the 99% confidence intervals for the population variance and standard deviation. Assume that the amount of soda in all 12-ounce cans has a normal distribution.

b. Test at the 2.5% significance level if the variance of the amounts of soda in all 12-ounce cans for this company is greater than .01 square ounces.

11.47 The two-inch-long bolts manufactured by a company must have a variance of .003 square inches or less for acceptance by a buyer. A random sample of 29 such bolts gave a variance of .0061 square inches.

a. Test at the 1% significance level if the variance of all such bolts is greater than .003 square inches. Assume that the lengths of all two-inch-long bolts manufactured by this company are (approximately) normally distributed.

b. Make the 98% confidence intervals for the population variance and standard deviation.

11.48 An auto manufacturing company wants to estimate the variance of miles per gallon for its auto model AST727. A random sample of 24 cars of this model showed that the variance of miles per gallon for these cars is .62.

a. Construct the 95% confidence intervals for the population variance and standard deviation. Assume that the miles per gallon for all such cars are (approximately) normally distributed.

b. Test at the 1% significance level if the sample result indicates that the population variance is different from .30.

11.49 The manufacturer of a certain brand of light bulbs claims that the variance of the lives of these bulbs is 4000 square hours. A consumer agency took a random sample of 25 such bulbs and tested them. The variance of the lives of these bulbs was found to be 4990 square hours. Assume that the lives of all such bulbs are (approximately) normally distributed.

a. Make the 99% confidence intervals for the variance and standard deviation of the lives of all such bulbs.

b. Test at the 5% significance level if the variance of such bulbs is different from 4000 square hours.

GLOSSARY

Chi-square distribution A distribution, with degrees of freedom as the only parameter, that is skewed to the right for small df and looks like a normal curve for large df.

Expected frequencies The frequencies for different categories of a multinomial experiment or for different cells of a contingency table that are expected to occur when a given null hypothesis is true.

Goodness-of-fit test A test of the null hypothesis that the observed frequencies for an experiment follow a certain pattern or theoretical distribution.

Multinomial experiment An experiment with n trials for which (1) the trials are identical, (2) there are more than two possible outcomes per trial, (3) the trials are independent, and (4) the probabilities of various outcomes remain constant for each trial.

Observed frequencies The frequencies actually obtained from the performance of an experiment.

Test of homogeneity A test of the null hypothesis that the proportions of elements that belong to different groups in two (or more) populations are similar.

Test of independence A test of the null hypothesis that two attributes of a population are not related.

KEY FORMULAS

1. **Expected frequency for a category for a goodness-of-fit test**

$$E = np$$

where n is the sample size and p is the probability that an element belongs to this category.

2. **Value of the test statistic χ^2 for a goodness-of-fit test and a test of independence or homogeneity**

$$\chi^2 = \sum \frac{(O - E)^2}{E}$$

where O and E are the observed and expected frequencies, respectively, for a category or cell.

3. **Degrees of freedom for a goodness-of-fit test**

$$df = k - 1$$

where k denotes the number of categories for the experiment.

4. **Degrees of freedom for a test of independence or homogeneity**

$$df = (R - 1)(C - 1)$$

where R and C are, respectively, the number of rows and columns for the contingency table.

5. **Expected frequency for a cell for an independence or homogeneity test**

$$E = \frac{(\text{Row total}) (\text{Column total})}{\text{Sample size}}$$

6. **The $(1 - \alpha)100\%$ confidence interval for population variance σ^2**

$$\frac{(n - 1) s^2}{\chi^2_{\alpha/2}} \quad \text{to} \quad \frac{(n - 1) s^2}{\chi^2_{1-\alpha/2}}$$

where $\chi^2_{\alpha/2}$ and $\chi^2_{1-\alpha/2}$ are obtained from the chi-square table for $\alpha/2$ and $1 - \alpha/2$ areas in the right tail of the chi-square distribution curve and $n - 1$ degrees of freedom.

7. **Value of the test statistic χ^2 in a hypothesis test about σ^2**

$$\chi^2 = \frac{(n - 1) s^2}{\sigma^2}$$

where s^2 is the sample variance, σ^2 is the hypothesized value of the population variance, and $n - 1$ are the degrees of freedom.

SUPPLEMENTARY EXERCISES

11.50 In a *Self* magazine readers' survey, women were asked about the greatest source of stress in their lives. The responses were work, 26%; finances, 22%; family, 17%; "significant other," 17%; and other, 18% (*USA Today*, March 30, 1995). Assume that these results hold true for the population of all U.S. women at the time of the survey. A random sample of 1000 women was taken recently, and these women were asked the same question. The results of this survey are given in the table below.

Source of stress	Work	Finances	Family	Significant Other	Other
Number of women	251	240	158	191	160

Test at the 1% significance level whether the current distribution of sources of stress for women differs from that of 1995.

11.51 One of the products produced by Branco Food Company is All-Bran Cereal, which competes with three other brands of similar all-bran cereals. The company's research office wants to investigate if the percentage of people who consume all-bran cereal is the same for each of these four brands. Let us denote the four brands of cereal by A, B, C, and D. A sample of 1000 persons who consume all-bran cereal was taken, and they were asked which brand they most often consume. Of the respondents, 212 said they usually consume Brand A, 284 consume Brand B, 259 consume Brand C, and 245 consume Brand D. Does the sample provide enough evidence to reject the null hypothesis that the percentage of people who consume all-bran cereal is the same for all four brands? Use $\alpha = .05$.

11.52 A 1994 survey by the U.S. Department of Health and Human Services collected information on injury-related visits to hospital emergency departments. The following table lists the classification of these visits according to the place the injury occurred (*Source:* National Center for Health Statistics, *Advance Data*, May 17, 1996).

Place of injury	Home	Street or Highway	Work	School or Day Care	Other or Unspecified
Percentage of patients	39.0	13.6	13.0	4.0	30.4

A recently taken random sample of 10,000 injury-related emergency visits yielded the results given in the following table.

Place of injury	Home	Street or Highway	Work	School or Day Care	Other or Unspecified
Number of patients	3810	1503	1195	506	2986

Does the sample information provide sufficient evidence to reject the null hypothesis that the current distribution of places of injury is the same as for 1994? Use $\alpha = .01$.

11.53   In a 1995 Bruskin/Goldring Research poll, U.S. taxpayers were asked to indicate the greatest source of difficulty they had in preparing their federal income tax returns. The sources named by these taxpayers were: understanding IRS jargon, 43%; knowing deductions, 29%; obtaining the correct forms, 10%; calculations, 8%; and do not know, 10% (*Source: USA Today*, February 22, 1995). Assume that these results hold true for the 1995 population of all U.S. taxpayers. A recent sample of 500 U.S. taxpayers produced the results listed in the following table.

Source of difficulty	IRS Jargon	Deductions	Correct Forms	Calculations	Unknown
Number of taxpayers	202	136	53	37	72

Test at the 5% significance level whether the current percentage distribution of sources of difficulty in preparing federal income tax returns differs from that of 1995.

11.54   The following table shows the number of persons in a random sample of 210 listed according to the day of the week on which they prefer to do their grocery shopping.

Day	Mon	Tue	Wed	Thu	Fri	Sat	Sun
Number of persons	9	15	12	26	38	69	41

Using the 2.5% significance level, test the null hypothesis that the proportion of persons who prefer to do their grocery shopping on a particular day is the same for all days of the week.

11.55   A randomly selected sample of 100 persons who suffer from allergies were asked during what season they suffer the most. The results of the survey are recorded in the following table.

Season	Fall	Winter	Spring	Summer
Persons allergic	18	13	31	38

Using the 1% significance level, test the null hypothesis that the proportions of all allergic persons are equally distributed over the four seasons.

11.56   The president of a bank selected a sample of 200 loan applications to check if the approval or rejection of an application depends on which one of the two loan officers, Thurow or Webber, handles that application. The information obtained from the sample is summarized in the following table.

	Approved	Rejected
Thurow	57	38
Webber	69	36

Test at the 2.5% significance level if the approval or rejection of a loan application depends on which loan officer handles the application.

11.57 A study of the hunting behavior of African wild dogs was conducted in the Selous Game Reserve in Tanzania between November 1991 and March 1994 (*Source:* Scott Creel and Nancy Marusha Creel, "Communal Hunting and Pack Size in African Wild Dogs, *Lycaon pictus,*" *Animal Behavior* 50[5], November 1995). The table below summarizes researchers' observations of 736 hunts by wild dogs. Each hunt is classified by species of prey and the outcome of the hunt.

		Outcome of the Hunt	
		Success	**Failure**
	Impala	188	105
	Wildebeest	100	166
Species	Warthog	31	57
	African hare	10	22
	Zebra	2	28
	Common duiker	16	11

Using the 1% significance level, can you conclude that the species and the outcomes of the hunts are related?

11.58 A random sample of 100 jurors was selected and asked whether or not each of them had ever been a victim of crime. The jurors were also asked whether they are strict, fair, or lenient regarding punishment for crime. The following table gives the results of the survey.

	Strict	**Fair**	**Lenient**
Have been a victim	20	8	3
Have never been a victim	22	38	9

Test at the 5% significance level if the two attributes for all jurors are dependent.

11.59 The recession and bad economic conditions have forced many people to hold more than one job to make ends meet. A sample of 500 persons who held more than one job produced the following two-way table.

	Single	**Married**	**Other**
Male	69	212	39
Female	33	102	45

Test at the 10% significance level if gender and marital status are related for all people who hold more than one job.

11.60 In the past few years the drug paclitaxel has been prominent in media coverage of new cancer treatments. Paclitaxel, originally made from the bark of the pacific yew tree but now manufactured synthetically, is also known as "taxol." This drug was featured in a recent study comparing two therapies for advanced ovarian cancer (*Source:* William P. McGuire, M.D., et al., "Cyclophosphamide and Cisplatin Compared with Paclitaxel and Cisplatin in Patients with Stage III and Stage IV Ovarian Cancer," *The New England Journal of Medicine* 334[1], January 4, 1996). A total of 216 women with ovarian cancer were randomly divided into two groups. Of them, 116 received the standard combination of cisplatin and cyclophosphamide, and the remaining 100 women were given cisplatin and paclitaxel. The data given in the table below indicate the number of these patients who experienced each of three levels of response (cure) for the two treatments.

		Cisplatin and Cyclophosphamide	Cisplatin and Paclitaxel
Clinical Response	Complete	36	51
	Partial	34	22
	None	46	27

Using the 2.5% significance level, test to find whether responses are similar for both treatments.

11.61   A random sample of 100 persons was selected from each of four regions in the United States. These people were asked whether or not they support a certain farm subsidy program. The results of the survey are summarized in the following table.

	Favor	Oppose	Uncertain
Northeast	56	33	11
Midwest	73	23	4
South	67	28	5
West	59	35	6

Using the 1% significance level, test the null hypothesis that the percentages of people with different opinions are similar for all four regions.

11.62   Construct the 98% confidence intervals for the population variance and standard deviation for the following data assuming that the respective populations are (approximately) normally distributed.

a. $n = 21$, $s^2 = 9.2$ b. $n = 17$, $s^2 = 1.7$

11.63   Construct the 95% confidence intervals for the population variance and standard deviation for the following data assuming that the respective populations are (approximately) normally distributed.

a. $n = 12$, $s^2 = 7.2$ b. $n = 19$, $s^2 = 14.8$

11.64   Refer to Exercise 11.62a. Test at the 5% significance level if the population variance is different from 6.5.

11.65   Refer to Exercise 11.62b. Test at the 2.5% significance level if the population variance is greater than 1.1.

11.66   Refer to Exercise 11.63a. Test at the 1% significance level if the population variance is greater than 4.2.

11.67   Refer to Exercise 11.63b. Test at the 5% significance level if the population variance is different from 10.4.

11.68   Usually people do not like waiting in line for service for a long time. A bank manager does not want the variance of the waiting times for her customers to be higher than 4.0 square minutes. A random sample of 25 customers taken from this bank gave the variance of the waiting times equal to 8.1 square minutes.

  a.   Test at the 1% significance level if the variance of the waiting times for all customers at this bank is higher than 4.0 square minutes. Assume that the waiting times for all customers are normally distributed.

  b.   Construct a 99% confidence interval for the population variance.

11.69   The variance of the SAT scores for all students who took that test this year is 5000. The variance of the SAT scores for a random sample of 20 students from one school is equal to 3225.

  a.   Test at the 2.5% significance level if the variance of the SAT scores for students from this school is lower than 5000. Assume that the SAT scores for all students at this school are (approximately) normally distributed.

b. Construct the 98% confidence intervals for the variance and the standard deviation of SAT scores for all students at this school.

11.70 A company manufactures ball bearings that it supplies to other companies. The machine that is used to manufacture these ball bearings produces them with a variance of diameters of .025 square millimeters or less. The quality control officer takes a sample of such ball bearings quite often and checks, using confidence intervals and tests of hypotheses, whether or not the variance of these bearings is within .025 square millimeters. If it is not, the machine is stopped and adjusted. A recently taken random sample of 23 ball bearings gave a variance of the diameters equal to .034 square millimeters.

a. Using the 5% significance level, can you conclude that the machine needs an adjustment? Assume that the diameters of all ball bearings have a normal distribution.

b. Construct a 95% confidence interval for the population variance.

11.71 A random sample of 25 students taken from a university gave the variance of their GPAs equal to .19.

a. Construct the 99% confidence intervals for the population variance and standard deviation. Assume that the GPAs of all students are (approximately) normally distributed.

b. The variance of GPAs of all students at this university was .13 two years ago. Test at the 1% significance level if the variance of GPAs now is different from .13.

11.72 A sample of seven students selected from a small college produced the following data on their heights (in inches).

$$65 \quad 74 \quad 68 \quad 62 \quad 60 \quad 68 \quad 66$$

a. Using the formula learned in Chapter 3, find the sample variance s^2 for these data.

b. Make the 98% confidence intervals for the population variance and standard deviation. Assume that the population from which this sample is selected is normally distributed.

c. Test at the 1% significance level if the population variance is greater than 15 square inches.

11.73 The following are the prices of the same brand of camcorder found at eight stores in Los Angeles.

$$\$749 \quad 815 \quad 789 \quad 799 \quad 732 \quad 825 \quad 799 \quad 769$$

a. Using the formula learned in Chapter 3, find the sample variance s^2 for these data.

b. Make the 95% confidence intervals for the population variance and standard deviation. Assume that the prices of this camcorder at all stores in Los Angeles follow a normal distribution.

c. Test at the 5% significance level if the population variance is different from 500 square dollars.

*11.74** A clinical trial was conducted to test the effectiveness of ''nicotine patch'' therapy combined with counseling for cessation of smoking (*Source:* Douglas E. Jorenby, Ph.D. et al., ''Varying Nicotine Patch Dose and Type of Smoking Cessation Counseling,'' *The Journal of the American Medical Association* 274[17], November 1, 1995). Each of a total of 252 smokers was given a 22-mg dose of transdermal nicotine (patch) therapy and then randomly assigned to one of three types of counseling groups. Of 252 smokers, 85 received minimal counseling, 80 received individual counseling, and 87 had group counseling. At the end of 26 weeks, the number of participants who had quit smoking was tabulated for each group. These numbers are given in the table below.

	Type of Counseling		
	Minimal	**Individual**	**Group**
Number of Smokers Who Quit Smoking	22	27	23

Using the 5% significance level, test the null hypothesis that all three types of counseling are similar in regard to the cessation rates.

*11.75   The author of an article published in the *CONNECTICUT Magazine* (February, 1996) interviewed 75 shoppers. Of them, 69% were females and 31% were males. Of all these shoppers, 31% indicated that they enjoy grocery shopping, while 69% said they do not. Also, 83% of the men and 63% of the women said they do not enjoy grocery shopping. Using $\alpha = .05$, can you conclude that the enjoyment of grocery shopping and gender are dependent? Note that all percentages mentioned here are rounded to the nearest whole numbers.

SELF-REVIEW TEST

1. The random variable χ^2 assumes only
    a. positive   **b.** nonnegative   **c.** nonpositive values

2. The parameter(s) of the chi-square distribution is(are)
    a. degrees of freedom   **b.** *df* and *n*   **c.** χ^2

3. Which of the following is *not* a characteristic of a multinomial experiment?
    a. It consists of *n* identical trials.
    b. There are *k* possible outcomes for each trial and $k > 2$.
    c. The trials are random.
    d. The trials are independent.
    e. The probabilities of outcomes remain constant for each trial.

4. The observed frequencies for a goodness-of-fit test are
    a. the frequencies obtained from the performance of an experiment
    b. the frequencies given by the product of *n* and *p*
    c. the frequencies obtained by adding the results of a and b

5. The expected frequencies for a goodness-of-fit test are
    a. the frequencies obtained from the performance of an experiment
    b. the frequencies given by the product of *n* and *p*
    c. the frequencies obtained by adding the results of a and b

6. The degrees of freedom for a goodness-of-fit test are
    a. $n - 1$ **b.** $k - 1$ **c.** $n + k - 1$

7. The chi-square goodness-of-fit test is always
    a. two-tailed   **b.** left-tailed   **c.** right-tailed

8. To apply a goodness-of-fit test, the expected frequency of each category must be at least
    a. 10   **b.** 5   **c.** 8

9. The degrees of freedom for a test of independence are
    a. $(R - 1)(C - 1)$ **b.** $n - 2$ **c.** $(n - 1)(k - 1)$

10. In 1994, U.S. residents aged 12 and older experienced approximately 42.4 million crimes, of which 73% were property crimes, 26% were violent crimes, and 1% were personal thefts such as purse snatching (*Source:* U.S. Bureau of Justice Statistics, April 1996). A recent random sample of 1000 U.S. residents aged 12 and older who had experienced one or more crimes yielded the following frequency distribution.

Type of Crime	Number of Crimes
Property crime	745
Violent crime	241
Personal theft	14

Test at the 1% significance level if the current distribution of types of crime is different from that of 1994.

11. The following table gives the two-way classification of 1000 persons who have been married at least once. They are classified by educational level and marital status.

	Educational Level			
	Less Than High School	High School Degree	Some College	College Degree
Divorced	173	158	95	53
Never divorced	162	126	116	117

Test at the 1% significance level if educational level and ever being divorced are dependent.

12. A researcher wanted to investigate if people belonging to different income groups are homogeneous with regard to playing lotteries. She took a sample of 600 people from the low-income group, another sample of 500 people from the middle-income group, and a third sample of 400 people from the high-income group. All these people were asked whether they play the lottery often, sometimes, or never. The results of the survey are summarized in the following table.

	Income Group		
	Low	Middle	High
Play often	170	160	90
Play sometimes	290	220	120
Never play	140	120	190

Using the 5% significance level, can you reject the null hypothesis that the percentages of people who play the lottery often, sometimes, and never are the same for each income group?

13. A cough syrup drug manufacturer requires that the variance for a chemical contained in the bottles of this drug should not exceed .03 square ounces. A sample of 25 such bottles gave the variance for this chemical as .06 square ounces.

    a. Construct the 99% confidence intervals for the population variance and the population standard deviation. Assume that the amount of this chemical in all such bottles is (approximately) normally distributed.

    b. Test at the 1% significance level if the variance of this chemical in all such bottles exceeds .03 square ounces.

USING MINITAB

In this section we first discuss how MINITAB is used to make a goodness-of-fit test and then how to make a test of independence.

A GOODNESS-OF-FIT TEST

MINITAB does not have a direct command to make a goodness-of-fit test. However, by combining a few commands, we can make such a test. Illustration M11–1 describes the procedure to perform a goodness-of-fit test using MINITAB. Note that in Illustration M11–1, we use the MINITAB COMMAND LANGUAGE to make the goodness-of-fit test because MINITAB FOR WINDOWS does not have a procedure to perform this test.

ILLUSTRATION M11–1 Refer to Example 11–4. In a 1995 American Express Financial Advisor sample survey, married women aged 45 and over were asked who makes the financial decisions in their households. The results of the survey are shown in the table below. Assume that these results hold true for the 1995 population of all such households.

Decisions made by	Wife	Husband	Both
Percentage of households	15	36	49

In a recent sample of 400 married women aged 45 and over, 68 indicated that in their households they make the financial decisions themselves, 124 said that their husbands make such decisions, and 208 stated that such decisions are made by both together. Test at the 2.5% significance level whether the current pattern of financial decision making is different from that for 1995.

Solution The null and alternative hypotheses are

H_0: The current percentage distribution of financial decision making in such households is the same as that for 1995

H_1: The current percentage distribution of financial decision making in such households is different from that for 1995

To make a goodness-of-fit test for this illustration, perform the following steps.

1. Enter the data on observed frequencies and probabilities in columns C1 and C2, respectively, using the **READ** command.
2. Find the sum of column C1 (the column of observed frequencies) and put it in K1. This gives the sample size.
3. Create column C3 by multiplying K1 by the probabilities listed in column C2. Column C3 lists the expected frequencies.
4. Create column C4 using the formula $(O - E)^2/E$. In MINITAB, this formula will be written as **LET C4 = (C1 − C3)**2/C3**.
5. Find the sum of the values listed in column C4, which gives the value of the test statistic χ^2.
6. Find the value of χ^2 for the given α and df from the chi-square distribution table and make a decision by comparing it with the value of the test statistic χ^2.

All these steps are shown in Figure 11.12, which presents the MINITAB input and output display.

Figure 11.12 MINITAB input and output for goodness-of-fit test.

```
MTB > Note: making a goodness-of-fit test for illustration m11-1

MTB > read c1 c2  ⟵  ⎰This command enters the data on observed frequencies and
DATA > 68    .15        ⎱probabilities in columns C1 and C2, respectively.
DATA > 124   .36
DATA > 208   .49
DATA > end
       3 rows read

MTB > sum c1 put in k1  ⟵  ⎰This command calculates the sum of the values
                            ⎱entered in column C1 and puts it in K1.
```

Column Sum

```
    Sum of C1  =   400.00  ⟵  This is the sample size.

MTB > let c3 = k1 * c2  ⟵  ⎰This command creates column C3 of expected frequencies
                            ⎱by multiplying K1 and column C2 values.

MTB > let c4 = (c1c3)**2/c3  ⟵  ⎧This command calculates
                                 ⎨$\dfrac{(O - E)^2}{E}$ for each category.
                                 ⎩

MTB > print c1 - c4
```

Data Display

```
Row      C1        C2        C3        C4
 1       68       0.15       60      1.06667 ⎫   ⎰Compare this table
 2      124       0.36      144      2.77778 ⎬ ⟵ ⎨with Table 11.4 of
 3      208       0.49      196      0.73469 ⎭   ⎱Example 11–4.

MTB > sum c4  ⟵  ⎰This command prints the sum of column C4,
                  ⎱which is the value of the test statistic $\chi^2$.
```

Column Sum

```
    Sum of C4 =   4.5791  ⟵  The value of the test statistic $\chi^2$
```

From the MINITAB solution given in Figure 11.12, the value of the test statistic χ^2 is 4.5791. For this illustration, $\alpha = .025$ and $df = 3 - 1 = 2$. From the chi-square distribution table, the critical value of χ^2 for $\alpha = .025$ and $df = 2$ is 7.378. (See Figure 11.5 of Example 11–4.) The value of the test statistic, $\chi^2 = 4.5791$, is smaller than the critical value of $\chi^2 = 7.378$, and it falls in the nonrejection region. Consequently, we fail to reject the null hypothesis and state that the current percentage distribution of financial decision making in households with married women aged 45 and over is not different from that for 1995. The difference between the observed frequencies and the expected frequencies seems to have occurred only because of sampling error.

A TEST OF INDEPENDENCE OR HOMOGENEITY

Illustration M11–2 describes the procedure used to make a chi-square test of independence. The procedure for a chi-square test of homogeneity is similar.

ILLUSTRATION M11–2 Refer to Example 11–5. Violence and lack of discipline have become major problems in schools in the United States. A random sample of 300 adults was selected, and they were asked if they favor giving more freedom to schoolteachers to punish students for violence and lack of discipline. The two-way classification of the responses of these adults is presented in the following table.

	In Favor (F)	Against (A)	No Opinion (N)
Men (M)	93	70	12
Women (W)	87	32	6

Does the sample provide sufficient evidence to conclude that the two attributes, gender and opinion, are dependent? Use $\alpha = .01$.

Solution   The null and alternative hypotheses are

$$H_0: \text{Gender and opinion are independent}$$

$$H_1: \text{Gender and opinion are dependent}$$

If you are using MINITAB FOR WINDOWS, perform the following steps to obtain the solution.

Step 1. Enter the data given in the three columns of the contingency table into columns C1, C2, and C3 of the Data window.

Step 2. Click the **Stat** pull-down menu at the top of the screen.

Step 3. Click **Tables** from the selections available in the **Stat** menu.

Step 4. Click **Chisquare Test** from the options available in the **Tables** menu.

Step 5. You will see a dialog box entitled **Chisquare Test** appear on the screen. Type **C1–C3** in the box below **Columns containing the table**.

Step 6. Click the **OK** button at the bottom of this dialog box. The MINITAB output will appear on the screen.

If you are using the MINITAB COMMAND LANGUAGE, enter the given data in columns C1, C2, and C3 using the **SET** or **READ** command. Then, type the following MINITAB command to obtain the solution.

```
MTB > CHISQUARE C1–C3
```

Whether you use MINITAB FOR WINDOWS or the MINITAB COMMAND LANGUAGE, you will obtain the MINITAB output given in Figure 11.13 for the data of Illustration M11–2.

Figure 11.13 MINITAB output for the test of independence of Illustration M11–2.

CHI-Square Test

Expected counts are printed below observed counts

```
             C1        C2        C3     Total
   1         93        70        12       175
          105.00     59.50     10.50 ←
                                            { These rows give the
   2         87        32         6       125 { expected frequencies.
           75.00     42.50      7.50 ←

Total      180       102        18       300


Chi-Sq = 1.371 + 1.853 + 0.214 +
         1.920 + 2.594 + 0.300 = 8.253    This is the value of the test statistic.
←

DF = 2, P-Value = 0.016
```

From the MINITAB output presented in Figure 11.13, the value of the test statistic is $\chi^2 = 8.253$. The critical value of χ^2 from Table IX of Appendix C for $\alpha = .01$ and

$df = (2 - 1)(3 - 1) = 2$ is 9.210. Because the value of the test statistic, $\chi^2 = 8.253$ is smaller than the critical value of $\chi^2 = 9.210$ and it falls in the nonrejection region, we fail to reject the null hypothesis (see Figure 11.6 of Example 11–6). Consequently, we conclude that the two characteristics, *gender* and *opinion*, are independent.

We can also use the *p*-value to make this decision. As we can observe from Figure 11.13, the *p*-value is .016, which is greater than $\alpha = .01$. Hence, we fail to reject the null hypothesis and conclude that the two characteristics, *gender* and *opinion*, are independent. ▬

COMPUTER ASSIGNMENTS

M11.1 The following table gives the percentage distribution by educational attainment of U.S. full-time workers 25 years of age and over for the first quarter of 1996 (*Source*: *U.S. Bureau of Labor Statistics News,* April 19, 1996.)

Educational Attainment	Percentage of Workers
Less than a high school diploma	9.9
High school graduate, no college	32.5
Some college or associate's degree	27.8
Bachelor's degree	19.7
Advanced degree	10.1

A recent sample of 1000 U.S. full-time workers in this age group showed that 112 have less than a high school diploma, 344 are high school graduates with no college, 252 have some college or an associate's degree, 193 hold a bachelor's degree, and 99 possess an advanced degree. Using MINITAB, test at the 1% significance level if the percentage distribution by educational attainment of U.S. full-time workers in this age group has changed since the first quarter of 1996.

M11.2 A sample of 4000 persons aged 18 and older produced the following two-way classification table.

	Men	Women
Single	531	357
Married	1375	1179
Widowed	55	195
Divorced	139	169

Using MINITAB, test at the 10% significance level if sex and marital status are dependent for all persons aged 18 and older.

M11.3 Two samples, one of 3000 students from urban high schools and another of 2000 students from rural high schools, were taken. These students were asked if they have ever smoked. The following table lists the summary of the results.

	Urban	Rural
Have never smoked	1448	1228
Have smoked	1552	772

Using the 5% significance level, test the null hypothesis that the proportions of urban and rural students who have smoked and who have never smoked are homogeneous.

12

ANALYSIS OF VARIANCE

Chapter 10 described the procedures that are used to test hypotheses about the difference between two population means using the normal and t distributions. Also described in that chapter were the hypothesis-testing procedures for the difference between two population proportions using the normal distribution. Then, Chapter 11 explained the procedures to test hypotheses about the equality of more than two population proportions using the chi-square distribution.

This chapter explains how to test the null hypothesis that the means of more than two populations are equal. For example, suppose that teachers at a school have devised three different methods to teach arithmetic. They want to find out if these three methods produce different mean scores. Let μ_1, μ_2, and μ_3 be the mean scores of all students who will be taught by methods I, II, and III, respectively. To test whether or not the three teaching methods produce different means, we test the null hypothesis

$$H_0: \mu_1 = \mu_2 = \mu_3 \qquad \text{(All three population means are equal)}$$

against the alternative hypothesis

$$H_1: \text{All three population means are not equal}$$

We use the analysis of variance procedure to perform such a test of hypothesis.

Note that the analysis of variance procedure can be used to compare two population means. However, the procedures learned in Chapter 10 are more efficient for performing tests of hypotheses about the difference between two population means, and the analysis of variance procedure, to be discussed in this chapter, is used to compare three or more population means.

An *analysis of variance* test is performed using the F distribution. First, the F distribution is described in Section 12.1 of this chapter. Then, Section 12.2 discusses the application of the one-way analysis of variance procedure to perform tests of hypotheses.

12.1 THE *F* DISTRIBUTION

Like the t and chi-square distributions, the shape of a particular **F distribution**[1] curve depends on the number of degrees of freedom. However, the F distribution has *two* numbers of degrees of freedom: *degrees of freedom for the numerator* and *degrees of freedom for the denominator*. These two numbers representing two types of degrees of freedom are the *parameters of the F distribution*. Each combination of degrees of freedom for the numerator and for the denominator gives a different F distribution curve. The units of an F distribution are denoted by F, which assumes only nonnegative values. Like the normal, t, and chi-square distributions, the F distribution is also a continuous distribution. The shape of an F distribution curve is skewed to the right, but the skewness decreases as the number of degrees of freedom increases.

[1]The F distribution is named after Sir Ronald Fisher.

THE *F* DISTRIBUTION

1. The *F distribution* is continuous and skewed to the right.
2. The *F* distribution has two numbers of degrees of freedom: *df* for the numerator and *df* for the denominator.
3. The units of an *F* distribution, denoted by *F*, are nonnegative.

For an *F* distribution, degrees of freedom for the numerator and degrees of freedom for the denominator are usually written as follows.

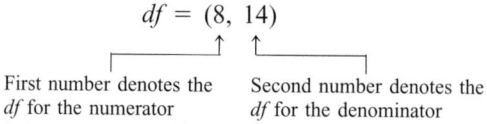

$$df = (8, 14)$$

First number denotes the   Second number denotes the
df for the numerator   *df* for the denominator

Figure 12.1 gives three *F* distribution curves for three sets of degrees of freedom for the numerator and for the denominator. In the figure, the first number gives the degrees of freedom associated with the numerator and the second number gives the degrees of freedom associated with the denominator. We can observe from this figure that as the degrees of freedom increase, the peak of the curve moves to the right, that is, the skewness decreases.

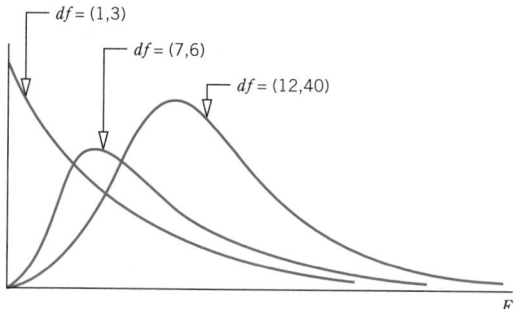

Figure 12.1   Three *F* distribution curves.

Table X in Appendix C lists the values of *F* for the *F* distribution. To read Table X, we need to know three quantities: the degrees of freedom for the numerator, the degrees of freedom for the denominator, and an area in the right tail of an *F* distribution curve. Note that the *F* distribution table (Table X) is read only for an area in the right tail of the *F* distribution curve. Also note that Table X has four parts. These four parts give the *F* values for an area of .01, .025, .05, and .10, respectively, in the right tail of the *F* distribution curve. Example 12–1 illustrates how to read Table X.

Reading the F distribution table.

EXAMPLE 12–1   Find the *F* value for 8 degrees of freedom for the numerator, 14 degrees of freedom for the denominator, and .05 area in the right tail of the *F* distribution curve.

Solution   To find the required value of *F*, we consult the portion of Table X of Appendix C that corresponds to .05 area in the right tail of the *F* distribution curve. The relevant portion of that table is shown here as Table 12.1. To find the required *F* value, we locate 8 in the row for degrees of freedom for the numerator (at the top of Table X) and 14 in the column for degrees of freedom for the denominator (the first column on the left side in Table X). The entry where the column for 8 and the row for 14 intersect gives the required F value. This value of F is 2.70, as shown in Table 12.1 and Figure 12.2. The F value taken from this table for a test of hypothesis is called the critical value of F.

Table 12.1

		Degrees of Freedom for the Numerator				
		1	2	...	8	...	100
	1	161.5	199.5	...	238.9	...	253.0
Degrees of Freedom for	2	18.51	19.00	...	19.37	...	19.49
the Denominator	.	...	...	...	...	...	...
	.	...	...	...	...	...	...
	14	4.60	3.74	...	2.70	...	2.19
	.	...	...	...	...	...	...
	.	...	...	...	...	...	...
	100	3.94	3.09	...	2.03	...	1.39

The F value for 8 df for the numerator, 14 df for the denominator, and .05 area in the right tail.

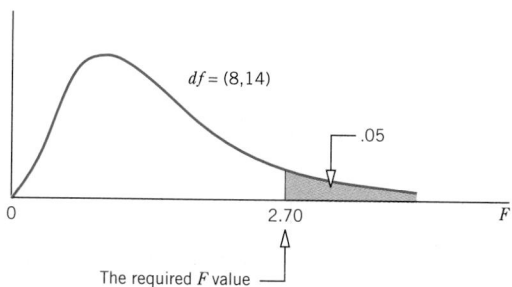

$df = (8,14)$

.05

0   2.70 F

The required F value

Figure 12.2   The critical value of F for 8 df for the numerator, 14 df for the denominator, and .05 area in the right tail.

EXERCISES

Concepts and Procedures

12.1   Describe the main characteristics of an F distribution.

12.2   Find the critical value of F for the following.
   a. $df = (5, 12)$ and area in the right tail $= .05$
   b. $df = (4, 18)$ and area in the right tail $= .05$
   c. $df = (12, 7)$ and area in the right tail $= .05$

12.3   Find the critical value of F for the following.
   a. $df = (5, 12)$ and area in the right tail $= .025$
   b. $df = (4, 18)$ and area in the right tail $= .025$
   c. $df = (12, 7)$ and area in the right tail $= .025$

12.4   Determine the critical value of F for the following.
   a. $df = (8, 11)$ and area in the right tail $= .01$

    b. $df = (6, 12)$ and area in the right tail $= .01$

    c. $df = (15, 5)$ and area in the right tail $= .01$

12.5 Determine the critical value of F for the following.

    a. $df = (3, 14)$ and area in the right tail $= .10$

    b. $df = (9, 10)$ and area in the right tail $= .10$

    c. $df = (11, 4)$ and area in the right tail $= .10$

12.6 Find the critical value of F for an F distribution with $df = (4, 12)$ and

    a. area in the right tail $= .05$

    b. area in the right tail $= .10$

12.7 Find the critical value of F for an F distribution with $df = (11, 6)$ and

    a. area in the right tail $= .01$

    b. area in the right tail $= .025$

12.8 Find the critical value of F for an F distribution with .025 area in the right tail and

    a. $df = (5, 11)$ **b.** $df = (15, 5)$

12.9 Find the critical value of F for an F distribution with .01 area in the right tail and

    a. $df = (10, 10)$ **b.** $df = (9, 25)$

12.2 ONE-WAY ANALYSIS OF VARIANCE

As mentioned in the beginning of this chapter, the analysis of variance procedure is used to test the null hypothesis that the means of three or more populations are the same against the alternative hypothesis that all population means are not the same. The analysis of variance procedure can be used to compare two population means. However, the procedures learned in Chapter 10 are more efficient for performing tests of hypotheses about the difference between two population means, and the analysis of variance procedure is used to compare three or more population means.

Reconsider the example of teachers at a school who have devised three different methods to teach arithmetic. They want to find out if these three methods produce different mean scores. Let μ_1, μ_2, and μ_3 be the mean scores of all students who are taught by methods I, II, and III, respectively. To test if the three teaching methods produce different means, we test the null hypothesis

$$H_0: \mu_1 = \mu_2 = \mu_3 \qquad \text{(All three population means are equal)}$$

against the alternative hypothesis

$$H_1: \text{All three population means are not equal}$$

One method to test such a hypothesis is to test the three hypotheses $H_0: \mu_1 = \mu_2$, $H_0: \mu_1 = \mu_3$, and $H_0: \mu_2 = \mu_3$ separately using the procedure discussed in Chapter 10. Besides being time consuming, such a procedure has other disadvantages. First, if we reject even one of these three hypotheses, then we must reject the null hypothesis $H_0: \mu_1 = \mu_2 = \mu_3$. Second, combining the Type I errors for the three tests (one for each test) will give a very large Type I error for the test $H_0: \mu_1 = \mu_2 = \mu_3$. Hence, we should prefer a procedure that can test the equality of three means in one test. The **ANOVA**, short for **analysis of variance**, provides such a procedure. It is used to compare three or more population means in a single test.

> **ANOVA**
>
> *ANOVA* is a procedure used to test the null hypothesis that the means of three or more populations are equal.

This section discusses the **one-way ANOVA** procedure to make tests comparing the means of several populations. By using a one-way ANOVA test, we analyze only one factor or variable. For instance, in the example of testing for the equality of mean arithmetic scores of students taught by each of the three different methods, we are considering only one factor, which is the effect of different teaching methods on the scores of students. Sometimes we may analyze the effects of two factors. For example, if different teachers teach arithmetic using these three methods, we can analyze the effects of teachers and teaching methods on the scores of students. This is done by using a two-way ANOVA. The procedure under discussion in this chapter is called the analysis of variance because the test is based on the analysis of variation in the data obtained from different samples. The application of one-way ANOVA requires that the following assumptions hold true.

> **ASSUMPTIONS OF ONE-WAY ANOVA**
>
> The following assumptions must hold true to use *one-way ANOVA*.
>
> **1.** The populations from which the samples are drawn are (approximately) normally distributed.
> **2.** The populations from which the samples are drawn have the same variance (or standard deviation).
> **3.** The samples drawn from different populations are random and independent.

For instance, in the example about three methods of teaching arithmetic, we first assume that the scores of all students taught by each method are (approximately) normally distributed. Second, the means of the distributions of scores for three teaching methods may or may not be the same, but all three distributions have the same variance σ^2. Third, when we take samples to make an ANOVA test, these samples are drawn independently and randomly from three different populations.

The ANOVA test is applied by calculating two estimates of the variance, σ^2, of population distributions: the **variance between samples** and the **variance within samples**. The variance between samples is also called the **mean square between samples** or **MSB**. The variance within samples is also called the **mean square within samples** or **MSW**.

The variance between samples, MSB, gives an estimate of σ^2 based on the variation among the means of samples taken from different populations. For the example of three teaching methods, MSB will be based on the values of the mean scores of three samples of students taught by three different methods. If the means of all populations under consideration are equal, the means of respective samples will still be different but the variation among them is expected to be small and, consequently, the value of MSB is expected to be small. However, if the means of populations under consideration are not all equal, the variation among the means of respective samples is expected to be large and, consequently, the value of MSB is expected to be large.

The variance within samples, MSW, gives an estimate of σ^2 based on the variation within the data of different samples. For the example of three teaching methods, MSW will be based on the scores of individual students included in the three samples taken from three populations. The concept of MSW is similar to the concept of the pooled standard deviation, s_p, for two samples discussed in Section 10.2 of Chapter 10.

The one-way ANOVA test is always right-tailed with the rejection region in the right tail of the F distribution curve. The hypothesis-testing procedure using ANOVA involves the same five steps that were used in earlier chapters. The next subsection explains how to calculate the value of the test statistic F for an ANOVA test.

12.2.1   CALCULATING THE VALUE OF THE TEST STATISTIC

The value of the test statistic F for a test of hypothesis using ANOVA is given by the ratio of two variances, the variance between samples (MSB) and the variance within samples (MSW).

TEST STATISTIC F FOR A ONE-WAY ANOVA TEST

The value of the *test statistic F* for an ANOVA test is calculated as

$$F = \frac{\text{Variance between samples}}{\text{Variance within samples}} \quad \text{or} \quad \frac{\text{MSB}}{\text{MSW}}$$

The calculation of MSB and MSW is explained in Example 12–2 below.

Example 12–2 describes the calculation of MSB, MSW, and the value of the test statistic F. Since the basic formulas are laborious to use, they are not presented here. We have used only the short-cut formulas to make calculations in this chapter.

Calculating the value of the test statistic F.

EXAMPLE 12–2   Fifteen fourth-grade students were randomly assigned to three groups in order to experiment with three different methods of teaching arithmetic. At the end of the semester, the same test was given to all 15 students. The table below gives the scores of students in the three groups.

Method I	Method II	Method III
48	55	84
73	85	68
51	70	95
65	69	74
87	90	67

Calculate the value of the test statistic F. Assume that all the required assumptions mentioned in Section 12.2 hold true.

Solution   In ANOVA terminology, the three methods used to teach arithmetic are called **treatments**. The table contains data on the scores of fourth-graders included in the three samples. Each sample of students is taught by a different method. Let

x = the score of a student

k = the number of different samples (or treatments)

n_i = the size of sample i

T_i = the sum of the values in sample i

n = the number of values in all samples = $n_1 + n_2 + n_3 + \ldots$

Σx = the sum of the values in all samples = $T_1 + T_2 + T_3 + \ldots$

Σx^2 = the sum of the squares of the values in all samples

To calculate MSB and MSW, we first compute the **between-samples sum of squares** denoted by **SSB** and the **within-samples sum of squares** denoted by **SSW**. The sum of SSB and SSW is called the **total sum of squares** and it is denoted by **SST**, that is,

$$SST = SSB + SSW$$

The values of SSB and SSW are calculated using the following formulas.

BETWEEN- AND WITHIN-SAMPLES SUM OF SQUARES

The *between-samples sum of squares,* denoted by SSB, is calculated as

$$SSB = \left(\frac{T_1^2}{n_1} + \frac{T_2^2}{n_2} + \frac{T_3^2}{n_3} + \ldots \right) - \frac{(\Sigma x)^2}{n}$$

The *within-samples sum of squares,* denoted by SSW, is calculated as

$$SSW = \Sigma x^2 - \left(\frac{T_1^2}{n_1} + \frac{T_2^2}{n_2} + \frac{T_3^2}{n_3} + \ldots \right)$$

Table 12.2 lists the scores of 15 students who were taught arithmetic by each of the three different methods, the values of T_1, T_2, and T_3, and the values of n_1, n_2, and n_3.

Table 12.2

Method I	Method II	Method III
48	55	84
73	85	68
51	70	95
65	69	74
87	90	67
$T_1 = 324$	$T_2 = 369$	$T_3 = 388$
$n_1 = 5$	$n_2 = 5$	$n_3 = 5$

In Table 12.2, T_1 is obtained by adding the five scores of the first sample. Thus, $T_1 = 48 + 73 + 51 + 65 + 87 = 324$. Similarly, the sum of the values in the second and third samples give $T_2 = 369$ and $T_3 = 388$, respectively. Because there are five observations in each sample, $n_1 = n_2 = n_3 = 5$. The values of Σx and n are

$$\Sigma x = T_1 + T_2 + T_3 = 324 + 369 + 388 = 1081$$

$$n = n_1 + n_2 + n_3 = 5 + 5 + 5 = 15$$

To calculate Σx^2, we square all scores included in all three samples and then add them. Thus,

$$\Sigma x^2 = (48)^2 + (73)^2 + (51)^2 + (65)^2 + (87)^2 + (55)^2 + (85)^2 + (70)^2$$
$$+ (69)^2 + (90)^2 + (84)^2 + (68)^2 + (95)^2 + (74)^2 + (67)^2$$
$$= 80{,}709$$

Substituting all the values in the formulas for SSB and SSW, we obtain the following values of SSB and SSW.

$$\text{SSB} = \left(\frac{(324)^2}{5} + \frac{(369)^2}{5} + \frac{(388)^2}{5} \right) - \frac{(1081)^2}{15} = 432.1333$$

$$\text{SSW} = 80{,}709 - \left(\frac{(324)^2}{5} + \frac{(369)^2}{5} + \frac{(388)^2}{5} \right) = 2372.8000$$

The value of SST is obtained by adding the values of SSB and SSW. Thus,

$$\text{SST} = 432.1333 + 2372.8000 = 2804.9333$$

The variance between samples (MSB) and the variance within samples (MSW) are calculated using the following formulas.

CALCULATING THE VALUES OF MSB AND MSW

The *MSB* and *MSW* are calculated as

$$\text{MSB} = \frac{\text{SSB}}{k - 1} \quad \text{and} \quad \text{MSW} = \frac{\text{SSW}}{n - k}$$

where $k - 1$ and $n - k$ are, respectively, the *df* for the numerator and the *df* for the denominator for the *F* distribution.

Consequently, the variance between samples is

$$\text{MSB} = \frac{\text{SSB}}{k - 1} = \frac{432.1333}{3 - 1} = 216.0667$$

The variance within samples is

$$\text{MSW} = \frac{\text{SSW}}{n - k} = \frac{2372.8000}{15 - 3} = 197.7333$$

The value of the test statistic F is given by the ratio of MSB and MSW. Therefore,

$$F = \frac{\text{MSB}}{\text{MSW}} = \frac{216.0667}{197.7333} = \mathbf{1.09}$$

For convenience, all these calculations are often recorded in a table called the *ANOVA table*. Table 12.3 gives the general form of an ANOVA table.

Table 12.3 ANOVA Table

Source of Variation	Degrees of Freedom	Sum of Squares	Mean Square	Value of the Test Statistic
Between	$k - 1$	SSB	MSB	
Within	$n - k$	SSW	MSW	$F = \dfrac{\text{MSB}}{\text{MSW}}$
Total	$n - 1$	SST		

Substituting the values of the various quantities in Table 12.3, we write an ANOVA table for our example as Table 12.4.

Table 12.4 ANOVA Table for Example 12–2

Source of Variation	Degrees of Freedom	Sum of Squares	Mean Square	Value of the Test Statistic
Between	2	432.1333	216.0667	
Within	12	2372.8000	197.7333	$F = \dfrac{216.0667}{197.7333} = 1.09$
Total	14	2804.9333		

12.2.2 ONE-WAY ANOVA TEST

Now suppose we want to test the null hypothesis that the mean scores are equal for all three groups of fourth-graders taught by three different methods of Example 12–2 against the alternative hypothesis that the mean scores of all three groups are not equal. Note that in a one-way ANOVA test, the null hypothesis is that the means for all populations are equal. The alternative hypothesis is that all population means are not equal. In other words, the alternative hypothesis states that at least one of the population means is different from the others. Example 12–3 demonstrates how we use the one-way ANOVA procedure to make such a test.

Performing a one-way ANOVA test: all samples of the same size.

EXAMPLE 12–3   Reconsider Example 12–2 about the scores of 15 fourth-grade students who were randomly assigned to three groups in order to experiment with three different methods of teaching arithmetic. At the 1% significance level, can we reject the null hypothesis that the mean arithmetic score of all fourth-grade students taught by each of these three methods is the same? Assume that all the assumptions required to apply the one-way ANOVA procedure hold true.

Solution   To make a test about the equality of means of three populations, we follow our standard procedure with five steps.

Step 1.   *State the null and alternative hypotheses*

Let μ_1, μ_2, and μ_3 be the mean arithmetic scores of all fourth-grade students who are taught, respectively, by methods I, II, and III. The null and alternative hypotheses are

H_0: $\mu_1 = \mu_2 = \mu_3$ (the mean scores of the three groups are equal)

H_1: All three means are not equal

Note that the alternative hypothesis states that at least one population mean is different from the other two.

Step 2. *Select the distribution to use*

Because we are comparing the means for three normally distributed populations, we use the F distribution to make this test.

Step 3. *Determine the rejection and nonrejection regions*

The significance level is .01. Because a one-way ANOVA test is always right-tailed, the area in the right tail of the F distribution curve is .01, which is the rejection region in Figure 12.3.

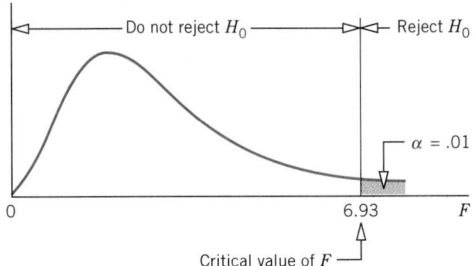

Figure 12.3   Critical value of F for $df = (2, 12)$ and $\alpha = .01$

Next we need to know the degrees of freedom for the numerator and the denominator. In our example, the students were assigned to three different methods. As mentioned earlier, these methods are called treatments. The number of treatments is denoted by k. The total number of observations in all samples taken together is denoted by n. Then the number of degrees of freedom for the numerator is equal to $k - 1$ and the number of degrees of freedom for the denominator is equal to $n - k$. In our example, there are 3 treatments (methods of teaching) and 15 total observations (total number of students) in all three samples. Thus,

$$\text{Degrees of freedom for the numerator} = k - 1 = 3 - 1 = 2$$

$$\text{Degrees of freedom for the denominator} = n - k = 15 - 3 = 12$$

From Table X of Appendix C, we find the critical value of F for 2 df for the numerator, 12 df for the denominator, and .01 area in the right tail of the F distribution curve. This value is shown in Figure 12.3. The required value of F is 6.93.

Thus, we will fail to reject H_0 if the calculated value of the test statistic F is less than 6.93 and we will reject H_0 if it is greater than 6.93.

Step 4. *Calculate the value of the test statistic*

We computed the value of the test statistic F for these data in Example 12–2. This value is

$$F = 1.09$$

Step 5. *Make a decision*

Because the value of the test statistic $F = 1.09$ is less than the critical value of $F = 6.93$, it falls in the nonrejection region. Hence, we fail to reject the null hypothesis and conclude that the means of the three populations are equal. In other words, the three different methods of teaching arithmetic do not seem to affect the mean scores of students. The difference in the three mean scores in the case of our three samples occurred only because of sampling error. ◼

In Example 12–3, the sample sizes were the same for all treatments. Example 12–4 describes a case in which the sample sizes are not the same for all treatments.

Performing a one-way ANOVA test: all samples not of the same size.

EXAMPLE 12–4 From time to time, unknown to its employees, the research department at Post Bank observes various employees for work productivity. Recently this department wanted to check whether the four tellers at a branch of this bank serve, on average, the same number of customers per hour. The research manager observed each of the four tellers for a certain number of hours. The following table gives the number of customers served by the four tellers during each of the observed hours.

Teller A	Teller B	Teller C	Teller D
19	14	11	24
21	16	14	19
26	14	21	21
24	13	13	26
18	17	16	20
	13	18	

At the 5% significance level, test the null hypothesis that the mean number of customers served per hour by each of these four tellers is the same. Assume that all the assumptions required to apply the one-way ANOVA procedure hold true.

Solution To make a test about the equality of means of four populations, we follow our standard procedure with five steps.

Step 1. *State the null and alternative hypotheses*

Let μ_1, μ_2, μ_3, and μ_4 be the mean number of customers served per hour by tellers A, B, C, and D, respectively. The null and alternative hypotheses are

H_0: $\mu_1 = \mu_2 = \mu_3 = \mu_4$ (the mean number of customers served per hour by each of the four tellers is the same)

H_1: All four population means are not equal

Step 2. *Select the distribution to use*

Because we are testing for the equality of four means for four normally distributed populations, we use the F distribution to make the test.

Step 3. *Determine the rejection and nonrejection regions*

The significance level is .05, which means the area in the right tail of the F distribution curve is .05.

In this example, there are four treatments (tellers) and 22 total observations in all four samples. Thus,

$$\text{Degrees of freedom for the numerator} = k - 1 = 4 - 1 = 3$$

$$\text{Degrees of freedom for the denominator} = n - k = 22 - 4 = 18$$

The critical value of F from Table X for 3 df for the numerator, 18 df for the denominator, and .05 area in the right tail of the F distribution curve is 3.16. This value is shown in Figure 12.4.

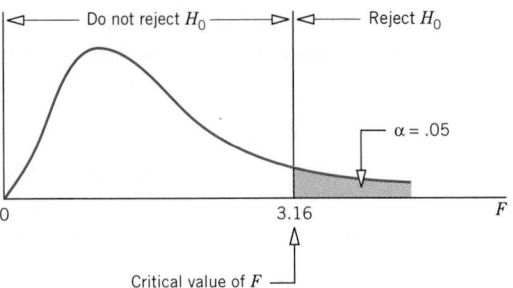

Figure 12.4 Critical value of F for $df = (3, 18)$ and $\alpha = .05$.

Step 4. *Calculate the value of the test statistic*

First we calculate SSB and SSW. Table 12.5 lists the number of customers served by the four tellers during the selected hours, the values of T_1, T_2, T_3, and T_4, and the values of n_1, n_2, n_3, and n_4.

Table 12.5

Teller A	Teller B	Teller C	Teller D
19	14	11	24
21	16	14	19
26	14	21	21
24	13	13	26
18	17	16	20
	13	18	
$T_1 = 108$	$T_2 = 87$	$T_3 = 93$	$T_4 = 110$
$n_1 = 5$	$n_2 = 6$	$n_3 = 6$	$n_4 = 5$

The values of Σx and n are

$$\Sigma x = T_1 + T_2 + T_3 + T_4 = 108 + 87 + 93 + 110 = 398$$

$$n = n_1 + n_2 + n_3 + n_4 = 5 + 6 + 6 + 5 = 22$$

The value of Σx^2 is calculated as follows.

$$\Sigma x^2 = (19)^2 + (21)^2 + (26)^2 + (24)^2 + (18)^2 + (14)^2 + (16)^2 + (14)^2$$
$$+ (13)^2 + (17)^2 + (13)^2 + (11)^2 + (14)^2 + (21)^2 + (13)^2$$
$$+ (16)^2 + (18)^2 + (24)^2 + (19)^2 + (21)^2 + (26)^2 + (20)^2$$

$$= 7614$$

Substituting all the values in the formulas for SSB and SSW, we obtain the following values of SSB and SSW.

$$SSB = \left(\frac{T_1^2}{n_1} + \frac{T_2^2}{n_2} + \frac{T_3^2}{n_3} + \frac{T_4^2}{n_4} \right) - \frac{(\Sigma x)^2}{n}$$

$$= \left(\frac{(108)^2}{5} + \frac{(87)^2}{6} + \frac{(93)^2}{6} + \frac{(110)^2}{5} \right) - \frac{(398)^2}{22} = 255.6182$$

$$SSW = \Sigma x^2 - \left(\frac{T_1^2}{n_1} + \frac{T_2^2}{n_2} + \frac{T_3^2}{n_3} + \frac{T_4^2}{n_4} \right)$$

$$= 7614 - \left(\frac{(108)^2}{5} + \frac{(87)^2}{6} + \frac{(93)^2}{6} + \frac{(110)^2}{5} \right) = 158.2000$$

Hence, the variance between samples MSB and the variance within samples MSW are

$$\text{MSB} = \frac{\text{SSB}}{k-1} = \frac{255.6182}{4-1} = 85.2061$$

$$\text{MSW} = \frac{\text{SSW}}{n-k} = \frac{158.2000}{22-4} = 8.7889$$

The value of the test statistic F is given by the ratio of MSB and MSW, which is

$$F = \frac{\text{MSB}}{\text{MSW}} = \frac{85.2061}{8.7889} = 9.69$$

Writing the values of the various quantities in the ANOVA table, we obtain Table 12.6.

Table 12.6 ANOVA Table for Example 12–4

Source of Variation	Degrees of Freedom	Sum of Squares	Mean Square	Value of the Test Statistic
Between	3	255.6182	85.2061	$F = \dfrac{85.2061}{8.7889} = 9.69$
Within	18	158.2000	8.7889	
Total	21	413.8182		

Step 5. *Make a decision*

Because the value of the test statistic $F = 9.69$ is greater than the critical value of $F = 3.16$, it falls in the rejection region. Consequently, we reject the null hypothesis and conclude that the mean number of customers served per hour by each of the four tellers is not the same. In other words, at least one of the four means is different from the other three. ■

EXERCISES

Concepts and Procedures

12.10 Briefly explain when a one-way ANOVA procedure is used to make a test of hypothesis.

12.11 Describe the assumptions that must hold true to apply the one-way analysis of variance procedure to test hypotheses.

12.12 Consider the following data obtained for two samples selected at random from two populations that are independent and normally distributed with equal variances.

Sample I	Sample II
32	27
26	37
31	33
23	40
27	38
34	31

a. Calculate the means and standard deviations for these samples using the formulas learned in Chapter 3.

b. Using the procedure learned in Section 10.2 of Chapter 10, test at the 1% significance level whether the means of the populations from which these samples are drawn are equal.

   c.   Using the one-way ANOVA procedure, test at the 1% significance level whether the
        means of the populations from which these samples are drawn are equal.
   d.   Are the conclusions reached in parts b and c the same?

12.13   Consider the following data obtained for two samples selected at random from two populations
that are independent and normally distributed with equal variances.

Sample I	Sample II
14	11
16	8
11	12
9	18
13	15
20	7
17	9

   a.   Calculate the means and standard deviations for these samples using the formulas learned
        in Chapter 3.
   b.   Using the procedure learned in Section 10.2 of Chapter 10, test at the 5% significance
        level whether the means of the populations from which these samples are drawn are
        equal.
   c.   Using the one-way ANOVA procedure, test at the 5% significance level whether the
        means of the populations from which these samples are drawn are equal.
   d.   Are the conclusions reached in parts b and c the same?

12.14   The following ANOVA table, based on information obtained for three samples selected from
three independent populations that are normally distributed with equal variances, has a few missing
values.

Source of Variation	Degrees of Freedom	Sum of Squares	Mean Square	Value of the Test Statistic
Between	2		19.2813	
Within		89.3677		$F = \underline{\quad\quad} =$
Total	12			

   a.   Find the missing values and complete the ANOVA table.
   b.   Using $\alpha = .01$, what is your conclusion for the test with the null hypothesis that the
        means of the three populations are all equal against the alternative hypothesis that the
        means of the three populations are not all equal?

12.15   The following ANOVA table, based on information obtained for four samples selected from
four independent populations that are normally distributed with equal variances, has a few missing
values.

Source of Variation	Degrees of Freedom	Sum of Squares	Mean Square	Value of the Test Statistic
Between				
Within	15		9.2154	$F = \underline{\quad\quad} = 4.07$
Total	18			

a. Find the missing values and complete the ANOVA table.

b. Using $\alpha = .05$, what is your conclusion for the test with the null hypothesis that the means of the four populations are all equal against the alternative hypothesis that the means of the four populations are not all equal?

Applications

For the following exercises assume that all the assumptions required to apply the one-way ANOVA procedure hold true.

12.16 A dietitian wanted to test three different diets to find out if the mean weight loss for each of these diets is the same. She randomly selected 21 overweight persons, randomly divided them into three groups, and put each group on one of the three diets. The following table records the weight (in pounds) lost by these persons after being on these diets for two months.

Diet I	Diet II	Diet III
15	11	9
8	16	17
17	9	11
7	16	8
22	24	11
12	17	6
8	19	14

a. We are to test if the mean weight lost by all persons on each of the three diets is the same. Write the null and alternative hypotheses.

b. Show the rejection and nonrejection regions on the F distribution curve for $\alpha = .025$.

c. Calculate SSB, SSW, and SST.

d. What are the degrees of freedom for the numerator and denominator, respectively?

e. Calculate the between-samples and within-samples variances.

f. What is the critical value of F for $\alpha = .025$?

g. What is the calculated value of the test statistic F?

h. Write the ANOVA table for this exercise.

i. Will you reject the null hypothesis stated in part a at a significance level of 2.5%?

12.17 The following table gives the number of classes missed during one semester by 25 randomly selected college students drawn from three different age groups.

Below 25	25 to 30	31 & above
19	9	5
12	6	8
25	11	4
10	14	3
19	8	10
4	9	9
15	3	
14	13	
16	18	
9		

a. We are to test if the mean number of classes missed during the semester by all students in each of these three age groups is the same. Write the null and alternative hypotheses.

b. Show the rejection and nonrejection regions on the F distribution curve for $\alpha = .01$.

c. Calculate SSB, SSW, and SST.

d. What are the degrees of freedom for the numerator and denominator, respectively?

 e. Calculate the between-samples and within-samples variances.
 f. What is the critical value of F for $\alpha = .01$?
 g. What is the calculated value of the test statistic F?
 h. Write the ANOVA table for this exercise.
 i. Will you reject the null hypothesis stated in part a at a significance level of 1%?

12.18 A consumer agency wanted to investigate if four insurance companies differed with regard to the premiums charged for auto insurance. The agency randomly selected a few auto drivers who were insured by each of these four companies and had similar driving records, autos, and insurance policies. The following table gives premiums paid per month by these drivers insured with these four insurance companies.

Company A	Company B	Company C	Company D
65	48	57	62
73	69	61	53
54	88	89	45
43		77	51
		69	

Using the 1% significance level, test the null hypothesis that the mean auto insurance premium paid per month by all drivers insured by each of these four companies is the same.

12.19 Three new brands of fertilizer that a farmer can use to grow crops came on the market recently. Before deciding which brand he should use permanently for all crops, a farmer decided to experiment for one season. To do so, he randomly assigned each fertilizer to eight one-acre tracts of land that he used to grow wheat. The following table gives the production of wheat (in bushels) for each acre for the three brands of fertilizer.

Fertilizer I	Fertilizer II	Fertilizer III
72	58	61
69	42	58
75	53	63
57	47	70
64	45	55
68	52	65
73	47	57
67	57	63

At the 5% significance level, can you conclude that the mean yield of wheat for each of these three brands of fertilizer is the same?

12.20 A consumer agency wanted to find out if the mean time it takes for each of three brands of medicines to provide relief from a headache is the same. The first drug was administered to six randomly selected patients, the second to four randomly selected patients, and the third to five randomly selected patients. The following table gives the time (in minutes) taken by each patient to get relief from a headache after taking the medicine.

Drug I	Drug II	Drug III
25	15	44
38	21	39
42	19	54
65	25	58
47		73
52		

At the 2.5% significance level, will you conclude that the mean time taken to provide relief from a headache is the same for all three drugs?

12.21 A large company buys thousands of light bulbs every year. The company is currently considering four brands of light bulbs to choose from. Before the company decides which light bulbs to buy, it wants to investigate if the mean life of the four types of light bulbs is the same. The company's research department randomly selected a few bulbs of each type and tested them. The following table lists the number of hours (in thousands) that each of the bulbs in each brand survived before being burned out.

Brand I	Brand II	Brand III	Brand IV
23	19	23	26
24	23	27	24
19	18	25	21
26	24	26	29
22	20	23	28
23	22	21	24
25	19	27	28

At the 2.5% significance level, test the null hypothesis that the mean life of bulbs for each of these four brands is the same.

GLOSSARY

Analysis of variance (ANOVA) A statistical technique used to test whether the means of three or more populations are equal.

***F* distribution** A continuous distribution that has two parameters: *df* for the numerator and *df* for the denominator.

Mean square between samples or **MSB** A measure of the variation among means of samples taken from different populations.

Mean square within samples or **MSW** A measure of the variation within data of all samples taken from different populations.

One-way ANOVA The analysis of variance technique that analyzes one variable only.

SSB The sum of squares between samples. Also called the sum of squares of the factor or treatment.

SST The total sum of squares given by the sum of SSB and SSW.

SSW The sum of squares within samples. Also called the sum of squares of errors.

KEY FORMULAS

Let

$$k = \text{the number of different samples (or treatments)}$$
$$n_i = \text{the size of sample } i$$
$$T_i = \text{the sum of the values in sample } i$$
$$n = \text{the number of values in all samples} = n_1 + n_2 + n_3 + \ldots$$
$$\Sigma x = \text{the sum of the values in all samples} = T_1 + T_2 + T_3 + \ldots$$
$$\Sigma x^2 = \text{the sum of the squares of values in all samples}$$

1. **Degrees of freedom for the F distribution**

$$\text{Degrees of freedom for the numerator} = k - 1$$
$$\text{Degrees of freedom for the denominator} = n - k$$

2. **Between-samples sum of squares**

$$\text{SSB} = \left(\frac{T_1^2}{n_1} + \frac{T_2^2}{n_2} + \frac{T_3^2}{n_3} + \dots \right) - \frac{(\Sigma x)^2}{n}$$

3. **Within-samples sum of squares**

$$\text{SSW} = \Sigma x^2 - \left(\frac{T_1^2}{n_1} + \frac{T_2^2}{n_2} + \frac{T_3^2}{n_3} + \dots \right)$$

4. **Total sum of squares**

$$\text{SST} = \text{SSB} + \text{SSW}$$

5. **Variance between samples**

$$\text{MSB} = \frac{\text{SSB}}{k - 1}$$

6. **Variance within samples**

$$\text{MSW} = \frac{\text{SSW}}{n - k}$$

7. **Value of the test statistic F**

$$F = \frac{\text{Variance between samples}}{\text{Variance within samples}} \quad \text{or} \quad \frac{\text{MSB}}{\text{MSW}}$$

SUPPLEMENTARY EXERCISES

For the following exercises, assume that all the assumptions required to apply the one-way ANOVA procedure hold true.

12.22 The following table lists the number of violent crimes reported to police on randomly selected days for this year. The data are taken from three large cities of about the same size.

City A	City B	City C
5	2	8
9	4	12
10	1	7
3	11	3
8	7	9
7	6	11
13		

Using the 5% significance level, test the null hypothesis that the mean number of violent crimes reported per day is the same for each of these three cities.

12.23 A consumer agency wants to check if the mean lives of four brands of auto batteries, which sell for nearly the same price, are the same. The agency randomly selected a few batteries of each brand and tested them. The following table gives the lives of these batteries in thousands of hours.

Brand A	Brand B	Brand C	Brand D
74	53	57	56
68	51	71	51
51	47	81	42
56	55	70	43
65		68	

a. At the 5% significance level, will you reject the null hypothesis that the mean life of each of these four brands of batteries is the same?

b. What is the Type I error in this case and what is the probability of committing such an error? Explain.

 12.24 A researcher wanted to investigate if recent college graduates with different majors who got jobs in San Francisco are commanding the same average salary. She selected a random sample of recent graduates in four areas—engineering, business, mathematics, and sociology. The following table gives the starting salaries (in thousands of dollars) for these samples.

Engineering	Business	Mathematics	Sociology
29.2	23.3	23.3	18.6
36.5	29.8	21.4	19.2
34.3	32.4	28.3	23.9
31.2	27.5	23.6	
32.1	30.6		

a. At the 1% significance level, test the null hypothesis that the mean starting salaries of all recent college graduates with these four majors who got jobs in San Francisco are equal.

b. What is the Type I error in this case and what is the probability of committing such an error? Explain.

12.25 A farmer wants to test three brands of weight-gain diets for chickens to determine if the mean weight gain for each of these brands is the same. He selected 15 chickens and randomly put each of them on one of these three brands of diet. The following table lists the weight (in pounds) gained by these chickens after a period of 1 month.

Brand A	Brand B	Brand C
.8	.6	1.2
1.3	1.3	.8
1.7	.6	.7
.9	.4	1.5
.6	.7	.9

a. At the 1% significance level, can you conclude that the mean weight gained by all chickens is the same for each of these three diets?

b. If you did not reject the null hypothesis in part a, explain the Type II error that you may have made in this case. Note that you cannot calculate the probability of committing a Type II error without additional information.

 12.26 The following table lists the prices of certain randomly selected college textbooks in statistics, psychology, economics, and business.

Statistics	Psychology	Economics	Business
57	68	67	64
51	75	54	75
59	80	80	56
62	71	72	60
51		62	82

a. Using the 5% significance level, test the null hypothesis that the mean prices of college textbooks in statistics, psychology, economics, and business are all equal.

b. If you did not reject the null hypothesis in part a, explain the Type II error that you may have made in this case. Note that you cannot calculate the probability of committing a Type II error without additional information.

12.27 A consumer agency that wanted to compare drying times for paints made by three companies tested a few samples of paints from each of these three companies. The following table records the drying times (in minutes) for these samples of paints.

Company A	Company B	Company C
42	57	45
53	63	45
43	61	51
47	54	58
40	49	44
51	60	41
56		47

a. Using the 5% significance level, test the null hypothesis that the mean drying times for paints of these three companies are equal.

b. What will your decision be if the probability of making a Type I error is zero? Explain.

12.28 According to a 1995 study by the U.S. auto industry, the average assembly time per vehicle for the three principal American auto manufacturers were: General Motors, 46 hours; Ford, 38 hours; and Chrysler, 43 hours (*Source: U.S. News & World Report*, June 10, 1996). Suppose that these means were calculated based on independent random samples of 14 vehicles from General Motors, 13 from Ford, and 16 from Chrysler. Also assume that the between-samples sum of squares was 438.70, the within-samples sum of squares was 571.60, and the assembly times for the three companies were normally distributed with equal variances. Using the 1% level of significance, can you conclude that the mean assembly times for the three companies are the same?

*12.29 Passive smoking (exposure to someone else's tobacco smoke) has been associated with heart disease and respiratory ailments, including lung cancer. A recent study by Dr. Celermajer and others was concerned with the relationship between passive smoking and heart disease (*Source:* D. S. Celermajer et al., "Passive Smoking and Impaired Endothelium-Dependent Arterial Dilatation in Healthy Young Adults," *The New England Journal of Medicine* 334[3], January 18, 1996). The study was based on 78 people who were divided into three groups: (1) a *control* group of 26 people who had never smoked nor been exposed to others' tobacco smoke at home or at work, (2) a *passive smoking* group of 26 people who had never smoked but were exposed to environmental tobacco smoke for at least one hour per day for at least three years, and (3) an *active smoking* group of 26 current smokers with smoking histories of at least two *pack years* (a *pack year* is 20 cigarettes per day for one year or the equivalent). For each of the 78 subjects, the "flow-mediated dilatation of the brachial artery" was measured. Low values of this dilatation are thought to be an early indication of arterial damage. The following table lists the various values based on flow-mediated dilatation results from the experiment.

Group	Mean	T_i
Control	8.2	213.2
Passive smoking	3.1	80.6
Active smoking	4.4	114.4

The calculations show that $\Sigma x^2 = 3164.21$. Using the 1% significance level, can you conclude that the mean dilatation is the same for all three groups? Assume that all assumptions required to apply ANOVA hold true.

*12.30   Suppose that you are a newspaper reporter whose editor has asked you to compare the hourly wages of carpenters, plumbers, electricians, and masons in your city. Since many of these workers are not union members, the wages vary considerably among individuals in the same trade.

    a.   What data should you gather and how would you collect it? What statistics would you present in your article and how would you calculate them? Assume that your newspaper is not intended for technical readers.

    b.   Suppose that you must submit your findings to a technical journal that will require statistical analysis of your data. If you want to determine whether or not the mean hourly wages are the same for all four trades, briefly describe how you would analyze the data. Assume that hourly wages in each trade are normally distributed and that the four variances are equal.

*12.31   The editor of an automotive magazine has asked you to compare the mean gas mileage in city driving of three makes of compact cars. The editor has made available to you one car of each of the three makes, three drivers, and a budget sufficient to buy gas and pay the drivers for approximately 500 miles of city driving for each car.

    a.   Explain how you would conduct an experiment and gather the data for a magazine article comparing the gas mileage.

    b.   Suppose you wish to test the null hypothesis that the mean gas mileages in city driving are the same for all three makes. Outline the procedure for using your data to conduct this test. Assume that the assumptions for applying analysis of variance are satisfied.

SELF-REVIEW TEST

1.   The F distribution is

    a.   continuous   **b.**   discrete   **c.**   neither

2.   The F distribution is always

    a.   symmetric   **b.**   skewed to the right   **c.**   skewed to the left

3.   The units of the F distribution, denoted by F, are always

    a.   nonpositive   **b.**   positive   **c.**   nonnegative

4.   The one-way ANOVA test analyzes only one

    a.   variable   **b.**   population   **c.**   sample

5.   The one-way ANOVA test is always

    a.   right-tailed   **b.**   left-tailed   **c.**   two-tailed

6. For a one-way ANOVA with k treatments and n observations in all samples taken together, the number of degrees of freedom for the numerator are

    a. $k - 1$ **b.** $n - k$ **c.** $n - 1$

7. For a one-way ANOVA with k treatments and n observations in all samples taken together, the number of degrees of freedom for the denominator are

    a. $k - 1$ **b.** $n - k$ **c.** $n - 1$

8. The ANOVA test can be applied to compare

    a. three or more population means
    b. more than four population means only
    c. more than three population means only

9. Briefly describe the assumptions that must hold true to apply the one-way ANOVA procedure as mentioned in this chapter.

10. The following table gives the hourly wage of computer programmers for samples taken from three cities.

New York	Boston	Los Angeles
$15.45	$23.50	$31.75
28.80	18.60	11.40
26.45	21.75	29.40
33.10	30.00	22.30
31.50	35.40	24.60
39.30	26.40	19.50
35.50		21.30

    a. Using the 1% significance level, test the null hypothesis that the mean hourly wage for all computer programmers in each of these three cities is the same.
    b. Is it a Type I error or a Type II error that may have been committed in part a? Explain.

USING MINITAB

Illustration M12–1 describes the use of MINITAB to perform a test of hypothesis using the one-way analysis of variance procedure.

ILLUSTRATION M12-1 According to Example 12–4, the research department at Post Bank wants to know if the mean number of customers served per hour by each of the four tellers at a branch of this bank is the same. The research manager observed each of the four tellers for a certain number of hours. The following table gives the number of customers served by each of the four tellers during each of the observed hours.

Teller A	Teller B	Teller C	Teller D
19	14	11	24
21	16	14	19
26	14	21	21
24	13	13	26
18	17	16	20
	13	18	

Using MINITAB, test the null hypothesis that the mean number of customers served per hour by each of these four tellers is the same. Use $\alpha = .05$ and assume that all the required assumptions to apply the one-way analysis of variance procedure hold true.

Solution Let μ_1, μ_2, μ_3, and μ_4 be the mean number of customers served per hour by each of the four tellers, respectively. Then the null and alternative hypotheses are

$$H_0: \mu_1 = \mu_2 = \mu_3 = \mu_4 \quad \text{(All four population means are equal)}$$

$$H_1: \text{All four population means are not equal}$$

If you are using MINITAB FOR WINDOWS, perform the following steps to obtain the ANOVA solution.

Step 1. Enter the given data in four columns C1 to C4 of the Data window.

Step 2. Click the **Stat** pull-down menu at the top of the screen.

Step 3. Click **ANOVA** from the selections available in the **Stat** menu.

Step 4. Click **Oneway (Unstacked)** from the options available in the **ANOVA** menu.

Step 5. You will see a dialog box entitled **Oneway Analysis of Variance** appear on the screen. Type **C1–C4** in the box below **Responses (in separate columns)**.

Step 6. Click the **OK** button at the bottom of this dialog box. The MINITAB output will appear on the screen.

If you are using the MINITAB COMMAND LANGUAGE, first enter the given data in four columns C1 to C4 using the **SET** command. Remember, you will enter data for one column at a time. Then, type the following MINITAB command to obtain the solution.

    MTB > AOVONEWAY C1–C4

In this MINITAB command, **AOVONEWAY**, **AOV** stands for analysis of variance and **ONEWAY** stands for one-way.

Whether you use MINITAB FOR WINDOWS or the MINITAB COMMAND LANGUAGE, you will obtain the MINITAB output given in Figure 12.5 for the data of Illustration M12–1.

Figure 12.5 MINITAB output for Illustration M12–1.

```
One-Way Analysis of Variance

Analysis of Variance
Source   DF       SS      MS      F       P  ⎫  ← Compare this table
Factor    3   255.62   85.21   9.69   0.000 ⎬    to the ANOVA Table
Error    18   158.20    8.79                 ⎭    12.6 of Example 12–4.
Total    21   413.82

                                   Individual 95% CIs for Mean
                                   Based on Pooled StDev
Level    N     Mean    StDev   ------+---------+---------+---------+
C1       5   21.600    3.362                       (-------*-------)
C2       6   14.500    1.643   (------*-------)
C3       6   15.500    3.619     (------*-------)
C4       5   22.000    2.915                   (------*-------)
                               ------+---------+---------+---------+
Pooled StDev =    2.965         14.0      17.5      21.0      24.5
```

Compare the analysis of variance table in the MINITAB solution of Figure 12.5 with Table 12.6 of Example 12–4. In Table 12.6, the two sources of variation were called the variations between- and within-samples in the column labeled *Source of Variation*. In the MINITAB solution of Figure 12.5 these two sources are called the *factor* and *error*, respectively. Also, MINITAB prints the *p*-value for the test.

The MINITAB solution also gives the following information.

1. The mean and standard deviation for data contained in each of the columns C1, C2, C3, and C4. Thus, for example, the mean and standard deviation for the data of column C1 (for teller A) are 21.6 and 3.362, respectively.

2. The 95% confidence interval for the mean of the population corresponding to each of the four samples, that is, the 95% confidence interval for the mean number of customers served per hour by each of the four tellers.

3. The pooled standard deviation, which is 2.965. This pooled standard deviation is nothing but the square root of what we called MSW in this chapter. The MSW calculated in Example 12–4 was 8.79. The square root of 8.79 is 2.965, which is printed as the pooled standard deviation in the MINITAB solution.

From the MINITAB printout of Figure 12.5, the value of the test statistic F is 9.69. The critical value of F from the F distribution table for $\alpha = .05$, df for the numerator $= 3$, and df for the denominator $= 18$ is 3.16 (see Figure 12.4 of Example 12–4). The value of the test statistic, $F = 9.69$, is larger than the critical value of $F = 3.16$ and it falls in the rejection region. Consequently, we reject the null hypothesis and conclude that the mean number of customers served per hour by each of the four tellers is not the same.

We can reach the same conclusion by considering the *p*-value. The *p*-value from the MINITAB solution is 0.000. Because this *p*-value is less than $\alpha = .05$, we reject the null hypothesis. ■

COMPUTER ASSIGNMENTS

M12.1 Refer to Exercise 12.18. Solve that exercise using MINITAB.

M12.2 Refer to Exercise 12.26. Solve that exercise using MINITAB.

1. Does the use of cellular telephones increase the risk of brain tumors? Suppose a manufacturer of cellular telephones hires you to answer this question because of concern about product liability suits. How would you conduct an experiment to address this question? Be specific. Explain who you would observe, what you would observe, how many observations you would take, and how you would analyze the data once you collect it. What are your null and alternative hypotheses? Would you want to use a high or a low significance level for the test? Explain.

2. Do rock music CDs and country music CDs give the consumer the same amount of music listening time? A sample of 12 randomly selected single rock music CDs and a sample of 14 randomly selected single country music CDs have the following total lengths (in minutes).

Rock Music	Country Music
43.0	45.3
44.3	40.2
63.8	42.8
32.8	33.0
54.2	33.5
51.3	37.7
64.8	36.8
36.1	34.6
33.9	33.4
51.7	36.5
36.5	43.3
59.7	31.7
	44.0
	42.7

Assume that the two populations are normally distributed with equal standard deviations.

    a. Compute the value of the test statistic t for testing the null hypothesis that the mean lengths of the rock and country music single CDs are the same against the alternative hypothesis that these mean lengths are not the same. Use the value of this t statistic to compute the (approximate) p-value.

    b. Compute the value of the (one-way ANOVA) test statistic F for performing the test of equality of the mean lengths of the rock and country music single CDs and use it to find the (approximate) p-value.

    c. How do the test statistics in parts a and b compare? How do the p-values computed in parts a and b compare? Do you think this is a coincidence or will this always happen?

3. It is desired to estimate the difference in the mean scores on a standardized test of students taught by Instructors A and B. The scores of all students taught by Instructor A have a normal distribution with standard deviation 15 and the scores of all students taught by Instructor B have a normal distribution with standard deviation 10. To estimate the difference in the two means, you decide the same number of students should be observed from each instructor.

    a. Assuming that the sample size is the same for each instructor, how large a sample should be taken from each instructor to estimate the difference between the mean scores of two populations to within 5 points with 90% confidence?

b. Suppose samples of the size computed in part *a* will be selected to test for the difference between the two population mean scores using a .05 level of significance. How large does the difference between the two sample means have to be to conclude that the two population means are different?

4. The weekly weight losses of all dieters on Diet I have a normal distribution with a mean of 1.3 pounds and a standard deviation of .4 pounds. The weekly weight losses of all dieters on Diet II have a normal distribution with a mean of 1.5 pounds and a standard deviation of .7 pounds. A random sample of 25 dieters on Diet I and another sample of 36 dieters on Diet II are observed.

a. What is the probability that the difference between the two sample means, $\bar{x}_1 - \bar{x}_2$, will be less than .15 pounds?

b. What is the probability that the average weight loss $\bar{x}_1$ for dieters on Diet I will be higher than the average weight loss $\bar{x}_2$ for dieters on Diet II?

c. If the average weight loss of the 25 dieters using Diet I is computed to be 2.0 pounds, what is the probability that the difference between the two sample means, $\bar{x}_1 - \bar{x}_2$, will be less than .15 pounds?

5. Sixty-five percent of all male voters and 40% of all female voters favor a particular candidate. A sample of 100 male voters and another sample of 100 female voters will be polled. What is the probability that at least 10 more male voters than female voters will favor this candidate?

6. Wheat crop yields with the fertilizer currently being used have a standard deviation of 3 bushels per acre. The producer of a new fertilizer claims that the wheat yields will have less variation when the new fertilizer is used than those with the fertilizer currently being used. If a sample of 20 acres of wheat crops using the new fertilizer are to be observed to test the claim at a .01 level of significance, what can be the largest value of the sample standard deviation that can allow us to conclude that the new fertilizer has less variation? Assume that the yields with the new fertilizer have a normal distribution.

7. A student who needs to pass elementary statistics wonders if it makes a difference with which of two possible instructors she takes the class. Observing the final grades given by each instructor in a recent elementary statistics course, she finds that Instructor I gave 48 passing grades in a class of 52 students and Instructor II gave 44 passing grades in a class of 54 students.

a. Compute the value of the standard normal test statistic z, of Section 10.5.3, for the data and use it to find the p-value when testing for a difference between the proportions of passing grades given by these instructors.

b. Construct a 2×2 contingency table for these data. Compute the value of the χ^2 test statistic for the test of homogeneity and use it to find the p-value.

c. How do the test statistics in parts *a* and *b* compare? How do the p-values for the tests in parts *a* and *b* compare? Do you think this is a coincidence or do you think this will always happen?

8. Each of five boxes contains a large (but unknown) number of red and green marbles. You have been asked to find if the proportions of red and green marbles are the same for each of the five boxes. You sample fifty times, with replacement, from each of the five boxes and observe 20, 14, 23, 30, and 18 red marbles, respectively. Can you conclude that all five boxes have the same proportions of red and green marbles? Use a .05 level of significance.

9. A dietician wanted to investigate if the mean weight loss for each of three diet plans is the same. She took a random sample of 12 persons who wanted to lose weight. Then she randomly assigned four of these persons to Diet A, four to Diet B, and four to Diet C. The following table gives the MINITAB output for a one-way ANOVA test based on the mean weight lost by persons on each diet plan at the end of a 3-week period. Find the missing entries.

ANALYSIS OF VARIANCE

SOURCE	DF	SS	MS	F	P
FACTOR	___	___	___	.65	.545
ERROR	___	___	56.92		
TOTAL	___	___			

LEVEL	N	MEAN	STDEV
C1	___	12.25	9.323
C2	___	11.75	5.560
C3	___	6.75	7.274

POOLED STDEV = ___

10. In Exercise 12.26 there was significant evidence that the mean prices of college textbooks in statistics, psychology, economics, and business were not all equal using a 5% significance level.

    a. After inspecting the sample means of the prices from the four disciplines, does it appear that the mean textbook price for each population is different from that of every other population or does it appear that some disciplines have the same mean textbook price? Which disciplines seem to have the same mean textbook price and which seem to have different mean textbook prices?

    b. One method of determining which population means differ significantly and which do not is the Least Significant Difference method. By using this method, μ_i and μ_j are concluded significantly different at α significance level if

$$\left| \overline{x}_i - \overline{x}_j \right| > t \sqrt{MSW} \sqrt{\frac{1}{n_i} + \frac{1}{n_j}}$$

where t is obtained from the t distribution table for degrees of freedom and the tail area of $\alpha/2$. Using this method, determine which disciplines have mean textbook prices that differ significantly. How do these conclusions compare with your suspicions in part a? Use a .05 significance level. Note that the degrees of freedom for this test will be $n - k$.

11.   a. In problem 10 you should have performed $\binom{4}{2} = 6$ tests, each using a .05 level of significance. If 6 tests of hypothesis are performed using a .05 significance level, what is the overall significance level (i.e., what is the probability of rejecting at least one H_0 if actually all null hypotheses are true)? Assume that the tests are independent.

    b. Suppose we perform pairwise tests for 10 population means for a problem using a .05 level of significance for each test. What would be the probability of concluding at least one pair of population means to be different if the population means are really all equal? Assume that the tests are independent.

13 | SIMPLE LINEAR REGRESSION

This chapter considers the relationship between two variables in two ways: (1) by using the regression analysis and (2) by computing the correlation coefficient. By using the regression model, we can evaluate the magnitude of change in one variable due to a certain change in another variable. For example, an economist can estimate the amount of change in food expenditure due to a certain change in the income of a household by using the regression model. A sociologist may want to estimate the increase in the crime rate due to a particular increase in the unemployment rate. Besides answering these questions, a regression model also helps to predict the value of one variable for a given value of another variable. For example, by using the regression line, we can predict the (approximate) food expenditure of a household with a given income.

The correlation coefficient, on the other hand, simply tells us how strongly two variables are related. It does not provide any information about the size of change in one variable as a result of a certain change in the other variable. For example, the correlation coefficient tells us how strongly income and food expenditure or crime rate and unemployment rate are related.

13.1 SIMPLE LINEAR REGRESSION MODEL

Only simple linear regression will be discussed in this chapter.[1] In the next two subsections the meaning of the words *simple* and *linear* as used in *simple linear regression* is explained.

13.1.1 SIMPLE REGRESSION

Let us return to the example of an economist investigating the relationship between food expenditure and income. What factors or variables does a household consider when deciding how much money should be spent on food every week or every month? Certainly, income of the household is one factor. However, many other variables also affect food expenditure. For instance, the assets owned by the household, the size of the household, the preferences and tastes of household members, and any special dietary needs of household members are some of the variables that will influence a household's decision about food expenditure. These variables are called **independent** or **explanatory variables** because they all vary independently, and they explain the variation in food expenditure among different households. In other words, these variables explain why different households spend different amounts of money on food. Food expenditure is called the **dependent variable** because it depends on the independent variables. Studying the effect of two or more independent variables on a dependent variable using regression analysis is called **multiple regression**. However, if we choose only one (usually the most important) independent variable and study the effect of that single variable on a dependent variable, it is called a **simple regression**. Thus, a simple regression includes only two variables: one independent and one dependent. Note that whether it is a simple or a multiple regression analysis, it always includes one and only one dependent variable. It is the number of independent variables that changes in simple and multiple regressions.

---

[1]The term *regression* was first used by Sir Francis Galton (1822–1911), who studied the relationship between the heights of children and the heights of their parents.

> **SIMPLE REGRESSION**
>
> A regression model is a mathematical equation that describes the relationship between two or more variables. A *simple regression* model includes only two variables: one independent and one dependent. The dependent variable is the one being explained and the independent variable is the one used to explain the variation in the dependent variable.

13.1.2 LINEAR REGRESSION

The relationship between two variables in a regression analysis is expressed by a mathematical equation called a **regression equation** or **model**. A regression equation, when plotted, may assume one of many possible shapes, including that of a straight line. A regression equation that gives a straight-line relationship between two variables is called a **linear regression model**; otherwise, it is called a **nonlinear regression model**. In this chapter, only linear regression models are studied.

> **LINEAR REGRESSION**
>
> A (simple) regression model that gives a straight-line relationship between two variables is called a *linear regression* model.

The two diagrams in Figure 13.1 show a linear and a nonlinear relationship between the dependent variable food expenditure and the independent variable income.

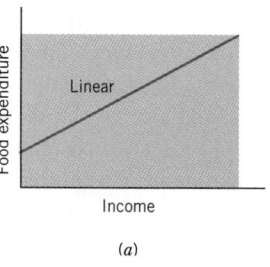
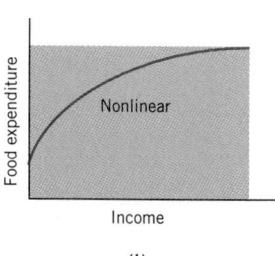

(a)   (b)

Figure 13.1   Relationship between food expenditure and income. (a) Linear relationship. (b) Nonlinear relationship.

A linear relationship between income and food expenditure, which is shown in Figure 13.1*a*, indicates that as income increases the food expenditure always increases at the same rate. However, a nonlinear relationship between income and food expenditure, as depicted in Figure 13.1*b*, shows that as income increases the food expenditure increases, although, after a point, the rate of increase in food expenditure is lower for every subsequent increase in income.

The **equation of a linear relationship** between two variables x and y is written as

$$y = a + bx$$

Each set of values of a and b gives a different straight line. For instance, when $a = 50$ and $b = 5$, then this equation becomes

$$y = 50 + 5x$$

To plot a straight line, we need to know two points that lie on that line. We can find two points on a line by assigning any two values to x and then calculating the corresponding values of y. For the equation $y = 50 + 5x$,

1. When $x = 0$, then $y = 50 + 5(0) = 50$
2. When $x = 10$, then $y = 50 + 5(10) = 100$

These two points are plotted in Figure 13.2. By joining these two points we obtain the line representing the equation $y = 50 + 5x$.

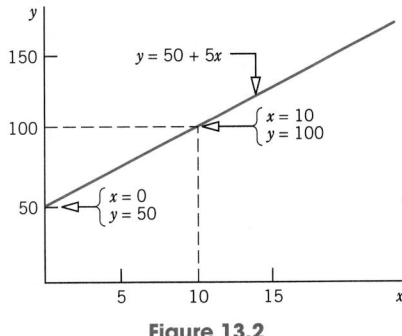

Figure 13.2

Note that in Figure 13.2 the line intersects the y (vertical) axis at 50. Consequently, 50 is called the **y-intercept**. The y-intercept is given by the constant term in the equation. It is the value of y when x is zero.

In the equation $y = 50 + 5x$, 5 is called the **coefficient of x** or the **slope** of the line. It gives the amount of change in y due to a change of one unit in x. For example,

$$\text{If } x = 10, \text{ then } y = 50 + 5(10) = 100$$

$$\text{If } x = 11, \text{ then } y = 50 + 5(11) = 105$$

Hence, as x increases by 1 unit (from 10 to 11), y increases by 5 units (from 100 to 105). This is true for any value of x. Such changes in x and y are shown in Figure 13.3.

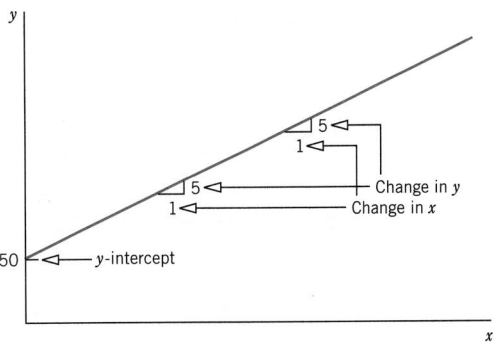

Figure 13.3

In general, when an equation is written in the form

$$y = a + bx$$

a gives the y-intercept and b represents the slope of the line. In other words, a represents the point where the line intersects the y-axis and b gives the amount of change in y due to a change of one unit in x. Note that b is also called the coefficient of x.

13.2   SIMPLE LINEAR REGRESSION ANALYSIS

In a regression model, the independent variable is usually denoted by x and the dependent variable is usually denoted by y. The x variable, with its coefficient, is written on the right side of the "=" sign, whereas the y variable is written on the left side of the "=" sign. The y-intercept and the slope, which we earlier denoted by a and b, respectively, can be represented by any of the many commonly used symbols. Let us denote the y-intercept (which is also called the *constant term*) by A, and the slope (or the coefficient of the x variable) by B. Then, our simple linear regression model is written as

Constant term or y-intercept⎤   ⎡Slope

$$y = A + Bx \tag{1}$$

Dependent variable   Independent variable

In model (1), A gives the value of y for $x = 0$, and B gives the change in y due to a change of one unit in x.

Model (1) is called a **deterministic model**. It gives an **exact relationship** between x and y. This model simply states that y is determined exactly by x and for a given value of x there is one and only one (unique) value of y.

However, in many cases the relationship between variables is not exact. For instance, if y is food expenditure and x is income, then model (1) would state that food expenditure is determined by income only and that all households with the same income spend the same amount on food. But as mentioned earlier, food expenditure is determined by many variables, only one of which is included in model (1). In reality, different households with the same income spend different amounts of money on food because of the differences in the size of the household, the assets they own, and their preferences and tastes. Hence, to take these variables into consideration and to make our model complete, we add another term to the right side of model (1). This term is called the **random error term**. It is denoted by ϵ (greek letter *epsilon*). The complete regression model is written as

$$y = A + Bx + \epsilon \tag{2}$$

Random error term

The regression model (2) is called a **probabilistic model** (or a **statistical relationship**.)

EQUATION OF A REGRESSION MODEL

In the *regression model* $y = A + Bx + \epsilon$, A is called the y-intercept or constant term, B is the slope, and ϵ is the random error term. The dependent and independent variables are y and x, respectively.

The random error term ϵ is included in the model to represent the following two phenomena.

1. *Missing or omitted variables.* As mentioned earlier, food expenditure is affected by many variables other than income. The random error term ϵ is included to capture the effect of all those missing or omitted variables that have not been included in the model.

2. *Random variation.* Human behavior is unpredictable. For example, a household may have many parties during one month and may spend more than usual on food during that month. The same household may spend less than usual during another month because it spent quite a bit of money to buy furniture. The variation in food expenditure for such reasons may be called random variation.

In model (2), A and B are the **population parameters**. The regression line obtained for model (2) by using the population data is called the **population regression line**. The values of A and B in the population regression line are called the **true values of the y-intercept and slope**.

However, population data are difficult to obtain. As a result, we almost always use sample data to estimate model (2). The values of the y-intercept and slope calculated from sample data on x and y are called the **estimated values of A and B and are denoted by a and b.** Using a and b we write the estimated regression model as

$$\hat{y} = a + bx \tag{3}$$

where $\hat{y}$ (read as *y hat*) is the **estimated** or **predicted value of y** for a given value of x. Equation (3) is called the **estimated regression model**; it gives the **regression of y on x.**

ESTIMATES OF *A* AND *B*

In the model $\hat{y} = a + bx$, a and b, which are calculated using sample data, are called the *estimates of A and B.*

13.2.1 SCATTER DIAGRAM

Suppose we take a sample of seven households and collect information on their incomes and food expenditures for the past month. The information obtained (in hundreds of dollars) is given in Table 13.1.

Table 13.1 Incomes and Food Expenditures of Seven Households

Income (hundreds of dollars)	Food Expenditure (hundreds of dollars)
35	9
49	15
21	7
39	11
15	5
28	8
25	9

In Table 13.1, we have a pair of observations for each of the seven households. Each pair consists of one observation on income and a second on food expenditure. For example, the first household's income for the past month was $3500 and its food expenditure was $900. By plotting all seven pairs of values, we obtain a **scatter diagram** or **scattergram**. Figure 13.4 gives the scatter diagram for the data of Table 13.1. Each dot in this diagram represents one household. A scatter diagram is helpful in detecting a relationship between two variables. For example, by looking at the scatter diagram of Figure 13.4, we can observe that there exists a strong linear relationship between food expenditure and income. If a straight line is drawn through the points, the points will be scattered closely around the line.

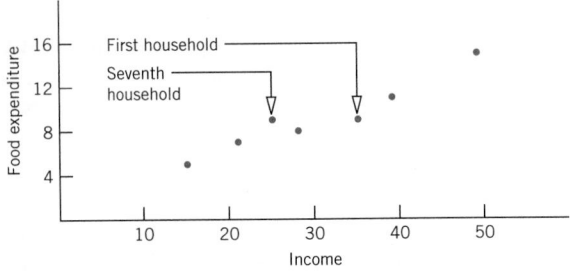

Figure 13.4   Scatter diagram.

SCATTER DIAGRAM

A plot of paired observations is called a *scatter diagram*.

As shown in Figure 13.5, a large number of straight lines can be drawn through the scatter diagram of Figure 13.4. Each of these lines will give different values for a and b of model (3).

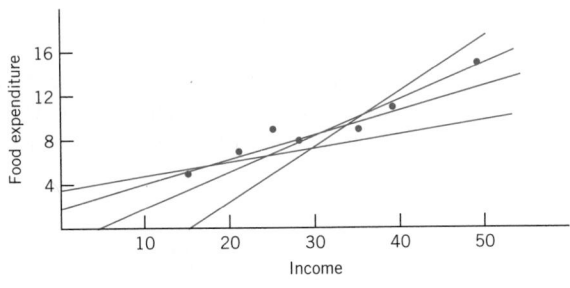

Figure 13.5

In regression analysis, we try to find a line that best fits the points in the scatter diagram. Such a line provides the best possible description of the relationship between the dependent and independent variables. The **least squares method**, discussed in the next section, gives such a line. The line obtained by using the least squares method is called the **least squares regression line**.

13.2.2 LEAST SQUARES LINE

The value of y obtained for a member from the survey is called the **observed or actual value of y**. As mentioned earlier in Section 13.2, the value of y, denoted by $\hat{y}$, obtained for a given x by using the regression line is called the **predicted value of y**. The random error ϵ denotes the difference between the actual value of y and the predicted value of y for population data. For example, for a given household, ϵ is the difference between what this household actually spent on food during the past month and what is predicted using the population regression line. The ϵ is also called the *residual*, as it measures the surplus (positive or negative) of actual food expenditure over what is predicted by using the regression model. If we estimate model (2) by using sample data, the difference between the actual y and predicted y based on this estimation cannot be denoted by ϵ. *The random error for the sample regression model is denoted by e.* Thus, e is an estimator of ϵ. If we estimate model (2) using sample data, then the value of e is given by

$$e = \text{Actual food expenditure} - \text{Predicted food expenditure} = y - \hat{y}$$

In Figure 13.6, e is the vertical distance between the actual position of a household and the point on the regression line. Note that in such a diagram, we always measure the dependent variable on the vertical axis and the independent variable on the horizontal axis.

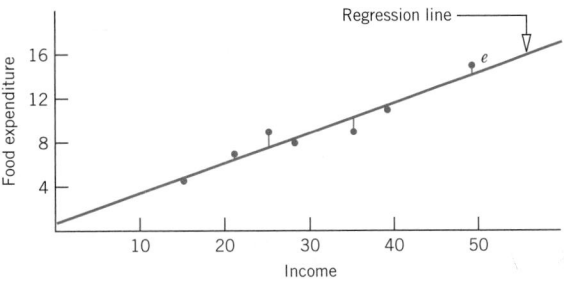

Figure 13.6

The value of an error is positive if the point that gives the actual food expenditure is above the regression line and negative if it is below the regression line. *The sum of these errors is always zero.* In other words, the sum of the actual food expenditures for seven households included in the sample will be the same as the sum of the food expenditures predicted from the regression model. Thus,

$$\Sigma e = \Sigma(y - \hat{y}) = 0$$

Hence, to find the line that best fits the scatter of points, we cannot minimize the sum of errors. Instead, we minimize the **error sum of squares**, denoted by **SSE**, which is obtained by adding the squares of errors. Thus,

$$\text{SSE} = \Sigma e^2 = \Sigma(y - \hat{y})^2$$

The least squares method gives the values of a and b for model (3) such that the sum of squared errors (SSE) is minimum.

ERROR SUM OF SQUARES (SSE)

The *error sum of squares*, denoted by SSE, is

$$SSE = \Sigma e^2 = \Sigma(y - \hat{y})^2$$

The values of a and b which give the minimum SSE are called the **least squares estimates** of A and B and the regression line obtained with these estimates is called the least squares line.

The least squares values of a and b are computed using the following formulas.

THE LEAST SQUARES LINE

For the least squares regression line $\hat{y} = a + bx$

$$b = \frac{SS_{xy}}{SS_{xx}} \quad \text{and} \quad a = \bar{y} - b\bar{x}$$

where $\quad SS_{xy} = \Sigma xy - \dfrac{(\Sigma x)(\Sigma y)}{n} \quad$ and $\quad SS_{xx} = \Sigma x^2 - \dfrac{(\Sigma x)^2}{n}$

and "SS" stands for "sum of squares."[2]

The least squares regression line $\hat{y} = a + bx$ is also called the regression of y on x.

The above formulas are for estimating a sample regression line. Suppose we have access to a population data set. We can find the population regression line by using the same formulas with a little adaptation. If we have access to population data, we replace a by A, b by B, n by N in these formulas, and use the values of Σx, Σy, Σxy, and Σx^2 calculated for population data to make the required computations. The population regression line is written as

$$\mu_{y|x} = A + Bx$$

where $\mu_{y|x}$ is read as *the mean value of y for a given x*. When plotted on a graph, the points on this population regression line give the average values of y for the corresponding values of x. These average values of y are denoted by $\mu_{y|x}$.

Example 13–1 illustrates how to estimate a regression line for sample data.

Estimating the least squares regression line.

EXAMPLE 13–1   Find the least squares regression line for the data on incomes and food expenditures of seven households given in Table 13.1. Use income as an independent variable and food expenditure as a dependent variable.

Solution   We are to find the values of a and b for the regression model $\hat{y} = a + bx$. Table 13.2 shows the calculations required for the computation of a and b. We denote the independent variable (income) by x and the dependent variable (food expenditure) by y.

---

[2]The values of SS_{xy} and SS_{xx} can also be obtained by using the following basic formulas.

$$SS_{xy} = \Sigma(x - \bar{x})(y - \bar{y}) \quad \text{and} \quad SS_{xx} = \Sigma(x - \bar{x})^2$$

However, these formulas usually take longer to make calculations.

Table 13.2

Income	Food Expenditure		
x	y	xy	x^2
35	9	315	1225
49	15	735	2401
21	7	147	441
39	11	429	1521
15	5	75	225
28	8	224	784
25	9	225	625
$\Sigma x = 212$	$\Sigma y = 64$	$\Sigma xy = 2150$	$\Sigma x^2 = 7222$

The following steps are performed to compute a and b.

Step 1. Compute Σx, Σy, $\bar{x}$, and $\bar{y}$.

$$\Sigma x = 212, \quad \Sigma y = 64$$
$$\bar{x} = \Sigma x/n = 212/7 = 30.2857$$
$$\bar{y} = \Sigma y/n = 64/7 = 9.1429$$

Step 2. Compute Σxy and Σx^2.

To calculate Σxy, we multiply the corresponding values of x and y. Then, we sum all the products. The products of x and y are recorded in the third column of Table 13.2. To compute Σx^2, we square each of the x values and then add them. The squared values of x are listed in the fourth column of Table 13.2. From these calculations,

$$\Sigma xy = 2150 \quad \text{and} \quad \Sigma x^2 = 7222$$

Step 3. Compute SS_{xy} and SS_{xx}.

$$SS_{xy} = \Sigma xy - \frac{(\Sigma x)(\Sigma y)}{n} = 2150 - \frac{(212)(64)}{7} = 211.7143$$

$$SS_{xx} = \Sigma x^2 - \frac{(\Sigma x)^2}{n} = 7222 - \frac{(212)^2}{7} = 801.4286$$

Step 4. Compute a and b.

$$b = \frac{SS_{xy}}{SS_{xx}} = \frac{211.7143}{801.4286} = .2642$$

$$a = \bar{y} - b\bar{x} = 9.1429 - (.2642)(30.2857) = 1.1414$$

Thus, our estimated regression model $\hat{y} = a + bx$ is

$$\hat{y} = 1.1414 + .2642\, x$$

This regression line is called the least squares regression line. It gives the *regression of food expenditure on income*.

Note that we have rounded all calculations to four decimal places. We can round the values of a and b in the regression equation to two decimal places, but it is not done here because we will use this regression equation for prediction and estimation purposes later on.

Using this estimated regression model, we can find the predicted value of y for any specific value of x. For instance, suppose we randomly select a household whose monthly income is \$3500 so that $x = 35$ (recall that x denotes income in hundreds of dollars). The predicted value of food expenditure for this household is

$$\hat{y} = 1.1414 + (.2642)\,(35) = \$10.3884 \text{ hundred} = \$1038.84$$

In other words, based on our regression line, we predict that a household with a monthly income of \$3500 is expected to spend \$1038.84 per month on food. This value of $\hat{y}$ can also be interpreted as a point estimator of the mean value of y for $x = 35$. Thus, we can state that, on average, all households with a monthly income of \$3500 spend about \$1038.84 per month on food.

In our data on seven households, there is one household whose income is \$3500. The actual food expenditure for that household is \$900 (see Table 13.1). The difference between the actual and predicted values gives the error of prediction. Thus, the error of prediction for this household, which is shown in Figure 13.7, is

$$e = y - \hat{y} = 9.00 - 10.3884 = -\$1.3884 \text{ hundreds} = -\$138.84$$

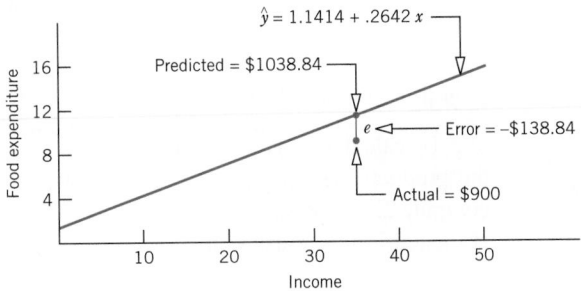

Figure 13.7 Error of prediction.

Therefore, the error of prediction is $-\$138.84$. The negative error indicates that the predicted value of y is greater than the actual value of y. Thus, if we use the regression model, this household's food expenditure is overestimated by \$138.84.

13.2.3 INTERPRETATION OF *a* AND *b*

How do we interpret $a = 1.1414$ and $b = .2642$ obtained in Example 13–1 for the regression of food expenditure on income? A brief explanation of the y-intercept and slope of a regression line was given in Section 13.1.2. The next two parts of this subsection explain the meaning of a and b in more detail.

Interpretation of *a*

Consider a household with zero income. Using the estimated regression line obtained in Example 13–1, the predicted value of y for $x = 0$ is

$$\hat{y} = 1.1414 + .2642\,(0) = \$1.1414 \text{ hundred} = \$114.14$$

Thus, we can state that a household with no income is expected to spend \$114.14 per month on food. Alternatively, we can also state that the point estimate of the average monthly food expenditure for all households with zero income is \$114.14. Note that here we have used

$\hat{y}$ as a point estimate of $\mu_{y|x}$. Thus, $a = 1.1414$ gives the predicted or mean value of y for $x = 0$ based on the regression model estimated for the sample data.

However, we should be very careful while making this interpretation of a. In our sample of seven households, the incomes vary from a minimum of \$1500 to a maximum of \$4900. (Note that in Table 13.1, the minimum value of x is 15 and the maximum value is 49.) Hence, our regression line is valid only for the values of x between 15 and 49. If we predict y for a value of x outside this range, the prediction usually will not hold true. Thus, since $x = 0$ is outside the range of household incomes that we have in the sample data, the prediction that a household with zero income spends \$114.14 per month on food does not carry much credibility. The same is true if we try to predict y for an income greater than \$4900, which is the maximum value of x in Table 13.1.

Interpretation of b

The value of b in a regression model gives the change in y (dependent variable) due to a change of one unit in x (independent variable). For example, by using the regression equation obtained in Example 13–1,

$$\text{when } x = 30, \quad \hat{y} = 1.1414 + .2642\,(30) = 9.0674$$

$$\text{when } x = 31, \quad \hat{y} = 1.1414 + .2642\,(31) = 9.3316$$

Hence, when x increased by one unit, from 30 to 31, $\hat{y}$ increased by $9.3316 - 9.0674 = .2642$, which is the value of b. Because our unit of measurement is hundreds of dollars, we can state that, on average, a \$100 increase in income will cause a \$26.42 increase in food expenditure. We can also state that, on average, a \$1 increase in income of a household will increase the food expenditure by \$.2642. Note the phrase ''on average'' in these statements. The regression line is seen as a measure of the mean value of y for a given value of x. If one household's income is increased by \$100, that household's food expenditure may or may not increase by \$26.42. However, if the incomes of all households are increased by \$100 each, the average increase in their food expenditures will be very close to \$26.42.

Note that when b is positive, an increase in x will lead to an increase in y and a decrease in x will lead to a decrease in y. In other words, when b is positive, the movements in x and y are in the same direction. Such a relationship between x and y is called a **positive linear relationship**. The regression line in this case slopes upward from left to right. On the other hand, if the value of b is negative, an increase in x will cause a decrease in y and a decrease in x will cause an increase in y. The changes in x and y in this case are in opposite directions. Such a relationship between x and y is called a **negative linear relationship**. The regression line in this case slopes downward from left to right. The two diagrams in Figure 13.8 show these two cases.

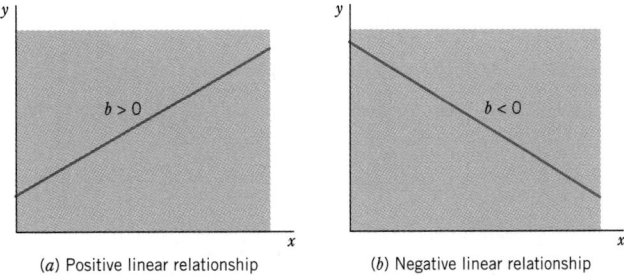

(a) Positive linear relationship   (b) Negative linear relationship

Figure 13.8 Positive and negative linear relationships between x and y.

☞ *Remember* For a regression model, b is computed as $b = SS_{xy}/SS_{xx}$. The value of SS_{xx} is always positive and that of SS_{xy} can be positive or negative. Hence, the sign of b depends on the sign of SS_{xy}. If SS_{xy} is positive (as in our example on incomes and food expenditures of seven households) then b will be positive, and if SS_{xy} is negative then b will be negative.

Case Study 13–1 illustrates the difference between the population regression line and a sample regression line.

CASE STUDY 13-1   REGRESSION OF HEIGHTS AND WEIGHTS OF NBA PLAYERS

Data Set III of Appendix B lists the heights and weights of all NBA (National Basketball Association) players who were on the rosters of all NBA teams at the beginning of the 1996–97 season. These data comprise the population of NBA players for that point in time. We postulate the following simple linear regression model for these data

$$y = A + Bx + \epsilon$$

where y is the weight (in pounds) and x is the height (in inches) of an NBA player.

Using the population data, we obtain the following regression line

$$\mu_{y|x} = -309.10 + 6.72\,x$$

This equation gives the population regression line because it is obtained by using the population data. (Note that in the population regression line we write $\mu_{y|x}$ instead of $\hat{y}$.) Thus, the true values of A and B are

$$A = -309.10 \quad \text{and} \quad B = 6.72$$

The value of B indicates that for every one-inch increase in the height of an NBA player, weight increases on average by 6.72 pounds. However, $A = -309.10$ does not make any sense. It states that the weight of a player with zero height is -309.10 pounds. (Recall from Section 13.2.3 that we cannot apply the regression equation to predict y for values of x outside the range of data used to find the regression line.) Figure 13.9, constructed using MINITAB, gives the scatter diagram for the heights and weights of all NBA players.

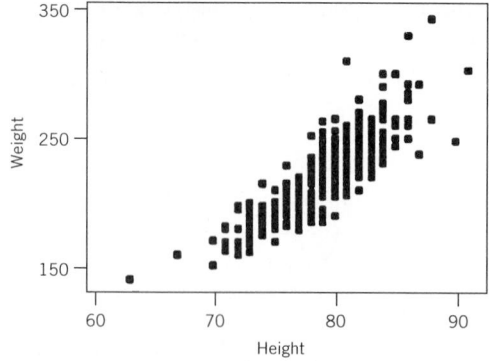

Figure 13.9 Scatter diagram for the data on heights and weights of all NBA players.

Next, we selected a random sample of 30 players and estimated the regression model for this sample. The estimated regression line for the sample is

$$\hat{y} = -295.29 + 6.56\,x$$

The values of a and b are $a = -295.29$ and $b = 6.56$. These values of a and b give the estimates of A and B based on sample data. The scatter diagram for the sample observations on heights and weights is given in Figure 13.10.

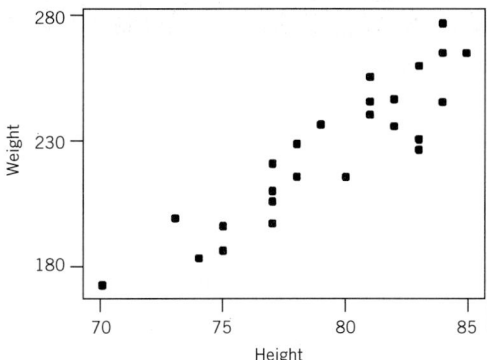

Figure 13.10   Scatter diagram for the data on heights and weights of 30 NBA players.

As we can observe from Figures 13.9 and 13.10, the scatter diagrams for population and sample data both show a (positive) linear relationship between heights and weights of NBA players.

13.2.4   ASSUMPTIONS OF THE REGRESSION MODEL

Like any other theory, the linear regression analysis is also based on certain assumptions. Consider the population regression model

$$y = A + Bx + \epsilon \tag{4}$$

There are four assumptions made about this model, which are described next. These assumptions are explained here with reference to the example regarding incomes and food expenditures of households. Note that these assumptions are made about the population regression model and not about the sample regression model.

Assumption 1:   The random error term ϵ has a mean equal to zero for each x. In other words, among all households with the same income, some spend more than the predicted food expenditure (and, hence, have positive errors) and others spend less than the predicted food expenditure (and, consequently, have negative errors). This assumption simply states that the sum of the positive errors is equal to the sum of the negative errors so that the mean of errors for all households with the same income is zero. Thus, when the mean value of ϵ is zero, the mean value of y for a given x is equal to $A + Bx$ and it is written as

$$\mu_{y|x} = A + Bx$$

As mentioned earlier in this chapter, $\mu_{y|x}$ is read as *the mean value of y for a given value of x*. When we find the values of A and B for model (4) using the population data, the points on the regression line give the average values of y, denoted by $\mu_{y|x}$, for the corresponding values of x.

Assumption 2: The errors associated with different observations are independent. According to this assumption, the errors for any two households in our example are independent. In other words, all households decide independently how much to spend on food.

Assumption 3: For any given x, the distribution of errors is normal. The corollary of this assumption is that the food expenditures for all households with the same income are normally distributed.

Assumption 4: The distribution of population errors for each x has the same (constant) standard deviation, which is denoted by σ_ϵ. This assumption indicates that the spread of points around the regression line is similar for all x values.

Figure 13.11 illustrates the meaning of the first, third, and fourth assumptions for households with incomes of $2000 and $3500 per month. The same assumptions hold true for any other income level. In the population of all households, there will be many households with a monthly income of $2000. Using the population regression line, if we calculate the errors for all these households and prepare the distribution of these errors, it will look like the distribution given in Figure 13.11a. Its standard deviation will be σ_ϵ. Similarly, Figure 13.11b gives the distribution of errors for all those households in the population whose monthly income is $3500. Its standard deviation is also σ_ϵ. Both these distributions are identical. Note that the mean of both of these distributions is $E(\epsilon) = 0$.

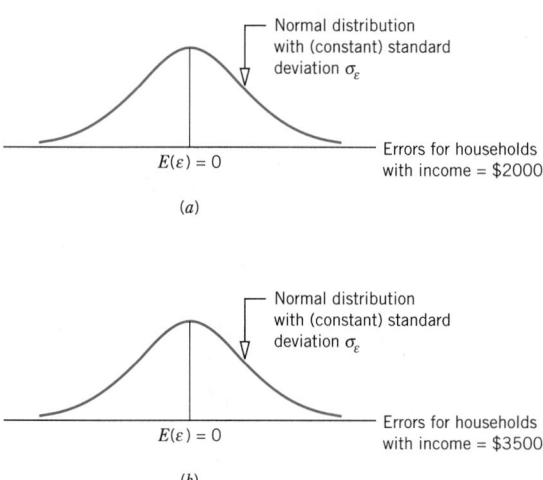

Figure 13.11 (a) Errors for households with an income of $2000 per month. (b) Errors for households with an income of $3500 per month.

Figure 13.12 shows how these distributions look when they are imposed on the same diagram with the population regression line. The points on the vertical line through $x = 20$ give the food expenditures for various households in the population, each of which has the same monthly income of \$2000. The same is true about the vertical line through $x = 35$ or any other vertical line for some other value of x.

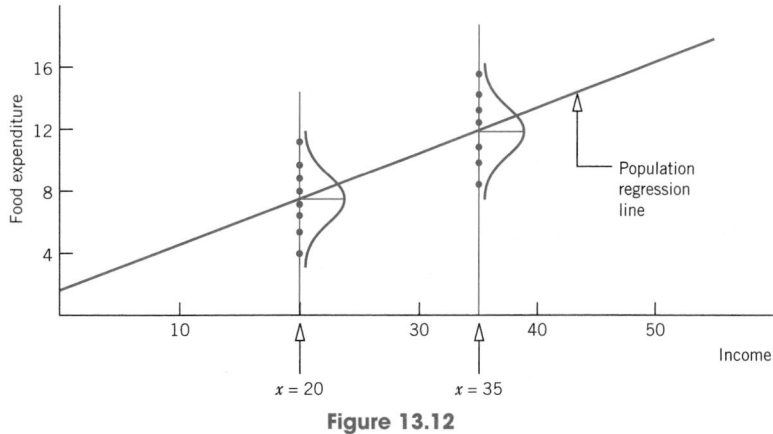

Figure 13.12

13.2.5 A NOTE ON THE USE OF SIMPLE LINEAR REGRESSION

We should apply linear regression with caution. When we use simple linear regression, we assume that the relationship between two variables is described by a straight line. In the real world, the relationship between variables may not be linear. Hence, before we use a simple linear regression, it is better to construct a scatter diagram and look at the plot of the data points. We should estimate a linear regression model only if the scatter diagram indicates such a relationship. The scatter diagrams of Figure 13.13 give two examples where the relationship between x and y is not linear. Consequently, fitting linear regression in such cases would be wrong.

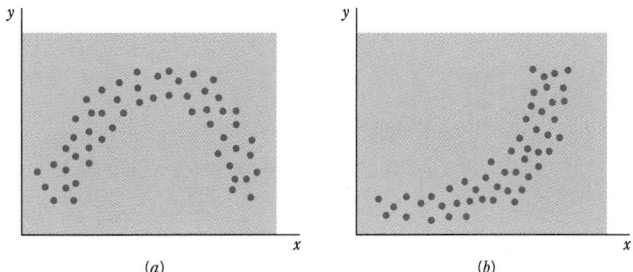

Figure 13.13 Nonlinear relationship between x and y.

EXERCISES

Concepts and Procedures

13.1 Explain the meaning of the words *simple* and *linear* as used in *simple linear regression*.

13.2 Explain the meaning of independent and dependent variables for a regression model.

13.3 Explain the difference between exact and nonexact relationships between two variables.

13.4 Explain the difference between linear and nonlinear relationships between two variables.

13.5 Explain the difference between a simple and a multiple regression model.

13.6 Briefly explain the difference between a deterministic and a probabilistic regression model.

13.7 Why is the random error term included in a regression model?

13.8 Explain the least squares method and least squares regression line. Why are they called by these names?

13.9 Explain the meaning and concept of SSE. You may use a graph for illustration purposes.

13.10 Explain the difference between y and $\hat{y}$.

13.11 Two variables x and y have a positive linear relationship. Explain what happens to the value of y when x increases.

13.12 Two variables x and y have a negative linear relationship. Explain what happens to the value of y when x increases.

13.13 Explain the following.
    a. Population regression line   **b.** Sample regression line
    c. True values of A and B
    d. Estimated values of A and B that are denoted by a and b, respectively

13.14 Briefly explain the assumptions of the population regression model.

13.15 Plot the following straight lines. Give the values of the y-intercept and slope for each of these lines and interpret them. Indicate whether each of the lines gives a positive or a negative relationship between x and y.
    a. $y = 100 + 5x$ **b.** $y = 400 - 4x$

13.16 Plot the following straight lines. Give the values of the y-intercept and slope for each of these lines and interpret them. Indicate whether each of the lines gives a positive or a negative relationship between x and y.
    a. $y = -60 + 8x$ **b.** $y = 300 - 6x$

13.17 A population data set produced the following information.

$$N = 250, \quad \Sigma x = 9880, \quad \Sigma y = 1456, \quad \Sigma xy = 85,080, \quad \Sigma x^2 = 485,870$$

Find the population regression line.

13.18 A population data set produced the following information.

$$N = 460, \quad \Sigma x = 3920, \quad \Sigma y = 2650, \quad \Sigma xy = 26,570, \quad \Sigma x^2 = 48,530$$

Find the population regression line.

13.19 The following information is obtained from a sample data set.

$$n = 10, \quad \Sigma x = 100, \quad \Sigma y = 220, \quad \Sigma xy = 3680, \quad \Sigma x^2 = 1140$$

Find the estimated regression line.

13.20 The following information is obtained from a sample data set.

$$n = 12, \quad \Sigma x = 66, \quad \Sigma y = 588, \quad \Sigma xy = 2244, \quad \Sigma x^2 = 396$$

Find the estimated regression line.

Applications

13.21 A car rental company charges $30 a day and 15 cents per mile for renting a car. Let y be the total rental charges (in dollars) for a car for one day and x be the miles driven. The equation for the relationship between x and y is

$$y = 30 + .15x$$

a. How much will a person pay who rents a car for one day and drives it 100 miles?
b. Suppose each of 20 persons rents a car from this agency for one day and drives it 100 miles. Will each of them pay the same amount for renting a car for a day or is each person expected to pay a different amount? Explain.
c. Is the relationship between x and y exact or nonexact?

13.22 Ben is an electrician who makes house calls for electrical repairs. He charges $30 to go to a house plus $22 per hour. Let y be the total amount (in dollars) paid by a household who uses Ben's services and x the number of hours Ben spends doing repairs in that household's home. The equation for the relationship between x and y is

$$y = 30 + 22x$$

a. Ben spent six hours doing repairs in Kristine's home. How much will he be paid?
b. Suppose seven persons called Ben for repairs during a week. Surprisingly, each of these jobs took six hours. Would each of these home owners pay the same amount for repairs or do you expect each to pay a different amount? Explain.
c. Is the relationship between x and y exact or nonexact?

13.23 A researcher took a sample of 25 electronics companies and found the following relationship between x and y where x is the amount of money (in millions of dollars) spent on advertising by a company in 1996 and y represents the total gross sales (in millions of dollars) of that company for 1996.

$$\hat{y} = 3.4 + 11.55x$$

a. An electronics company spent $2 million on advertising in 1996. What are its expected gross sales for 1996?
b. Suppose four electronics companies spent $2 million each on advertising in 1996. Do you expect these four companies to have the same actual gross sales for 1996? Explain.
c. Is the relationship between x and y exact or nonexact?

13.24 A researcher took a sample of 10 years and found the following relationship between x and y where x is the number of major natural calamities (such as tornadoes, hurricanes, earthquakes, floods, etc.) that occurred during a year and y represents the average total profits (in millions of dollars) of all insurance companies in the United States.

$$\hat{y} = 212.6 - 1.90x$$

a. A randomly selected year had 24 major calamities. What are the expected average profits of U.S. insurance companies for that year?
b. Suppose the number of major calamities was the same for each of three years. Do you expect the average profits for all U.S. insurance companies to be the same for these three years? Explain.
c. Is the relationship between x and y exact or nonexact?

13.25 An auto manufacturing company wanted to investigate how the price of one of its car models depreciates with age. The research department at the company took a sample of eight cars of this model and collected the following information on the ages (in years) and prices (in hundreds of dollars) of these cars.

Age	8	3	6	9	2	5	6	3
Price	16	74	40	19	124	36	33	89

a. Construct a scatter diagram for these data. Does the scatter diagram exhibit a linear relationship between ages and prices of cars?
b. Find the regression line with price as a dependent variable and age as an independent variable.
c. Give a brief interpretation of the values of a and b calculated in part b.

  d. Plot the regression line on the scatter diagram of part a and show the errors by drawing vertical lines between scatter points and the regression line.

  e. Predict the price of a 7-year-old car of this model.

  f. Estimate the price of an 18-year-old car of this model. Comment on this finding.

13.26 Seven students were tested for stress before a mathematics test. The following table gives the stress scores (on a scale of 1 to 10) of these students and their scores on the math test.

Stress score	6.5	4.0	2.5	7.2	8.1	3.4	5.5
Test score	81	96	93	68	63	84	71

  a. Construct a scatter diagram for these data. Does the scatter diagram exhibit a linear relationship between stress scores and test scores?

  b. Find the regression of test scores on stress scores.

  c. Give a brief interpretation of the values of a and b calculated in part b.

  d. Plot the regression line on the scatter diagram of part a and show the errors by drawing vertical lines between scatter points and the regression line.

  e. Predict the test score of a student with a 7.5 stress score before a math test.

  f. Estimate the test score of a student with a stress score of 9.5 before a math test. Comment on this finding.

13.27 An insurance company wants to know how the amount of life insurance depends on the income of persons. The research department at the company collected information on six persons. The following table lists the annual incomes (in thousands of dollars) and amounts (in thousands of dollars) of life insurance policies for these six persons.

Annual income	47	54	27	37	62	21
Life insurance	250	300	100	150	500	75

  a. Construct a scatter diagram for these data. Does the scatter diagram show a linear relationship between annual incomes and amounts of life insurance policies?

  b. Find the regression line $\hat{y} = a + bx$ with annual income as an independent variable and amount of life insurance policy as a dependent variable.

  c. Give a brief interpretation of the values of a and b calculated in part b.

  d. Plot the regression line on the scatter diagram of part a and show the errors by drawing vertical lines between the scatter points and the regression line.

  e. What is the estimated value of life insurance for a person with an annual income of $45,000?

  f. One of the persons in our sample has an annual income of $54,000 and $300,000 of life insurance. What is the predicted value of life insurance for this person? Find the error for this observation.

13.28 A consumer welfare agency wants to investigate the relationship between the sizes of houses and rents paid by tenants in a small city. The agency collected the following information on the sizes (in hundreds of square feet) of six houses and the monthly rents (in dollars) paid by tenants.

Size of the house	21	13	19	27	34	23
Monthly rent	700	580	720	850	1050	800

  a. Construct a scatter diagram for these data. Does the scatter diagram show a linear relationship between the sizes of houses and monthly rents?

  b. Find the regression line $\hat{y} = a + bx$ with the size of a house as an independent variable and monthly rent as a dependent variable.

   c.   Give a brief interpretation of the values of a and b calculated in part b.

   d.   Plot the regression line on the scatter diagram of part a and show the errors by drawing vertical lines between the scatter points and the regression line.

   e.   Predict the monthly rent for a house with 2500 square feet.

   f.   One of the houses in our sample is 2700 square feet and its rent is \$850. What is the predicted rent for this house? Find the error of estimation for this observation.

13.29   The following table gives the total 1996 payroll (rounded to the nearest million dollars) and the percentage of games won during the 1996 season by each of the National League baseball teams.

Team	Total Payroll (millions of dollars)	Percentage of Games Won
Atlanta Braves	54	59
Chicago Cubs	28	47
Cincinnati Reds	42	50
Colorado Rockies	38	51
Florida Marlins	25	49
Houston Astros	24	51
Los Angeles Dodgers	37	56
Montreal Expos	17	54
New York Mets	25	44
Philadelphia Phillies	29	41
Pittsburgh Pirates	17	45
St. Louis Cardinals	37	54
San Diego Padres	33	56
San Francisco Giants	34	42

   a.   Find the least squares regression line with total payroll as an independent variable and percentage of games won as a dependent variable.

   b.   Is the regression line obtained in part a the population regression line? Why or why not? Do the values of the y-intercept and the slope of the regression line give A and B or a and b?

   c.   Give a brief interpretation of the values of the y-intercept and the slope.

   d.   Predict the percentage of games won for a team with a total payroll of \$35 million.

13.30   The following table gives the total 1996 payroll (rounded to the nearest million dollars) and the percentage of games won during the 1996 season by each of the American League baseball teams.

Team	Total Payroll (millions of dollars)	Percentage of Games Won
Baltimore Orioles	53	54
Boston Red Sox	38	53
California Angels	24	44
Chicago White Sox	45	53
Cleveland Indians	48	62
Detroit Tigers	17	33
Kansas City Royals	20	47
Milwaukee Brewers	11	49
Minnesota Twins	21	48
New York Yankees	61	57
Oakland A's	19	48
Seattle Mariners	42	53
Texas Rangers	41	56
Toronto Blue Jays	28	46

a.   Find the least squares regression line with total payroll as an independent variable and percentage of games won as a dependent variable.
b.   Is the regression line obtained in part a the population regression line? Why or why not? Do the values of the y-intercept and the slope in the regression line give A and B or a and b?
c.   Give a brief interpretation of the values of the y-intercept and the slope.
d.   Predict the percentage of games won for a team with a total payroll of $38 million.

13.3   STANDARD DEVIATION OF RANDOM ERRORS

When we consider income and food expenditures, all households with the same income are expected to spend different amounts on food. Consequently, the random error ϵ will assume different values for these households. The standard deviation σ_ϵ measures the spread of these errors around the population regression line. The **standard deviation of errors** tells us how widely the errors and, hence, the values of y are spread for a given x. In Figure 13.12, which is reproduced below as Figure 13.14, the points on the vertical line through $x = 20$ give the monthly food expenditures for all households with a monthly income of $2000. The distance of each dot from the point on the regression line gives the value of the corresponding error. The standard deviation of errors σ_ϵ measures the spread of such points around the population regression line. The same is true for $x = 35$ or any other value of x.

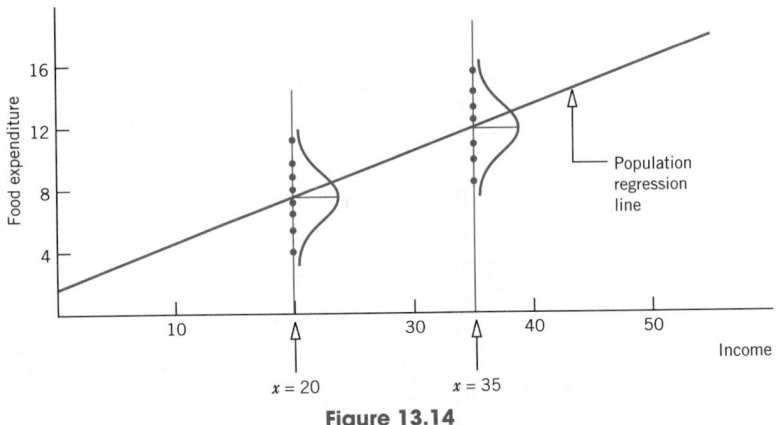

Figure 13.14

Note that σ_ϵ denotes the standard deviation of errors for the population. However, usually σ_ϵ is unknown. In such cases, it is estimated by s_e, which is the standard deviation of errors for the sample data. The following is the basic formula to calculate s_e.

$$s_e = \sqrt{\frac{\text{SSE}}{n - 2}} \qquad \text{where} \qquad \text{SSE} = \Sigma(y - \hat{y})^2$$

In the above formula, $n - 2$ represents the **degrees of freedom** for the regression model. The reason that $df = n - 2$ is that we lose one degree of freedom to calculate $\bar{x}$ and one for $\bar{y}$.

DEGREES OF FREEDOM FOR A SIMPLE LINEAR REGRESSION MODEL

The *degrees of freedom for a simple linear regression model* are

$$df = n - 2$$

For computational purposes, it is more convenient to use the following formula to calculate the standard deviation of errors s_e.

STANDARD DEVIATION OF ERRORS

The *standard deviation of errors* is calculated as[3]

$$s_e = \sqrt{\frac{SS_{yy} - b\, SS_{xy}}{n - 2}}$$

where

$$SS_{yy} = \Sigma y^2 - \frac{(\Sigma y)^2}{n}$$

The calculation of SS_{xy} was discussed earlier in this chapter.[4]

Like the value of SS_{xx}, the value of SS_{yy} is always positive.

Example 13–2 illustrates the calculation of the standard deviation of errors for the data of Table 13.1.

Calculating the standard deviation of errors.

EXAMPLE 13-2 Compute the standard deviation of errors s_e for the data on monthly incomes and food expenditures of seven households given in Table 13.1.

Solution To compute s_e, we need to know the values of SS_{yy}, SS_{xy}, and b. Earlier in Example 13–1, we computed SS_{xy} and b. These values are

$$SS_{xy} = 211.7143 \qquad \text{and} \qquad b = .2642$$

To compute SS_{yy}, we calculate Σy^2 as shown in Table 13.3.

---

[3]If we have access to population data, the value of σ_ϵ is calculated using the formula

$$\sigma_\epsilon = \sqrt{\frac{SS_{yy} - B\, SS_{xy}}{N}}$$

[4]The basic formula to calculate SS_{yy} is $SS_{yy} = \Sigma(y - \bar{y})^2$.

Table 13.3

Income	Food Expenditure	
x	y	y^2
35	9	81
49	15	225
21	7	49
39	11	121
15	5	25
28	8	64
25	9	81
$\Sigma x = 212$	$\Sigma y = 64$	$\Sigma y^2 = 646$

The value of SS_{yy} is

$$SS_{yy} = \Sigma y^2 - \frac{(\Sigma y)^2}{n} = 646 - \frac{(64)^2}{7} = 60.8571$$

Hence, the standard deviation of errors is

$$s_e = \sqrt{\frac{SS_{yy} - b \, SS_{xy}}{n - 2}} = \sqrt{\frac{60.8571 - .2642 \, (211.7143)}{7 - 2}} = \textbf{.9922}$$

13.4 COEFFICIENT OF DETERMINATION

We may ask the question, ''How good is the regression model?'' In other words, ''How well does the independent variable explain the dependent variable in the regression model?'' The *coefficient of determination* is one concept that answers this question.

For a moment, assume that we possess information only on food expenditures of households and not on their incomes. Hence, in this case, we cannot use the regression line to predict the food expenditure for any household. As we did in earlier chapters, in the absence of a regression model, we use $\bar{y}$ to estimate or predict every household's food expenditure. Consequently, the error of prediction for each household is now given by $y - \bar{y}$, which is the difference between the actual food expenditure of a household and the mean food expenditure. If we calculate such errors for all households, then square and add them, the resulting sum is called the **total sum of squares** and is denoted by **SST**. Actually SST is the same as SS_{yy} and is defined as

$$SST = SS_{yy} = \Sigma(y - \bar{y})^2$$

However, for computational purposes, SST is calculated using the following formula.

TOTAL SUM OF SQUARES (SST)

The *total sum of squares*, denoted by SST, is

$$SST = \Sigma y^2 - \frac{(\Sigma y)^2}{n}$$

Note that this is the same formula that we used to calculate SS_{yy}.

The value of SS_{yy}, which is 60.8571, was calculated in Example 13–2. Consequently, the value of SST is

$$SST = 60.8571$$

From Example 13–1, $\bar{y} = 9.1429$. Figure 13.15 shows the total errors for each of the seven households in our sample using the scatter diagram of Figure 13.4.

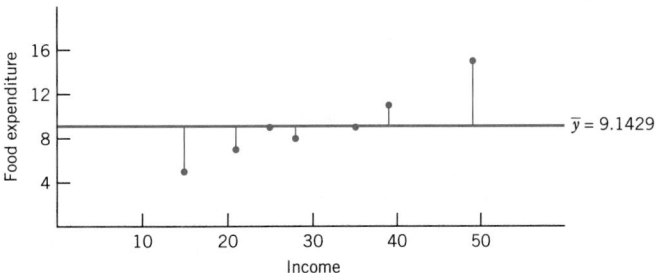

Figure 13.15 Total errors.

Now suppose we use the simple linear regression model to predict the food expenditure of each of the seven households in our sample. In this case, we predict each household's food expenditure by using the regression line we estimated earlier in Example 13–1, which is

$$\hat{y} = 1.1414 + .2642x$$

The predicted food expenditures, denoted by $\hat{y}$, for the seven households are shown in Table 13.4. Also shown are the errors and error squares.

Table 13.4

x	y	$\hat{y} = 1.1414 + .2642x$	$e = y - \hat{y}$	$e^2 = (y - \hat{y})^2$
35	9	10.3884	−1.3884	1.9277
49	15	14.0872	.9128	.8332
21	7	6.6896	.3104	.0963
39	11	11.4452	−.4452	.1982
15	5	5.1044	−.1044	.0109
28	8	8.5390	−.5390	.2905
25	9	7.7464	1.2536	1.5715
				$\Sigma e^2 = \Sigma(y - \hat{y})^2 = 4.9283$

We calculate the values of $\hat{y}$ (given in the third column of Table 13.4) by substituting the values of x in the estimated regression model. For example, the value of x for the first household is 35. Substituting this value of x in the regression equation, we obtain

$$\hat{y} = 1.1414 + .2642 (35) = 10.3884$$

Similarly we find the other values of $\hat{y}$.

The error sum of squares SSE is given by the sum of the fifth column in Table 13.4. Thus,

$$SSE = \Sigma(y - \hat{y})^2 = 4.9283$$

The errors of prediction for the regression model for the seven households are shown in Figure 13.16.

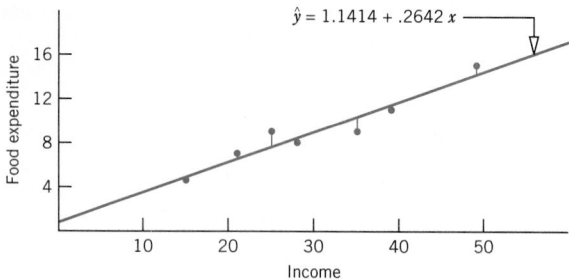

Figure 13.16   Errors of prediction when regression model is used.

Thus, from the foregoing calculations,

$$\text{SST} = 60.8571 \quad \text{and} \quad \text{SSE} = 4.9283$$

These values indicate that the sum of squared errors decreased from 60.8571 to 4.9283 when we used $\hat{y}$ in place of $\bar{y}$ to predict food expenditures. This reduction in squared errors is called the **regression sum of squares** and is denoted by **SSR**. Thus,

$$\text{SSR} = \text{SST} - \text{SSE} = 60.8571 - 4.9283 = 55.9288$$

The value of SSR can also be computed by using the formula

$$\text{SSR} = \Sigma(\hat{y} - \bar{y})^2$$

REGRESSION SUM OF SQUARES (SSR)

The *regression sum of squares*, denoted by SSR, is

$$\text{SSR} = \text{SST} - \text{SSE}$$

Thus, SSR is the portion of SST that is explained by the use of the regression model, and SSE is the portion of SST that is not explained by the use of the regression model. The sum of SSR and SSE is always equal to SST. Thus,

$$\text{SST} = \text{SSR} + \text{SSE}$$

The ratio of SSR to SST gives the **coefficient of determination**. The coefficient of determination calculated for population data is denoted by ρ^2 (ρ is the Greek letter *rho*) and the one calculated for sample data is denoted by r^2. The coefficient of determination gives the proportion of SST that is explained by the use of the regression model. The value of the coefficient of determination always lies in the range zero to 1. The coefficient of determination can be calculated by using the formula

$$r^2 = \frac{\text{SSR}}{\text{SST}} \quad \text{or} \quad \frac{\text{SST} - \text{SSE}}{\text{SST}}$$

However, for computational purposes, the formula given below is more efficient to use to calculate the coefficient of determination.

COEFFICIENT OF DETERMINATION

The *coefficient of determination*, denoted by r^2, represents the proportion of SST that is explained by the use of the regression model. The computational formula for r^2 is[5]

$$r^2 = \frac{b \, SS_{xy}}{SS_{yy}}$$

and $$0 \le r^2 \le 1$$

Example 13–3 illustrates the calculation of the coefficient of determination for a sample data set.

Calculating the coefficient of determination.

EXAMPLE 13–3 For the data of Table 13.1 on monthly incomes and food expenditures of seven households, calculate the coefficient of determination.

Solution From earlier calculations made in Examples 13–1 and 13–2,

$$b = .2642, \qquad SS_{xy} = 211.7143, \qquad \text{and} \qquad SS_{yy} = 60.8571$$

Hence,

$$r^2 = \frac{b \, SS_{xy}}{SS_{yy}} = \frac{(.2642)\,(211.7143)}{60.8571} = \mathbf{.92}$$

Thus, we can state that SST is reduced by approximately 92% (from 60.8571 to 4.9283) when we use $\hat{y}$, instead of $\bar{y}$, to predict the food expenditures of households. Note that r^2 is usually rounded to two decimal places. ■

The total sum of squares SST is a measure of the total variation in food expenditures, the regression sum of squares SSR is the portion of total variation explained by the regression model (or by income), and the error sum of squares SSE is the portion of total variation not explained by the regression model. Hence, for Example 13–3 we can state that 92% of the total variation in food expenditures of households occurs because of the variation in their incomes, and the remaining 8% is due to randomness and other variables.

Usually, the higher the value of r^2, the better the regression model. This is so because if r^2 is larger, a greater portion of the total errors is explained by the included independent variable and a smaller portion of errors is attributed to other variables and randomness.

---

[5]If we have access to population data, the value of ρ^2 is calculated using the formula

$$\rho^2 = \frac{B \, SS_{xy}}{SS_{yy}}$$

The values of SS_{xy} and SS_{yy} used here are calculated for the population data set.

EXERCISES

Concepts and Procedures

13.31 What are the degrees of freedom for a simple linear regression model?

13.32 Explain the meaning of coefficient of determination.

13.33 Explain the meaning of SST and SSR. You may use graphs for illustration purposes.

13.34 A population data set produced the following information.

$$N = 250, \quad \Sigma x = 9880, \quad \Sigma y = 1456, \quad \Sigma xy = 85{,}080,$$
$$\Sigma x^2 = 485{,}870, \quad \text{and} \quad \Sigma y^2 = 135{,}675$$

Find the values of σ_ϵ and ρ^2.

13.35 A population data set produced the following information.

$$N = 460, \quad \Sigma x = 3920, \quad \Sigma y = 2650, \quad \Sigma xy = 26{,}570,$$
$$\Sigma x^2 = 48{,}530, \quad \text{and} \quad \Sigma y^2 = 39{,}347$$

Find the values of σ_ϵ and ρ^2.

13.36 The following information is obtained from a sample data set.

$$n = 10, \quad \Sigma x = 100, \quad \Sigma y = 220, \quad \Sigma xy = 3680,$$
$$\Sigma x^2 = 1140, \quad \text{and} \quad \Sigma y^2 = 25{,}272$$

Find the values of s_e and r^2.

13.37 The following information is obtained from a sample data set.

$$n = 12, \quad \Sigma x = 66, \quad \Sigma y = 588, \quad \Sigma xy = 2244,$$
$$\Sigma x^2 = 396, \quad \text{and} \quad \Sigma y^2 = 58{,}734$$

Find the values of s_e and r^2.

Applications

13.38 The following table gives information on the monthly incomes (in hundreds of dollars) and monthly telephone bills (in dollars) for a random sample of 10 households.

Income	16	45	36	32	30	13	41	15	36	40
Phone bill	35	142	175	70	95	26	160	42	79	97

Find the following.

a. SS_{xx}, SS_{yy}, and SS_{xy} b. Standard deviation of errors
c. SST, SSE, and SSR d. Coefficient of determination

13.39 The following table gives information on the average saturated fat (in grams) consumed per day and the cholesterol level (in milligrams per hundred milliliters) for eight males.

Fat consumption	55	65	50	34	43	58	72	36
Cholesterol level	180	215	195	165	170	204	235	150

Compute the following.

a. SS_{xx}, SS_{yy}, and SS_{xy} b. Standard deviation of errors
c. SST, SSE, and SSR d. Coefficient of determination

13.40 Refer to Exercise 13.25. The following table, which gives the ages (in years) and prices (in hundreds of dollars) of eight cars of a specific model, is reproduced from that exercise.

Age	8	3	6	9	2	5	6	3
Price	16	74	40	19	124	36	33	89

   a. Calculate the standard deviation of errors.
   b. Compute the coefficient of determination and give a brief interpretation of it.

13.41 The following data on the stress scores before a math test and the math test scores for seven students are reproduced from Exercise 13.26.

Stress score	6.5	4.0	2.5	7.2	8.1	3.4	5.5
Test score	81	96	93	68	63	84	71

   a. Determine the standard deviation of errors.
   b. Find the coefficient of determination and give a brief interpretation of it.

13.42 The following data set on annual incomes (in thousands of dollars) and amounts (in thousands of dollars) of life insurance policies for six persons is reproduced from Exercise 13.27.

Annual income	47	54	27	37	62	21
Life insurance	250	300	100	150	500	75

   a. Find the standard deviation of errors.
   b. Compute the coefficient of determination. What percentage of the variation in life insurance amounts is explained by the annual incomes? What percentage of this variation is not explained?

13.43 The following table, reproduced from Exercise 13.28, lists the sizes of six houses (in hundreds of square feet) and the monthly rents (in dollars) paid by tenants for those houses.

Size of the house	21	13	19	27	34	23
Monthly rent	700	580	720	850	1050	800

   a. Compute the standard deviation of errors.
   b. Calculate the coefficient of determination. What percentage of the variation in monthly rents is explained by the sizes of the houses? What percentage of this variation is not explained?

13.44 Refer to data given in Exercise 13.29 on the total 1996 payroll and the percentage of games won during the 1996 season by each of the National League baseball teams.
   a. Find the standard deviation of errors, σ_ϵ. (Note that this data set belongs to a population.)
   b. Compute the coefficient of determination, ρ^2.

13.45 Refer to data given in Exercise 13.30 on the total 1996 payroll and the percentage of games won during the 1996 season by each of the American League baseball teams.
   a. Find the standard deviation of errors, σ_ϵ. (Note that this data set belongs to a population.)
   b. Compute the coefficient of determination, ρ^2.

13.5   INFERENCES ABOUT *B*

This section is concerned with estimation and tests of hypotheses about the population regression slope *B*. We can also make confidence intervals and test hypotheses about the *y*-intercept *A* of the population regression line. However, making inferences about *A* is beyond the scope of this text.

13.5.1   SAMPLING DISTRIBUTION OF *b*

One of the main purposes for determining a regression line is to find the true value of the slope *B* of the population regression line. However, in almost all cases, the regression line is estimated using sample data. Then, based on the sample regression line, inferences are made about the population regression line. The slope *b* of a sample regression line is a point estimator of the slope *B* of the population regression line. The different sample regression lines estimated for different samples taken from the same population will give different values of *b*. If only one sample is taken and the regression line for that sample is estimated, the value of *b* will depend on which elements are included in the sample. Thus, *b* is a random variable and it possesses a probability distribution that is more commonly called its sampling distribution. The shape of the sampling distribution of *b*, its mean, and standard deviation are given below.

> **MEAN, STANDARD DEVIATION, AND SAMPLING DISTRIBUTION OF *b***
>
> Because of the assumption of normally distributed random errors, the sampling distribution of *b* is normal. The mean and standard deviation of *b*, denoted by μ_b and σ_b respectively, are
>
> $$\mu_b = B \quad \text{and} \quad \sigma_b = \frac{\sigma_\epsilon}{\sqrt{SS_{xx}}}$$

However, usually the standard deviation of population errors σ_ϵ is not known. Hence, the sample standard deviation of errors s_e is used to estimate σ_ϵ. In such a case, when σ_ϵ is unknown, the standard deviation of *b* is estimated by s_b, which is calculated as

$$s_b = \frac{s_e}{\sqrt{SS_{xx}}}$$

If σ_ϵ is not known and the sample size is large ($n \geq 30$), the normal distribution can be used to make inferences about *B*. However, if σ_ϵ is not known and the sample size is small ($n < 30$), the normal distribution is replaced by the *t* distribution to make inferences about *B*.

13.5.2   ESTIMATION OF *B*

The value of *b* obtained from the sample regression line is a point estimate of the slope *B* of the population regression line. As mentioned in Section 13.5.1, if σ_ϵ is not known and the sample size is small, the *t* distribution is used to make a confidence interval for *B*.

CONFIDENCE INTERVAL FOR *B*

The $(1 - \alpha)100\%$ *confidence interval for B* is given by

$$b \pm ts_b$$

where $\qquad s_b = \dfrac{s_e}{\sqrt{SS_{xx}}}$

and the value of t is obtained from the t distribution table for $\alpha/2$ area in the right tail of the t distribution and $n - 2$ degrees of freedom.

Example 13–4 describes the procedure for making a confidence interval for *B*.

Constructing a confidence interval for B.

EXAMPLE 13–4 Construct a 95% confidence interval for *B* for the data on incomes and food expenditures of seven households given in Table 13.1.

Solution From the given information and earlier calculations in Examples 13–1 and 13–2,

$$n = 7, \qquad b = .2642, \qquad SS_{xx} = 801.4286, \qquad \text{and} \qquad s_e = .9922$$

The confidence level is 95%.

$$s_b = \frac{s_e}{\sqrt{SS_{xx}}} = \frac{.9922}{\sqrt{801.4286}} = .0350$$

$$df = n - 2 = 7 - 2 = 5$$

$$\alpha/2 = .5 - (.95/2) = .025$$

From the t distribution table, the value of t for 5 *df* and .025 area in the right tail of the t distribution curve is 2.571. The 95% confidence interval for *B* is

$$b \pm ts_b = .2642 \pm 2.571(.0350) = .2642 \pm .0900 = \textbf{.17 to .35}$$

Thus, we are 95% confident that the slope *B* of the population regression line is between .17 and .35. ◼

13.5.3 HYPOTHESIS TESTING ABOUT *B*

Testing a hypothesis about *B* when the null hypothesis is $B = 0$ (that is, the slope of the regression line is zero) is equivalent to testing that x does not determine y and that the regression line is of no use in predicting y for a given x. However, we should remember that we are testing for a linear relationship between x and y. It is possible that x may determine y nonlinearly. Hence, a nonlinear relationship may exist between x and y.

To test the hypothesis that x does not determine y linearly, we will test the null hypothesis that the slope of the regression line is zero, that is, $B = 0$. The alternative hypothesis can be: (1) x determines y, that is, $B \neq 0$, (2) x determines y positively, that is, $B > 0$, or (3) x determines y negatively, that is, $B < 0$.

The procedure used to make a hypothesis test about *B* is similar to the one used in earlier chapters. It involves the same five steps.

TEST STATISTIC FOR *b*

The value of the *test statistic t* for *b* is calculated as

$$t = \frac{b - B}{s_b}$$

The value of *B* is substituted from the null hypothesis.

Example 13–5 illustrates the procedure for testing a hypothesis about *B*.

Conducting a test of
hypothesis about B.

EXAMPLE 13-5   Test at the 1% significance level if the slope of the regression line for the example on incomes and food expenditures of seven households is positive.

Solution   From the given information and earlier calculations in Examples 13–1 and 13–4,

$$n = 7, \quad b = .2642, \quad \text{and} \quad s_b = .0350$$

Step 1.   *State the null and alternative hypotheses*

We are to test whether or not the slope *B* of the population regression line is positive. Hence, the two hypotheses are

$$H_0: B = 0 \quad \text{(slope is zero)}$$
$$H_1: B > 0 \quad \text{(slope is positive)}$$

Note that we can also write the null hypothesis as $H_0: B \leq 0$, which states that the slope is either zero or negative.

Step 2.   *Select the distribution to use*

The sample size is small ($n < 30$) and σ_ϵ is not known. Hence, we use the *t* distribution to make the test about *B*.

Step 3.   *Determine the rejection and nonrejection regions*

The significance level is .01. The $>$ sign in the alternative hypothesis indicates that the test is right-tailed. Therefore,

Area in the right tail of the *t* distribution $= \alpha = .01$

$$df = n - 2 = 7 - 2 = 5$$

From the *t* distribution table, the critical value of *t* for 5 *df* and .01 area in the right tail of the *t* distribution is 3.365, as shown in Figure 13.17.

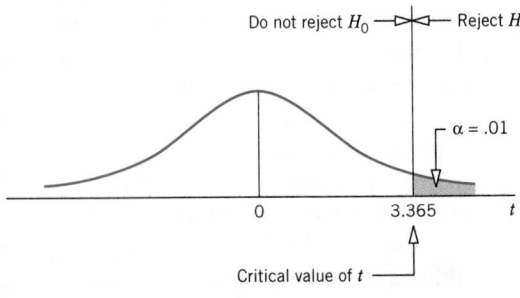

Figure 13.17

Step 4. *Calculate the value of the test statistic*

The value of the test statistic t for b is calculated as follows.

$$t = \frac{b - B}{s_b} = \frac{.2642 - \overset{\text{From } H_0}{0}}{.0350} = 7.549$$

Step 5. *Make a decision*

The value of the test statistic $t = 7.549$ is greater than the critical value of $t = 3.365$, and it falls in the rejection region. Hence, we reject the null hypothesis and conclude that x (income) determines y (food expenditure) positively. That is, food expenditure increases with an increase in income and it decreases with a decrease in income. ■

Note that the null hypothesis does not always have to be $B = 0$. We may test the null hypothesis that B is equal to a certain value. See Exercises 13.47 to 13.50, 13.57, and 13.58 for such cases.

EXERCISES

Concepts and Procedures

13.46 Describe the mean, standard deviation, and shape of the sampling distribution of the slope b of the simple linear regression model.

13.47 The following information is obtained for a sample of 16 observations taken from a population.

$$SS_{xx} = 340.700, \qquad s_e = 1.951, \qquad \text{and} \qquad \hat{y} = 12.45 + 6.32x$$

   a. Make a 99% confidence interval for B.
   b. Using a significance level of .025, can you conclude that B is positive?
   c. Using a significance level of .01, can you conclude that B is different from zero?
   d. Using a significance level of .02, test whether B is different from 4.50. (*Hint:* The null hypothesis here will be H_0: $B = 4.50$, and the alternative hypothesis will be H_1: $B \neq 4.50$. Notice that the value of $B = 4.50$ will be used to calculate the value of the test statistic t.)

13.48 The following information is obtained for a sample of 25 observations taken from a population.

$$SS_{xx} = 274.600, \qquad s_e = .932, \qquad \text{and} \qquad \hat{y} = 280.56 - 3.77x$$

   a. Make a 95% confidence interval for B.
   b. Using a significance level of .01, test whether B is negative.
   c. Testing at the 5% significance level, can you conclude that B is different from zero?
   d. Test if B is different from -5.20. Use $\alpha = .01$.

13.49 The following information is obtained for a sample of 100 observations taken from a population. (Note that because $n > 30$, we can use the normal distribution to make a confidence interval and test a hypothesis about B.)

$$SS_{xx} = 524.884, \qquad s_e = 1.464, \qquad \text{and} \qquad \hat{y} = 5.48 + 2.50x$$

   a. Make a 98% confidence interval for B.
   b. Test at the 2% significance level whether B is positive.
   c. Can you conclude that B is different from zero? Use $\alpha = .01$.
   d. Using a significance level of .01, test whether B is greater than 1.75.

13.50   The following information is obtained for a sample of 80 observations taken from a population.

$$SS_{xx} = 380.592, \qquad s_e = .961, \qquad \text{and} \qquad \hat{y} = 160.24 - 2.70x$$

a.   Make a 97% confidence interval for B.
b.   Test at the 1% significance level whether B is negative.
c.   Can you conclude that B is different from zero? Use $\alpha = .01$.
d.   Using a significance level of .02, test whether B is less than -1.25.

Applications

13.51   Refer to Exercise 13.25. The data on ages (in years) and prices (in hundreds of dollars) for eight cars of a specific model are reproduced below from that exercise.

Age	8	3	6	9	2	5	6	3
Price	16	74	40	19	124	36	33	89

a.   Construct a 95% confidence interval for B. You can use results obtained in Exercises 13.25 and 13.40 here.
b.   Test at the 5% significance level if B is negative.

13.52   The data given in the table below give the midterm scores in a course for a sample of 10 students and the scores of student evaluations of the instructor. (In the instructor evaluation scores, 1 is the lowest and 4 is the highest score.)

Instructor score	3	2	3	1	2	4	3	4	4	2
Midterm score	90	75	97	64	47	99	75	88	93	81

a.   Find the regression of instructor scores on midterm scores.
b.   Construct a 99% confidence interval for B.
c.   Test at the 1% significance level if B is positive.

13.53   The following data give the experience (in years) and monthly salaries (in hundreds of dollars) of nine randomly selected secretaries.

Experience	14	3	5	6	4	9	18	5	16
Monthly salary	26	15	18	20	18	22	29	16	32

a.   Find the least squares regression line with experience as an independent variable and monthly salary as a dependent variable.
b.   Construct a 98% confidence interval for B.
c.   Test at the 2.5% significance level whether B is greater than zero.

13.54   The following data on the stress scores before a math test and the math test scores for seven students are reproduced from Exercise 13.26.

Stress score	6.5	4.0	2.5	7.2	8.1	3.4	5.5
Test score	81	96	93	68	63	84	71

a.   Find the regression line $\hat{y} = a + bx$ where x is the stress score and y is the test score. You can use the calculations made in Exercises 13.26 and 13.41 here.
b.   Make a 95% confidence interval for B.
c.   Test at the 2.5% significance level if B is negative.

13.55 The following data set on annual incomes (in thousands of dollars) and amounts (in thousands of dollars) of life insurance policies for six persons is reproduced from Exercise 13.27.

Annual income	47	54	27	37	62	21
Life insurance	250	300	100	150	500	75

    a. Construct a 99% confidence interval for B. You can use the calculations made in Exercises 13.27 and 13.42 here.

    b. Test at the 1% significance level if B is different from zero.

13.56 The data on the sizes of six houses (in hundreds of square feet) and the monthly rents (in dollars) paid by tenants for those houses are reproduced below from Exercise 13.28.

Size of the house	21	13	19	27	34	23
Monthly rent	700	580	720	850	1050	800

    a. Construct a 98% confidence interval for B. You can use the calculations made in Exercises 13.28 and 13.43 here.

    b. Testing at the 5% significance level, can you conclude that B is different from zero?

13.57 Refer to Exercise 13.38. The following table gives information on the monthly incomes (in hundreds of dollars) and monthly telephone bills (in dollars) for a random sample of 10 households.

Income	16	45	36	32	30	13	41	15	36	40
Phone bill	35	142	175	70	95	26	160	42	79	97

    a. Find the least squares regression line with monthly income as an independent variable and monthly telephone bill as a dependent variable.

    b. Make a 95% confidence interval for B. You can use the results obtained in Exercise 13.38 here.

    c. An earlier study claims that for the relationship between income and amount of phone bills $B = 1.50$. Test at the 1% significance level if B is greater than 1.50. [*Hint:* Here the null hypothesis will be H_0: $B = 1.50$, and the alternative hypothesis will be H_1: $B > 1.50$. Notice that the value of $B = 1.50$ will be used to calculate the value of the test statistic t.]

13.58 The following table, reproduced from Exercise 13.39, gives information on the average saturated fat (in grams) consumed per day and the cholesterol level (in milligrams per hundred milliliters) of eight males.

Fat consumption	55	65	50	34	43	58	72	36
Cholesterol level	180	215	195	165	170	204	235	150

    a. Find the regression line $\hat{y} = a + bx$ where x is the fat consumption and y is the cholesterol level. You can use the results obtained in Exercise 13.39 here.

    b. Construct a 90% confidence interval for B.

    c. An earlier study claims that B is 1.75. Test at the 5% significance level if B is different from 1.75.

13.6 LINEAR CORRELATION

Another measure of the relationship between two variables is the correlation coefficient. This section describes the simple linear correlation, for short **linear correlation**, which measures the strength of the linear association between two variables. In other words, the linear correlation coefficient measures how closely the points in a scatter diagram are spread around the regression line. The correlation coefficient calculated for the population data is denoted by ρ (Greek letter *rho*) and the one calculated for sample data is denoted by r. (Note that the square of the correlation coefficient is equal to the coefficient of determination.)

VALUE OF THE CORRELATION COEFFICIENT

The *value of the correlation coefficient* always lies in the range -1 to 1, that is,

$$-1 \leq \rho \leq 1 \qquad \text{and} \qquad -1 \leq r \leq 1$$

Although we can explain the linear correlation using the population correlation coefficient ρ, we will do so using the sample correlation coefficient r.

If $r = 1$, it is said to be a case of *perfect positive linear correlation*. In such a case, all points in the scatter diagram lie on a straight line that slopes upward from left to right, as shown in Figure 13.18a. If $r = -1$, the correlation is said to be a *perfect negative linear correlation*. In this case, all points in the scatter diagram fall on a straight line that slopes downward from left to right, as shown in Figure 13.18b. If the points are scattered all over the diagram, as shown in Figure 13.18c, then there is *no linear correlation* between the two variables and consequently r is close to 0.

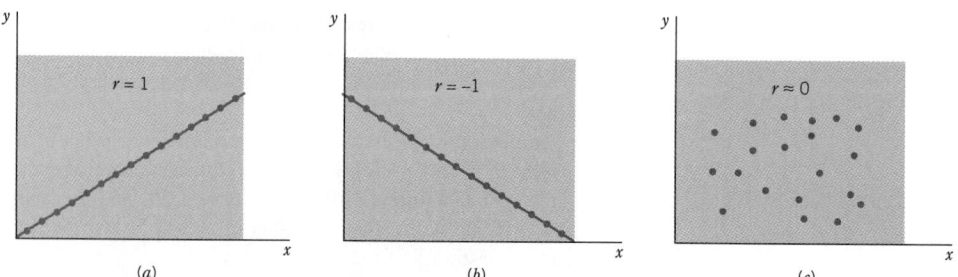

Figure 13.18   Linear correlation between two variables. (a) Perfect positive linear correlation, $r = 1$. (b) Perfect negative linear correlation, $r = -1$. (c) No linear correlation, $r \approx 0$.

We do not usually encounter an example with perfect positive or perfect negative correlation. What we observe in real-world problems is either a positive linear correlation with $0 < r < 1$ (that is, the correlation coefficient is greater than zero but less than 1) or a negative linear correlation with $-1 < r < 0$ (that is, the correlation coefficient is greater than -1 but less than zero).

If the correlation between two variables is positive and close to 1, we say that the variables have a *strong positive linear correlation*. If the correlation between two variables is positive but close to zero, then the variables have a *weak positive linear correlation*. On the other hand, if the correlation between two variables is negative and close to -1, then the variables are said to have a *strong negative linear correlation*. Also, if the correlation between variables is negative but close to zero, there exists a *weak negative linear correlation* between the variables. Graphically, a strong correlation indicates that the points in the scatter diagram are very close to the regression line and a weak correlation indicates that the points in the scatter diagram are widely spread around the regression line. These four cases are shown in Figure 13.19a–d.

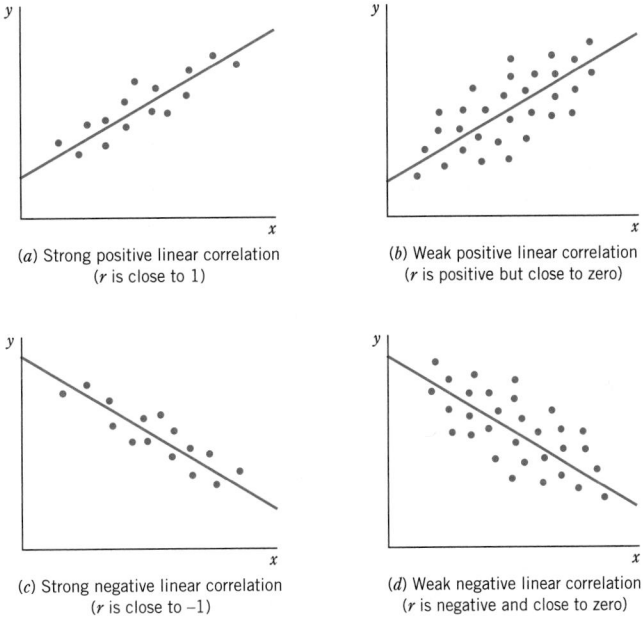

(a) Strong positive linear correlation
(*r* is close to 1)

(b) Weak positive linear correlation
(*r* is positive but close to zero)

(c) Strong negative linear correlation
(*r* is close to −1)

(d) Weak negative linear correlation
(*r* is negative and close to zero)

Figure 13.19 Linear correlation between variables.

The linear correlation coefficient is calculated by using the following formula. (This correlation coefficient is also called the *Pearson product moment correlation coefficient*.)

LINEAR CORRELATION COEFFICIENT

The *simple linear correlation*, denoted by r, measures the strength of the linear relationship between two variables for a sample and is calculated as[6]

$$r = \frac{\text{SS}_{xy}}{\sqrt{\text{SS}_{xx}\,\text{SS}_{yy}}}$$

As both SS_{xx} and SS_{yy} are always positive, the sign of the correlation coefficient r depends on the sign of SS_{xy}. If SS_{xy} is positive then r will be positive and if SS_{xy} is negative then r will be negative. Another important observation to remember is that r *and* b,

---

[6]If we have access to population data, the value of ρ is calculated using the formula

$$\rho = \frac{SS_{xy}}{\sqrt{SS_{xx}\,SS_{yy}}}$$

Here the values of SS_{xy}, SS_{xx}, and SS_{yy} are calculated using the population data.

calculated for the same sample, will always have the same sign. That is, both r and b are either positive or negative. This is so because both r and b provide information about the relationship between x and y. Likewise, the corresponding population parameters ρ and B will always have the same sign.

Example 13–6 illustrates the calculation of the linear correlation coefficient r.

Calculating the linear correlation coefficient.

EXAMPLE 13–6 Calculate the correlation coefficient for the example on incomes and food expenditures of seven households.

Solution From earlier calculations made in Examples 13–1 and 13–2,

$$SS_{xy} = 211.7143, \quad SS_{xx} = 801.4286, \quad \text{and} \quad SS_{yy} = 60.8571$$

Substituting these values in the formula for r, we obtain

$$r = \frac{SS_{xy}}{\sqrt{SS_{xx}\ SS_{yy}}} = \frac{211.7143}{\sqrt{(801.4286)(60.8571)}} = .96$$

Thus, the linear correlation coefficient is .96. The correlation coefficient is usually rounded to two decimal places.

The linear correlation coefficient simply tells us how strongly the two variables are (linearly) related. The correlation coefficient of .96 for incomes and food expenditures of seven households indicates that income and food expenditure are very strongly and positively correlated. This correlation coefficient does not, however, provide us with any more information.

The square of the correlation coefficient gives the coefficient of determination, which was explained in Section 13.4. Thus, $(.96)^2$ is .92, which is the value of r^2 calculated in Example 13–3.

Sometimes the calculated value of r may indicate that the two variables are very strongly linearly correlated but in reality they are not. For example, if we calculate the correlation coefficient between the price of Coke and the size of families in the United States using data for the past 30 years, we will find a strong negative linear correlation. Over time, the price of Coke has increased and the size of families has decreased. This finding does not mean that family size and price of Coke are related. As a result, before we calculate the correlation coefficient, we must seek help from a theory or from common sense to postulate whether or not the two variables have a causal relationship.

Another point to note is that in a simple regression model one of the two variables is categorized as an independent variable and the other is classified as a dependent variable. However, no such distinction is made between the two variables when the correlation coefficient is calculated.

EXERCISES

Concepts and Procedures

13.59 What does a linear correlation coefficient tell you about the relationship between two variables? Within what range can a correlation coefficient assume a value?

13.60 What is the difference between ρ and r? Explain.

13.61 Explain each of the following concepts. You may use graphs to illustrate each concept.

a. Perfect positive linear correlation
b. Perfect negative linear correlation
c. Strong positive linear correlation
d. Strong negative linear correlation
e. Weak positive linear correlation
f. Weak negative linear correlation
g. No linear correlation

13.62 Can the values of B and ρ calculated for the same population data have different signs? Explain.

13.63 For a sample data set, the linear correlation coefficient r has a positive value. Which of the following is true about the slope b of the regression line estimated for the same sample data?

a. The value of b will be positive
b. The value of b will be negative
c. The value of b can be positive or negative

13.64 For a sample data set, the slope b of the regression line has a positive value. Which of the following is true about the linear correlation coefficient r calculated for the same sample data?

a. The value of r will be positive
b. The value of r will be negative
c. The value of r can be positive or negative

13.65 For a sample data set on two variables, the value of the linear correlation coefficient is zero. Does this mean that these variables are not related? Explain.

13.66 Will you expect a positive, zero, or negative correlation between the two variables for each of the following examples?

a. Grade of a student and hours spent studying
b. Incomes and entertainment expenditures of households
c. Ages of women and makeup expenses per month
d. Price of a computer and consumption of Coke
e. Price and consumption of wine

13.67 Will you expect a positive, zero, or negative correlation between the two variables for each of the following examples?

a. SAT scores and GPAs of students
b. Stress level and blood pressure of individuals
c. Amount of fertilizer used and yield of corn per acre
d. Ages and prices of houses
e. Heights of husbands and incomes of their wives

13.68 A population data set produced the following information.

$$N = 250, \quad \Sigma x = 9880, \quad \Sigma y = 1456, \quad \Sigma xy = 85,080,$$

$$\Sigma x^2 = 485,870, \quad \text{and} \quad \Sigma y^2 = 135,675$$

Find the linear correlation coefficient ρ.

13.69 A population data set produced the following information.

$$N = 460, \quad \Sigma x = 3920, \quad \Sigma y = 2650, \quad \Sigma xy = 26,570,$$

$$\Sigma x^2 = 48,530, \quad \text{and} \quad \Sigma y^2 = 39,347$$

Find the linear correlation coefficient ρ.

13.70 A sample data set produced the following information.

$$n = 10, \quad \Sigma x = 100, \quad \Sigma y = 220, \quad \Sigma xy = 3680,$$

$$\Sigma x^2 = 1140, \quad \text{and} \quad \Sigma y^2 = 25,272$$

Calculate the linear correlation coefficient r.

13.71   A sample data set produced the following information.

$$n = 12, \quad \Sigma x = 66, \quad \Sigma y = 588, \quad \Sigma xy = 2244,$$

$$\Sigma x^2 = 396, \quad \text{and} \quad \Sigma y^2 = 58{,}734$$

Calculate the linear correlation coefficient r.

Applications

13.72   Refer to Exercise 13.25. The data on ages (in years) and prices (in hundreds of dollars) for eight cars of a specific model are reproduced below from that exercise.

Age	8	3	6	9	2	5	6	3
Price	16	74	40	19	124	36	33	89

   a.   Do you expect the ages and prices of cars to be positively or negatively related? Explain.
   b.   Calculate the correlation coefficient.

13.73   The following table, reproduced from Exercise 13.53, gives the experience (in years) and monthly salaries (in hundreds of dollars) of nine randomly selected secretaries.

Experience	14	3	5	6	4	9	18	5	16
Monthly salary	26	15	18	20	18	22	29	16	32

   a.   Do you expect the experience and monthly salaries to be positively or negatively related? Explain.
   b.   Compute the correlation coefficient.

13.74   The following table lists the midterm and final exam scores for seven students in a statistics class.

Midterm score	79	95	81	66	87	94	59
Final exam score	85	97	78	76	94	84	67

   a.   Do you expect the midterm and final exam scores to be positively or negatively related?
   b.   Plot a scatter diagram. By looking at the scatter diagram, do you expect the correlation coefficient between these two variables to be close to zero, 1, or -1?
   c.   Find the correlation coefficient. Is the value of r consistent with what you expected in parts a and b?

13.75   The following data give the ages of husbands and wives for six couples.

Husband's age	43	57	28	19	35	39
Wife's age	37	51	32	20	33	38

   a.   Do you expect the ages of husbands and wives to be positively or negatively related?
   b.   Plot a scatter diagram. By looking at the scatter diagram, do you expect the correlation coefficient between these two variables to be close to zero, 1, or -1?
   c.   Find the correlation coefficient. Is the value of r consistent with what you expected in parts a and b?

13.76   The following data on stress scores before a math test and math test scores for seven students are reproduced from Exercise 13.26.

Stress score	6.5	4.0	2.5	7.2	8.1	3.4	5.5
Test score	81	96	93	68	63	84	71

Find the correlation coefficient. Is the sign of the correlation coefficient the same as that of b calculated in Exercise 13.26?

 13.77 The following table, reproduced from Exercise 13.58, gives information on average saturated fat (in grams) consumed per day and cholesterol level (in milligrams per hundred milliliters) of eight males.

Fat consumption	55	65	50	34	43	58	72	36
Cholesterol level	180	215	195	165	170	204	235	150

Find the correlation coefficient. Is the sign of the correlation coefficient the same as that of b calculated in Exercise 13.58?

13.78 Refer to data given in Exercise 13.29 on the total 1996 payroll and the percentage of games won during the 1996 season by each of the National League baseball teams. Compute the linear correlation coefficient, ρ.

13.79 Refer to data given in Exercise 13.30 on the total 1996 payroll and the percentage of games won during the 1996 season by each of the American League baseball teams. Compute the linear correlation coefficient, ρ.

13.7 REGRESSION ANALYSIS: A COMPLETE EXAMPLE

This section works out an example that includes all the topics we have discussed so far in this chapter.

A complete example of regression analysis.

EXAMPLE 13–7 A random sample of eight auto drivers insured with a company and having similar auto insurance policies was selected. The following table lists their driving experience (in years) and the monthly auto insurance premium (in dollars).

Driving Experience (years)	Monthly Auto Insurance Premium (dollars)
5	64
2	87
12	50
9	71
15	44
6	56
25	42
16	60

(a) Does the insurance premium depend on driving experience or does the driving experience depend on insurance premium? Do you expect a positive or a negative relationship between these two variables?

(b) Compute SS_{xx}, SS_{yy}, and SS_{xy}.

(c) Find the least squares regression line by choosing appropriate dependent and independent variables based on your answer in part a.

(d) Interpret the meaning of the values of a and b calculated in part c.

(e) Plot the scatter diagram and the regression line.

(f) Calculate r and r^2 and explain what they mean.

(g) Predict the monthly auto insurance premium for a driver with 10 years of driving experience.

(h) Compute the standard deviation of errors.

(i) Construct a 90% confidence interval for B.

(j) Test at the 5% significance level if B is negative.

Solution

(a) Based on theory and intuition, we expect the insurance premium to depend on driving experience. Consequently, the insurance premium would be a dependent variable and driving experience an independent variable in the regression model. A new driver is considered a high risk by the insurance companies, and he or she has to pay a higher premium for auto insurance. On average, the insurance premium is expected to decrease with an increase in the years of driving experience. Therefore, we expect a negative relationship between these two variables. In other words, both the population correlation coefficient ρ and the population regression slope B are expected to be negative.

(b) Table 13.5 shows the calculation of Σx, Σy, Σxy, Σx^2, and Σy^2.

Table 13.5

Experience x	Premium y	xy	x^2	y^2
5	64	320	25	4096
2	87	174	4	7569
12	50	600	144	2500
9	71	639	81	5041
15	44	660	225	1936
6	56	336	36	3136
25	42	1050	625	1764
16	60	960	256	3600
$\Sigma x = 90$	$\Sigma y = 474$	$\Sigma xy = 4739$	$\Sigma x^2 = 1396$	$\Sigma y^2 = 29{,}642$

The values of $\bar{x}$ and $\bar{y}$ are

$$\bar{x} = \Sigma x/n = 90/8 = 11.25$$

$$\bar{y} = \Sigma y/n = 474/8 = 59.25$$

The values of SS_{xy}, SS_{xx}, and SS_{yy} are computed as follows.

$$SS_{xy} = \Sigma xy - \frac{(\Sigma x)(\Sigma y)}{n} = 4739 - \frac{(90)(474)}{8} = -593.5000$$

$$SS_{xx} = \Sigma x^2 - \frac{(\Sigma x)^2}{n} = 1396 - \frac{(90)^2}{8} = 383.5000$$

$$SS_{yy} = \Sigma y^2 - \frac{(\Sigma y)^2}{n} = 29{,}642 - \frac{(474)^2}{8} = 1557.5000$$

(c)   To find the regression line, we calculate a and b as follows.

$$b = \frac{SS_{xy}}{SS_{xx}} = \frac{-593.5000}{383.5000} = -1.5476$$

$$a = \bar{y} - b\bar{x} = 59.25 - (-1.5476)(11.25) = 76.6605$$

Thus, our estimated regression line $\hat{y} = a + bx$ is

$$\hat{y} = 76.6605 - 1.5476x$$

(d)   The value of $a = 76.6605$ gives the value of $\hat{y}$ for $x = 0$, that is, it gives the monthly auto insurance premium for a driver with no driving experience. However, as mentioned earlier in this chapter, we should not attach much importance to this statement because the sample contains drivers with only two or more years of experience.

The value of b gives the change in $\hat{y}$ due to a change of one unit in x. Thus, $b = -1.5476$ indicates that, on average, for every extra year of driving experience the monthly auto insurance premium decreases by \$1.55. Note that when b is negative, y decreases as x increases.

(e)   Figure 13.20 shows the scatter diagram and the regression line for the data on eight auto drivers. Note that the regression line slopes downward from left to right. This result is consistent with the negative relationship we anticipated between driving experience and insurance premium.

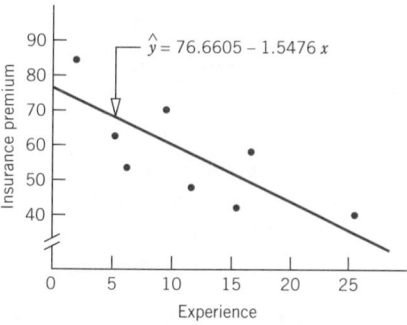

Figure 13.20

(f)   The values of r and r^2 are computed as follows.

$$r = \frac{SS_{xy}}{\sqrt{SS_{xx}\ SS_{yy}}} = \frac{-593.5000}{\sqrt{(383.5000)(1557.5000)}} = -.77$$

$$r^2 = \frac{b\ SS_{xy}}{SS_{yy}} = \frac{(-1.5476)(-593.5000)}{1557.5000} = .59$$

The value of $r = -.77$ indicates that the driving experience and the monthly auto insurance premium are negatively related. The (linear) relationship is strong but not very strong.

The value of $r^2 = .59$ states that 59% of the total variation in insurance premiums is explained by years of driving experience and 41% is not. The low value of r^2 indicates that there may be many other important variables that contribute to the determination of auto insurance premiums. For example, the premium is expected to depend on the driving record of a driver and the type and age of the car.

(g) Using the estimated regression line, the predicted value of y for $x = 10$ is

$$\hat{y} = 76.6605 - 1.5476x = 76.6605 - 1.5476\,(10) = \mathbf{\$61.18}$$

Thus, we expect the monthly auto insurance premium of a driver with 10 years of driving experience to be $61.18.

(h) The standard deviation of errors is

$$s_e = \sqrt{\frac{\text{SS}_{yy} - b\,\text{SS}_{xy}}{n-2}} = \sqrt{\frac{1557.5000 - (-1.5476)\,(-593.5000)}{8-2}} = \mathbf{10.3199}$$

(i) To construct a 90% confidence interval for B, first we calculate the standard deviation of b.

$$s_b = \frac{s_e}{\sqrt{\text{SS}_{xx}}} = \frac{10.3199}{\sqrt{383.5000}} = .5270$$

For a 90% confidence level, the area in each tail of the t distribution is

$$\alpha/2 = .5 - (.90/2) = .05$$

The degrees of freedom are

$$df = n - 2 = 8 - 2 = 6$$

From the t distribution table, the t value for .05 area in the right tail of the t distribution and 6 df is 1.943. The 90% confidence interval for B is

$$b \pm ts_b = -1.5476 \pm 1.943\,(.5270)$$

$$= -1.5476 \pm 1.0240 = \mathbf{-2.57 \text{ to } -.52}$$

Thus, we can state with 90% confidence that B lies in the interval -2.57 to $-.52$. That is, on average, the monthly auto insurance premium of a driver decreases by an amount between $.52 and $2.57 for every extra year of driving experience.

(j) We perform the following five steps to test the hypothesis about B.

Step 1. *State the null and alternative hypotheses*

The null and alternative hypotheses are written as follows.

$$H_0\!: B = 0 \qquad (B \text{ is not negative})$$
$$H_1\!: B < 0 \qquad (B \text{ is negative})$$

Note that the null hypothesis can also be written as $H_0\!: B \geq 0$.

Step 2. *Select the distribution to use*

As the sample size is small ($n < 30$) and σ_ϵ is not known, we use the t distribution to make the hypothesis test.

Step 3. *Determine the rejection and nonrejection regions*

The significance level is .05. The $<$ sign in the alternative hypothesis indicates that it is a left-tailed test.

$$\text{Area in the left tail of the } t \text{ distribution} = \alpha = .05$$
$$df = n - 2 = 8 - 2 = 6$$

From the t distribution table, the critical value of t for .05 area in the left tail of the t distribution and 6 df is -1.943, as shown in Figure 13.21.

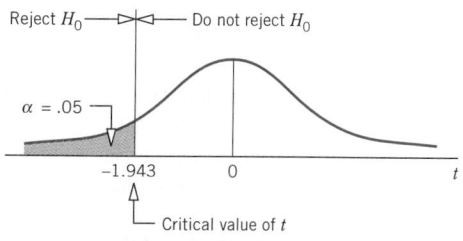

Figure 13.21

Step 4. *Calculate the value of the test statistic*

The value of the test statistic t for b is calculated as follows.

$$t = \frac{b - B}{s_b} = \frac{-1.5476 - 0}{.5270} = -2.937$$

From H_0

Step 5. *Make a decision*

The value of the test statistic $t = -2.937$ falls in the rejection region. Hence, we reject the null hypothesis and conclude that B is negative. That is, the monthly auto insurance premium decreases with an increase in years of driving experience. ■

EXERCISES

Applications

13.80 Eight students, randomly selected from a large class, were asked to keep a record of the hours they spent studying before the midterm examination. The following table gives the number of hours these eight students studied before the midterm and their scores on the midterm.

Hours studied	15	7	12	8	18	6	9	11
Midterm score	97	78	87	92	89	57	74	69

a. Do the midterm scores depend on hours studied or do hours studied depend on the midterm scores? Do you expect a positive or a negative relationship between these two variables?

b. Taking hours studied as an independent variable and midterm scores as a dependent variable, compute SS_{xx}, SS_{yy}, and SS_{xy}.

c. Find the least squares regression line.

d. Interpret the meaning of the values of a and b calculated in part c.

e. Plot the scatter diagram and the regression line.

f. Calculate r and r^2 and briefly explain what they mean.

g. Predict the midterm score of a student who studied for 14 hours.

h. Compute the standard deviation of errors.

i. Construct a 99% confidence interval for B.

j. Test at the 1% significance level if B is positive.

13.81 The following table gives information on ages and cholesterol levels for a random sample of 10 men.

Age	58	69	43	39	63	52	47	31	74	36
Cholesterol level	189	235	193	177	154	191	213	175	198	181

    a. Taking age as an independent variable and cholesterol level as a dependent variable, compute SS_{xx}, SS_{yy}, and SS_{xy}.
    b. Find the regression of cholesterol level on age.
    c. Briefly explain the meaning of the values of a and b calculated in part b.
    d. Calculate r and r^2 and explain what they mean.
    e. Plot the scatter diagram and the regression line.
    f. Predict the cholesterol level of a 60-year-old man.
    g. Compute the standard deviation of errors.
    h. Construct a 95% confidence interval for B.
    i. Test at the 5% significance level if B is positive.

13.82 A farmer wanted to find the relationship between the amount of fertilizer used and the yield of corn. He selected seven acres of his land on which he used different amounts of fertilizer to grow corn. The following table gives the amount (in pounds) of fertilizer used and the yield (in bushels) of corn for each of the seven acres.

Fertilizer Used	Yield of Corn
120	138
80	112
100	129
70	96
88	119
75	104
110	134

    a. With the amount of fertilizer used as an independent variable and yield of corn as a dependent variable, compute SS_{xx}, SS_{yy}, and SS_{xy}.
    b. Find the least squares regression line.
    c. Interpret the meaning of the values of a and b calculated in part b.
    d. Calculate r and r^2 and explain what they mean.
    e. Compute the standard deviation of errors.
    f. Predict the yield of corn per acre for $x = 105$.
    g. Construct a 98% confidence interval for B.
    h. Test at the 5% significance level if B is different from zero.

13.83 The following table gives information on the incomes (in thousands of dollars) and charitable contributions (in hundreds of dollars) for the past year for a random sample of 10 households.

Income (thousands of dollars)	Charitable Contributions (hundreds of dollars)
33	10
23	4
82	29
47	23
26	3
71	28
28	8
39	16
58	18
17	1

a. With income as an independent variable and charitable contributions as a dependent variable, compute SS_{xx}, SS_{yy}, and SS_{xy}.
b. Find the regression of charitable contributions on income.
c. Briefly explain the meaning of the values of a and b.
d. Calculate r and r^2 and briefly explain what they mean.
e. Compute the standard deviation of errors.
f. Construct a 99% confidence interval for B.
g. Test at the 1% significance level if B is positive.

 13.84 The following data give information on the lowest cost ticket price (in dollars) and the average attendance (rounded to the nearest thousand) for the past year for six football teams.

Ticket price	12.50	9.50	10.00	14.50	16.00	12.00
Attendance	56	65	71	69	55	42

a. Taking ticket price as an independent variable and attendance as a dependent variable, compute SS_{xx}, SS_{yy}, and SS_{xy}.
b. Find the least squares regression line.
c. Briefly explain the meaning of the values of a and b calculated in part b.
d. Calculate r and r^2 and briefly explain what they mean.
e. Compute the standard deviation of errors.
f. Construct a 90% confidence interval for B.
g. Test at the 2.5% significance level if B is negative.

13.85 The following table gives information on GPAs (grade point averages) and starting salaries (rounded to the nearest thousand dollars) of seven recent college graduates.

GPA	2.90	3.81	3.20	2.42	3.94	2.05	2.25
Starting salary	23	28	23	21	32	19	22

a. With GPA as an independent variable and starting salary as a dependent variable, compute SS_{xx}, SS_{yy}, and SS_{xy}.
b. Find the least squares regression line.
c. Interpret the meaning of the values of a and b calculated in part b.
d. Calculate r and r^2 and briefly explain what they mean.
e. Compute the standard deviation of errors.
f. Construct a 95% confidence interval for B.
g. Test at the 1% significance level if B is different from zero.

13.8 USING THE REGRESSION MODEL

Let us return to the example on incomes and food expenditures to discuss two major uses of a regression model. These two uses are

1. Estimating the mean value of y for a given value of x. For instance, we can use our food expenditure regression model to estimate the mean food expenditure of all households with a specific income (say $3500 per month).

2. Predicting a particular value of y for a given value of x. For instance, we can determine the expected food expenditure of a randomly selected household with a particular monthly income (say $3500) using our food expenditure regression model.

13.8.1   USING THE REGRESSION MODEL FOR ESTIMATING THE MEAN VALUE OF _y_

Our population regression model is

$$y = A + Bx + \epsilon$$

As mentioned earlier in this chapter, the mean value of y for a given x is denoted by $\mu_{y|x}$, read as "the mean value of y for a given value of x." Because of the assumption that the mean value of ϵ is zero, the mean value of y is given by

$$\mu_{y|x} = A + Bx$$

Our objective is to estimate this mean value. The value of $\hat{y}$, obtained from the sample regression line by substituting the value of x, is the *point estimate of* $\mu_{y|x}$ for that x.

For our example on incomes and food expenditures, the estimated sample regression line is (see Example 13–1)

$$\hat{y} = 1.1414 + .2642x$$

Suppose we want to estimate the mean food expenditure for all households with a monthly income of \$3500. We will denote this population mean by $\mu_{y|x=35}$ or $\mu_{y|35}$. Note that we have written $x = 35$ and not $x = 3500$ in $\mu_{y|35}$ because the units of measurement for the data used to estimate the above regression line in Example 13–1 were hundreds of dollars. Using the regression line, we find that the point estimate of $\mu_{y|35}$ is

$$\hat{y} = 1.1414 + .2642 \,(35) = \$10.3884 \text{ hundreds}$$

Thus, based on the sample regression line, the point estimate for the mean food expenditure $\mu_{y|35}$ for all households with a monthly income of \$3500 is \$1038.84 per month.

However, suppose we take a second sample of seven households from the same population and estimate the regression line for this sample. The point estimate of $\mu_{y|35}$ obtained from the regression line for the second sample is expected to be different. All possible samples of the same size taken from the same population will give different regression lines as shown in Figure 13.22, and, consequently, a different point estimate of $\mu_{y|x}$. Therefore, a confidence interval constructed for $\mu_{y|x}$ based on one sample will give a more reliable estimate of $\mu_{y|x}$ than will a point estimate.

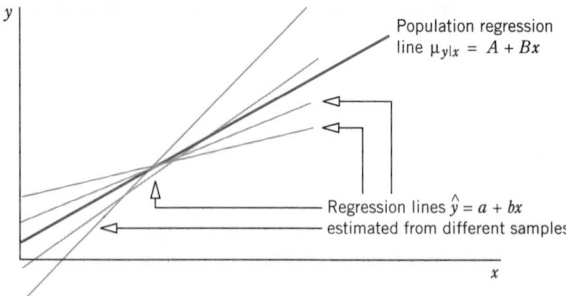

Figure 13.22   Population and sample regression lines.

To construct a confidence interval for $\mu_{y|x}$, we must know the mean, the standard deviation, and the shape of the sampling distribution of its point estimator $\hat{y}$.

The point estimator $\hat{y}$ of $\mu_{y|x}$ is normally distributed with a mean of $A + Bx$ and a standard deviation of

$$\sigma_{\hat{y}_m} = \sigma_\epsilon \sqrt{\frac{1}{n} + \frac{(x_0 - \bar{x})^2}{SS_{xx}}}$$

where $\sigma_{\hat{y}_m}$ is the standard deviation of $\hat{y}$ when it is used to estimate $\mu_{y|x}$, x_0 is the value of x for which we are estimating $\mu_{y|x}$, and σ_ϵ is the population standard deviation of ϵ.

However, usually σ_ϵ is not known. Rather, it is estimated by the standard deviation of sample errors s_e. In this case, we replace σ_ϵ by s_e and $\sigma_{\hat{y}_m}$ by $s_{\hat{y}_m}$ in the foregoing expression. To make a confidence interval for $\mu_{y|x}$, we use the normal distribution when the sample size is large ($n \geq 30$) and the t distribution when the sample size is small ($n < 30$).

CONFIDENCE INTERVAL FOR $\mu_{y|x}$

The $(1 - \alpha)100\%$ *confidence interval for* $\mu_{y|x}$ for $x = x_0$ is

$$\hat{y} \pm t s_{\hat{y}_m}$$

where the value of t is obtained from the t distribution table for $\alpha/2$ area in the right tail of the t distribution curve and $df = n - 2$. The value of $s_{\hat{y}_m}$ is calculated as follows.

$$s_{\hat{y}_m} = s_e \sqrt{\frac{1}{n} + \frac{(x_0 - \bar{x})^2}{SS_{xx}}}$$

Example 13–8 illustrates how to make a confidence interval for the mean value of y, $\mu_{y|x}$.

Constructing a confidence interval for the mean value of y.

EXAMPLE 13-8 Refer to Example 13–1 on incomes and food expenditures. Find a 99% confidence interval for the mean food expenditure for all households with a monthly income of $3500.

Solution Using the regression line estimated in Example 13–1, the point estimate of the mean food expenditure for $x = 35$ is

$$\hat{y} = 1.1414 + .2642\,(35) = \$10.3884 \text{ hundreds}$$

The confidence level is 99%. Hence, the area in each tail of the t distribution is

$$\alpha/2 = .5 - (.99/2) = .005$$

The degrees of freedom are

$$df = n - 2 = 7 - 2 = 5$$

From the t distribution table, the t value for .005 area in the right tail of the t distribution and 5 df is 4.032. From calculations in Examples 13–1 and 13–2, we know that

$$s_e = .9922, \qquad \bar{x} = 30.2857, \qquad \text{and} \qquad SS_{xx} = 801.4286$$

The standard deviation of $\hat{y}$ as an estimate of $\mu_{y|x}$ for $x = 35$ is calculated as follows.

$$s_{\hat{y}_m} = s_e \sqrt{\frac{1}{n} + \frac{(x_0 - \bar{x})^2}{SS_{xx}}} = (.9922) \sqrt{\frac{1}{7} + \frac{(35 - 30.2857)^2}{801.4286}} = .4098$$

Hence, the 99% confidence interval for $\mu_{y|35}$ is

$$\hat{y} \pm ts_{\hat{y}_m} = 10.3884 \pm 4.032 \,(.4098)$$

$$= 10.3884 \pm 1.6523 = \textbf{8.7361 to 12.0407}$$

Thus, with 99% confidence we can state that the mean food expenditure for all households with a monthly income of \$3500 is between \$873.61 and \$1204.07.

13.8.2 USING THE REGRESSION MODEL FOR PREDICTING A PARTICULAR VALUE OF y

The second major use of a regression model is to predict a particular value of y for a given value of x, say x_0. For example, we may want to predict the food expenditure of a randomly selected household with a monthly income of \$3500. In this case, we are not interested in the mean food expenditure of all households with a monthly income of \$3500 but in the food expenditure of one particular household with a monthly income of \$3500. This predicted value of y is denoted by y_p. Again, to predict a single value of y for $x = x_0$ from the estimated sample regression line, we use the value of $\hat{y}$ *as a point estimate of y_p.* Using the estimated regression line, we find that $\hat{y}$ for $x = 35$ is

$$\hat{y} = 1.1414 + .2642(35) = \$10.3884 \text{ hundreds}$$

Thus, based on our regression line, the point estimate for the food expenditure of a given household with a monthly income of \$3500 is \$1038.84 per month. Note that $\hat{y} = 1038.84$ is the point estimate for the mean food expenditure for all households with $x = 35$ as well as for the predicted value of food expenditure of one household with $x = 35$.

Different regression lines estimated by using different samples of seven households each taken from the same population will give different values of the point estimator for the predicted value of y for $x = 35$. Hence, a confidence interval constructed for y_p based on one sample will give a more reliable estimate of y_p than will a point estimate. The confidence interval constructed for y_p is more commonly called a **prediction interval**.

The procedure to construct a prediction interval for y_p is similar to that for constructing a confidence interval for $\mu_{y|x}$ except that the standard deviation of $\hat{y}$ is larger when we predict a single value of y than when we estimate $\mu_{y|x}$.

The point estimator $\hat{y}$ of y_p is normally distributed with a mean of $A + Bx$ and a standard deviation of

$$\sigma_{\hat{y}_p} = \sigma_\epsilon \sqrt{1 + \frac{1}{n} + \frac{(x_0 - \bar{x})^2}{SS_{xx}}}$$

where $\sigma_{\hat{y}_p}$ is the standard deviation of the predicted value of y, x_0 is the value of x for which we are predicting y, and σ_ϵ is the population standard deviation of ϵ.

However, usually σ_ϵ is not known. In this case, we replace σ_ϵ by s_e and $\sigma_{\hat{y}_p}$ by $s_{\hat{y}_p}$ in the foregoing expression. To make a prediction interval for y_p, we use the normal distribution when the sample size is large ($n \geq 30$) and the t distribution when the sample size is small ($n < 30$).

> **PREDICTION INTERVAL FOR y_p**
>
> The $(1 - \alpha)100\%$ *prediction interval* for the predicted value of y, denoted by y_p, for $x = x_0$ is
>
> $$\hat{y} \pm t s_{\hat{y}_p}$$
>
> where the value of t is obtained from the t distribution table for $\alpha/2$ area in the right tail of the t distribution curve and $df = n - 2$. The value of $s_{\hat{y}_p}$ is calculated as follows
>
> $$s_{\hat{y}_p} = s_e \sqrt{1 + \frac{1}{n} + \frac{(x_0 - \bar{x})^2}{\text{SS}_{xx}}}$$

Example 13–9 illustrates the procedure to make a prediction interval for a particular value of y.

Making a prediction interval for a particular value of y.

EXAMPLE 13-9   Refer to Example 13–1 on incomes and food expenditures. Find a 99% prediction interval for the predicted food expenditure for a randomly selected household with a monthly income of $3500.

Solution   Using the regression line estimated in Example 13–1, the point estimate of the predicted food expenditure for $x = 35$ is given by

$$\hat{y} = 1.1414 + .2642 (35) = \$10.3884 \text{ hundreds}$$

The area in each tail of the t distribution for a 99% prediction level is

$$\alpha/2 = .5 - (.99/2) = .005$$

The degrees of freedom are

$$df = n - 2 = 7 - 2 = 5$$

From the t distribution table, the t value for .005 area in the right tail of the t distribution curve and 5 df is 4.032. From calculations in Examples 13–1 and 13–2,

$$s_e = .9922, \quad \bar{x} = 30.2857, \quad \text{and} \quad \text{SS}_{xx} = 801.4286$$

The standard deviation of $\hat{y}$ as an estimator of y_p for $x = 35$ is calculated as follows.

$$s_{\hat{y}_p} = s_e \sqrt{1 + \frac{1}{n} + \frac{(x_0 - \bar{x})^2}{\text{SS}_{xx}}}$$

$$= (.9922) \sqrt{1 + \frac{1}{7} + \frac{(35 - 30.2857)^2}{801.4286}} = 1.0735$$

Hence, the 99% prediction interval for y_p for $x = 35$ is

$$\hat{y} \pm t s_{\hat{y}_p} = 10.3884 \pm 4.032 (1.0735)$$

$$= 10.3884 \pm 4.3284 = \mathbf{6.0600 \text{ to } 14.7168}$$

Thus, with 99% confidence we can state that the predicted food expenditure of a household with a monthly income of $3500 is between $606.00 and $1471.68. ∎

As we can observe, this interval is much wider than the one for the mean value of y for $x = 35$ calculated in Example 13–8, which was $873.61 to $1204.07. This is always true. The prediction interval for predicting a single value of y is always larger than the confidence interval for estimating the mean value of y for a certain value of x.

13.9 CAUTIONS IN USING REGRESSION

When carefully applied, regression is a very helpful technique for making predictions and estimations about one variable for a certain value of another variable. However, we need to be cautious while using the regression analysis, for it can give us misleading results and predictions. The following are the two most important points to remember while using regression.

Extrapolation

The regression line estimated for the sample data is true only for the range of x values observed in the sample. For example, the values of x in our example on incomes and food expenditures vary from a minimum of 15 to a maximum of 49. Hence, our estimated regression line is only applicable for values of x between 15 and 49, that is, we should use this regression line to estimate the mean food expenditure or to predict the food expenditure of a single household only for income levels between $1500 and $4900. If we estimate or predict y for a value of x either less than 15 or greater than 49, it is called *extrapolation*. This does not mean that we should never use the regression line for extrapolation. Instead, we should interpret such predictions cautiously and not attach much value to them.

Similarly, if the data used for the regression estimation are time-series data, the predicted values of y for periods outside the time interval used for the estimation of the regression line should be interpreted very cautiously. When using the estimated regression line for extrapolation, we are assuming that the same linear relationship between the two variables holds true for values of x outside the given range. It is possible that the relationship between the two variables may not be linear outside that range. Nonetheless, even if it is linear, adding a few more observations at either end will probably give a new estimation of the regression line.

Causality

The regression line does not prove causality between two variables. That is, it does not predict that a change in y is caused by a change in x. The information about causality is based on theory or common sense. A regression line only describes whether or not a significant quantitative relationship between x and y exists. Significant relationship means that we reject the null hypothesis H_0: $B = 0$ at a given significance level. The estimated regression line gives the change in y due to a change of one unit in x. Note that it does not indicate that the reason y has changed is that x has changed. In our example on incomes and food expenditures, it is economic theory and common sense, not the regression line, that tells us that food expenditure depends on income. The regression analysis simply helps to determine whether or not this dependence is significant.

EXERCISES

Concepts and Procedures

13.86 Briefly explain the difference between estimating the mean value of y and predicting a particular value of y using a regression model.

13.87 Construct a 99% confidence interval for the mean value of y and a 99% prediction interval for the predicted value of y for the following.

  a. $\hat{y} = 3.25 + .80x$ for $x = 15$ given $s_e = .954$, $\bar{x} = 18.52$, $SS_{xx} = 144.65$, and $n = 10$
  b. $\hat{y} = -27 + 7.67x$ for $x = 12$ given $s_e = 2.46$, $\bar{x} = 13.43$, $SS_{xx} = 369.77$, and $n = 10$

13.88 Construct a 95% confidence interval for the mean value of y and a 95% prediction interval for the predicted value of y for the following.

  a. $\hat{y} = 13.40 + 2.58x$ for $x = 8$ given $s_e = 1.29$, $\bar{x} = 11.30$, $SS_{xx} = 210.45$, and $n = 12$
  b. $\hat{y} = -8.6 + 3.72x$ for $x = 24$ given $s_e = 1.89$, $\bar{x} = 19.70$, $SS_{xx} = 315.40$, and $n = 10$

Applications

13.89 Refer to Exercise 13.53. Construct a 90% confidence interval for the mean monthly salary of all secretaries with 10 years of experience. Construct a 90% prediction interval for the monthly salary of a randomly selected secretary with 10 years of experience.

13.90 Refer to the data on hours studied and the midterm scores of eight students given in Exercise 13.80. Construct a 99% confidence interval for $\mu_{y|x}$ for $x = 13$ and a 99% prediction interval for y_p for $x = 13$.

13.91 Refer to Exercise 13.82. Construct a 99% confidence interval for the mean yield of corn per acre for all acres on which 90 pounds of fertilizer are used. Determine a 99% prediction interval for the yield of corn for a randomly selected acre on which 90 pounds of fertilizer are used.

13.92 Using the data on ages and cholesterol levels of 10 men given in Exercise 13.81, find a 95% confidence interval for the mean cholesterol level for all 53-year-old men. Make a 95% prediction interval for the cholesterol level for a randomly selected 53-year-old man.

13.93 Refer to Exercise 13.83. Construct a 95% confidence interval for the mean charitable contributions made by all households with an income of $64,000. Make a 95% prediction interval for the charitable contributions made by a randomly selected household with an income of $64,000.

13.94 Refer to Exercise 13.85. Construct a 98% confidence interval for the mean starting salary of recent college graduates with a GPA of 3.15. Construct a 98% prediction interval for the starting salary of a randomly selected recent college graduate with a GPA of 3.15.

GLOSSARY

Coefficient of determination A measure that gives the proportion (or percentage) of the total variation in a dependent variable that is explained by a given independent variable.

Degrees of freedom for a simple linear regression model Sample size minus 2, that is, $n - 2$.

Dependent variable The variable to be predicted or explained.

Deterministic model A model in which the independent variable determines the dependent variable exactly. Such a model gives an **exact relationship** between two variables.

Estimated or **predicted value of y** The value of the dependent variable, denoted by $\hat{y}$, that is calculated for a given value of x using the estimated regression model.

Independent or **explanatory variable** The variable included in a model to explain the variation in the dependent variable.

Least squares estimates of A and B The values of a and b that are calculated by using the sample data.

Least squares method The method used to fit a regression line through a scatter diagram such that the error sum of squares is minimum.

Least squares regression line A regression line obtained by using the least squares method.

Linear correlation coefficient A measure of the strength of the linear relationship between two variables.

Linear regression model A regression model that gives a straight-line relationship between two variables.

Multiple regression model A regression model that contains two or more independent variables.

Negative relationship between two variables The value of the slope in the regression line and the correlation coefficient between two variables are both negative.

Nonlinear (simple) regression model A regression model that does not give a straight-line relationship between two variables.

Population parameters for a simple regression model The values of A and B for the regression model $y = A + Bx + \epsilon$ that are obtained by using population data.

Positive relationship between two variables The value of the slope in the regression line and the correlation coefficient between two variables are both positive.

Prediction interval The confidence interval for a particular value of y for a given value of x.

Probabilistic or **statistical model** A model in which the independent variable does not determine the dependent variable exactly.

Random error term (ϵ) The difference between the actual and predicted values of y.

Scatter diagram or **scattergram** A plot of the paired observations of x and y.

Simple linear regression A regression model with one dependent and one independent variable that assumes a straight-line relationship.

Slope The coefficient of x in a regression model that gives the change in y for a change of one unit in x.

SSE (error sum of squares) The sum of the squared differences between the actual and predicted values of y. It is the portion of the SST that is not explained by the regression model.

SSR (regression sum of squares) The portion of the SST that is explained by the regression model.

SST (total sum of squares) The sum of the squared differences between actual y values and $\bar{y}$.

Standard deviation of errors A measure of spread for the random errors.

y-intercept The point at which the regression line intersects the vertical axis on which the dependent variable is marked. It is the value of y when x is zero.

KEY FORMULAS

1. **Simple linear regression model**

$$y = A + Bx + \epsilon$$

2. **Estimated simple linear regression model**

$$\hat{y} = a + bx$$

3. **Least squares estimates of A and B**

$$b = \frac{\text{SS}_{xy}}{\text{SS}_{xx}} \quad \text{and} \quad a = \bar{y} - b\bar{x}$$

where

$$\text{SS}_{xy} = \Sigma xy - \frac{(\Sigma x)\,(\Sigma y)}{n}$$

$$\text{SS}_{xx} = \Sigma x^2 - \frac{(\Sigma x)^2}{n}$$

4. **Standard deviation of the sample errors**

$$s_e = \sqrt{\frac{SS_{yy} - b\, SS_{xy}}{n - 2}}$$

where

$$SS_{yy} = \Sigma y^2 - \frac{(\Sigma y)^2}{n}$$

5. **Error sum of squares**

$$SSE = \Sigma e^2 = \Sigma(y - \hat{y})^2$$

6. **Total sum of squares**

$$SST = \Sigma y^2 - \frac{(\Sigma y)^2}{n}$$

7. **Regression sum of squares**

$$SSR = SST - SSE$$

8. **Coefficient of determination**

$$r^2 = \frac{b\, SS_{xy}}{SS_{yy}}$$

9. **Standard deviation of b**

$$s_b = \frac{s_e}{\sqrt{SS_{xx}}}$$

10. **The $(1 - \alpha)100\%$ confidence interval for B**

$$b \pm t s_b$$

11. **Value of the test statistic t for b**

$$t = \frac{b - B}{s_b}$$

12. **Linear correlation coefficient**

$$r = \frac{SS_{xy}}{\sqrt{SS_{xx}\, SS_{yy}}}$$

13. **The $(1 - \alpha)100\%$ confidence interval for $\mu_{y|x}$**

$$\hat{y} \pm t s_{\hat{y}_m}$$

where

$$s_{\hat{y}_m} = s_e \sqrt{\frac{1}{n} + \frac{(x_0 - \bar{x})^2}{SS_{xx}}}$$

14. **The $(1 - \alpha)100\%$ prediction interval for y_p**

$$\hat{y} \pm t s_{\hat{y}_p}$$

where

$$s_{\hat{y}_p} = s_e \sqrt{1 + \frac{1}{n} + \frac{(x_0 - \bar{x})^2}{SS_{xx}}}$$

SUPPLEMENTARY EXERCISES

13.95 The following data give information on the ages (in years) and the number of breakdowns during the past month for a sample of seven machines at a large company.

Age	12	7	2	8	13	9	4
Number of breakdowns	9	5	1	4	11	7	2

   a. Taking age as an independent variable and number of breakdowns as a dependent variable, what is your hypothesis about the sign of B in the regression line? (In other words, do you expect B to be positive or negative?)

   b. Find the least squares regression line. Is the sign of b the same as you hypothesized for B in part a?

   c. Give a brief interpretation of the values of a and b calculated in part b.

   d. Compute r and r^2 and explain what they mean.

   e. Compute the standard deviation of errors.

   f. Construct a 99% confidence interval for B.

   g. Test at the 2.5% significance level if B is positive.

13.96 The following table gives information on the number of hours that eight bank loan officers slept the previous night and the number of loan applications that they processed the next day.

Number of hours slept	8	5	7	6	4	8	6	5
Number of applications processed	14	11	16	13	8	17	10	8

   a. Taking the number of hours slept as an independent variable and the number of applications processed as a dependent variable, do you expect B to be positive or negative in the regression model $y = A + Bx + \epsilon$?

   b. Find the least squares regression line. Is the sign of b the same as you hypothesized for B in part a?

   c. Compute r and r^2 and explain what they mean.

   d. Compute the standard deviation of errors.

   e. Construct a 90% confidence interval for B.

   f. Test at the 5% significance level if B is positive.

13.97 The management of a supermarket wants to find if there is a relationship between the number of times a specific product is promoted on the intercom system in the store and the number of units of that product sold. To experiment, the management selected a product and promoted it on the intercom system for seven days. The following table gives the number of times this product was promoted each day and the number of units sold.

Number of Promotions per Day	Number of Units Sold per Day (hundreds)
15	11
22	22
42	30
30	24
18	17
12	15
38	21

a. With the number of promotions as an independent variable and the number of units sold as a dependent variable, what do you expect the sign of B in the regression line $y = A + Bx + \epsilon$ will be?

b. Find the least squares regression line $\hat{y} = a + bx$. Is the sign of b the same as you hypothesized for B in part a?

c. Give a brief interpretation of the values of a and b calculated in part b.

d. Compute r and r^2 and explain what they mean.

e. Predict the number of units of this product sold on a day with 35 promotions.

f. Compute the standard deviation of errors.

g. Construct a 98% confidence interval for B.

h. Testing at the 1% significance level, can you conclude that B is positive?

 13.98 The following table gives information on the temperature in a city and the volume of ice cream (in pounds) sold at an ice cream parlor for a random sample of eight days during the summer of 1996.

Temperature	93	86	77	89	98	102	87	79
Ice cream sold	202	175	123	198	232	267	158	117

a. Find the least squares regression line $\hat{y} = a + bx$. Take temperature as an independent variable and volume of ice cream sold as a dependent variable.

b. Give a brief interpretation of the values of a and b.

c. Compute r and r^2 and explain what they mean.

d. Predict the amount of ice cream sold on a day with a temperature of 95°.

e. Compute the standard deviation of errors.

f. Construct a 99% confidence interval for B.

g. Testing at the 1% significance level, can you conclude that B is different from zero?

13.99 The following table gives the milk production (rounded to billions of pounds) in the United States for the years 1984 to 1995. (*Source*: U.S. Department of Agriculture.)

Year	Milk Production
1984	135
1985	143
1986	144
1987	143
1988	145
1989	144
1990	148
1991	148
1992	151
1993	151
1994	154
1995	156

a. Assign a value of 0 to 1984, 1 to 1985, 2 to 1986, and so on. Call this new variable *time*. Write a new table with the variables *time* and *milk production*.

b. With time as an independent variable and milk production as a dependent variable, compute SS_{xx}, SS_{yy}, and SS_{xy}.

c. Construct a scatter diagram for these data. Does the scatter diagram exhibit a linear positive relationship between time and milk production?

d. Find the least squares regression line $\hat{y} = a + bx$.

   e.   Give a brief interpretation of the values of *a* and *b* calculated in part d.
   f.   Compute the correlation coefficient *r*.
   g.   Predict the milk production for 1999. Comment on this prediction.

13.100   The following table gives the gross national product (in hundreds of billions of dollars) for the United States for the years 1984 to 1994.

Year	Gross National Product
1984	38.02
1985	40.54
1986	42.78
1987	45.45
1988	49.08
1989	52.67
1990	55.68
1991	57.41
1992	60.26
1993	63.48
1994	67.27

   a.   Assign a value of 0 to 1984, 1 to 1985, 2 to 1986, and so on. Call this new variable *time*. Write a new table with the variables *time* and *gross national product*.
   b.   With time as an independent variable and gross national product as a dependent variable, compute SS_{xx}, SS_{yy}, and SS_{xy}.
   c.   Construct a scatter diagram for these data. Does the scatter diagram exhibit a linear positive relationship between time and gross national product?
   d.   Find the least squares regression line $\hat{y} = a + bx$.
   e.   Give a brief interpretation of the values of *a* and *b* calculated in part d.
   f.   Compute the correlation coefficient *r*.
   g.   Predict the gross national product for $x = 15$. Comment on this prediction.

13.101   Refer to the data on ages and number of breakdowns for seven machines given in Exercise 13.95. Construct a 99% confidence interval for the mean number of breakdowns per month for all machines with an age of eight years. Find a 99% prediction interval for the number of breakdowns per month for a randomly selected machine with an age of eight years.

13.102   Refer to the data on the number of hours slept and the number of loan applications processed by bank loan officers given in Exercise 13.96. Determine a 95% confidence interval for the mean number of applications processed by bank loan officers who slept for six hours. Make a 95% prediction interval for the number of applications processed by a bank loan officer who slept for six hours.

13.103   Refer to the data given in Exercise 13.97 on the number of times a specific product is promoted on the intercom system in a supermarket and the number of units of that product sold. Make a 90% confidence interval for the mean number of units of that product sold on days with 35 promotions. Construct a 90% prediction interval for the number of units of that product sold on a randomly selected day with 35 promotions.

13.104   Refer to the data given in Exercise 13.98 on temperatures and the volumes of ice cream sold at an ice cream parlor for a sample of eight days. Construct a 98% confidence interval for the mean volume of ice cream sold at this parlor on all days with a temperature of 95°. Determine a 98% prediction interval for the volume of ice cream sold at this parlor on a randomly selected day with a temperature of 95°.

*13.105   Consider the data given in the following table.

x	10	20	30	40	50	60
y	12	15	19	21	25	30

a. Find the least squares regression line and the linear correlation coefficient, r.

b. Suppose that each value of y given in the table is increased by 5, leaving the x-values unchanged. Would you expect r to increase, decrease, or remain the same? How do you expect the least squares regression line to change?

c. Increase each value of y given in the table by 5 and find the new least squares regression line and the correlation coefficient r. Do these results agree with your expectation in part b?

*13.106   A few recent studies seem to indicate that moderate consumption of wine is beneficial to your heart. A recent article discussed this relationship between these two variables (*Source:* Laura Shapiro, ''Food to Your Health?'' *Newsweek*, January 22, 1996). Let x be the annual consumption of wine in liters per capita and y the annual death rate due to heart disease per 100,000 population. The following table, adapted from the above mentioned article, gives data on x and y for 10 countries.

Country	x	y
France	63.5	61.1
Italy	58.0	94.1
Switzerland	46.0	106.4
Australia	15.7	173.0
Britain	12.2	199.7
United States	8.9	176.0
Russia	2.7	373.6
Czech Republic	1.7	283.7
Japan	1.0	34.7
Mexico	0.2	36.4

a. Compute the correlation coefficient between x and y.

b. Does your correlation coefficient indicate a positive or a negative association between wine consumption and death rate due to heart disease for these countries? Is this association strong, moderate, or weak?

c. Does your answer to part a demonstrate a cause-and-effect relationship between wine consumption and heart disease? What other explanations might there be for the association between the two variables found in part a?

*13.107   Recently a group of researchers studied the diving behavior of Antarctic fur seals (*Source:* I. L. Boyd et al. ''Swimming Speed and Allocation of Time during the Dive Cycle in Antarctic Fur Seals,'' *Animal Behavior* 50[3], September 1995). Ten lactating female fur seals were captured, weighed, fitted with a time-depth recorder, and released. Each time a seal dived, the depth, duration, and other information about the dive were stored in the recorder's memory. The seals were recaptured 3 to 10 days later, and the recorders were recovered. Let x be the mass of a seal (in kilograms) and y the median dive duration (in seconds). The following table gives the data on x and y obtained from this study.

x	29.0	34.5	33.0	31.0	32.5	39.5	29.5	29.5	37.5	35.0
y	45	45	85	70	25	65	15	50	40	35

a. Compute SS_{xx}, SS_{yy}, and SS_{xy}.

b. Find the regression of y on x.

c. Briefly explain the meaning of the values of a and b calculated in part b.

d. Calculate r and r^2 and explain what they mean.

e. Compute the standard deviation of errors.

f. Test at the 5% significance level if the coefficient of x is different from zero.

g. Does the result of your hypothesis test in part f suggest that a seal's mass would be a useful predictor of her median dive duration?

*13.108 Suppose that you work part-time at a bowling alley that is open daily from twelve noon to twelve midnight. Although business is usually slow from twelve noon to 6:00 P.M., the owner has noticed that it is better on hotter days during the summer, perhaps because the premises are comfortably air-conditioned. The owner shows you some data that she gathered last summer. This data set includes the maximum temperature and the number of lines bowled between twelve noon and 6:00 P.M. for each of 20 days. (The maximum temperatures ranged from 77° to 95° Fahrenheit during this period.) The owner would like to know if she can estimate tomorrow's business from twelve noon to 6:00 P.M. by looking at tomorrow's weather forecast. She asks you to analyze the data. Let x be the maximum temperature for a day and y the number of lines bowled between twelve noon and 6:00 P.M. on that day. The computer output based on the data for 20 days provided the following results.

$$\hat{y} = -432 + 7.7x, \qquad s_e = 28.17, \qquad SS_{xx} = 607, \qquad \bar{x} = 87.5$$

Assume that the weather forecasts are reasonably accurate.

a. Does the maximum temperature seem to be a useful predictor of bowling activity between twelve noon and 6:00 P.M.? Use an appropriate statistical procedure based on the information given. Use $\alpha = .05$.

b. The owner wants to know how many lines of bowling she can expect, on average, for days with a maximum temperature of 90°. Answer using a 95% confidence level.

c. The owner has seen tomorrow's weather forecast, which predicts a high of 90°. About how many lines of bowling can she expect? Answer using a 95% confidence level.

d. Give a brief commonsense explanation to the owner for the difference in the interval estimates of parts b and c.

e. The owner asks you how many lines of bowling she could expect if the high temperature were 100°. Give a point estimate, together with an appropriate warning to the owner.

*13.109 An economist wanted to study the relationship between incomes of 30-year-old persons employed full-time and the incomes of their fathers. Let x be the mean of the three highest annual incomes for a 30-year-old person employed full-time and y the mean of the three highest annual incomes for the person's father. Suppose that after collecting data for a random sample of 300 such 30-year-olds and their fathers, the economist finds a linear correlation coefficient of .60. A friend of yours, who has read about this research, asks you several questions, such as: Does the positive value of the correlation coefficient suggest that the 30-year-olds tend to earn more than their fathers? Does the correlation coefficient reveal anything at all about the difference between the incomes of 30-year-olds and their fathers? If not, what other information would we need from this study? What does the correlation coefficient tell us about the relationship between the two variables in this example? Write a short note to your friend answering these questions.

SELF-REVIEW TEST

1. A simple regression is a regression model that contains
    a. only one independent variable
    b. only one dependent variable
    c. more than one independent variable
    d. both a and b

2. The relationship between independent and dependent variables represented by the (simple) linear regression is that of

    a. a straight line   **b.** a curve   **c.** both a and b

3. A deterministic regression model is a model that

    a. contains the random error term
    b. does not contain the random error term
    c. gives a nonlinear relationship

4. A probabilistic regression model is a model that

    a. contains the random error term
    b. does not contain the random error term
    c. shows an exact relationship

5. The least squares regression line minimizes the sum of

    a. errors   **b.** squared errors   **c.** predictions

6. The degrees of freedom for a simple regression model are

    a. $n - 1$ **b.** $n - 2$ **c.** $n - 5$

7. Indicate if the following statement is true or false.

> The coefficient of determination gives the proportion of total squared errors (SST) that is explained by the use of the regression model.

8. Indicate if the following statement is true or false.

> The linear correlation coefficient measures the strength of the linear association between two variables.

9. The value of the coefficient of determination is always in the range

    a. 0 to 1   **b.** -1 to 1   **c.** -1 to 0

10. The value of the correlation coefficient is always in the range

    a. 0 to 1   **b.** -1 to 1   **c.** -1 to 0

11. Explain why the random error term ϵ is added to the regression model.

12. Explain the difference between A and a and between B and b for a regression model.

13. Briefly explain the assumptions of a regression model.

14. Briefly explain the difference between the population regression line and a sample regression line.

15. The following table gives the experience (in years) and the number of computers sold during the previous three months by seven salespersons.

Experience	4	12	9	6	10	16	7
Computers sold	19	42	28	33	39	35	23

    a. Do you think experience depends on the number of computers sold or the number of computers sold depends on experience?
    b. With experience as an independent variable and the number of computers sold as a dependent variable, what is your hypothesis about the sign of B in the regression model?
    c. Construct a scatter diagram for these data. Does the scatter diagram exhibit a linear relationship between the two variables?

d. Find the least squares regression line. Is the sign of b the same as the one you hypothesized for B in part b?

e. Give a brief interpretation of the values of the y-intercept and slope calculated in part d.

f. Compute r and r^2 and explain what they mean.

g. Predict the number of computers sold during the past three months by a salesperson with 11 years of experience.

h. Compute the standard deviation of errors.

i. Construct a 99% confidence interval for B.

j. Testing at the 1% significance level, can you conclude that B is positive?

k. Construct a 95% confidence interval for the mean number of computers sold in a three-month period by all salespersons with eight years of experience.

l. Make a 95% prediction interval for the number of computers sold in a three-month period by a randomly selected salesperson with eight years of experience.

USING MINITAB

In this section we describe the use of MINITAB for doing regression analysis. We use Illustration M13–1 to show how MINITAB FOR WINDOWS and the MINITAB COMMAND LANGUAGE can be used to obtain a computer solution.

ILLUSTRATION M13–1   Refer to the data of Example 13–7 on driving experience (in years) and monthly auto insurance premiums (in dollars) for eight auto drivers. That data set is reproduced below.

Driving Experience (years)	Monthly Auto Insurance Premium (dollars)
5	64
2	87
12	50
9	71
15	44
6	56
25	42
16	60

Use MINITAB to answer the following.

(a) Construct a scatter diagram for these data.

(b) Find the correlation between these two variables.

(c) Find the regression line with experience as an independent variable and premium as a dependent variable.

(d) Make a 90% confidence interval for B.

(e) Test at the 5% significance level if B is negative.

(f) Make a 95% confidence interval for the mean monthly auto insurance premium for all drivers with 10 years of driving experience. Construct a 95% prediction interval for the monthly auto insurance premium for a randomly selected driver with 10 years of driving experience.

Solution

(a) Note that in this example experience is the independent variable, and premium is the dependent variable. If you are using MINITAB FOR WINDOWS, perform the following steps to obtain the scatter diagram.

Step 1.   Enter the given data in columns C1 and C2 of the Data window. You can name these columns EXPER and PREMIUM. Note that the data entered in column C1 represent the x variable and those entered in column C2 represent the y variable.

Step 2.   Click the **Graph** pull-down menu at the top of the screen.

Step 3.   Click **Plot** from the selections available in the **Graph** menu.

Step 4. You will see a dialog box entitled **Plot** appear on the screen. Type **C2** or **Premium** in the box below **Y** and **C1** or **Exper** in the box below **X**.

Step 5. Click the **OK** button at the bottom of this dialog box. The scatter diagram will appear on the screen.

If you are using the MINITAB COMMAND LANGUAGE, first enter the given data in columns C1 and C2 using the **SET** or **READ** command. Then, type the following MINITAB command to obtain the scatter diagram.

```
MTB > PLOT C2 * C1
```

If you named the two columns EXPER and PREMIUM, then you can type the following command.

```
MTB > PLOT 'PREMIUM' * 'EXPER'
```

Whether you use MINITAB FOR WINDOWS or the MINITAB COMMAND LANGUAGE, you will obtain the scatter diagram given in Figure 13.23.

Figure 13.23   Scatter diagram for Illustration M13–1.

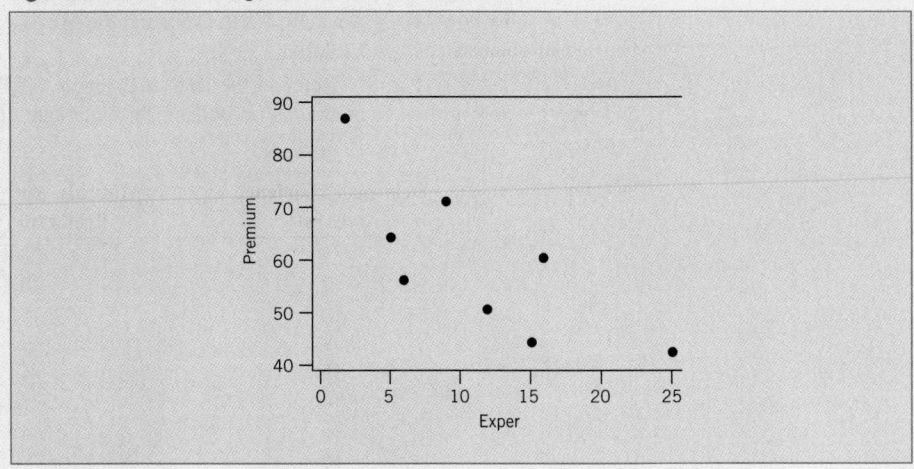

To do the remaining parts of this illustration, we assume that you have already entered the given data in columns C1 and C2. We also assume that you have named these columns EXPER and PREMIUM. If you did not, then replace EXPER by C1 and PREMIUM by C2 in the MINITAB commands in sections b through f below.

(b) If you are using MINITAB FOR WINDOWS, perform the following steps to obtain the correlation coefficient.

Step 1. Click the **Stat** pull-down menu at the top of the screen.

Step 2. Click **Basic Statistics** from the selections available in the **Stat** menu.

Step 3. Click **Correlation** from the options available in the **Basic Statistics** menu.

Step 4. You will see a dialog box entitled **Correlation** appear on the screen. Type **C1 C2** or **EXPER PREMIUM** in the box below **variables**.

Step 5. Click the **OK** button at the bottom of this dialog box. The MINITAB output will appear on the screen.

If you are using the MINITAB COMMAND LANGUAGE, type the following MINITAB command to obtain the correlation coefficient.

```
MTB > CORRELATION 'EXPER' * 'PREMIUM'
```

Whether you use MINITAB FOR WINDOWS or MINITAB COMMAND LANGUAGE, you will obtain the MINITAB output shown in Figure 13.24.

Figure 13.24 Correlation coefficient.

> ## Correlations (Pearson)
>
> Correlation of EXPER and PREMIUM = −0.768

Hence, the linear correlation coefficient between experience and premium is

$$r = -.768$$

(c) If you are using MINITAB FOR WINDOWS, perform the following steps to obtain the regression line.

Step 1. Click the **Stat** pull-down menu at the top of the screen.

Step 2. Click **Regression** from the selections available in the **Stat** menu.

Step 3. Click **Regression** again from the options available in the **Regression** menu.

Step 4. You will see a dialog box entitled **Regression** appear on the screen. Type **C2** or **PREMIUM** in the box next to **Response** and **C1** or **EXPER** in the box next to **Predictors**.

Step 5. Click the **OK** button at the bottom of this dialog box. The MINITAB output will appear on the screen.

If you are using the MINITAB COMMAND LANGUAGE, type the following MINITAB command to obtain the regression line.

MTB > REGRESS 'PREMIUM' 1 'EXPER'

In this MINITAB command, 1 indicates that there is only one independent variable.

Whether you use MINITAB FOR WINDOWS or the MINITAB COMMAND LANGUAGE, you will obtain the MINITAB output given in Figure 13.25.

Figure 13.25 MINITAB regression output.

> ## Regression Analysis
>
> The regression equation is
> PREMIUM = 76.7 − 1.55 EXPER
>
> | Predictor | Coef | StDev | T | P |
> |-----------|---------|--------|-------|-------|
> | Constant | 76.660 | 6.961 | 11.01 | 0.000 |
> | EXPER | −1.5476 | 0.5270 | −2.94 | 0.026 |
>
> S = 10.32   R-Sq = 59.0%   R-sq(adj) = 52.1%
>
> Analysis of Variance
>
> | Source | DF | SS | MS | F | P |
> |------------|----|--------|-------|------|-------|
> | Regression | 1 | 918.5 | 918.5 | 8.62 | 0.026 |
> | Error | 6 | 639.0 | 106.5 | | |
> | Total | 7 | 1557.5 | | | |

The MINITAB output given in Figure 13.25 has four main parts. The first part gives the regression equation, which is

$$\text{PREMIUM} = 76.7 - 1.55 \text{ EXPER}$$

Thus, $a = 76.7$ and $b = -1.55$

The second part of the MINITAB output gives the values of a and b, their standard deviations, the values of the test statistic t for the hypothesis tests about A and B, and the p-values for these two tests. Figure 13.26 explains this part of the output.

Figure 13.26

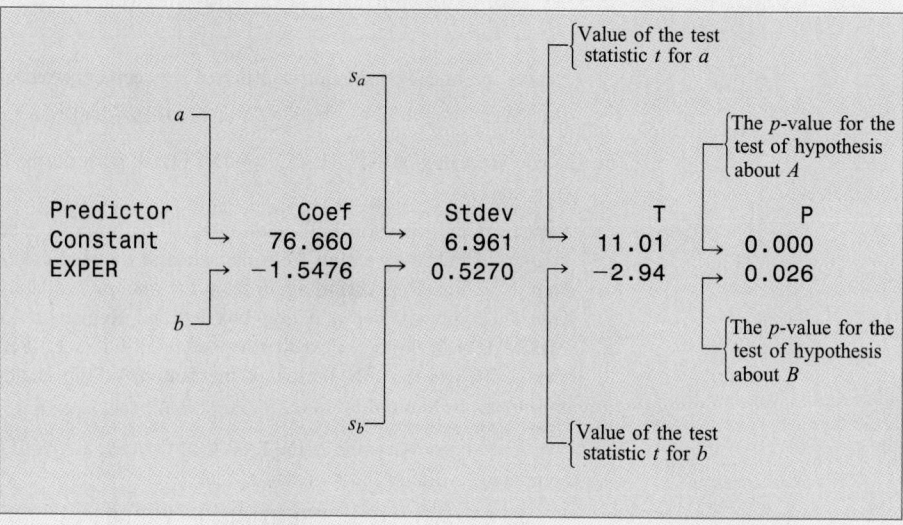

Note that we have not discussed the standard deviation s_a of a and the test of hypothesis about a (and, hence, the value of the test statistic t for a and the p-value for the test of hypothesis about a) in this chapter. Therefore, in the row for Constant in Figure 13.26, all the values except the coefficient value are irrelevant for this chapter. If we compare the various values in this figure with the ones calculated in Example 13–7, we notice a slight difference due to rounding.

The third part of the MINITAB output in Figure 13.25 gives the standard deviation of errors s_e, the coefficient of determination r^2, and the adjusted r^2 (the value of r^2 adjusted for the degrees of freedom). However, we have not discussed the concept of adjusted r^2 in this chapter. The explanation of this part of the output appears in Figure 13.27.

Figure 13.27

S = 10.32	R-Sq = 59.0%	R-Sq(adj) = 52.1%
↑	↑	↑
s_e	r^2	r^2 adjusted for df

The fourth part of the MINITAB output gives the analysis of variance table. We have not discussed such a table in this chapter except for the SS column. However, this table is similar to the analysis of variance table discussed in Chapter 12. The column of SS gives the values of SSR, SSE, and SST as shown in Figure 13.28.

Figure 13.28

Source	SS	
Regression	918.5	←——SSR
Error	639.0	←——SSE
Total	1557.5	←——SST

(d) The 90% confidence interval for B is given by the formula

$$b \pm ts_b$$

From the row corresponding to *EXPER* in Figure 13.26,

$$b = -1.5476 \quad \text{and} \quad s_b = 0.5270$$

For a 90% confidence interval, $\alpha = .10$ and $\alpha/2 = .05$. The degrees of freedom are $n - 2 = 8 - 2 = 6$. From the t distribution table, the t value for .05 area in the right tail of the t distribution curve and 6 df is 1.943. The 90% confidence interval for B is

$$b \pm ts_b = -1.5476 \pm 1.943 \, (.5270) = -1.5476 \pm 1.0240 = -2.57 \text{ to } -.52$$

(e) The null and alternative hypotheses are

$$H_0: B = 0 \quad \text{(B is not negative)}$$

$$H_1: B < 0 \quad \text{(B is negative)}$$

The significance level is .05. The degrees of freedom are 6. The test is left-tailed. From the t distribution table, the critical value of t for $\alpha = .05$ and $df = 6$ is -1.943.

We again use the information given in the row corresponding to *EXPER* in Figure 13.26. From that row of information,

$$\text{Value of the test statistic } t \text{ for } b = -2.94$$

Because the value of the test statistic, $t = -2.94$, is less than the critical value of $t = -1.943$, it falls in the rejection region, which is in the left tail of the t distribution curve (see Figure 13.21 of Example 13–7). Consequently, we reject the null hypothesis and state that our sample information supports the alternative hypothesis that B is negative.

(f) To make a 95% confidence interval for $\mu_{y|x}$ and a 95% prediction interval for y_p for EXPER = 10, we cannot use MINITAB FOR WINDOWS. In this case we use the MINITAB COMMAND LANGUAGE as follows.

```
MTB > REGRESS 'PREMIUM' 1 'EXPER';
SUBC > PREDICT 10.
```

Note that when we use the above command and subcommand, MINITAB will give all the output we listed in Figure 13.25 and the output given in Figure 13.29. We have omitted the output of Figure 13.25 from Figure 13.29.

Figure 13.29

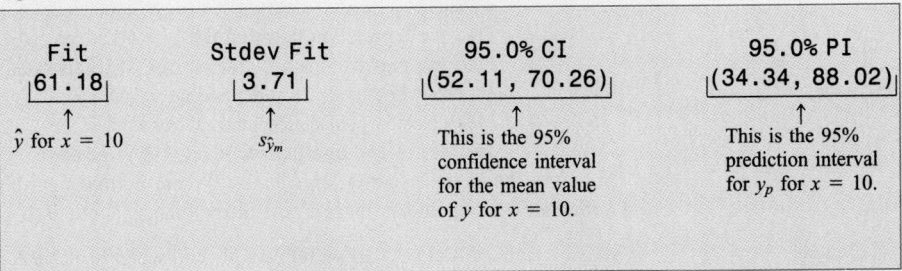

Thus, from Figure 13.29, the value of $\hat{y}$ for $x = 10$ is 61.18. This can be considered as the point estimate of the mean value of y as well as the point estimate of the predicted value of y for $x = 10$. The interval 52.11 to 70.26 in the printout gives a 95% confidence interval for the mean value $\mu_{y|10}$ of y for $x = 10$ as was discussed in Section 13.8.1. The interval 34.34 to 88.02 gives a 95% prediction interval for y_p for $x = 10$ as was discussed in Section 13.8.2.

COMPUTER ASSIGNMENTS

M13.1 Professor Hamid Zangenehzadeh studied the relationship between student evaluations of a teacher and the expected grades of students in a course. The following table lists the (average of the) ratings of teachers by students and (the average of the) students' expected grades for 39 faculty members of the three departments of the School of Management at Widener University as reported in this study. (*Note:* The ratings of teachers reported in this table are what the author called the *unadjusted ratings of teachers.*)

Ratings of Teachers	Students' Expected Grades	Ratings of Teachers	Students' Expected Grades
3.833	3.500	2.739	3.000
3.769	3.769	2.543	2.829
3.642	3.214	2.286	3.143
3.625	3.250	2.278	2.833
3.529	3.529	2.133	2.800
3.500	3.300	2.103	2.620
3.500	3.500	2.053	2.368
3.409	3.864	2.043	2.696
3.380	3.048	1.944	2.944
3.333	3.200	1.923	2.846
3.294	3.059	1.800	2.800
3.267	3.000	1.800	3.000
3.263	3.368	1.692	2.769
3.120	3.440	1.692	3.462
3.045	2.909	1.688	3.125
3.000	3.500	1.667	4.000
3.000	3.500	1.625	2.375
2.923	2.538	1.333	3.555
2.826	3.086	0.521	2.652
2.778	3.111		

Source: Hamid Zangenehzadeh, "Grade Inflation: A Way Out." *Journal of Economic Education,* summer 1988, 217–226. Reprinted with permission of the Helen Dwight Reid Educational Foundation. Published by Heldref Publications, Washington, D.C. Copyright © 1988.

Use MINITAB to answer the following.
  a. Construct a scatter diagram for these data.
  b. Find the correlation between the two variables.
  c. Find the regression line with ratings of teachers as a dependent variable and students' expected grades as an independent variable.
  d. Make a 95% confidence interval for B.
  e. Test at the 1% significance level if B is positive.
(*Note:* As the sample size is large, $n = 39$, the normal distribution can be applied to construct a confidence interval and to make a test of hypothesis about B in parts d and e if we so desire.)

M13.2 Refer to Data Set III on the heights and weights of NBA players given in Appendix B. Select a random sample of 25 players from that population. Use MINITAB to answer the following.
  a. Construct a scatter diagram for these data.
  b. Find the correlation between these two variables.
  c. Find the regression line with weight as a dependent variable and height as an independent variable.
  d. Make a 90% confidence interval for B.
  e. Test at the 5% significance level if B is positive.

     f. Make a 95% confidence interval for the mean weight of all NBA players who are 78 inches tall. Construct a 95% prediction interval for the weight of a randomly selected NBA player with a height of 78 inches.

M13.3 Refer to the data on the ages and the number of breakdowns for a sample of seven machines given in Exercise 13.95. Using MINITAB,

     a. Construct a scatter diagram for these data.

     b. Find the least squares regression line with age as an independent variable and the number of breakdowns as a dependent variable.

     c. Compute the correlation coefficient.

     d. Construct a 99% confidence interval for B.

     e. Test at the 2.5% significance level if B is positive.

MORE CHALLENGING
EXERCISES (Optional)
CHAPTER 13

1. For the past 25 years Burton Hodge has been keeping track of how many times he mows his lawn and the average size of the ears of corn in his garden. Hearing about the Pearson correlation coefficient from a statistician buddy of his, Burton decides to substantiate his suspicion that the more often he mows his lawn the bigger the ears of corn are by computing the correlation coefficient. Lo and behold, Burton computes a .93 coefficient of correlation! Elated, he calls his friend the statistician to thank him and announce that next year he will have prize-winning ears of corn because he plans to mow his lawn every day. Do you think Burton's logic is correct? If not, how would you explain to Burton the mistake he is making in his presumption (without eroding the usefulness he now feels statistics has)? Suggest what Burton could do next year to make the ears of corn large and relate this to the Pearson correlation coefficient.

2. It seems reasonable that the more hours per week a full-time college student works at a job, the less time he/she will have to study and, consequently, the lower his/her GPA would be.

    a. Assuming a linear relationship, suggest specifically what the equation relating x and y would be where x is the average number of hours a student works per week and y represents a student's GPA. Try several values of x and see if your equation gives reasonable values of y.

    b. Compute the regression equation using the following observations taken from 10 randomly selected students and compare it to yours.

Student	1	2	3	4	5	6	7	8	9	10
Average number of hours worked	20	28	10	35	5	14	0	40	8	23
GPA	2.8	2.5	3.1	2.1	3.4	3.3	2.8	2.5	3.6	1.8

3. You are hired by the local telephone company to study the monthly telephone bills at residences of college students serviced by a single phone line. The phone company wants you to relate the number of students living at the address with the size of the phone bill.

    a. Before taking any observations, suggest a specific relation between these two variables. Try some values to see if your relation seems reasonable.

    b. A random sample of 15 residences of students revealed the following monthly phone bills (in dollars) and number of students living at those addresses.

Phone Bill	Number of Students
67.40	2
171.12	5
53.67	1
101.07	2
84.26	3
152.71	3
123.22	5
168.21	4
82.55	3
126.98	4
41.85	1
239.01	6
75.18	2
114.38	4
82.27	3

Compute the regression equation based on these data and compare it to your answer in part a.

4. A researcher observed measurements on intoxication and reaction times for a sample of 25 intoxicated drivers. According to her calculations, the intoxication measures have a mean of .5 and a standard deviation of .2, and the reaction times have a mean of 20 seconds and a standard deviation of 6 seconds. She also found a correlation coefficient of .8 between the intoxication measures and reaction times.

    a. Find the appropriate regression line. Define the variables you use and interpret all the parameter estimates you find.

    b. What proportion of variation in reaction time is explained by the intoxication measure?

    c. Find the standard deviation of errors.

    d. Test if B is positive using a .025 level of significance.

    e. What would you predict the reaction time to be of a person with a .3 intoxication measure? Based on your conclusion in part d, how valid do you think this predicted value is? Find a 95% prediction interval for this reaction time.

5. A bowling establishment owner interested in the relation between the price she charges for a game of bowling and the number of games bowled per day collects data on the number of games bowled per day at 15 different prices. Fill in the missing entries in the following MINITAB output that was obtained for these data. In this output, X represents the price of a game of bowling and Y is the number of games bowled per day.

```
THE REGRESSION EQUATION IS

Y=____ ____ X


PREDICTOR     COEF     STDEV     T-RATIO     P
CONSTANT      691.02   21.70     ____        .000
X             -141.30  ____      -16.83      .000

S=____        R-SQ=____          R-SQ(ADJ)=95.3%


ANALYSIS OF VARIANCE

SOURCE        DF       SS        MS       F          P
REGRESSION    ____     148484.0  ____     283.13     .000
ERROR         ____     ____      ____
TOTAL         ____     ____
```

APPENDIX A

SAMPLE SURVEYS, SAMPLING TECHNIQUES, AND DESIGN OF EXPERIMENTS

In Chapter 1 we briefly explained a few terms and concepts such as sources of data, sample surveys, census, reasons for conducting a sample survey instead of a census, and simple random sample. In Chapter 7 we discussed sampling and nonsampling errors. However, nonsampling errors were not discussed in detail in Chapter 7.

This appendix will explain these concepts in more detail. In addition, this appendix discusses four sampling techniques, the use of a table of random numbers to select a sample, and the design of experiments.

A.1 SOURCES OF DATA

The availability of accurate data is essential for deriving reliable results and making accurate decisions. As the truism "garbage in, garbage out" (GIGO) indicates, policy decisions based on the results of poor data may prove to be disastrous.

Data sources can be divided into three categories: (1) internal sources, (2) external sources, and (3) surveys and experiments.

1. INTERNAL SOURCES

Many times data come from **internal sources**, such as a company's own personnel files or accounting records. A company that wants to forecast the future sales of its products might use data from its own records for previous periods. A police department might use data that exist in its own records to analyze changes in the nature of crimes over a period of time.

2. EXTERNAL SOURCES

All needed data may not be available from internal sources. Hence, to obtain data we may have to depend on sources outside the company. Such sources of data are called **external sources**. Data obtained from external sources may be primary or secondary data. Data obtained from the organization that originally collected them are called **primary data**. If we obtain data from the Bureau of Labor Statistics that were collected by this organization, then these are primary data. Data obtained from a source that did not originally collect them are called **secondary data**. For example, data originally collected by the Bureau of Labor Statistics and published in the *Statistical Abstract of the United States* are secondary data.

3. SURVEYS AND EXPERIMENTS

Sometimes the data we need may not be available from internal or external sources. In such cases, we may have to obtain data by conducting our own survey or experiment.

I. Surveys

In a **survey** we do not exercise any control over the factors when collecting information.

SURVEY

In a *survey*, data are collected from the members of a population or sample without exercising any particular control over the factors that may affect the characteristic of interest or the results of the survey.

For example, if we want to collect data on money spent last month on clothes by various families, we will ask each of the families included in the survey how much it spent last month on clothes. Then we will record this information.

A survey may be a census or a sample survey.

Census

A **census** includes every member of the population of interest, which is called the **target population**.

CENSUS

A survey that includes every member of the population is called a *census*.

In practice, a census is rarely taken because it is very expensive and time consuming. Furthermore, in many cases it is impossible to identify each member of the target population. We will discuss these reasons in more detail in Section A.2.1.

Sample Survey

Usually, to conduct research, we select a portion of the target population. This portion of the population is called a sample. Then we collect the required information from the elements included in the sample.

SAMPLE SURVEY

The technique of collecting information from a portion of the population is called a *sample survey*.

A survey can be conducted by personal interviews, by telephone, or by mail. The personal interview technique has the advantage of having a high response rate and a high quality of answers obtained. However, it is the most expensive and time consuming technique. The telephone survey also gives a high response rate. Unlike personal interviews, it is less expensive and less time consuming. Nonetheless, a problem with this technique is that many people do not like to be called at home, and those who do not have a phone are left out of the survey. A survey conducted by mail is the least expensive method, but the response rate is usually very low. Many people included in such a survey do not return the questionnaires.

Conducting a survey with accurate and reliable results is not an easy task. To quote Warren Mitofsky, director of Elections and Surveys for CBS News, ''Any damn fool with 10 phones and a typewriter thinks he can conduct a poll.''[1] The preparation of a questionnaire is probably the most difficult part of a survey. The way a question is phrased can affect the results of the survey. Case Study A-1, which is excerpted from an article published in *Psychology Today*, shows that writing questions for a questionnaire is a much more complex task than is usually thought.

---

[1]''The Numbers Racket: How Polls and Statistics Lie,'' *U.S. News & World Report,* July 11, 1988.

CASE STUDY A-1 IS IT A SIMPLE QUESTION?

Even the seemingly simplest of questions can yield complex answers. "Do you own a car?" asks Stanley Presser, a sociologist at the National Science Foundation in Washington, D.C. "That sounds like an awfully simple question. But is it really? What does 'you' mean? Suppose a wife is answering the poll, and the car is registered in her husband's name. How is she supposed to answer? What does 'own' mean? What if the car is on a long-term lease? What does 'car' mean? What if they have one of those new little vans, or a four-wheel-drive vehicle? My God, that sounds like a simple question! You can imagine how diverse the factors become in a more complicated one."

Suppose, however, that the question about car ownership had been preceded by a series of related questions: "Are you married? Does your spouse drive an automotive vehicle? Is it a car, a van or some other sort of vehicle? Is it leased, or does your spouse own it? Now about you—do you own a car?" Such a series of questions would serve to clarify the intended meaning of the one about car ownership.

Source: Rich Jaroslovsky, "What's on Your Mind, America?" *Psychology Today*, July–August 1988, 54–59. Copyright © 1988 Sussex Publishers, Inc. Reprinted with permission.

Section A.2 discusses sample surveys and sampling techniques in detail.

II. Experiments

In an **experiment**, we exercise control over some factors when collecting information.

> **EXPERIMENT**
>
> In an *experiment*, data are collected from members of a population or sample by exercising control over some or all of the factors that may affect the characteristic of interest or the results of the experiment.

For example, how is a new drug to be tested to find out whether or not it cures a disease? It is done by designing an experiment in which the patients under study are divided into two groups as follows.

1. The **treatment group**—the members of this group receive the actual drug.
2. The **control group**—the members of this group do not receive the actual drug but are given a substitute (called a placebo) that appears to be the actual drug.

The two groups are formed in such a way that the patients in one group are similar to the patients in the other group. This is done by making random assignments of patients to two groups. Neither the doctors nor the patients know to which group a patient belongs. Such an experiment is called a **double-blind experiment**. Then, after a comparison of the percentage of patients cured in each of the two groups, a decision is made about the effectiveness or noneffectiveness of the new drug. For more on experiments, refer to Section A.3 of this appendix on design of experiments.

A.2   SAMPLE SURVEYS AND SAMPLING TECHNIQUES

In this section we will discuss the reasons sample surveys are preferred over a census, a representative sample, random and nonrandom samples, sampling and nonsampling errors, and random sampling techniques.

A.2.1   WHY SAMPLE?

As mentioned in the previous section, most of the time surveys are conducted by using samples and not a census of the population. Some of the main reasons for conducting a sample survey instead of a census are as follows.

1. Time

In most cases, the size of the population is quite large. Consequently, conducting a census will take a long time, whereas a sample survey can be conducted very quickly. It will be time consuming to interview or contact hundreds of thousands or even millions of members of a population. On the other hand, a survey of a sample of a few hundred elements may be completed in little time. In fact, because of the amount of time needed to conduct a census, by the time the census is completed the results may be obsolete.

2. Cost

The cost of collecting information from all members of a population may easily fall outside the limited budget of most, if not all, surveys. Consequently, to stay within the available resources, conducting a sample survey may be the best approach.

3. Impossibility of Conducting a Census

Sometimes it is impossible to conduct a census. First, it may not be possible to identify and access each member of the population. For example, if a researcher wants to conduct a survey about homeless people, it will not be possible to locate each member of the population and include him/her in the survey. Second, sometimes conducting a survey means destroying the items included in the survey. For example, to estimate the mean life of light bulbs would necessitate burning out all the bulbs included in the survey. The same is true about finding the average life of batteries. In such cases, only a portion of the population can be selected for the survey.

A.2.2   RANDOM AND NONRANDOM SAMPLES

Depending on how a sample is drawn, it may be a **random sample** or a **nonrandom sample**.

RANDOM AND NONRANDOM SAMPLES

A *random sample* is a sample drawn in such a way that each member of the population has some chance of being selected in the sample. In a *nonrandom sample*, some members of the population may not have any chance of being selected in the sample.

Suppose we have a list of 100 students and we want to select 10 of them. If we write the names of all 100 students on pieces of paper, put them in a hat, mix them, and then draw 10 names, the result will be a random sample of 10 students. However, if we arrange the names of these 100 students alphabetically and pick the first 10 names, it will be a nonrandom sample because the students who are not among the first 10 have no chance of being selected in the sample.

A random sample is usually a representative sample. Note that for a random sample, each member of the population may or may not have the same chance of being included in the sample. Four types of random samples are discussed in Section A.2.4 of this appendix.

Two types of nonrandom samples are a *convenience sample* and a *judgment sample*. In a **convenience sample**, the most accessible members of the population are selected to obtain the results quickly. For example, an opinion poll may be conducted in a few hours by collecting information from certain shoppers at a single shopping mall. In a **judgment sample**, the members are selected from the population based on the judgment and prior knowledge of an expert. Although such a sample may happen to be a representative sample, the chances of it being so are small. If the population is large, it is not an easy task to select a representative sample based on judgment.

The so-called *pseudo polls* are examples of nonrepresentative samples. For instance, a survey conducted by a magazine that includes only its readers does not usually involve a representative sample. Similarly, a poll conducted by a television station giving two separate 900 telephone numbers for *yes* and *no* votes is not based on a representative sample. In these two examples, respondents will be only those people who read that magazine or watch that television station, who do not mind paying the postage and telephone charges, or who feel emotionally compelled to respond. To quote Larry King on this subject:

> All over the board ... The 900 telephone number is very popular these days, but viewers should be warned that in the case of political polling it has absolutely no basis in fact. Poor people in the audience can't contribute to the survey, so it's faulty to begin with. ... So next time you see a poll based on 900 numbers, treat it as some sort of middle-class amusement and forget about it. (''Larry King's People, News and Views,'' *USA Today*, July 17, 1989. Copyright © 1989, *USA Today*. Reprinted with permission.)

Another kind of sample is the **quota sample**. To draw such a sample we divide the target population into different subpopulations based on certain characteristics. Then a subsample is selected from each subpopulation in such a way that each subpopulation is represented in the sample in exactly the same proportion as in the target population. As an example of a quota sample, suppose we want to select a sample of 1000 persons from a city whose population is comprised of 48% males and 52% females. To select a quota sample, we choose 480 males from the male population and 520 females from the female population. The sample selected in this way will contain exactly 48% males and 52% females. Another way to select a quota sample is to select from the population one person at a time until we have exactly 480 males and 520 females.

Until the 1948 presidential election in the United States, quota sampling was the most commonly used sampling procedure to conduct opinion polls. The voters included in the samples were selected in such a way that they represented the population proportions of voters based on age, sex, education, income, race, etc. However, this procedure was abandoned after the 1948 presidential election in which the underdog, Harry Truman, defeated Thomas E. Dewey, who was heavily favored based on the opinion polls. First, the quota samples failed to be representative because the interviewers were allowed to fill their quotas by choosing voters based on their own judgments. This caused selection of more upper-income and highly educated people who happened to be Republicans. Thus, the quota samples were unrepresentative of the population because Republicans were overrepresented in these samples. Second, the results of the opinion polls based on quota sampling happened to be false because a large number of factors differentiate voters, and the pollsters considered only a few of those factors. A quota sample based on a few factors will skew the results. A random sample (that is not based on quotas) has a much better chance of being representative of the population of all voters than a quota sample based on a few factors.

A.2.3 SAMPLING AND NONSAMPLING ERRORS

The results obtained from a sample survey may contain two types of errors: the sampling and nonsampling errors. The sampling error is also called the chance error, and the nonsampling errors are also called the systematic errors.

1. Sampling or Chance Error

Usually, all samples taken from the same population will give different results because they contain different elements of the population. Moreover, the results obtained from any one sample will not be exactly the same as the ones obtained from a census. The difference between a sample result and the result we would have obtained by conducting a census is called the **sampling error**, assuming that the sample is random and no nonsampling error has been made.

SAMPLING ERROR

The *sampling error* is the difference between the result obtained from a sample survey and the result that would have been obtained if the whole population had been included in the survey.

The sampling error occurs because of chance, and it cannot be avoided. A sampling error can occur only in a sample survey. It does not occur in a census. Sampling error is discussed in detail in Section 7.2 of Chapter 7, and an example of it is given there.

2. Nonsampling or Systematic Errors

Nonsampling errors can occur both in a sample survey and in a census. Such errors occur because of human mistakes and not chance.

NONSAMPLING ERRORS

The errors that occur in the collection, recording, and tabulation of data are called *nonsampling errors*.

Nonsampling errors occur because of human mistakes and not chance. Nonsampling errors can be minimized if questions are prepared carefully and data are handled cautiously. There are many types of systematic errors or biases that can occur in a survey—including selection error, nonresponse error, response error, and voluntary response error. The following chart shows the types of errors.

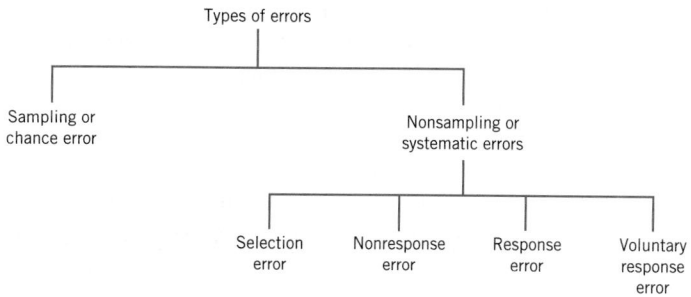

I. Selection Error

When we need to select a sample, we use a list of elements from which we draw a sample, and this list usually does not include many members of the target population. Most of the time it is not feasible to include every member of the target population in this list. This list of members of the population that is used to select a sample is called the **sampling frame**. For example, if we use a telephone directory to select a sample, the list of names that appears in this directory makes the sampling frame. In this case we will miss the people who are not listed in the telephone directory. The people we miss, for example, will be poor people (including homeless people) who do not have telephones and people who do not want to be listed in the directory. Thus, the sampling frame that is used to select a sample may not be representative of the population. This may cause the sample results to be different from the population results. The error that occurs because the sampling frame is not representative of the population is called the **selection error**.

SELECTION ERROR

The list of members of the target population that is used to select a sample is called the sampling frame. The error that occurs because the sampling frame is not representative of the population is called the *selection error*.

If a sample is nonrandom (and, hence, nonrepresentative), the sample results may be quite different from the census results.

II. Nonresponse Error

Even if our sampling frame and, consequently, the sample is representative of the population, the **nonresponse error** may occur because many of the people included in the sample did not respond to the survey.

NONRESPONSE ERROR

The error that occurs because many of the people included in the sample do not respond to a survey is called the *nonresponse error*.

This type of error occurs especially when a survey is conducted by mail. A lot of people do not return the questionnaires. It has been observed that families with low and high incomes do not respond to surveys by mail. Consequently, such surveys overrepresent middle-income families. This kind of error occurs in other types of surveys, too. For instance, in a face-to-face survey where the interviewer interviews people at their homes, many people may not be home when the interviewer visits their homes. The people who are home at the time the interviewer visits their homes and the ones who are not home at that time may differ in many respects, causing a bias in the survey results. This kind of error may also occur in a telephone survey. Many people may not be home when the interviewer calls. This may distort the results. To avoid the nonresponse error, every effort should be made to contact all people included in the survey.

III. Response Error

The **response error** occurs when the answer given by a person included in the survey is not correct. This may happen for many reasons. One reason is that the respondent may not have understood the question. Thus, the wording of the question may have caused the respondent to answer incorrectly. It has been observed that when the same question is worded differently, many people do not respond the same way. Usually such an error on the part of respondents is not intentional.

RESPONSE ERROR

The *response error* occurs when people included in the survey do not provide correct answers.

Sometimes the respondents do not want to give correct information when answering a question. For example, many respondents will not disclose their true incomes on questionnaires or in interviews. When information on income is provided, it is almost always biased in the upward direction.

Sometimes the race of the interviewer may affect the answers of respondents. This is especially true if the questions asked are about race relations. The answers given by respondents will differ depending on whether the interviewer is white or nonwhite.

IV. Voluntary Response Error

Another source of systematic error is a survey based on a voluntary response sample.

VOLUNTARY RESPONSE ERROR

Voluntary response error occurs when a survey is not conducted on a randomly selected sample but a questionnaire is published in a magazine or newspaper and people are invited to respond to that questionnaire.

The polls conducted based on samples of readers of magazines and newspapers suffer from **voluntary response error** or **bias**. Usually only those readers who have very strong opinions about the issues involved respond to such surveys. Surveys in which the respondents are required to call 900 telephone numbers also suffer from this type of error. Here, in order to participate, a respondent must pay for the call, and many people do not want to bear this cost. Consequently, the sample is usually neither random nor representative of the target population because participation is voluntary.

A.2.4 RANDOM SAMPLING TECHNIQUES

There are many ways to select a random sample. Four of these techniques are discussed below.

1. Simple Random Sampling

A sample that assigns the same probability of being selected to each member of the population is called a **simple random sample**.

SIMPLE RANDOM SAMPLE

A *simple random sample* is a sample that is selected in such a way that each member of the population has the same chance of being included in the sample.

One way to select a simple random sample is by a lottery or drawing. For example, if we need to select five students from a class of 50, we write each of the 50 names on separate pieces of paper. Then, we place all 50 names in a hat and mix them thoroughly. Next, we draw one name randomly from the hat. We repeat this experiment four more times. The five drawn names comprise a simple random sample.

The second procedure to select a simple random sample is to use a table of random numbers. Table I in Appendix C lists random numbers. These numbers are generated by a random process. Suppose we have a group of 400 persons and we need to select 30 persons randomly from this group. To select a simple random sample, we arrange the names of all 400 persons in alphabetic order and assign a three-digit number, from 001 to 400, to each person.

Next, we use the table of random numbers to select 30 persons. The random numbers in Table I are recorded in blocks of five digits. To use this table, we can start anywhere. One way to do so is to close our eyes and put a finger anywhere on the page and start at that point. From there, we can move in any direction. We need to pick three-digit numbers from the table because we have assigned three-digit numbers to the 400 persons in our population.

Suppose we start at the first block of the 31st row from the top of Table I. The five rows starting with the 31st row from that table are reproduced as Table A.1 here. The first block of five numbers in Table A.1 is 13049. We use the first three digits of this block to select the first person from the population. Hence, the first person selected is the one with the number 130. Suppose we move along the row to the right to make the next selection. The second block of five numbers in Table A.1 is 85293. The first three digits of this block give 852. However, we have only 400 persons in the population with assigned numbers of 001 to 400. Consequently, we cannot use 852 to select a person. Therefore, we move to the next block of five numbers without making a selection. The third block of numbers is 32747. The first three digits of this block are 327. Consequently, the second person selected is the one with the number 327. We continue this process until all 30 required persons are selected. This gives us a simple random sample of 30 persons.

Table A.1

13049	85293	32747	17728	50495	34617	73707	33976	86177
86544	52703	74990	98288	61833	48803	75258	83382	79099
77295	70694	97326	35430	53881	94007	70471	66815	73042
54637	32831	59063	72353	87365	15322	33156	40331	93942
50938	12004	18585	23896	62559	44470	27701	66780	56157

Although the table of random numbers given in Appendix C contains only 1485 blocks of five-digit numbers, we can easily construct a table of as many random numbers as we want using a computer software package such as MINITAB.

If we have access to a computer, we can use a statistical package, such as MINITAB, to select a simple random sample. The MINITAB sections at the end of this chapter and Chapters 1 and 7 explain and illustrate how we can draw such a sample by using MINITAB.

2. Systematic Random Sampling

The simple random sampling procedure will become very tedious if the size of the population is large. For example, if we need to select 150 households from a list of 45,000, it will be very time consuming either to write the 45,000 names on pieces of paper and then select 150 households or to assign a five-digit number to each of the 45,000 households and then select 150 households using the table of random numbers. In such cases, it is more convenient to use **systematic random sampling**.

The procedure to select a systematic random sample is as follows. In the example just mentioned, we would arrange all 45,000 households alphabetically (or based on some other characteristic). Since the sample size should equal 150, the ratio of population to sample size is $45,000/150 = 300$. Using this ratio, we randomly select one household from the first 300 households in the arranged list either by using the lottery system or by using a table of random numbers. Suppose by using either of these methods, we select the 210th household. We then select every 210th household from every 300 households in the list. In other words, our sample includes the households with numbers 210, 510, 810, 1110, 1410, 1710, and so on.

SYSTEMATIC RANDOM SAMPLE

In *systematic random sampling*, we first randomly select one member from the first k units. Then every kth member, starting with the first selected member, is included in the sample.

Note that systematic random sampling does not give a simple random sample because we cannot select two adjacent elements. Hence, every member of the population does not have the same probability of being selected.

3. Stratified Random Sampling

Suppose we need to select a sample from the population of a city and we want households with different income levels to be equally represented in the sample. In this case, instead of selecting a simple random sample or a systematic random sample, we may prefer to apply a different technique. First, we divide the whole population into different groups based on income levels. For example, we may form three groups of low-, medium-, and high-income households. We will now have three *subpopulations*, which are usually called **strata**.

We then select one sample from each subpopulation or stratum. The collection of all three samples selected from three strata gives the required sample, called the **stratified random sample**. Usually, the sizes of the samples selected from different strata are proportionate to the sizes of the subpopulations in these strata. Note that the elements of each stratum are identical with regard to the possession of a characteristic.

STRATIFIED RANDOM SAMPLE

In a *stratified random sample*, we first divide the population into subpopulations, which are called strata. Then, one sample is selected from each of these strata. The collection of all samples from all strata gives the stratified random sample.

Thus, whenever we observe that a population differs widely in the possession of a characteristic, we may prefer to divide it into different strata and then select one sample from each stratum. We can divide the population on the basis of any characteristic, such as income, expenditure, sex, education, race, employment, or family size.

4. Cluster Sampling

Sometimes the target population is scattered over a wide geographical area. Consequently, if a simple random sample is selected, it may be costly to contact each member of the sample. In such a case, we divide the population into different geographical groups or clusters and as a first step select a random sample of certain clusters from all clusters. We then take a random sample of certain elements from each selected cluster. For example, suppose we are to conduct a survey of households in the state of New York. First, we divide the whole state of New York into, say, 40 regions, which will be called **clusters** or **primary units**. We make sure that all clusters are similar and, hence, representative of the population. We then select at random, say, 5 clusters from 40. Next, we randomly select certain households from each of these 5 clusters and conduct a survey of these selected households. This is called **cluster sampling**. Note that all clusters must be representative of the population.

CLUSTER SAMPLING

In *cluster sampling*, the whole population is first divided into (geographical) groups called clusters. Each cluster is representative of the population. Then a random sample of clusters is selected. Finally, a random sample of elements from each of the selected clusters is selected.

A.3   DESIGN OF EXPERIMENTS

As mentioned earlier, to use statistical methods to make decisions, we need access to data. Consider the following examples about decision making.

1. A government agency wants to find the average income of households in the United States.
2. A company wants to find the percentage of defective items produced on a machine.

3. A researcher wants to know if there is an association between eating unhealthy food and cholesterol level.

4. A pharmaceutical company has invented a new medicine for a disease and it wants to check if this medicine cures the disease.

All of these cases relate to decision making. We cannot reach a conclusion in these examples unless we have access to data. Data can be obtained from observational studies, experiments, or surveys. This section is devoted mainly to controlled experiments. However, it also explains observational studies and how they differ from surveys.

Suppose two diets, Diet 1 and Diet 2, are being promoted by two different companies and each of these companies claims that its diet is successful in reducing weight. A research nutritionist wants to compare these diets with regard to their effectiveness for losing weight. Following are the two alternatives for the researcher to conduct this research.

1. The researcher contacts the persons who are using these diets and collects information on their weight loss. The researcher may contact as many persons as she has the time and financial resources for. Based on this information, the researcher makes a decision about the comparative effectiveness of these diets.

2. The researcher selects a sample of persons who want to lose weight, divides them randomly into two groups, and assigns each group to one of the two diets. Then she compares these two groups with regard to the effectiveness of these diets.

The first alternative is an example of an **observational study**, and the second alternative is an example of a **controlled experiment**.

TREATMENT

A condition (or a set of conditions) that is imposed on a group of elements by the experimenter is called a *treatment*.

In an observational study the investigator does not impose a treatment on subjects or elements included in the study. For instance, in the first alternative mentioned above, the researcher simply collects information from the persons who are currently using these diets. In this case, the persons were not assigned to the two diets at random; instead, they chose the diets voluntarily. In this situation the researcher's conclusion about the comparative effectiveness of the two diets may not be valid because the effects of the diets will be **confounded** with many other factors or variables. When the effects of one factor cannot be separated from the effects of some other factors, the effects are said to be confounded. The persons who chose Diet 1 may be completely different with regard to age, gender, and eating and exercise habits from the persons who chose Diet 2. Thus, the weight loss may not be due entirely to the diet but to other factors or variables as well. Persons in one group may aggressively manage both diet and exercise, for example, whereas the persons in the second group may depend entirely on diet. Thus, the effects of these other variables will get mixed up (confounded) with the effect of the diets.

Under the second alternative the researcher selects a group of people, say 100, and randomly assigns them to two diets. One way to make random assignments is to write the name of each of these persons on a piece of paper, put them in a hat, and then randomly draw 50 names from this hat. These 50 persons will be assigned to one of the two diets, say Diet 1. The remaining 50 persons will be assigned to the second diet, Diet 2. This procedure is called **randomization**. Note that random assignments can also be made by using the table of random numbers.

RANDOMIZATION

The procedure under which elements are assigned to different groups at random is called *randomization*.

When people are assigned to one or the other of two diets at random, the other differences among people in the two groups almost disappear. In this case these groups will not differ very much with regard to such factors as age, gender, and eating and exercise habits. The two groups will be very similar to each other. By using the random process to assign people to one or the other of two diets, we have *controlled* the other factors that can affect the weights of people. Consequently, this is an example of a **designed experiment**.

As mentioned earlier, a condition (or a set of conditions) that is imposed on a group of elements by the experimenter is called a treatment. In the example on diets, each of the two diet types is called a treatment. The experimenter randomly assigns the elements to these two treatments. Again, in such cases the study is called a designed experiment.

DESIGNED EXPERIMENT AND OBSERVATIONAL STUDY

When the experimenter controls the assignment of elements to different treatment groups, the study is said to be a *designed experiment*. On the other hand, in an *observational study* the assignment of elements to different treatments is voluntary and the experimenter simply observes the results of the study.

The group of people who receive a treatment is called the **treatment group**, and the group of people who do not receive a treatment is called the **control group**. In our example on diets, both groups are treatment groups because each group is assigned to one of the two types of diets. That example does not contain a control group.

TREATMENT AND CONTROL GROUPS

The group of elements that receives a treatment is called the *treatment group*, and the group of elements that does not receive a treatment is called the *control group*.

An example of an observational study.

EXAMPLE A-1   Suppose a pharmaceutical company has invented a new medicine to cure a disease. To see whether or not this medicine is effective in curing this disease, it will have to be tested on a group of humans. Suppose there are 100 persons who have this disease; 50 of them voluntarily decide to take this medicine and the remaining 50 decide not to take it. The researcher then compares the cure rates for the two groups of patients. Is this an example of a designed experiment or an observational study?

Solution   This is an example of an observational study because 50 patients voluntarily joined the treatment group; they were not randomly selected. In this case, the results of the study may not be valid because the effects of the medicine will be confounded with other variables. All of the patients who decided to take the medicine may not be similar to the ones who decided not to take it. It is possible that the persons who decided to take the medicine are in the advanced stages of the disease. Consequently, they do not have much to lose by being in the treatment group. The patients in the two groups may also differ with regard to other factors such as age, gender, etc.   ■

An example of a designed experiment.

EXAMPLE A-2   Reconsider Example A-1. Now suppose that out of the 100 people who have this disease 50 are selected at random. These 50 people comprise one group, and the remaining 50 belong to the second group. One of these groups is the treatment group, and the second is the control group. The researcher then compares the cure rates for the two groups of patients. Is this an example of a designed experiment or an observational study?

Solution   In this case, the two groups will be very similar to each other. Note that we do not expect the two groups to be exactly identical. However, when randomization is used, the two groups will be very close to being exactly similar. After these two groups have been formed, one group will be given the actual medicine. This group is called the treatment group. The other group will be administered a placebo (a dummy medicine that looks exactly like the actual medicine). This group is called the control group. This is an example of a designed experiment because the patients are assigned to one of two groups—the treatment or the control group—randomly.   ■

Usually in an experiment like the one mentioned in Example A-2, patients do not know which group they belong to. Most of the time even the experimenters do not know which group a patient belongs to. This is done to avoid any bias or distortion in the results of the experiment. When neither patients nor experimenters know who is taking the real medicine and who is taking the placebo, it is called a **double-blind experiment**. For the results of the study to be unbiased and valid, an experiment must be a double-blind designed experiment. Note that if either doctors, patients, or both have access to information regarding which patients belong to treatment or control groups, it will no longer be a double-blind experiment.

The use of placebos in medical experiments is very important. A placebo is just a dummy pill that looks exactly like the real medicine. Often, patients respond to any kind of medicine. Many studies have shown that even when the patients were given sugar pills (and patients did not know it), many of them indicated a decrease in pain. Patients respond to placebos because they have confidence in their physicians and medicines. This is called the **placebo effect**.

Note that there can be more than two groups of elements in an experiment. For example, an investigator may need to compare three diets with regard to weight gain for chickens. Here, in a designed experiment, the chickens will be randomly assigned to one of the three diets, which are the three treatments.

In some instances we have to base our research on observational studies because it is not feasible to conduct a designed experiment. For example, suppose a researcher wants to compare the starting salaries of business and psychology majors. The researcher, in this case, will have to depend on an observational study. She will select two samples, one of recent business majors and another of recent psychology majors. Based on the starting salaries of these two groups, the researcher will make a decision. Note that, here, the effects of the majors on the starting salaries of the two groups of graduates will be confounded with other variables. One of these other factors is that the business and psychology majors may be different in regard to intelligence level, which may affect their salaries. However, the researcher cannot conduct a designed experiment in this case. She cannot select a group of persons randomly and ask them to major in business and select another group and ask them to major in psychology. Instead, persons voluntarily choose their majors.

In a survey we do not exercise any control over the factors when collecting information. This characteristic of a survey makes it very close to an observational study. However, a survey may be based on a probability sample, which differentiates it from an observational study.

If an observational study or a survey indicates that two variables are related, it does not mean that there is a cause-and-effect relationship between them. For example, if an economist takes a sample of families, collects data on the incomes and rents paid by these families, and establishes an association between these two variables, it does not necessarily mean that families with higher incomes pay higher rents. Here the effects of many variables on rents are confounded. A family may pay a higher rent not because of higher income but because of various other factors, such as family size, preferences, place of residence, etc. We cannot make a statement about the cause-and-effect relationship between incomes and rents paid by families unless we control for these other variables. The association between incomes and rents paid by families may fit any of the following three scenarios.

1. These two variables have a cause-and-effect relationship. Families that have higher incomes do pay higher rents. A change in incomes of families causes a change in rents paid.

2. The incomes and rents paid by families do not have a cause-and-effect relationship. Both of these variables have a cause-and-effect relationship with a third variable. Whenever that third variable changes, these two variables change.

3. The effect of income on rent is confounded with other variables, and this indicates that income affects rent paid by families.

If our purpose in a study is to establish a cause-and-effect relationship between two variables, we must control the effects of other variables. In other words, we must conduct a designed study.

EXERCISES

A.1 Briefly describe the various sources of data.

A.2 What is the difference between internal and external sources of data? Explain.

A.3 Explain the difference between a sample survey and a census. Why is a sample survey usually preferred over a census?

A.4 What is the difference between a survey and an experiment? Explain.

A.5 Explain the following.
  a. Random sample   b. Nonrandom sample
  c. Convenience sample   d. Judgment sample   e. Quota sample

A.6 Explain briefly the following four sampling techniques.
  a. Simple random sampling   b. Systematic random sampling
  c. Stratified random sampling   d. Cluster sampling

A.7 Under what sampling technique do all elements of a population have the same chance of being selected in a sample?

A.8 A statistics professor wanted to find out the average GPA (grade point average) for all students at her university. She used all students enrolled in her statistics class as a sample and collected information on their GPAs to find the average GPA.

a.   Is this sample a random or a nonrandom sample? Explain.

b.   What kind of sample is it? In other words, is it a simple random sample, a systematic sample, a stratified sample, a cluster sample, a convenience sample, a judgment sample, or a quota sample? Explain.

c.   What kind of systematic error, if any, will be made in this kind of sample? Explain.

A.9   A professor wanted to select 20 students from his class of 300 students to collect detailed information on the profiles of his students. He used his knowledge and expertise to select these 20 students.

a.   Is this sample a random or a nonrandom sample? Explain.

b.   What kind of sample is it? In other words, is it a simple random sample, a systematic sample, a stratified sample, a cluster sample, a convenience sample, a judgment sample, or a quota sample? Explain.

c.   What kind of systematic error, if any, will be made in this kind of sample? Explain.

A.10   Refer to Exercise A.8. Suppose the professor obtains a list of all students enrolled at the university from the registrar's office and then selects 150 students at random from this list using a statistical software package such as Minitab.

a.   Is this sample a random or a nonrandom sample? Explain.

b.   What kind of sample is it? In other words, is it a simple random sample, a systematic sample, a stratified sample, a cluster sample, a convenience sample, a judgment sample, or a quota sample? Explain.

c.   Do you think any systematic error will be made in this case? Explain.

A.11   Refer to Exercise A.9. Suppose the professor enters the names of all students enrolled in his class on a computer. He then selects a sample of 20 students at random using a statistical software package such as Minitab.

a.   Is this sample a random or a nonrandom sample? Explain.

b.   What kind of sample is it? In other words, is it a simple random sample, a systematic sample, a stratified sample, a cluster sample, a convenience sample, a judgment sample, or a quota sample? Explain.

c.   Do you think any systematic error will be made in this case? Explain.

A.12   A company has 1000 employees of whom 58% are males and 42% are females. The research department at the company wanted to conduct a quick survey by selecting a sample of 50 employees and asking them about their opinions on an issue. They divided the population of employees into two groups, males and females, and then selected 29 males and 21 females from these respective groups. The interviewers were free to choose any 29 men and 21 women they wanted. What kind of sample is it? Explain.

A.13   Many magazines regularly publish questionnaires and ask their readers to send their responses by mail. The tallied answers are published in the magazine at a later date. What kind of systematic error, if any, is made in this kind of a sample survey? Explain.

A.14   A researcher wanted to conduct a survey of major companies to find out what benefits are offered to their employees. She mailed questionnaires to 2500 companies and received questionnaires back from 493 companies. What kind of systematic error does this survey suffer from? Explain.

A.15   An opinion poll agency conducted a survey based on a random sample in which the interviewers called the parents included in the sample and asked them the following questions.

i.   Do you believe in spanking children?

ii.   Have you ever spanked your children?

iii.   If the answer to the second question is yes, how often?

What kind of systematic error, if any, does this survey suffer from? Explain.

A.16   A survey, based on a random sample taken from a borough of New York City, showed that 65% of the people living there would prefer to live somewhere other than New York City if they had the opportunity to do so. Based on this result, can the researcher say that 65% of people living in New York City would prefer to live somewhere else if they had the opportunity to do so? Explain.

A.17 A study conducted by the Steering Committee of the Physicians' Health Study Research Group in 1988 showed that taking one adult-size aspirin every other day reduces the risk of heart attack. A group of physicians at a research university wanted to investigate whether or not this claim is true. They chose 500 volunteers who offered to be included in the study. Of these 500 persons, 300 volunteered to take one adult-size aspirin every other day. The remaining 200 made up the control group. After two years the physicians compared the heart attack rates for the two groups.

    a. Is this an observational study or a designed experiment? Explain.

    b. Is this study a double-blind study? Explain.

A.18 Refer to Exercise A.17. Suppose the group of physicians randomly selected 500 persons to be included in the study. Then, of these 500 persons, 300 were randomly selected to make up the treatment group. The remaining 200 made up the control group. The persons in the treatment group were given one adult-size aspirin every other day and the persons in the control group were given a placebo. The patients did not know what group they belonged to but doctors had access to this information.

    a. Is this an observational study or a designed experiment? Explain.

    b. Is this study a double-blind study? Explain.

A.19 Refer to Exercise A.18. Now suppose that neither patients nor doctors knew what group patients belonged to.

    a. Is this an observational study or a designed experiment? Explain.

    b. Is this study a double-blind study? Explain.

A.20 A federal government think tank wanted to investigate whether a job training program helps the families who are on welfare to get off the welfare program. The researchers at this agency selected 5000 volunteer families who were on welfare and offered the adults in those families free job training. The researchers selected another group of 5000 volunteer families who were on welfare and did not offer them such job training. After three years the two groups were compared in regard to the percentage of families who got off welfare. Is this an observational study or a designed experiment? Explain.

A.21 Refer to Exercise A.20. Now suppose the agency selected 10,000 families at random from the list of all families that were on welfare. Of these 10,000 families, the agency randomly selected 5000 families and offered them free job training. The remaining 5000 families were not offered such job training. After three years the two groups were compared in regard to the percentage of families who got off welfare. Is this an observational study or a designed experiment? Explain.

A.22 Refer to Exercise A.20. Based on that study the researchers concluded that the job training program causes (helps) families who are on welfare to get off the welfare program. Do you agree with this conclusion? Explain.

A.23 Refer to Exercise A.21. Based on that study the researchers concluded that the job training program causes (helps) families who are on welfare to get off the welfare program. Do you agree with this conclusion? Explain.

A.24 A researcher advertised for volunteers to study the relationship between the amount of meat consumed and cholesterol level. In response to this advertisement, 3476 persons volunteered. The researcher collected information on the meat consumption and cholesterol level of each of these persons. Based on these data, the researcher concluded that there is a very strong positive association between these two variables.

    a. Is this an observational study or a designed experiment? Explain.

    b. Based on this study, can the researcher conclude that consumption of meat increases cholesterol level? Explain why or why not.

A.25 A pharmaceutical company invented a new medicine for compulsive behavior. To test this medicine on humans, the company advertised for volunteers who were suffering from this disease and wanted to participate in the study. As a result, 1820 persons responded. Using their own judgment, the group of physicians who were conducting this study assigned 910 of these patients to the treatment group and the remaining 910 to the control group. The patients in the treatment group were administered the actual medicine, and the patients in the control group were given a placebo. Six months later the conditions of the patients in the two groups were examined and compared. Based on this comparison, the physicians concluded that this medicine improves the condition of patients suffering from compulsive behavior.

    a. Comment on this study and its conclusion.
    b. Is this an observational study or a designed experiment? Explain.
    c. Is this a double-blind study? Explain.

A.26 Refer to Exercise A.25. Suppose the physicians conducting this study obtained a list of all patients suffering from compulsive behavior who were being treated by doctors in all hospitals in the country. Further, assume that this list is representative of the population of all such patients. The physicians then randomly selected 1820 patients from this list. Of these 1820, a randomly selected group of 910 patients was assigned to the treatment group, and the remaining 910 patients were assigned to the control group. The patients did not know which group they belonged to, but the doctors had access to such information. Six months later the conditions of the patients in the two groups were examined and compared. Based on this comparison, the physicians concluded that this medicine improves the condition of patients suffering from compulsive behavior.

    a. Comment on this study and its conclusion.
    b. Is this an observational study or a designed experiment? Explain.
    c. Is this a double-blind study? Explain.

A.27 Refer to Exercise A.26. Now suppose that neither patients nor doctors knew what group the patients belonged to.

    a. Is this an observational study or a designed experiment? Explain.
    b. Is this a double-blind study? Explain.

A.28 An electronics company that has chain stores in many states hired 50 new salespersons six months ago. Of these 50 salespersons, 20 are business/economics majors, and the remaining 30 hold a degree in one or more other areas. Based on the data for six months, the company president announced in the meeting of the board of directors that business/economics majors are better salespersons.

    a. Comment on this study and its conclusion.
    b. Is this an observational study or a designed experiment? Explain.

A.29 A psychologist needs 10 piglets for a study of the intelligence of pigs. She goes to a pig farm where there are 40 young pigs in a large pen. Assume that these pigs are representative of the population of all pigs. She selects the first 10 pigs she can catch and uses them for her study.

    a. Do these 10 pigs make a random sample?
    b. Are these 10 pigs likely to be representative of the entire population? Why or why not?
    c. If these 10 pigs do not form a random sample, what type of sample is it?
    d. Can you suggest a better procedure for selecting a sample of 10 from the 40 pigs in the pen?

A.30 A newspaper wants to conduct a poll to estimate the percentage of its readers who favor a gambling casino in their city. People register their opinions by placing a phone call that costs them $1.

    a. Is this method likely to produce a random sample?
    b. Which, if any, of the types of biases listed in this appendix are likely to be present and why?

*A.31 Computer crime is becoming an increasingly serious problem for business in the United States. In the summer of 1995, David Carter, a professor of criminal justice at Michigan State University, conducted a mail-in survey of 500 major corporations. Only 150 corporations responded, of which 98.7% had experienced computer crimes. Of the corporations that had experienced such crimes, 43.34% had suffered 25 or more incidents. Write a brief article for a business magazine summarizing the results of the survey and cautioning the readers about possible bias in the results. Indicate which types of bias are likely to be present, how they could arise, and whether the percentages given above are likely to underestimate or overestimate the true percentages of all U.S. corporations that have experienced such crimes.

GLOSSARY

Census A survey conducted by including every element of the population.

Control group The group on which no condition is imposed.

Cluster A subgroup (usually geographical) of the population that is representative of the population.

Cluster sampling The sampling technique under which the population is divided into clusters and a sample is chosen from one or a few clusters.

Convenience sample A sample that includes the most accessible members of the population.

Designed experiment A study in which the experimenter controls the assignment of elements to different treatment groups.

Double-blind experiment Experiment in which neither the doctors (or researchers) nor the patients (or members) know to which group a patient (or member) belongs.

Experiment Method of collecting data by controlling some or all factors.

Judgment sample A sample that includes the elements of the population selected based on the judgment and prior knowledge of an expert.

Nonresponse error The error that occurs because many of the people included in the sample do not respond.

Nonsampling or systematic errors The errors that occur in the collection, recording, and tabulation of data.

Observational study A study in which the assignment of elements to different treatments is voluntary and the researcher simply observes the results of the study.

Quota sample A sample selected in such a way that each group or subpopulation is represented in the sample in exactly the same proportion as in the target population.

Random sample A sample that assigns some chance of being selected in the sample to each member of the population.

Randomization The procedure under which elements are assigned to different (treatment and control) groups at random.

Representative sample A sample that contains the characteristics of the population as closely as possible.

Response error The error that occurs because people included in the survey do not provide correct answers.

Sample A portion of the population of interest.

Sample survey A survey that includes elements of a sample.

Sampling frame The list of elements of the target population that is used to select a sample.

Sampling or chance error The difference between the result obtained from a sample survey and the result that would be obtained from the census.

Selection error The error that occurs because the sampling frame is not representative of the population.

Simple random sample A sample chosen in such a way that each element of the population has the same probability of being included in the sample.

Stratified random sampling The sampling technique under which the population is divided into different strata and a sample is chosen from each stratum.

Stratum   A subgroup of the population whose members are identical with regard to the possession of a characteristic.

Survey   Collecting data from the elements of a population or sample.

Systematic random sampling   Sampling method used to choose a sample by selecting every kth unit from the list.

Target population   The collection of all subjects of interest.

Treatment   A condition (or a set of conditions) that is imposed on a group of elements by the experimenter. This group is called the **treatment group**.

Voluntary response error   The error that occurs because a survey is not conducted on a randomly selected sample but people are invited to respond voluntarily to the survey.

USING MINITAB

This MINITAB section describes how to generate random numbers and select a sample from a population using MINITAB.

GENERATING RANDOM NUMBERS

Illustration MA–1 describes the procedure to generate random numbers using MINITAB.

ILLUSTRATION MA-1   Using MINITAB, generate eight three-digit random numbers.

Solution   Note that 100 is the smallest three-digit number and 999 is the largest three-digit number. If you are using MINITAB FOR WINDOWS, perform the following steps to generate eight three-digit random numbers.

Step 1.   Click the **Calc** pull-down menu at the top of the screen.

Step 2.   Click **Random Data** from the selections available in the **Calc** menu.

Step 3.   Click **Integer** from the selections available in the **Random Data** menu.

Step 4.   You will see a dialog box entitled **Integer Distribution** appear on the screen. Type **8** in the box next to **Generate**. This tells MINITAB to generate eight rows of data. Next, enter **C1** in the box below **Store in column(s)**. Since the smallest and the largest values of three-digit numbers are 100 and 999, respectively, enter **100** in the box next to **Minimum value** and **999** in the box next to **Maximum value**.

Step 5.   Click the **OK** button at the bottom of this dialog box. Eight three-digit random numbers will appear in column C1 of the Data window.

If you are using the MINITAB COMMAND LANGUAGE, type the following MINITAB commands to obtain eight three-digit random numbers.

```
MTB  > RANDOM 8 OBSERVATIONS IN C1;
SUBC > INTEGERS BETWEEN 100 AND 999.
MTB  > PRINT C1
```

Each time you use the above set of MINITAB commands or the MINITAB FOR WINDOWS steps, you will obtain a different set of eight three-digit random numbers. One such MINITAB output is given in Figure A.1.

Figure A.1   MINITAB output for eight three-digit random numbers.

```
Data Display

C1
    781   819   183   589   469   853   198   401
```

Suppose you want to generate two-digit random numbers. Because 10 is the smallest two-digit number and 99 is the largest two-digit number, replace 100 by 10 and 999 by 99 in step 4 of the MINITAB FOR WINDOWS steps and in the MINITAB COMMAND LANGUAGE subcommand. Similarly, to generate four-digit random numbers, replace 100 by 1000 and 999 by 9999.

SELECTING A SAMPLE

MINITAB can be used to select a sample from a population. Suppose you have some population data entered in column C1 and want to select a sample of 16 observations from this population and store the sample data in column C2. If you are using MINITAB FOR WINDOWS, perform the following steps to select a sample.

Step 1.   Click the **Calc** pull-down menu at the top of the screen.
Step 2.   Click **Random Data** from the selections available in the **Calc** menu.
Step 3.   Click **Sample From Columns** from the selections available in the **Random Data** menu.
Step 4.   You will see a dialog box entitled **Sample From Columns** appear on the screen. Type **16** in the box next to **Sample** and **C1** in the box below **Samples**. This tells MINITAB to select a sample of 16 rows (observations) from the data entered in column C1. Then type **C2** in the box below **Store samples in**. If you want the sample to be selected with replacement, then check the box next to **Sample with replacement**. Otherwise make sure this box is not checked.
Step 5.   Click the **OK** button at the bottom of this dialog box. A sample of 16 observations will appear in column C2 of the Data window.

If you are using the MINITAB COMMAND LANGUAGE, type the following MINITAB commands to obtain a sample of 16 observations from the population data of column C1 and store the sample data in column C2.

```
MTB > SAMPLE 16 FROM C1 PUT IN C2
MTB > PRINT C2
```

Each time you use the above MINITAB COMMAND LANGUAGE or MINITAB FOR WINDOWS steps, you will obtain a different sample of 16 observations.

Now suppose the population data contains a large number of observations on three variables and these data are entered in columns C1 to C3. Further suppose that you want to select a sample of 12 observations, without replacement, from these data and store the sample data in columns C4 to C6. If you are using MINITAB FOR WINDOWS, Step 4 in the above procedure will be changed as follows: Type **12** in the box next to **Sample** and **C1–C3** in the box below **Samples**. Then type **C4–C6** in the box below **Store samples in**. Make sure the box next to **Sample with replacement** is not checked. This will give you a sample of 12 observations; store those data in columns C4–C6.

If you are using the MINITAB COMMAND LANGUAGE, type the following MINITAB commands to obtain a sample of 12 observations from the population data of columns C1–C3 and store the sample data in columns C4–C6.

```
MTB > SAMPLE 12 FROM C1–C3 PUT IN C4–C6
MTB > PRINT C4–C6
```

The MINITAB section at the end of Chapter 1 also shows how we can select a sample from a population using MINITAB.

COMPUTER ASSIGNMENTS

MA.1   Using MINITAB, construct a table of 100 two-digit random numbers.

MA.2   Using MINITAB, construct a table of 60 four-digit random numbers.

MA.3   Using MINITAB, construct a table of 40 five-digit random numbers.

MA.4   Using MINITAB, take a sample of 15 observations from Data Set III (NBA data) of Appendix B.

MA.5   Using MINITAB, take a sample of 15 observations from Data Set I (City Data) of Appendix B.

DATA SETS[1]

[1]These data sets are available on a floppy disk in MINITAB and ASCII format. To obtain this disk, contact either John Wiley's College Division or your area representative of John Wiley. The disk contains the following files:
1. CITYDATA (This file contains Data Set I.)
2. STATDATA (This file contains Data Set II.)
3. NBA (This file contains Data Set III.)
4. RRACESAM (This file contains Data Set IV.)
5. ROADRACE (This file contains the population data for Data Set IV.)
The extension ''MTW'' indicates that the file is in MINITAB format, and the extension ''DAT'' refers to the ASCII format file.

DATA SET I: CITY DATA[2]

Data on prices of selected products for selected cities across the country

Explanation of Columns

C1   Name of the city

C2   Price of T-bone steak per pound

C3   Price of half-gallon carton of whole milk

C4   Price of 2-liter Coca-Cola bottle, excluding any deposit

C5   Monthly rent of an unfurnished two-bedroom apartment (excluding all utilities except water), 1-1/2 or 2 baths, approximately 950 square feet

C6   Purchase price of 1800-square-foot living area new house, on 8000-square-foot lot in urban area with all utilities

C7   Monthly telephone charges for a private residential line (customer owns instruments)

C8   Price of one gallon regular unleaded gas, national brand, including all taxes; cash price at self-service pump if available

C9   Average cost per day of a semiprivate room in a hospital

C10   Price for woman's shampoo, trim, and blow dry

C11   Price of dry cleaning, man's two-piece suit

C12   Price of Gallo Chablis blanc wine, 1.5-liter bottle

C1	C2	C3	C4	C5	C6	C7	C8	C9	C10	C11	C12
ALABAMA											
1. Birmingham	5.64	1.58	1.39	556	129750	24.02	1.171	436.33	21.17	6.10	5.39
2. Decatur-Hartselle	5.97	1.46	1.16	410	123450	22.06	1.287	293.67	21.90	6.23	5.69
3. Gadsden	4.46	1.53	1.22	410	107342	23.10	1.263	415.75	21.49	5.47	4.96
4. Huntsville	5.98	1.35	1.05	468	113920	23.07	1.227	352.50	16.40	5.75	6.16
5. Marshall County	5.67	1.42	1.11	426	116125	17.87	1.222	290.00	18.50	6.10	5.59
6. Mobile	6.06	1.45	1.11	459	102433	22.85	1.227	216.80	21.00	5.50	5.69
7. Montgomery	5.83	1.49	1.16	506	110000	18.43	1.171	415.75	20.20	5.52	5.49
ALASKA											
8. Anchorage	6.42	2.20	1.99	750	186061	14.43	1.262	684.00	26.60	9.48	5.99
9. Fairbanks	5.81	2.02	1.73	786	175088	17.53	1.319	503.00	23.29	9.44	6.19
10. Juneau	5.87	2.00	1.90	1000	200900	14.86	1.326	400.00	24.00	9.03	6.05
11. Kodiak	6.09	2.49	2.49	1050	195000	21.71	1.587	400.00	25.67	9.75	6.84
ARIZONA											
12. Flagstaff	4.96	1.65	1.35	708	169063	18.84	1.389	411.00	21.40	7.91	4.26
13. Lake Havasu	4.96	1.68	1.34	555	114000	17.38	1.343	398.00	19.25	7.50	4.39
14. Phoenix	4.73	1.67	1.30	625	126566	19.03	1.343	472.14	22.50	8.33	4.43
15. Prescott	4.91	1.60	1.35	522	169137	20.57	1.323	498.00	26.60	6.63	4.48
16. Scottsdale	5.02	1.66	1.17	644	149720	18.66	1.311	460.67	26.40	7.53	4.61
17. Yuma	4.31	1.57	1.14	532	99125	19.28	1.499	545.00	25.00	7.18	4.56

[2]Data Set I is excerpted from *ACCRA Cost of Living Index*, 29(2): Second Quarter 1996. Copyright © 1996 ACCRA, 4232 King Street, Alexandria, VA 22302-1507. Reproduced with permission of ACCRA.

C1	C2	C3	C4	C5	C6	C7	C8	C9	C10	C11	C12
ARKANSAS											
18. Fayetteville	5.56	1.37	1.11	462	112100	26.44	1.098	190.00	19.10	6.00	5.99
19. Fort Smith	5.12	1.32	1.12	393	100800	22.32	1.049	200.00	15.00	5.25	6.53
20. Jonesboro	4.78	1.25	1.16	447	98667	21.34	1.182	220.00	18.33	6.25	5.49
21. Little Rock	5.73	1.41	1.05	430	93875	23.63	1.189	221.80	20.40	5.10	4.93
CALIFORNIA											
22. Bakersfield	5.77	1.54	1.03	553	120145	15.35	1.293	516.25	24.40	8.97	4.13
23. Fresno	5.59	1.59	1.00	545	133288	15.76	1.301	442.50	25.70	7.16	4.69
24. Los Angeles-Long Beach	6.26	1.86	0.91	720	170210	17.58	1.353	632.00	23.30	7.00	4.34
25. Marin County	5.91	1.71	1.37	991	346450	20.75	1.455	865.00	33.67	8.61	4.86
26. Palm Springs	6.07	1.87	0.93	677	160000	20.75	1.327	605.00	31.80	6.75	4.31
27. Riverside City	5.65	1.70	0.99	566	151573	16.31	1.347	596.33	25.39	7.05	3.89
28. San Diego	5.74	1.92	0.97	821	204951	15.64	1.391	614.36	25.44	6.35	4.39
29. San Francisco	6.04	1.78	1.32	1121	386791	20.75	1.436	1120.00	29.50	6.89	4.86
30. Santa Rosa	5.59	1.67	1.19	755	244700	20.75	1.527	680.00	28.00	8.50	4.53
COLORADO											
31. Boulder	5.35	1.72	1.07	801	222882	21.34	1.295	400.00	24.20	9.46	7.29
32. Colorado Springs	5.05	1.82	1.02	682	158313	20.18	1.263	542.50	20.13	5.19	4.08
33. Denver	5.18	1.60	1.06	699	154868	20.51	1.258	484.24	21.50	7.09	4.47
34. Fort Collins	5.24	1.69	1.03	727	160000	20.98	1.311	500.00	19.80	6.73	4.39
35. Glenwood Springs	5.29	1.49	1.49	667	199350	20.97	1.349	420.00	20.50	8.00	4.99
36. Loveland	5.19	1.81	1.13	517	126475	21.12	1.257	484.00	12.25	7.09	4.25
37. Pueblo	5.21	1.89	1.22	459	103374	20.77	1.249	342.50	16.18	6.44	4.12
DELAWARE											
38. Dover	6.91	1.29	0.93	567	124167	16.00	1.219	447.00	19.75	6.92	5.48
39. Wilmington	6.31	1.44	0.91	569	163133	16.90	1.239	491.83	18.75	6.10	5.37
FLORIDA											
40. Fort Myers-Cape Coral	5.83	1.53	1.21	601	122094	15.34	1.268	365.67	24.95	7.59	5.08
41. Gainesville	5.98	1.60	1.12	578	121674	15.72	1.239	433.33	23.60	7.47	5.16
42. Jacksonville	5.39	1.55	1.09	671	103333	16.10	1.333	331.60	24.36	6.82	5.09
43. Miami–Dade County	5.63	1.65	1.09	703	140700	16.27	1.315	476.60	26.00	7.80	4.89
44. Orlando	6.06	1.62	1.21	525	123966	19.65	1.251	469.10	27.60	6.86	4.99
45. Panama City	6.21	1.45	1.21	495	109750	14.28	1.229	357.50	17.80	7.15	5.11
46. Tallahassee	5.39	1.53	1.11	586	133916	17.49	1.267	414.50	22.40	7.12	5.21
GEORGIA											
47. Americus	5.59	1.38	1.16	455	96900	17.28	1.135	257.00	17.00	4.87	6.34
48. Atlanta	6.24	1.41	1.08	575	125200	22.75	1.083	319.20	26.00	6.67	6.19
49. Augusta	5.15	1.41	1.21	485	110000	20.38	1.093	320.60	24.00	6.17	5.81
50. Bainbridge	4.98	1.54	0.86	425	108000	19.01	1.139	203.00	18.75	5.63	6.35
IDAHO											
51. Boise	4.73	1.35	1.33	698	135219	16.78	1.226	447.50	22.26	7.10	4.95
52. Idaho Falls	4.66	1.69	1.19	408	125400	18.40	1.266	525.00	19.00	6.58	4.96
ILLINOIS											
53. Champaign-Urbana	5.24	1.56	1.21	536	135880	20.00	1.322	544.50	23.70	7.47	4.19
54. Decatur	6.24	1.59	1.26	447	102250	17.32	1.284	332.50	17.60	7.94	5.09

C1	C2	C3	C4	C5	C6	C7	C8	C9	C10	C11	C12
55. DeKalb	5.74	1.54	1.18	494	125775	37.65	1.299	395.00	18.00	6.42	4.49
56. Freeport	4.44	1.38	1.33	459	128857	28.05	1.332	308.00	18.75	6.50	5.01
57. Peoria	5.69	1.69	1.13	586	139567	26.02	1.311	314.00	26.60	7.38	4.81
58. Quad-Cities	5.84	1.51	1.21	415	130698	22.07	1.285	311.67	16.30	7.27	4.11
59. Quincy	4.61	1.54	1.05	448	136500	21.13	1.233	285.00	18.70	7.23	4.27
60. Springfield	5.91	1.63	0.95	497	100350	21.83	1.219	366.00	16.80	6.80	3.91
INDIANA											
61. Anderson	5.10	1.42	1.09	494	131000	18.92	1.227	339.50	19.20	5.41	4.25
62. Bloomington	4.55	1.47	1.21	607	123426	15.31	1.207	493.00	18.50	6.97	4.11
63. Elkhart-Goshen	5.58	1.43	1.31	559	108000	20.89	1.221	396.00	17.60	6.79	4.45
64. Fort Wayne	5.33	1.17	0.99	489	107850	23.83	1.229	396.33	20.79	6.72	4.46
65. Indianapolis	5.52	1.60	1.16	578	119625	19.66	1.231	374.47	21.00	7.34	4.30
66. Muncie	5.84	1.29	1.13	580	128412	16.95	1.234	351.00	20.75	6.53	4.28
67. South Bend	5.95	1.41	1.35	581	110540	19.18	1.167	529.87	25.20	6.54	4.01
IOWA											
68. Cedar Rapids	5.22	1.39	1.31	526	137500	20.20	1.243	343.00	18.40	7.38	5.14
69. Des Moines	4.63	1.29	1.06	510	119840	19.76	1.205	393.00	22.79	7.07	5.83
70. Dubuque	5.73	1.45	1.31	575	166000	18.74	1.229	265.00	17.40	5.85	5.33
71. Mason City	5.13	1.24	1.31	380	117000	18.85	1.199	345.00	13.38	6.40	5.51
KANSAS											
72. Garden City	4.55	1.38	0.99	444	128200	17.68	1.259	340.00	19.80	6.12	5.99
73. Lawrence	4.91	1.46	1.02	525	140673	17.77	1.182	318.00	16.25	7.75	5.69
74. Manhattan	4.92	1.47	1.28	517	139000	18.68	1.119	438.50	14.32	6.97	5.70
75. Wichita	4.54	1.42	0.99	437	118108	18.76	1.179	504.75	20.19	6.96	5.12
KENTUCKY											
76. Bowling Green	5.62	1.34	1.29	460	125000	19.83	1.110	337.00	20.00	5.50	5.18
77. Covington	6.82	1.54	1.17	558	104351	20.38	1.173	376.00	16.80	6.00	5.11
78. Hopkinsville	5.71	1.43	1.19	419	109450	20.16	1.119	316.00	21.09	5.98	6.65
79. Lexington	5.79	1.50	1.01	643	121240	27.38	1.223	374.20	23.20	6.43	5.53
80. Louisville	5.95	1.38	1.19	506	108400	25.31	1.149	345.47	17.20	6.78	6.05
LOUISIANA											
81. Alexandria	4.83	1.30	1.14	483	115400	18.69	1.173	230.62	23.80	7.14	4.61
82. Baton Rouge	6.09	1.52	1.23	519	132708	21.15	1.205	363.00	22.40	6.70	5.02
83. Lafayette	5.16	1.49	1.14	575	115250	19.56	1.221	232.00	27.00	6.75	4.99
84. Lake Charles	4.85	1.86	1.26	492	118170	19.82	1.235	330.00	20.30	7.29	5.41
85. Monroe	5.01	1.32	1.19	471	109150	19.06	1.199	242.00	16.00	7.62	5.09
86. New Orleans	6.85	1.33	1.07	535	105335	23.23	1.161	366.00	17.53	5.69	4.66
MARYLAND											
87. Baltimore	5.79	1.48	0.93	505	135968	22.01	1.225	500.20	22.30	6.04	5.29
88. Cumberland	5.91	1.19	1.15	559	142300	26.83	1.207	304.33	17.20	6.17	5.40
89. Hagerstown	5.38	0.98	1.02	488	115905	20.43	1.212	398.00	19.64	6.31	5.28
90. Worcester County	5.16	1.36	1.01	560	136500	16.95	1.262	377.00	19.67	6.98	6.99
MASSACHUSETTS											
91. Boston	6.53	1.37	1.17	1001	243800	22.77	1.243	619.00	21.50	6.56	6.21

C1	C2	C3	C4	C5	C6	C7	C8	C9	C10	C11	C12
MICHIGAN											
92. Ann Arbor	5.89	1.57	1.21	807	155000	22.53	1.279	535.00	28.20	8.00	5.49
93. Grand Rapids	6.19	1.54	1.20	548	148213	18.53	1.256	397.00	25.12	7.74	5.89
94. Holland	5.89	1.64	1.14	487	151800	15.98	1.232	242.00	15.90	7.76	5.66
95. Lansing	5.54	1.59	1.22	590	164083	18.00	1.284	462.75	21.62	6.32	5.87
MINNESOTA											
96. Minneapolis	6.22	1.46	1.21	600	124463	22.28	1.273	536.80	21.20	7.62	5.21
97. Rochester	5.43	1.47	1.28	577	115144	23.88	1.317	441.67	18.83	7.43	5.00
98. St. Cloud	5.55	1.52	1.32	464	116130	22.35	1.289	349.00	19.17	6.91	3.99
99. St. Paul	6.31	1.53	1.18	634	116517	20.63	1.299	685.43	22.60	7.48	4.35
MISSISSIPPI											
100. Hattiesburg	6.19	1.69	1.01	479	98657	23.18	1.075	219.00	16.56	5.69	5.21
101. Jackson	5.51	1.43	0.99	548	109664	28.26	1.171	224.62	19.70	5.44	6.10
MISSOURI											
102. Columbia	5.89	1.35	1.23	406	110500	14.30	1.164	404.33	23.70	6.59	5.32
103. Joplin	4.95	1.31	1.08	395	112400	14.38	1.189	413.00	16.40	6.07	4.81
104. Kansas City	5.87	1.39	1.00	638	114816	17.73	1.168	514.50	20.56	7.08	5.15
105. Kennett	4.80	1.41	1.14	435	89100	16.45	1.179	330.00	22.00	6.58	6.22
106. Kirksville	5.26	1.12	1.13	438	154667	14.42	1.191	366.50	16.67	5.63	5.24
107. St. Joseph	5.21	1.27	0.97	574	117098	15.10	1.162	390.00	14.54	5.75	5.08
108. St. Louis	5.62	1.68	1.01	628	121317	19.04	1.104	463.50	22.85	6.78	4.94
109. Springfield	5.36	1.16	1.02	482	112875	16.29	1.112	355.00	21.25	6.94	5.46
MONTANA											
110. Billings	5.49	1.48	1.28	425	142000	21.31	1.338	414.00	19.40	7.82	5.09
111. Bozeman	5.01	1.42	1.28	583	135467	22.08	1.299	395.00	18.00	7.57	5.16
112. Great Falls	5.47	1.42	1.12	486	142000	23.37	1.322	396.00	19.62	7.17	5.18
113. Helena	5.12	1.37	1.12	450	107910	19.77	1.449	410.00	20.17	7.50	4.89
114. Missoula	5.37	1.52	1.27	586	135263	18.22	1.339	408.00	20.80	7.36	5.46
NEBRASKA											
115. Grand Island	5.21	1.28	1.09	438	133475	23.32	1.269	314.00	18.20	7.98	5.29
116. Hastings	4.94	1.33	1.07	438	113037	17.67	1.241	300.00	17.20	8.35	5.09
117. Kearney	5.22	1.32	1.36	475	113563	15.44	1.257	312.00	13.50	8.50	5.13
118. Lincoln	4.61	1.28	1.02	460	99438	16.58	1.219	355.00	24.80	6.30	4.60
119. Omaha	5.36	1.30	1.07	487	125050	21.02	1.275	303.40	17.99	6.53	4.66
NEVADA											
120. Las Vegas	5.91	1.54	0.99	709	135045	10.86	1.342	341.91	22.80	7.68	4.64
121. Reno-Sparks	5.59	1.39	1.29	716	170462	13.58	1.466	546.33	21.40	7.59	4.41
NEW HAMPSHIRE											
122. Manchester	5.49	1.20	1.06	610	143950	20.09	1.219	456.00	16.67	7.09	4.99
NEW MEXICO											
123. Albuquerque	4.87	1.67	1.19	645	132000	22.20	1.250	358.25	20.90	6.51	5.54
124. Carlsbad	5.09	1.39	1.76	414	115000	20.74	1.296	365.00	21.00	5.08	5.62
125. Clovis-Portales	4.79	1.54	1.76	560	117455	18.44	1.279	230.00	19.33	5.92	5.74
126. Farmington	5.33	1.66	1.44	519	122441	20.40	1.339	350.00	18.60	6.34	6.61

C1	C2	C3	C4	C5	C6	C7	C8	C9	C10	C11	C12
127. Hobbs	4.66	1.39	1.82	420	99000	21.24	1.299	345.00	23.33	5.88	5.79
128. Las Cruces	4.76	1.50	1.15	484	135350	20.52	1.216	369.00	21.73	6.88	5.24
129. Santa Fe	4.89	1.37	1.49	699	167125	20.37	1.339	305.00	25.20	7.29	5.55
NEW YORK											
130. Albany	6.29	1.21	1.07	660	145000	30.25	1.257	334.00	17.30	7.45	5.43
131. Buffalo	6.65	1.27	1.00	464	111380	25.17	1.231	377.00	15.00	5.44	5.25
132. Glens Falls	5.79	1.19	1.15	508	112475	25.03	1.239	318.00	18.90	6.90	5.84
133. New York	10.19	1.77	1.57	2740	545400	24.06	1.409	1284.00	38.00	7.90	6.29
134. Rochester	6.31	1.31	1.69	484	140000	20.68	1.309	494.00	15.66	7.79	6.66
135. Syracuse	5.97	1.37	1.21	536	120064	34.75	1.207	493.00	15.80	7.35	5.27
NORTH CAROLINA											
136. Burlington	5.32	1.41	1.05	453	116750	16.64	1.139	370.00	19.00	6.50	5.08
137. Charlotte	5.55	1.42	1.09	486	131600	17.51	1.203	349.33	24.00	5.99	4.74
138. Fayetteville	6.14	1.40	1.21	451	112000	17.77	1.202	432.00	20.40	6.68	5.21
139. Greenville	5.76	1.48	1.05	473	117234	17.66	1.193	305.00	20.00	7.38	5.17
140. Hickory	5.99	1.39	1.09	500	112125	17.45	1.275	282.50	19.25	5.87	4.94
141. Raleigh-Durham	5.24	1.60	1.15	590	142836	19.00	1.193	308.00	20.86	6.99	5.11
142. Winston-Salem	5.79	1.46	1.03	458	141840	16.30	1.191	218.50	27.00	7.16	5.17
NORTH DAKOTA											
143. Fargo	5.39	1.53	1.19	522	135001	19.69	1.299	375.00	21.50	7.15	5.58
144. Minot	5.12	1.58	1.06	450	116796	10.39	1.299	378.00	13.50	5.78	5.13
OHIO											
145. Akron	6.26	1.30	1.04	534	119025	22.45	1.246	494.40	19.40	6.31	5.19
146. Canton	5.51	1.17	1.01	519	135620	21.62	1.237	242.00	20.90	6.70	5.42
147. Cincinnati	6.93	1.59	1.23	562	128839	20.51	1.233	375.00	20.99	7.16	4.99
148. Toledo	5.95	1.38	1.07	418	120925	20.10	1.259	303.20	22.60	7.45	5.49
149. Youngstown	6.83	1.36	0.97	438	134200	19.44	1.209	309.75	13.00	5.97	5.11
OKLAHOMA											
150. Oklahoma City	4.90	1.34	1.13	503	98965	20.58	1.174	266.20	20.60	6.13	6.29
151. Stillwater	5.28	1.36	1.04	467	105242	18.84	1.170	250.00	20.14	6.33	5.92
152. Tulsa	5.09	1.39	1.10	453	103375	19.69	1.119	314.40	21.60	6.37	5.39
OREGON											
153. Portland	4.99	1.44	1.19	620	160500	23.34	1.385	484.93	25.80	8.05	4.81
154. Salem	4.98	1.36	1.16	513	153072	19.82	1.317	380.00	22.90	8.00	4.72
PENNSYLVANIA											
155. Hanover	4.58	1.24	1.14	535	137400	16.06	1.171	340.00	17.00	7.28	5.99
156. Harrisburg	5.33	1.24	1.13	582	118878	20.47	1.179	503.75	21.00	7.10	5.99
157. Lancaster	4.99	1.25	1.02	577	137033	17.63	1.197	309.50	23.20	7.61	5.99
158. Philadelphia	5.94	1.29	1.31	721	191990	16.59	1.269	451.00	23.46	7.49	5.99
159. York County	5.69	1.24	1.13	510	121169	22.56	1.201	347.75	17.80	7.52	5.99
SOUTH CAROLINA											
160. Charleston	5.99	1.43	1.01	564	112295	24.79	1.171	379.75	24.33	7.31	5.55
161. Columbia	4.95	1.40	1.05	554	116116	23.05	1.169	370.25	16.79	6.29	5.25
162. Greenville	5.41	1.47	0.99	576	137900	21.73	1.181	308.33	32.20	6.95	5.39

C1	C2	C3	C4	C5	C6	C7	C8	C9	C10	C11	C12
163. Myrtle Beach	5.25	1.47	1.01	520	133855	22.55	1.168	393.00	22.50	6.44	5.69
164. Spartanburg	5.87	1.48	0.99	468	120750	22.35	1.093	277.00	21.40	6.70	5.39
165. Sumter	5.45	1.45	1.11	416	105834	21.13	1.159	383.00	18.00	6.46	5.23
SOUTH DAKOTA											
166. Sioux Falls	5.11	1.39	1.15	595	114950	22.99	1.257	387.00	16.80	6.57	5.31
167. Vermillion	4.64	1.28	1.24	470	139500	21.23	1.199	296.00	16.87	7.25	5.49
TENNESSEE											
168. Chattanooga	5.75	1.48	1.25	556	106650	20.74	1.139	333.20	28.10	5.66	5.63
169. Dyersburg	4.64	1.41	1.01	423	97666	17.63	1.279	242.00	15.83	7.50	8.05
170. Johnson City	5.43	1.31	0.95	540	127733	21.60	1.244	271.00	18.00	6.15	6.89
171. Knoxville	5.81	1.29	0.99	545	107225	19.47	1.151	332.60	23.40	6.48	5.29
172. Memphis	5.59	1.47	1.03	646	119216	20.37	1.303	269.80	22.00	7.60	5.31
173. Morristown	5.66	1.48	1.01	455	124400	17.69	1.184	199.00	15.25	6.07	7.27
TEXAS											
174. Amarillo	5.51	1.30	1.74	492	113475	14.54	1.213	309.67	21.20	5.76	5.69
175. Dallas	6.25	1.51	1.25	701	114001	16.33	1.225	436.60	29.10	6.76	5.43
176. El Paso	4.77	1.59	1.15	514	112900	14.58	1.289	340.60	28.40	6.30	5.80
177. Fort Worth	5.46	1.70	1.29	574	101572	16.62	1.242	325.00	16.73	7.04	5.41
178. Georgetown	5.34	1.26	1.29	603	120667	13.73	1.232	325.00	25.00	7.97	5.39
179. Houston	5.47	1.40	0.99	624	108863	17.52	1.240	395.81	28.00	4.63	5.38
180. San Antonio	5.37	1.20	1.36	561	106000	16.38	1.239	350.20	26.10	6.46	5.73
181. Waco	4.55	1.23	1.04	582	92980	17.56	1.163	293.50	17.50	5.36	5.00
182. Wichita Falls	5.33	1.51	1.69	483	100602	15.34	1.249	350.75	13.75	5.52	5.09
UTAH											
183. Cedar City	4.62	1.74	1.21	438	112500	13.52	1.239	390.00	15.67	7.62	5.45
184. St. George	4.26	1.61	1.20	548	150375	13.67	1.287	371.50	21.00	8.00	5.45
185. Salt Lake City	4.71	1.48	1.07	517	124840	18.77	1.341	542.00	18.20	6.43	5.45
VERMONT											
186. Barre-Montpelier	5.99	1.22	1.16	582	134500	33.77	1.212	631.00	16.75	7.40	7.29
VIRGINIA											
187. Lynchburg	5.74	1.45	0.99	446	115950	18.53	1.178	325.67	20.57	5.93	5.30
188. Prince William	6.21	1.44	0.99	658	176360	23.51	1.235	292.50	19.17	6.51	5.24
189. Richmond	5.59	1.46	1.01	668	132279	19.86	1.227	400.00	28.00	5.67	5.01
WASHINGTON											
190. Spokane	5.19	1.58	1.00	605	167688	15.94	1.373	507.50	19.20	7.76	4.19
191. Tacoma	5.23	1.47	1.11	602	131500	20.08	1.313	417.20	24.40	7.19	4.35
192. Yakima	5.50	1.56	1.22	563	169167	18.95	1.255	368.50	18.70	6.55	3.99
WEST VIRGINIA											
193. Charleston	6.39	1.39	1.13	517	144560	29.93	1.277	275.40	20.00	6.44	5.19
194. Martinsburg/Berkeley County	5.22	1.12	0.96	523	114160	22.17	1.239	243.00	14.92	5.95	4.94

C1	C2	C3	C4	C5	C6	C7	C8	C9	C10	C11	C12
WISCONSIN											
195. Eau Claire	5.15	1.34	1.05	579	133500	15.80	1.299	300.00	16.31	7.06	4.57
196. Fond du Lac	5.86	1.56	0.93	565	150000	16.31	1.246	360.00	18.17	7.82	4.65
197. Green Bay	5.65	1.42	1.15	547	129500	15.94	1.299	318.00	17.29	7.65	4.63
198. Marshfield	4.79	1.44	1.14	550	135000	26.53	1.324	280.00	15.67	6.58	4.62
199. Wausau	5.47	1.15	1.21	561	170000	23.44	1.289	275.00	18.10	7.28	4.83
WYOMING											
200. Gillette	5.09	1.39	1.01	379	106813	22.42	1.207	349.00	14.60	5.88	6.31

DATA SET II: DATA ON STATES

Information on different variables for 50 states

Explanation of columns

C1 Name of the state

C2 Average tuition and fees (in dollars) for in-state residents at four-year public colleges, 1995–96 (Source: General Accounting Office)

C3 Female labor force participation rate (in percent), 1995 (Source: U.S. Bureau of Labor Statistics)

C4 Estimated 1996 spending (in millions of dollars) by states (Source: Governor's Association Survey)

C5 Per capita income (in dollars), 1995 (Source: U.S. Bureau of Economic Analysis)

C6 Per capita energy expenditure (in dollars), 1993 (Source: U.S. Energy Information Administration)

C7 Infant mortality rate (in percent), 1993 (Source: U.S. National Center for Health Statistics)

C1	C2	C3	C4	C5	C6	C7
1. Alabama	2234	55.3	4235	19,181	2092	10.3
2. Alaska	2502	66.4	2483	24,002	3085	8.2
3. Arizona	1943	59.6	4562	20,489	1871	7.6
4. Arkansas	2062	58.4	2583	18,101	2100	10.0
5. California	2918	56.5	44,246	24,073	1564	6.8
6. Colorado	2458	67.2	4152	23,961	1648	7.9
7. Connecticut	3828	60.7	8914	31,776	2015	7.1
8. Delaware	3962	62.9	1714	26,273	2141	8.8
9. Florida	1790	54.5	14,797	23,061	1553	8.6
10. Georgia	2076	59.2	10,358	21,741	2004	10.4
11. Hawaii	1524	61.2	3154	24,590	1693	7.2
12. Idaho	1714	62.1	1331	18,906	1892	7.2
13. Illinois	3388	60.3	18,121	25,225	1934	9.9
14. Indiana	3040	64.2	7180	21,433	2159	9.2
15. Iowa	2565	66.6	3783	20,921	1971	6.9
16. Kansas	2110	63.4	3474	21,841	2129	8.8
17. Kentucky	2160	56.0	5443	18,849	2044	8.2
18. Louisiana	2139	53.6	4852	18,981	3052	10.8
19. Maine	3582	61.6	1732	20,105	2152	6.8
20. Maryland	3572	64.5	7442	26,333	1731	9.8
21. Massachusetts	4178	60.8	16,304	28,021	1882	6.2
22. Michigan	3789	57.9	8439	23,915	1854	9.5
23. Minnesota	3108	69.6	9257	23,971	1831	7.5
24. Mississippi	2443	55.8	2776	16,683	1936	11.5
25. Missouri	3007	65.0	5758	21,819	1873	8.4
26. Montana	2346	59.6	999	18,445	2201	7.4
27. Nebraska	2294	68.4	1810	21,477	1968	9.1
28. Nevada	1830	60.5	1233	24,390	1987	6.7
29. New Hampshire	4537	65.3	845	25,587	1861	5.6
30. New Jersey	3649	58.7	15,758	29,848	2088	8.3
31. New Mexico	1938	55.6	2781	18,206	1975	8.4
32. New York	3697	52.8	32,748	27,678	1673	8.4

C1	C2	C3	C4	C5	C6	C7
33. North Carolina	1622	59.8	10,032	21,103	1959	10.5
34. North Dakota	2211	64.9	639	18,625	2564	7.9
35. Ohio	3664	58.3	16,286	22,514	2034	9.2
36. Oklahoma	1741	55.7	3547	18,580	1988	8.8
37. Oregon	3241	60.8	3563	21,611	1768	7.2
38. Pennsylvania	4693	55.0	16,220	23,558	1911	8.6
39. Rhode Island	3619	58.4	1651	23,844	1926	7.3
40. South Carolina	3103	59.9	4404	18,998	1955	10.1
41. South Dakota	2549	65.8	622	19,576	1943	9.5
42. Tennessee	2001	60.3	5474	21,038	1942	9.4
43. Texas	1832	60.0	21,836	21,206	2524	7.5
44. Utah	2007	61.2	2612	18,232	1571	6.0
45. Vermont	5521	65.3	703	21,231	2026	6.7
46. Virginia	3965	63.0	7666	23,974	1825	8.7
47. Washington	2726	60.6	8645	23,774	1641	6.4
48. West Virginia	1997	46.3	2422	17,687	2160	8.6
49. Wisconsin	2555	68.2	8216	22,261	1719	7.9
50. Wyoming	2005	64.1	491	20,684	3546	7.9

DATA SET III: NBA DATA[3]

The heights and weights of NBA players who were on the rosters of National Basketball Association teams at the beginning of 1996–1997 season

Explanation of data columns

C1 Name of a player

C2 Height (in inches) of a player

C3 Weight (in pounds) of a player

C1	C2	C3	C1	C2	C3
1. Abdul-Rauf, Mahmoud	73	162	38. Brown, Marcus	75	185
2. Abdur-Rahim, Shareef	81	225	39. Brown, P.J.	83	240
3. Addison, Rafael	80	241	40. Brown, Randy	74	191
4. Alexander, Cory	73	190	41. Bryant, Kobe	78	200
5. Allen, Jerome	76	184	42. Bryant, Mark	81	245
6. Allen, Ray	77	205	43. Buechler, Jud	78	228
7. Amaya, Ashraf	80	250	44. Bullard, Matt	82	235
8. Anderson, Greg	82	250	45. Burrell, Scott	79	226
9. Anderson, Kenny	73	168	46. Butler, Mitchell	77	210
10. Anderson, Nick	78	228	47. Caldwell, Adrian	80	265
11. Anderson, Rod	78	208	48. Caffey, Jason	80	256
12. Anthony, Greg	73	176	49. Cage, Michael	81	248
13. Armstrong, B.J.	74	185	50. Camby, Marcus	83	220
14. Armstrong, Darrell	73	180	51. Campbell, Elden	83	250
15. Askew, Vincent	78	235	52. Carr, Antoine	81	255
16. Askins, Keith	80	224	53. Carr, Chris	78	220
17. Augmon, Stacey	80	205	54. Cassell, Sam	75	185
18. Austin, Isaac	82	270	55. Causwell, Duane	84	240
19. Baker, Vin	83	245	56. Ceballos, Cedric	79	225
20. Barkley, Charles	78	252	57. Chaney, Calbert	79	215
21. Barros, Dana	71	163	58. Chapman, Rex	76	195
22. Barry, Brent	78	185	59. Chilcutt, Pete	83	235
23. Barry, Jon	76	194	60. Childress, Randolph	74	188
24. Benjamin, Benoit	84	265	61. Childs, Chris	75	195
25. Benoit, David	80	220	62. Christie, Doug	78	205
26. Best, Travis	71	182	63. Coleman, Derrick	82	260
27. Blaylock, Mookie	73	185	64. Coles, Bimbo	74	182
28. Blount, Corie	82	242	65. Corbin, Tyrone	78	225
29. Bogues, Muggsy	63	141	66. Curley, Bill	81	245
30. Boyce, Donnie	77	196	67. Curry, Dell	77	200
31. Bradley, Shawn	90	248	68. Curry, Michael	77	210
32. Bradtke, Mark	82	265	69. Dampier, Erick	83	265
33. Brandon, Terrell	71	180	70. Danilovic, Predrag	78	200
34. Brickowski, Frank	81	248	71. Dare, Yinka	84	265
35. Brooks, Scott	71	165	72. Davis, Antonio	81	230
36. Brown, Chucky	80	215	73. Davis, Ben	81	240
37. Brown, Dee	74	192	74. Davis, Dale	83	230

[3]Data Set III, representing the 1996–97 NBA opening day team rosters is excerpted from the *USA Today*, November 1, 1996. Copyright © 1996, *USA Today*. Excerpted and reproduced with permission. The list includes the players on injury lists also.

C1	C2	C3	C1	C2	C3
75. Davis, Emanual	77	195	127. Grant, Harvey	81	225
76. Davis, Hubert	77	183	128. Grant, Horace	82	245
77. Davis, Mark	79	210	129. Green, A.C.	81	225
78. Day, Todd	78	188	130. Green, Litterial	73	195
79. DeClercq, Andrew	82	230	131. Gugliotta, Tom	82	240
80. Dehere, Terry	76	190	132. Ham, Darvin	79	220
81. Delk, Tony	74	193	133. Hamer, Steve	84	245
82. Del Negro, Vinny	76	200	134. Hamilton, Thomas	86	330
83. Divac, Vlade	85	250	135. Hammonds, Tom	81	225
84. Djordjevic, Aleksandar	74	198	136. Hancock, Darrin	79	205
85. Douglas, Sherman	73	198	137. Hardaway, Anfernee	79	215
86. Dreiling, Greg	85	265	138. Hardaway, Tim	72	195
87. Drexler, Clyde	79	222	139. Harper, Derek	76	206
88. Duckworth, Kevin	84	300	140. Harper, Ron	78	216
89. Dudley, Chris	83	240	141. Harrington, Othella	81	235
90. Dumars, Joe	75	195	142. Harris, Anthony	74	215
91. Dumas, Tony	78	190	143. Harris, Lucious	77	205
92. Earl, Acie	82	240	144. Hawkins, Hersey	75	190
93. Edney, Tyus	70	152	145. Heal, Shane	72	180
94. Edwards, Blue	76	229	146. Henderson, Alan	81	235
95. Edwards, Doug	79	235	147. Hendrickson, Mark	81	220
96. Edwards, Kevin	75	210	148. Herrera, Carl	81	225
97. Ehlo, Craig	79	205	149. Hill, Grant	80	225
98. Eisley, Howard	74	177	150. Hill, Tyrone	81	245
99. Elie, Mario	77	210	151. Hoiberg, Fred	76	203
100. Elliott, Sean	80	220	152. Hornacek, Jeff	76	190
101. Ellis, Dale	79	215	153. Horry, Robert	82	220
102. Ellis, LaPhonso	80	240	154. Houston, Allan	78	200
103. Ellison, Pervis	82	242	155. Howard, Juwan	81	250
104. Evans, Brian	80	220	156. Hughes, Mark	80	235
105. Ewing, Patrick	84	240	157. Hunter, Lindsey	74	195
106. Farmer, Tony	81	245	158. Hurley, Bobby	72	165
107. Ferrell, Duane	79	215	159. Ilgauskas, Zydrunas	87	238
108. Ferry, Danny	82	235	160. Iverson, Allen	72	165
109. Finley, Michael	79	215	161. Jackson, Jaren	78	200
110. Fish, Matt	83	235	162. Jackson, Jim	78	215
111. Fisher, Derek	73	200	163. Jackson, Mark	75	185
112. Foster, Greg	83	240	164. Jent, Chris	79	220
113. Fox, Rick	79	249	165. Jennings, Keith	67	160
114. Fuller, Todd	83	255	166. Johnson, Avery	71	180
115. Gamble, Kevin	78	225	167. Johnson, Eddie	79	215
116. Garnett, Kevin	83	220	168. Johnson, Ervin	83	245
117. Garrett, Dean	83	250	169. Johnson, Kevin	73	190
118. Gatling, Chris	82	230	170. Johnson, Larry	79	263
119. Geary, Reggie	74	187	171. Jones, Charles	81	215
120. Geiger, Matt	84	245	172. Jones, Dontae	80	220
121. Gill, Kendall	77	216	173. Jones, Eddie	78	190
122. Gilliam, Armon	81	245	174. Jones, Popeye	80	250
123. Goldwire, Anthony	74	182	175. Jordan, Michael	78	216
124. Graham, Greg	76	182	176. Jordan, Reggie	76	195
125. Grant, Brian	81	254	177. Keefe, Adam	81	241
126. Grant, Gary	75	185	178. Kemp, Shawn	82	256

C1	C2	C3	C1	C2	C3
179. Kempton, Tim	82	265	231. Miller, Oliver	81	310
180. Kerr, Steve	75	181	232. Miller, Reggie	79	185
181. Kersey, Jerome	79	240	233. Mills, Chris	78	216
182. Kidd, Jason	76	208	234. Mills, Terry	82	250
183. King, Rich	86	265	235. Minor, Greg	78	230
184. Kittles, Kerry	77	179	236. Mitchell, Sam	79	215
185. Kleine, Joe	84	271	237. Mobley, Eric	83	257
186. Knight, Travis	84	235	238. Montross, Eric	84	270
187. Koncak, Jon	84	255	239. Moore, Tracy	76	200
188. Kukoc, Toni	83	232	240. Morris, Chris	80	220
189. Laettner, Christian	83	235	241. Moten, Lawrence	77	186
190. Lenard, Voshon	76	205	242. Mourning, Alonzo	82	261
191. Lang, Andrew	83	250	243. Mullin, Chris	79	215
192. Lang, Antonio	80	230	244. Murdock, Eric	73	200
193. Lauderdale, Priest	88	343	245. Muresan, Gheorghe	91	303
194. Leckner, Eric	83	265	246. Murray, Lamond	79	236
195. Legler, Tim	76	200	247. Murray, Tracy	79	228
196. Lister, Alton	84	245	248. Mutombo, Dikembe	86	250
197. Livingston, Randy	76	209	249. Muursepp, Martin	81	235
198. Long, Grant	81	248	250. Nash, Steve	75	195
199. Longley, Luc	86	292	251. Newbill, Ivano	83	245
200. Lynch, George	80	223	252. Newman, Johnny	79	205
201. Mack, Sam	79	220	253. Norman, Ken	80	228
202. MacLean, Don	82	235	254. Norris, Moochie	73	175
203. Mahorn, Rick	82	260	255. Nwosu, Julius	82	255
204. Majerle, Dan	78	220	256. O'Bannon, Ed	80	222
205. Malone, Karl	81	256	257. O'Neal, Jermaine	83	226
206. Maloney, Matt	75	200	258. O'Neal, Shaquille	85	300
207. Mann, Marcus	80	245	259. Oakley, Charles	81	245
208. Manning, Danny	82	234	260. Olajuwon, Hakeem	84	255
209. Manning, Rich	83	260	261. Ostertag, Greg	86	280
210. Marbury, Stephon	74	180	262. Outlaw, Charles	80	210
211. Marciulionis, Sarunas	77	215	263. Overton, Doug	75	190
212. Marshall, Donny	79	230	264. Owens, Billy	81	225
213. Marshall, Donyell	81	230	265. Owes, Ray	81	224
214. Martin, Darrick	71	170	266. Pack, Robert	74	190
215. Mashburn, Jamal	80	250	267. Parish, Robert	85	244
216. Mason, Anthony	79	250	268. Parks, Cherokee	83	240
217. Massenberg, Tony	81	210	269. Payton, Gary	76	190
218. Maxwell, Vernon	76	190	270. Peeler, Anthony	76	212
219. Mayberry, Lee	73	172	271. Peplowski, Mike	82	270
220. McCarty, Walter	82	230	272. Perdue, Will	84	240
221. McCaskill, Amal	83	235	273. Perkins, Sam	81	255
222. McCloud, George	80	225	274. Perry, Elliot	72	160
223. McDaniel, Xavier	79	218	275. Person, Chuck	80	235
224. McDyess, Antonio	81	220	276. Person, Wesley	78	195
225. McInnis, Jeff	76	190	277. Phills, Bobby	77	220
226. McKey, Derrick	82	225	278. Piatkowski, Eric	79	215
227. McKie, Aaron	77	209	279. Pierce, Ricky	76	215
228. McIlvaine, Jim	85	260	280. Pinckney, Ed	81	240
229. McMillan, Nate	77	200	281. Pippen, Scottie	79	228
230. Meyer, Loren	82	260	282. Polynice, Olden	84	250

C1	C2	C3	C1	C2	C3
283. Porter, Terry	75	195	335. Smith, Joe	82	225
284. Potapenko, Vitaly	82	280	336. Smith, Kenny	75	170
285. Praskevicius, Virginius	80	230	337. Smith, Michael	80	230
286. Price, Brent	73	185	338. Smith, Steve	80	215
287. Price, Mark	72	180	339. Smith, Tony	76	205
288. Radja, Dino	83	255	340. Smits, Rik	88	265
289. Ratliff, Theo	82	225	341. Snow, Eric	75	200
290. Recasner, Eldridge	75	190	342. Spencer, Elmore	84	270
291. Reeves, Bryant	84	275	343. Spencer, Felton	84	265
292. Reeves, Khalid	75	201	344. Sprewell, Latrell	77	190
293. Reid, Don	80	250	345. Stackhouse, Jerry	78	218
294. Respert, Shawn	74	195	346. Starks, John	77	185
295. Rice, Glen	80	214	347. Stewart, Larry	80	230
296. Richardson, Pooh	73	180	348. Stith, Bryant	77	208
297. Richmond, Mitch	77	215	349. Stockton, John	73	175
298. Rider, Isaiah	77	215	350. Stoudamire, Damon	70	171
299. Roberts, Fred	82	210	351. Strickland, Erick	75	210
300. Roberts, Stanley	84	290	352. Strickland, Mark	81	220
301. Robinson, Chris	77	205	353. Strickland, Rod	75	185
302. Robinson, Clifford	82	225	354. Strong, Carlos	80	235
303. Robinson, David	85	250	355. Strong, Derek	81	240
304. Robinson, Glenn	79	240	356. Sura, Bob	77	200
305. Robinson, James	74	180	357. Szabo, Brett	83	230
306. Robinson, Rumeal	74	195	358. Tabak, Zan	84	245
307. Rodman, Dennis	78	220	359. Thomas, Carl	76	195
308. Roe, Lou	79	220	360. Thomas, Kurt	81	230
309. Rogers, Carlos	83	220	361. Thompson, Brooks	76	193
310. Rogers, Rodney	79	255	362. Thompson, LaSalle	82	260
311. Rogers, Roy	82	235	363. Thorpe, Otis	82	246
312. Rooks, Sean	82	260	364. Tisdale, Wayman	81	260
313. Rose, Jalen	80	210	365. Tower, Keith	82	260
314. Rose, Malik	79	250	366. Trent, Gary	80	250
315. Royal, Donald	80	218	367. Van Exel, Nick	73	183
316. Rozier, Clifford	83	255	368. Vaughn, David	81	240
317. Russell, Bryon	79	225	369. Vaught, Loy	81	240
318. Sabonis, Arvydas	87	292	370. Vrankovic, Stoyko	86	260
319. Salvadori, Kevin	84	231	371. Wallace, Ben	80	238
320. Schayes, Danny	83	260	372. Wallace, John	81	225
321. Scheffler, Steve	81	250	373. Wallace, Rasheed	82	225
322. Schintzius, Dwayne	86	285	374. Walker, Antoine	81	225
323. Schrempf, Detlef	82	235	375. Walker, Samaki	81	240
324. Scott, Brent	82	250	376. Walters, Rex	76	190
325. Scott, Byron	76	205	377. Ward, Charlie	74	190
326. Scott, Dennis	80	235	378. Watson, Jamie	79	190
327. Scott, James	78	195	379. Weatherspoon, Clarence	79	240
328. Scott, Shawnelle	83	250	380. Webber, Chris	82	250
329. Sealy, Malik	80	190	381. Wennington, Bill	84	277
330. Seikaly, Rony	83	253	382. Wesley, David	72	198
331. Shaw, Brian	78	200	383. West, Doug	78	220
332. Simmons, Lionel	79	210	384. West, Mark	82	246
333. Simpkins, Dickey	82	264	385. Whiteside, Donald	71	170
334. Smith, Charles	82	245	386. Whitney, Chris	72	170

C1	C2	C3	C1	C2	C3
387. Wilkins, Dominique	80	230	400. Williams, Walt	80	230
388. Wilkins, Gerald	78	225	401. Williamson, Corliss	79	245
389. Williams, Buck	80	225	402. Willis, Kevin	84	245
390. Williams, Eric	80	220	403. Wilson, Trevor	80	222
391. Williams, Herb	83	260	404. Wingate, David	77	185
392. Williams, Jayson	82	245	405. Wingfield, Dontonio	80	256
393. Williams, Jerome	81	206	406. Wolf, Joe	83	230
394. Williams, John	83	245	407. Wood, David	81	230
395. Williams, Lorenzo	81	230	408. Workman, Haywoode	75	180
396. Williams, Micheal	74	175	409. Wright, Lorenzen	83	225
397. Williams, Monty	80	225	410. Wright, Sharone	83	260
398. Williams, Reggie	79	195	411. Zidek, George	84	266
399. Williams, Scott	82	230			

DATA SET IV: **SAMPLE OF 500 OBSERVATIONS SELECTED FROM MANCHESTER (CONNECTICUT) ROAD RACE DATA**

(This data set represents a random sample of 500 observations selected from the data on the Sixtieth Road Race held on Thanksgiving day, 28th of November 1996, in Manchester (Connecticut). The total distance for this race is 4.748 miles. The data set represents time (in minutes) taken by 500 participants to complete this race. The complete data set, which includes 9070 participants who completed that race, is available on a floppy disk, in MINITAB and ASCII format, from the publisher. The data were published in *The Hartford Courant* of December 2, 1996 and are reproduced here with the permission of *The Hartford Courant*.)

23.59	24.09	24.21	25.35	26.05	26.07	26.26	27.00	27.39	27.40	27.43	27.59	28.07
28.51	29.01	29.04	29.18	29.39	29.58	30.01	30.04	30.12	30.21	31.20	31.36	31.39
31.42	31.45	31.46	31.47	31.48	31.50	31.53	32.12	32.15	32.15	32.17	32.21	32.39
32.40	32.42	32.46	32.52	32.58	33.13	33.15	33.18	33.28	33.35	33.41	33.44	33.47
33.50	33.58	34.08	34.29	34.31	34.39	34.46	34.46	34.47	34.49	34.55	35.03	35.04
35.18	35.20	35.21	35.23	35.43	35.47	35.47	35.54	35.58	36.00	36.04	36.04	36.07
36.08	36.11	36.20	36.21	36.24	36.24	36.26	36.28	36.29	36.33	36.34	36.35	36.39
36.39	36.49	36.50	36.55	36.56	36.59	37.03	37.05	37.09	37.10	37.11	37.11	37.13
37.13	37.26	37.29	37.32	37.40	37.41	37.48	37.50	37.55	37.58	38.04	38.06	38.19
38.22	38.28	38.29	38.40	38.41	38.56	39.00	39.01	39.03	39.09	39.10	39.11	39.11
39.12	39.14	39.18	39.18	39.19	39.24	39.24	39.34	39.36	39.37	39.41	39.43	39.43
39.43	39.44	39.44	39.50	39.54	40.00	40.00	40.18	40.21	40.26	40.31	40.36	40.37
40.37	40.39	40.42	40.44	40.47	40.48	40.51	40.52	40.54	41.09	41.13	41.14	41.15
41.16	41.18	41.23	41.26	41.26	41.26	41.28	41.31	41.36	41.42	41.46	41.57	41.57
41.57	41.58	41.58	41.58	42.00	42.07	42.09	42.11	42.20	42.22	42.24	42.24	42.28
42.33	42.38	42.40	42.40	42.53	43.06	43.20	43.31	43.34	43.35	43.38	43.39	43.41
43.43	43.44	43.50	43.50	43.51	43.55	43.58	44.09	44.11	44.13	44.18	44.23	44.26
44.27	44.31	44.35	44.36	44.38	44.39	44.43	44.43	44.47	44.49	44.51	44.52	44.57
44.58	44.59	45.08	45.18	45.22	45.25	45.28	45.35	45.35	45.36	45.45	45.48	45.49
45.50	45.55	46.02	46.03	46.04	46.04	46.09	46.15	46.17	46.20	46.29	46.29	46.31
46.33	46.43	46.51	46.52	46.52	46.55	47.03	47.05	47.09	47.10	47.11	47.12	47.12
47.14	47.22	47.31	47.34	47.39	47.40	47.41	47.41	47.45	47.46	47.48	47.52	47.53
47.54	47.55	47.56	47.57	48.02	48.02	48.03	48.03	48.03	48.03	48.04	48.06	48.18
48.20	48.21	48.22	48.24	48.24	48.27	48.29	48.29	48.35	48.37	48.38	48.39	48.40
48.47	49.03	49.06	49.09	49.16	49.17	49.23	49.30	49.31	49.32	49.38	49.42	49.44
49.47	49.52	49.53	49.58	50.05	50.08	50.09	50.19	50.26	50.38	50.44	50.49	50.50
50.50	50.52	50.52	50.52	50.55	51.11	51.17	51.22	51.25	51.27	51.31	51.31	51.35
51.37	51.37	51.38	51.44	51.45	51.52	51.54	52.00	52.05	52.17	52.24	52.28	52.28
52.29	52.36	52.47	52.56	53.00	53.01	53.06	53.08	53.11	53.11	53.15	53.18	53.20
53.22	53.23	53.25	53.29	53.38	53.39	53.40	53.45	53.58	54.09	54.15	54.15	54.26
54.32	54.34	54.40	54.53	54.58	55.16	55.44	55.45	55.49	55.50	55.50	55.53	56.10
56.25	56.26	56.27	56.27	56.31	56.33	56.38	56.45	56.47	57.00	57.07	57.09	57.14
57.27	57.31	58.11	58.22	58.28	58.47	58.50	58.53	58.59	59.07	59.09	59.48	59.49
59.56	60.27	60.39	60.59	60.59	61.04	61.04	61.06	61.13	61.54	62.17	62.19	62.22
62.31	62.38	62.47	63.14	63.36	64.08	66.46	66.47	67.18	67.46	68.11	68.21	68.33
68.44	68.47	69.11	69.11	70.19	70.25	70.51	71.00	71.37	71.47	72.19	72.20	72.37
72.49	73.57	73.58	74.14	74.37	76.16	77.35	77.37	78.01	78.18	78.49	79.22	79.29
79.36	79.47	80.08	82.03	83.47	84.08	84.22	85.05	85.15	85.22	85.42	86.01	87.36
89.56	90.45	92.50	94.14	96.13	96.27							

APPENDIX C

STATISTICAL TABLES[1]

[1]All tables included in this appendix were prepared by the author. Tables I, IV, VI, and VII were constructed using MINITAB, and Tables VIII through X were made using SAS statistical software package. Tables II, III, and V were prepared by using a computer program.

TABLE I   RANDOM NUMBERS

57728	16308	27337	53884	60742	61693	39887	81779	36354
63962	45765	75060	46767	28844	32354	91463	25057	91907
51041	22252	38447	71567	95103	11124	34960	35710	91098
84048	53578	67379	42605	59122	39415	82869	86971	64817
17736	34458	67227	97041	77846	20338	52372	34645	56563
82238	83763	45464	18493	98489	72138	38942	97661	95788
28853	61793	44664	69427	68144	71949	57192	25592	49835
22251	73098	68108	87626	76724	56495	87357	83065	95316
66236	46591	69225	29867	60815	51931	40507	52568	47097
50006	91666	86406	92778	51232	38761	21861	98596	42673
68328	12840	61206	64298	27378	61452	13349	27223	79637
83039	25015	95983	82835	67268	23355	44647	25542	10536
53158	82329	81756	81429	54366	97530	51447	11324	49939
46802	61720	97508	73339	29277	17964	35421	39880	38180
25162	78468	44303	14425	42587	37212	58866	39008	91938
65957	15171	22417	95571	90679	54774	43979	71017	49647
10876	36062	91375	90128	14906	81447	49158	14703	89517
35354	66633	62311	58185	67310	95474	21878	89101	38299
70822	69983	23726	97422	46713	20340	42807	10859	26897
64299	12987	60370	70165	43306	14417	79261	53891	72816
74007	61658	86698	31571	75098	11676	35867	39764	47504
70909	68300	55074	42093	55745	80364	18488	47981	18702
67898	98830	97705	10723	82370	45586	19013	60915	84961
59386	25440	92441	14265	26123	85453	57326	72790	55243
71469	49833	95737	84195	78444	32104	89917	88361	35344
34064	12993	23818	28197	33755	96438	84223	10400	36797
86492	25367	65712	81581	89579	31759	56108	24476	47696
86914	87565	20344	39027	98338	95171	75562	54283	35342
88418	58064	13624	32978	90704	56218	84064	69990	45354
87948	83451	96217	40534	40775	74376	43157	74856	13950
13049	85293	32747	17728	50495	34617	73707	33976	86177
86544	52703	74990	98288	61833	48803	75258	83382	79099
77295	70694	97326	35430	53881	94007	70471	66815	73042
54637	32831	59063	72353	87365	15322	33156	40331	93942
50938	12004	18585	23896	62559	44470	27701	66780	56157
80999	49724	76745	25232	74291	74184	91055	58903	18172
71303	36255	77310	95847	30282	77207	34439	47763	99697
79264	16901	55814	89734	30255	87209	31629	19328	42532
30235	69368	38685	32790	58980	42159	88577	18427	73504
59110	69783	93713	29151	34933	95745	72271	38684	15426
28094	19560	27259	82736	49700	37876	52322	69562	75837
40341	20666	26662	16422	76351	70520	36890	86559	89160
30117	68850	28319	44992	68110	47007	22243	72813	60934
62287	44957	47690	79484	69449	27981	34770	34228	81686
96976	77830	61746	67846	15584	28070	79200	12663	63273
82584	34789	33494	55533	25040	84187	14479	26286	10665
35728	87881	70271	13115	35745	99145	92717	74357	16716

TABLE I   RANDOM NUMBERS

737

TABLE I   (*Continued*)

88458	63625	59577	92037	99012	40836	58817	30757	37934
49789	20873	53858	91356	11387	75208	33643	88210	42440
49131	34078	45396	56884	81416	46292	36012	30806	65220
96256	82566	34796	88012	43066	35786	93715	15550	16690
43742	97487	68089	69887	23737	71136	21108	85204	60726
34527	87490	81183	95864	59430	19473	57978	39853	47877
58906	37390	88924	80917	58840	29907	99098	33761	50335
30438	12056	12104	61012	44674	49815	85298	94129	59542
79149	98261	48599	54336	71894	82889	51219	70291	60922
48703	25290	13835	35695	15440	52533	82849	25504	81623
96050	74505	18706	10572	66015	53509	48115	87578	86099
98859	48791	15048	73300	48045	33559	98939	98003	39453
32758	55597	11686	18385	31103	87621	39659	81413	68625
17238	73653	15557	79374	60965	75564	15872	34611	86497
79748	21687	94964	43348	26957	27085	81760	29099	23553
89199	75213	37815	99891	60990	37062	80331	54009	26812
87491	62544	51229	13028	81370	16309	28493	98555	24278
87338	17647	40018	48386	49992	44304	23330	38730	21601
48635	73063	37450	65403	65134	83119	16341	95766	83949
32197	94930	50586	88559	48025	97023	15372	18847	97168
83421	68819	69623	45088	54839	70855	86714	38202	98163
66167	84791	40631	33428	78200	41145	57816	86795	31646
26555	70521	69140	93495	38179	43253	78172	44239	60701
47786	37539	17452	88719	24423	59201	24979	51019	35458
47775	42564	15665	92454	98345	87963	81142	34356	41518
49414	83761	74309	82620	53677	34575	81871	76615	27653
75918	39825	60958	96584	26872	15379	84080	40371	35019
13440	85096	85668	24896	65261	83757	68388	29797	48376
39614	53926	97122	85279	15622	29329	59579	60250	73895
40067	48944	98882	39023	31677	41118	52818	29586	43848
65350	11148	63012	59418	54688	83692	95840	91627	84057
98902	62170	49281	29406	63143	43722	35838	98979	67024
10529	81048	29639	50740	93253	77339	80328	88580	30970
99123	48497	35247	33488	63781	19388	98534	27479	44269
65147	42913	50654	64220	13950	74293	53489	39014	86040
86886	15231	43834	88205	87159	30789	10959	81631	15575
17264	57846	52347	96649	69212	28053	62290	93328	98520
36110	87509	95913	66687	67149	81500	44107	27546	94868
58288	91109	66433	75388	80441	95720	64891	63049	68237
67834	18606	88840	39705	17329	15690	90382	35725	21362
67746	23016	87357	89427	98266	39452	58011	86665	70716
32196	36633	63350	73154	47699	15479	63905	81186	67181
33646	35175	41141	75793	58908	80681	88974	84611	46634
66973	98812	21094	45209	52503	51038	96306	75653	32482
74819	41419	91296	31736	99727	68791	93588	99566	98413
76495	37282	43051	17275	30370	76105	55926	98910	84767
39350	85262	59225	47343	63449	47004	94970	77067	16857
29657	66820	47420	37404	80296	94070	54249	60378	54670

TABLE I (*Continued*)

98059	31868	86468	80389	66521	23304	99582	48791	74154
78131	47852	62735	79575	48757	25712	22468	66035	29237
36071	71312	33098	38558	57088	26162	32752	24827	95562
23742	21969	82378	28923	19944	91024	63237	36022	76979
48418	36759	92342	93571	86923	26627	46138	86343	21083
74678	64188	39402	65189	30854	65086	43052	54042	79127
45693	86700	39667	53646	11663	86785	84727	83728	21758
68539	65113	68955	71627	38626	57160	63171	41707	51634
56807	61373	21941	71481	88523	72157	92088	41244	75735
29320	38387	89881	59789	50099	64811	31131	74334	65674
63399	73318	61578	28141	21655	65378	56261	69795	67096
33316	23627	55609	48463	92502	64287	99853	54497	85985
15403	81891	20190	72235	85636	16745	99483	43583	89137
76057	62447	71848	60035	76280	38017	26998	82690	16512
93885	29489	26222	77121	83244	74614	62527	36019	31265
66312	17182	96913	10736	52184	57082	58901	87749	34684
88960	52088	92432	18463	25562	20674	48988	41829	98681
38903	23457	87215	47089	38395	21929	94929	59489	79066
88281	90912	22965	16428	32289	99354	87068	55884	69518
98627	47123	32667	69196	70158	19828	42793	54593	53682
13183	37010	53184	53434	53631	96983	21201	10236	90134
69177	18284	72840	68433	88300	85396	10298	15680	13859
79501	74784	50483	37213	67077	59481	43976	35404	90683
52734	93419	47519	85203	27665	97179	47002	41258	39219
70256	58003	11565	67432	56505	18468	10293	46490	98191
47603	62142	37636	43374	19773	10538	27243	20800	50383
41507	43884	18253	81908	22803	84840	38968	70176	59393
56464	84865	65387	97484	95349	55548	54214	86814	20654
54523	35676	93542	20744	23942	22935	19794	53413	12979
86965	67669	42284	68532	24766	41411	97597	34998	70248
80707	35351	29958	12270	76227	31529	26105	94145	30469
57351	44045	67826	18191	75712	86420	36234	93377	20205
56860	89252	53362	82306	24114	91538	49114	67506	73489
87844	42168	83234	59134	17403	20418	65647	14702	59080
47157	27318	46686	59507	31598	16152	41184	95641	58835
63530	51286	46562	43739	51259	39836	72962	96998	89257
74595	35004	61728	28879	60412	81320	99003	20824	47086
26649	55512	58180	87954	91885	22660	31132	15752	11807
63596	44068	12648	91827	35448	59307	64466	68502	36292
66621	89136	46721	43322	78706	60249	90841	79917	18000
98128	99125	86432	87068	88376	65121	64402	55931	45748
85968	99264	64582	85694	29027	62883	53615	26692	73490
93011	71694	78514	63842	33754	84577	78698	38667	54673
36994	29619	36095	44782	85794	28498	25870	83655	97905
58857	32343	61392	65331	66939	51145	77060	85743	85278
97430	82854	28720	52153	37246	87152	95563	51769	79320
64642	69774	67582	95955	91433	95515	35211	39734	82631
56789	90056	28697	88922	39250	66008	55324	39129	63408

TABLE I   RANDOM NUMBERS

739

TABLE I   (*Continued*)

78707	36317	69939	63529	88044	66897	16846	67664	99997
20752	71605	38186	18221	79499	14660	86115	80339	34321
20794	82021	37432	97568	85812	97016	15655	40601	39475
84832	45347	60186	66673	62148	37683	79034	46572	69243
39960	63046	99657	28301	19953	84261	30215	52274	91374
51835	19676	40685	45677	57150	73208	59526	76240	24209
88213	83367	21935	72494	87548	43000	72275	81974	54718
39746	38989	28721	38803	89668	57496	97127	59364	83335
95915	51291	14163	40972	33163	85169	66522	72010	53429
43937	78760	47672	69700	87058	19072	89435	13390	72315
28633	29330	20463	89033	16968	62815	65802	53006	70674
27415	32278	61924	61670	38880	13911	85037	93738	94913
10157	74513	43054	44601	35689	54559	91660	60035	83733
18041	40798	39274	72760	83644	48960	52193	95674	22516
22679	12792	60046	80515	12962	57351	36431	52277	50567
81468	99534	30455	17430	92600	85813	90223	15335	97102
50636	87932	25489	29395	87683	84579	10396	38276	33729
60635	13409	81824	77150	51472	65915	62520	33839	52209
68336	29892	94343	37822	55260	97321	20488	50172	45199
79273	96036	89979	78654	38959	36250	91126	90337	91381
71942	89335	75664	75278	40445	12818	24033	11809	44129
87842	60665	73523	55824	61257	22080	74425	54851	84786

TABLE II   FACTORIALS

n	$n!$
0	1
1	1
2	2
3	6
4	24
5	120
6	720
7	5,040
8	40,320
9	362,880
10	3,628,800
11	39,916,800
12	479,001,600
13	6,227,020,800
14	87,178,291,200
15	1,307,674,368,000
16	20,922,789,888,000
17	355,687,428,096,000
18	6,402,373,705,728,000
19	121,645,100,408,832,000
20	2,432,902,008,176,640,000
21	51,090,942,171,709,440,000
22	1,124,000,727,777,607,680,000
23	25,852,016,738,884,976,640,000
24	620,448,401,733,239,439,360,000
25	15,511,210,043,330,985,984,000,000

TABLE III   VALUES OF $\binom{n}{x}$ (COMBINATION)

n \ x	0	1	2	3	4	5	6	7	8	9	10	11	12	13	14	15	16	17	18	19	20
1	1	1																			
2	1	2	1																		
3	1	3	3	1																	
4	1	4	6	4	1																
5	1	5	10	10	5	1															
6	1	6	15	20	15	6	1														
7	1	7	21	35	35	21	7	1													
8	1	8	28	56	70	56	28	8	1												
9	1	9	36	84	126	126	84	36	9	1											
10	1	10	45	120	210	252	210	120	45	10	1										
11	1	11	55	165	330	462	462	330	165	55	11	1									
12	1	12	66	220	495	792	924	792	495	220	66	12	1								
13	1	13	78	286	715	1,287	1,716	1,716	1,287	715	286	78	13	1							
14	1	14	91	364	1,001	2,002	3,003	3,432	3,003	2,002	1,001	364	91	14	1						
15	1	15	105	455	1,365	3,003	5,005	6,435	6,435	5,005	3,003	1,365	455	105	15	1					
16	1	16	120	560	1,820	4,368	8,008	11,440	12,870	11,440	8,008	4,368	1,820	560	120	16	1				
17	1	17	136	680	2,380	6,188	12,376	19,448	24,310	24,310	19,448	12,376	6,188	2,380	680	136	17	1			
18	1	18	153	816	3,060	8,568	18,564	31,824	43,758	48,620	43,758	31,824	18,564	8,568	3,060	816	153	18	1		
19	1	19	171	969	3,876	11,628	27,132	50,388	75,582	92,378	92,378	75,582	50,388	27,132	11,628	3,876	969	171	19	1	
20	1	20	190	1,140	4,845	15,504	38,760	77,520	125,970	167,960	184,756	167,960	125,970	77,520	38,760	15,504	4,845	1,140	190	20	1
21	1	21	210	1,330	5,985	20,349	54,264	116,280	203,490	293,930	352,716	352,716	293,930	203,490	116,280	54,264	20,349	5,985	1,330	210	21
22	1	22	231	1,540	7,315	26,334	74,613	170,544	319,770	497,420	646,646	705,432	646,646	497,420	319,770	170,544	74,613	26,334	7,315	1,540	231
23	1	23	253	1,771	8,855	33,649	100,947	245,157	490,314	817,190	1,144,066	1,352,078	1,352,078	1,144,066	817,190	490,314	245,157	100,947	33,649	8,855	1,771
24	1	24	276	2,024	10,626	42,504	134,596	346,104	735,471	1,307,504	1,961,256	2,496,144	2,704,156	2,496,144	1,961,256	1,307,504	735,471	346,104	134,596	42,504	10,626
25	1	25	300	2,300	12,650	53,130	177,100	480,700	1,081,575	2,042,975	3,268,760	4,457,400	5,200,300	5,200,300	4,457,400	3,268,760	2,042,975	1,081,575	480,700	177,100	53,130

TABLE IV   **TABLE OF BINOMIAL PROBABILITIES**

n	x	.05	.10	.20	.30	.40	.50	.60	.70	.80	.90	.95
1	0	.9500	.9000	.8000	.7000	.6000	.5000	.4000	.3000	.2000	.1000	.0500
	1	.0500	.1000	.2000	.3000	.4000	.5000	.6000	.7000	.8000	.9000	.9500
2	0	.9025	.8100	.6400	.4900	.3600	.2500	.1600	.0900	.0400	.0100	.0025
	1	.0950	.1800	.3200	.4200	.4800	.5000	.4800	.4200	.3200	.1800	.0950
	2	.0025	.0100	.0400	.0900	.1600	.2500	.3600	.4900	.6400	.8100	.9025
3	0	.8574	.7290	.5120	.3430	.2160	.1250	.0640	.0270	.0080	.0010	.0001
	1	.1354	.2430	.3840	.4410	.4320	.3750	.2880	.1890	.0960	.0270	.0071
	2	.0071	.0270	.0960	.1890	.2880	.3750	.4320	.4410	.3840	.2430	.1354
	3	.0001	.0010	.0080	.0270	.0640	.1250	.2160	.3430	.5120	.7290	.8574
4	0	.8145	.6561	.4096	.2401	.1296	.0625	.0256	.0081	.0016	.0001	.0000
	1	.1715	.2916	.4096	.4116	.3456	.2500	.1536	.0756	.0256	.0036	.0005
	2	.0135	.0486	.1536	.2646	.3456	.3750	.3456	.2646	.1536	.0486	.0135
	3	.0005	.0036	.0256	.0756	.1536	.2500	.3456	.4116	.4096	.2916	.1715
	4	.0000	.0001	.0016	.0081	.0256	.0625	.1296	.2401	.4096	.6561	.8145
5	0	.7738	.5905	.3277	.1681	.0778	.0312	.0102	.0024	.0003	.0000	.0000
	1	.2036	.3280	.4096	.3602	.2592	.1562	.0768	.0284	.0064	.0005	.0000
	2	.0214	.0729	.2048	.3087	.3456	.3125	.2304	.1323	.0512	.0081	.0011
	3	.0011	.0081	.0512	.1323	.2304	.3125	.3456	.3087	.2048	.0729	.0214
	4	.0000	.0004	.0064	.0283	.0768	.1562	.2592	.3601	.4096	.3281	.2036
	5	.0000	.0000	.0003	.0024	.0102	.0312	.0778	.1681	.3277	.5905	.7738
6	0	.7351	.5314	.2621	.1176	.0467	.0156	.0041	.0007	.0001	.0000	.0000
	1	.2321	.3543	.3932	.3025	.1866	.0937	.0369	.0102	.0015	.0001	.0000
	2	.0305	.0984	.2458	.3241	.3110	.2344	.1382	.0595	.0154	.0012	.0001
	3	.0021	.0146	.0819	.1852	.2765	.3125	.2765	.1852	.0819	.0146	.0021
	4	.0001	.0012	.0154	.0595	.1382	.2344	.3110	.3241	.2458	.0984	.0305
	5	.0000	.0001	.0015	.0102	.0369	.0937	.1866	.3025	.3932	.3543	.2321
	6	.0000	.0000	.0001	.0007	.0041	.0156	.0467	.1176	.2621	.5314	.7351
7	0	.6983	.4783	.2097	.0824	.0280	.0078	.0016	.0002	.0000	.0000	.0000
	1	.2573	.3720	.3670	.2471	.1306	.0547	.0172	.0036	.0004	.0000	.0000
	2	.0406	.1240	.2753	.3177	.2613	.1641	.0774	.0250	.0043	.0002	.0000
	3	.0036	.0230	.1147	.2269	.2903	.2734	.1935	.0972	.0287	.0026	.0002
	4	.0002	.0026	.0287	.0972	.1935	.2734	.2903	.2269	.1147	.0230	.0036
	5	.0000	.0002	.0043	.0250	.0774	.1641	.2613	.3177	.2753	.1240	.0406
	6	.0000	.0000	.0004	.0036	.0172	.0547	.1306	.2471	.3670	.3720	.2573
	7	.0000	.0000	.0000	.0002	.0016	.0078	.0280	.0824	.2097	.4783	.6983
8	0	.6634	.4305	.1678	.0576	.0168	.0039	.0007	.0001	.0000	.0000	.0000
	1	.2793	.3826	.3355	.1977	.0896	.0312	.0079	.0012	.0001	.0000	.0000
	2	.0515	.1488	.2936	.2965	.2090	.1094	.0413	.0100	.0011	.0000	.0000
	3	.0054	.0331	.1468	.2541	.2787	.2187	.1239	.0467	.0092	.0004	.0000
	4	.0004	.0046	.0459	.1361	.2322	.2734	.2322	.1361	.0459	.0046	.0004

TABLE IV   TABLE OF BINOMIAL PROBABILITIES

743

TABLE IV   (*Continued*)

n	x	.05	.10	.20	.30	.40	.50	.60	.70	.80	.90	.95
	5	.0000	.0004	.0092	.0467	.1239	.2187	.2787	.2541	.1468	.0331	.0054
	6	.0000	.0000	.0011	.0100	.0413	.1094	.2090	.2965	.2936	.1488	.0515
	7	.0000	.0000	.0001	.0012	.0079	.0312	.0896	.1977	.3355	.3826	.2793
	8	.0000	.0000	.0000	.0001	.0007	.0039	.0168	.0576	.1678	.4305	.6634
9	0	.6302	.3874	.1342	.0404	.0101	.0020	.0003	.0000	.0000	.0000	.0000
	1	.2985	.3874	.3020	.1556	.0605	.0176	.0035	.0004	.0000	.0000	.0000
	2	.0629	.1722	.3020	.2668	.1612	.0703	.0212	.0039	.0003	.0000	.0000
	3	.0077	.0446	.1762	.2668	.2508	.1641	.0743	.0210	.0028	.0001	.0000
	4	.0006	.0074	.0661	.1715	.2508	.2461	.1672	.0735	.0165	.0008	.0000
	5	.0000	.0008	.0165	.0735	.1672	.2461	.2508	.1715	.0661	.0074	.0006
	6	.0000	.0001	.0028	.0210	.0743	.1641	.2508	.2668	.1762	.0446	.0077
	7	.0000	.0000	.0003	.0039	.0212	.0703	.1612	.2668	.3020	.1722	.0629
	8	.0000	.0000	.0000	.0004	.0035	.0176	.0605	.1556	.3020	.3874	.2985
	9	.0000	.0000	.0000	.0000	.0003	.0020	.0101	.0404	.1342	.3874	.6302
10	0	.5987	.3487	.1074	.0282	.0060	.0010	.0001	.0000	.0000	.0000	.0000
	1	.3151	.3874	.2684	.1211	.0403	.0098	.0016	.0001	.0000	.0000	.0000
	2	.0746	.1937	.3020	.2335	.1209	.0439	.0106	.0014	.0001	.0000	.0000
	3	.0105	.0574	.2013	.2668	.2150	.1172	.0425	.0090	.0008	.0000	.0000
	4	.0010	.0112	.0881	.2001	.2508	.2051	.1115	.0368	.0055	.0001	.0000
	5	.0001	.0015	.0264	.1029	.2007	.2461	.2007	.1029	.0264	.0015	.0001
	6	.0000	.0001	.0055	.0368	.1115	.2051	.2508	.2001	.0881	.0112	.0010
	7	.0000	.0000	.0008	.0090	.0425	.1172	.2150	.2668	.2013	.0574	.0105
	8	.0000	.0000	.0001	.0014	.0106	.0439	.1209	.2335	.3020	.1937	.0746
	9	.0000	.0000	.0000	.0001	.0016	.0098	.0403	.1211	.2684	.3874	.3151
	10	.0000	.0000	.0000	.0000	.0001	.0010	.0060	.0282	.1074	.3487	.5987
11	0	.5688	.3138	.0859	.0198	.0036	.0005	.0000	.0000	.0000	.0000	.0000
	1	.3293	.3835	.2362	.0932	.0266	.0054	.0007	.0000	.0000	.0000	.0000
	2	.0867	.2131	.2953	.1998	.0887	.0269	.0052	.0005	.0000	.0000	.0000
	3	.0137	.0710	.2215	.2568	.1774	.0806	.0234	.0037	.0002	.0000	.0000
	4	.0014	.0158	.1107	.2201	.2365	.1611	.0701	.0173	.0017	.0000	.0000
	5	.0001	.0025	.0388	.1321	.2207	.2256	.1471	.0566	.0097	.0003	.0000
	6	.0000	.0003	.0097	.0566	.1471	.2256	.2207	.1321	.0388	.0025	.0001
	7	.0000	.0000	.0017	.0173	.0701	.1611	.2365	.2201	.1107	.0158	.0014
	8	.0000	.0000	.0002	.0037	.0234	.0806	.1774	.2568	.2215	.0710	.0137
	9	.0000	.0000	.0000	.0005	.0052	.0269	.0887	.1998	.2953	.2131	.0867
	10	.0000	.0000	.0000	.0000	.0007	.0054	.0266	.0932	.2362	.3835	.3293
	11	.0000	.0000	.0000	.0000	.0000	.0005	.0036	.0198	.0859	.3138	.5688
12	0	.5404	.2824	.0687	.0138	.0022	.0002	.0000	.0000	.0000	.0000	.0000
	1	.3413	.3766	.2062	.0712	.0174	.0029	.0003	.0000	.0000	.0000	.0000
	2	.0988	.2301	.2835	.1678	.0639	.0161	.0025	.0002	.0000	.0000	.0000
	3	.0173	.0852	.2362	.2397	.1419	.0537	.0125	.0015	.0001	.0000	.0000
	4	.0021	.0213	.1329	.2311	.2128	.1208	.0420	.0078	.0005	.0000	.0000
	5	.0002	.0038	.0532	.1585	.2270	.1934	.1009	.0291	.0033	.0000	.0000
	6	.0000	.0005	.0155	.0792	.1766	.2256	.1766	.0792	.0155	.0005	.0000
	7	.0000	.0000	.0033	.0291	.1009	.1934	.2270	.1585	.0532	.0038	.0002
	8	.0000	.0000	.0005	.0078	.0420	.1208	.2128	.2311	.1329	.0213	.0021
	9	.0000	.0000	.0001	.0015	.0125	.0537	.1419	.2397	.2362	.0852	.0173

TABLE IV   (*Continued*)

n	x	.05	.10	.20	.30	.40	.50	.60	.70	.80	.90	.95
	10	.0000	.0000	.0000	.0002	.0025	.0161	.0639	.1678	.2835	.2301	.0988
	11	.0000	.0000	.0000	.0000	.0003	.0029	.0174	.0712	.2062	.3766	.3413
	12	.0000	.0000	.0000	.0000	.0000	.0002	.0022	.0138	.0687	.2824	.5404
13	0	.5133	.2542	.0550	.0097	.0013	.0001	.0000	.0000	.0000	.0000	.0000
	1	.3512	.3672	.1787	.0540	.0113	.0016	.0001	.0000	.0000	.0000	.0000
	2	.1109	.2448	.2680	.1388	.0453	.0095	.0012	.0001	.0000	.0000	.0000
	3	.0214	.0997	.2457	.2181	.1107	.0349	.0065	.0006	.0000	.0000	.0000
	4	.0028	.0277	.1535	.2337	.1845	.0873	.0243	.0034	.0001	.0000	.0000
	5	.0003	.0055	.0691	.1803	.2214	.1571	.0656	.0142	.0011	.0000	.0000
	6	.0000	.0008	.0230	.1030	.1968	.2095	.1312	.0442	.0058	.0001	.0000
	7	.0000	.0001	.0058	.0442	.1312	.2095	.1968	.1030	.0230	.0008	.0000
	8	.0000	.0000	.0011	.0142	.0656	.1571	.2214	.1803	.0691	.0055	.0003
	9	.0000	.0000	.0001	.0034	.0243	.0873	.1845	.2337	.1535	.0277	.0028
	10	.0000	.0000	.0000	.0006	.0065	.0349	.1107	.2181	.2457	.0997	.0214
	11	.0000	.0000	.0000	.0001	.0012	.0095	.0453	.1388	.2680	.2448	.1109
	12	.0000	.0000	.0000	.0000	.0001	.0016	.0113	.0540	.1787	.3672	.3512
	13	.0000	.0000	.0000	.0000	.0000	.0001	.0013	.0097	.0550	.2542	.5133
14	0	.4877	.2288	.0440	.0068	.0008	.0001	.0000	.0000	.0000	.0000	.0000
	1	.3593	.3559	.1539	.0407	.0073	.0009	.0001	.0000	.0000	.0000	.0000
	2	.1229	.2570	.2501	.1134	.0317	.0056	.0005	.0000	.0000	.0000	.0000
	3	.0259	.1142	.2501	.1943	.0845	.0222	.0033	.0002	.0000	.0000	.0000
	4	.0037	.0349	.1720	.2290	.1549	.0611	.0136	.0014	.0000	.0000	.0000
	5	.0004	.0078	.0860	.1963	.2066	.1222	.0408	.0066	.0003	.0000	.0000
	6	.0000	.0013	.0322	.1262	.2066	.1833	.0918	.0232	.0020	.0000	.0000
	7	.0000	.0002	.0092	.0618	.1574	.2095	.1574	.0618	.0092	.0002	.0000
	8	.0000	.0000	.0020	.0232	.0918	.1833	.2066	.1262	.0322	.0013	.0000
	9	.0000	.0000	.0003	.0066	.0408	.1222	.2066	.1963	.0860	.0078	.0004
	10	.0000	.0000	.0000	.0014	.0136	.0611	.1549	.2290	.1720	.0349	.0037
	11	.0000	.0000	.0000	.0002	.0033	.0222	.0845	.1943	.2501	.1142	.0259
	12	.0000	.0000	.0000	.0000	.0005	.0056	.0317	.1134	.2501	.2570	.1229
	13	.0000	.0000	.0000	.0000	.0001	.0009	.0073	.0407	.1539	.3559	.3593
	14	.0000	.0000	.0000	.0000	.0000	.0001	.0008	.0068	.0440	.2288	.4877
15	0	.4633	.2059	.0352	.0047	.0005	.0000	.0000	.0000	.0000	.0000	.0000
	1	.3658	.3432	.1319	.0305	.0047	.0005	.0000	.0000	.0000	.0000	.0000
	2	.1348	.2669	.2309	.0916	.0219	.0032	.0003	.0000	.0000	.0000	.0000
	3	.0307	.1285	.2501	.1700	.0634	.0139	.0016	.0001	.0000	.0000	.0000
	4	.0049	.0428	.1876	.2186	.1268	.0417	.0074	.0006	.0000	.0000	.0000
	5	.0006	.0105	.1032	.2061	.1859	.0916	.0245	.0030	.0001	.0000	.0000
	6	.0000	.0019	.0430	.1472	.2066	.1527	.0612	.0116	.0007	.0000	.0000
	7	.0000	.0003	.0138	.0811	.1771	.1964	.1181	.0348	.0035	.0000	.0000
	8	.0000	.0000	.0035	.0348	.1181	.1964	.1771	.0811	.0138	.0003	.0000
	9	.0000	.0000	.0007	.0116	.0612	.1527	.2066	.1472	.0430	.0019	.0000
	10	.0000	.0000	.0001	.0030	.0245	.0916	.1859	.2061	.1032	.0105	.0006
	11	.0000	.0000	.0000	.0006	.0074	.0417	.1268	.2186	.1876	.0428	.0049
	12	.0000	.0000	.0000	.0001	.0016	.0139	.0634	.1700	.2501	.1285	.0307
	13	.0000	.0000	.0000	.0000	.0003	.0032	.0219	.0916	.2309	.2669	.1348
	14	.0000	.0000	.0000	.0000	.0000	.0005	.0047	.0305	.1319	.3432	.3658
	15	.0000	.0000	.0000	.0000	.0000	.0000	.0005	.0047	.0352	.2059	.4633

TABLE IV   TABLE OF BINOMIAL PROBABILITIES   745

TABLE IV   (*Continued*)

						p						
n	x	**.05**	**.10**	**.20**	**.30**	**.40**	**.50**	**.60**	**.70**	**.80**	**.90**	**.95**
16	0	.4401	.1853	.0281	.0033	.0003	.0000	.0000	.0000	.0000	.0000	.0000
	1	.3706	.3294	.1126	.0228	.0030	.0002	.0000	.0000	.0000	.0000	.0000
	2	.1463	.2745	.2111	.0732	.0150	.0018	.0001	.0000	.0000	.0000	.0000
	3	.0359	.1423	.2463	.1465	.0468	.0085	.0008	.0000	.0000	.0000	.0000
	4	.0061	.0514	.2001	.2040	.1014	.0278	.0040	.0002	.0000	.0000	.0000
	5	.0008	.0137	.1201	.2099	.1623	.0667	.0142	.0013	.0000	.0000	.0000
	6	.0001	.0028	.0550	.1649	.1983	.1222	.0392	.0056	.0002	.0000	.0000
	7	.0000	.0004	.0197	.1010	.1889	.1746	.0840	.0185	.0012	.0000	.0000
	8	.0000	.0001	.0055	.0487	.1417	.1964	.1417	.0487	.0055	.0001	.0000
	9	.0000	.0000	.0012	.0185	.0840	.1746	.1889	.1010	.0197	.0004	.0000
	10	.0000	.0000	.0002	.0056	.0392	.1222	.1983	.1649	.0550	.0028	.0001
	11	.0000	.0000	.0000	.0013	.0142	.0666	.1623	.2099	.1201	.0137	.0008
	12	.0000	.0000	.0000	.0002	.0040	.0278	.1014	.2040	.2001	.0514	.0061
	13	.0000	.0000	.0000	.0000	.0008	.0085	.0468	.1465	.2463	.1423	.0359
	14	.0000	.0000	.0000	.0000	.0001	.0018	.0150	.0732	.2111	.2745	.1463
	15	.0000	.0000	.0000	.0000	.0000	.0002	.0030	.0228	.1126	.3294	.3706
	16	.0000	.0000	.0000	.0000	.0000	.0000	.0003	.0033	.0281	.1853	.4401
17	0	.4181	.1668	.0225	.0023	.0002	.0000	.0000	.0000	.0000	.0000	.0000
	1	.3741	.3150	.0957	.0169	.0019	.0001	.0000	.0000	.0000	.0000	.0000
	2	.1575	.2800	.1914	.0581	.0102	.0010	.0001	.0000	.0000	.0000	.0000
	3	.0415	.1556	.2393	.1245	.0341	.0052	.0004	.0000	.0000	.0000	.0000
	4	.0076	.0605	.2093	.1868	.0796	.0182	.0021	.0001	.0000	.0000	.0000
	5	.0010	.0175	.1361	.2081	.1379	.0472	.0081	.0006	.0000	.0000	.0000
	6	.0001	.0039	.0680	.1784	.1839	.0944	.0242	.0026	.0001	.0000	.0000
	7	.0000	.0007	.0267	.1201	.1927	.1484	.0571	.0095	.0004	.0000	.0000
	8	.0000	.0001	.0084	.0644	.1606	.1855	.1070	.0276	.0021	.0000	.0000
	9	.0000	.0000	.0021	.0276	.1070	.1855	.1606	.0644	.0084	.0001	.0000
	10	.0000	.0000	.0004	.0095	.0571	.1484	.1927	.1201	.0267	.0007	.0000
	11	.0000	.0000	.0001	.0026	.0242	.0944	.1839	.1784	.0680	.0039	.0001
	12	.0000	.0000	.0000	.0006	.0081	.0472	.1379	.2081	.1361	.0175	.0010
	13	.0000	.0000	.0000	.0001	.0021	.0182	.0796	.1868	.2093	.0605	.0076
	14	.0000	.0000	.0000	.0000	.0004	.0052	.0341	.1245	.2393	.1556	.0415
	15	.0000	.0000	.0000	.0000	.0001	.0010	.0102	.0581	.1914	.2800	.1575
	16	.0000	.0000	.0000	.0000	.0000	.0001	.0019	.0169	.0957	.3150	.3741
	17	.0000	.0000	.0000	.0000	.0000	.0000	.0002	.0023	.0225	.1668	.4181
18	0	.3972	.1501	.0180	.0016	.0001	.0000	.0000	.0000	.0000	.0000	.0000
	1	.3763	.3002	.0811	.0126	.0012	.0001	.0000	.0000	.0000	.0000	.0000
	2	.1683	.2835	.1723	.0458	.0069	.0006	.0000	.0000	.0000	.0000	.0000
	3	.0473	.1680	.2297	.1046	.0246	.0031	.0002	.0000	.0000	.0000	.0000
	4	.0093	.0700	.2153	.1681	.0614	.0117	.0011	.0000	.0000	.0000	.0000
	5	.0014	.0218	.1507	.2017	.1146	.0327	.0045	.0002	.0000	.0000	.0000
	6	.0002	.0052	.0816	.1873	.1655	.0708	.0145	.0012	.0000	.0000	.0000
	7	.0000	.0010	.0350	.1376	.1892	.1214	.0374	.0046	.0001	.0000	.0000
	8	.0000	.0002	.0120	.0811	.1734	.1669	.0771	.0149	.0008	.0000	.0000
	9	.0000	.0000	.0033	.0386	.1284	.1855	.1284	.0386	.0033	.0000	.0000
	10	.0000	.0000	.0008	.0149	.0771	.1669	.1734	.0811	.0120	.0002	.0000
	11	.0000	.0000	.0001	.0046	.0374	.1214	.1892	.1376	.0350	.0010	.0000
	12	.0000	.0000	.0000	.0012	.0145	.0708	.1655	.1873	.0816	.0052	.0002

TABLE IV   (*Continued*)

n	x	.05	.10	.20	.30	.40	.50	.60	.70	.80	.90	.95
	13	.0000	.0000	.0000	.0002	.0045	.0327	.1146	.2017	.1507	.0218	.0014
	14	.0000	.0000	.0000	.0000	.0011	.0117	.0614	.1681	.2153	.0700	.0093
	15	.0000	.0000	.0000	.0000	.0002	.0031	.0246	.1046	.2297	.1680	.0473
	16	.0000	.0000	.0000	.0000	.0000	.0006	.0069	.0458	.1723	.2835	.1683
	17	.0000	.0000	.0000	.0000	.0000	.0001	.0012	.0126	.0811	.3002	.3763
	18	.0000	.0000	.0000	.0000	.0000	.0000	.0001	.0016	.0180	.1501	.3972
19	0	.3774	.1351	.0144	.0011	.0001	.0000	.0000	.0000	.0000	.0000	.0000
	1	.3774	.2852	.0685	.0093	.0008	.0000	.0000	.0000	.0000	.0000	.0000
	2	.1787	.2852	.1540	.0358	.0046	.0003	.0000	.0000	.0000	.0000	.0000
	3	.0533	.1796	.2182	.0869	.0175	.0018	.0001	.0000	.0000	.0000	.0000
	4	.0112	.0798	.2182	.1491	.0467	.0074	.0005	.0000	.0000	.0000	.0000
	5	.0018	.0266	.1636	.1916	.0933	.0222	.0024	.0001	.0000	.0000	.0000
	6	.0002	.0069	.0955	.1916	.1451	.0518	.0085	.0005	.0000	.0000	.0000
	7	.0000	.0014	.0443	.1525	.1797	.0961	.0237	.0022	.0000	.0000	.0000
	8	.0000	.0002	.0166	.0981	.1797	.1442	.0532	.0077	.0003	.0000	.0000
	9	.0000	.0000	.0051	.0514	.1464	.1762	.0976	.0220	.0013	.0000	.0000
	10	.0000	.0000	.0013	.0220	.0976	.1762	.1464	.0514	.0051	.0000	.0000
	11	.0000	.0000	.0003	.0077	.0532	.1442	.1797	.0981	.0166	.0002	.0000
	12	.0000	.0000	.0000	.0022	.0237	.0961	.1797	.1525	.0443	.0014	.0000
	13	.0000	.0000	.0000	.0005	.0085	.0518	.1451	.1916	.0955	.0069	.0002
	14	.0000	.0000	.0000	.0001	.0024	.0222	.0933	.1916	.1636	.0266	.0018
	15	.0000	.0000	.0000	.0000	.0005	.0074	.0467	.1491	.2182	.0798	.0112
	16	.0000	.0000	.0000	.0000	.0001	.0018	.0175	.0869	.2182	.1796	.0533
	17	.0000	.0000	.0000	.0000	.0000	.0003	.0046	.0358	.1540	.2852	.1787
	18	.0000	.0000	.0000	.0000	.0000	.0000	.0008	.0093	.0685	.2852	.3774
	19	.0000	.0000	.0000	.0000	.0000	.0000	.0001	.0011	.0144	.1351	.3774
20	0	.3585	.1216	.0115	.0008	.0000	.0000	.0000	.0000	.0000	.0000	.0000
	1	.3774	.2702	.0576	.0068	.0005	.0000	.0000	.0000	.0000	.0000	.0000
	2	.1887	.2852	.1369	.0278	.0031	.0002	.0000	.0000	.0000	.0000	.0000
	3	.0596	.1901	.2054	.0716	.0123	.0011	.0000	.0000	.0000	.0000	.0000
	4	.0133	.0898	.2182	.1304	.0350	.0046	.0003	.0000	.0000	.0000	.0000
	5	.0022	.0319	.1746	.1789	.0746	.0148	.0013	.0000	.0000	.0000	.0000
	6	.0003	.0089	.1091	.1916	.1244	.0370	.0049	.0002	.0000	.0000	.0000
	7	.0000	.0020	.0545	.1643	.1659	.0739	.0146	.0010	.0000	.0000	.0000
	8	.0000	.0004	.0222	.1144	.1797	.1201	.0355	.0039	.0001	.0000	.0000
	9	.0000	.0001	.0074	.0654	.1597	.1602	.0710	.0120	.0005	.0000	.0000
	10	.0000	.0000	.0020	.0308	.1171	.1762	.1171	.0308	.0020	.0000	.0000
	11	.0000	.0000	.0005	.0120	.0710	.1602	.1597	.0654	.0074	.0001	.0000
	12	.0000	.0000	.0001	.0039	.0355	.1201	.1797	.1144	.0222	.0004	.0000
	13	.0000	.0000	.0000	.0010	.0146	.0739	.1659	.1643	.0545	.0020	.0000
	14	.0000	.0000	.0000	.0002	.0049	.0370	.1244	.1916	.1091	.0089	.0003
	15	.0000	.0000	.0000	.0000	.0013	.0148	.0746	.1789	.1746	.0319	.0022
	16	.0000	.0000	.0000	.0000	.0003	.0046	.0350	.1304	.2182	.0898	.0133
	17	.0000	.0000	.0000	.0000	.0000	.0011	.0123	.0716	.2054	.1901	.0596
	18	.0000	.0000	.0000	.0000	.0000	.0002	.0031	.0278	.1369	.2852	.1887
	19	.0000	.0000	.0000	.0000	.0000	.0000	.0005	.0068	.0576	.2702	.3774
	20	.0000	.0000	.0000	.0000	.0000	.0000	.0000	.0008	.0115	.1216	.3585

TABLE IV   TABLE OF BINOMIAL PROBABILITIES   **747**

TABLE IV   (*Continued*)

							p					
n	*x*	.05	.10	.20	.30	.40	.50	.60	.70	.80	.90	.95
21	0	.3406	.1094	.0092	.0006	.0000	.0000	.0000	.0000	.0000	.0000	.0000
	1	.3764	.2553	.0484	.0050	.0003	.0000	.0000	.0000	.0000	.0000	.0000
	2	.1981	.2837	.1211	.0215	.0020	.0001	.0000	.0000	.0000	.0000	.0000
	3	.0660	.1996	.1917	.0585	.0086	.0006	.0000	.0000	.0000	.0000	.0000
	4	.0156	.0998	.2156	.1128	.0259	.0029	.0001	.0000	.0000	.0000	.0000
	5	.0028	.0377	.1833	.1643	.0588	.0097	.0007	.0000	.0000	.0000	.0000
	6	.0004	.0112	.1222	.1878	.1045	.0259	.0027	.0001	.0000	.0000	.0000
	7	.0000	.0027	.0655	.1725	.1493	.0554	.0087	.0005	.0000	.0000	.0000
	8	.0000	.0005	.0286	.1294	.1742	.0970	.0229	.0019	.0000	.0000	.0000
	9	.0000	.0001	.0103	.0801	.1677	.1402	.0497	.0063	.0002	.0000	.0000
	10	.0000	.0000	.0031	.0412	.1342	.1682	.0895	.0176	.0008	.0000	.0000
	11	.0000	.0000	.0008	.0176	.0895	.1682	.1342	.0412	.0031	.0000	.0000
	12	.0000	.0000	.0002	.0063	.0497	.1402	.1677	.0801	.0103	.0001	.0000
	13	.0000	.0000	.0000	.0019	.0229	.0970	.1742	.1294	.0286	.0005	.0000
	14	.0000	.0000	.0000	.0005	.0087	.0554	.1493	.1725	.0655	.0027	.0000
	15	.0000	.0000	.0000	.0001	.0027	.0259	.1045	.1878	.1222	.0112	.0004
	16	.0000	.0000	.0000	.0000	.0007	.0097	.0588	.1643	.1833	.0377	.0028
	17	.0000	.0000	.0000	.0000	.0001	.0029	.0259	.1128	.2156	.0998	.0156
	18	.0000	.0000	.0000	.0000	.0000	.0006	.0086	.0585	.1917	.1996	.0660
	19	.0000	.0000	.0000	.0000	.0000	.0001	.0020	.0215	.1211	.2837	.1981
	20	.0000	.0000	.0000	.0000	.0000	.0000	.0003	.0050	.0484	.2553	.3764
	21	.0000	.0000	.0000	.0000	.0000	.0000	.0000	.0006	.0092	.1094	.3406
22	0	.3235	.0985	.0074	.0004	.0000	.0000	.0000	.0000	.0000	.0000	.0000
	1	.3746	.2407	.0406	.0037	.0002	.0000	.0000	.0000	.0000	.0000	.0000
	2	.2070	.2808	.1065	.0166	.0014	.0001	.0000	.0000	.0000	.0000	.0000
	3	.0726	.2080	.1775	.0474	.0060	.0004	.0000	.0000	.0000	.0000	.0000
	4	.0182	.1098	.2108	.0965	.0190	.0017	.0001	.0000	.0000	.0000	.0000
	5	.0034	.0439	.1898	.1489	.0456	.0063	.0004	.0000	.0000	.0000	.0000
	6	.0005	.0138	.1344	.1808	.0862	.0178	.0015	.0000	.0000	.0000	.0000
	7	.0001	.0035	.0768	.1771	.1314	.0407	.0051	.0002	.0000	.0000	.0000
	8	.0000	.0007	.0360	.1423	.1642	.0762	.0144	.0009	.0000	.0000	.0000
	9	.0000	.0001	.0140	.0949	.1703	.1186	.0336	.0032	.0001	.0000	.0000
	10	.0000	.0000	.0046	.0529	.1476	.1542	.0656	.0097	.0003	.0000	.0000
	11	.0000	.0000	.0012	.0247	.1073	.1682	.1073	.0247	.0012	.0000	.0000
	12	.0000	.0000	.0003	.0097	.0656	.1542	.1476	.0529	.0046	.0000	.0000
	13	.0000	.0000	.0001	.0032	.0336	.1186	.1703	.0949	.0140	.0001	.0000
	14	.0000	.0000	.0000	.0009	.0144	.0762	.1642	.1423	.0360	.0007	.0000
	15	.0000	.0000	.0000	.0002	.0051	.0407	.1314	.1771	.0768	.0035	.0001
	16	.0000	.0000	.0000	.0000	.0015	.0178	.0862	.1808	.1344	.0138	.0005
	17	.0000	.0000	.0000	.0000	.0004	.0063	.0456	.1489	.1898	.0439	.0034
	18	.0000	.0000	.0000	.0000	.0001	.0017	.0190	.0965	.2108	.1098	.0182
	19	.0000	.0000	.0000	.0000	.0000	.0004	.0060	.0474	.1775	.2080	.0726
	20	.0000	.0000	.0000	.0000	.0000	.0001	.0014	.0166	.1065	.2808	.2070
	21	.0000	.0000	.0000	.0000	.0000	.0000	.0002	.0037	.0406	.2407	.3746
	22	.0000	.0000	.0000	.0000	.0000	.0000	.0000	.0004	.0074	.0985	.3235
23	0	.3074	.0886	.0059	.0003	.0000	.0000	.0000	.0000	.0000	.0000	.0000
	1	.3721	.2265	.0339	.0027	.0001	.0000	.0000	.0000	.0000	.0000	.0000
	2	.2154	.2768	.0933	.0127	.0009	.0000	.0000	.0000	.0000	.0000	.0000
	3	.0794	.2153	.1633	.0382	.0041	.0002	.0000	.0000	.0000	.0000	.0000

TABLE IV (*Continued*)

n	x	.05	.10	.20	.30	.40	.50	.60	.70	.80	.90	.95
	4	.0209	.1196	.2042	.0818	.0138	.0011	.0000	.0000	.0000	.0000	.0000
	5	.0042	.0505	.1940	.1332	.0350	.0040	.0002	.0000	.0000	.0000	.0000
	6	.0007	.0168	.1455	.1712	.0700	.0120	.0008	.0000	.0000	.0000	.0000
	7	.0001	.0045	.0883	.1782	.1133	.0292	.0029	.0001	.0000	.0000	.0000
	8	.0000	.0010	.0442	.1527	.1511	.0584	.0088	.0004	.0000	.0000	.0000
	9	.0000	.0002	.0184	.1091	.1679	.0974	.0221	.0016	.0000	.0000	.0000
	10	.0000	.0000	.0064	.0655	.1567	.1364	.0464	.0052	.0001	.0000	.0000
	11	.0000	.0000	.0019	.0332	.1234	.1612	.0823	.0142	.0005	.0000	.0000
	12	.0000	.0000	.0005	.0142	.0823	.1612	.1234	.0332	.0019	.0000	.0000
	13	.0000	.0000	.0001	.0052	.0464	.1364	.1567	.0655	.0064	.0000	.0000
	14	.0000	.0000	.0000	.0016	.0221	.0974	.1679	.1091	.0184	.0002	.0000
	15	.0000	.0000	.0000	.0004	.0088	.0584	.1511	.1527	.0442	.0010	.0000
	16	.0000	.0000	.0000	.0001	.0029	.0292	.1133	.1782	.0883	.0045	.0001
	17	.0000	.0000	.0000	.0000	.0008	.0120	.0700	.1712	.1455	.0168	.0007
	18	.0000	.0000	.0000	.0000	.0002	.0040	.0350	.1332	.1940	.0505	.0042
	19	.0000	.0000	.0000	.0000	.0000	.0011	.0138	.0818	.2042	.1196	.0209
	20	.0000	.0000	.0000	.0000	.0000	.0002	.0041	.0382	.1633	.2153	.0794
	21	.0000	.0000	.0000	.0000	.0000	.0000	.0009	.0127	.0933	.2768	.2154
	22	.0000	.0000	.0000	.0000	.0000	.0000	.0001	.0027	.0339	.2265	.3721
	23	.0000	.0000	.0000	.0000	.0000	.0000	.0000	.0003	.0059	.0886	.3074
24	0	.2920	.0798	.0047	.0002	.0000	.0000	.0000	.0000	.0000	.0000	.0000
	1	.3688	.2127	.0283	.0020	.0001	.0000	.0000	.0000	.0000	.0000	.0000
	2	.2232	.2718	.0815	.0097	.0006	.0000	.0000	.0000	.0000	.0000	.0000
	3	.0862	.2215	.1493	.0305	.0028	.0001	.0000	.0000	.0000	.0000	.0000
	4	.0238	.1292	.1960	.0687	.0099	.0006	.0000	.0000	.0000	.0000	.0000
	5	.0050	.0574	.1960	.1177	.0265	.0025	.0001	.0000	.0000	.0000	.0000
	6	.0008	.0202	.1552	.1598	.0560	.0080	.0004	.0000	.0000	.0000	.0000
	7	.0001	.0058	.0998	.1761	.0960	.0206	.0017	.0000	.0000	.0000	.0000
	8	.0000	.0014	.0530	.1604	.1360	.0438	.0053	.0002	.0000	.0000	.0000
	9	.0000	.0003	.0236	.1222	.1612	.0779	.0141	.0008	.0000	.0000	.0000
	10	.0000	.0000	.0088	.0785	.1612	.1169	.0318	.0026	.0000	.0000	.0000
	11	.0000	.0000	.0028	.0428	.1367	.1488	.0608	.0079	.0002	.0000	.0000
	12	.0000	.0000	.0008	.0199	.0988	.1612	.0988	.0199	.0008	.0000	.0000
	13	.0000	.0000	.0002	.0079	.0608	.1488	.1367	.0428	.0028	.0000	.0000
	14	.0000	.0000	.0000	.0026	.0318	.1169	.1612	.0785	.0088	.0000	.0000
	15	.0000	.0000	.0000	.0008	.0141	.0779	.1612	.1222	.0236	.0003	.0000
	16	.0000	.0000	.0000	.0002	.0053	.0438	.1360	.1604	.0530	.0014	.0000
	17	.0000	.0000	.0000	.0000	.0017	.0206	.0960	.1761	.0998	.0058	.0001
	18	.0000	.0000	.0000	.0000	.0004	.0080	.0560	.1598	.1552	.0202	.0008
	19	.0000	.0000	.0000	.0000	.0001	.0025	.0265	.1177	.1960	.0574	.0050
	20	.0000	.0000	.0000	.0000	.0000	.0006	.0099	.0687	.1960	.1292	.0238
	21	.0000	.0000	.0000	.0000	.0000	.0001	.0028	.0305	.1493	.2215	.0862
	22	.0000	.0000	.0000	.0000	.0000	.0000	.0006	.0097	.0815	.2718	.2232
	23	.0000	.0000	.0000	.0000	.0000	.0000	.0001	.0020	.0283	.2127	.3688
	24	.0000	.0000	.0000	.0000	.0000	.0000	.0000	.0002	.0047	.0798	.2920
25	0	.2774	.0718	.0038	.0001	.0000	.0000	.0000	.0000	.0000	.0000	.0000
	1	.3650	.1994	.0236	.0014	.0000	.0000	.0000	.0000	.0000	.0000	.0000
	2	.2305	.2659	.0708	.0074	.0004	.0000	.0000	.0000	.0000	.0000	.0000
	3	.0930	.2265	.1358	.0243	.0019	.0001	.0000	.0000	.0000	.0000	.0000

TABLE IV   TABLE OF BINOMIAL PROBABILITIES

749

TABLE IV   (*Continued*)

							p					
n	x	.05	.10	.20	.30	.40	.50	.60	.70	.80	.90	.95
	4	.0269	.1384	.1867	.0572	.0071	.0004	.0000	.0000	.0000	.0000	.0000
	5	.0060	.0646	.1960	.1030	.0199	.0016	.0000	.0000	.0000	.0000	.0000
	6	.0010	.0239	.1633	.1472	.0442	.0053	.0002	.0000	.0000	.0000	.0000
	7	.0001	.0072	.1108	.1712	.0800	.0143	.0009	.0000	.0000	.0000	.0000
	8	.0000	.0018	.0623	.1651	.1200	.0322	.0031	.0001	.0000	.0000	.0000
	9	.0000	.0004	.0294	.1336	.1511	.0609	.0088	.0004	.0000	.0000	.0000
	10	.0000	.0001	.0118	.0916	.1612	.0974	.0212	.0013	.0000	.0000	.0000
	11	.0000	.0000	.0040	.0536	.1465	.1328	.0434	.0042	.0001	.0000	.0000
	12	.0000	.0000	.0012	.0268	.1140	.1550	.0760	.0115	.0003	.0000	.0000
	13	.0000	.0000	.0003	.0115	.0760	.1550	.1140	.0268	.0012	.0000	.0000
	14	.0000	.0000	.0001	.0042	.0434	.1328	.1465	.0536	.0040	.0000	.0000
	15	.0000	.0000	.0000	.0013	.0212	.0974	.1612	.0916	.0118	.0001	.0000
	16	.0000	.0000	.0000	.0004	.0088	.0609	.1511	.1336	.0294	.0004	.0000
	17	.0000	.0000	.0000	.0001	.0031	.0322	.1200	.1651	.0623	.0018	.0000
	18	.0000	.0000	.0000	.0000	.0009	.0143	.0800	.1712	.1108	.0072	.0001
	19	.0000	.0000	.0000	.0000	.0002	.0053	.0442	.1472	.1633	.0239	.0010
	20	.0000	.0000	.0000	.0000	.0000	.0016	.0199	.1030	.1960	.0646	.0060
	21	.0000	.0000	.0000	.0000	.0000	.0004	.0071	.0572	.1867	.1384	.0269
	22	.0000	.0000	.0000	.0000	.0000	.0001	.0019	.0243	.1358	.2265	.0930
	23	.0000	.0000	.0000	.0000	.0000	.0000	.0004	.0074	.0708	.2659	.2305
	24	.0000	.0000	.0000	.0000	.0000	.0000	.0000	.0014	.0236	.1994	.3650
	25	.0000	.0000	.0000	.0000	.0000	.0000	.0000	.0001	.0038	.0718	.2774

TABLE V   VALUES OF $e^{-\lambda}$

λ	$e^{-\lambda}$	λ	$e^{-\lambda}$
0.0	1.00000000	5.5	.00408677
0.1	.90483742	5.6	.00369786
0.2	.81873075	5.7	.00334597
0.3	.74081822	5.8	.00302755
0.4	.67032005	5.9	.00273944
0.5	.60653066	6.0	.00247875
0.6	.54881164	6.1	.00224287
0.7	.49658530	6.2	.00202943
0.8	.44932896	6.3	.00183630
0.9	.40656966	6.4	.00166156
1.0	.36787944	6.5	.00150344
1.1	.33287108	6.6	.00136037
1.2	.30119421	6.7	.00123091
1.3	.27253179	6.8	.00111378
1.4	.24659696	6.9	.00100779
1.5	.22313016	7.0	.00091188
1.6	.20189652	7.1	.00082510
1.7	.18268352	7.2	.00074659
1.8	.16529889	7.3	.00067554
1.9	.14956862	7.4	.00061125
2.0	.13533528	7.5	.00055308
2.1	.12245643	7.6	.00050045
2.2	.11080316	7.7	.00045283
2.3	.10025884	7.8	.00040973
2.4	.09071795	7.9	.00037074
2.5	.08208500	8.0	.00033546
2.6	.07427358	8.1	.00030354
2.7	.06720551	8.2	.00027465
2.8	.06081006	8.3	.00024852
2.9	.05502322	8.4	.00022487
3.0	.04978707	8.5	.00020347
3.1	.04504920	8.6	.00018411
3.2	.04076220	8.7	.00016659
3.3	.03688317	8.8	.00015073
3.4	.03337327	8.9	.00013639
3.5	.03019738	9.0	.00012341
3.6	.02732372	9.1	.00011167
3.7	.02472353	9.2	.00010104
3.8	.02237077	9.3	.00009142
3.9	.02024191	9.4	.00008272
4.0	.01831564	9.5	.00007485
4.1	.01657268	9.6	.00006773
4.2	.01499558	9.7	.00006128
4.3	.01356856	9.8	.00005545
4.4	.01227734	9.9	.00005017
4.5	.01110900	10.0	.00004540
4.6	.01005184	11.0	.00001670
4.7	.00909528	12.0	.00000614
4.8	.00822975	13.0	.00000226
4.9	.00744658	14.0	.00000083
5.0	.00673795	15.0	.00000031
5.1	.00609675	16.0	.00000011
5.2	.00551656	17.0	.00000004
5.3	.00499159	18.0	.000000015
5.4	.00451658	19.0	.000000006
		20.0	.000000002

TABLE VI   TABLE OF POISSON PROBABILITIES

751

TABLE VI   TABLE OF POISSON PROBABILITIES

x	λ 0.1	0.2	0.3	0.4	0.5	0.6	0.7	0.8	0.9	1.0
0	.9048	.8187	.7408	.6703	.6065	.5488	.4966	.4493	.4066	.3679
1	.0905	.1637	.2222	.2681	.3033	.3293	.3476	.3595	.3659	.3679
2	.0045	.0164	.0333	.0536	.0758	.0988	.1217	.1438	.1647	.1839
3	.0002	.0011	.0033	.0072	.0126	.0198	.0284	.0383	.0494	.0613
4	.0000	.0001	.0003	.0007	.0016	.0030	.0050	.0077	.0111	.0153
5	.0000	.0000	.0000	.0001	.0002	.0004	.0007	.0012	.0020	.0031
6	.0000	.0000	.0000	.0000	.0000	.0000	.0001	.0002	.0003	.0005
7	.0000	.0000	.0000	.0000	.0000	.0000	.0000	.0000	.0000	.0001

x	λ 1.1	1.2	1.3	1.4	1.5	1.6	1.7	1.8	1.9	2.0
0	.3329	.3012	.2725	.2466	.2231	.2019	.1827	.1653	.1496	.1353
1	.3662	.3614	.3543	.3452	.3347	.3230	.3106	.2975	.2842	.2707
2	.2014	.2169	.2303	.2417	.2510	.2584	.2640	.2678	.2700	.2707
3	.0738	.0867	.0998	.1128	.1255	.1378	.1496	.1607	.1710	.1804
4	.0203	.0260	.0324	.0395	.0471	.0551	.0636	.0723	.0812	.0902
5	.0045	.0062	.0084	.0111	.0141	.0176	.0216	.0260	.0309	.0361
6	.0008	.0012	.0018	.0026	.0035	.0047	.0061	.0078	.0098	.0120
7	.0001	.0002	.0003	.0005	.0008	.0011	.0015	.0020	.0027	.0034
8	.0000	.0000	.0001	.0001	.0001	.0002	.0003	.0005	.0006	.0009
9	.0000	.0000	.0000	.0000	.0000	.0000	.0001	.0001	.0001	.0002

x	λ 2.1	2.2	2.3	2.4	2.5	2.6	2.7	2.8	2.9	3.0
0	.1225	.1108	.1003	.0907	.0821	.0743	.0672	.0608	.0550	.0498
1	.2572	.2438	.2306	.2177	.2052	.1931	.1815	.1703	.1596	.1494
2	.2700	.2681	.2652	.2613	.2565	.2510	.2450	.2384	.2314	.2240
3	.1890	.1966	.2033	.2090	.2138	.2176	.2205	.2225	.2237	.2240
4	.0992	.1082	.1169	.1254	.1336	.1414	.1488	.1557	.1622	.1680
5	.0417	.0476	.0538	.0602	.0668	.0735	.0804	.0872	.0940	.1008
6	.0146	.0174	.0206	.0241	.0278	.0319	.0362	.0407	.0455	.0504
7	.0044	.0055	.0068	.0083	.0099	.0118	.0139	.0163	.0188	.0216
8	.0011	.0015	.0019	.0025	.0031	.0038	.0047	.0057	.0068	.0081
9	.0003	.0004	.0005	.0007	.0009	.0011	.0014	.0018	.0022	.0027
10	.0001	.0001	.0001	.0002	.0002	.0003	.0004	.0005	.0006	.0008
11	.0000	.0000	.0000	.0000	.0000	.0001	.0001	.0001	.0002	.0002
12	.0000	.0000	.0000	.0000	.0000	.0000	.0000	.0000	.0000	.0001

x	λ 3.1	3.2	3.3	3.4	3.5	3.6	3.7	3.8	3.9	4.0
0	.0450	.0408	.0369	.0334	.0302	.0273	.0247	.0224	.0202	.0183
1	.1397	.1304	.1217	.1135	.1057	.0984	.0915	.0850	.0789	.0733
2	.2165	.2087	.2008	.1929	.1850	.1771	.1692	.1615	.1539	.1465

TABLE VI   (*Continued*)

					λ					
x	3.1	3.2	3.3	3.4	3.5	3.6	3.7	3.8	3.9	4.0
3	.2237	.2226	.2209	.2186	.2158	.2125	.2087	.2046	.2001	.1954
4	.1733	.1781	.1823	.1858	.1888	.1912	.1931	.1944	.1951	.1954
5	.1075	.1140	.1203	.1264	.1322	.1377	.1429	.1477	.1522	.1563
6	.0555	.0608	.0662	.0716	.0771	.0826	.0881	.0936	.0989	.1042
7	.0246	.0278	.0312	.0348	.0385	.0425	.0466	.0508	.0551	.0595
8	.0095	.0111	.0129	.0148	.0169	.0191	.0215	.0241	.0269	.0298
9	.0033	.0040	.0047	.0056	.0066	.0076	.0089	.0102	.0116	.0132
10	.0010	.0013	.0016	.0019	.0023	.0028	.0033	.0039	.0045	.0053
11	.0003	.0004	.0005	.0006	.0007	.0009	.0011	.0013	.0016	.0019
12	.0001	.0001	.0001	.0002	.0002	.0003	.0003	.0004	.0005	.0006
13	.0000	.0000	.0000	.0000	.0001	.0001	.0001	.0001	.0002	.0002
14	.0000	.0000	.0000	.0000	.0000	.0000	.0000	.0000	.0000	.0001

					λ					
x	4.1	4.2	4.3	4.4	4.5	4.6	4.7	4.8	4.9	5.0
0	.0166	.0150	.0136	.0123	.0111	.0101	.0091	.0082	.0074	.0067
1	.0679	.0630	.0583	.0540	.0500	.0462	.0427	.0395	.0365	.0337
2	.1393	.1323	.1254	.1188	.1125	.1063	.1005	.0948	.0894	.0842
3	.1904	.1852	.1798	.1743	.1687	.1631	.1574	.1517	.1460	.1404
4	.1951	.1944	.1933	.1917	.1898	.1875	.1849	.1820	.1789	.1755
5	.1600	.1633	.1662	.1687	.1708	.1725	.1738	.1747	.1753	.1755
6	.1093	.1143	.1191	.1237	.1281	.1323	.1362	.1398	.1432	.1462
7	.0640	.0686	.0732	.0778	.0824	.0869	.0914	.0959	.1002	.1044
8	.0328	.0360	.0393	.0428	.0463	.0500	.0537	.0575	.0614	.0653
9	.0150	.0168	.0188	.0209	.0232	.0255	.0281	.0307	.0334	.0363
10	.0061	.0071	.0081	.0092	.0104	.0118	.0132	.0147	.0164	.0181
11	.0023	.0027	.0032	.0037	.0043	.0049	.0056	.0064	.0073	.0082
12	.0008	.0009	.0011	.0014	.0016	.0019	.0022	.0026	.0030	.0034
13	.0002	.0003	.0004	.0005	.0006	.0007	.0008	.0009	.0011	.0013
14	.0001	.0001	.0001	.0001	.0002	.0002	.0003	.0003	.0004	.0005
15	.0000	.0000	.0000	.0000	.0001	.0001	.0001	.0001	.0001	.0002

					λ					
x	5.1	5.2	5.3	5.4	5.5	5.6	5.7	5.8	5.9	6.0
0	.0061	.0055	.0050	.0045	.0041	.0037	.0033	.0030	.0027	.0025
1	.0311	.0287	.0265	.0244	.0225	.0207	.0191	.0176	.0162	.0149
2	.0793	.0746	.0701	.0659	.0618	.0580	.0544	.0509	.0477	.0446
3	.1348	.1293	.1239	.1185	.1133	.1082	.1033	.0985	.0938	.0892
4	.1719	.1681	.1641	.1600	.1558	.1515	.1472	.1428	.1383	.1339
5	.1753	.1748	.1740	.1728	.1714	.1697	.1678	.1656	.1632	.1606
6	.1490	.1515	.1537	.1555	.1571	.1584	.1594	.1601	.1605	.1606
7	.1086	.1125	.1163	.1200	.1234	.1267	.1298	.1326	.1353	.1377

TABLE VI   TABLE OF POISSON PROBABILITIES   **753**

TABLE VI   (*Continued*)

					λ					
x	5.1	5.2	5.3	5.4	5.5	5.6	5.7	5.8	5.9	6.0
8	.0692	.0731	.0771	.0810	.0849	.0887	.0925	.0962	.0998	.1033
9	.0392	.0423	.0454	.0486	.0519	.0552	.0586	.0620	.0654	.0688
10	.0200	.0220	.0241	.0262	.0285	.0309	.0334	.0359	.0386	.0413
11	.0093	.0104	.0116	.0129	.0143	.0157	.0173	.0190	.0207	.0225
12	.0039	.0045	.0051	.0058	.0065	.0073	.0082	.0092	.0102	.0113
13	.0015	.0018	.0021	.0024	.0028	.0032	.0036	.0041	.0046	.0052
14	.0006	.0007	.0008	.0009	.0011	.0013	.0015	.0017	.0019	.0022
15	.0002	.0002	.0003	.0003	.0004	.0005	.0006	.0007	.0008	.0009
16	.0001	.0001	.0001	.0001	.0001	.0002	.0002	.0002	.0003	.0003
17	.0000	.0000	.0000	.0000	.0000	.0001	.0001	.0001	.0001	.0001

					λ					
x	6.1	6.2	6.3	6.4	6.5	6.6	6.7	6.8	6.9	7.0
0	.0022	.0020	.0018	.0017	.0015	.0014	.0012	.0011	.0010	.0009
1	.0137	.0126	.0116	.0106	.0098	.0090	.0082	.0076	.0070	.0064
2	.0417	.0390	.0364	.0340	.0318	.0296	.0276	.0258	.0240	.0223
3	.0848	.0806	.0765	.0726	.0688	.0652	.0617	.0584	.0552	.0521
4	.1294	.1249	.1205	.1162	.1118	.1076	.1034	.0992	.0952	.0912
5	.1579	.1549	.1519	.1487	.1454	.1420	.1385	.1349	.1314	.1277
6	.1605	.1601	.1595	.1586	.1575	.1562	.1546	.1529	.1511	.1490
7	.1399	.1418	.1435	.1450	.1462	.1472	.1480	.1486	.1489	.1490
8	.1066	.1099	.1130	.1160	.1188	.1215	.1240	.1263	.1284	.1304
9	.0723	.0757	.0791	.0825	.0858	.0891	.0923	.0954	.0985	.1014
10	.0441	.0469	.0498	.0528	.0558	.0588	.0618	.0649	.0679	.0710
11	.0244	.0265	.0285	.0307	.0330	.0353	.0377	.0401	.0426	.0452
12	.0124	.0137	.0150	.0164	.0179	.0194	.0210	.0227	.0245	.0263
13	.0058	.0065	.0073	.0081	.0089	.0099	.0108	.0119	.0130	.0142
14	.0025	.0029	.0033	.0037	.0041	.0046	.0052	.0058	.0064	.0071
15	.0010	.0012	.0014	.0016	.0018	.0020	.0023	.0026	.0029	.0033
16	.0004	.0005	.0005	.0006	.0007	.0008	.0010	.0011	.0013	.0014
17	.0001	.0002	.0002	.0002	.0003	.0003	.0004	.0004	.0005	.0006
18	.0000	.0001	.0001	.0001	.0001	.0001	.0001	.0002	.0002	.0002
19	.0000	.0000	.0000	.0000	.0000	.0000	.0001	.0001	.0001	.0001

					λ					
x	7.1	7.2	7.3	7.4	7.5	7.6	7.7	7.8	7.9	8.0
0	.0008	.0007	.0007	.0006	.0006	.0005	.0005	.0004	.0004	.0003
1	.0059	.0054	.0049	.0045	.0041	.0038	.0035	.0032	.0029	.0027
2	.0208	.0194	.0180	.0167	.0156	.0145	.0134	.0125	.0116	.0107
3	.0492	.0464	.0438	.0413	.0389	.0366	.0345	.0324	.0305	.0286
4	.0874	.0836	.0799	.0764	.0729	.0696	.0663	.0632	.0602	.0573
5	.1241	.1204	.1167	.1130	.1094	.1057	.1021	.0986	.0951	.0916
6	.1468	.1445	.1420	.1394	.1367	.1339	.1311	.1282	.1252	.1221

TABLE VI   (*Continued*)

					λ					
x	7.1	7.2	7.3	7.4	7.5	7.6	7.7	7.8	7.9	8.0
7	.1489	.1486	.1481	.1474	.1465	.1454	.1442	.1428	.1413	.1396
8	.1321	.1337	.1351	.1363	.1373	.1381	.1388	.1392	.1395	.1396
9	.1042	.1070	.1096	.1121	.1144	.1167	.1187	.1207	.1224	.1241
10	.0740	.0770	.0800	.0829	.0858	.0887	.0914	.0941	.0967	.0993
11	.0478	.0504	.0531	.0558	.0585	.0613	.0640	.0667	.0695	.0722
12	.0283	.0303	.0323	.0344	.0366	.0388	.0411	.0434	.0457	.0481
13	.0154	.0168	.0181	.0196	.0211	.0227	.0243	.0260	.0278	.0296
14	.0078	.0086	.0095	.0104	.0113	.0123	.0134	.0145	.0157	.0169
15	.0037	.0041	.0046	.0051	.0057	.0062	.0069	.0075	.0083	.0090
16	.0016	.0019	.0021	.0024	.0026	.0030	.0033	.0037	.0041	.0045
17	.0007	.0008	.0009	.0010	.0012	.0013	.0015	.0017	.0019	.0021
18	.0003	.0003	.0004	.0004	.0005	.0006	.0006	.0007	.0008	.0009
19	.0001	.0001	.0001	.0002	.0002	.0002	.0003	.0003	.0003	.0004
20	.0000	.0000	.0001	.0001	.0001	.0001	.0001	.0001	.0001	.0002
21	.0000	.0000	.0000	.0000	.0000	.0000	.0000	.0000	.0001	.0001

					λ					
x	8.1	8.2	8.3	8.4	8.5	8.6	8.7	8.8	8.9	9.0
0	.0003	.0003	.0002	.0002	.0002	.0002	.0002	.0002	.0001	.0001
1	.0025	.0023	.0021	.0019	.0017	.0016	.0014	.0013	.0012	.0011
2	.0100	.0092	.0086	.0079	.0074	.0068	.0063	.0058	.0054	.0050
3	.0269	.0252	.0237	.0222	.0208	.0195	.0183	.0171	.0160	.0150
4	.0544	.0517	.0491	.0466	.0443	.0420	.0398	.0377	.0357	.0337
5	.0882	.0849	.0816	.0784	.0752	.0722	.0692	.0663	.0635	.0607
6	.1191	.1160	.1128	.1097	.1066	.1034	.1003	.0972	.0941	.0911
7	.1378	.1358	.1338	.1317	.1294	.1271	.1247	.1222	.1197	.1171
8	.1395	.1392	.1388	.1382	.1375	.1366	.1356	.1344	.1332	.1318
9	.1255	.1269	.1280	.1290	.1299	.1306	.1311	.1315	.1317	.1318
10	.1017	.1040	.1063	.1084	.1104	.1123	.1140	.1157	.1172	.1186
11	.0749	.0775	.0802	.0828	.0853	.0878	.0902	.0925	.0948	.0970
12	.0505	.0530	.0555	.0579	.0604	.0629	.0654	.0679	.0703	.0728
13	.0315	.0334	.0354	.0374	.0395	.0416	.0438	.0459	.0481	.0504
14	.0182	.0196	.0210	.0225	.0240	.0256	.0272	.0289	.0306	.0324
15	.0098	.0107	.0116	.0126	.0136	.0147	.0158	.0169	.0182	.0194
16	.0050	.0055	.0060	.0066	.0072	.0079	.0086	.0093	.0101	.0109
17	.0024	.0026	.0029	.0033	.0036	.0040	.0044	.0048	.0053	.0058
18	.0011	.0012	.0014	.0015	.0017	.0019	.0021	.0024	.0026	.0029
19	.0005	.0005	.0006	.0007	.0008	.0009	.0010	.0011	.0012	.0014
20	.0002	.0002	.0002	.0003	.0003	.0004	.0004	.0005	.0005	.0006
21	.0001	.0001	.0001	.0001	.0001	.0002	.0002	.0002	.0002	.0003
22	.0000	.0000	.0000	.0000	.0001	.0001	.0001	.0001	.0001	.0001

TABLE VI   TABLE OF POISSON PROBABILITIES   **755**

TABLE VI   (*Continued*)

x	9.1	9.2	9.3	9.4	λ 9.5	9.6	9.7	9.8	9.9	10
0	.0001	.0001	.0001	.0001	.0001	.0001	.0001	.0001	.0001	.0000
1	.0010	.0009	.0009	.0008	.0007	.0007	.0006	.0005	.0005	.0005
2	.0046	.0043	.0040	.0037	.0034	.0031	.0029	.0027	.0025	.0023
3	.0140	.0131	.0123	.0115	.0107	.0100	.0093	.0087	.0081	.0076
4	.0319	.0302	.0285	.0269	.0254	.0240	.0226	.0213	.0201	.0189
5	.0581	.0555	.0530	.0506	.0483	.0460	.0439	.0418	.0398	.0378
6	.0881	.0851	.0822	.0793	.0764	.0736	.0709	.0682	.0656	.0631
7	.1145	.1118	.1091	.1064	.1037	.1010	.0982	.0955	.0928	.0901
8	.1302	.1286	.1269	.1251	.1232	.1212	.1191	.1170	.1148	.1126
9	.1317	.1315	.1311	.1306	.1300	.1293	.1284	.1274	.1263	.1251
10	.1198	.1209	.1219	.1228	.1235	.1241	.1245	.1249	.1250	.1251
11	.0991	.1012	.1031	.1049	.1067	.1083	.1098	.1112	.1125	.1137
12	.0752	.0776	.0799	.0822	.0844	.0866	.0888	.0908	.0928	.0948
13	.0526	.0549	.0572	.0594	.0617	.0640	.0662	.0685	.0707	.0729
14	.0342	.0361	.0380	.0399	.0419	.0439	.0459	.0479	.0500	.0521
15	.0208	.0221	.0235	.0250	.0265	.0281	.0297	.0313	.0330	.0347
16	.0118	.0127	.0137	.0147	.0157	.0168	.0180	.0192	.0204	.0217
17	.0063	.0069	.0075	.0081	.0088	.0095	.0103	.0111	.0119	.0128
18	.0032	.0035	.0039	.0042	.0046	.0051	.0055	.0060	.0065	.0071
19	.0015	.0017	.0019	.0021	.0023	.0026	.0028	.0031	.0034	.0037
20	.0007	.0008	.0009	.0010	.0011	.0012	.0014	.0015	.0017	.0019
21	.0003	.0003	.0004	.0004	.0005	.0006	.0006	.0007	.0008	.0009
22	.0001	.0001	.0002	.0002	.0002	.0002	.0003	.0003	.0004	.0004
23	.0000	.0001	.0001	.0001	.0001	.0001	.0001	.0001	.0002	.0002
24	.0000	.0000	.0000	.0000	.0000	.0000	.0000	.0001	.0001	.0001

x	11	12	13	14	λ 15	16	17	18	19	20
0	.0000	.0000	.0000	.0000	.0000	.0000	.0000	.0000	.0000	.0000
1	.0002	.0001	.0000	.0000	.0000	.0000	.0000	.0000	.0000	.0000
2	.0010	.0004	.0002	.0001	.0000	.0000	.0000	.0000	.0000	.0000
3	.0037	.0018	.0008	.0004	.0002	.0001	.0000	.0000	.0000	.0000
4	.0102	.0053	.0027	.0013	.0006	.0003	.0001	.0001	.0000	.0000
5	.0224	.0127	.0070	.0037	.0019	.0010	.0005	.0002	.0001	.0001
6	.0411	.0255	.0152	.0087	.0048	.0026	.0014	.0007	.0004	.0002
7	.0646	.0437	.0281	.0174	.0104	.0060	.0034	.0019	.0010	.0005
8	.0888	.0655	.0457	.0304	.0194	.0120	.0072	.0042	.0024	.0013
9	.1085	.0874	.0661	.0473	.0324	.0213	.0135	.0083	.0050	.0029
10	.1194	.1048	.0859	.0663	.0486	.0341	.0230	.0150	.0095	.0058
11	.1194	.1144	.1015	.0844	.0663	.0496	.0355	.0245	.0164	.0106
12	.1094	.1144	.1099	.0984	.0829	.0661	.0504	.0368	.0259	.0176
13	.0926	.1056	.1099	.1060	.0956	.0814	.0658	.0509	.0378	.0271
14	.0728	.0905	.1021	.1060	.1024	.0930	.0800	.0655	.0514	.0387

TABLE VI   (*Continued*)

x	11	12	13	14	λ 15	16	17	18	19	20
15	.0534	.0724	.0885	.0989	.1024	.0992	.0906	.0786	.0650	.0516
16	.0367	.0543	.0719	.0866	.0960	.0992	.0963	.0884	.0772	.0646
17	.0237	.0383	.0550	.0713	.0847	.0934	.0963	.0936	.0863	.0760
18	.0145	.0255	.0397	.0554	.0706	.0830	.0909	.0936	.0911	.0844
19	.0084	.0161	.0272	.0409	.0557	.0699	.0814	.0887	.0911	.0888
20	.0046	.0097	.0177	.0286	.0418	.0559	.0692	.0798	.0866	.0888
21	.0024	.0055	.0109	.0191	.0299	.0426	.0560	.0684	.0783	.0846
22	.0012	.0030	.0065	.0121	.0204	.0310	.0433	.0560	.0676	.0769
23	.0006	.0016	.0037	.0074	.0133	.0216	.0320	.0438	.0559	.0669
24	.0003	.0008	.0020	.0043	.0083	.0144	.0226	.0328	.0442	.0557
25	.0001	.0004	.0010	.0024	.0050	.0092	.0154	.0237	.0336	.0446
26	.0000	.0002	.0005	.0013	.0029	.0057	.0101	.0164	.0246	.0343
27	.0000	.0001	.0002	.0007	.0016	.0034	.0063	.0109	.0173	.0254
28	.0000	.0000	.0001	.0003	.0009	.0019	.0038	.0070	.0117	.0181
29	.0000	.0000	.0001	.0002	.0004	.0011	.0023	.0044	.0077	.0125
30	.0000	.0000	.0000	.0001	.0002	.0006	.0013	.0026	.0049	.0083
31	.0000	.0000	.0000	.0000	.0001	.0003	.0007	.0015	.0030	.0054
32	.0000	.0000	.0000	.0000	.0001	.0001	.0004	.0009	.0018	.0034
33	.0000	.0000	.0000	.0000	.0000	.0001	.0002	.0005	.0010	.0020
34	.0000	.0000	.0000	.0000	.0000	.0000	.0001	.0002	.0006	.0012
35	.0000	.0000	.0000	.0000	.0000	.0000	.0000	.0001	.0003	.0007
36	.0000	.0000	.0000	.0000	.0000	.0000	.0000	.0001	.0002	.0004
37	.0000	.0000	.0000	.0000	.0000	.0000	.0000	.0000	.0001	.0002
38	.0000	.0000	.0000	.0000	.0000	.0000	.0000	.0000	.0000	.0001
39	.0000	.0000	.0000	.0000	.0000	.0000	.0000	.0000	.0000	.0001

TABLE VII   STANDARD NORMAL DISTRIBUTION TABLE

757

TABLE VII   **STANDARD NORMAL DISTRIBUTION TABLE**

The entries in this table give the areas under
the standard normal curve from 0 to z.

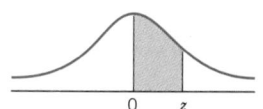

z	.00	.01	.02	.03	.04	.05	.06	.07	.08	.09
0.0	.0000	.0040	.0080	.0120	.0160	.0199	.0239	.0279	.0319	.0359
0.1	.0398	.0438	.0478	.0517	.0557	.0596	.0636	.0675	.0714	.0753
0.2	.0793	.0832	.0871	.0910	.0948	.0987	.1026	.1064	.1103	.1141
0.3	.1179	.1217	.1255	.1293	.1331	.1368	.1406	.1443	.1480	.1517
0.4	.1554	.1591	.1628	.1664	.1700	.1736	.1772	.1808	.1844	.1879
0.5	.1915	.1950	.1985	.2019	.2054	.2088	.2123	.2157	.2190	.2224
0.6	.2257	.2291	.2324	.2357	.2389	.2422	.2454	.2486	.2517	.2549
0.7	.2580	.2611	.2642	.2673	.2704	.2734	.2764	.2794	.2823	.2852
0.8	.2881	.2910	.2939	.2967	.2995	.3023	.3051	.3078	.3106	.3133
0.9	.3159	.3186	.3212	.3238	.3264	.3289	.3315	.3340	.3365	.3389
1.0	.3413	.3438	.3461	.3485	.3508	.3531	.3554	.3577	.3599	.3621
1.1	.3643	.3665	.3686	.3708	.3729	.3749	.3770	.3790	.3810	.3830
1.2	.3849	.3869	.3888	.3907	.3925	.3944	.3962	.3980	.3997	.4015
1.3	.4032	.4049	.4066	.4082	.4099	.4115	.4131	.4147	.4162	.4177
1.4	.4192	.4207	.4222	.4236	.4251	.4265	.4279	.4292	.4306	.4319
1.5	.4332	.4345	.4357	.4370	.4382	.4394	.4406	.4418	.4429	.4441
1.6	.4452	.4463	.4474	.4484	.4495	.4505	.4515	.4525	.4535	.4545
1.7	.4554	.4564	.4573	.4582	.4591	.4599	.4608	.4616	.4625	.4633
1.8	.4641	.4649	.4656	.4664	.4671	.4678	.4686	.4693	.4699	.4706
1.9	.4713	.4719	.4726	.4732	.4738	.4744	.4750	.4756	.4761	.4767
2.0	.4772	.4778	.4783	.4788	.4793	.4798	.4803	.4808	.4812	.4817
2.1	.4821	.4826	.4830	.4834	.4838	.4842	.4846	.4850	.4854	.4857
2.2	.4861	.4864	.4868	.4871	.4875	.4878	.4881	.4884	.4887	.4890
2.3	.4893	.4896	.4898	.4901	.4904	.4906	.4909	.4911	.4913	.4916
2.4	.4918	.4920	.4922	.4925	.4927	.4929	.4931	.4932	.4934	.4936
2.5	.4938	.4940	.4941	.4943	.4945	.4946	.4948	.4949	.4951	.4952
2.6	.4953	.4955	.4956	.4957	.4959	.4960	.4961	.4962	.4963	.4964
2.7	.4965	.4966	.4967	.4968	.4969	.4970	.4971	.4972	.4973	.4974
2.8	.4974	.4975	.4976	.4977	.4977	.4978	.4979	.4979	.4980	.4981
2.9	.4981	.4982	.4982	.4983	.4984	.4984	.4985	.4985	.4986	.4986
3.0	.4987	.4987	.4987	.4988	.4988	.4989	.4989	.4989	.4990	.4990

TABLE VIII   THE *t* DISTRIBUTION TABLE

The entries in this table give the critical values
of *t* for the specified number of degrees
of freedom and areas in the right tail.

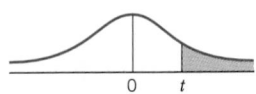

	Area in the Right Tail under the *t* Distribution Curve					
df	.10	.05	.025	.01	.005	.001
1	3.078	6.314	12.706	31.821	63.657	318.309
2	1.886	2.920	4.303	6.965	9.925	22.327
3	1.638	2.353	3.182	4.541	5.841	10.215
4	1.533	2.132	2.776	3.747	4.604	7.173
5	1.476	2.015	2.571	3.365	4.032	5.893
6	1.440	1.943	2.447	3.143	3.707	5.208
7	1.415	1.895	2.365	2.998	3.499	4.785
8	1.397	1.860	2.306	2.896	3.355	4.501
9	1.383	1.833	2.262	2.821	3.250	4.297
10	1.372	1.812	2.228	2.764	3.169	4.144
11	1.363	1.796	2.201	2.718	3.106	4.025
12	1.356	1.782	2.179	2.681	3.055	3.930
13	1.350	1.771	2.160	2.650	3.012	3.852
14	1.345	1.761	2.145	2.624	2.977	3.787
15	1.341	1.753	2.131	2.602	2.947	3.733
16	1.337	1.746	2.120	2.583	2.921	3.686
17	1.333	1.740	2.110	2.567	2.898	3.646
18	1.330	1.734	2.101	2.552	2.878	3.610
19	1.328	1.729	2.093	2.539	2.861	3.579
20	1.325	1.725	2.086	2.528	2.845	3.552
21	1.323	1.721	2.080	2.518	2.831	3.527
22	1.321	1.717	2.074	2.508	2.819	3.505
23	1.319	1.714	2.069	2.500	2.807	3.485
24	1.318	1.711	2.064	2.492	2.797	3.467
25	1.316	1.708	2.060	2.485	2.787	3.450
26	1.315	1.706	2.056	2.479	2.779	3.435
27	1.314	1.703	2.052	2.473	2.771	3.421
28	1.313	1.701	2.048	2.467	2.763	3.408
29	1.311	1.699	2.045	2.462	2.756	3.396
30	1.310	1.697	2.042	2.457	2.750	3.385
31	1.309	1.696	2.040	2.453	2.744	3.375
32	1.309	1.694	2.037	2.449	2.738	3.365
33	1.308	1.692	2.035	2.445	2.733	3.356
34	1.307	1.691	2.032	2.441	2.728	3.348
35	1.306	1.690	2.030	2.438	2.724	3.340
36	1.306	1.688	2.028	2.434	2.719	3.333
37	1.305	1.687	2.026	2.431	2.715	3.326

TABLE VIII   THE *t* DISTRIBUTION TABLE   **759**

TABLE VIII   (*Continued*)

df	Area in the Right Tail under the *t* Distribution Curve					
	.10	**.05**	**.025**	**.01**	**.005**	**.001**
38	1.304	1.686	2.024	2.429	2.712	3.319
39	1.304	1.685	2.023	2.426	2.708	3.313
40	1.303	1.684	2.021	2.423	2.704	3.307
41	1.303	1.683	2.020	2.421	2.701	3.301
42	1.302	1.682	2.018	2.418	2.698	3.296
43	1.302	1.681	2.017	2.416	2.695	3.291
44	1.301	1.680	2.015	2.414	2.692	3.286
45	1.301	1.679	2.014	2.412	2.690	3.281
46	1.300	1.679	2.013	2.410	2.687	3.277
47	1.300	1.678	2.012	2.408	2.685	3.273
48	1.299	1.677	2.011	2.407	2.682	3.269
49	1.299	1.677	2.010	2.405	2.680	3.265
50	1.299	1.676	2.009	2.403	2.678	3.261
51	1.298	1.675	2.008	2.402	2.676	3.258
52	1.298	1.675	2.007	2.400	2.674	3.255
53	1.298	1.674	2.006	2.399	2.672	3.251
54	1.297	1.674	2.005	2.397	2.670	3.248
55	1.297	1.673	2.004	2.396	2.668	3.245
56	1.297	1.673	2.003	2.395	2.667	3.242
57	1.297	1.672	2.002	2.394	2.665	3.239
58	1.296	1.672	2.002	2.392	2.663	3.237
59	1.296	1.671	2.001	2.391	2.662	3.234
60	1.296	1.671	2.000	2.390	2.660	3.232
61	1.296	1.670	2.000	2.389	2.659	3.229
62	1.295	1.670	1.999	2.388	2.657	3.227
63	1.295	1.669	1.998	2.387	2.656	3.225
64	1.295	1.669	1.998	2.386	2.655	3.223
65	1.295	1.669	1.997	2.385	2.654	3.220
66	1.295	1.668	1.997	2.384	2.652	3.218
67	1.294	1.668	1.996	2.383	2.651	3.216
68	1.294	1.668	1.995	2.382	2.650	3.214
69	1.294	1.667	1.995	2.382	2.649	3.213
70	1.294	1.667	1.994	2.381	2.648	3.211
71	1.294	1.667	1.994	2.380	2.647	3.209
72	1.293	1.666	1.993	2.379	2.646	3.207
73	1.293	1.666	1.993	2.379	2.645	3.206
74	1.293	1.666	1.993	2.378	2.644	3.204
75	1.293	1.665	1.992	2.377	2.643	3.202
∞	1.282	1.645	1.960	2.326	2.576	3.090

TABLE IX   CHI-SQUARE DISTRIBUTION TABLE

The entries in this table give
the critical values of χ^2 for the
specified number of degrees of freedom
and areas in the right tail.

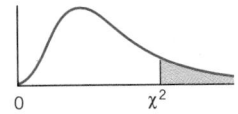

df	Area in the Right Tail under the Chi-square Distribution Curve									
	.995	.990	.975	.950	.900	.100	.050	.025	.010	.005
1	0.000	0.000	0.001	0.004	0.016	2.706	3.841	5.024	6.635	7.879
2	0.010	0.020	0.051	0.103	0.211	4.605	5.991	7.378	9.210	10.597
3	0.072	0.115	0.216	0.352	0.584	6.251	7.815	9.348	11.345	12.838
4	0.207	0.297	0.484	0.711	1.064	7.779	9.488	11.143	13.277	14.860
5	0.412	0.554	0.831	1.145	1.610	9.236	11.070	12.833	15.086	16.750
6	0.676	0.872	1.237	1.635	2.204	10.645	12.592	14.449	16.812	18.548
7	0.989	1.239	1.690	2.167	2.833	12.017	14.067	16.013	18.475	20.278
8	1.344	1.646	2.180	2.733	3.490	13.362	15.507	17.535	20.090	21.955
9	1.735	2.088	2.700	3.325	4.168	14.684	16.919	19.023	21.666	23.589
10	2.156	2.558	3.247	3.940	4.865	15.987	18.307	20.483	23.209	25.188
11	2.603	3.053	3.816	4.575	5.578	17.275	19.675	21.920	24.725	26.757
12	3.074	3.571	4.404	5.226	6.304	18.549	21.026	23.337	26.217	28.300
13	3.565	4.107	5.009	5.892	7.042	19.812	22.362	24.736	27.688	29.819
14	4.075	4.660	5.629	6.571	7.790	21.064	23.685	26.119	29.141	31.319
15	4.601	5.229	6.262	7.261	8.547	22.307	24.996	27.488	30.578	32.801
16	5.142	5.812	6.908	7.962	9.312	23.542	26.296	28.845	32.000	34.267
17	5.697	6.408	7.564	8.672	10.085	24.769	27.587	30.191	33.409	35.718
18	6.265	7.015	8.231	9.390	10.865	25.989	28.869	31.526	34.805	37.156
19	6.844	7.633	8.907	10.117	11.651	27.204	30.144	32.852	36.191	38.582
20	7.434	8.260	9.591	10.851	12.443	28.412	31.410	34.170	37.566	39.997
21	8.034	8.897	10.283	11.591	13.240	29.615	32.671	35.479	38.932	41.401
22	8.643	9.542	10.982	12.338	14.041	30.813	33.924	36.781	40.289	42.796
23	9.260	10.196	11.689	13.091	14.848	32.007	35.172	38.076	41.638	44.181
24	9.886	10.856	12.401	13.848	15.659	33.196	36.415	39.364	42.980	45.559
25	10.520	11.524	13.120	14.611	16.473	34.382	37.652	40.646	44.314	46.928
26	11.160	12.198	13.844	15.379	17.292	35.563	38.885	41.923	45.642	48.290
27	11.808	12.879	14.573	16.151	18.114	36.741	40.113	43.195	46.963	49.645
28	12.461	13.565	15.308	16.928	18.939	37.916	41.337	44.461	48.278	50.993
29	13.121	14.256	16.047	17.708	19.768	39.087	42.557	45.722	49.588	52.336
30	13.787	14.953	16.791	18.493	20.599	40.256	43.773	46.979	50.892	53.672
40	20.707	22.164	24.433	26.509	29.051	51.805	55.758	59.342	63.691	66.766
50	27.991	29.707	32.357	34.764	37.689	63.167	67.505	71.420	76.154	79.490
60	35.534	37.485	40.482	43.188	46.459	74.397	79.082	83.298	88.379	91.952
70	43.275	45.442	48.758	51.739	55.329	85.527	90.531	95.023	100.425	104.215
80	51.172	53.540	57.153	60.391	64.278	96.578	101.879	106.629	112.329	116.321
90	59.196	61.754	65.647	69.126	73.291	107.565	113.145	118.136	124.116	128.299
100	67.328	70.065	74.222	77.929	82.358	118.498	124.342	129.561	135.807	140.169

TABLE X   THE *F* DISTRIBUTION TABLE

761

TABLE X   THE *F* DISTRIBUTION TABLE

Area in the Right Tail under the *F* Distribution Curve = .01

Degrees of Freedom for the Denominator	Degrees of Freedom for the Numerator																		
	1	2	3	4	5	6	7	8	9	10	11	12	15	20	25	30	40	50	100
1	4052	5000	5403	5625	5764	5859	5928	5981	6022	6056	6083	6106	6157	6209	6240	6261	6287	6303	6334
2	98.50	99.00	99.17	99.25	99.30	99.33	99.36	99.37	99.39	99.40	99.41	99.42	99.43	99.45	99.46	99.47	99.47	99.48	99.49
3	34.12	30.82	29.46	28.71	28.24	27.91	27.67	27.49	27.35	27.23	27.13	27.05	26.87	26.69	26.58	26.50	26.41	26.35	26.24
4	21.20	18.00	16.69	15.98	15.52	15.21	14.98	14.80	14.66	14.55	14.45	14.37	14.20	14.02	13.91	13.84	13.75	13.69	13.58
5	16.26	13.27	12.06	11.39	10.97	10.67	10.46	10.29	10.16	10.05	9.96	9.89	9.72	9.55	9.45	9.38	9.29	9.24	9.13
6	13.75	10.92	9.78	9.15	8.75	8.47	8.26	8.10	7.98	7.87	7.79	7.72	7.56	7.40	7.30	7.23	7.14	7.09	6.99
7	12.25	9.55	8.45	7.85	7.46	7.19	6.99	6.84	6.72	6.62	6.54	6.47	6.31	6.16	6.06	5.99	5.91	5.86	5.75
8	11.26	8.65	7.59	7.01	6.63	6.37	6.18	6.03	5.91	5.81	5.73	5.67	5.52	5.36	5.26	5.20	5.12	5.07	4.96
9	10.56	8.02	6.99	6.42	6.06	5.80	5.61	5.47	5.35	5.26	5.18	5.11	4.96	4.81	4.71	4.65	4.57	4.52	4.41
10	10.04	7.56	6.55	5.99	5.64	5.39	5.20	5.06	4.94	4.85	4.77	4.71	4.56	4.41	4.31	4.25	4.17	4.12	4.01
11	9.65	7.21	6.22	5.67	5.32	5.07	4.89	4.74	4.63	4.54	4.46	4.40	4.25	4.10	4.01	3.94	3.86	3.81	3.71
12	9.33	6.93	5.95	5.41	5.06	4.82	4.64	4.50	4.39	4.30	4.22	4.16	4.01	3.86	3.76	3.70	3.62	3.57	3.47
13	9.07	6.70	5.74	5.21	4.86	4.62	4.44	4.30	4.19	4.10	4.02	3.96	3.82	3.66	3.57	3.51	3.43	3.38	3.27
14	8.86	6.51	5.56	5.04	4.69	4.46	4.28	4.14	4.03	3.94	3.86	3.80	3.66	3.51	3.41	3.35	3.27	3.22	3.11
15	8.68	6.36	5.42	4.89	4.56	4.32	4.14	4.00	3.89	3.80	3.73	3.67	3.52	3.37	3.28	3.21	3.13	3.08	2.98
16	8.53	6.23	5.29	4.77	4.44	4.20	4.03	3.89	3.78	3.69	3.62	3.55	3.41	3.26	3.16	3.10	3.02	2.97	2.86
17	8.40	6.11	5.18	4.67	4.34	4.10	3.93	3.79	3.68	3.59	3.52	3.46	3.31	3.16	3.07	3.00	2.92	2.87	2.76
18	8.29	6.01	5.09	4.58	4.25	4.01	3.84	3.71	3.60	3.51	3.43	3.37	3.23	3.08	2.98	2.92	2.84	2.78	2.68
19	8.18	5.93	5.01	4.50	4.17	3.94	3.77	3.63	3.52	3.43	3.36	3.30	3.15	3.00	2.91	2.84	2.76	2.71	2.60
20	8.10	5.85	4.94	4.43	4.10	3.87	3.70	3.56	3.46	3.37	3.29	3.23	3.09	2.94	2.84	2.78	2.69	2.64	2.54
21	8.02	5.78	4.87	4.37	4.04	3.81	3.64	3.51	3.40	3.31	3.24	3.17	3.03	2.88	2.79	2.72	2.64	2.58	2.48
22	7.95	5.72	4.82	4.31	3.99	3.76	3.59	3.45	3.35	3.26	3.18	3.12	2.98	2.83	2.73	2.67	2.58	2.53	2.42
23	7.88	5.66	4.76	4.26	3.94	3.71	3.54	3.41	3.30	3.21	3.14	3.07	2.93	2.78	2.69	2.62	2.54	2.48	2.37
24	7.82	5.61	4.72	4.22	3.90	3.67	3.50	3.36	3.26	3.17	3.09	3.03	2.89	2.74	2.64	2.58	2.49	2.44	2.33
25	7.77	5.57	4.68	4.18	3.85	3.63	3.46	3.32	3.22	3.13	3.06	2.99	2.85	2.70	2.60	2.54	2.45	2.40	2.29
30	7.56	5.39	4.51	4.02	3.70	3.47	3.30	3.17	3.07	2.98	2.91	2.84	2.70	2.55	2.45	2.39	2.30	2.25	2.13
40	7.31	5.18	4.31	3.83	3.51	3.29	3.12	2.99	2.89	2.80	2.73	2.66	2.52	2.37	2.27	2.20	2.11	2.06	1.94
50	7.17	5.06	4.20	3.72	3.41	3.19	3.02	2.89	2.78	2.70	2.63	2.56	2.42	2.27	2.17	2.10	2.01	1.95	1.82
100	6.90	4.82	3.98	3.51	3.21	2.99	2.82	2.69	2.59	2.50	2.43	2.37	2.22	2.07	1.97	1.89	1.80	1.74	1.60

Degrees of Freedom for the Denominator

TABLE X (*Continued*)

Area in the Right Tail under the *F* Distribution Curve = .025

	Degrees of Freedom for the Numerator																		
Denom. df	1	2	3	4	5	6	7	8	9	10	11	12	15	20	25	30	40	50	100
1	647.8	799.5	864.2	899.6	921.8	937.1	948.2	956.7	963.3	968.6	973.0	976.7	984.9	993.1	998.1	1001	1006	1008	1013
2	38.51	39.00	39.17	39.25	39.30	39.33	39.36	39.37	39.39	39.40	39.41	39.41	39.43	39.45	39.46	39.46	39.47	39.48	39.49
3	17.44	16.04	15.44	15.10	14.88	14.73	14.62	14.54	14.47	14.42	14.37	14.34	14.25	14.17	14.12	14.08	14.04	14.01	13.96
4	12.22	10.65	9.98	9.61	9.36	9.20	9.07	8.98	8.90	8.84	8.79	8.75	8.66	8.56	8.50	8.46	8.41	8.38	8.32
5	10.01	8.43	7.76	7.39	7.15	6.98	6.85	6.76	6.68	6.62	6.57	6.52	6.43	6.33	6.27	6.23	6.18	6.14	6.08
6	8.81	7.26	6.60	6.23	5.99	5.82	5.70	5.60	5.52	5.46	5.41	5.37	5.27	5.17	5.11	5.07	5.01	4.98	4.92
7	8.07	6.54	5.89	5.52	5.29	5.12	4.99	4.90	4.82	4.76	4.71	4.67	4.57	4.47	4.40	4.36	4.31	4.28	4.21
8	7.57	6.06	5.42	5.05	4.82	4.65	4.53	4.43	4.36	4.30	4.24	4.20	4.10	4.00	3.94	3.89	3.84	3.81	3.74
9	7.21	5.72	5.08	4.72	4.48	4.32	4.20	4.10	4.03	3.96	3.91	3.87	3.77	3.67	3.60	3.56	3.51	3.47	3.40
10	6.94	5.46	4.83	4.47	4.24	4.07	3.95	3.85	3.78	3.72	3.66	3.62	3.52	3.42	3.35	3.31	3.26	3.22	3.15
11	6.72	5.26	4.63	4.28	4.04	3.88	3.76	3.66	3.59	3.53	3.47	3.43	3.33	3.23	3.16	3.12	3.06	3.03	2.96
12	6.55	5.10	4.47	4.12	3.89	3.73	3.61	3.51	3.44	3.37	3.32	3.28	3.18	3.07	3.01	2.96	2.91	2.87	2.80
13	6.41	4.97	4.35	4.00	3.77	3.60	3.48	3.39	3.31	3.25	3.20	3.15	3.05	2.95	2.88	2.84	2.78	2.74	2.67
14	6.30	4.86	4.24	3.89	3.66	3.50	3.38	3.29	3.21	3.15	3.09	3.05	2.95	2.84	2.78	2.73	2.67	2.64	2.56
15	6.20	4.77	4.15	3.80	3.58	3.41	3.29	3.20	3.12	3.06	3.01	2.96	2.86	2.76	2.69	2.64	2.59	2.55	2.47
16	6.12	4.69	4.08	3.73	3.50	3.34	3.22	3.12	3.05	2.99	2.93	2.89	2.79	2.68	2.61	2.57	2.51	2.47	2.40
17	6.04	4.62	4.01	3.66	3.44	3.28	3.16	3.06	2.98	2.92	2.87	2.82	2.72	2.62	2.55	2.50	2.44	2.41	2.33
18	5.98	4.56	3.95	3.61	3.38	3.22	3.10	3.01	2.93	2.87	2.81	2.77	2.67	2.56	2.49	2.44	2.38	2.35	2.27
19	5.92	4.51	3.90	3.56	3.33	3.17	3.05	2.96	2.88	2.82	2.76	2.72	2.62	2.51	2.44	2.39	2.33	2.30	2.22
20	5.87	4.46	3.86	3.51	3.29	3.13	3.01	2.91	2.84	2.77	2.72	2.68	2.57	2.46	2.40	2.35	2.29	2.25	2.17
21	5.83	4.42	3.82	3.48	3.25	3.09	2.97	2.87	2.80	2.73	2.68	2.64	2.53	2.42	2.36	2.31	2.25	2.21	2.13
22	5.79	4.38	3.78	3.44	3.22	3.05	2.93	2.84	2.76	2.70	2.65	2.60	2.50	2.39	2.32	2.27	2.21	2.17	2.09
23	5.75	4.35	3.75	3.41	3.18	3.02	2.90	2.81	2.73	2.67	2.62	2.57	2.47	2.36	2.29	2.24	2.18	2.14	2.06
24	5.72	4.32	3.72	3.38	3.15	2.99	2.87	2.78	2.70	2.64	2.59	2.54	2.44	2.33	2.26	2.21	2.15	2.11	2.02
25	5.69	4.29	3.69	3.35	3.13	2.97	2.85	2.75	2.68	2.61	2.56	2.51	2.41	2.30	2.23	2.18	2.12	2.08	2.00
30	5.57	4.18	3.59	3.25	3.03	2.87	2.75	2.65	2.57	2.51	2.46	2.41	2.31	2.20	2.12	2.07	2.01	1.97	1.88
40	5.42	4.05	3.46	3.13	2.90	2.74	2.62	2.53	2.45	2.39	2.33	2.29	2.18	2.07	1.99	1.94	1.88	1.83	1.74
50	5.34	3.97	3.39	3.05	2.83	2.67	2.55	2.46	2.38	2.32	2.26	2.22	2.11	1.99	1.92	1.87	1.80	1.75	1.66
100	5.18	3.83	3.25	2.92	2.70	2.54	2.42	2.32	2.24	2.18	2.12	2.08	1.97	1.85	1.77	1.71	1.64	1.59	1.48

Degrees of Freedom for the Denominator

TABLE X (*Continued*)

TABLE X   THE *F* DISTRIBUTION TABLE

Area in the Right Tail under the *F* Distribution Curve = .05

Degrees of Freedom for the Denominator

							Degrees of Freedom for the Numerator												
	1	**2**	**3**	**4**	**5**	**6**	**7**	**8**	**9**	**10**	**11**	**12**	**15**	**20**	**25**	**30**	**40**	**50**	**100**
1	161.5	199.5	215.7	224.6	230.2	234.0	236.8	238.9	240.5	241.9	243.0	243.9	246.0	248.0	249.3	250.1	251.1	251.8	253.0
2	18.51	19.00	19.16	19.25	19.30	19.33	19.35	19.37	19.38	19.40	19.40	19.41	19.43	19.45	19.46	19.46	19.47	19.48	19.49
3	10.13	9.55	9.28	9.12	9.01	8.94	8.89	8.85	8.81	8.79	8.76	8.74	8.70	8.66	8.63	8.62	8.59	8.58	8.55
4	7.71	6.94	6.59	6.39	6.26	6.16	6.09	6.04	6.00	5.96	5.94	5.91	5.86	5.80	5.77	5.75	5.72	5.70	5.66
5	6.61	5.79	5.41	5.19	5.05	4.95	4.88	4.82	4.77	4.74	4.70	4.68	4.62	4.56	4.52	4.50	4.46	4.44	4.41
6	5.99	5.14	4.76	4.53	4.39	4.28	4.21	4.15	4.10	4.06	4.03	4.00	3.94	3.87	3.83	3.81	3.77	3.75	3.71
7	5.59	4.74	4.35	4.12	3.97	3.87	3.79	3.73	3.68	3.64	3.60	3.57	3.51	3.44	3.40	3.38	3.34	3.32	3.27
8	5.32	4.46	4.07	3.84	3.69	3.58	3.50	3.44	3.39	3.35	3.31	3.28	3.22	3.15	3.11	3.08	3.04	3.02	2.97
9	5.12	4.26	3.86	3.63	3.48	3.37	3.29	3.23	3.18	3.14	3.10	3.07	3.01	2.94	2.89	2.86	2.83	2.80	2.76
10	4.96	4.10	3.71	3.48	3.33	3.22	3.14	3.07	3.02	2.98	2.94	2.91	2.85	2.77	2.73	2.70	2.66	2.64	2.59
11	4.84	3.98	3.59	3.36	3.20	3.09	3.01	2.95	2.90	2.85	2.82	2.79	2.72	2.65	2.60	2.57	2.53	2.51	2.46
12	4.75	3.89	3.49	3.26	3.11	3.00	2.91	2.85	2.80	2.75	2.72	2.69	2.62	2.54	2.50	2.47	2.43	2.40	2.35
13	4.67	3.81	3.41	3.18	3.03	2.92	2.83	2.77	2.71	2.67	2.63	2.60	2.53	2.46	2.41	2.38	2.34	2.31	2.26
14	4.60	3.74	3.34	3.11	2.96	2.85	2.76	2.70	2.65	2.60	2.57	2.53	2.46	2.39	2.34	2.31	2.27	2.24	2.19
15	4.54	3.68	3.29	3.06	2.90	2.79	2.71	2.64	2.59	2.54	2.51	2.48	2.40	2.33	2.28	2.25	2.20	2.18	2.12
16	4.49	3.63	3.24	3.01	2.85	2.74	2.66	2.59	2.54	2.49	2.46	2.42	2.35	2.28	2.23	2.19	2.15	2.12	2.07
17	4.45	3.59	3.20	2.96	2.81	2.70	2.61	2.55	2.49	2.45	2.41	2.38	2.31	2.23	2.18	2.15	2.10	2.08	2.02
18	4.41	3.55	3.16	2.93	2.77	2.66	2.58	2.51	2.46	2.41	2.37	2.34	2.27	2.19	2.14	2.11	2.06	2.04	1.98
19	4.38	3.52	3.13	2.90	2.74	2.63	2.54	2.48	2.42	2.38	2.34	2.31	2.23	2.16	2.11	2.07	2.03	2.00	1.94
20	4.35	3.49	3.10	2.87	2.71	2.60	2.51	2.45	2.39	2.35	2.31	2.28	2.20	2.12	2.07	2.04	1.99	1.97	1.91
21	4.32	3.47	3.07	2.84	2.68	2.57	2.49	2.42	2.37	2.32	2.28	2.25	2.18	2.10	2.05	2.01	1.96	1.94	1.88
22	4.30	3.44	3.05	2.82	2.66	2.55	2.46	2.40	2.34	2.30	2.26	2.23	2.15	2.07	2.02	1.97	1.94	1.91	1.85
23	4.28	3.42	3.03	2.80	2.64	2.53	2.44	2.37	2.32	2.27	2.24	2.20	2.13	2.05	2.00	1.96	1.91	1.88	1.82
24	4.26	3.40	3.01	2.78	2.62	2.51	2.42	2.36	2.30	2.25	2.22	2.18	2.11	2.03	1.97	1.94	1.89	1.86	1.80
25	4.24	3.39	2.99	2.76	2.60	2.49	2.40	2.34	2.28	2.24	2.20	2.16	2.09	2.01	1.96	1.92	1.87	1.84	1.78
30	4.17	3.32	2.92	2.69	2.53	2.42	2.33	2.27	2.21	2.16	2.13	2.09	2.01	1.93	1.88	1.84	1.79	1.76	1.70
40	4.08	3.23	2.84	2.61	2.45	2.34	2.25	2.18	2.12	2.08	2.04	2.00	1.92	1.84	1.78	1.74	1.69	1.66	1.59
50	4.03	3.18	2.79	2.56	2.40	2.29	2.20	2.13	2.07	2.03	1.99	1.95	1.87	1.78	1.73	1.69	1.63	1.60	1.52
100	3.94	3.09	2.70	2.46	2.31	2.19	2.10	2.03	1.97	1.93	1.89	1.85	1.77	1.68	1.62	1.57	1.52	1.48	1.39

764

TABLE X (*Continued*)

Area in the Right Tail under the *F* Distribution Curve = .10

Degrees of Freedom for the Numerator

	1	2	3	4	5	6	7	8	9	10	11	12	15	20	25	30	40	50	100
1	39.86	49.50	53.59	55.83	57.24	58.20	58.91	59.44	59.86	60.19	60.47	60.71	61.22	61.74	62.05	62.26	62.53	62.69	63.01
2	8.53	9.00	9.16	9.24	9.29	9.33	9.35	9.37	9.38	9.39	9.40	9.41	9.42	9.44	9.45	9.46	9.47	9.47	9.48
3	5.54	5.46	5.39	5.34	5.31	5.28	5.27	5.25	5.24	5.23	5.22	5.22	5.20	5.18	5.17	5.17	5.16	5.15	5.14
4	4.54	4.32	4.19	4.11	4.05	4.01	3.98	3.95	3.94	3.92	3.91	3.90	3.87	3.84	3.83	3.82	3.80	3.80	3.78
5	4.06	3.78	3.62	3.52	3.45	3.40	3.37	3.34	3.32	3.30	3.28	3.27	3.24	3.21	3.19	3.17	3.16	3.15	3.13
6	3.78	3.46	3.29	3.18	3.11	3.05	3.01	2.98	2.96	2.94	2.92	2.90	2.87	2.84	2.81	2.80	2.78	2.77	2.75
7	3.59	3.26	3.07	2.96	2.88	2.83	2.78	2.75	2.72	2.70	2.68	2.67	2.63	2.59	2.57	2.56	2.54	2.52	2.50
8	3.46	3.11	2.92	2.81	2.73	2.67	2.62	2.59	2.56	2.54	2.52	2.50	2.46	2.42	2.40	2.38	2.36	2.35	2.32
9	3.36	3.01	2.81	2.69	2.61	2.55	2.51	2.47	2.44	2.42	2.40	2.38	2.34	2.30	2.27	2.25	2.23	2.22	2.19
10	3.29	2.92	2.73	2.61	2.52	2.46	2.41	2.38	2.35	2.32	2.30	2.28	2.24	2.20	2.17	2.16	2.13	2.12	2.09
11	3.23	2.86	2.66	2.54	2.45	2.39	2.34	2.30	2.27	2.25	2.23	2.21	2.17	2.12	2.10	2.08	2.05	2.04	2.01
12	3.18	2.81	2.61	2.48	2.39	2.33	2.28	2.24	2.21	2.19	2.17	2.15	2.10	2.06	2.03	2.01	1.99	1.97	1.94
13	3.14	2.76	2.56	2.43	2.35	2.28	2.23	2.20	2.16	2.14	2.12	2.10	2.05	2.01	1.98	1.96	1.93	1.92	1.88
14	3.10	2.73	2.52	2.39	2.31	2.24	2.19	2.15	2.12	2.10	2.07	2.05	2.01	1.96	1.93	1.91	1.89	1.87	1.83
15	3.07	2.70	2.49	2.36	2.27	2.21	2.16	2.12	2.09	2.06	2.04	2.02	1.97	1.92	1.89	1.87	1.85	1.83	1.79
16	3.05	2.67	2.46	2.33	2.24	2.18	2.13	2.09	2.06	2.03	2.01	1.99	1.94	1.89	1.86	1.84	1.81	1.79	1.76
17	3.03	2.64	2.44	2.31	2.22	2.15	2.10	2.06	2.03	2.00	1.98	1.96	1.91	1.86	1.83	1.81	1.78	1.76	1.73
18	3.01	2.62	2.42	2.29	2.20	2.13	2.08	2.04	2.00	1.98	1.95	1.93	1.89	1.84	1.80	1.78	1.75	1.74	1.70
19	2.99	2.61	2.40	2.27	2.18	2.11	2.06	2.02	1.98	1.96	1.93	1.91	1.86	1.81	1.78	1.76	1.73	1.71	1.67
20	2.97	2.59	2.38	2.25	2.16	2.09	2.04	2.00	1.96	1.94	1.91	1.89	1.84	1.79	1.76	1.74	1.71	1.69	1.65
21	2.96	2.57	2.36	2.23	2.14	2.08	2.02	1.98	1.95	1.92	1.90	1.87	1.83	1.78	1.74	1.72	1.69	1.67	1.63
22	2.95	2.56	2.35	2.22	2.13	2.06	2.01	1.97	1.93	1.90	1.88	1.86	1.81	1.76	1.73	1.70	1.67	1.65	1.61
23	2.94	2.55	2.34	2.21	2.11	2.05	1.99	1.95	1.92	1.89	1.87	1.84	1.80	1.74	1.71	1.69	1.66	1.64	1.59
24	2.93	2.54	2.33	2.19	2.10	2.04	1.98	1.94	1.91	1.88	1.85	1.83	1.78	1.73	1.70	1.67	1.64	1.62	1.58
25	2.92	2.53	2.32	2.18	2.09	2.02	1.97	1.93	1.89	1.87	1.84	1.82	1.77	1.72	1.68	1.66	1.63	1.61	1.56
30	2.88	2.49	2.28	2.14	2.05	1.98	1.93	1.88	1.85	1.82	1.79	1.77	1.72	1.67	1.63	1.61	1.57	1.55	1.51
40	2.84	2.44	2.23	2.09	2.00	1.93	1.87	1.83	1.79	1.76	1.74	1.71	1.66	1.61	1.57	1.54	1.51	1.48	1.43
50	2.81	2.41	2.20	2.06	1.97	1.90	1.84	1.80	1.76	1.73	1.70	1.68	1.63	1.57	1.53	1.50	1.46	1.44	1.39
100	2.76	2.36	2.14	2.00	1.91	1.83	1.78	1.73	1.69	1.66	1.64	1.61	1.56	1.49	1.45	1.42	1.38	1.35	1.29

Degrees of Freedom for the Denominator

ANSWERS TO SELECTED ODD-NUMBERED EXERCISES AND SELF-REVIEW TESTS

(*Note:* Due to differences in rounding, the answers obtained by readers may differ slightly from the ones given in this Appendix).

CHAPTER 1

1.7 **a.** population   **b.** sample   **c.** population   **d.** sample   **e.** population

1.11 **a.** score   **b.** five observations   **c.** five elements

1.15 **a.** quantitative   **b.** quantitative   **c.** qualitative   **d.** quantitative   **e.** qualitative

1.17 **a.** discrete   **b.** continuous   **d.** continuous

1.21 **a.** cross-section data   **b.** cross-section data   **c.** time-series data   **d.** time-series data

1.23 **a.** $\Sigma f = 59$ **b.** $\Sigma m^2 = 819$ **c.** $\Sigma mf = 576$ **d.** $\Sigma m^2 f = 7614$

1.25 **a.** $\Sigma x = 88$ **b.** $\Sigma y = 58$ **c.** $\Sigma xy = 855$ **d.** $\Sigma x^2 = 1590$ **e.** $\Sigma y^2 = 622$

1.27 **a.** $\Sigma x = 13$ **b.** $(\Sigma x)^2 = 169$ **c.** $\Sigma x^2 = 35$

1.29 **a.** $\Sigma x = 61$ **b.** $(\Sigma x)^2 = 3721$ **c.** $\Sigma x^2 = 555$

1.33 **a.** sample   **b.** population for the week   **c.** sample   **d.** population

1.35 **a.** sampling without replacement   **b.** sampling with replacement

1.37 **a.** $\Sigma x = 56$ **b.** $(\Sigma x)^2 = 3136$ **c.** $\Sigma x^2 = 586$

1.39 **a.** $\Sigma m = 59$ **b.** $\Sigma f^2 = 2662$ **c.** $\Sigma mf = 1508$ **d.** $\Sigma m^2 f = 24{,}884$ **e.** $\Sigma m^2 = 867$

Self-Review Test

1. b   **2.** c

3. **a.** sampling without replacement   **b.** sampling with replacement

4. **a.** qualitative   **b.** quantitative (discrete)   **c.** quantitative (continuous)

6. **a.** $\Sigma x = 76$ **b.** $\Sigma x^2 = 1348$ **c.** $(\Sigma x)^2 = 5776$

7. **a.** $\Sigma m = 45$ **b.** $\Sigma f = 112$ **c.** $\Sigma m^2 = 495$ **d.** $\Sigma mf = 975$ **e.** $\Sigma m^2 f = 9855$ **f.** $\Sigma f^2 = 2994$

CHAPTER 2

2.3 **c.** 30%   **d.** 70%   **2.5** **c.** 52%   **2.7** **c.** 46.7%   **2.15** **d.** 58%

2.17 **a.** class limits: \$1–\$4, \$5–\$8, \$9–\$12, \$13–\$16, \$17–\$20
   b. class boundaries: \$0.5, \$4.5, \$8.5, \$12.5, \$16.5, \$20.5; width = \$4
   c. class midpoints: \$2.5, \$6.5, \$10.5, \$14.5, \$18.5

2.19 **d.** 30%   **2.29** **c.** 12   **2.35** **c.** 42%   **e.** about 50%

2.47 218, 245, 256, 329, 367, 383, 397, 404, 427, 433, 471, 523, 537, 551, 563, 581, 592, 622, 636, 647, 655, 678, 689, 810, 841

2.57 **d.** 27.5%   **2.59** **c.** 56.7%   **2.61** **c.** 56.7%

2.63 **d.** Boundaries of the fourth class are 4200.5 and 5600.5; width = 1400.

2.73 **b.** Widths of class intervals are unequal; readers may associate frequencies with areas rather than heights.
   c. Using the original data, make class intervals equal in width.

2.75 **b.** No, because the percentages in the table do not indicate what proportion of the labor force falls into each category (mining, etc.); they merely indicate the proportion unemployed within each category.
   c. No. We would have to know the number of workers in each category, construction and manufacturing.

Self-Review Test

2. **a.** 5   **b.** 5   **c.** 12   **d.** 4.5   **e.** 9   **f.** 100   **g.** .08

4. **c.** 30%   **5.** **c.** 37.5%   **6.** About 60%

8. 30, 33, 37, 42, 44, 46, 47, 49, 51, 53, 53, 56, 60, 67, 67, 71, 79

CHAPTER 3

3.5   mode   **3.9**   mean = 2.50;   median = 3.50;   no mode   **3.11**   mean = 598.83;   median = 608.50

3.13   mean = 21.38;   median = 20.30;   no mode   **3.15**   mean = 4.90;   median = 4.73

3.17   mean = 3.82;   median = 4.60;   no mode

3.19   mean = 5.70 hours;   median = 6 hours;   mode = 0 and 7 hours

3.21   mean = 27.50;   median = 27.50;   mode = 23

3.23   a.   mean = 46.80;   median = 37
          b.   outlier = 129;   when the outlier is dropped: mean = 37.67; median = 34;   mean changes by a larger amount
          c.   median

3.25   combined mean = $41.89   **3.27**   total = $855   **3.29**   age of the sixth person = 48 years

3.31   mean for data set I = 24.60; mean for data set II = 31.60
          The mean of the second data set is equal to the mean of the first data set plus 7.

3.33   10% trimmed mean = 38.25 years   **3.39**   range = 21; $\sigma^2 = 50.250$; $\sigma = 7.09$

3.41   a. $\bar{x} = 105$; deviations from the mean: $-23, 11, -40, 65, -13$. The sum of these deviations is zero.
          b.   range = 105; $s^2 = 1661.000$; $s = 40.76$

3.43   range = 11; $s^2 = 10.788$; $s = 3.28$ **3.45**   range = 26 cars; $s^2 = 72.278$; $s = 8.50$ cars

3.47   range = 20 pounds; $s^2 = 32.971$; $s = 5.74$ pounds   **3.49**   range = 30; $s^2 = 107.357$; $s = 10.36$

3.51   range = 6.5; $s^2 = 6.701$; $s = 2.59$ **3.53** $s = 0$

3.55   CV for salaries = 9.02%;   CV for years of schooling = 13.33%;
          The relative variation in salaries is lower.

3.57 $s = 14.64$ for both data sets   **3.61** $\mu = 10.30$; $\sigma^2 = 23.680$; $\sigma = 4.87$

3.63 $\bar{x} = 9.40$; $s^2 = 37.711$; $s = 6.14$ **3.65** $\bar{x} = 27.20$; $s^2 = 176.694$; $s = 13.29$

3.67 $\mu = 7.8$ pounds; $\sigma^2 = 3.596$; $\sigma = 1.90$ pounds

3.69 $\bar{x} = 24.65$ computers; $s^2 = 114.438$; $s^2 = 10.70$ computers

3.71 $\bar{x} = 36.80$ minutes; $s^2 = 597.714$; $s = 24.45$ minutes   **3.73** $\bar{x} = 1.74$ sets; $s^2 = .930$; $s = .96$ sets

3.77   at least 75%;   at least 84%;   at least 88.89%   **3.79**   68%;   95%;   99.7%

3.81   a.   at least 75%;   b.   at least 84%;   c.   at least 88.89%

3.83   a.   i.   at least 75%   ii.   at least 88.89%   b. $765 to $1965

3.85   a.   95%   b.   68%   c.   99.7%

3.87   a.   i.   68%   ii.   99.7%   b.   3 to 11 years old

3.93   a. $Q_1 = 66.5$; $Q_2 = 69$; $Q_3 = 74$; $IQR = 7.5$ b. $P_{35} = 67$ c.   53.85%

3.95   a. $Q_1 = 35$; $Q_2 = 40$; $Q_3 = 43$; $IQR = 8$ b. $P_{79} = 45$ c.   36.67%

3.97   a. $Q_1 = 24$; $Q_2 = 27.5$; $Q_3 = 31$; $IQR = 7$ b. $P_{65} = 29$ c.   23.33%

3.99   a. $Q_1 = 69$; $Q_2 = 78$; $Q_3 = 84$; $IQR = 15$ b. $P_{93} = 97$ c.   57.89%   **3.101**   no outlier

3.111   a.   mean = 1716.57;   median = 1724
           b.   outliers = 233 and 3293;   when the outliers are dropped: mean = 1698;   median = 1724; mean changes by a larger
                amount
           c.   median

3.113   a.   mean = 26.6;   median = 19;   mode = 17, 19, and 22   b.   range = 50; $s^2 = 292.044$; $s = 17.09$

3.115 $\bar{x} = 5.16$ inches; $s^2 = 6.994$; $s = 2.64$ inches

3.117   a.   i.   at least 75%   ii.   at least 88.89%   b.   160 to 240 minutes

3.119   a.   i.   68%   ii.   95%   b.   140 to 260 minutes

3.121   a. $Q_1 = 17$; $Q_2 = 19$; $Q_3 = 22$; $IQR = 5$ b. $P_{70} = 22$ c.   30%

3.123   The data set is skewed slightly to the right.   **3.125**   The minimum score is 169.

3.127   a.   new mean = 76.4 inches;   new median = 78 inches;   new range = 13 inches   b.   new mean = 75.2 inches

3.129   mean = $22.31 per barrel   **3.131**   a.   trimmed mean = 9.5   b.   14.29%

3.133   a.   low-risk: rate for A = .01;   rate for B = .02

    b. high-risk: rate for A = .05;   rate for B = .08

    c. overall: rate for A = .04;   rate for B = .03

    d. A was used in more of the high-risk surgeries, for which both anesthetics had higher death rates.

3.135   **a.** $k = 1.41$ **b.** $k = 2.24$

Self-Review Test

1. b   **2.** a and d   **3.** c   **4.** c   **5.** b   **6.** b   **7.** a

8. a   **9.** b   **10.** a   **11.** b   **12.** c   **13.** a   **14.** a

15. mean = 9.8; median = 8; mode = 6; range = 20; $s^2 = 42.178$; $s = 6.49$

19. **b.** $\bar{x} = 18.98$; $s^2 = 44.760$; $s = 6.69$

20. **a.** **i.** at least 84%   **ii.** at least 88.89%   **b.** 2.9 to 11.7 years

21. **a.** **i.** 68%   **ii.** 99.7%   **b.** 2.9 to 11.7 years

22. **a.** $Q_1 = 6$; $Q_2 = 10.5$; $Q_3 = 16.5$; $IQR = 10.5$ **b.** $P_{68} = 15$ **c.** 68.75%

23. Data are skewed slightly to the right.   **24.** combined mean = \$466.43   **25.** GPA of fifth student = 3.17

26. 10% trimmed mean = 197.75; trimmed mean is a better measure

27. **a.** mean for data set I = 19.75; mean for data set II = 16.75. The mean of the second data set is equal to the mean of the first data set minus 3.

    b. $s = 11.32$ for both data sets.

CHAPTER 4

4.3 $S = \{AB, AC, BA, BC, CA, CB\}$ **4.5**   four possible outcomes; $S = \{RR, RG, GR, GG\}$

4.7   four possible outcomes; $S = \{DD, DG, GD, GG\}$ **4.9** $S = \{HHH, HHT, HTH, HTT, THH, THT, TTH, TTT\}$

4.11   **a.** {RG and GR}; a compound event   **b.** { RG, GR, and RR}; a compound event

    c. {RG, GR, and RR}; a compound event   **d.** {GR}; a simple event

4.13   **a.** {DG, GD, and GG}; a compound event   **b.** {DG and GD}; a compound event

    c. {GD}; a simple event   **d.** {DD, DG, and GD}; a compound event

4.19 $-.35, 1.56, 5/3, -2/7$ **4.21**   not equally likely events; use relative frequency approach

4.23   subjective probability   **4.25** **a.** .4000   **b.** .6000   **4.27** .3000   **4.29** .5400

4.31   **a.** .2000   **b.** .8000   **4.33** .3333; .6667   **4.35** .3000; .7000

4.37   **a.** .3760   **b.** .1170   **4.39** use relative frequency approach   **4.45** 1296

4.47   **a.** no   **b.** no   **c.** $\bar{A} = \{1, 3, 4, 6, 8\}$; $\bar{B} = \{1, 3, 5, 6, 7\}$; $P(\bar{A}) = .6250$; $P(\bar{B}) = .6250$ **4.49**   15

4.51   960

4.53   **a.** **i.** .3929   **ii.** .7262   **iii.** .3137   **iv.** .4262

    b. Events "smoker" and "college graduate" are not mutually exclusive. Events "smoker" and "nonsmoker" are mutually exclusive.

    c. Events "smoker" and "not a college graduate" are dependent.

4.55   **a.** **i.** .5575   **ii.** .4425   **iii.** .5636   **iv.** .4576

    b. Events "male" and "in favor" are not mutually exclusive. Events "in favor" and "against" are mutually exclusive.

    c. Events "female" and "in favor" are dependent.

4.57   **a.** **i.** .1725   **ii.** .4685   **iii.** .6337   **iv.** .1550

    b. Events "man" and "commutes for more than one hour" are not mutually exclusive. Events "less than 30 minutes" and "more than one hour" are mutually exclusive.

    c. Events "woman" and "commutes for 30 minutes to one hour" are dependent.

4.59   Events "female" and "pediatrician" are dependent but not mutually exclusive.

4.61   Events "female" and "business major" are dependent but not mutually exclusive.

4.63 $P(A) = .3333$; $P(\bar{A}) = .6667$ **4.65** .35   **4.71**   **a.** .4543   **b.** .0420

4.73   **a.** .1720   **b.** .1824   **4.75**   **a.** .1960   **b.** .0990   **4.77** .6923   **4.79** .7250

4.81　a.　i.　.1958　　ii.　.2028　　b.　.0000　　4.83　a.　i.　.3095　　ii.　.1905

4.85　a.　i.　.2250　　ii.　.0350　　b.　.0000　　4.87　.2381　　4.89　.2667　　4.91　.0576

4.93　a.　.0025　　b.　.9025　　4.95　.4219　　4.97　.5278　　4.99　.4727

4.105　a.　.5400　　b.　.8600　　4.107　a.　.4200　　b.　.4700　　4.109　a.　.4825　　b.　.9161

4.111　a.　.5833　　b.　.6905　　4.113　a.　.7800　　b.　.5500　　c.　.7900　　4.115　.9100

4.117　.8100　　4.119　.8386　　4.121　.8800　　4.123　.9220　　4.125　a.　.2571　　b.　.1429

4.127　a.　i.　.4360　　ii.　.4800　　iii.　.3462　　iv.　.6809　　v.　.3400　　vi.　.6600

　　　b.　Events "female" and "prefers watching sports" are dependent but not mutually exclusive.

4.129　a.　i.　.7500　　ii.　.7000　　iii.　.2250　　iv.　.7750

　　　b.　Events "student athlete" and "should be paid" are dependent but not mutually exclusive.

4.131　a.　.1211　　b.　.8789　　4.133　.0605　　4.135　.0048　　4.137　a.　17,576,000　　b.　2600

4.139　b.　.4000　　c.　.6000

4.141　a.　.5000　　b.　.3333

　　　c.　No; the sixth toss is independent of the first five tosses. Equivalent to part a.

4.143　.0300

Self-Review Test

1.　a　　2.　b　　3.　c　　4.　a　　5.　a　　6.　b

7.　c　　8.　b　　9.　b　　10.　c　　11.　b　　12.　36

13.　a.　.3333　　b.　.6667

14.　a.　Events "male" and "married" are dependent but not mutually exclusive.　　b.　i.　.4000　　ii.　.6786

15.　.7800　　16.　.1595　　17.　.4225　　18.　.6000　　19.　a.　.1680　　b.　.6880

20.　a.　i.　.3600　　ii.　.2750　　iii.　.3500　　iv.　.6500

　　　b.　Events "male" and "yes" are dependent but not mutually exclusive.

CHAPTER 5

5.3　a.　discrete random variable　　b.　continuous random variable　　c.　continuous random variable

　　　d.　discrete random variable　　e.　discrete random variable　　f.　continuous random variable

5.5　discrete random variable

5.9　a.　not a valid probability distribution　　b.　not a valid probability distribution　　c.　a valid probability distribution

5.11　a.　.13　　b.　.16　　c.　.62　　d.　.38　　e.　.38　　f.　.31　　g.　.72

5.13　b.　i.　.14　　ii.　.50　　iii.　.50　　iv.　.44

5.15　a.

x	2	3	4	5	6
$P(x)$	.12	.21	.34	.19	.14

　　　b.　approximate　　c.　i.　.21　　ii.　.88　　iii.　.67　　　.33

5.17

x	0	1	2
$P(x)$	.0676	.3848	.5476

5.19

x	0	1	2
$P(x)$	.3844	.4712	.1444

5.21

x	0	1	2
$P(x)$	.1474	.5052	.3474

5.23　a.　$\mu = 1.670$; $\sigma = .906$　　b.　$\mu = 7.190$; $\sigma = 1.102$　　5.25　$\mu = .440$ errors; $\sigma = .852$ errors

5.27　$\mu = 2.860$ camcorders; $\sigma = 1.435$ camcorders　　5.29　$\mu = 1.000$ head; $\sigma = .707$ heads

5.31　$\mu = 1.896$ sets; $\sigma = 1.079$ sets　　5.33　$\mu = .100$ cars; $\sigma = .308$ cars

5.35　$\mu = \$38,250$; $\sigma = \$18,660$　　5.37　$\mu = 1.400$ parts; $\sigma = .611$ parts

5.39 $3! = 6$; $(7-3)! = 24$; $9! = 362,880$; $(14-12)! = 2$;

$\binom{5}{3} = 10$; $\binom{7}{4} = 35$; $\binom{9}{3} = 84$; $\binom{6}{0} = 1$; $\binom{3}{3} = 1$

5.41 105 5.43 220 5.45 15,504 5.47 125,970

5.51 a. not a binomial experiment b. a binomial experiment c. a binomial experiment

5.53 a. .2787 b. .0756 c. .2458 **5.55** b. $\mu = 2.100$; $\sigma = 1.212$

5.59 a. 0, 1, 2, 3, 4, 5, 6, 7, 8, 9, 10, 11, 12 b. .0836 5.61 a. .7215 b. .7419 c. .0501

5.63 a. .1878 b. .0006 c. .0554 5.65 a. .2725 b. .0839

5.67 a. i. .1115 ii. .0060 b. i. .3670 ii. .6330 iii. .7780

5.69 a. $\mu = 5.600$ customers; $\sigma = 1.296$ customers b. .0467 5.73 a. .1991 b. .0771

5.75 a. $\mu = .600$; $\sigma = .775$ b. $\mu = 1.800$; $\sigma = 1.342$ 5.77 .0982 5.79 .0012

5.81 a. .0998 b. i. .3301 ii. .0430 iii. .8571

5.83 a. .0682 b. i. .2388 ii. .1898 iii. .3225

5.85 a. .0203 b. i. .0257 ii. .9743

5.87 a. .5488 c. $\mu = .600$ accidents; $\sigma = .775$ accidents

5.89 a. .0418 b. i. .0697 ii. .5942 iii. .1249 5.91 $\mu = 2.040$ trips; $\sigma = 1.174$ trips

5.93 a.

x	\$350	$-\$99,650$
$P(x)$	.998	.002

b. $\mu = \$150$; $\sigma = \$4467.66$

5.95 a. .0393 b. .3781 c. .4186 5.97 a. $\mu = 1.500$ returns; $\sigma = 1.162$ returns b. .0105

5.99 a. .1239 b. i. .0084 ii. .2255 iii. .6858 5.101 a. .0383 5.103 $p = .3214$

5.105 a. .1840 b. assume that conditions for binomial random variable are satisfied

5.107 a. 36.5 b. no, his expected score would decrease c. no, her expected score would be unchanged

Self-Review Test

2. probability distribution table 3. a 4. b 5. a 7. b 8. a 9. b

10. a 11. c 12. a 14. $\mu = 1.980$ sales; $\sigma = 1.386$ sales

15. a. i. .1678 ii. .8821 iii. .0017 b. $\mu = 8.400$; $\sigma = 1.587$

16. a. i. .0361 ii. .9473 iii. .0526

CHAPTER 6

6.11 .8664 6.13 .9876 6.15 a. .4713 b. .4599 c. .0967 d. .0603 e. .9429

6.17 a. .0594 b. .0244 c. .9798 d. .9686

6.19 a. .5 approximately b. .5 approximately c. .00 approximately d. .00 approximately

6.21 a. .9626 b. .4830 c. .4706 d. .0838

6.23 a. .0207 b. .2430 c. .1841 d. .9564

6.25 a. .7823 b. .8553 c. .5 approximately d. .5 approximately e. .00 approximately
     f. .00 approximately

6.27 a. 1.40 b. -2.20 c. -1.40 d. 2.80 6.29 a. .4599 b. .1210 c. .2223

6.31 a. .3336 b. .9564 c. .9564 d. .00 approximately 6.33 a. .2178 b. .5997

6.35 a. .7967 b. .3372 c. .0475 d. .7734 6.37 a. .0162 b. .3069

6.39 a. .2005 b. .5410 6.41 a. 18.39% b. 79.65% 6.43 a. .0548 b. .2037

6.45 a. .6808 b. 12.83% 6.47 a. .2578 b. .2764

6.49 a. 13.35% b. 67.00% c. 22.36%

6.51   2.64%   **6.53**   **a.**   2.00   **b.** -2.02 approximately   **c.** $-.37$ approximately   **d.**   1.02 approximately

6.55   **a.**   1.65   **b.** -1.96 **c.** -2.33 approximately   **d.**   2.58 approximately

6.57   **a.**   208.50 approximately   **b.**   241.25 approximately   **c.**   178.50   **d.**   145.75   **e.**   158.25
  f.   251.25 approximately

6.59   19 minutes approximately   **6.61**   2060 kilowatt hours approximately   **6.63** $62.02 approximately

6.65 $np > 5$ and $nq > 5$ **6.67**   **a.**   .6928   **b.**   .6922 ; difference is .0006

6.69   **a.** $\mu = 72$; $\sigma = 5.367$ **b.**   .3897   **c.**   .3102   **6.71**   **a.**   .0764   **b.**   .6793   **c.**   .8413   **d.**   .8238

6.73   .3354   **6.75**   **a.**   .0383   **b.**   .4483   **c.**   .5401   **6.77**   **a.**   .0484   **b.**   .0192   **c.**   .1525

6.79   **a.**   .7549   **b.**   .2451

6.81   **a.**   .0359   **b.**   2.35%   **c.**   29.02%   **d.**   It is possible, but its probability is close to zero.

6.83   1.24%   **6.85**   **a.**   848 hours   **b.**   792 hours approximately   **6.87**   16.23 ounces

6.89   **a.**   .0151   **b.**   .0465   **c.**   .8340   **d.**   .2540   **6.91**   .0039

6.93   84% of Plant B's pints are satisfactory, compared to 68.26% of Plant A's.   **6.95**   8:10 AM approximately

6.97   **a.**   95.24 approximately   **b.**   41.68%

6.99   **b.**   The probability is .4866 for single-number bets; .3974 for color bets.

Self-Review Test

1. a   **2.** a   **3.** d   **4.** b   **5.** a   **6.** c   **7.** b   **8.** b

9. **a.** .1823   **b.** .9264   **c.** .1170   **d.** .7611

10. **a.** -1.28 approximately   **b.** .61   **c.** 1.65 approximately   **d.** -1.07 approximately

11. **a.** .7492   **b.** .2033   **c.** .3372   **d.** .2897

12. **a.** 5.02 hours approximately   **b.** 8.54 hours approximately

13. **a.** .0461   **b.** .2072   **c.** .0901   **d.** .0869   **e.** .7451   **f.** 1.00 approximately   **g.** .00 approximately

Chapter 7

7.5   **a.**   15.90   **b.**   sampling error $= -.34$ **c.**   sampling error $= -.34$;   nonsampling error $= 1.11$
  d. $\bar{x}_1 = 15.44$; $\bar{x}_2 = 14.89$; $\bar{x}_3 = 16.22$; $\bar{x}_4 = 15.56$; $\bar{x}_5 = 16.67$; $\bar{x}_6 = 16$; $\bar{x}_7 = 16.44$; $\bar{x}_8 = 16.89$;
  $\bar{x}_9 = 15.78$; $\bar{x}_{10} = 15.11$;   sampling errors: $-.46$; -1.01; .32; $-.34$; .77; .10; .54; .99; $-.12$; $-.79$

7.7   **b.** $\bar{x}_1 = 28.4$; $\bar{x}_2 = 28.8$; $\bar{x}_3 = 33.8$; $\bar{x}_4 = 34.4$; $\bar{x}_5 = 35.2$; $\bar{x}_6 = 36.4$;   **c.** $\mu = 32.83$

7.13   **a.** $\mu_{\bar{x}} = 60$; $\sigma_{\bar{x}} = 2.828$ **b.** $\mu_{\bar{x}} = 60$; $\sigma_{\bar{x}} = 1.265$ **7.15**   **a.** $\sigma_{\bar{x}} = 1.120$ **b.** $\sigma_{\bar{x}} = 2.000$

7.17   **a.** $n = 25$ **b.** $n = 64$ **7.19** $\mu_{\bar{x}} = 34.2$; $\sigma_{\bar{x}} = .632$ **7.21** $\mu_{\bar{x}} = \$51,400$; $\sigma_{\bar{x}} = \$1850$

7.23 $n = 400$ **7.25**   **a.** $\mu_{\bar{x}} = 80.600$ **b.** $\sigma_{\bar{x}} = 3.302$ **d.** $\sigma_{\bar{x}} = 3.302$

7.33 $\mu_{\bar{x}} = 68$ miles per hour; $\sigma_{\bar{x}} = .671$ miles per hour;   the normal distribution

7.35 $\mu_{\bar{x}} = 3.02$; $\sigma_{\bar{x}} = .042$;   approximately normal distribution

7.37 $\mu_{\bar{x}} = \$70$; $\sigma_{\bar{x}} = \$2.635$;   approximately normal distribution

7.39 $\mu_{\bar{x}} = \$50,980$; $\sigma_{\bar{x}} = \$704.361$;   approximately normal distribution   **7.41**   86.64%

7.43   **a.** $z = 2.44$ **b.** $z = -7.25$ **c.** $z = -3.65$ **d.** $z = 5.82$ **7.45**   **a.**   .1940   **b.**   .8749

7.47   **a.**   .00 approximately   **b.**   .9505   **7.49**   **a.**   .1093   **b.**   .0322   **c.**   .7776

7.51   **a.**   .1588   **b.**   .7498   **c.**   .0618   **7.53**   **a.**   .8664   **b.**   .0668

7.55   **a.**   .2373   **b.**   .9624   **c.**   .0418   **7.57**   **a.**   .3085   **b.**   .8543   **c.**   .8664   **d.**   .1056

7.59   .0124   **7.61** $p = .12$; $\hat{p} = .15$ **7.63**   7125 subjects in the population; 312 subjects in the sample

7.65   sampling error $= -.06$ **7.71**   **a.** $\mu_{\hat{p}} = .18$; $\sigma_{\hat{p}} = .019$ **b.** $\mu_{\hat{p}} = .18$; $\sigma_{\hat{p}} = .014$

7.73   **a.** $\sigma_{\hat{p}} = .051$;   **b.** $\sigma_{\hat{p}} = .070$ **7.75**   in cases a and c

7.77   **a.** $p = .667$ **b.**   6   **d.**   sampling errors: $-.067, -.067, .133, .133, -.067, -.067$

7.79 $\mu_{\hat{p}} = .73$; $\sigma_{\hat{p}} = .020$; approximately normal distribution

7.81 $\mu_{\hat{p}} = .20$; $\sigma_{\hat{p}} = .057$;   approximately normal distribution   **7.83**   95.44%

7.85   a. $z = -.82$ b. $z = 1.63$ c. $z = -1.43$ d. $z = 1.02$ 7.87   a.   .9402   b.   .1949

7.89   a.   .4122   b.   .7422   7.91   .2005

7.93 $\mu_{\bar{x}} = 750$ hours; $\sigma_{\bar{x}} = 10$ hours;   the normal distribution

7.95   a.   .9332   b.   .1525   c.   .8664   d.   .0228

7.97   a.   .0125   b.   .8861   c.   .7850   d.   .0064

7.99 $\mu_{\hat{p}} = .10$; $\sigma_{\hat{p}} = .034$;   approximately normal distribution

7.101   a.   i.   .0146   ii.   .0907   b.   .9912   c.   .0146

7.103   .9544   7.105   10 approximately   7.107   a.   .8023   b.   754 approximately

Self-Review Test

1. b   2. b   3. a   4. a   5. b   6. b   7. c   8. a   9. b   10. a   11. c   12. a

14.   a. $\mu_{\bar{x}} = 145$ pounds; $\sigma_{\bar{x}} = 3.600$ pounds;   approximately normal distribution
     b. $\mu_{\bar{x}} = 145$ pounds; $\sigma_{\bar{x}} = 1.800$ pounds;   approximately normal distribution

15.   a. $\mu_{\bar{x}} = 47$ minutes; $\sigma_{\bar{x}} = 1.878$ minutes;   unknown distribution
     b. $\mu_{\bar{x}} = 47$ minutes; $\sigma_{\bar{x}} = 1.004$ minutes;   approximately normal distribution

16.   a.   .2162   b.   .0179   c.   .8413   d.   .8709

17.   a.   i.   .1203   ii.   .1335   iii.   .7486   b.   .9736   c.   .0013

18.   a. $\mu_{\hat{p}} = .92$; $\sigma_{\hat{p}} = .038$;   approximately normal distribution
     b. $\mu_{\hat{p}} = .92$; $\sigma_{\hat{p}} = .019$;   approximately normal distribution
     c. $\mu_{\hat{p}} = .92$; $\sigma_{\hat{p}} = .009$;   approximately normal distribution

19.   a.   i.   .2005   ii.   .8525   iii.   .7995   iv.   .1831   b.   .9652   c.   .0455   d.   .1020

CHAPTER 8

8.11   a.   22.5   b. $\pm.83$ c.   21.40 to 23.60   d.   1.10

8.13   a.   71.33 to 78.27   b.   70.68 to 78.92   c.   69.38 to 80.22   d.   yes

8.15   a.   76.19 to 81.61   b.   77.09 to 80.71   c.   77.27 to 80.53   d.   yes

8.17   a.   53.93 to 56.71   b.   56.16 to 58.64   c.   54.95 to 57.55
     d.   Confidence intervals of parts a and c cover μ, but the confidence interval of part b does not cover μ.

8.19   a.   38.20   b. $\pm.81$ c.   37.24 to 39.16   d.   .96   8.21 $1024.61 to $1127.39

8.23   8.13 to 8.27 hours

8.25   a.   5.28 kilograms; margin of error $= \pm .07$ kilograms   b.   5.20 to 5.36 kilograms

8.27   a.   4.65 to 4.95 nights   8.29   35.98 to 36.06 ounces; the machine needs to be adjusted

8.31   a. $206.04 to $239.96   8.39   a.   1.771   b. -2.080 c. -3.646 d.   2.787

8.41   a.   .025; right tail   b.   .001; left tail   c.   .01; left tail   d.   .005; right tail

8.43   a.   2.921   b.   2.069   c.   1.729   8.45   a.   12.25   b.   9.71 to 14.79   c.   2.54

8.47   a.   60.92 to 70.08   b.   61.73 to 69.27   c.   61.95 to 69.05   8.49   16.69 to 21.31 minutes

8.51   164.27 to 185.73 beats per minute   8.53 $106.04 to $133.96

8.55   a.   24.71 to 28.09 miles per gallon   8.57   66.72 to 72.68 mph

8.63   a.   yes, sample size is large   b.   no, sample size is not large   c.   yes, sample size is large
     d.   no, sample size is not large

8.65   a.   .583 to .677   b.   .542 to .638   c.   .624 to .716
     d.   confidence intervals of parts a and c cover .65, but the confidence interval of part b does not.

8.67   a.   .687 to .753   b.   .681 to .759   c.   .668 to .772   d.   yes

8.69   a.   .549 to .791   b.   .620 to .720   c.   .639 to .701   d.   yes

8.71   a.   .49; margin of error $= \pm .036$ b.   .443 to .537

8.73   a.   .92; margin of error $= \pm .027$ b.   .893 to .947

8.75   **a.**   .719 to .961   **8.77**   **a.**   .671 to .789

8.79   **a.**   .60;   margin of error $= \pm$ .248   **b.**   35.2% to 84.8%   **8.83**   **a.** $n = 118$ **b.** $n = 46$

8.85   **a.** $n = 299$ **b.** $n = 126$ **c.** $n = 61$ **8.87** $n = 84$ **8.89** $n = 221$

8.91   **a.** $n = 1046$ **b.** $n = 1359$ **8.93**   **a.** $n = 1849$ **b.** $n = 601$ **c.** $n = 6807$

8.95 $n = 4161$ **8.97** $n = 1350$ **8.99**   **a.**   \$6.75; margin of error $= \pm$ \$.16   **b.**   \$6.59 to \$6.91

8.101   23.989 to 24.041; the machine needs to be adjusted

8.103   **a.**   \$161.97; margin of error $= \pm$ \$15.95   **b.**   \$143.01 to \$180.93

8.105   \$385.32 to \$454.68   **8.107**   9.40 to 11.10 hours   **8.109**   1.92 to 2.54 hours

8.111   **a.**   .12; margin of error $= \pm$ .090   **b.**   .001 to .239   **8.113**   .117 to .683   **8.115** $n = 42$

8.117 $n = 1068$ **8.119**   **a.** $\bar{x} = \$12.90$ **b.**   \$10.99 to \$14.81

8.121   .793 to .867;   .593 to .687;   .471 to .569;   .079 to .141

Self-Review Test

1. **a.**   population parameter; sample statistic   **b.**   sample statistic; population parameter
    c.   sample statistic; population parameter

2. b   3. a   4. a   5. d   6. b   7. **a.**   \$2.85; margin of error $= \pm$ \$.20   **b.**   \$2.59 to \$3.11

8. \$266,478.30 to \$328,249.70   9. **a.**   .58; margin of error $= \pm$ .041   **b.**   .539 to .621

10. $n = 33$ 11. $n = 2663$ 12. $n = 1125$

CHAPTER 9

9.5   **a.**   a left-tailed test   **b.**   a right-tailed test   **c.**   a two-tailed test

9.7   **a.**   Type II error   **b.**   Type I error

9.9   **a.** $H_0:\mu = \$143,000$; $H_1:\mu > \$143,000$; a right-tailed test   **b.** $H_0:\mu = 15$ hours; $H_1:\mu \neq 15$ hours; a two-tailed test
    c. $H_0:\mu = 45$ months; $H_1:\mu < 45$ months; a left-tailed test   **d.** $H_0:\mu = 5$ hours; $H_1:\mu \neq 5$ hours; a two-tailed test
    e. $H_0:\mu = 24$ years; $H_1:\mu \neq 24$ years; a two-tailed test

9.15   **a.**   rejection region is to the left of -1.96 and to the right of 1.96; nonrejection region is between -1.96 and 1.96
    b.   rejection region is to the left of -2.33; nonrejection region is to the right of -2.33
    c.   rejection region is to the right of 2.05; nonrejection region is to the left of 2.05

9.17   statistically significant   **9.19**   **a.**   .025   **b.**   .05   **c.**   .01

9.21   **a.**   observed value of z is 6.67; critical value of z is 1.65
    b.   observed value of z is 6.67; critical values of z are -1.96 and 1.96

9.23   **a.**   reject H_0 if $z < -1.65$ **b.**   reject H_0 if $z < -1.96$ or $z > 1.96$ **c.**   reject H_0 if $z > 1.65$

9.25   **a.**   critical values: $z = -2.58$ and 2.58; test statistic: $z = -1.33$; do not reject H_0
    b.   critical values: $z = -2.58$ and 2.58; test statistic: $z = 3.20$; reject H_0

9.27   **a.**   critical values: $z = -2.58$ and 2.58; test statistic: $z = 9.00$; reject H_0
    b.   critical value: $z = -1.65$; test statistic: $z = -1.49$; do not reject H_0
    c.   critical value: $z = 1.28$; test statistic: $z = 8.57$; reject H_0

9.29 $H_0:\mu = \$220$; $H_1:\mu > \$220$; critical value: $z = 1.96$; test statistic: $z = 2.89$; reject H_0

9.31 $H_0:\mu = \$53,459$; $H_1:\mu > \$53,459$; critical value: $z = 2.05$; test statistic: $z = 2.96$; reject H_0

9.33   **a.** $H_0:\mu = \$15.66$; $H_1:\mu \neq \$15.66$; critical values : $z = -2.58$ and 2.58; test statistic: $z = -4.13$; reject H_0
    b. P (Type I error) $= .01$

9.35   **a.** $H_0:\mu = 11$ hours; $H_1:\mu < 11$ hours; critical value: $z = -2.33$; test statistic: $z = -6.82$; reject H_0
    b.   do not reject H_0

9.37   **a.** $H_0:\mu = 45$ months; $H_1:\mu < 45$ months; critical value: $z = -1.96$; test statistic: $z = -1.88$; do not reject H_0
    b.   critical value: $z = -1.65$; test statistic: $z = -1.88$; reject H_0

9.39   critical values: $z = -2.58$ and 2.58; test statistic: $z = 1.90$; do not reject H_0; the machine does not need adjustment

9.45 **a.** p-value = .0250   **b.** p-value = .0082   **c.** p-value = .0113

9.47 **a.** p-value = .0192   **b.** no, do not reject H_0 **c.** yes, reject H_0

9.49 $H_0: \mu = 2.4$ days; $H_1: \mu < 2.4$ days; test statistic: $z = -2.72$; p-value = .0033

9.51 $H_0: \mu \geq 14$ hours; $H_1: \mu < 14$ hours; test statistic: $z = -2.05$; p-value = .0202

9.53 **a.** $H_0: \mu = 10$ minutes; $H_1: \mu < 10$ minutes; test statistic: $z = -1.92$; p-value = .0274
   b. no, do not reject H_0 **c.** yes, reject H_0

9.55 **a.** test statistic: $z = -3.94$; p-value = .00 approximately   **b.** yes, adjust for $\alpha = .01$; yes, adjust for $\alpha = .05$

9.57 **a.** reject H_0 if $t < -2.539$ or $t > 2.539$ **b.** reject H_0 if $t < -2.602$ **c.** reject H_0 if $t > 1.740$

9.59 **a.** critical value: $t = 2.492$; observed value: $t = 2.333$
   b. critical values: $t = -2.797$ and 2.797; observed value: $t = 2.333$

9.61 **a.** reject H_0 if $t < -2.539$ **b.** reject H_0 if $t < -2.861$ or $t > 2.861$ **c.** reject H_0 if $t > 2.539$

9.63 **a.** critical values: $t = -2.797$ and 2.797; test statistic: $t = -1.875$; do not reject H_0
   b. critical values: $t = -2.797$ and 2.797; test statistic: $t = 5.000$; reject H_0

9.65 **a.** critical values: $t = -2.797$ and 2.797; test statistic: $t = 4.082$; reject H_0
   b. critical value: $t = -2.131$; test statistic: $t = -1.818$; do not reject H_0
   c. critical value: $t = 1.328$; test statistic: $t = 2.236$; reject H_0

9.67 $H_0: \mu = 69.5$ inches; $H_1: \mu \neq 69.5$ inches; critical values: $t = -2.797$ and 2.797; test statistic: $t = 1.667$; do not reject H_0

9.69 $H_0: \mu \leq 7$ hours; $H_1: \mu > 7$ hours; critical value: $t = 2.093$; test statistic: $t = 7.194$;   reject H_0

9.71 $H_0: \mu \leq 30$ calories; $H_1: \mu > 30$ calories; critical value: $t = 1.753$; test statistic: $t = 2.400$; reject H_0

9.73 **a.** $H_0: \mu \leq 45$ minutes; $H_1: \mu > 45$ minutes; critical value: $t = 2.539$; test statistic: $t = 7.454$; reject H_0
   b. P(Type I error) = .01

9.75 **a.** yes, the claim is true
   b. $H_0: \mu \geq 1200$; $H_1: \mu < 1200$; critical value: $t = -1.711$; test statistic: $t = -4.118$; reject H_0

9.77 $H_0: \mu = 18$ hours; $H_1: \mu \neq 18$ hours; critical values: $t = -2.262$ and 2.262; test statistic: $t = 2.692$; reject H_0

9.83 **a.** large enough   **b.** not large enough   **c.** not large enough   **d.** large enough

9.85 **a.** reject H_0 if $z < -1.96$ or $z > 1.96$ **b.** reject H_0 if $z < -2.05$ **c.** reject H_0 if $z > 1.96$

9.87 **a.** critical value: $z = -2.33$; observed value: $z = -1.17$
   b. critical values: $z = -2.58$ and 2.58; observed value: $z = -1.17$

9.89 **a.** reject H_0 if $z < -2.33$ **b.** reject H_0 if $z < -2.58$ or $z > 2.58$ **c.** reject H_0 if $z > 2.33$

9.91 **a.** critical value: $z = -1.96$; test statistic: $z = -1.61$; do not reject H_0
   b. critical value: $z = -1.96$; test statistic: $z = -2.21$; reject H_0

9.93 **a.** critical values: $z = -1.96$ and 1.96; test statistic: $z = -3.43$; reject H_0
   b. critical value: $z = -2.33$; test statistic: $z = -1.82$; do not reject H_0
   c. critical value: $z = 1.96$; test statistic: $z = .86$; do not reject H_0

9.95 $H_0: p = .49$; $H_1: p < .49$; critical value: $z = -2.33$; test statistic: $z = -2.00$; do not reject H_0

9.97 $H_0: p = .59$; $H_1: p \neq .59$; critical values: $z = -1.96$ and 1.96; test statistic: $z = 2.57$; reject H_0

9.99 **a.** $H_0: p = .63$; $H_1: p \neq .63$; critical values: $z = -1.96$ and 1.96; test statistic: $z = -1.43$; do not reject H_0
   b. P(Type I error) = .05

9.101 **a.** $H_0: p \geq .60$; $H_1: p < .60$; critical value: $z = -2.33$; test statistic: $z = -2.86$; reject H_0 **b.** do not reject H_0

9.103 **a.** critical value: $z = 2.05$; test statistic: $z = 2.22$; reject H_0; adjust machine
   b. critical value: $z = 2.33$; test statistic: $z = 2.22$; do not reject H_0; do not adjust the machine

9.107 **a.** critical values: $z = -2.33$ and 2.33; test statistic: $z = -2.13$; do not reject H_0
   b. P(Type I error) = .02   **c.** p-value = .0332; do not reject H_0 if $\alpha = .01$; reject H_0 if $\alpha = .05$

9.109 **a.** critical value: $z = -2.05$; test statistic: $z = -2.14$; reject H_0 **b.** P(Type I error) = .02
   c. p-value = .0162; do not reject H_0 if $\alpha = .01$; reject H_0 if $\alpha = .025$

9.111 **a.** $H_0: \mu = \$2865$; $H_1: \mu > \$2865$; critical value: $z = 2.33$; test statistic: $z = 2.85$; reject H_0
   b. P(Type I error) = .01   **c.** yes

9.113 **a.** $H_0: \mu = 111.9$ pounds; $H_1: \mu \neq 111.9$ pounds; critical values: $z = -1.96$ and 1.96; test statistic: $z = -2.04$; reject H_0
   b. do not reject H_0 if $\alpha = .01$

9.115 $H_0: \mu = 8$ minutes; $H_1: \mu < 8$ minutes; critical value: $z = -1.96$; test statistic: $z = -1.62$; do not reject H_0

9.117 $H_0: \mu = 1245$ cubic feet; $H_1: \mu < 1245$ cubic feet; test statistic: $z = -2.80$; p-value $= .0026$

9.119 $H_0: \mu = 25$ minutes; $H_1: \mu \neq 25$ minutes; critical values: $t = -2.947$ and 2.947; test statistic: $t = 2.083$; do not reject H_0

9.121 a. $H_0: \mu = 114$ minutes; $H_1: \mu < 114$ minutes; critical value: $t = -2.492$; test statistic: $t = -2.273$; do not reject H_0
  b. do not reject H_0

9.123 a. $H_0: \mu \leq 150$ calories; $H_1: \mu > 150$ calories; critical value: $t = 2.262$; test statistic: $t = 1.157$; do not reject H_0

9.125 a. $H_0: p = .11$; $H_1: p > .11$; critical value: $z = 2.33$; test statistic: $z = 1.96$; do not reject H_0
  b. P(Type I error) $= .01$

9.127 $H_0: p = .78$; $H_1: p \neq .78$; critical values: $z = -1.96$ and 1.96; test statistic: $z = 1.62$; do not reject H_0

9.129 a. $H_0: p \geq .90$; $H_1: p < .90$; critical value: $z = -2.05$; test statistic: $z = -2.11$; reject H_0 b. do not reject H_0

9.131 $H_0: p \geq .90$; $H_1: p < .90$; p-value $= .0174$; reject H_0 if $\alpha = .05$; do not reject H_0 if $\alpha = .01$

9.133 8 approximately

Self-Review Test

1. a   2. b   3. a   4. b   5. a   6. a   7. a   8. b
9. c   10. a   11. c   12. b   13. d   14. c   15. a   16. b

17. a. $H_0: \mu = \$1365$; $H_1: \mu \neq \$1365$; critical values: $z = -2.58$ and 2.58; test statistic: $z = 4.46$; reject H_0
   b. P(Type I error) $= .01$ c. do not reject H_0

18. a. $H_0: \mu \geq 30$ months; $H_1: \mu < 30$ months; critical value: $t = -2.131$; test statistic: $t = -2.778$; reject H_0
   b. P(Type I error) $= .025$ c. critical value: $t = -3.733$; do not reject H_0

19. a. $H_0: p = .23$; $H_1: p < .23$; critical value: $z = -1.65$; test statistic: $z = -2.47$; reject H_0
   b. P(Type I error) $= .05$ c. do not reject H_0

20. a. $H_0: \mu = 336$ minutes; $H_0: \mu \neq 336$ minutes; test statistic: $z = 2.10$; p-value $= .0358$
   b. do not reject H_0 if $\alpha = .01$; reject H_0 if $\alpha = .05$

21. a. p-value $= .0068$ b. reject H_0 if $\alpha = .01$; do not reject H_0 if $\alpha = .005$

CHAPTER 10

10.3 a. .76   b. .29 to 1.23

10.5 $H_0: \mu_1 - \mu_2 = 0$; $H_1: \mu_1 - \mu_2 \neq 0$; critical values: $z = -1.96$ and 1.96; test statistic: $z = 4.19$; reject H_0

10.7 $H_0: \mu_1 - \mu_2 = 0$; $H_1: \mu_1 - \mu_2 > 0$; critical value: $z = 2.33$; test statistic: $z = 4.19$; reject H_0

10.9 a. $\$2432$ b. $-\$3514.54$ to $\$8378.54$
   c. $H_0: \mu_1 - \mu_2 = 0$; $H_1: \mu_1 - \mu_2 \neq 0$; critical values: $z = -1.96$ and 1.96; test statistic: $z = 1.06$; do not reject H_0

10.11 a. $\$.10$ b. $\$.08$ to $\$.12$
   c. $H_0: \mu_1 - \mu_2 = 0$; $H_1: \mu_1 - \mu_2 > 0$; critical value: $z = 2.33$; test statistic: $z = 7.00$; reject H_0

10.13 a. -3.64 to -2.16 days
   b. $H_0: \mu_1 - \mu_2 = 0$; $H_1: \mu_1 - \mu_2 < 0$; critical value: $z = -1.96$; test statistic: $z = -9.15$; reject H_0
   c. P(Type I error) $= .025$

10.15 a. $\$1.65$ to $\$1.93$ b. $H_0: \mu_1 - \mu_2 = 0$; $H_1: \mu_1 - \mu_2 > 0$; critical value: $z = -1.96$; test statistic: $z = 25.17$; reject H_0
   c. do not reject H_0

10.17 a. $-.51$ to $.01$ minutes
   b. $H_0: \mu_1 - \mu_2 = 0$; $H_1: \mu_1 - \mu_2 < 0$; critical value: $z = -1.96$; test statistic: $z = -2.06$; reject H_0
   c. p-value $= .0197$; do not reject H_0 if $\alpha = .01$; reject H_0 if $\alpha = .05$

10.21 a. 5.25   b. 1.21 to 9.29

10.23 $H_0: \mu_1 - \mu_2 = 0$; $H_1: \mu_1 - \mu_2 \neq 0$; critical values: $t = -2.701$ and 2.701; test statistic: $t = 3.514$; reject H_0

10.25 $H_0: \mu_1 - \mu_2 = 0$; $H_1: \mu_1 - \mu_2 > 0$; critical value: $t = 1.683$; test statistic: $t = 3.514$; reject H_0

10.27 a. -5.32 b. -8.33 to -2.31
   c. $H_0: \mu_1 - \mu_2 = 0$; $H_1: \mu_1 - \mu_2 < 0$; critical value: $t = -2.080$; test statistic: $t = -4.996$; reject H_0

10.29 **a.** 1.50 to 16.50 items
      b. $H_0:\mu_1 - \mu_2 = 0$; $H_1:\mu_1 - \mu_2 > 0$; critical value: $t = 2.120$; test statistic: $t = 2.543$; reject H_0

10.31 **a.** $-\$4302.15$ to $-\$1917.85$
      b. $H_0:\mu_1 - \mu_2 = 0$; $H_1:\mu_1 - \mu_2 \neq 0$; critical values: $t = -2.035$ and 2.035; test statistic: $t = -4.414$; reject H_0

10.33 **a.** -14.52 to 4.52 minutes
      b. $H_0:\mu_1 - \mu_2 = 0$; $H_1:\mu_1 - \mu_2 < 0$; critical value: $t = -2.412$; test statistic: $t = -1.413$; do not reject H_0

10.35 **a.** $\$6167.65$ to $\$9332.35$
      b. $H_0:\mu_1 - \mu_2 = 0$; $H_1:\mu_1 - \mu_2 > 0$; critical value: $t = 2.400$; test statistic: $t = 8.204$; reject H_0

10.37 4.49 to 13.23

10.39 $H_0:\mu_1 - \mu_2 = 0$; $H_1:\mu_1 - \mu_2 \neq 0$; critical values: $t = -2.756$ and 2.756; test statistic: $t = 5.588$; reject H_0

10.41 $H_0:\mu_1 - \mu_2 = 0$; $H_1:\mu_1 - \mu_2 > 0$; critical value: $t = 1.699$; test statistic: $t = 5.588$; reject H_0

10.43 **a.** .85 to 17.15 items
      b. $H_0:\mu_1 - \mu_2 = 0$; $H_1:\mu_1 - \mu_2 > 0$; critical value: $t = 2.201$; test statistic: $t = 2.429$; reject H_0

10.45 **a.** $-\$4323.84$ to $-\$1896.16$
      b. $H_0:\mu_1 - \mu_2 = 0$; $H_1:\mu_1 - \mu_2 \neq 0$; critical values: $t = -2.045$ and 2.045; test statistic: $t = -4.353$; reject H_0

10.47 **a.** -14.43 to 4.43 minutes
      b. $H_0:\mu_1 - \mu_2 = 0$; $H_1:\mu_1 - \mu_2 < 0$; critical value: $t = -2.414$; test statistic: $t = -1.428$; do not reject H_0

10.49 **a.** $\$6176.24$ to $\$9323.76$
      b. $H_0:\mu_1 - \mu_2 = 0$; $H_1:\mu_1 - \mu_2 > 0$; critical value: $t = 2.402$; test statistic: $t = 8.249$; reject H_0

10.51 **a.** 10.30 to 40.50   **b.** 11.26 to 15.14   **c.** 28.53 to 40.67

10.53 **a.** critical values: $t = -1.860$ and 1.860; test statistic: $t = 8.040$; reject H_0
      b. critical value: $t = 1.721$; test statistic: $t = 10.847$; reject H_0
      c. critical value: $t = -2.583$; test statistic: $t = -7.989$; reject H_0

10.55 **a.** -2.93 to $.07$ **b.** $H_0:\mu_d = 0$; $H_1:\mu_d < 0$; critical value: $t = -3.143$; test statistic: $t = -2.338$; do not reject H_0

10.57 **a.** -16.55 to -3.21 **b.** $H_0:\mu_d = 0$; $H_1:\mu_d < 0$; critical value: $t = -1.895$; test statistic: $t = -2.809$; reject H_0

10.59 **a.** -4.07 to $.63$ miles per gallon
      b. $H_0:\mu_d = 0$; $H_1:\mu_d < 0$; critical value: $t = -2.571$; test statistic: $t = -2.951$; reject H_0

10.63 $-.177$ to $.057$

10.65 $H_0:p_1 - p_2 = 0$; $H_1:p_1 - p_2 \neq 0$; critical values: $z = -2.58$ and 2.58; test statistic: $z = -1.32$; do not reject H_0

10.67 $H_0:p_1 - p_2 = 0$; $H_1:p_1 - p_2 < 0$; critical value: $z = -2.33$; test statistic: $z = -1.32$; do not reject H_0

10.69 **a.** .04   **b.** $-.024$ to $.104$
      c. rejection region to the right of $z = 1.96$; non-rejection region to the left of $z = 1.96$
      d. test statistic: $z = 1.35$ **e.** do not reject H_0

10.71 **a.** $-.052$ to $-.002$
      b. $H_0:p_1 - p_2 = 0$; $H_1:p_1 - p_2 < 0$; critical value: $z = -2.33$; test statistic: $z = -2.82$; reject H_0

10.73 **a.** $-.139$ to $-.001$
      b. $H_0:p_1 - p_2 = 0$; $H_1:p_1 - p_2 < 0$; critical value: $z = -2.33$; test statistic: $z = -2.327$; do not reject H_0

10.75 **a.** .03   **b.** $-.023$ to $.083$
      c. $H_0:p_1 - p_2 = 0$; $H_1:p_1 - p_2 \neq 0$; critical values: $z = -1.96$ and 1.96; test statistic: $z = 1.10$; do not reject H_0

10.77 **a.** $-.070$ to $.010$
      b. $H_0:p_1 - p_2 = 0$; $H_1:p_1 - p_2 < 0$; critical value: $z = -1.96$; test statistic: $z = -1.57$; do not reject H_0
      c. p-value $= .0582$

10.79 **a.** .53 to .91 hours   **b.** $H_0:\mu_1 - \mu_2 = 0$; $H_1:\mu_1 - \mu_2 > 0$; critical value: $z = 2.33$; test statistic: $z = 9.89$; reject H_0
      c. do not reject H_0

10.81 **a.** .60 to 2.40 policies
      b. $H_0:\mu_1 - \mu_2 = 0$; $H_1:\mu_1 - \mu_2 > 0$; critical value: $z = 2.33$; test statistic: $z = 4.30$; reject H_0

10.83 **a.** $H_0:\mu_1 - \mu_2 = 0$; $H_1:\mu_1 - \mu_2 \neq 0$; critical values: $t = -2.009$ and 2.009; test statistic: $t = -1.058$; do not reject H_0
      b. $-.31$ to $.07$

10.85 **a.** $-\$3.63$ to $\$17.63$
      b. $H_0:\mu_1 - \mu_2 = 0$; $H_1:\mu_1 - \mu_2 > 0$; critical value: $t = 2.416$; test statistic: $t = 1.774$; do not reject H_0

10.87 a. $H_0: \mu_1 - \mu_2 = 0$; $H_1: \mu_1 - \mu_2 \neq 0$; critical values: $t = -2.010$ and 2.010; test statistic: $t = -1.068$; do not reject H_0 b. $-.31$ to $.07$

10.89 a. $-\$3.46$ to $\$17.46$
   b. $H_0: \mu_1 - \mu_2 = 0$; $H_1: \mu_1 - \mu_2 > 0$; critical value: $t = 2.418$; test statistic: $t = 1.805$; do not reject H_0

10.91 a. $-.64$ to $.14$ b. $H_0: \mu_d = 0$; $H_1: \mu_d < 0$; critical value: $t = -2.365$; test statistic: $t = -2.236$; do not reject H_0

10.93 a. $-.109$ to $-.031$
   b. $H_0: p_1 - p_2 = 0$; $H_1: p_1 - p_2 \neq 0$; critical values: $z = -2.33$ and 2.33; test statistic: $z = -3.55$; reject H_0

10.95 a. $.064$ to $.196$
   b. $H_0: p_1 - p_2 = 0$; $H_1: p_1 - p_2 \neq 0$; critical values: $z = -2.58$ and 2.58; test statistic: $z = 3.81$; reject H_0

10.97 a. 35   b. 11.31 to 58.69   **10.99**   .2611

10.101 a. 27.88 to 32.12 beats per minute
   b. $H_0: \mu_d = 0$; $H_1: \mu_d > 0$; critical value: $z = 2.33$; test statistic: $z = 36.51$; reject H_0

10.103 The following should be included in the report.
   Number of contacts: 90% confidence interval for μ_d: -2.38 to 5.24 contacts; $H_0: \mu_d = 0$; $H_1: \mu_d > 0$; critical value: $t = 1.943$; test statistic: $t = .729$; do not reject H_0
   Gas Mileage: 90% confidence interval for μ_d: -3.89 to $-.97$ MPG;
   $H_0: \mu_d = 0$; $H_1: \mu_d < 0$; critical value: $t = -1.943$; test statistic: $t = -3.234$; reject H_0

Self-Review Test

1. a.

3. a. 1.62 to 2.78
   b. $H_0: \mu_1 - \mu_2 = 0$; $H_1: \mu_1 - \mu_2 > 0$; critical value: $z = 1.96$; test statistic: $z = 9.86$; reject H_0

4. a. -2.72 to -1.88 hours
   b. $H_0: \mu_1 - \mu_2 = 0$; $H_1: \mu_1 - \mu_2 < 0$; critical value: $t = -2.416$; test statistic: $t = -10.997$; reject H_0

5. a. -2.70 to -1.90 hours
   b. $H_0: \mu_1 - \mu_2 = 0$; $H_1: \mu_1 - \mu_2 < 0$; critical value: $t = -2.421$; test statistic: $t = -11.474$; reject H_0

6. a. -3.43 to 1.43 items
   b. $H_0: \mu_d = 0$; $H_1: \mu_d \neq 0$; critical values: $t = -2.447$ and 2.447; test statistic: $t = -1.528$; do not reject H_0

7. a. $-.052$ to $.092$
   b. $H_0: p_1 - p_2 = 0$; $H_1: p_1 - p_2 \neq 0$; critical values: $z = -2.58$ and 2.58; test statistic: $z = .60$; do not reject H_0

CHAPTER 11

11.3 $\chi^2 = 43.773$ **11.5** $\chi^2 = 36.191$ **11.7** a. $\chi^2 = 5.009$ b. $\chi^2 = 3.565$

11.13 critical value: $\chi^2 = 11.070$; test statistic: $\chi^2 = 5.200$; do not reject H_0

11.15 critical value: $\chi^2 = 11.345$; test statistic: $\chi^2 = 4.073$; do not reject H_0

11.17 critical value: $\chi^2 = 9.488$; test statistic: $\chi^2 = 7.876$; do not reject H_0

11.19 critical value: $\chi^2 = 17.275$; test statistic: $\chi^2 = 22.853$; reject H_0

11.21 critical value: $\chi^2 = 9.348$; test statistic: $\chi^2 = 68.663$; reject H_0

11.27 a. H_0: the proportion in each row is the same for all four populations
      H_1: the proportion in each row is not the same for all four populations
   c. critical value: $\chi^2 = 14.449$ d. test statistic: $\chi^2 = 54.840$ e. reject H_0

11.29 critical value: $\chi^2 = 5.024$; test statistic: $\chi^2 = .253$; do not reject H_0

11.31 critical value: $\chi^2 = 15.507$; test statistic: $\chi^2 = 25.390$; reject H_0

11.33 critical value: $\chi^2 = 9.210$; test statistic: $\chi^2 = 49.287$; reject H_0

11.35 critical value: $\chi^2 = 6.635$; test statistic: $\chi^2 = 10.445$; reject H_0

11.37 critical value: $\chi^2 = 6.251$; test statistic: $\chi^2 = 28.006$; reject H_0

11.39 critical value: $\chi^2 = 7.378$; test statistic: $\chi^2 = 1.748$; do not reject H_0

11.41   a.   20.5448 to 94.6793   b.   23.7781 to 75.4778   c.   25.7037 to 67.5910

11.43   a. $H_0:\sigma^2 = .80$; $H_1:\sigma^2 > .80$ b.   reject if $\chi^2 > 30.578$ c.   test statistic: $\chi^2 = 21.188$ d.   do not reject H_0

11.45   a. $H_0:\sigma^2 = 2.2$; $H_1:\sigma^2 \neq 2.2$ b.   reject H_0 if $\chi^2 < 5.009$ or $\chi^2 > 24.736$
      c.   test statistic: $\chi^2 = 27.182$ d.   reject H_0

11.47   a. $H_0:\sigma^2 = .003$; $H_1:\sigma^2 > .003$; critical value: $\chi^2 = 48.278$; test statistic: $\chi^2 = 56.933$; reject H_0
      b.   .0035 to .0126; .059 to .112

11.49   a.   2628.6793 to 12,114.1007; 51.271 to 110.064
      b. $H_0:\sigma^2 = 4000$; $H_1:\sigma^2 \neq 4000$; critical values: $\chi^2 = 12.401$ and 39.364; test statistic: $\chi^2 = 29.940$; do not reject H_0

11.51   critical value: $\chi^2 = 7.815$; test statistic: $\chi^2 = 10.824$; reject H_0

11.53   critical value: $\chi^2 = 9.488$; test statistic: $\chi^2 = 11.430$; reject H_0

11.55   critical value: $\chi^2 = 11.345$; test statistic: $\chi^2 = 15.920$; reject H_0

11.57   critical value: $\chi^2 = 15.086$; test statistic: $\chi^2 = 73.362$; reject H_0

11.59   critical value: $\chi^2 = 4.605$; test statistic: $\chi^2 = 13.529$; reject H_0

11.61   critical value: $\chi^2 = 16.812$; test statistic: $\chi^2 = 10.181$; do not reject H_0

11.63   a.   3.6131 to 20.7547; 1.901 to 4.556   b.   8.4502 to 32.3654; 2.907 to 5.689

11.65 $H_0:\sigma^2 = 1.1$; $H_1:\sigma^2 > 1.1$; critical value: $\chi^2 = 28.845$; test statistic: $\chi^2 = 24.727$; reject H_0

11.67 $H_0:\sigma^2 = 10.4$; $H_1:\sigma^2 \neq 10.4$; critical values: $\chi^2 = 8.231$ and 31.526; test statistic: $\chi^2 = 25.615$; do not reject H_0

11.69   a. $H_0:\sigma^2 = 5000$; $H_1:\sigma^2 < 5000$; critical value: $\chi^2 = 8.907$; test statistic: $\chi^2 = 12.255$; do not reject H_0
      b.   1693.1005 to 8027.6431; 41.147 to 89.597

11.71   a.   .1001 to .4613; .316 to .679
      b. $H_0:\sigma^2 = .13$; $H_1:\sigma^2 \neq .13$; critical values: $\chi^2 = 9.886$ and 45.559; test statistic: $\chi^2 = 35.077$; do not reject H_0

11.73   a. $s^2 = 1038.2679$ b.   453.8734 to 4300.5179; 21.304 to 65.578
      c. $H_0:\sigma^2 = 500$; $H_1:\sigma^2 \neq 500$; critical values: $\chi^2 = 1.690$ and 16.013; test statistic: $\chi^2 = 14.536$; do not reject H_0

11.75   critical value: $\chi^2 = 3.841$; test statistic: $\chi^2 = 2.744$; do not reject H_0

Self-Review Test

1. b   2. a   3. c   4. a   5. b   6. b   7. c   8. b   9. a

10.   critical value: $\chi^2 = 9.210$; test statistic: $\chi^2 = 3.296$; do not reject H_0

11.   critical value: $\chi^2 = 11.345$; test statistic: $\chi^2 = 28.435$; reject H_0

12.   critical value: $\chi^2 = 9.488$; test statistic: $\chi^2 = 82.844$; reject H_0

13.   a.   .0316 to .1457; .178 to .382
      b. $H_0:\sigma^2 = .03$; $H_1:\sigma^2 > .03$; critical value: $\chi^2 = 42.980$; test statistic: $\chi^2 = 48.000$; reject H_0

CHAPTER 12

12.3   a.   3.89   b.   3.61   c.   4.67   **12.5**   a.   2.52   b.   2.35   c.   3.91

12.7   a.   7.79   b.   5.41   **12.9**   a.   4.85   b.   3.22

12.13   a. $\bar{x}_1 = 14.286$; $\bar{x}_2 = 11.429$; $s_1 = 3.7289$; $s_2 = 3.9521$
      b. $H_0:\mu_1 = \mu_2$; $H_1:\mu_1 \neq \mu_2$; critical values: $t = -2.179$ and 2.179; test statistic: $t = 1.391$; do not reject H_0
      c.   critical value: $F = 4.75$; test statistic: $F = 1.94$; do not reject H_0 d.   conclusions are the same

12.15   b.   critical value: $F = 3.29$; test statistic: $F = 4.07$; reject H_0

12.17   a. $H_0:\mu_1 = \mu_2 = \mu_3$; H_1: all three population means are not equal
      b.   reject H_0 if $F > 5.72$ c. $SSB = 237.3511$; $SSW = 522.4889$; $SST = 759.8400$
      d.   numerator: $df = 2$; denominator: $df = 22$ e. $MSB = 118.6756$; $MSW = 23.7495$
      f.   critical value: $F = 5.72$ g.   test statistic: $F = 5.00$ i.   do not reject H_0

12.19   critical value: $F = 3.47$; test statistic: $F = 22.25$; reject H_0

12.21   critical value: $F = 3.72$; test statistic: $F = 5.44$; reject H_0

12.23   a.   critical value: $F = 3.34$; test statistic: $F = 7.66$; reject H_0 b.   .05

12.25 **a.** critical value: $F = 6.93$; test statistic: $F = 1.24$; do not reject H_0

12.27 **a.** critical value: $F = 3.59$; test statistic: $F = 6.53$; reject H_0 **b.** do not reject H_0

12.29 critical value: $F = 4.95$ approximately; test statistic: $F = 20.66$; reject H_0

Self-Review Test

1. a   **2.** b   **3.** c   **4.** a   **5.** a   **6.** a   **7.** b   **8.** a

10. **a.** critical value: $F = 6.11$; test statistic: $F = 1.88$; do not reject H_0 **b.** Type II error

CHAPTER 13

13.15 **a.** y-intercept = 100; slope = 5; positive relationship   **b.** y-intercept = 400; slope = −4; negative relationship

13.17 $\mu_{y|x} = -5.582 + .289x$ **13.19** $\hat{y} = -83.714 + 10.571x$

13.21 **a.** $45.00   **b.** the same amount   **c.** exact relationship

13.23 **a.** $26.50 million   **b.** different amounts   **c.** nonexact relationship

13.25 **b.** $\hat{y} = 127.3451 - 13.9943x$ **e.** $2938.50   **f.** −$12,455.23

13.27 **b.** $\hat{y} = -163.4005 + 9.4976x$ **e.** $263,991.50   **f.** $349,469.90; error = −$49,469.90

13.29 **a.** $\mu_{y|x} = 41.7917 + .2589x$
  b. population regression line because data set includes all 14 National League teams; values of A and B
  d. 50.85%

13.35 $\sigma_\epsilon = 7.0756$; $\rho^2 = .04$ **13.37** $s_e = 4.7117$; $r^2 = .99$

13.39 **a.** $SS_{xx} = 1277.8750$; $SS_{yy} = 5591.5000$; $SS_{xy} = 2536.7500$ **b.** $s_e = 9.6246$
  c. SST = 5591.5000; SSE = 555.7175; SSR = 5035.7825   **d.** $r^2 = .90$

13.41 **a.** $s_e = 7.0269$ **b.** $r^2 = .74$ **13.43** **a.** $s_e = 30.3886$ **b.** $r^2 = .97$

13.45 **a.** $\sigma_\epsilon = 4.4628$ **b.** $\rho^2 = .56$

13.47 **a.** 6.01 to 6.63   **b.** $H_0:B = 0$; $H_1: B > 0$; critical value: $t = 2.145$; test statistic: $t = 59.792$; reject H_0
  c. $H_0:B = 0$; $H_1:B \neq 0$; critical values: $t = -2.977$ and 2.977; test statistic: $t = 59.792$; reject H_0
  d. $H_0:B = 4.50$; $H_1:B \neq 4.50$; critical values: $t = -2.624$ and 2.624; test statistic: $t = 17.219$; reject H_0

13.49 **a.** 2.35 to 2.65   **b.** $H_0:B = 0$; $H_1:B > 0$; critical value: $z = 2.05$; test statistic: $z = 39.12$; reject H_0
  c. $H_0:B = 0$; $H_1:B \neq 0$; critical values: $z = -2.58$ and 2.58; test statistic: $z = 39.12$; reject H_0
  d. $H_0:B = 1.75$; $H_1:B > 1.75$; critical value: $z = 2.33$; test statistic: $z = 11.74$; reject H_0

13.51 **a.** −20.1204 to −7.8682   **b.** $H_0:B = 0$; $H_1:B < 0$; critical value: $t = -1.943$; test statistic: $t = -5.590$; reject H_0

13.53 **a.** $\hat{y} = 12.7547 + 1.0151x$ **b.** .69 to 1.34
  c. $H_0:B = 0$; $H_1:B > 0$; critical value: $t = 2.365$; test statistic: $t = 9.425$; reject H_0

13.55 **a.** 2.36 to 16.64   **b.** $H_0:B = 0$; $H_1:B \neq 0$; critical values: $t = -4.604$ and 4.604; test statistic: $t = 6.122$; reject H_0

13.57 **a.** $\hat{y} = -21.3954 + 3.7334x$ **b.** 1.70 to 5.77
  c. $H_0:B = 1.50$; $H_1:B > 1.50$; critical value: $t = 2.896$; test statistic: $t = 2.532$; do not reject H_0

13.63 a   **13.67** **a.** positive   **b.** positive   **c.** positive   **d.** negative   **e.** zero

13.69 $\rho = .21$ **13.71** $r = -.996$ **13.73** **a.** positively   **b.** $r = .96$

13.75 **a.** positively   **b.** close to 1   **c.** $r = .97$ **13.77** $r = .95$ **13.79** $\rho = .75$

13.81 **a.** $SS_{xx} = 1895.6000$; $SS_{yy} = 4396.4000$; $SS_{xy} = 1029.8000$ **b.** $\hat{y} = 162.7830 + .5433x$ **d.** $r = .36$; $r^2 = .13$
  f. 195.3810   **g.** $s_e = 21.9001$ **h.** −.62 to 1.70
  i. $H_0:B = 0$; $H_1:B > 0$; critical value: $t = 1.860$; test statistic: $t = 1.080$; do not reject H_0

13.83 **a.** $SS_{xx} = 4248.4000$; $SS_{yy} = 964$; $SS_{xy} = 1920$ **b.** $\hat{y} = -5.1606 + .4519x$ **d.** $r = .95$; $r^2 = .90$
  e. $s_e = 3.4704$ **f.** .27 to .63   **g.** $H_0:B = 0$; $H_1:B > 0$; critical value: $t = 2.896$; test statistic: $t = 8.494$; reject H_0

13.85 **a.** $SS_{xx} = 3.3647$; $SS_{yy} = 120.0000$; $SS_{xy} = 18.6500$ **b.** $\hat{y} = 7.7119 + 5.5428x$ **d.** $r = .93$; $r^2 = .86$
  e. $s_e = 1.8236$ **f.** 2.99 to 8.10
  g. $H_0:B = 0$; $H_1:B \neq 0$; critical values: $t = -4.032$ and 4.032; test statistic: $t = 5.575$; reject H_0

13.87   a. 13.8708 to 16.6292; 11.7648 to 18.7352   b. 62.3590 to 67.7210; 56.3623 to 73.7177

13.89   21.7926 to 24.0188; 19.4518 to 26.3596   **13.91**   111.5117 to 123.1553; 100.9487 to 133.7183

13.93 $2009.52 to $2742.68; $1495.85 to $3256.35

13.95   a. positive relationship   b. $\hat{y} = -1.4340 + .8916x$ d. $r = .97$; $r^2 = .95$ e. $s_e = .9285$
      f. .51 to 1.28   g. $H_0{:}B = 0$; $H_1{:}B > 0$; critical value: $t = 2.571$; test statistic: $t = 9.356$; reject H_0

13.97   a. positive   b. $\hat{y} = 8.3483 + .4608x$ d. $r = .85$; $r^2 = .73$ e. 2448   f. $s_e = 3.5811$
      g. .04 to .88   h. $H_0{:}B = 0$; $H_1{:}B > 0$; critical value: $t = 3.365$; test statistic: $t = 3.660$; reject H_0

13.99   b. $SS_{xx} = 143.0000$; $SS_{yy} = 361.6667$; $SS_{xy} = 216.0000$ d. $\hat{y} = 138.5256 + 1.5105x$ f. $r = .95$
      g. 161.18 billion pounds

13.101   4.2828 to 7.1148; 1.6962 to 9.7014   **13.103**   20.8007 to 28.1519; 16.3782 to 32.5744

13.105   a. $\hat{y} = 8.1323 + .3486x$; $r = .99$ c. $\hat{y} = 13.1323 + .3486x$; $r = .99$

13.107 $SS_{xx} = 113.4000$; $SS_{yy} = 4012.5000$; $SS_{xy} = 132.5000$ b. $\hat{y} = 8.8260 + 1.1684x$ d. $r = .20$; $r^2 = .04$
      e. $s_e = 21.9593$ f. $H_0{:}B = 0$; $H_1{:}B \neq 0$; critical values: $t = -2.306$ and 2.306; test statistic: $t = .567$;
      do not reject H_0 g. no

Self-Review Test

1. d   2. a   3. b   4. a   5. b   6. b   7. true   8. true
9. a   10. b

15.   a. The number of computers sold depends on experience.   b. positive   d. $\hat{y} = 18.1912 + 1.4322x$
     f. $r = .69$; $r^2 = .47$ g. 33.95 computers   h. $s_e = 6.6748$ i. -1.3023 to 4.1667
     j. $H_0{:}B = 0$; $H_1{:}B > 0$; critical value: $t = 3.365$; test statistic: $t = 2.112$; do not reject H_0 k. 22.8634 to 36.4342
     l. 11.1952 to 48.1024

APPENDIX A

A.7   simple random sample

A.9   a. nonrandom sample   b. judgment sample   c. selection error

A.11   a. random sample   b. simple random sample   c. no

A.13   voluntary response error   **A.15**   response error

A.17   a. observational study   b. not a double-blind study

A.19   a. designed experiment   b. double-blind study

A.21   designed experiment   **A.23**   yes

A.25   b. observational study   c. not a double-blind study

A.27   a. designed experiment   b. double-blind study

A.29   a. no   b. no   c. convenience sample

INDEX